# ANNUAL REVIEW OF
# NEUROSCIENCE

# ANNUAL REVIEW OF NEUROSCIENCE

## VOLUME 12, 1989

W. MAXWELL COWAN, *Editor*
Howard Hughes Medical Institute

ERIC M. SHOOTER, *Associate Editor*
Stanford University School of Medicine

CHARLES F. STEVENS, *Associate Editor*
Yale University School of Medicine

RICHARD F. THOMPSON, *Associate Editor*
University of Southern California

ANNUAL REVIEWS INC.   4139 EL CAMINO WAY   P.O. BOX 10139   PALO ALTO, CALIFORNIA 94303-0897

# ANNUAL REVIEWS INC.
Palo Alto, California, USA

*International Standard Serial Number : 0147-006X*
*International Standard Book Number : 0-8243-2412-9*

Annual Review and publication titles are registered trademarks of Annual Reviews Inc.

∞ The paper used in this publication meets the minimum requirements of American National Standard for Information Sciences—Permanence of Paper for Printed Library Materials, ANSI Z39.48-1984.

Annual Reviews Inc. and the Editors of its publications assume no responsibility for the statements expressed by the contributors to this *Review*.

TYPESET BY AUP TYPESETTERS (GLASGOW) LTD., SCOTLAND
PRINTED AND BOUND IN THE UNITED STATES OF AMERICA

Annual Review of Neuroscience
Volume 12, 1989

# CONTENTS

(*Continued*)   v

vi   CONTENTS *(Continued)*

# SOME RELATED ARTICLES IN OTHER *ANNUAL REVIEWS*

From the *Annual Review of Biochemistry*, Volume 58 (1989):

*The Protein Kinase C Family: Heterogeneity and Its Implications*,
U. Kikkawa, A. Kishimoto, and Y. Nishizuka

From the *Annual Review of Cell Biology*, Volume 4 (1988):

*Development of the Peripheral Nervous System from the Neural Crest*,
N. M. LeDouarin and J. Smith

*Growth Cone Motility and Guidance*, D. Bray and P. J. Hollenbeck

*Hair Cells: Transduction, Tuning, and Transmission in the Inner Ear*,
W. M. Roberts, J. Howard, and A. J. Hudspeth

*Regulation of Adenylyl Cyclase-Coupled β-Adrenergic Receptors*,
J. L. Benovic, M. Bouvier, M. G. Caron, and R. J. Lefkowitz

From the *Annual Review of Entomology*, Volume 34 (1989):

*Structure and Function of the Deutocerebrum in Insects*, U. Homberg,
T. A. Christensen, and J. G. Hildebrand

From the *Annual Review of Pharmacology and Toxicology*, Volume
29 (1989):

*Bernard B. Brodie and the Rise of Chemical Pharmacology*, E. Costa,
A. G. Karczmar, and E. S. Vesell

*The Biochemical Pharmacology of Atrial Peptides*, P. Needleman,
E. H. Blaine, J. E. Greenwald, M. L. Michener, C. B. Saper, P. T. Stockman,
and H. Eser Tolunay

*Tissue Kallikreins and Kinins: Regulation and Roles in Hypertensive and
Diabetic Disease*, H. S. Margolius

*The Excitatory Amino Acid Receptors*, D. T. Monaghan, R. J. Bridges, and
C. W. Cotman

*Modulation of Glutamate Receptors: Molecular Mechanisms and Functional
Implications*, J. T. Wroblewski and W. Danysz

*Chemical Coding of Neurons and Plurichemical Transmission*, J. B. Furness,
J. L. Morris, I. L. Gibbins, and M. Costa

*Neurotransmitter Receptors and Phosphoinositide Turnover*, D.-M. Chuang

*Peripheral Type Benzodiazepine Receptors*, A. Verma and S. H. Snyder

From the *Annual Review of Physiology*, Volume 51 (1989):

From the *Annual Review of Psychology*, Volume 40 (1989):

Viktor Hamburger

*Ann. Rev. Neurosci. 1989. 12 : 1–12*

# THE JOURNEY OF A NEUROEMBRYOLOGIST

*Viktor Hamburger*

Department of Biology, Washington University, St. Louis, Missouri 63130

How did one become an experimental neuroembryologist in the 1920s? As an American, one would go to Ross G. Harrison at Yale, or to one of his students. Harrison had pioneered the field of neuroembryology at the beginning of the century and had made his mark by designing the first tissue culture experiment, in order to provide indisputable evidence for the axon outgrowth theory. By the adoption of the limb transplantation experiment in amphibians he had provided us with a paradigmatic model for the analysis of the formation of nerve patterns. The handful of neuro-embryologists who were active at that time were mostly his students. He had turned to other problems, but retained a strong interest in the field (Harrison 1935). As a student of Hans Spemann in Germany, whose domain was the analysis of determination and inductions in early amphi-bian development, I entered the field through the back door, so to speak. When Spemann accepted me as a PhD candidate, he assigned me a topic unrelated to his turf, because he wanted to assure me of an independent career from the start. But I have never found out what prompted him to have me repeat an experiment that had given rather questionable results. B. Duerken (1913) had done eye extirpations on young frog tadpoles and had found leg abnormalities in a fairly high percentage of cases. He had interpreted them as neurogenic, in the sense that the operation would cause a primary defect in the midbrain that would cause secondary deficiencies in the motor centers of the spinal cord, resulting in deficient leg innervation. Spemann was skeptical of these findings, but I surprised him by actually obtaining some slightly deformed legs, though at a low percentage (1925).[1] However, when I repeated the experiments the following year on another 1000 tadpoles, I obtained 2000 entirely normal legs (1927). Spemann was

---

[1] References that contain only the year of publication refer to my own publications.

0147–006X/89/0301–0001$02.00

convinced that Duerken's results were spurious, but I decided to do the crucial experiment of creating nerveless legs by extirpating the lumbar part of the spinal cord in early stages preceding nerve outgrowth. Such legs developed normally in every respect, though the musculature atrophied and eventually degenerated (1928). This, my first original contribution, seems to have settled the issue, since to my knowledge the experiment has never been challenged.

Hidden behind my clear-cut result was an uphill fight against the resolute efforts of the embryo to regenerate the defect I had inflicted on its spinal cord. In fact, I found myself in the possession of only a few nerveless legs but an abundance of cases with partial or poor innervation. I exploited them for information on nerve pattern formation. Among other observations, I found that even tiny nerves would manage to form a typical pattern, though they would not reach their targets. I concluded that the signals that guide nerves to their destinations are distinct from those responsible for the appropriate terminal connections (1929).

By that time I became aware that the developing nervous system provides the analytically minded embryologist with almost unlimited challenges to keep him busy for a life-time. In my first review article, entitled "Developmental-physiological Correlations Between the Amphibian Limbs and Their Innervation" (1927), I identified three major areas: the influence of innervation on limb development (which I considered as settled); the guidance of nerves to their targets; and a possible effect of the growing limb on the nerve centers that innervate it. This section was the shortest—less than half a page. I mentioned in passing a largely-forgotten study by one M. Shorey without even indicating that it had been done on chick embryos. She had removed the wing buds of chick embryos by electrocautery and found that the lateral motor columns and lumbar spinal ganglia (DRG) were severely defective (Shorey 1909). The only other experiments dealing with this topic had been done by Harrison's student, S. Detwiler, on salamander embryos. Following forelimb extirpation he had found a hypoplasia of the brachial DRG and, following the transplantation of a forelimb primordium to the flank, a hyperplasia of the "overloaded" thoracic DRG (Detwiler 1920). Strangely enough, he had not observed any changes in the motor region of the spinal cord. These experiments are of great historical significance. They initiated a trend of thought that led eventually to the discovery of the nerve growth factor.

While I realized that neurogenesis offered many opportunities to a budding experimental embryologist, my own role was not clear. The problem of the influence of the target on its nerve centers was in the hands of Detwiler, who worked assiduously in this field (Detwiler 1936), and there was then no obvious way of advancing the problem of nerve guidance. I turned to another, very different field that had fascinated me: developmental genetics (which at that time did not even have a name). It was clear to me—though not to many geneticists—that genes "produce" a particular phenotype by controlling the developmental processes that lead to it. It occurred to me that one might gain insight into the way genes do this by combining hybridization with the method of transplantation. Since there were then no known mutants in salamanders, I undertook hybridization of two local species and studied in the reciprocal hybrids the development of some characteristics in which the species differed markedly, such as the development of toes. The project was terminated before I got around to transplantations, when I moved to the United States. The results were published in English (1936). As fate would have it, one set of species hybrids showed leg abnormalities which again eluded an explanation (1935).

My penchant for stage series dates back to the earliest days; a limited one was created for the frog species that gave the material for my PhD thesis. I assigned the tedious and not particularly inspiring task of designing a stage series of the two *Triturus* species used for hybridization to the only PhD candidate I had in Germany, Salome Gluecksohn (1931). The only lasting result of her effort was that it started her career as an eminent developmental geneticist. My friend Salome Gluecksohn Waelsch has held the chair of Genetics at Albert Einstein Medical School for many years. Incidentally, I returned briefly to developmental genetics later, using the creeper mutant of the chick embryo. Reciprocal transplantations of limb and eye primordia between normal and mutant embryos gave some interesting results (1942).

My life took a fateful turn in 1932. I received a fellowship from the Rockefeller Foundation to spend a year at the Zoological Laboratory of the University of Chicago; its Director, Dr. Frank Lillie, was a friend of Spemann. The change of continents became permanent; it saved me from Nazi persecution, and at the same time I changed my allegiance from the amphibian to the chick embryo, and, fortuitously, all these changes brought about the return to my destiny: experimental neurogenesis. How did this happen? Lillie's classic book, *The Development of the Chick* (1908), had introduced the chick embryo to the classroom and to experimental research. When I had my first meeting with Dr. Lillie, in which my plans were discussed, he reminded me that M. Shorey had done her work in his laboratory, at his suggestion. Would it not be expedient for me to try out Spemann's glass neddle technique on the chick embryo by repeating her experiment? I would then have the chance to resolve the discrepancy between her and Detwiler's results, with respect to the effect of the ablation on the motor system. Dr. Lillie was at that time the Dean of the Bio-Medical Sciences. I was entrusted to his former student, B. H. Willier, then Professor of Embryology in the Zoology Department. He and his exceptionally skillful research associate, Dr. Mary Rawles, initiated me in the art of handling chick embryos. During the winter of 1932–1933, using Spemann's glass needle technique, I learned how to extirpate limb buds and to transplant them to the flank. I was thrilled to see them develop and grow to almost normal size and to quite advanced stages (1938). What a difference compared to the amphibian experimentation, to have results in a few days and not have to fight any longer the high mortality rates. Lillie and Willier and Mary Rawles followed my adventure with great interest. We all realized that since the Spemann technique worked beautifully on the chick embryo, it would now become available for a wide range of experimentation; but I did not anticipate that in neuroembryology the chick embryo would eventually play a more important role than the amphibian embryo. It should be mentioned that up to then the manipulation of the embryo itself had not been feasible. The embryos had been used, however, for growing embryonic tissues on their chorio-allantoic membrane, but this substrate did not permit the normal morphogenesis of structures such as limb or eye primordia.

By the end of the 1940s, the chick embryo had conquered wide areas of experimental research, far beyond the boundaries of experimental embryology. Now there arose the need for a precise characterization of the different stages of development. Up to then, embryos had been identified by their age in terms of hours or days of incubation. But the embryological staging was inadequate, partly because conditions of incubation varied widely, and partly due to individual differences in rate of development. A three-day embryo can show a wide range of morphological differences. In 1948, I saw my colleague Howard Hamilton of the University of Iowa at a meeting in Chapel Hill, North Carolina. We discussed a project he was engaged in: a revised edition of Lillie's *Development of the Chick*. We commented on the inadequate description of stages in the earlier editions, and I suggested that we produce a stage series that would meet the highest standards of precision. The series would have to fulfill two critical requirements: each stage had to be characterized unequivocally by a set of easily identifiable morphological features; and the continuum of development had to be transected into an arbitrary sequence of stages that would be so close to each other that they would indicate the smallest detectable changes. We ended up with 45 stages, presented by photographs and descriptions (Hamburger & Hamilton 1951). Obviously, we were very successful. The stage series soon became indispensable; it still is one of the most frequently quoted publications. Without a doubt it contributed materially to the precision whereby data obtained with chick embryos could be processed.

Returning to the wing extirpations, my results confirmed those of M.

Shorey. Both the dorsal root ganglia (DRG) and the lateral motor column (lmc) were distinctly hypoplastic. I became aware that I was dealing with a fundamental issue and that its analysis could be carried much further than Shorey and Detwiler had done previously. The clear demarcation of the lmc in the chick embryo, which one does not find in the salamander embryo, played into my hands, as did my imperfection as a beginner in my dealings with the chick embryo. It was easy enough to cut off the wing bud neatly; but I did not realize that the primordia of the massive pectoral muscles that were also supplied by the brachial nerves were situated beneath the surface. So I obtained a wide range of muscle deficiencies, which was reflected in the range of motor hypoplasia. When I quantified the data, I found a close correlation between the two sets of deficiencies. On the other hand, the hypoplasia of the DRG amounted to 50% with rather little variation. This I attributed to a fairly uniform loss of skin. I drew the conclusion that "every structure within the growing limb, muscles, as well as sense organs, send stimuli to the nervous system. Each part of the peripheral field controls directly its own nervous center, i.e. the limb muscles affect the lateral motor centers and the sensory fields control the ganglia" (1934, p. 470). From this conceptualization it follows logically that targets and their centers must be connected directly, and I suggested "that the nerve fibers themselves serve as mediators between the two links of the correlation" (1934, p. 475). To remove any lingering doubt that the neuronal deficiencies might have been caused by a nonspecific trauma inflicted by the operation, I made use of my extensive material of limb transplantations. I found that the most strongly innervated transplants were wings placed immediately behind the normal wing and supplied by the brachial plexus, and leg buds placed immediately in front of the normal leg and supplied by the lumbar plexus. A quantitative analysis of individual segments showed clearly that only those lmc segments and DRG which contributed to the innervation of the transplants were hyperplastic (1939). The model of growth-controlling agents produced by the targets and transported retrogradely to their respective nerve centers has stood the test of time well. It is often forgotten that concepts that are now taken for granted, at one time had to pass the test of authenticity.

The ground was now prepared for the pursuit of two specific problems: the nature of the growth-regulating agents, and their specific mode of action on the nerve centers. The first question did not seem to be manageable at that time. Several answers to the second question were proposed. Detwiler had suggested that the targets might regulate the size of centers by controlling proliferation. This option was ruled out for the lmc in the chick by mitotic counts in the spinal cord in cases of wing extirpation and transplantation that showed no difference between the left and right sides

(Hamburger & Keefe 1944). Then I formulated a "recruitment hypothesis," in which I postulated that the first pathfinder nerves that reached the target would explore its size and recruit from a pool of (hypothetical) undifferentiated neuroblasts the appropriate number that would saturate the target. The recruitment I imagined would take the form of an induction of undifferentiated neuroblasts by the mature pathfinder neurons (1934; Hamburger & Keefe 1944). The hypothesis had the merit that it could account for both hypo- and hyperplasia.

Soon after the end of the war, I became acquainted with the publications of two Italian investigators, Rita Levi-Montalcini and Guiseppe Levi (1942, 1944), who had proposed an entirely different explanation of hypoplasia in DRGs. They had repeated the limb bud extirpation and counted neurons in successive stages from 6 to 19 days of incubation. The counts provided evidence that the hypoplasia resulted from the degeneration of fully differentiated neurons rather than from some form of interference with progressive differentiation. This spelled the doom of my brain child, but it did not alter the underlying basic conception of the control of the nerve centers by their targets via the nerve fiber route. After I invited Dr. Levi-Montalcini to join me in St. Louis in 1947, we repeated both limb ablation and transplantation experiments. The former fully substantiated her claim, since large numbers of degenerating neurons were found in the brachial DRG. In the same material she made another observation: cervical and thoracic DRG that were not affected by the operation also displayed numerous degenerating cells. This was the seminal discovery of naturally occurring neuronal death. As far as the hyperplasia in the transplantation experiments was concerned, mitotic and cell counts seemed to indicate that we were dealing with the enhancement of proliferation; however, in this matter we were on the wrong track (Hamburger & Levi-Montalcini 1949).

Conceptually, our new data gave my model of the target-nerve center relationship a new slant, and we suggested that "substances necessary for . . . neuroblast growth and maintenance would not be provided in adequate quantities, when the limb bud is removed" (Hamburger & Levi-Montalcini 1949, p. 493). The hypothetical agent was thus identified as a *maintenance* factor. When we reflected on the cause of naturally occurring death, we were struck by the fact that the presence of degenerating cells was limited to a short period of a few days and that it occurred at exactly the same stages in the experimentally affected brachial DRG and the adjacent normal cervical and thoracic DRG. This strongly suggested a common denominator for both types of neuronal death. "In both instances, the early differentiating VL cells [a special subpopulation in DRG] are affected after their neurites have reached the periphery, and in both instances the breakdown occurs between 5 and 7 days. . . . In the experimental situation the reduction of the peripheral area is definitely responsible for degeneration. It is possible that the same mechanism operates in the case of the cervical and thoracic ganglia. This would imply that in early stages cervical and thoracic VL cells send out more neurites than the periphery can support. The excess of neurons would break down at the stage at which the VL cells are highly susceptible to environmental conditions" (Hamburger & Levi-Montalcini 1949, pp. 495–96). This was clearly an anticipation of the competition hypothesis.

In retrospect, we made the right choice not to pursue this topic further but to concentrate on the analysis of hyperplasia. The exuberant vitality of the DRG displayed in their overgrowth, far beyond their normal performance, seemed more worthy of our attention than the morbid story of neuronal death. But how to proceed? We realized that limb transplantation would yield no further insights. We needed a more homogeneous tissue with growth-stimulating activity. I then remembered that my former student, Elmer Bueker, with a similar idea in mind, had already provided us with an ideal solution to our quest, namely to use tumor tissue. Applying an experimental design that he had devised for his PhD thesis Bueker (1943) implanted fragments of mouse sarcoma 180 in the place of the leg bud, which amounted practically to a coelomic graft. The tumor had grown well and had been invaded by sensory axons but was avoided by motor fibers. The corresponding DRG showed a considerable hyperplasia. Bueker concluded "that sarcoma 180, because of intrinsic physicochemical properties and mechanics of growth, selectively causes the enlargement of spinal ganglia" (Bueker 1948, p. 382). This ingenious experiment was a crucial link to the discovery of NGF. It provided us with a growth-promoting, fast-growing tissue that was specific for a particular type of neuron and available in large quantities for future biochemical analysis. With Bueker's consent we repeated the experiment. The results far surpassed our expectations. The tumors were profusely invaded by fibers that could be traced not only to DRG but also to sympathetic chain ganglia. The former were enlarged up to $2\frac{1}{2}$ times and the latter up to 6 times their normal size. The enlargement did not begin until the fibers had spread in the tumor. Now for the first time the prospect of giving concrete reality to the paradigm that had guided me since the 1930s, namely that specific nerve-growth regulating agents were produced by the targets and transported retrogradely to the centers, seemed to be within reach. Indeed, a series of brilliant experimental masterstrokes by Dr. Levi-Montalcini, in collaboration with a superb biochemist, Dr. Stanley Cohen, whom I invited to join my laboratory in 1953, led within a few years to the discovery of the nerve growth factor and its identification as a protein. I actively participated in the early phases of this work and in the preparation of the first two publications (Levi-Montalcini & Hamburger 1951, 1953) but withdrew from the project in 1953 to pursue other interests. I later returned to the exploration of the role of NGF in naturally occurring cell death (see below).

Ever since I became acquainted with the chick embryo, I had noticed its stirrings, but had paid little attention to them. However, I was familiar with the general topic of embryonic motility and with the sharp controversy over this issue that had been fought in the 1920s and 1930s. The behaviorists who then dominated the scene contended that local reflexes are the primary units of embryonic behavior and that they become gradually integrated into coordinated performances. Among their opponents, the neuroembryologist G. E. Coghill stood out as an astute observer and an eminent theorist. He had done pioneer work on the correlation of neurogenesis and genesis of behavior. On the basis of these investigations he had developed the theory that motility is an integrated performance from the first bending of the head in the early embryo to the complex behavior patterns of locomotion, feeding, etc. Local reflexes would originate secondarily by "emancipation" from the total pattern (Coghill 1929). The controversy had ended in an impasse and not much had been done about it.

It occurred to me that since this was a fundamental issue, it could perhaps be revived by approaching it with the experimental method. So far observation and stimulation had been the standard implements. The project that I started in the late 1950s kept me and a group of very able co-workers busy for over a decade. My associates were M. Balaban, A. Bekoff, J. Decker, C. H. Narayanan, R. Oppenheim, R. Provine, S. Sharma, and E. Wenger. We realized immediately that the chick embryo does not comply with either one of the old paradigms. The main characteristic of its performance is that it is spontaneous, that is, nonreflexogenic. This is apparent from its two main features: the movements are entirely uncoordinated and they are performed intermittently, activity phases alternating with inactivity phases. They are performed in the absence of any apparent stimulation. By uncoordinated I mean that at any moment the head, trunk, wings, legs, and later beak and eyelids, can be active in any combination. The movements are jerky, convulsive-like, and they appear completely aimless. The activity phases get gradually longer and the inactivity phases shorter, until from day 13 of incubation on the embryo is almost continuously active.

We confirmed an important discovery that an eminent physiological psychologist of the last century, W. Preyer, had reported in his classic book *Spezielle Physiologie des Embryo* (1885). He had observed that in the chick embryo, motility begins at about 4 days of incubation, but responses to stimulation cannot be elicited until about day 8. Hence, there exists a prereflexogenic period of spontaneous motility. To determine whether sensory input plays a role at later stages we performed complete deafferentations of the leg level. We did this by removing the dorsal part of the lumbar spinal cord including the neural crest, which gives rise to the DRG, and by producing a gap in the thoracic cord to exclude descending inputs from more rostral levels. Overall activity and the periodicity pattern were the same in experimental embryos of different stages and in controls with thoracic gaps only. Thus we had established that spontaneous motility is the basic form of behavior in the chick embryo. This implies that any form of stimulation, including self-stimulation by the brushing of the legs against the head, which had been postulated by behaviorists as an essential ingredient, plays a very minor role, if any. Later on, we found that rat fetuses display the same uncoordinated, intermittent, jerky motility as the chick embryos, and we believe that this type of embryonic and fetal motility is paradigmatic for all warm-blooded forms.

If the motility is not elicited by stimulation, it must be generated by endogenous bioelectrical activity. This aspect was investigated in the second phase of our project. The main result of extensive extracellular recordings from the ventral regions of the spinal cord can be summarized briefly: The periods of motility and the duration of activity phases were reflected precisely in polyneuronal burst patterns. We could not have wished for a better correlation between overt motility and electrophysiological activity. This holds for all stages from the beginning of motility to near-hatching stages. One can consider the spontaneous, intermittent, self-generated motility of warm-blooded embryos and fetuses as the earliest manifestation of innate behavior patterns. This is a novel paradigm; it has created a solid foundation for all modern studies of fetal activity, including the human fetus (reviews in Hamburger 1963, 1970, 1973).

How does the embryo manage to extricate itself from the shell? Certainly not by the uncoordinated, jerky movements that it has displayed so far. When I searched the literature, I found to my surprise that my question pointed to unexplored territory. In the mid-1960s Ron Oppenheim and I embarked on the project to detect the embryo's way of solving a major problem in its life-history. We spent many months observing the embryos through enlarged windows in the shell and became witness to most remarkable events. They begin 3–4 days before the actual hatching occurs. While the uncoordinated random movements are temporarily suspended, the embryo performs a series of well-coordinated movements by which it maneuvers itself into what we have called the "hatching position." The embryo is oriented lengthwise in the shell, with the head near the blunt end. The head is tucked under the right wing and the beak points obliquely toward the shell, almost touching it. This position is attained by coordinated head, trunk, and leg movements with a rotatory component; they are synchronized with wing lifting. This positioning takes many hours and much trial and error and it is interrupted by long rest periods. Next, a hole is pierced in the shell by vigorous back thrusts of head and beak; this is referred to as "pipping." About a day later (on the twenty-first day of incubation), the hatching act begins rather suddenly with repetitions of the powerful back thrusts of head and beak, but this time they are combined with the slow rotation of the whole body. In this way, the shell is cracked along a circle at some distance from the blunt end. When pieces of shell are chipped off along about 2/3 of the circumference, the cap of the shell is loosened, and it soon breaks open and the chicken escapes. All this takes between 1/2 to 1 hour. Having watched the aimless fidgeting of embryos for several years, viewing this smooth, goal-directed performance held its fascination. I may mention that this was the only investigation of mine that did not involve experimentation (apart from the designing of the stage series, which was not a problem-solving project).

In the meantime, the nerve growth factor had become a cause célèbre, but interest in neuronal death had faded. I had brought the lateral motor column in line with the DRG by demonstrating that limb bud extirpation likewise causes massive degeneration of differentiated motor neurons, which had sent their axons to the target area (1958). But it was not until after my excursion to the field of embryonic behavior that I returned once more to my old favorite theme, the dialogue between nerve centers and their partners. While natural neuronal death had been established as an integral part of neurogenesis, a rationale for this strange phenomenon was not yet in evidence. When it was first discovered in 1949, we had suggested a possible explanation: that the target might produce a maintenance factor in insufficient amounts and that the neurons that lose in the competition for it might be the ones that die. Now, in the mid-1970s, I had the idea that the old stand-by, the limb transplantation experiment, could be used to test the competition hypothesis. An enlargement of the target area should result in the rescue of neurons that would otherwise die. The leg buds that M. Hollyday transplanted skillfully close to the normal leg buds developed to normal size and became innervated by nerves from the lumbar plexus. We succeeded in salvaging an average of 30% motor neurons, and in one case 53% neurons (Hollyday & Hamburger 1976).[2] Later on, others,

---

[2] The figures reported in our publication are considerably lower than those mentioned here. We had underestimated the gain by comparing the percentage difference of the surviving cells between the left (control) side and the right side, at the end of the degeneration period (day 12). The figures stated above represent a comparison between the total numbers of neurons before the onset of the degeneration period at day 6 and after its termination at day 12. I owe this rescue of our success to my friend J. Sanes.

using different experimental designs, managed to rescue up to 100% neurons that would have died. These results provide conclusive evidence that neuronal death in vertebrates is probabilistic and not programmed, as in some invertebrates.

Finally, we tried to link the two seemingly disparate events in neurogenesis: neuronal death and neuron-growth promotion by NGF. Is NGF identical with the hypothetical maintenance factor for DRG, produced by limb tissue? The best we could do with the techniques at our disposal was to provide indirect evidence. First, we showed that if labeled NGF is injected into the leg during the critical period of neuron degeneration, it is transported retrogradely and selectively to the lumbar DRG (Brunso-Bechtold & Hamburger 1979).

Next, we were able to reduce normally occurring neuronal death in both brachial and thoracic DRG by daily injections of NGF into the yolk sac from early stages on (Hamburger et al 1981). Finally, we subjected the competition hypothesis to the most severe test: we combined wing bud extirpations, which normally cause a near-total breakdown of the brachial DRG, with daily injections of NGF. With the dosage we used, we reduced the death rate in the subpopulation of large, early differentiating neurons (VL) by 50% and rescued nearly all neurons in the subpopulation of small, late-differentiating cells (DM) (Hamburger & Yip 1984). The fact that NGF is an effective substitute for the hypothetical trophic maintenance factor for DRG strengthened our belief that NGF is identical with this factor.

Some investigators are adventurous, and enjoy exploring uncharted territory. They make bold forays and get their deepest satisfaction from the unforeseen surprises that await them when they reach a clearing in the forest or a mountain top. The best are guided by an unfailing magnet or endowed with a green thumb. They are often rewarded by momentous discoveries. Others, with a different temperament, start with some general idea that may be nothing more than a dim hunch. They conquer their territory by patient step-by-step analysis, guided by the inner logic of their pursuit. They get their rewards when the story unfolds and has a happy ending, and when an unexpected discovery comes their way. Obviously, I belong to the second type; undoubtedly, inherent tendencies were reinforced by my mentor Spemann. But I had the good fortune to count among my close friends some very successful personifications of the first type, who have enriched my life.

At the end of my experimental exploits I realize that I have come full circle to my original ideas of 1934 about the influence of targets on the nerve centers that innervate them. With the book, *The Heritage of Experimental Embryology* (1988), in which the concept of embryonic induction plays a major role, I have closed another circle. It took me back to my student days in Spemann's laboratory where I was exposed to the major issues in experimental embryology, before I started my journey as an experimental neuroembryologist.

## Literature Cited

Brunso-Bechtold, J., Hamburger, V. 1979. *Proc. Natl. Acad. Sci. USA* 76: 1494–96

Bueker, E. 1943. *J. Exp. Zool.* 93: 99–129

Bueker, E. 1948. *Anat. Rec.* 102: 369–90

Coghill, G. E. 1929. *Anatomy and the Problem of Behaviour.* London: Cambridge Univ. Press

Detwiler, S. R. 1920. *Proc. Natl. Acad. Sci. USA* 6: 96–101

Detwiler, S. R. 1936. *Neuroembryology.* New York: Macmillan

Duerken, B. 1913. *Z. Wiss. Zool.* 105: 191–242

Glucksohn, S. 1931. *Roux' Arch. Entw. Mech.* 125: 341–405

Hamburger, V. 1925. *Roux' Arch. Entw. Mech.* 105: 149–201

Hamburger, V. 1927. *Naturwissenschaften* 15: 657–81

Hamburger, V. 1928. *Roux' Arch. Entw. Mech.* 114: 272–362

Hamburger, V. 1929. *Roux' Arch. Entw. Mech.* 119: 47–99

Hamburger, V. 1934. *J. Exp. Zool.* 68: 449–94

Hamburger, V. 1935. *J. Exp. Zool.* 70: 43–54

Hamburger, V. 1936. *J. Exp. Zool.* 73: 319–73

Hamburger, V. 1938. *J. Exp. Zool.* 77: 379–97

Hamburger, V. 1939. *J. Exp. Zool.* 80: 347–89

Hamburger, V. 1942. *Biol. Symp.* 6: 311–34

Hamburger, V. 1958. *Am. J. Anat.* 102: 365–410

Hamburger, V. 1963. *Q. Rev. Biol.* 38: 342–65

Hamburger, V. 1970. *The Neurosciences Second Study Program,* ed. F. O. Schmitt, pp. 141–51

Hamburger, V. 1973. In *Studies on the Development of Behavior and the Nervous System,* ed. G. Gottlieb, 1: 51–76. New York/London: Academic

Hamburger, V. 1988. *The Heritage of Experimental Embryology. Hans Spemann and the Organizer.* New York/Oxford: Oxford Univ. Press

Hamburger, V., Hamilton, H. 1951. *J. Morphol.* 88: 49–92

Hamburger, V., Keefe, E. L. 1944. *J. Exp. Zool.* 96: 223–42

Hamburger, V., Levi-Montalcini, R. 1949. *J. Exp. Zool.* 111: 457–502

Hamburger, V., Brunso-Bechtold, J. K., Yip, J. 1981. *J. Neurosci.* 1: 60–71

Hamburger, V., Yip, J. 1984. *J. Neurosci.* 4: 767–74

Harrison, R. G. 1935. *Proc. R. Soc. London Ser. B* 118: 155–96

Hollyday, M., Hamburger, V. 1976. *J. Comp. Neurol.* 170: 311–20

Levi-Montalcini, R., Hamburger, V. 1951. *J. Exp. Zool.* 116: 321–62

Levi-Montalcini, R., Hamburger, V. 1953. *J. Exp. Zool.* 123: 233–88

Levi-Montalcini, R., Levi, G. 1942. *Arch. Biol.* 53: 537–45

Levi-Montalcini, R., Levi, G. 1944. *Comment. Pontific. Acad. Sci.* 8: 627–68

Lillie, F. R. 1908. *The Development of the Chick.* New York: Holt

Preyer, W. 1885. *Specielle Physiologie des Embryo.* Leipzig: Grieben

Shorey, M. L. 1909. *J. Exp. Zool.* 7: 25–63

*Ann. Rev. Neurosci. 1989. 12 : 13–31*

# SHORT-TERM SYNAPTIC PLASTICITY

*Robert S. Zucker*

Department of Physiology-Anatomy, University of California, Berkeley, California 94720

## INTRODUCTION

Chemical synapses are not static. Postsynaptic potentials (PSPs) wax and wane, depending on the recent history of presynaptic activity. At some synapses PSPs grow during repetitive stimulation to many times the size of an isolated PSP. When this growth occurs within one second or less, and decays after a tetanus equally rapidly, it is called *synaptic facilitation*. A gradual rise of PSP amplitude during tens of seconds of stimulation is called *potentiation*; its slow decay after stimulation is *post-tetanic potentiation* (PTP). Enhanced synaptic transmission with an intermediate lifetime of a few seconds is sometimes called *augmentation*. Potentiated responses lasting for hours or days are called *long-term potentiation*. This latter process, not usually regarded as short-term, is the subject of a separate review (Brown et al 1989, this volume).

Other chemical synapses are subject to fatigue or depression. Sustained presynaptic activity results in a progressive decline in PSP amplitude. Most synapses display a mixture of these dynamic characteristics (Figure 1). During a tetanus, or train of action potentials, transmission may rise briefly due to facilitation before it is overwhelmed by depression (Hubbard 1963). If depression is not too severe, augmentation and potentiation lead to a partial recovery of transmission during the tetanus. Following the tetanus, facilitation decays rapidly, leaving depressed responses which recover to the potentiated level, causing what appears as a delayed post-tetanic potentiation (Magleby 1973b). Finally, PTP decays and PSPs return to the same amplitude as that elicited by an isolated presynaptic spike.

Short-term synaptic plasticity often determines the information pro-

13

FREQUENCY DEPENDENT CHANGES IN SYNAPTIC EFFICACY

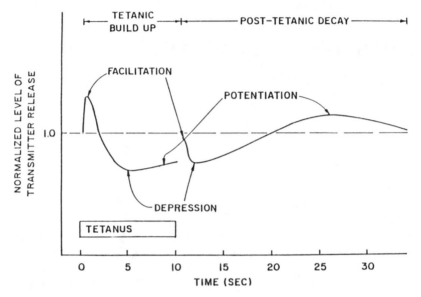

*Figure 1*  The effects of simultaneous facilitation, depression, and potentiation on transmitter release by each spike in a tetanus, and by single spikes as a function of time after the end of the tetanus.

cessing and response molding functions of neural circuits. In fish and insects, synaptic depression in visual and auditory pathways causes sensory adaptation and alteration in receptive fields of higher order sensory cells (O'Shea & Rowell 1976, Furukawa et al 1982). In *Aplysia*, depression at sensory to motor neuron synapses is responsible for habituation of gill withdrawal responses (Castellucci et al 1970). Synaptic depression at sensory terminals in fish, crustacea, and insects leads to habituation of escape responses to repeated stimuli (Auerbach & Bennett 1969, Zucker 1972, Zilber-Gachelin & Chartier 1973). And neuromuscular depression can weaken responses such as tail flicks in crayfish (Larimer et al 1971). In contrast, highly facilitating synapses respond effectively only to high frequency inputs. This shapes the frequency response characteristic of mammalian neurosecretory and sympathetic neurons and crustacean and amphibian peripheral synapses (Bittner 1968, Landau & Lass 1973, Dutton & Dyball 1979, Birks et al 1981). The important integrative consequences of synaptic plasticity motivate efforts to understand the underlying physiological mechanisms.

# SYNAPTIC DEPRESSION

At some synapses depression is the dominant effect of repetitive stimulation. Quantal analysis at neuromuscular junctions demonstrates that depression is due to a presynaptic reduction in the number of quanta of transmitter released by impulses (Del Castillo & Katz 1954). Depression can often be relieved by reducing the level of transmitter release, for example by reducing the external calcium concentration or adding magnesium to block calcium influx at the nerve terminal (Thies 1965). The dependence of depression on initial level of transmission suggests that it is due to a limited store of releasable transmitter, which is depleted by a train of stimuli and not instantaneously replenished. Development of depression during a train and subsequent recovery are roughly exponential (Takeuchi 1958, Mallart & Martin 1968, Betz 1970), suggesting a first order process for renewing the releasable store within seconds.

## Depletion Model

These characteristics of depression are consistent with a simple model (Figure 2) which has each action potential liberating a constant fraction of an immediately releasable store with subsequent refilling (or mobilization of replacement quanta) from a larger depot (Liley & North 1953). Only minor deviations from the predictions of this model have been observed:

1. The fraction of the store released by each impulse, as indicated by the fractional reduction in successive PSP amplitudes during a tetanus, may decline during depression (Betz 1970). Perhaps the most easily released quanta are secreted first, while those remaining are less easily released.
2. Depression in a train of impulses may be less severe than predicted from the decline of the first few PSPs (Kusano & Landau 1975), suggesting that replenishment of the releasable store is boosted (subject to extra nonlinear mobilization) by excessive release of transmitter.
3. Stimulation for several minutes often results in a second slow phase

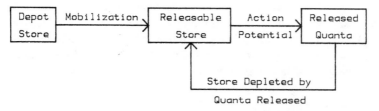

*Figure 2*    The depletion model of synaptic depression.

of depression (Birks & MacIntosh 1961, Elmqvist & Quastel 1965, Rosenthal 1969, Lass et al 1973), from which recovery also requires minutes. This may represent gradual depletion of the depot store of transmitter from which releasable quanta are mobilized.

One might expect the store of releasable quanta of transmitter to correspond to synaptic vesicles, or perhaps to those near the presynaptic membrane at release sites. However, synaptic depression develops faster and exceeds the reduction in vesicle number (Ceccarelli & Hurlbut 1980), leaving still unclear the identification of the structural correlate of the releasable store.

## Release Statistics

Transmitter release is a statistical process, in which a variable number of quanta are released by repeated action potentials (Martin 1966). The quantal number is usually well described as a binomial random variable, characterized by $n$ releasable quanta, each secreted by a spike with probability $p$ (Johnson & Wernig 1971, Bennett & Florin 1974, McLachlan 1975a, Miyamoto 1975, Wernig 1975, Furukawa et al 1978, Korn et al 1982). Synaptic depression is sometimes associated with a drop in $p$ (McLachlan 1975b, Korn et al 1982, 1984), but more often with a drop in $n$ (Barrett & Stevens 1972a, Bennett & Florin 1974, McLachlan 1975b, Furukawa & Matsuura 1978, Glavinović 1979, Smith 1983).

Interpretation of these results depends on the physiological or structural meaning assigned to the parameters $n$ and $p$. In one view (Bennett & Fisher 1977, Glavinović 1979), $n$ is thought to be a measure of the releasable store of quanta, and $p$ the fraction of this store released by a spike. Then reduction in $n$ would be expected if depression is due to depletion of the releasable store. However, since $n$ would then be reduced by depletion after each action potential, and would recover by mobilization from the depot store in the interval until the next spike, $n$ itself would be a fluctuating random variable, and would not correspond to $n$ of binomial release statistics (Vere-Jones 1966).

In another view (Zucker 1973, Wernig 1975, Bennett & Lavidis 1979, Furukawa et al 1982, Korn et al 1982, Neale et al 1983, Smith 1983), binomial release statistics are thought to arise from a fixed number of release sites ($n$), whereas $p$ is the probability that a site releases a quantum. This notion is based on a correspondence between $n$ and the number of morphological release sites observed at the same synapse. In this view, the store of releasable quanta corresponds to the fraction of release sites loaded with a quantum (probably a vesicle—see Oorschot & Jones 1987). This fraction drops during depletion, whereas $n$ remains constant. The probability $p$ that a release site releases a quantum depends both on its probability of being filled ($p_f$) and its probability of being activated by an action potential ($p_a$) (Zucker 1973), according to $p = p_f p_a$. The binomial parameter $p$ would then be less than the fraction of the releasable store activated by a spike, $p_a$. This has been observed experimentally (Christensen & Martin 1970). And only $p$ should drop during depression.

Although appealing, this simple view is not supported by evidence cited above that depression is often accompanied by a reduction in $n$. This discrepancy may arise in part from the assumptions underlying the estimation of $n$ and $p$. In particular, $p$ is assumed to be uniform across release sites (or releasable quanta). This is an extremely unlikely assumption, and is actually controverted by experimental evidence (Hatt & Smith 1976b, Bennett & Lavidis 1979, Jack et al 1981). Moreover, any variance in $p$ causes overestimation of average $p$, underestimation of $n$, and real changes in the values of $p$ to be mirrored or even overshadowed by apparent changes in $n$ (Zucker 1973, Brown et al 1976, Barton & Cohen 1977). These and other considerations (Zucker 1977) make accurate estimation of $n$ and $p$, and their direct association with structures or physiological processes, difficult at best. Thus reductions in $n$ are often thought to be loosely associated with a reduction in the proportion of release sites effectively activated by an action potential (Furukawa et al 1982, Smith 1983), due either to a depletion of quanta available to load the sites, or reduced activation of sites by partially blocked action potentials. The latter mechanism, although not usually a prominent factor in synaptic depression, has been found to be important during prolonged stimulation at some crustacean neuromuscular junctions (Parnas 1972, Hatt & Smith 1976a).

## Other Mechanisms

At some central and peripheral synapses, depression is less dependent on the level of transmission and develops with a different time course during a tetanus than predicted by depletion models (Zucker & Bruner 1977, Byrne 1982). In *Aplysia*, habituation of gill withdrawal is due to presynaptically generated depression at synapses formed by sensory neurons (Castellucci & Kandel 1974). This depression is temporally correlated with a long-lasting inactivation of presynaptic calcium current measured in the cell body (Klein et al 1980). A similar correlation has been observed at synapses between cultured spinal cord neurons (Jia & Nelson 1986). This contrasts sharply with the squid giant synapse, where synaptic depression occurs in the clear absence of calcium current inactivation (Charlton et al 1982). A recent analysis indicates that this inactivation in *Aplysia* is insufficient to account for synaptic depression. A new model (Gingrich

& Byrne 1985) proposes that transmitter depletion also contributes to depression, and postulates a calcium-dependent mobilization of transmitter to counterbalance the change in transmitter release. This could result in the independence of short-term depression of the level of transmission when calcium levels are altered.

A long-lasting form of depression at these synapses underlies the long-lasting gill withdrawal habituation to trials of stimuli repeated for several days (Castellucci et al 1978). This depression is accompanied by a reduction in number and size of transmitter release sites and the number of synaptic vesicles each contains (Bailey & Chen 1983). How short-term depression is consolidated into long-lasting morphological changes is still unknown.

Although depression normally involves only a reduction in the number of quanta released, prolonged stimulation at central synapses in fishes and at neuromuscular junctions in frogs results also in a reduction in the size of quanta released (Bennett et al 1975, Glavinović 1987). It appears that after releasable vesicles or activated release sites are strongly depleted, reloaded sites or newly formed vesicles are not entirely refilled between stimuli in a tetanus.

Finally, at some multi-action synapses, depression arises from postsynaptic desensitization of neurotransmitter receptors. In *Aplysia*, a cholinergic interneuron in the abdominal ganglion binds to excitatory and inhibitory receptors on a motoneuron to elicit a diphasic excitatory-inhibitory PSP. The excitatory receptor is subject to desensitization, so that repeated activation results in a brief excitation followed by tonic inhibition (Wachtel & Kandel 1971). Iontophoresing acetylcholine onto the postsynaptic cell has the same effect. Just the opposite situation is seen at a buccal ganglion synapse. Here it is the inhibitory cholinergic receptors that are subject to desensitization, so that the synaptic effect changes from inhibition to excitation during repeated activation (Gardner & Kandel 1977).

These examples illustrate the multifaceted nature of *short-term depression*, caused by a variety of physiological processes at different synapses, and often having interesting consequences for information processing and behavior. A more prolonged form of depression, called *long-term depression*, is treated in a separate chapter (Ito 1989).

## FACILITATION AND AUGMENTATION

Most synapses display a short-term facilitation, in which successive spikes at high frequency evoke PSPs of increasing amplitude. Depression may mask facilitation, which will then be evident only when depression is relieved by reducing the amount of transmitter released by spikes. At numerous synapses, a quantal analysis indicates that facilitation is presynaptic in origin, reflecting increasing numbers of transmitter quanta released per spike (reviewed in Zucker 1973).

## Early Theories of Facilitation

Early theories of facilitation invoked increased spike invasion of presynaptic terminals or effects of afterpotentials in nerve terminals (for reviews see Atwood 1976, Zucker 1977, Atwood & Wojtowicz 1986). The operation of such mechanisms has been refuted at central neurons (Charlton & Bittner 1978), peripheral neurons (Martin & Pilar 1964), and neuromuscular junctions (Hubbard 1963, Braun & Schmidt 1966, Zucker 1974a,c). Another hypothesis holds that spike broadening in nerve terminals, due to inactivation of potassium currents (Aldrich et al 1979), causes facilitation by increasing the calcium influx to successive action potentials (Gainer 1978, Andrew & Dudek 1985, Cooke 1985). Surprisingly, however, spike broadening in molluscan neurons is not accompanied by a measurable increase in calcium influx (Smith & Zucker 1980), and it is not involved in facilitation at crayfish neuromuscular junctions (Zucker & Lara-Estrella 1979, Bittner & Baxter 1983). Finally, synaptic facilitation could arise from a facilitated activation of calcium channels (Zucker 1974b), as has been observed in chromaffin cells (Hoshi et al 1984). However, calcium channels in *Aplysia* neurons (Smith & Zucker 1980) and at presynaptic terminals of squid synapses (Charlton et al 1982) exhibit no such facilitation to repeated depolarization.

## Residual Calcium Hypothesis

At present, the residual calcium hypothesis of Katz & Miledi (Katz & Miledi 1968, Miledi & Thies 1971, H. Parnas et al 1982) enjoys the greatest popularity among synaptic physiologists. They propose that facilitation is the natural consequence of a nonlinear dependence of transmitter release upon intracellular calcium activity and the probability that after a presynaptic action potential some residual calcium will persist at sites of transmitter release (Figure 3).

To be more specific, transmitter release varies with about the fourth power of external calcium concentration at several synapses (Dodge & Rahamimoff 1967, Hubbard et al 1968, Katz & Miledi 1970, Dudel 1981). It has been argued that this measure will underestimate the cooperativity of calcium action (Parnas & Segel 1981, Barton et al 1983), so we will assume that transmitter release is determined by the fifth power of calcium concentration at release sites. Perhaps vesicle exocytosis requires the binding of several calcium ions to sites on the vesicular or plasma membrane.

RESIDUAL CALCIUM MODEL
OF SYNAPTIC FACILITATION

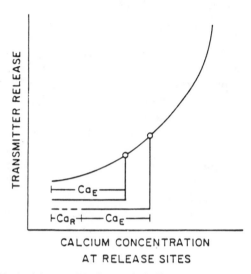

CALCIUM CONCENTRATION
AT RELEASE SITES

*Figure 3*   The residual calcium model of synaptic facilitation. Calcium entering in a spike ($Ca_E$) summates with residual calcium from prior activity ($Ca_R$) to release more transmitter than in the absence of prior activity. The nonlinear dependence of release on calcium causes $Ca_R$ alone to release little transmitter.

Let the peak calcium concentration at release sites reach one unit during an action potential. Imagine that 10 ms later the calcium concentration has dropped to 0.05 unit. This residual calcium should release transmitter at a rate of $(0.05)^5$ or one three-millionth the rate of transmitter release during the spike. At frog neuromuscular junctions in low calcium solution, a spike releases about 1 quantum in 1 ms, so the residual calcium 10 ms after the spike should increase spontaneous release about $3 \times 10^{-7}$ times 1000/s, or about 1 quantum/hr. A second action potential at this time will generate a peak calcium concentration at release sites of 1.05, which when raised to the fifth power will release 28% more quanta than did the first spike. Once having worked through such calculations, it is difficult to imagine that residual calcium would not lead to facilitation in this way.

## Experimental Support

Calculations like those in the preceding paragraph show that after a tetanus in which residual calcium at release sites may reach 20% of its peak in the first spike, a facilitation of 660% will occur in the presence of an acceleration of miniature PSP frequency (spontaneous release of quanta) of

31/s. Such a correlation between facilitation of PSP amplitude and increase in miniature PSP frequency has been observed in several (Miledi & Thies 1971, Barrett & Stevens 1972b, Zucker & Lara-Estrella 1983) experiments using brief tetani. When the normal calcium gradient across the presynaptic membrane is reversed by removing all extracellular calcium, similar tetani cause a fall in miniature PSP frequency, presumably because of a drop in internal calcium as calcium exits through open calcium channels (Erulkar & Rahamimoff 1978).

The residual calcium hypothesis receives more direct support from three other sets of experiments:

1. Calcium is required for facilitation: Katz & Miledi (1968) showed that not only transmitter release but also facilitation requires calcium in the external medium. When they raised calcium after a conditioning impulse but before a test impulse, the first spike released no transmitter and also caused no facilitation of release to the second spike. One might conclude that the first spike must release transmitter in order to facilitate release to subsequent spikes. However, transmitter release fluctuates from spike to spike, and sometimes failures (releases of zero quanta) occur. Spikes releasing no transmitter cause as much facilitation as spikes that do release transmitter (Del Castillo & Katz 1954, Dudel & Kuffler 1961). Apparently, calcium entry during the first spike causes facilitation whether or not transmitter is released by the first spike.

2. Calcium elicits facilitation: Raising presynaptic calcium by fusing calcium-containing liposomes with presynaptic terminals (Rahamimoff et al 1978), poisoning calcium sequestering organelles (Alnaes & Rahamimoff 1975), or injecting calcium directly into terminals (Charlton et al 1982) facilitates transmitter release by action potentials.

3. Residual calcium accumulates during repeated activity: Calcium concentration in presynaptic terminals is seen to increase about ten-fold during a tetanus of 50 spikes, when it is measured spectrophotometrically with the indicator dye arsenazo III (Miledi & Parker 1981, Charlton et al 1982).

## Residual Calcium Kinetics

Augmentation appears to be a longer lasting form of facilitation arising from similar mechanisms. It has been observed at neuromuscular junctions in frogs and synapses in sympathetic ganglia in rabbits, cerebral cortex in rats, and at central synapses in *Aplysia* (Magleby & Zengel 1976, Zengel et al 1980, Kretz et al 1982, Racine & Milgram 1983). A slow phase in increased miniature PSP frequency is also seen that corresponds to this phase of increased evoked transmitter release (Zengel & Magleby 1981). Like facilitation, augmentation requires calcium entry, since tetani in calcium-free media do not elicit this increase in miniature PSP frequency of duration intermediate between facilitation and potentiation (Erulkar & Rahamimoff 1978).

The time course of the growth of facilitation and augmentation in a tetanus and its subsequent decline have received much attention. It was originally proposed that each impulse in a train added a constant increment of facilitation that decayed with two or more exponential components (Mallart & Martin 1967). This description is inadequate except for very brief tetani (Magleby 1973a, Linder 1974, Zucker 1974b, Bittner & Sewell 1976). A better fit to facilitation and augmentation is obtained by assuming that each impulse contributes an equal increment of residual calcium to a presynaptic compartment regulating transmitter release, that calcium is removed from this compartment by processes approximated as the sum of three exponentials (two for facilitation and one for augmentation), and that transmitter release is proportional to the fourth or higher power of calcium concentration in this compartment (Zengel & Magleby 1982).

## Physical Models of Residual Calcium Kinetics

Recent attempts have been made to formulate physical models to explain the magnitude and time course of residual calcium at release sites necessary to account for facilitation and augmentation. Calcium crosses the presynaptic membrane into nerve terminals during action potentials (Llinás et al 1981, 1982) and acts at the surface to release transmitter. Calcium is bound to axoplasmic proteins (Alemà et al 1973, Brinley 1978) and diffuses toward the interior of the terminal after each spike, where it can no longer affect transmitter release. Finally, calcium is taken up into organelles (Blaustein et al 1978) and extruded by surface membrane pumps (Requena & Mullins 1979, I. Parnas et al 1982). The diffusion equation may be solved in cylindrical coordinates with boundary conditions imposed by measured rates of influx, binding, uptake, and extrusion (Alemà et al 1973, Blaustein et al 1978, Brinley 1978, Requena & Mullins 1979) to predict the magnitude and time course of intracellular calcium gradients during and after nervous activity. Transmitter release may be calculated from a power-law dependence upon calcium concentration at release sites.

The first simulations of these physical constraints used a one-dimensional model of radial calcium diffusion away from the surface and assumed uniform calcium influx across the membrane (Zucker & Stockbridge 1983, Stockbridge & Moore 1984). The time course and magnitude of facilitation following one spike at squid giant synapses and frog neuromuscular junctions were predicted reasonably accurately, as well as the tetanic accumulation of calcium and its decay as measured spectrophotometrically and the time course of spike-evoked transmitter release as measured electrophysiologically (Zucker & Stockbridge 1983, Fogelson & Zucker 1985). However, these simulations predicted too high a post-tetanic residual calcium compared to peak submembrane calcium in a single spike (Fogelson & Zucker 1985).

This defect was remedied in a subsequent model (Fogelson & Zucker 1985) in which calcium enters through an array of discrete channels and releases transmitter from release sites near these channels. The brief synaptic delay from calcium influx to transmitter release (0.2 ms) requires that transmitter release occur near calcium channels before calcium equilibrates at the surface (Simon & Llinás 1985), when distinct clouds of calcium ions still surround each open channel. After a spike, calcium diffuses in three dimensions away from each channel, and away from the clusters of channels, vesicles, and release sites called *active zones* (Pumplin et al 1981). The peak calcium concentration at release sites in active zones in such a model is much higher than in the simpler, one-dimensional diffusion model, and even after a tetanus the residual calcium never reaches this level. Simulations using this model provide a quantitatively better, although still imperfect, fit to data on phasic transmitter release, accumulation of presynaptic calcium, and facilitation and augmentation at squid synapses and neuromuscular junctions.

These simulations demonstrate that diffusion of calcium away from release sites will resemble a multi-exponential time course. This is because diffusion follows a second-order differential equation. Therefore, the existence of multiple exponentials in descriptions of the kinetics of facilitation and augmentation does not indicate that these necessarily reflect independent processes. Changes in a single parameter, such as cytoplasmic calcium binding, have unequal effects on the different apparent exponential components of facilitation. However, it is true that changes in cytoplasmic binding affect mainly the fast process of facilitation through effects on diffusion, whereas changes in calcium uptake or extrusion affect mainly the slower process of augmentation in these simulations. Substituting strontium for calcium prolongs mainly the slow component of facilitation, while addition of barium accentuates augmentation (Zengel & Magleby 1980). It is possible that strontium binds differently than calcium to cytoplasmic proteins, while barium interferes with extrusion or uptake pumps.

## Release Statistics

As with synaptic depression, facilitation and potentiation are accompanied by changes in the binomial release parameters $n$ and $p$. In different preparations, apparent increases are observed mainly in $p$ (Zucker 1973, Hirst et al 1981), mainly in $n$ (Bennett & Florin 1974, McLachlan 1975a,

Branisteanu et al 1976, Wojtowicz & Atwood 1986), and in both $p$ and $n$ (Wernig 1972, Smith 1983). These results are all consistent with transmitter release occurring at release sites with nonuniform probabilities of activation. Both facilitation and potentiation might cause release sites to be more effectively activated by spikes. Whether this will be expressed mainly as an increase in $n$ or in $p$ depends on the exact form of the distribution of the values of $p$ among release sites.

## POTENTIATION

Potentiation is an increase in efficacy of transmission requiring minutes for its development and decay at synapses in sympathetic ganglia, olfactory and hippocampal cortex, and *Aplysia* ganglia (Waziri et al 1969, Richards 1972, Magleby & Zengel 1975, Atwood 1976, Schlapfer et al 1976, Zengel et al 1980, Racine & Milgram 1983). At crustacean neuromuscular junctions, quantal analysis shows potentiation to be presynaptic in origin (Baxter et al 1985, Wojtowicz & Atwood 1986). Unlike facilitation and augmentation, post-tetanic potentiation decays more slowly following tetani of longer duration or higher frequency (Magleby & Zengel 1975, Schlapfer et al 1976).

Potentiation appears to arise from two sources. It is reduced but not abolished by stimulation in a calcium-free medium (Rosenthal 1969, Weinreich 1971, Erulkar & Rahamimoff 1978). This suggests that potentiation is partly due to slow phases of removal of calcium that entered through calcium channels. Perhaps calcium pumps become saturated, or energy stores are limiting, in high calcium loads. The decay of PTP resembles that of post-tetanic calcium-activated potassium current and spectrophotometrically measured presynaptic calcium activity in *Aplysia* neurons (Kretz et al 1982, Connor et al 1986), a finding again suggesting that PTP reflects a late component in removal of residual calcium. The existence of a transition temperature in the decay kinetics of PTP (Schlapfer et al 1975) and the influence of alcohol on this decay rate (Woodson et al 1976) suggest that potentiation depends on some membrane process, such as calcium uptake into endoplasmic reticulum or its extrusion by surface pumps.

At neuromuscular junctions, part of potentiation is independent of calcium entry during a tetanus. This part is enhanced by treatments that augment sodium loading of nerve terminals, such as blocking the sodium pump with ouabain, and is reduced when sodium loading is minimized in low sodium media (Birks & Cohen 1968a,b, Atwood 1976). Transmitter release can be potentiated by exposing junctions to sodium-containing liposomes (Rahamimoff et al 1978), introducing sodium with ionophores

(Meiri et al 1981, Atwood et al 1983), and injecting sodium into nerve terminals (Charlton & Atwood 1977, Wojtowicz & Atwood 1985). It has been proposed that sodium that accumulates presynaptically during a tetanus potentiates transmitter release by displacing calcium from intracellular stores (Rahamimoff et al 1980) or reducing calcium extrusion by Na/Ca exchange (Misler & Hurlbut 1983). Lithium and rubidium Ringers enhance potentiation, presumably by blocking Na/Ca exchange (Misler et al 1987).

These results suggest that potentiation might be viewed as another consequence of increased residual calcium, dependent in part upon sodium accumulation. However, if this were the whole story, potentiation would summate with facilitation and augmentation. When this point has been examined, however, the interaction of potentiation with facilitation has appeared more multiplicative than additive (Landau et al 1973, Magleby & Zengel 1982). This suggests that another site of action of presynaptic calcium may also be involved in potentiation (Figure 4). Recently, Llinás et al (1985) have found that a calcium-dependent phosphorylation of presynaptic synapsin I, a synaptic vesicle protein, can potentiate transmitter release. It is possible that a calcium-dependent mobilization of transmitter mediated by this protein plays a role in PTP.

## CONCLUSION

This concludes my brief survey of processes involved in short-term synaptic plasticity. Synaptic efficacy is a highly plastic variable, subject to numerous

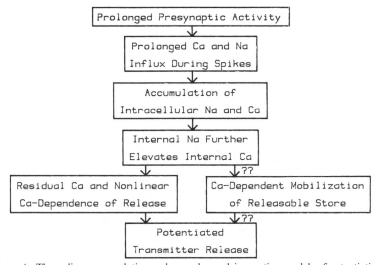

*Figure 4*   The sodium accumulation and secondary calcium action models of potentiation.

pre- and postsynaptic modulations affected by prior activity. These processes shape dramatically the pattern selectivity of synapses and the information transfer they mediate. Sensory phenomena such as adaptation and dynamic versus static sensitivity often arise from synaptic processes like depression and facilitation. These synaptic qualities are also expressed behaviorally as habituation and in the recruitment of elements in a pool of target neurons. Longer lasting processes such as long-term potentiation or depression build on these shorter processes to span the gap between synaptic plasticity and permanent structural changes involved in long-term memory. As processes providing clues to the basic mechanisms underlying synaptic transmission, the various forms of short-term synaptic plasticity promise to remain popular topics of intensive research.

ACKNOWLEDGMENT

My research is supported by National Institutes of Health Grant NS 15114.

*Literature Cited*

Aldrich, R. W. Jr., Getting, P. A., Thompson, S. H. 1979. Mechanism of frequency-dependent broadening of molluscan neurone soma spikes. *J. Physiol.* 291: 531–44

Alemà, S., Calissano, P., Rusca, G., Giuditta, A. 1973. Identification of a calcium-binding, brain specific protein in the axoplasm of squid giant axons. *J. Neurochem.* 20: 681–89

Alnaes, E., Rahamimoff, R. 1975. On the role of mitochondria in transmitter release from motor nerve terminals. *J. Physiol.* 248: 285–306

Andrew, R. D., Dudek, F. E. 1985. Spike broadening in magnocellular neuroendocrine cells of rat hypothalamic slices. *Brain Res.* 334: 176–79

Atwood, H. L. 1976. Organization and synaptic physiology of crustacean neuromuscular systems. *Prog. Neurobiol.* 7: 291–391

Atwood, H. L., Charlton, M. P., Thompson, C. S. 1983. Neuromuscular transmission in crustaceans is enhanced by a sodium ionophore, monensin, and by prolonged stimulation. *J. Physiol.* 335: 179–95

Atwood, H. L., Wojtowicz, J. M. 1986. Short-term and long-term plasticity and physiological differentiation of crustacean motor synapses. *Int. Rev. Neurobiol.* 28: 275–362

Auerbach, A. A., Bennett, M. V. L. 1969. Chemically mediated transmission at a giant fiber synapse in the central nervous system of a vertebrate. *J. Gen. Physiol.* 53: 183–210

Bailey, C. H., Chen, M. 1983. Morphological basis of long-term habituation and sensitization in *Aplysia. Science* 220: 91–93

Barrett, E. F., Stevens, C. F. 1972a. Quantal independence and uniformity of presynaptic release kinetics at the frog neuromuscular junction. *J. Physiol.* 227: 665–89

Barrett, E. F., Stevens, C. F. 1972b. The kinetics of transmitter release at the frog neuromuscular junction. *J. Physiol.* 227: 691–708

Barton, S. B., Cohen, I. S. 1977. Are transmitter release statistics meaningful? *Nature* 268: 267–68

Barton, S. B., Cohen, I. S., van der Kloot, W. 1983. The calcium dependence of spontaneous and evoked quantal release at the frog neuromuscular junction. *J. Physiol.* 337: 735–51

Baxter, D. A., Bittner, G. D., Brown, T. H. 1985. Quantal mechanism of long-term synaptic potentiation. *Proc. Natl. Acad. Sci. USA* 82: 5978–82

Bennett, M. R., Fisher, C. 1977. The effect of calcium ions on the binomial parameters that control acetylcholine release during trains of nerve impulses at amphibian neuromuscular synapses. *J. Physiol.* 271: 673–98

Bennett, M. R., Florin, T. 1974. A statistical analysis of the release of acetylcholine at newly formed synapses in striated muscle. *J. Physiol.* 238: 93–107

Bennett, M. R., Lavidis, N. A. 1979. The effect of calcium ions on the secretion of quanta evoked by an impulse at nerve terminal release sites. *J. Gen. Physiol.* 74: 429–56

Bennett, M. V. L., Model, P. G., Highstein, S. M. 1975. Stimulation-induced depletion of vesicles, fatigue of transmission and recovery processes at a vertebrate central synapse. *Cold Spring Harbor Symp. Quant. Biol.* 40: 25–35

Betz, W. J. 1970. Depression of transmitter release at the neuromuscular junction of the frog. *J. Physiol.* 206: 629–44

Birks, R. I., Cohen, M. W. 1968a. The action of sodium pump inhibitors on neuromuscular transmission. *Proc. R. Soc. London. Ser. B* 170: 381–99

Birks, R. I., Cohen, M. W. 1968b. The influence of internal sodium on the behaviour of motor nerve endings. *Proc. R. Soc. London Ser. B* 170: 401–21

Birks, R. I., Laskey, W., Polosa, C. 1981. The effect of burst patterning of preganglionic input on the efficacy of transmission at the cat stellate ganglion. *J. Physiol.* 318: 531–39

Birks, R. I., MacIntosh, F. C. 1961. Acetylcholine metabolism of a sympathetic ganglion. *Can. J. Biochem. Physiol.* 39: 787–827

Bittner, G. D. 1968. Differentiation of nerve terminals in the crayfish opener muscle and its functional significance. *J. Gen. Physiol.* 51: 731–58

Bittner, G. D., Baxter, D. A. 1983. Intracellular recordings from synaptic terminals during facilitation of transmitter release. *Soc. Neurosci. Abstr.* 9: 883

Bittner, G. D., Sewell, V. L. 1976. Facilitation at crayfish neuromuscular junctions. *J. Comp. Physiol.* 109: 287–308

Blaustein, M. P., Ratzlaff, R. W., Schweitzer, E. S. 1978. Calcium buffering in presynaptic nerve terminals. II. Kinetic properties of the nonmitochondrial Ca sequestration mechanism. *J. Gen. Physiol.* 72: 43–66

Branisteanu, D. D., Miyamoto, M. D., Volle, R. L. 1976. Effects of physiologic alterations on binomial transmitter release at magnesium-depressed neuromuscular junctions. *J. Physiol.* 254: 19–37

Braun, M., Schmidt, R. F. 1966. Potential changes recorded from the frog motor nerve terminal during its activation. *Pflügers Arch.* 287: 56–80

Brinley, F. J. Jr. 1978. Calcium buffering in squid axons. *Ann. Rev. Biophys. Bioeng.* 7: 363–92

Brown, T. H., Ganong, A. H., Kairiss, E. W.,

Keenan, C. L. 1989. Hebbian synapses— Computations and biophysical mechanisms. *Ann. Rev. Neurosci.* 13: Submitted

Brown, T. H., Perkel, D. H., Feldman, M. W. 1976. Evoked neurotransmitter release: Statistical effects of nonuniformity and nonstationarity. *Proc. Natl. Acad. Sci. USA* 73: 2913–17

Byrne, J. H. 1982. Analysis of synaptic depression contribution to habituation of gill-withdrawal reflex in *Aplysia californica. J. Neurophysiol.* 48: 431–38

Castellucci, V. F., Carew, T. J., Kandel, E. R. 1978. Cellular analysis of long-term habituation of the gill-withdrawal reflex of *Aplysia californica. Science* 202: 1306–8

Castellucci, V. F., Kandel, E. R. 1974. A quantal analysis of the synaptic depression underlying habituation of the gill-withdrawal reflex in *Aplysia. Proc. Natl. Acad. Sci. USA* 71: 5004–8

Castellucci, V., Pinsker, H., Kupfermann, I., Kandel, E. R. 1970. Neuronal mechanisms of habituation and dishabituation of the gill-withdrawal reflex in Aplysia. *Science* 167: 1745–48

Ceccarelli, B., Hurlbut, W. P. 1980. Vesicle hypothesis of the release of quanta of acetylcholine. *Physiol. Rev.* 60: 396–441

Charlton, M. P., Atwood, H. L. 1977. Modulation of transmitter release by intracellular sodium in squid giant synapse. *Brain Res.* 134: 367–71

Charlton, M. P., Bittner, G. D. 1978. Presynaptic potentials and facilitation of transmitter release in the squid giant synapse. *J. Gen. Physiol.* 72: 487–511

Charlton, M. P., Smith, S. J., Zucker, R. S. 1982. Role of presynaptic calcium ions and channels in synaptic facilitation and depression at the squid giant synapse. *J. Physiol.* 323: 173–93

Christensen, B. N., Martin, A. R. 1970. Estimates of probability of transmitter release at the mammalian neuromuscular junction. *J. Physiol.* 210: 933–45

Connor, J. A., Kretz, R., Shapiro, E. 1986. Calcium levels measured in a presynaptic neurone of *Aplysia* under conditions that modulate transmitter release. *J. Physiol.* 375: 625–42

Cooke, I. M. 1985. Electrophysiological characterization of peptidergic neurosecretory terminals. *J. Exp. Biol.* 118: 1–35

Del Castillo, J., Katz, B. 1954. Statistical factors involved in neuromuscular facilitation and depression. *J. Physiol.* 124: 574–85

Dodge, F. A. Jr., Rahamimoff, R. 1967. Cooperative action of calcium ions in transmitter release at the neuromuscular junction. *J. Physiol.* 193: 419–32

Dudel, J. 1981. The effect of reduced calcium on quantal unit current and release at the crayfish neuromuscular junction. *Pflügers Arch.* 391: 35–40

Dudel, J., Kuffler, S. W. 1961. Mechanism of facilitation at the crayfish neuromuscular junction. *J. Physiol.* 155: 530–42

Dutton, A., Dyball, R. E. J. 1979. Phasic firing enhances vasopressin release from the rat neurohypophysis. *J. Physiol.* 290: 433–40

Elmqvist, D., Quastel, D. M. J. 1965. A quantitative study of end-plate potentials in isolated human muscle. *J. Physiol.* 178: 505–29

Erulkar, S. D., Rahamimoff, R. 1978. The role of calcium ions in tetanic and post-tetanic increase of miniature end-plate potential frequency. *J. Physiol.* 278: 501–11

Fogelson, A. L., Zucker, R. S. 1985. Presynaptic calcium diffusion from various arrays of single channels. Implications for transmitter release and synaptic facilitation. *Biophys. J.* 48: 1003–17

Furukawa, T., Hayashida, Y., Matsuura, S. 1978. Quantal analysis of the size of excitatory post-synaptic potentials at synapses between hair cells and afferent nerve fibres in goldfish. *J. Physiol.* 276: 211–26

Furukawa, T., Kuno, M., Matsuura, S. 1982. Quantal analysis of a decremental response at hair cell-afferent fibre synapses in the goldfish sacculus. *J. Physiol.* 322: 181–95

Furukawa, T., Matsuura, S. 1978. Adaptive rundown of excitatory post-synaptic potentials at synapses between hair cells and eighth nerve fibres in the goldfish. *J. Physiol.* 276: 193–209

Gainer, H. 1978. Input-output relations of neurosecretory cells. In *Comparative Endocrinology*, ed. P. J. Gaillard, H. H. Boar, pp. 293–304. Amsterdam: Elsevier

Gardner, D., Kandel, E. R. 1977. Physiological and kinetic properties of cholinergic receptors activated by multiaction interneurons in buccal ganglia of *Aplysia*. *J. Neurophysiol.* 40: 333–48

Gingrich, K. J., Byrne, J. H. 1985. Simulation of synaptic depression, posttetanic potentiation, and presynaptic facilitation of synaptic potentials from sensory neurons mediating gill-withdrawal reflex in *Aplysia*. *J. Neurophysiol.* 53: 652–69

Glavinović, M. I. 1979. Change of statistical parameters of transmitter release during various kinetic tests in unparalysed voltage-clamped rat diaphragm. *J. Physiol.* 290: 481–97

Glavinović, M. I. 1987. Synaptic depression in frog neuromuscular junction. *J. Neurophysiol.* 58: 230–46

Hatt, H., Smith, D. O. 1976a. Synaptic depression related to presynaptic axon conduction block. *J. Physiol.* 259: 367–93

Hatt, H., Smith, D. O. 1976b. Non-uniform probabilities of quantal release at the crayfish neuromuscular junction. *J. Physiol.* 259: 395–404

Hirst, G. D. S., Redman, S. J., Wong, K. 1981. Post-tetanic potentiation and facilitation of synaptic potentials evoked in cat spinal motoneurones. *J. Physiol.* 321: 97–109

Hoshi, T., Rothlein, J., Smith, S. J. 1984. Facilitation of $Ca^{2+}$-channel currents in bovine adrenal chromaffin cells. *Proc. Natl. Acad. Sci. USA* 81: 5871–75

Hubbard, J. I. 1963. Repetitive stimulation at the neuromuscular junction, and the mobilization of transmitter. *J. Physiol.* 169: 641–62

Hubbard, J. I., Jones, S. F., Landau, E. M. 1968. On the mechanism by which calcium and magnesium affect the release of transmitter by nerve impulses. *J. Physiol.* 196: 75–87

Ito, M. 1989. Long-term depression. *Ann. Rev. Neurosci.* 12: 85–102

Jack, J. J. B., Redman, S. J., Wong, K. 1981. The components of synaptic potentials evoked in cat spinal motoneurones by impulses in single group Ia afferents. *J. Physiol.* 321: 65–96

Jia, M., Nelson, P. G. 1986. Calcium currents and transmitter output in cultured spinal cord and dorsal root ganglion neurons. *J. Neurophysiol.* 56: 1257–67

Johnson, E. W., Wernig, A. 1971. The binomial nature of transmitter release at the crayfish neuromuscular junction. *J. Physiol.* 218: 757–67

Katz, B., Miledi, R. 1968. The role of calcium in neuromuscular facilitation. *J. Physiol.* 195: 481–92

Katz, B., Miledi, R. 1970. Further study of the role of calcium in synaptic transmission. *J. Physiol.* 207: 789–801

Klein, M., Shapiro, E., Kandel, E. R. 1980. Synaptic plasticity and the modulation of the $Ca^{2+}$ current. *J. Exp. Biol.* 89: 117–57

Korn, H., Faber, D. S., Burnod, Y., Triller, A. 1984. Regulation of efficacy at central synapses. *J. Neurosci.* 4: 125–30

Korn, H., Mallet, A., Triller, A., Faber, D. S. 1982. Transmission at a central inhibitory synapse. II. Quantal description of release, with a physical correlate for binomial n. *J. Neurophysiol.* 48: 679–707

Kretz, R., Shapiro, E., Kandel, E. R. 1982. Post-tetanic potentiation at an identified synapse in *Aplysia* is correlated with a $Ca^{2+}$-activated $K^+$ current in the presynaptic neuron: evidence for $Ca^{2+}$

accumulation. *Proc. Natl. Acad. Sci. USA* 79: 5430–34

Kusano, K., Landau, E. M. 1975. Depression and recovery of transmission at the squid giant synapse. *J. Physiol.* 245: 13–32

Landau, E. M., Lass, Y. 1973. Synaptic frequency response: The influence of sinusoidal changes in stimulation frequency on the amplitude of the end-plate potential. *J. Physiol.* 228: 27–40

Landau, E. M., Smolinsky, A., Lass, Y. 1973. Post-tetanic potentiation and facilitation do not share a common calcium-dependent mechanism. *Nature New Biol.* 244: 155–57

Lass, Y., Halevi, Y., Landau, E. M., Gitter, S. 1973. A new model for transmitter mobilization in the frog neuromuscular junction. *Pflügers Arch.* 343: 157–63

Larimer, J. L., Eggleston, A. C., Masukawa, L. M., Kennedy, D. 1971. The different connections and motor outputs of lateral and medial giant fibres in the crayfish. *J. Exp. Biol.* 54: 391–402

Liley, A. W., North, K. A. K. 1953. An electrical investigation of effects of repetitive stimulation on mammalian neuromuscular junction. *J. Neurophysiol.* 16: 509–27

Linder, T. M. 1974. The accumulative properties of facilitation at crayfish neuromuscular synapses. *J. Physiol.* 238: 223–34

Llinás, R., McGuinness, T. L., Leonard, C. S., Sugimori, M., Greengard, P. 1985. Intraterminal injection of synapsin I or calcium/calmodulin-dependent protein kinase II alters neurotransmitter release at the squid giant synapse. *Proc. Natl. Acad. Sci. USA* 82: 3035–39

Llinás, R., Steinberg, I. Z., Walton, K. 1981. Relationship between presynaptic calcium current and postsynaptic potential in squid giant synapse. *Biophys. J.* 33: 323–52

Llinás, R., Sugimori, M., Simon, S. M. 1982. Transmission by presynaptic spike-like depolarization in the squid giant synapse. *Proc. Natl. Acad. Sci. USA* 79: 2415–19

Magleby, K. L. 1973a. The effect of repetitive stimulation on facilitation of transmitter release at the frog neuromuscular junction. *J. Physiol.* 234: 327–52

Magleby, K. L. 1973b. The effect of tetanic and post-tetanic potentiation on facilitation of transmitter release at the frog neuromuscular junction. *J. Physiol.* 234: 353–71

Magleby, K. L., Zengel, J. E. 1975. A quantitative description of tetanic and post-tetanic potentiation of transmitter release

at the frog neuromuscular junction. *J. Physiol.* 245: 183–208

Magleby, K. L., Zengel, J. E. 1976. Augmentation: A process that acts to increase transmitter release at the frog neuromuscular junction. *J. Physiol.* 257: 449–70

Magleby, K. L., Zengel, J. E. 1982. A quantitative description of stimulation-induced changes in transmitter release at the frog neuromuscular junction. *J. Gen. Physiol.* 30: 613–38

Mallart, A., Martin, A. R. 1967. An analysis of facilitation of transmitter release at the neuromuscular junction of the frog. *J. Physiol.* 193: 679–94

Mallart, A., Martin, A. R. 1968. The relation between quantum content and facilitation at the neuromuscular junction of the frog. *J. Physiol.* 196: 593–604

Martin, A. R. 1966. Quantal nature of synaptic transmission. *Physiol. Rev.* 46: 51–66

Martin, A. R., Pilar, G. 1964. Presynaptic and post-synaptic events during post-tetanic potentiation and facilitation in the avian ciliary ganglion. *J. Physiol.* 175: 17–30

Meiri, H., Erulkar, S. D., Lerman, T., Rahamimoff, R. 1981. The action of the sodium ionophore, monensin, on transmitter release at the frog neuromuscular junction. *Brain Res.* 204: 204–8

McLachlan, E. M. 1975a. An analysis of the release of acetylcholine from preganglionic nerve terminals. *J. Physiol.* 245: 447–66

McLachlan, E. M. 1975b. Changes in statistical release parameters during prolonged stimulation of preganglionic nerve terminals. *J. Physiol.* 253: 477–91

Miledi, R., Parker, I. 1981. Calcium transients recorded with arsenazo III in the presynaptic terminal of the squid giant synapse. *Proc. R,. Soc. London Ser. B* 212: 197–211

Miledi, R., Thies, R. 1971. Tetanic and post-tetanic rise in frequency of miniature end-plate potentials in low-calcium solutions. *J. Physiol.* 212: 245–57

Misler, S., Falke, L., Martin, S. 1987. Cation dependence of posttetanic potentiation of neuromuscular transmission. *Amer. J. Physiol.* 252: C55–C62

Misler, S., Hurlbut, W. P. 1983. Post-tetanic potentiation of acetylcholine release at the frog neuromuscular junction develops after stimulation in $Ca^{2+}$-free solutions. *Proc. Natl. Acad. Sci. USA* 80: 315–19

Miyamoto, M. D. 1975. Binomial analysis of quantal transmitter release at glycerol treated frog neuromuscular junctions. *J. Physiol.* 250: 121–42

Neale, E. A., Nelson, P. G., Macdonald, R.

L., Christian, C. N., Bowers, L. M. 1983. Synaptic interactions between mammalian central neurons in cell culture. III. Morphophysiological correlates of quantal synaptic transmission. *J. Neurophysiol.* 49: 1459–68

Oorschot, D. E., Jones, D. G. 1987. The vesicle hypothesis and its alternatives: A critical assessment. *Curr. Top. Res. Synapses* 4: 85–153

O'Shea, M., Rowell, C. H. F. 1976. The neuronal basis of a sensory analyser, the acridid movement detector system. II. Response decrement, convergence, and the nature of the excitatory afferents to the fan-like dendrites of the LGMD. *J. Exp. Biol.* 65: 289–308

Parnas, H., Dudel, J., Parnas, I. 1982. Neurotransmitter release and its facilitation in crayfish. I. Saturation kinetics of release, and of entry and removal of calcium. *Pflügers Arch.* 393: 1–14

Parnas, H., Segel, L. A. 1981. A theoretical study of calcium entry in nerve terminals, with application to neurotransmitter release. *J. Theoret. Biol.* 91: 125–169

Parnas, I. 1972. Differential block at high frequency of branches of a single axon innervating two muscles. *J. Neurophysiol.* 35: 903–14

Parnas, I., Parnas, H., Dudel, J. 1982. Neurotransmitter release and its facilitation in crayfish. II. Duration of facilitation and removal processes of calcium from the terminal. *Pflügers Arch.* 393: 232–36

Pumplin, D. W., Reese, T. S., Llinás, R. 1981. Are the presynaptic membrane particles the calcium channels? *Proc. Natl. Acad. Sci. USA* 78: 7210–13

Racine, R. J., Milgram, N. W. 1983. Short-term potentiation phenomena in the rat limbic forebrain. *Brain Res.* 260: 201–16

Rahamimoff, R., Lev-tov, A., Meiri, H. 1980. Primary and secondary regulation of quantal transmitter release: Calcium and sodium. *J. Exp. Biol.* 89: 5–18

Rahamimoff, R., Meiri, H., Erulkar, S. D., Barenholz, Y. 1978. Changes in transmitter release induced by ion containing liposomes. *Proc. Natl. Acad. Sci. USA* 75: 5214–16

Requena, J., Mullins, L. J. 1979. Calcium movement in nerve fibres. *Q. Rev. Biophys.* 12: 371–460

Richards, C. D. 1972. Potentiation and depression of synaptic transmission in the olfactory cortex of the guinea-pig. *J. Physiol.* 222: 209–31

Rosenthal, J. 1969. Post-tetanic potentiation at the neuromuscular junction of the frog. *J. Physiol.* 203: 121–33

Schlapfer, W. T., Tremblay, J. P., Woodson,

P. B. J., Barondes, S. H. 1976. Frequency facilitation and post-tetanic potentiation of a unitary synaptic potential in *Aplysia californica* are limited by different processes. *Brain Res.* 109: 1–20

Schlapfer, W. T., Woodson, P. B. J., Smith, G. A., Tremblay, J. P., Barondes, S. H. 1975. Marked prolongation of post-tetanic potentiation at a transition temperature and its adaptation. *Nature* 258: 623–25

Simon, S. M., Llinás, R. R. 1985. Compartmentalization of the submembrane calcium activity during calcium influx and its significance in transmitter release. *Biophys. J.* 48: 485–98

Smith, S. J., Zucker, R. S. 1980. Aequorin response facilitation and intracellular calcium accumulation in molluscan neurones. *J. Physiol.* 300: 167–96

Smith, D. O. 1983. Variable activation of synaptic release sites at the neuromuscular junction. *Exp. Neurol.* 80: 520–28

Stockbridge, N., Moore, J. W. 1984. Dynamics of intracellular calcium and its possible relationship to phasic transmitter release and facilitation at the frog neuromuscular junction. *J. Neurosci.* 4: 803–11

Takeuchi, A. 1958. The long-lasting depression in neuromuscular transmission of frog. *Jpn. J. Physiol.* 8: 102–13

Thies, R. E. 1965. Neuromuscular depression and the apparent depletion of transmitter in mammalian muscle. *J. Neurophysiol.* 28: 427–42

Vere-Jones, D. 1966. Simple stochastic models for the release of quanta of transmitter from a nerve terminal. *Aust. J. Statist.* 8: 53–63

Wachtel, H., Kandel, E. R. 1971. Conversion of synaptic excitation to inhibition at a dual chemical synapse. *J. Neurophysiol.* 34: 56–68

Waziri, R., Kandel, E. R., Frazier, W. T. 1969. Organization of inhibition in abdominal ganglion of *Aplysia*. II. Post-tetanic potentiation, heterosynaptic depression, and increments in frequency of inhibitory postsynaptic potentials. *J. Neurophysiol.* 32: 509–15

Weinreich, D. 1971. Ionic mechanism of post-tetanic potentiation at the neuromuscular junction of the frog. *J. Physiol.* 212: 431–46

Wernig, A. 1972. The effects of calcium and magnesium on statistical release parameters at the crayfish neuromuscular junction. *J. Physiol.* 226: 761–68

Wernig, A. 1975. Estimates of statistical release parameters from crayfish and frog neuromuscular junctions. *J. Physiol.* 244: 207–21

Wojtowicz, J. M., Atwood, H. L. 1985. Correlation of presynaptic and postsynaptic events during establishment of long-term facilitation at crayfish neuromuscular junction. *J. Neurophysiol.* 54: 220–30

Wojtowicz, J. M., Atwood, H. L. 1986. Long-term facilitation alters transmitter releasing properties at the crayfish neuromuscular junction. *J. Neurophysiol.* 55: 484–98

Woodson, P. B. J., Traynor, M. E., Schlapfer, W. T., Barondes, S. H. 1976. Increased membrane fluidity implicated in acceleration of decay of post-tetanic potentiation by alcohols. *Nature* 260: 797–99

Zengel, J. E., Magleby, K. L. 1980. Differential effects of $Ba^{2+}$, $Sr^{2+}$, and $Ca^{2+}$ on stimulation-induced changes in transmitter release at the frog neuromuscular junction. *J. Gen. Physiol.* 76: 175–211

Zengel, J. E., Magleby, K. L. 1981. Changes in miniature endplate potential frequency during repetitive nerve stimulation in the presence of $Ca^{2+}$, $Ba^{2+}$, and $Sr^{2+}$ at the frog neuromuscular junction. *J. Gen. Physiol.* 77: 503–29

Zengel, J. E., Magleby, K. L. 1982. Augmentation and facilitation of transmitter release. A quantitative description at the frog neuromuscular junction. *J. Gen. Physiol.* 80: 583–611

Zengel, J. E., Magleby, K. L., Horn, J. P., McAfee, D. A., Yarowsky, P. J. 1980. Facilitation, augmentation, and potentiation of synaptic transmission at the superior cervical ganglion of the rabbit. *J. Gen. Physiol.* 76: 213–31

Zilber-Gachelin, N. F., Chartier, M. P. 1973. Modification of the motor reflex responses due to repetition of the peripheral stimulus in the cockroach. I. Habituation at the level of an isolated abdominal ganglion. *J. Exp. Biol.* 59: 359–82

Zucker, R. S. 1972. Crayfish escape behavior and central synapses. II. Physiological mechanisms underlying behavioral habituation. *J. Neurophysiol.* 35: 621–37

Zucker, R. S. 1973. Changes in the statistics of transmitter release during facilitation. *J. Physiol.* 229: 787–810

Zucker, R. S. 1974a. Crayfish neuromuscular facilitation activated by constant presynaptic acition potentials and depolarizing pulses. *J. Physiol.* 241: 69–89

Zucker, R. S. 1974b. Characteristics of crayfish neuromuscular facilitation and their calcium dependence. *J. Physiol.* 241: 91–110

Zucker, R. S. 1974c. Excitability changes in crayfish motor neurone terminals. *J. Physiol.* 241: 111–26

Zucker, R. S. 1977. Synaptic plasticity at crayfish neuromuscular junctions. In *Identified Neurons and Behavior of Arthropods*, ed. G. Hoyle, pp. 49–65. New York: Plenum

Zucker, R. S., Bruner, J. 1977. Long-lasting depression and the depletion hypothesis at crayfish neuromuscular junctions. *J. Comp. Physiol.* 121: 223–40

Zucker, R. S., Lara-Estrella, L. O. 1979. Is synaptic facilitation caused by presynaptic spike broadening? *Nature* 278: 57–59

Zucker, R. S., Lara-Estrella, L. O. 1983. Post-tetanic decay of evoked and spontaneous transmitter release and a residual-calcium model of synaptic facilitation at crayfish neuromuscular junctions. *J. Gen. Physiol.* 81: 355–72

Zucker, R. S., Stockbridge, N. 1983. Presynaptic calcium diffusion and the time courses of transmitter release and synaptic facilitation at the squid giant synapse. *J. Neurosci.* 3: 1263–69

*Ann. Rev. Neurosci. 1989. 12 : 33–45*

# INTEGRATING WITH NEURONS

## D. A. Robinson

Department of Ophthalmology and Biomedical Engineering, The Johns Hopkins University, School of Medicine, Baltimore, Maryland 21205

## INTRODUCTION

Integration has two meanings in neurophysiology: One indicates some vaguely specified combination as in ". . . an integration of visual and vestibular signals may lead to a perception of . . ."; the other comes from calculus. If $x(t)$ is one variable in time and $y(t)$ another, then

$$y(t) = \int_0^t x(\tau) \, d\tau \qquad \qquad 1.$$

says that $y$ is the time integral of $x$. This mathematical operation occurs in the central nervous system and is the subject of this review. The review concentrates on the integrator of the vestibulo-ocular reflex as a prominent example, offers a model of how integration might be done by neurons, and speculates about the extent to which neural integrators occur elsewhere in motor control.

Integration describes physical processes all around us: The volume of a fluid (blood) in a container (ventricle) is the integral of the inflow (venous return); the position of the shaft of a d.c. motor is the integral of the current applied to its armature. These examples, however, are just statements of physics. They are not examples of devices deliberately constructed by nature or technology to integrate a signal to achieve some desired end. Such devices are not very visible in our world. One exception is attached to the back of our houses, it measures the energy we use by integrating our power consumption, but most integrators hide in boxes that operate cranes, fly airplanes, orient satellites, and so on.

In these examples, the integrators are usually located in negative feedback control systems. Their value there, to oversimplify, is that integrators have very large gains at low frequencies, making controllers very accurate

0147–006X/89/0301–0033$02.00

in the steady state. On the other hand, integrators have low gains at high frequencies, helping to prevent oscillations. If a control system does not contain an integrator naturally (such as a motor), the design engineer will probably add one, if not two, to the system to get the desired performance.

Consequently, when engineers became interested in biological control systems, they took it for granted that integrators were everywhere—how else could all these control systems possibly work? The oculomotor system offers an especially clear example: The retina senses the error between the eye (fovea) and the target, and the system turns the eye until the error is zero—a simple negative feedback scheme (e.g. Young & Stark 1963). Moreover, when the goal is reached, a constant eye deviation (output) is maintained while the error (input) is zero. But that is just what an integrator does—it holds signals in the absence of new information. Indeed, the only way its output can be constant is if the input error is zero—the desired condition. So obviously the oculomotor system had an integrator and to the bioengineer this idea was so obvious as to be trivial.

To the neurophysiologist a neural integrator seemed exotic, but, so long as it only appeared in top-down, black-box models, it could be relegated to "higher centers" and ignored. This was not, however, the case for the vestibulo-ocular reflex. By the early 1960s, it was clear that the signal from the semicircular canals, coded in the rate modulation of the primary afferents, was head velocity. The discharge rate of motoneurons, on the other hand, largely determined eye position. If a constant head velocity is to make the eyes move at a constant velocity, the motoneurons must respond to the time integral of the canal signal. This integrator was not hidden under a feedback loop—it was the major signal-processing element in a short, well-defined, pontine reflex. It could not be ignored without also ignoring the main function of the reflex.

Still, it was not until 1968 that this obvious observation first appeared, however briefly, in print (Robinson 1968). Soon thereafter neurophysiological evidence appeared. Cohen & Komatsuzaki (1972) found that electrical stimulation of the reticular formation caused the eyes of monkeys to move at a constant velocity—the time integral of the step of excitation. We know now that they were stimulating an input to the integrator (from saccadic burst neurons), not the integrator itself, but no matter; integration was clearly occurring. It was also confirmed that motoneurons were responsible for determining the position of the eye (Robinson 1970), and by recording from them during sinusoidal rotations of monkeys we demonstrated the requisite 90 deg phase lag between vestibular and oculomotor motoneurons created by the integrator (Skavenski & Robinson 1973).

Although the integrator's existence was not open to question simply on theoretical grounds, these findings lent a sort of respectability to the idea and helped in making clear the integrator's essential role. The concept was readily accepted in neuro-ophthalmology; after all, it is the neural integrator that generates the slow phases of nystagmus and holds the eye eccentrically after a saccade. Disorders of these basic operations could now be attributed to the integrator.

# THE NEURAL INTEGRATOR IN THE VESTIBULO-OCULAR REFLEX

Figure 1 shows the signal processing involved in the reflex. On the right, the canal produces, in the frequency range of physiological head movements, a signal proportional to head velocity, $\dot{H}(t)$, coded in the discharge-rate modulation, $R_{v1}$, of primary vestibular afferents. The background rate (90 spikes/sec) and sensitivity [0.4 (spikes/sec)/(deg/sec)] are taken from Fernandez & Goldberg (1971). On the left is shown the well-established relationship between the modulation of motoneuron discharge rate, $\Delta R_m$

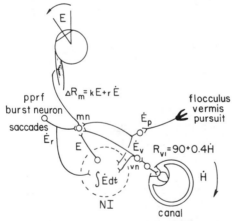

*Figure 1*   The final common integrator. On the *right* the canals transduce head velocity, $\dot{H}$, and report it, coded as the modulation of the discharge rate, $R_{v1}$, of primary vestibular afferents to the vestibular nucleus, vn. This signal becomes an eye velocity command for vestibular movements, $\dot{E}_v$, which is sent directly to the motoneurons, mn, and to the neural integrator, NI, to provide the needed position signal $E$. These signals provide those needed by the motoneurons modulating by $\Delta R_m$. The pursuit and saccadic signals also arrive as eye velocity commands, $\dot{E}_p$ and $\dot{E}_r$, from the cerebellum (flocculus, vermis) and paramedian pontine reticular formation (pprf), respectively, and are also sent to the motoneurons and the neural integrator. The latter is contained in the nucleus prepositus hypoglossi and the vestibular nuclei (*dashed lines*).

(around a background rate of typically 100 spikes/sec), and eye position and velocity. The coefficient $k$ is the neural reflection of the elasticity of the orbital tissues while $r$ represents orbital viscosity. Typical values are 4 (spikes/sec)/deg and 1.0 (spikes/sec)/(deg/sec), respectively.

For a vestibular command, the motoneurons need a signal proportional to desired eye velocity ($\dot{E}_v$ in this case) and desired eye position $E(t)$. The former can obviously be obtained directly from the vestibular nucleus by the direct path shown. This agrees, fortunately, with the anatomical fact that the neurons in the vestibular nucleus project directly to the motoneurons to form the well-known three-neuron arc.

The signal $E$, on the other hand, is the integral of $\dot{E}$, and, no matter how it is done, the process can be given its mathematical name: integration, labeled NI in Figure 1. Since this operation occurs in the dark when no other sense modality can help, it would appear that a network of neurons in the pons must perform this mathematical operation.

Interestingly, bioengineers themselves felt uncomfortable with the idea of integrating just with neurons and suggested an alternative familiar to engineers—velocity feedback. If the output of a control system is differentiated before being fed back, the overall system behaves like an integrator. So it was proposed that velocity feedback from muscle proprioception could do the trick (e.g. Fender & Nye 1961). We were able to eliminate this hypothesis by showing that there was no mono- or paucisynaptic stretch reflex for the eye muscles of the rhesus monkey (Keller & Robinson 1971). This indicated that integration was done somehow by a network of neurons.

Figure 1 is greatly simplified to emphasize the signal processing. The integrator must be a bilateral structure with its halves coupled across the midline. It receives a push-pull signal from a pair of canals, one canal modulation decreasing the other increasing. It sends a push-pull signal to the motoneurons to modulate the agonist and antagonist muscles in reciprocal innervation.

The integrator does its job well. Figure 1 shows, as discussed below, that a single integrator is shared by all the conjugate oculomotor subsystems. When, for example, the burst neurons create a saccade by sending a pulse of activity directly to the motoneurons, the eye is held in its new position by the step of innervation produced by integrating the pulse. The integrator is not perfect; it leaks and, in the dark, the eye begins sliding back toward the center with a time constant, $T_n$, of about 25 sec (Becker & Klein 1973). This is so much longer than the interval between most normal eye movements that this imperfection can be largely ignored. Lesions can greatly decrease $T_n$ and create a failure of gaze-holding called gaze-paretic nystagmus.

# LOCATION OF THE INTEGRATOR

For a long time the neural integrator was thought to lie in the paramedian pontine reticular formation because lesions there affected eye movements profoundly and the reticular formation seemed a good place for anything mysterious. This idea was finally disproved when Henn and his colleagues made neurotoxin lesions there. Ipsilateral saccades were abolished but not gaze holding (Henn et al 1984). The cerebellum is important in minimizing leak rate. Total cerebellectomy (Robinson 1974), ablation of the flocculus in particular (Zee et al 1981), reduces the time constant $T_n$ to about 1.3 sec. This might tempt one to put the integrator entirely in the cerebellum (Carpenter 1972) except that 1.3 sec is not negligible. During vestibular nystagmus, the integrator need only integrate well from one quick phase to the next (roughly 0.3 sec), and even after total cerebellectomy only close examination will detect such leakiness in nystagmus recordings. Moreover, during ice-water stimulation of one ear, simulating a vestibular lesion, $T_n$ in humans is deliberately decreased to about 2.4 sec within 80 sec of the onset of the inappropriate nystagmus (Robinson et al 1984). The cerebellum may well be responsible for parametric adjustments with this time scale. It is shown below that lesions of the vestibulo-prepositus complex abolish *all* integrator action, so the current thinking is that the integrator is basically in this complex, but the cerebellum has a powerful influence in adjusting its time constant.

Studies of cells in the vestibular nuclei of alert monkeys discovered that a large proportion of cells in the superior and rostral-medial subdivisions carried, among others, the eye position signal $E$. Many cells are purely oculomotor in that their activity reflects eye movements whether or not of vestibular origin. Thus, significant subdivisions of these nuclei form an eye-movement nucleus. This led Tomlinson & Robinson (1984) to propose this region as the site of the neural integrator. Meanwhile, the nucleus prepositus hypoglossi, right next door, had also been suggested because its cells also carried the eye-position signal and projected directly to moto-neurons. Finally, Cheron et al (1986) showed integrator failure after electro-lytic lesions of either region in the cat, and Cannon & Robinson (1987) showed in the monkey total loss of the neural integrator after bilateral neurotoxin lesions of both the prepositus and medial vestibular nuclei.

The latter study showed that, as one would predict from Figure 1, without the integrator the eye velocity commands would pass directly to the motoneurons and produce an eye *position* that was proportional to desired eye *velocity*. For a step of head velocity, for example, the step of $\dot{E}_v$, in the absence of the ramp normally generated by the integrator, simply causes a step change in eye position without the usual slow phases of nystagmus. Similar results occurred with pursuit and optokinetic stimulation. For the saccadic pulse, the eyes moved quickly to one side but, without the usual step from the integrator, the eyes returned rapidly to straight ahead with the time constant of the orbital mechanics (about 200 msec). Consequently, the time constant of the integrator, if any, was significantly less than 0.2 sec. Thus, in addition to locating the integrator, this study showed, as had long been proposed, that a single integrator was used by all conjugate oculomotor systems.

## MODELS OF NEURAL INTEGRATORS

If an engineer wants to build an integrator, either rate feedback, mentioned above, or positive feedback are the two usual choices. The former is risky: It requires large gains around the feedback loop so that if, by a lesion, the loop is broken, the output would try to rise to very large values. Positive feedback is more failsafe. One starts with a very leaky integrator with a time constant $\tau$. Positive feedback of gain $k$ around it raises the effective time constant to $\tau/(1-k)$. The integrator becomes perfect when $k$ is 1.0; no large gains are involved, and if feedback is lost, the integrator simply returns to being very leaky. Put another way, if cells excite their neighbors and are excited by them, then cells excite themselves and this perseverates activity, once started, by a "system of reverberating collaterals." To date, this has been the only model seriously considered.

On the other hand, one could, for example, hypothesize a neurotransmitter we could call integratide. When released into the subsynaptic cleft, it binds to the subsynaptic cell membrane and depolarizes it (so the cell fires faster) indefinitely until integratase is released into the cleft to inactivate the integratide and decrease depolarization and firing rate, again indefinitely, until a new signal comes along. One cannot object to this idea because the time constant of the neural integrator can be decreased from 25 to 2.4 sec within 80 sec of caloric stimulation; one need only hypothesize another neuromodulator that breaks down both integratide and integratase with the same suitable rate constant. Unfortunately, such neuromodulatory behavior has not yet been observed. Long-term synaptic changes have been observed in *Aplysia* (Frost et al 1985) but these are changes in sensitivity or gain, a multiplicative operation, quite unlike the linear operation of integration.

Returning to reverberating collaterals or positive feedback, we find two problems arising. The first is that all the signals to be integrated (except the saccadic pulse) ride on a background discharge rate. From Figure 1, the vestibular background rate is 90 spikes/sec. Recordings from other neurons in the region of the integrator (e.g. Tomlinson & Robinson 1984) show that 100 spikes/sec is typical. We don't want to integrate this background rate, that would be disastrous, just the modulation riding on it. Shamma & Cannon hit upon lateral inhibition as an exceedingly simple way to solve this problem (Cannon et al 1983). In Figure 2A, each neuron inhibits its neighbor with strength $w$ and is inhibited by it. Thus, each cell excites itself by disinhibition—positive feedback. In this scheme, the initial lag $\tau$ is the membrane time constant of an individual neuron, about 5 msec. Analysis shows that when both inputs, $u_1(t)$ and $u_2(t)$, change together, the outputs, $x_1(t)$ and $x_2(t)$, respond with a time constant of $\tau/(1+w)$ or, since $w$ is close to 1.0, about 2.5 msec. Thus, the background rates are simply passed through the system unchanged, as shown initially in Figure 2A. But when $u_1$ and $u_2$ change in push-pull, by $\Delta u$ in Figure 2A, $x_1$ and $x_2$ differ, the feedback loop starts to operate, and the cells respond with a time constant of $\tau/(1-w)$. If $w$ is very close to 1.0, this time constant can be very large, and integration takes place. It is remarkable that lateral inhibition, a work horse of neural network modelers, solves this problem as well.

The second problem is that to increase the time constant from 5 msec to 20 sec, the value of $w$ must be 0.99975. Of course, if the membrane time constant of these particular neurons was 50 instead of 5 msec, the value of $w$ becomes 0.9975, but this is still hardly robust. On the other hand, no one supposes that two neurons are adequate. Cannon et al (1983) examined a ring of 32 neurons connected as in Figure 2B. Figure 2C shows that the Bode diagram (log gain vs log frequency, $\omega$, if the input were a sinusoid) depends on the spatial frequency, $P$, of the inputs. If all the cells received the same signal (e.g. the background rate), its spatial frequency would be zero and the network would act like a wide-band, low-gain system with a time constant of 2.5 msec (*front edge* in Figure 2C). If the inputs from left and right canals (or pursuit cells or saccadic burst neurons) are interleaved as shown, one has the highest spatial frequency ($\pi/2$ or one half cycle/neuron), and the system has the Bode diagram of an integrator with a time constant of 20 sec (*back edge* in Figure 2C).

We were able to show that adding spatial noise to such a model (changing synaptic strengths randomly throughout) did little to perturb its integrating behavior. Nothing depended heavily on only a few synapses. Killing one cell (out of 32, 3%) caused the impulse response (a saccade) to drift back quickly by about 15%, because the dead neuron broke local feedback loops with nearby cells, but then to recover and drift toward zero at a rate appropriate to a 20 sec time constant. This result indicated that the major property of the network did not depend critically on any one parameter in the model. The model was robust.

Integration still requires a critical adjustment in a global sense. If $X(s, P)$

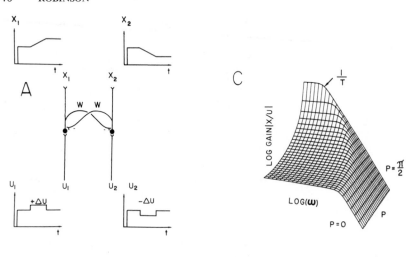

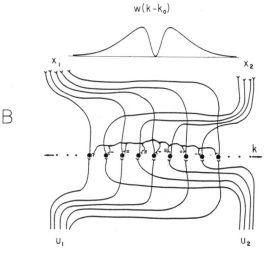

*Figure 2*  A model of the neural integrator. A: Lateral inhibition. Each cell excites itself (positive feedback) by self-disinhibition through its neighbor. A common input, $u_1$ and $u_2$, is not integrated but is simply repeated by the outputs $x_1$ and $x_2$. A differential input, on the other hand, such as $\Delta u$, is integrated as shown. B: A fuller model with 32 neurons. The spatial variable $k$ runs from 1 to 32. The neuron at $k_0$ inhibits itself and neighboring cells with the strength $w(k-k_0)$ shown by the double Gaussian curve at the *top*. The inputs enter from push-pull sources and alternate on the neurons to achieve the highest spatial frequency: $\pi/2$ or one half cycle per neuron. C: The Bode diagram of the gain $X/U$ of this network with the spatial frequency of the input $P$ as a third dimension. As $P$ approaches $\pi/2$, the system behaves like an integrator with a time constant $T$ of 20 sec.

and $U(s, P)$ are the transforms of the outputs and inputs, where $s$ is the Laplace transform frequency in the temporal frequency domain and $P$ is the Fourier transform frequency in the spatial frequency domain, then

$$\frac{X(s, P)}{U(s, P)} = \frac{1}{s\tau + [1 - W(P)]} \qquad 2.$$

where $W(P)$ is the Fourier transform of the spatial function of the pattern of lateral inhibition $w(k - k_0)$ shown in Figure 2B. The effective network time constant is $\tau/[1 - W(P)]$ when $P$ is near the highest spatial frequency of one half cycle/neuron. Thus, $W(P)$, encompassing hundreds of cells and hundreds of thousands of synapses, must have a global value of 0.99975.

Of course, any network must be able to have its function monitored and its parameters adjusted. In the case of this integrator one thinks of the cerebellum. Integrator failure results in eye motion during attempted, eccentric fixation. The resulting image motion is sensed by the retina and reported to the inferior olive by way of brainstem visual pathways via the nucleus of the optic tract. The olive reports this signal to the cerebellum, particularly the flocculus, via climbing fibers. The flocculus in turn is well connected to the prepositus-vestibular complex and is known to be involved in several ways in decreasing retinal image motion (Zee et al 1981). Thus, a reasonable hypothesis is that the flocculus monitors the error (retinal slip) and adjusts the connectivity of the integrator network to fine tune its ability to hold eye position. How the latter might be done is quite unknown. The main point here is that in the model of Figure 2, such visual feedback and parametric adaptation should be able to adjust and maintain the time constant of the integrator. Note that moderate changes of many individual synapses still contribute only infinitesimal changes in the global property $W(P)$.

The model in Figure 2B is, of course, much too simple. All the neurons are identical and inhibitory and all the inputs are identical. To show that these conditions could be relaxed, Cannon & Robinson (1985) extended the model. The background discharge rates of the incoming fibers and cells of the network could be allowed to vary over the population in a realistic manner. Most important is that almost all the cells in the prepositus-vestibular complex carry not just the eye position signal $E$ but a variety of combinations of $\dot{E}_r$, $\dot{E}_v$, and $\dot{E}_p$ as well (Figure 1). The model in Figure 2 is not dynamically rich enough to do this; it integrates too well and all the cells just carry $E$. To provide velocity terms, we used a double-layer model; a row of excitatory and a row of inhibitory cells that talk, in certain ways, to both types of neighbors. This is, of course, a minimal step in the right direction since the integrator output to downstream cells, especially the motoneurons, must be excitatory as well as inhibitory. Finally, we showed that the integrated signal could spread almost instantly through the whole network, even if the velocity input entered only part of it, so that neurons could carry $E$ while carrying $\dot{E}_r$, $\dot{E}_v$, and $\dot{E}_p$ at quite different strengths, as observed experimentally.

The time is ripe, now that we know the location of the integrator, to test the hypothesis in Figure 2 and to modify it or replace it with a better one. For example, it depends heavily on fibers crossing between the bilateral prepositus-vestibular complexes, similar to a proposal by Galiana & Outerbridge (1984). Preliminary results by T. J. Anastasio and myself, trying to make midline lesions to interrupt these fibers, have met with peculiar results. In only one monkey so far, electrolytic lesions in the midline, in the region 0 to 4 mm caudal to the abducens nuclei and 4 mm deep, did decrease the integrator time constant to 0.6 sec after each lesion, but recovery was significant in just a few hours and more so after two days. Two nagging questions arise: What is the anatomical course of these crossing fibers—did we interrupt them? What pathways and mechanisms mediate this incredible capacity for repeated recuperation?

Equally important, new models should be proposed by theoreticians to provoke further experimental tests. The operation we are considering is not trivial.

## INTEGRATORS ELSEWHERE

The integrator we have considered so far is for horizontal, conjugate eye movements. Vertical eye movements are probably combinations of movements created by two integrators, in part in the caudal mesencephalon, with planes of action near those of the vertical semicircular canals and the vertical recti and oblique muscles. Current findings in the vergence system suggest another integrator there (Mays et al 1986). That's four. Current models of gaze saccades suggest integrating the head velocity signal, $\dot{H}$, to obtain head position, $H$, needed for internal calculations of eye in space, as opposed to eye in head (e.g. Laurutis & Robinson 1986). That could add three more, one for each degree of freedom. Thus, in the oculomotor system, we can see that there is nothing special about the integrator shown in Figure 1; integrators crop up everywhere.

Can we then conclude that integrators are everywhere in motor control? Before extrapolating too quickly, it is important to appreciate some unusual features of the oculomotor system. Phylogenetically, the vestibulo-ocular reflex has dominated its development. In lateral-eyed, afoveate animals, this reflex and its visual extension, the optokinetic system, represent almost their entire oculomotor capability. In frontal-eyed, foveate animals, the saccadic, pursuit, and vergence systems were added. The vestibulo-ocular reflex is dominated by the semicircular canals, which are unique sense organs. Although each contains a sheet of sensory neurons (the hair cells), all the receptors report the same signal to the brainstem— that component of the head rotation vector that is perpendicular to the plane of that canal. There are no maps as in the visual, auditory, and somatosensory systems—the signal is entirely one-dimensional. Moreover, the canal signal rides on a high, stable, resting background rate so that modulation both up and down, in push-pull, can occur, thereby avoiding the most severe nonlinearity in the nervous system: Discharge rate cannot be negative. The result is linearity. It has also been shown that the canals have even imposed their planes upon the pulling directions of the extraocular muscles, even in frontal-eyed animals. Thus, the canals have imposed on the oculomotor system (*a*) the pulling directions of the muscles, (*b*) a high background discharge rate, (*c*) push-pull operation (expressed at the motor end by strict reciprocal innervation), (*d*) linearity. These features are exhibited not only by the motoneurons (Figure 1) but by all the interneurons in this reflex.

In the spinal cord, on the other hand, high background rates, reciprocal innervation, and especially linearity, are not the order of the day. Because of cocontraction and the need to handle a wide variety of loads utilizing proprioceptive feedback (unnecessary and not found in the eye), one would not even hope to write an equation, as in Figure 1, between motoneuron discharge rate and load position that was even unique, let alone linear. Nevertheless, this does not mean that it is too soon to think about integrators in the spinal cord. If I point straight ahead with extended arm and hand, what keeps my arm from succumbing to gravity? To say that "tonic signals" form "higher centers" are responsible is simply to avoid the question.

In the cord one can, of course, use position feedback via proprioception to hold a limb still, turning it on only during limb fixation. This is, however, not an integrator; a tonic, central signal is still needed, if only to the $\gamma$ motoneurons, and where does that come from? One could try to build an integrator by proprioceptive rate feedback, as mentioned above. The major problem with all these speculations is that because recording from single units in the spinal cords of alert, behaving animals is technically difficult, we have almost no idea of the signal processing that goes on there. Working out anatomical pathways and some properties of basic spinal reflexes in immobile animals helps almost not at all in fathoming what the cord does to the signals we see descending from the cerebellum and motor cortex before they reach the motoneurons. Without seeing the signals in the incredibly rich internuncial networks in the cord, it is difficult to even guess about premotor signal processing.

In the saccadic system, the signal most often seen in supramesencephalic structures is transient, indicating an impending change in eye position, more resembling movement velocity than position. Only when these signals descend to the caudal pons (for horizontal movements) are they converted by an immediately premotor network—the integrator—from velocity to position commands. Similarly, the majority of intracranial signals related to limb movements also seem to be phasic or velocity-related, although this is certainly not always so. If there is any analogy to the saccadic (and pursuit system as well), these phasic signals would be converted to limb position commands at the last minute, namely in the cord close to the motor nuclei. Of course, identification of such a process will be considerably complicated by the wealth of proprioceptive feedback signals, but for those who see value in the bottom-up approach, we will never even start to unravel such mysteries unless we record from premotor neurons in behaving animals and find out what they are telling the motoneurons.

*Literature Cited*

Becker, W., Klein, H. M. 1973. Accuracy of saccadic eye movements and maintenance of eccentric eye positions in the dark. *Vision Res.* 13: 1021–34

Cannon, S. C., Robinson, D. A. 1985. An improved neural-network model for the neural integrator of the oculomotor system: More realistic neuron behavior. *Biol. Cybern.* 53: 93–108

Cannon, S. C., Robinson, D. A. 1987. Loss of the neural integrator of the oculomotor system from brain stem lesions in monkey. *J. Neurophysiol.* 57: 1383–1409

Cannon, S. C., Robinson, D. A., Shamma, S. 1983. A proposed neural network for the integrator of the oculomotor system. *Biol. Cybern.* 49: 127–36

Carpenter, R. H. S. 1972. Cerebellectomy and the transfer function of the vestibulo-ocular reflex in the decerebrate cat. *Proc. R. Soc. London Ser. B* 181: 353–74

Cheron, G., Godaux, E., Laune, J. M., vanDerkelen, B. 1986. Lesions in the cat prepositus: Effects on the vestibulo-ocular reflex and saccades. *J. Physiol. London* 372: 75–94

Cohen, B., Komatsuzaki, A. 1972. Eye movements induced by stimulation of the pontine reticular formation: Evidence for integration in oculomotor pathways. *Exp. Neurol.* 36: 101–7

Fender, D. H., Nye, P. W. 1961. An investigation of the mechanisms of eye movement control. *Kybernetik* 1: 81–88

Fernandez, C., Goldberg, J. M. 1971. Physiology of peripheral neurons innervating semicircular canals of the squirrel monkey. II. Response to sinusoidal stimulation and dynamics of peripheral vestibular system. *J. Neurophysiol.* 34: 661–75

Frost, W. N., Castellucci, R. D., Hawkins, R. D., Kandel, E. R. 1985. The monosynaptic connections from the sensory neurons participate in the storage of long-term memory for sensitization of the gill- and siphon-withdrawal reflex in *Aplysia*. *Proc. Natl. Acad. Sci. USA* 82: 8266–69

Galiana, H. L., Outerbridge, J. S. 1984. A bilateral model for central neural pathways in vestibuloocular reflex. *J. Neurophysiol.* 51: 210–41

Henn, V., Lang, W., Hepp, K., Reisine, H. 1984. Experimental gaze palsies in monkeys and their relation to human pathology. *Brain* 107: 619–36

Keller, E. L., Robinson, D. A. 1971. Absence of a stretch reflex in extraocular muscles of the monkey. *J. Neurophysiol.* 34: 908–19

Laurutis, V. P., Robinson, D. A. 1986. The vestibulo-ocular reflex during human saccadic eye movements. *J. Physiol. London* 373: 209–33

Mays, L. E., Porter, J. D., Gamlin, P. D. R., Tello, C. A. 1986. Neural control of vergence eye movements: Neurons encoding vergence velocity. *J. Neurophysiol.* 56: 1007–21

Robinson, D. A. 1968. Eye movement control in primates. *Science* 161: 1219–24

Robinson, D. A. 1970. Oculomotor unit behavior in the monkey. *J. Neurophysiol.* 33: 393–404

Robinson, D. A. 1974. The effect of cerebellectomy on the cat's vestibulo-ocular integrator. *Brain Res.* 71: 195–207

Robinson, D. A., Zee, D. S., Hain, T. C., Holmes, A., Rosenberg, L. F. 1984. Alexander's law: Its behavior and origin in the human vestibulo-ocular reflex. *Ann. Neurol.* 16: 714–22

Skavenski, A. A., Robinson, D. A. 1973. Role of abducens neurons in the vestibuloocular reflex. *J. Neurophysiol.* 36: 724–38

Tomlinson, R. D., Robinson, D. A. 1984. Signals in vestibular nucleus mediating vertical eye movements in the monkey. *J. Neurophysiol.* 51: 1121–36

Young, L. R., Stark, L. 1963. Variable feedback experiments testing a sampled data model for eye tracking movements. *IEEE Trans. Prof. Tech. Group on Human Factors Electron.* HFE-4: 38–51

Zee, D. S., Yamazaki, A., Butler, P., Gücer, G. 1981. Effect of ablation of flocculus and paraflocculus on eye movement in primate. *J. Neurophysiol.* 46: 878–99

*Ann. Rev. Neurosci. 1989. 12:47–65*

# IMMORTALIZATION OF NEURAL CELLS VIA RETROVIRUS-MEDIATED ONCOGENE TRANSDUCTION

## Constance L. Cepko

Department of Genetics, Harvard Medical School, Boston, Massachusetts 02115

## INTRODUCTION

Techniques that enable neurobiologists to prepare large, pure populations of specific neural cell types would be very powerful. The diversity and postmitotic nature of neurons, however, impose severe restrictions on the number of cells that can be obtained using classical techniques developed for non-neural tissue. A lack of markers and methods for the determination of purity or uniformity of a preparation similarly limit many experiments. The establishment of mitotic, immortal cell lines, then, could offer a potential solution to this problem.

Neural cell lines have been established by using several strategies, some of them quite familiar. These include cultivation of primary tumor tissue, fusion of primary cells with tumor cells, carcinogen-induced transformation in vivo and in vitro, and spontaneous transformation (i.e. outgrowth of cells from primary cultures without any deliberate genetic manipulations). Such techniques have contributed some useful lines (e.g. PC-12 cells), but have inherent limitations. Tumor cells are irreversibly transformed and have an ill-defined history, and they cannot be used as transplants into an animal, due to tumorigenicity. Another restriction concerns an apparent limitation in the nature of the target cell as most neural tumors are human glioblastomas. A technique that allowed the establishment of lines from diverse developmental times, locations, and genetic backgrounds, then, would be ideal.

47

0147–006X/89/0301–0047$02.00

The availability of oncogenes and an understanding of some aspects of oncogenesis, along with the advent of efficient gene transfer systems, have allowed for an approach to the establishment of neural cell lines that may offer some advantages over the classical techniques. This approach is the stable introduction of oncogenes into primary neural cells and the subsequent cultivation of neural lines under non-transforming conditions. The current status of this approach is similar in many ways to that of monoclonal antibodies. It is labor-intensive, calling for diligent cell culture and the simultaneous characterization of many colonies. Several groups have shown that it is possible to introduce oncogenes into primary neural cells and readily obtain cell lines. To date, most of the neuronal lines, or progenitor lines that produce neurons, appear to produce "immature" neurons. In some cases, glial lines may comprise more mature cells. Because this approach holds such promise, it seems worth the effort for many laboratories to develop it. Our understanding of the nature of the lines is limited by our general lack of knowledge of progenitors and growth or induction conditions so that progress in this area relies on advances in cellular and developmental neurobiology. I hope that this will be a synergistic process, with the cell lines providing reagents and assay systems.

## RETROVIRUSES AND ONCOGENES

Retroviruses are RNA-containing viruses composed of a protein shell surrounded by a lipid bilayer (for background see Weiss et al 1984, 1985). The first retrovirus was isolated and identified by Peyton Rous as a leukemogenic agent of chickens in 1911. Interest in retroviruses has remained high for many years because of their oncogenic potential and their utility as agents of gene transfer. Many different oncogenes were first brought to light in the context of a retrovirus. An acutely oncogenic retrovirus is generated by stable transduction and alteration of a cellular "proto-oncogene." Cellular proto-oncogenes (i.e. oncogenes in their native state in uninfected cells) have been observed to suffer alterations in their coding regions and transcription regulation by transduction in retroviruses. Alterations reveal themselves when retroviruses cause malignant diseases, usually in chicken flocks or mouse colonies. Isolation of oncogenic agents from sick animals has allowed both characterization of oncogenes and preparation of transforming agents that can be used for in vitro experiments, such as the establishment of cell lines. The first retrovirus so isolated, the Rous sarcoma virus (RSV), offered the *src* gene and a replication competent virus capable of transforming a wide range of cells in vitro. As is discussed below, RSV has been an archetypical virus for the immortalization of many types of cells.

Oncogenes have been identified and/or isolated by using several techniques and systems (for review see Bishop 1985, Adamson 1987). Naturally occurring oncogenic viruses (other than retroviruses), such as the DNA tumor viruses (e.g. SV40, Adenovirus), encode unique oncogenes not recently derived from the host genome. Transfection of tumor DNA and karyotypic abnormalities of tumors are two other methods that reveal oncogenes. The list of nonviral and viral oncogenes now includes some 60 genes and is still growing. An understanding of oncogene function remains elusive, although progress in cell biology, transcription regulation, and development may soon offer answers concerning the function of some of these genes.

In attempts to structure our thinking about this large array of genes, several classification schemes for oncogenes are useful. One helpful distinction is between the immortalizing and the transforming oncogenes. *Immortalization* is a term that is used to describe the establishment of cell division well beyond the normal level encountered when a cell is not transduced with an oncogene. *Transformation* is an overused term with different meanings in different circles. In the present context, *transformed* refers to the acquisition of growth properties that lead to loss of contact inhibition and anchorage independence (i.e. formation of foci in monolayer culture and growth in suspension in semisolid media). Land et al (1983) and Ruley (1983) performed experiments that distinguished among oncogenes having these two properties. For example, two oncogenes that have been used to establish neural cell lines, *myc* and Adenovirus *E1a*, were shown to have immortalization capability, but not to possess transformation activity. *Ras* was demonstrated to have transformation activity, but not immortalization capacity. Other oncogenes may have both properties, as is debated for SV40 large T antigen. Which category should be used? Unfortunately (or fortunately, depending upon your point of view), a rational argument can be made for either, and neural cell lines have been established with both types of genes.

## RETROVIRUS VECTORS

Since their ignominious beginning, retroviruses have taken center stage as agents of gene transfer. They owe their popularity to a unique feature of their lifecycle, the stable, efficient integration of the viral genome. The retroviral genome can suffer molecular manipulations, both in nature and in the laboratory, with remarkable resilience. Deletions, additions, and rearrangements of its structural genes can still allow the viral genome to become encapsidated in a viral coat, integrate into a host cell, and direct transcription of viral-encoded genes. The integration feature has been exploited to transduce nonviral genes from virtually any source. Heterologous genes can be cloned into a viral vector and transmitted to host cells by using the same mechanisms as normally employed by the wild type virus. A compilation of retrovirus vectors burgeons as the list of transduced genes and applications grows at a rapid rate. All retrovirus vectors nonetheless employ the same basic strategy for transmission as a virus. Subsequent expression of the genes encoded by the viral genome is the point of divergence among the various designs. Suffice it to note that genes can be transcribed from the viral promoter (LTR), located at the 5' end of the viral genome, or from an internal (usually) nonretroviral promoter. For a full description of retroviral vectors, see Weiss et al 1985, Mulligan 1983, Brown & Scott 1987, Cepko et al 1984, Cepko 1988.

Retrovirus particles (*virions*) are the agents of transduction, or stable gene transfer into recipient cells. A key aspect of the virion surface is a glycoprotein found in the virion envelope. This glycoprotein, a product of the viral *env* gene, mediates entry of the virion into the host cytoplasm through interaction with a host receptor. There are several classes of *env* genes and cognate host receptors. Viral vectors that comprise two of the murine *env* classes are readily produced by "packaging" lines. Packaging lines are stable lines derived from mouse fibroblasts and contain all of the structural protein genes (*gag*, *pol*, and *env*) necessary for production of retrovirus particles.

The most commonly used class of viral glycoprotein is the ecotropic class. Ecotropic viral glycoproteins allow entry only into rat and mouse cells via the ecotropic receptor on these species. Ecotropic retroviruses do not infect humans and are considered relatively safe for gene transfer experiments. One packaging line that uses the ecotropic *env* gene, $\psi 2$, makes the highest titers of vectors, relative to other packaging lines (for unknown reasons) (Mann et al 1983). The *env* gene of the amphotropic class endows the virus with a very broad host range, including mouse, human, chicken, dog, cat, and mink. There are three packaging lines for the production of vectors with this coat: $\psi$am (Cone & Mulligan 1984), PA12 (Miller et al 1985), and PA317 (Miller & Buttimore 1986). There is also a packaging line, D17-C3, based on the avian virus, spleen necrosis virus (Watanabe & Temin 1983). This line also produces virus with a broad host range, which includes birds and mammals, and will encapsidate murine vectors (Embretson & Temin 1987).

Viral titers produced by the packaging lines vary widely, from $10^4$ to $10^8$ colony forming units per milliliter (CFU/ml), and depend on a combination of the vector, the inserted gene, and the packaging line. Amphibians and fish cannot be infected with the current packaging line products. Some researchers are trying to generate a packaging line, or a series of lines, that will allow infection of a very broad host range, including fish and amphibians. Cliff Tabin (Harvard Medical School, Department of Genetics and Massachusetts General Hospital, Department of Molecular Biology) and Richard Mulligan's laboratory (Whitehead Institute, MIT) are attempting to use other viral glycoproteins, derived from viruses with a very broad host range, in place of the retroviral *env* products to allow the retrovirus to enter cells of virtually any species.

Integration of viral genes into the host genome is required for the stable association of the viral genome with the host cell. Integration requires that the host cell undergo at least one round of DNA synthesis (an S phase). Target cells are thus limited to mitotic neural progenitors or glial cells. No integration into postmitotic cells has been reported, with the possible exception of integration of a murine neurotropic virus (Gardner 1978). This naturally occurring virus has been reported to integrate into post-mitotic motoneurons, and if it actually does it may be useful to pro-duce vectors for postmitotic cells. Vectors based on other neurotropic viruses, such as Herpes virus, are being developed for transduction into neurons. Such vectors may not solve the problem of immortalization of postmitotic neurons, however, as it is unclear whether the level of dif-ferentiation of postmitotic cells can be maintained if they are forced to divide. Beyond the limitation to mitotic cells, no predictable host range limitations exist. Most cell types appear to support the entry, synthesis, and integration of the viral genome, although immature stem cells (e.g. preimplantation mouse embryos (Jaenisch & Berns 1977), embryonic stem cell lines (Teich et al 1977, Gorman et al 1985), Stage X–XII avian embryos (Mitrani et al 1987), and hematopoetic stem cells (Williams et al 1986, Magli et al 1987) do not support expression from the viral LTR promoter.

## NON-NEURAL CELL LINES

Results in the hematopoetic system regarding oncogenic immortalization and differentiation offer some perspectives for applications in neuro-biology. Naturally occurring oncogenic retroviruses quite often induce hematopoetic malignancies. Many studies aimed at understanding the basis of these diseases have led to insights into some aspects of development and oncogene action (for review see Graf & Beug 1978). In addition, the isolation of useful hematopoietic cell lines is an encouraging result. Murine erythroleukemia lines (MEL) produced by the action of the Friend retro-virus can be induced to differentiate and form postmitotic erythrocytes (Marks & Rifkind 1978). Although the differentiation is not always perfect and complete, MEL cells have allowed investigation of commitment to maturation, identification of *cis*-acting sequences that are required for globin gene expression, and isolation of factors that may mediate some of these responses. Abelson murine leukemia virus can immortalize early B lymphocytes (Rosenberg & Baltimore 1976). These lines can undergo immunoglobulin gene rearrangement in vitro and may aid in the isolation of enzymes and factors that permit this activity (Alt et al 1981). In addition, investigations into the nature of the in vivo target cells of leukemogenic retroviruses have contributed to the definition of hematopoetic lineages as well as in vitro culture conditions for several cell types. For example, infection of bone marrow cells with avian erythroblastosis virus (AEV) can yield erythroblasts arrested in an immature state of differentiation. If the cells are infected by a temperature-sensitive (*ts*) version of AEV and shifted to the nonpermissive temperature in the presence of erythropoetin, they differentiate into fully mature erythrocytes (Beug et al 1982). In the absence of erythropoetin, full differentiation is not achieved, even at the nonpermissive temperature. If this result can be extrapolated to neural cell lines, it may lead to the discovery and/or definition of factors that are needed for full differentiation of neural cells. Another feature of AEV infection of hematopoetic cells that can be considered when designing strategies for neural cells concerns an effect of culture conditions (Beug et al 1987). Factor-dependent, immature myelomonocytic cells, as opposed to erythroblasts, can be induced to proliferate after infection of immature hematopoetic cells, if the culture conditions favor this lineage.

Nonhematopoetic cell lines, including epithelial, melanoblast, fibroblast, chondroblast, and muscle, have also been successfully established by retrovirus transduction of oncogenes (reviewed by Holtzer et al 1981). Transduction of *src* into muscle progenitors has generated cultures that have been used to investigate myotube formation (Holtzer et al 1975) and may enable investigations of acetycholine receptor clustering (Anthony et al 1984). Primary cells that are refractory to cell culture can also be induced to proliferate and differentiate in vitro after infection with oncogenic retroviruses. For example, embryonic chicken epithelial cells ordinarily do not proliferate in vitro. After infection with a *ts* mutant of RSV, these cells proliferated at the permissive temperature and differentiated at the nonpermissive temperature (Yoshimura et al 1981). Generalizations concerning the success of this approach are difficult to make as many lines do not express a full set of antigens or differentiate upon demand. The results in the "best" cases, however, do indicate that the approach can be successful. Success may depend on availability of a well-defined system (e.g. well-defined markers and growth conditions) and a realistic set of goals.

# PARAMETERS FOR IMMORTALIZATION OF NEURAL CELLS

## Oncogene and Vector

The oncogene and vector for the establishment of neural lines must be chosen empirically, although progress in the study of oncogenes and experience in immortalization of neural cells may yield rational methods for making this choice in the future. We have used several oncogenes of the immortalizing type (Ryder et al 1987, 1988, Hen et al 1987, Roberts et al 1985, Jat et al 1986, Land et al 1986) with the hope that the cells will retain more of their nontumorigenic properties and be more likely to differentiate than they would if they were transduced with a transforming oncogene. However, at least some counter-examples to this logic exist (Noda et al 1983) and no general statements can be made. Trying several different oncogenes and expressing them at different levels seems to be the best current strategy because critical differences in the behavior of a variety of cell types in response to a given oncogene have been observed (e.g. *src* does not transform all cell types (Lipsich et al 1984).

As mentioned above, $\psi$2-produced vectors appear to enter, integrate, and express in most mitotic mouse and rat cells. Many constructs that encode oncogenes have been made, or occur naturally. The "best" choice of the oncogene and vector may then be the constructs that are available and tested. Most available vectors employ promoters for oncogenes, such as the LTR or SV40 early promoter, that are not specific for cell type, but cell type-specific promoters can be used to direct transcription of the oncogene and thus restrict the target cell of immortalization. Two examples of cell type-specific transcription via retrovirus vectors are the $\beta$-globin (Cone et al 1987b, Karlsson et al 1987, Soriano et al 1986) and immuno-globulin ($\lambda$ light chain, Cone et al 1987a) promoters. If the target cells are nonrodent, then the number of constructs and the hazards associated with their use are much more limiting. If human cells are the target, the vectors are by definition capable of infecting the experimenter and great care must be exercised. As the full host range of the amphotropic virus has not been investigated as thoroughly as that of the ecotropic virus ($\psi$2 virus), it is not clear how effectively vectors in an amphotropic coat will infect nonrodent neural progenitor cells. It is likely, however, that the vectors produced by amphotropic packaging lines will have a broad host range.

The ability to control expression of an oncogene may constitute an advantage over other methods of establishing neural lines (e.g. cultivation of tumor tissue). Control of an oncogene's activity can be exerted at several levels. Inducible promoters can be used to turn on the transcription of a gene. Although several inducible promoters have been described and isolated, no retrovirus vectors have been constructed in which such inducible promoters function properly. Several constructs have been tested with this aim in mind, but have failed to yield consistent results. Construction of such vectors should, however, be possible. Another level of control is through the use of conditional mutants, such as temperature sensitive (*ts*) mutants, as described above for AEV and RSV. The oncogene is active at the permissive temperature, allowing proliferation, and inactive at the nonpermissive temperature, perhaps allowing differentiation. Temperature-sensitive alleles of several oncogenes exist, and a few have been cloned into retrovirus vectors or remain in the parent retrovirus (Eckhart 1969, Wyke 1973, Tegtmeyer 1975, Graf et al 1978, Weizacker et al 1986).

## Infection and Cultivation

A straightforward strategy for transduction of oncogenes is to make primary, dissociated cell cultures and infect them by incubation with a packaging line supernatant that contains an oncogenic vector. Choices of dissociation method and culture conditions should optimize for mitotic capability. Conditions that may be described for maintenance or growth of postmitotic, differentiated cells may not be appropriate. Inclusion of growth factors and use of substrata that promote adhesion (poly-*l*-lysine, laminin, etc) are recommended, however. We have used Dulbecco's Modified Eagle Medium plus 10 % fetal calf serum and poly-*l*-lysine coated plastic dishes for cells from several sources (olfactory bulb, cerebellum, cerebral cortex) (Ryder et al 1987, 1988).

If a specific cell type is the target for immortalization, it may be advantageous to select in some way for that cell prior to the infection. A major problem often encountered is recognition of the desired progenitor among many immortalized colonies. If a feature of the progenitor can be identified, it may be possible to sort out the progenitors, or perhaps directly select them out, prior to the infection. Useful reagents for this purpose are monoclonal antibodies or ligands that recognize surface molecules. The fluorescence-activated cell sorter can greatly enrich the target cell population, or alternatively, a culture dish can be coated with the antibody, and selection by differential adhesion, or "panning," can be effective (Aruffo & Seed 1987). The ability to select specific target cells will greatly reduce the amount of work encountered at later stages.

Several days after infection, cells are cultured in a drug that selects only those cells that have successfully integrated the vector. Most oncogene-containing vectors encode a dominant drug resistance gene (e.g. *neo*) for this purpose. During the process of drug selection, well-separated colonies will form, allowing for a first round of cloning. Although drug selection may not be absolutely required, since the oncogene may select for infected cells, it is extremely useful because glial, fibroblast, meningeal, or endo-thelial cell growth may sometimes occur in the absence of infection. This unintended cell proliferation can obscure, or inhibit, the growth of desired cell types during the first few weeks in culture. Colonies grow at widely varying rates, and changes in morphology can take place within a colony over the period of a week. Without the ability to observe individual colonies, such changes would not be appreciated. In cases where a pre-selection was made and a well-defined population is in culture, the drug selection may not be necessary, although subculturing the cells at low dilution during drug selection is a good way to clone a line. The number of drug-resistant colonies obtained greatly depends on the titer of the virus and the mitotic index of the cells at the time of infection.

Virus can also be delivered to the mitotic population in situ. This may be the method of choice when the mitotic zone is defined and accessible and culture conditions are unknown. Although the culture conditions must be established eventually, integration and expression of the oncogene prior to explantation of the tissue may facilitate growth in vitro. This may be the only way to immortalize cells when the conditions for mitosis in vitro (in the absence of an oncogene) are unknown. If an in situ infection protocol is used, some time (several days) for integration and expression of the vector genome should be allowed prior to cultivation and selection in the appropriate drugs.

## Characterization

When attempting to screen a large number of colonies, it can be valuable to choose those with the morphology of the desired target, although morphology is difficult to predict for many progenitors. When morphology is not predictive, or is unknown, several types of colonies should be selected for further characterization. Antibodies that react with surface molecules can be particularly valuable in the early stages of characterization. Prior to colony isolation, such antibodies can be used to identify desirable cell types on plates containing many colonies. A dish of colonies can be incubated with a primary antibody and then a second antibody, coupled to red blood cells, can be used to visualize the positive colonies (Seed & Aruffo 1987). As it does not harm the cells, this procedure can be used to identify colonies for isolation. One can also use a replica plating technique to create copies of the colonies by using a filter that supports the growth of cells (Raetz et al 1982). One copy of the colonies can be screened by a noxious method, such as radioactive ligand binding and recovery of viable, positive colonies can proceed by using the master plate. After the colonies are picked, they can be screened for biochemical and physiological prop-erties (e.g. enzyme activity, channels) or for a defined function. For example, astrocytes are known to promote the growth of neurons. This assay has been used in characterization of astrocytic lines, in conjunction with antigenic criteria (Moura Neto et al 1986, Evrard et al 1986).

Perhaps the most rigorous assay of a cell's ability to develop and differentiate is an in vivo assay. Since the retroviral marking technique allows one to tag cells with an indelible marker, any clone can be marked via infection with a second retrovirus, such as a β galactosidase virus (Sanes et al 1986, Price et al 1987, Turner & Cepko 1987, Cepko 1988). Marked cells can be introduced into animals and later assayed by microscopy or isolation, and can be later isolated and characterized by biochemical or functional assays. Alternatively, the viral oncogene may provide an antigenic tag for the cells that can be identified with an anti-oncogene antibody (e.g. SV40 T antigen, Frederiksen et al 1987). (A further discussion of in vivo possibilities is given below, under "Future Directions.")

# INTERACTIONS OF NATURALLY OCCURRING ONCOGENIC RETROVIRUSES AND NEURAL CELLS

Over the past 15 years, several groups have used naturally occurring oncogenic retroviruses to stimulate proliferation in neural cells and to establish neural cell lines. Early experiments were concerned with defining oncogene function in different cell types. Pessac & Calothy (1974), Calothy et al (1980), and Bechade et al (1985) investigated oncogene-induced proliferation in chick neural retinal cells with RSV, AEV, and another acute oncogenic avian retrovirus, MH-2. Embryonic day 7 chick neural retinal cells did not proliferate for more than two or three doublings in vitro without the action of the oncogenes encoded by these viruses. Oncogene-induced proliferation resulted in 15–20 doublings. The retinal cells continued to express several neural markers, such as tetanus-toxin binding, choline acetyl transferase (CAT), and glutamic acid decarboxylase (GAD); and synapse formation was observed (Crisanti-Combes et al 1982). Keane et al (1984) also investigated the transient effect of a *ts* RSV on early chick neural development. They infected cell cultures from prestreak presumptive neural plate (Stage 1 embryos) and definitive-neural streak cells (Stage 4 embryos). At the permissive temperature the cells were rounded and appeared transformed. By shifting to the nonpermissive temperature, the definitive-neural plate cells initiated expression of several differentiation-specific neural markers, whereas the prestreak cells did not. Since nontransformed prestreak cells do not express these markers, commitment to the many neural lineages occurs between Stages 1 and 4. The faithful maintenance of the commitment state and retention of the ability to differentiate, despite several generations under the influence of *src*, indicated that *src* did not abolish the normal programming of the cells.

These two examples of the "short-term" effect of an oncogene on the behavior of neural cells in vitro suggested that the presence of an oncogene did not irreversibly block differentiation. In both cases, the full extent of differentiation was not assessed. The characterization of a stable line is far easier. Giotta et al (1980) and Giotta & Cohn (1981) produced stable rat cerebellar lines by using *ts* RSV. They characterized two lines, WC5 and WC17a, at the permissive and nonpermissive temperatures. WC5 exhibited glial characteristics and contact-inhibition at the nonpermissive temperature, unlike its behavior at the permissive temperature. Interestingly, the cells could be shifted back to the permissive temperature, where they resumed their transformed appearance and growth. The WC17a line had two neuronal features in that veratridine stimulated $Na^+$ uptake and the morphology included round cell bodies and processes. Both characteristics were enhanced at the nonpermissive temperature. This line was not able to reverse the effects of a shift to the nonpermissive temperature, as if commitment to a later stage of neuronal differentiation at the nonpermissive temperature induced an irreversible postmitotic state. The WC5 line, and another *ts* RSV line, R2, also were shown to have ten times higher neural cell adhesion molecule (NCAM) levels at the nonpermissive temperature (Greenberg et al 1984). These studies suggest the utility of *ts* oncogenes in the establishment of neural lines. Further characterization of the "cerebellar" nature of the lines would be of value. As discussed below, a newer generation of cerebellar lines have been recently established by using SV40 large T antigen and avian *myc* (*avmyc*).

Pessac et al (1983) were unable to select continuous cell lines from their infected chick cultures, yet were able to select quail retinal lines by using the same conditions and a *ts* RSV (Pessac et al 1983). One such quail line, QNR/D, exhibited action potentials, bound tetanus toxin, reacted with several neural-specific antibodies, and contained GAD activity. In vivo, amacrine and retinal ganglion cell types exhibit GAD and stain with one of the monoclonal antibodies that was positive on QNR/D. The properties of the line were exhibited at both the permissive temperature (36°C) and at 39°C, close to the nonpermissive temperature of 42°C, although there was some enhancement of the differentiated properties of 39°C. (The cells were unable to tolerate the true nonpermissive temperature of 42°C). Further tests of the specific nature of this line, using physiological, immunological, and behavioral assays, would be useful.

Neural lines have been established by two non-retroviral transduction techniques. SV40 is a DNA virus that kills and lyses monkey cells, its normal host. In nonsimian species it occasionally integrates and expresses large T antigen, an immortalizing oncogene. De Vitry et al (1974) used SV40 virus to establish a line of primitive neurosecretory cells from embryonic mouse hypothalamus. These cells contained many secretory vesicles and other ultrastructural features of secretory cells. Neurophysin and vasopressin, two normal products of the hypothalamus, were found to be secreted. Moura Neto et al (1986) report isolation of glial lines from the embryonic mouse brain, also by using SV40 virus. They were interested in the different arborization patterns exhibited by dopaminergic neurons cultured on glial cells from mesencephalic and striatal regions. They made lines in order to study the biochemistry of surface molecules that contribute to this phenomenon. Although their glial lines supported attachment and sprouting of dopaminergic neurons, only a marginal difference in the ability to recapitulate the arborization patterns observed in primary cultures was observed (Mallat et al 1986, Autillo-Toutati et al 1986). The morphology was that of immature astrocytes, although maturation into the morphology of more mature astrocytes was observed after treatment with mitomycin C and dibutyryl cAMP (dBcAMP). Evrard et al (1986) also established mouse glial lines. They used a nonviral DNA transduction technique (CaPO$_4$ transfection) and bacterial plasmids encoding Adenovirus *E1a* and polyoma virus large T antigen. A large number of lines were isolated and also found to support the growth of neurons. Some of their lines expressed glial fibrillary acidic protein (GFAP) after induction with dBcAMP, while all expressed GFAP after induction with an astroglial growth factor (AGF 2). The response to this growth factor illustrates a potential use of neural cell lines as assay systems for growth factors. Lines that do not exhibit differentiation to the desired level in vitro could be used for assays of factors, supplied in supernatants from other lines that do differentiate, or from primary cells or tissue extracts.

## IMMORTALIZATION PRODUCED BY RETROVIRAL VECTOR TRANSDUCTION

A new generation of lines has been produced by several groups using retrovirus vectors and a slightly different strategy. Retrovirus vectors, unlike naturally occurring retroviruses, usually have been constructed to encode a drug-resistance marker. Any infected cell can thus be selected by growth in the appropriate drug. As the vectors are usually efficient at generating drug-resistant colonies, even in the absence of an oncogene, it is easy to generate many clones of widely varying growth rates and very different properties in one experiment. Whether the spectrum of neural lines isolated with this strategy will ultimately and significantly differ from those described above is not yet clear. Although clones can be isolated in the absence of an oncogene, the presence of an oncogene is necessary for more than one passage (Ryder et al 1987, 1988). We found that all oncogenes tested (*avmyc, p53, N-myc, SV40* and polyoma large T antigen, Adenovirus *E1a 12S* and *13S*) by infection of primary cultures of neonatal mouse and rat cerebral cortex, olfactory bulb, and cerebellum were able to induce the ability to passage in vitro at least several times. The majority of our colonies could passage indefinitely, but did not exhibit reactivity with informative neural-specific antibodies, or have an interesting morphology. A minority of the colonies proved to be of interest and are described below.

The clones produced by several groups fall into three categories. Ryder et al (1987, 1988) have found that rat neonatal cerebral cortex cells transduced with *avmyc* (Land et al 1986) can result in clones with expanded, but limited, growth potential. These mortal clones initially appeared as "holes" in an old dish of many overlapping, drug-resistant colonies. When isolated and subsequently cloned (by dilution), they continued to divide, up to about 30 doublings, before completely stopping all cell division. When stained with a battery of neural-specific antibodies, they were found to be negative for the neuronal marker, neurofilament, but positive with antibodies to the oligodendrocyte marker, galactocerebroside (galC), and GFAP. After several additional generations, or after freeze-thawing, the cells became 100% GFAP-positive and galC-negative, until they stopped growing. Such lines may be quite useful for screening of growth factors that may prolong their mitotic ability. Lines such as these may be similar to the chick retinal cultures of Pessac & Calothy (1974) described above.

Another category of lines (which overlaps with lines discussed in the preceding section) includes immortal cells that exhibit commitment to a single neural lineage. Several examples of lines produced by different groups and using different oncogenes have generated lines in this category. Trotter et al (1987) infected cultures of rat optic nerve cells with a retrovirus vector encoding *src* in an attempt to immortalize cells of the oligodendrocyte-type 2 astrocyte (O2A) lineage. Lines isolated by this strategy exhibit some of the antigens found on immature cells in this pathway. As no individual markers are specific to this pathway, the conclusion that they are O2A lines will await the demonstration of full differentiation into oligodendrocytes and astrocytes. Ryder et al (1987, 1988) isolated mouse olfactory bulb lines using an *avmyc* retrovirus vector. Two lines stably express neurofilament, bind tetanus-toxin, and show reactivity with several monoclonal antibodies that are consistent (although not specific) with a neuronal identity. Almost every cell in these lines shows antineurofilament staining and has neuronal morphology. These cells are currently being screened for "mature" neuronal markers, such as the presence of glutamate receptors. It should be noted that these lines are actively mitotic and divide approximately once a day.

The third category of lines may be the most interesting for developmental studies. This category is one of lines that can express a heterogeneous set of markers. Clonality has been demonstrated (Ryder et al 1988, Fredericksen et al 1987, Bartlett et al 1988) by examination of the viral integration site, a standard technique when using retroviruses, as each viral integration is essentially unique. The integration site can be characterized by restriction enzyme digestion and Southern blots. These clones exhibit characteristics of more than one lineage, as demonstrated by reactivity with antibodies that classically define neurons, astrocytes, and oligodendrocytes (e.g. neurofilament, GFAP, galC, respectively). Fredericksen et al (1987) used a *ts* allele of SV40 large T antigen to immortalize rat cerebellar cells. At the permissive temperature, their lines exhibited reactivity with rat 401, an antibody that they characterized as a marker of immature neurepithelial cells. Some of their lines were similar to the WC5 line of Giotta et al (1980), as they became GFAP-positive upon shifting to the nonpermissive temperature, but were negative at the permissive temperature. Interestingly, Fredericksen et al were able to use cocultivation with primary cells to induce the same line to form neurofilament-positive cells with neuronal morphology, and a few galC-positive cells. The ability to control lineage could prove to be extremely advantageous and facilitate the isolation of factors that contribute to commitment decisions.

Bartlett et al [1988 and Ora Bernard (personal communication)] produced neurepithelial clones by infection of E10 mouse neurepithelium cultures with retroviruses encoding *c-myc* or *N-myc*. Due to their early embryonic origin, such lines would be expected to be multipotential. Bartlett et al were able to induce the *c-myc*-derived lines to produce differentiation-specific antigens by using fibroblast growth factor (FGF), an inducer of uninfected neurepithelial cells. Within three days of induction, one line exhibited cells with neurofilament antigens and A2B5 (a surface ganglioside antigen found on several types of neural cells). After seven days, some cells of the same line showed reactivity with GFAP antisera, thus demonstrating that at least one line (2.3D) was multipotent and could give rise to neurons and astrocytes. The morphology of the two populations appeared to differ and thus presumably these distinct antigenic markers are on different cells, although double staining experiments should confirm this hypothesis. In contrast, the *N-myc*-derived lines were able to produce neuronal cells without the addition of FGF. The lines are in the early stages of characterization and have not yet been observed to produce cells of other lineages. It is not clear whether this difference in lines induced with the two oncogenes is due to intrinsic differences between *c-myc* and

*N-myc* activity. Bernard noticed a difference in the amount of oncogene mRNA in cells infected by the two different retroviruses, perhaps due to a variation in vector design in the two constructs. Is the level or type of oncogene the critical factor? The answer should prove interesting and instructive.

Some of our clonal lines from the olfactory bulb and cerebellum, immortalized by *avmyc*, also fall into the third category of multipotential lines (Ryder et al 1987, 1988). Some clones maintain a low percentage of cells showing reactivity with a variety of monoclonal antibodies specific for neural cells. Others show antigens characteristic of the neuronal and glial lineages (e.g. neurofilament, GFAP, and galC). In double-staining experiments we have not observed the same cell showing markers of more than one lineage. Rather, individual cells at the same time, under the same conditions, show markers of different lineages. In some cases this situation developed from negative cells during routine subculturing, or after thawing a frozen, early passage of the line. We have also observed changes from one lineage to another. For example, galC-positive, neurofilament-negative lines have become uniformly neurofilament-positive, without deliberate manipulation to effect the switch. It is never clear when a switch occurs whether this is due to outgrowth of a different or uncommitted population. Subcloning lines also can result in the appearance of a subclone with properties different from the parent. This may be caused by a dilution effect when subcloning, which triggers a switch, or selection of a minor percentage of cells within the parent clone. A multipotential parent clone that gives rise to neuronal and glial subclones also has been reported for a neuroblastoma line, RT4 (Droms & Sueoka 1987).

Multipotential clones may be an expected result if the findings of recent lineage mapping experiments are considered. For example, Turner & Cepko (1987) report that complex clones (e.g. containing rods, Muller glial cells, and bipolar cells) can result after infection of postnatal rat retina with a retrovirus vector designed to mark cells histochemically. It is interesting to note that the complex retinal clones occurred at the end of retinal histogenesis, and some of the cell lines discussed above (e.g. olfactory bulb and cerebellar lines) were made from postnatal tissue. The heterogeneous lines may reliably indicate the multipotential states of normal in vivo progenitors. Alternatively, these lines may be prone to artifacts, and it is too soon to know the difference. More in situ lineage experiments and a greater understanding of progenitor behavior are needed to interpret the behavior of these lines.

## FUTURE DIRECTIONS

Many questions remain unanswered concerning cell lines and their representation of in vivo counterparts. The methods and results outlined here provide the tools for investigation of this crucial point. Particularly when the behavior of lines is either unstable or surprising, it is critical to determine the contribution of the oncogene and the in vitro environment in the behavior. Assaying lines in vivo, by marking the cells genetically and transplanting them to different sites at different developmental times, should provide the most rigorous test of their authenticity. That transplantation will lead to tumors is not expected in most cases. The in vivo environment has suppressed tumorigenicity of lines in the past, perhaps most dramatically in the generation of chimeras by using teratocarcinoma cells. These multipotential mouse stem cell lines can be delivered to blastocysts, where they participate in the full development of an apparently normal mouse (for review, see LeDouarin & McLaren 1984). Transgenic mice that undergo complete development have been generated where an oncogene was expressed in an unregulated manner (Stewart et al 1984). Moreover, *ts* alleles of oncogenes typically have reduced activity at normal body temperatures. Lines immortalized with *ts* alleles should behave as though the oncogene product were absent in vivo. If the behavior of lines in the in vivo environment indicates that they have the potential to differentiate fully, the conditions that constitute such a permissive or supportive environment in vitro will be very important to determine. The use of lines will be limited if such conditions cannot be obtained.

If the results with the current set of lines prove to be encouraging, innumerable applications can be proposed. Lines can be used as a source of molecules that are currently difficult to obtain and of antigens for the production of monoclonal antibodies. Such antibodies could greatly further identification of normal progenitors and of the lines themselves. Cells from a variety of genetic backgrounds, including individuals with mutations, could be immortalized. Such lines could provide assay systems for poorly understood defects and provide critical reagents. They can act as recipients in gene transfer experiments that attempt to define a mutant cDNA or gene by complementation from a nonmutant source. Lines made from nonmutant animals can also be used for complementation by transplantation into mutant mice. Embryonic human tissue could be used directly for transplants into people who suffer from neurological diseases or as a source of therapeutic factors.

*Literature Cited*

Adamson, E. D. 1987. Oncogenes in development. *Development* 101: 449–71

Alt, F., Rosenberg, N., Lewis, S., Thomas, E., Baltimore, D. 1981. Organization and reorganization of immunoglobulin genes in A-MuLV-transformed cells: Rearrange- ment of heavy but not light chain genes. *Cell* 27: 381–90

Anthony, D. T., Schuetze, S. M., Rubin, L. L. 1984. Transformation by Rous sarcoma virus prevents acetylcholine receptor clustering on cultured muscle fibers. *Proc.*

*Natl. Acad. Sci. USA* 81: 2265–69

Aruffo, A., Seed, B. 1987. Molecular cloning of two CD7 (T-cell leukemia antigen)-cDNAs by a COS cell expression system. *EMBO J.* 6: 3313–16

Autillo-Touati, A., Mallat, M., Araud, D., Moura Neto, V., Vuillet, J., Glowinski, J., Seite, R., Prochiantz, A. 1986. Two simian virus 40(SV40)-transformed cell lines from the mouse striatum and mesencephalon presenting astrocytic characters. III. A light and electron microscopic study. *Dev. Brain Res.* 26: 33–47

Bartlett, P. F., Reid, H. H., Bailey, K. A., Bernard, O. 1988. Immortalization of mouse neural precursor cells by the *c-myc* oncogene. *Proc. Natl. Acad. Sci. USA* 85: 3255–59

Bechade, C., Calothy, G., Pessac, B. 1985. Induction of proliferation or transformation of neuroretina cells by the *mil* and *myc* viral oncogens. *Nature* 316: 559–62

Beug, H., Blundell, P. A., Graf, T. 1987. Reversibility of differentiation and proliferative capacity in avian myelomonocytic cells transformed by *ts*E26 leukemia virus. *Genes Dev.* 1: 277–86

Beug, H., Palmiere, S., Freudenstein, C., Zentgraf, H., Graf, T. 1982. Hormone-dependent terminal differentiation in vitro of chicken erythroleukemia cells transformed by *ts* mutants of avian erythroclastosis virus. *Cell* 28: 907–19

Bishop, J. M. 1985. Viral oncogenes. *Cell* 42: 23–38

Brown, A. M. C., Scott, M. R. D. 1987. Retroviral vectors. In *DNA Cloning: A Practical Approach*, ed. D. M. Glover, Vol. 3, Ch. 9, pp. 189–212. Oxford/Washington: IRL Press

Calothy, G., Poirer, F., Dambrine, G., Mignatti, P., Combes, P., Pessac, B. 1980. Expression of viral oncogenes in differentiating chick neuroretinal cells infected with avian tumor viruses. *Cold Spring Harbor Symp.* 44: 983–90

Cepko, C. L. 1988. Retroviral vectors and their applications to neurobiology. *Neuron* 1: 345–53

Cepko, C. L., Roberts, B. E., Mulligan, R. E. 1984. Construction and applications of a highly transmissible murine retrovirus shuttle vector. *Cell* 37: 1053–62

Cone, R. D., Mulligan, R. C. 1984. High-efficiency gene transfer into mammalian cells: Generation of helper-free recombinant retrovirus with broad mammalian host range. *Proc. Natl. Acad. Sci. USA* 81: 6349–53

Cone, R. D., Reilly, E. B., Eisen, H. N., Mulligan, R. C. 1987a. Tissue-specific expression of functionally rearranged λ1 Ig gene through a retrovirus vector. *Science* 236: 954–57

Cone, R. D., Weber-Benarous, A., Baorto, D., Mulligan, R. C. 1987b. Regulated expression of a complete human β-globin gene encoded by a transmissible retrovirus vector. *Mol. Cell. Biol.* 7: 887–97

Crisanti-Combes, P., Lorinet, A.-M., Girard, A., Pessac, B., Wasseff, M., Calothy, G. 1982. Expression of neuronal markers in chick and quail embryo neuroretinal cultures infected with Rous sarcoma virus. *Cell Differentiation* 11: 45–54

De Vitry, F., Camier, M., Czernichow, P., Benda, Ph., Cohen, P., Tixier-Vidal, A. 1974. Establishment of a clone of mouse hypothalamic neurosecretory cells synthesizing neurophysin and vasopressin. *Proc. Natl. Acad. Sci. USA* 71: 3575–79

Droms, K., Sueoka, N. 1987. Cell-type-specific responses of RT4 neural cell lines to dibutyryl-cAMP: Branch determination versus maturation. *Proc. Natl. Acad. Sci. USA* 84: 1309–13

Eckhart, W. 1969. Complementation and transformation by temperature sensitive mutants of polyoma virus. *Virology* 38: 120–25

Embretson, J. E., Temin, H. M. 1987. Lack of competition results in efficient packaging of heterologous murine retroviral RNAs and reticuloendotheliosis virus encapsidation-minus RNAs by the reticuloendotheliosis virus helper cell line. *J. Virol.* 61: 2675–83

Evrard, C., Galiana, E., Rouget, P. 1986. Establishment of "normal" nervous cell lines after transfer of polyoma virus and adenovirus early genes into murine brain cells. *EMBO J.* 5: 3157–62

Fredericksen, K., Jat, P.-J., Valtz, N., McKay, R. 1987. Immortal neuroepithelial precursor cells lines. *Abstr. Soc. Neurosci.* 13: 182

Gardner, M. B. 1978. Type C viruses of wild mice: Characterization and natural history of amphotropic, ecotropic, and xenotropic MuLV. *Curr. Top. Microbiol. Immunol.* 79: 215–39

Giotta, G. J., Cohn, M. 1981. The expression of glial fibrillary acidic protein in a rat cerebellar cell line. *J. Cell. Physiol.* 107: 219–30

Giotta, G. J., Heitzmann, J., Cohn, M. 1980. Properties of two temperature-sensitive Rous Sarcoma virus transformed cerebellar cell lines. *Brain Res.* 202: 445–58

Gorman, C. M., Rigby, P. W. J., Lane, D. P. 1985. Negative regulation of viral enhancers in undifferentiated embryonic stem cells. *Cell* 42: 519–26

Graf, T., Ade, N., Beug, H. 1978. Temperature-sensitive mutant of avian erythroblastosis virus suggests a block of differentiation as mechanism of leukaemogenesis. *Nature* 275: 496–501

Graf, T., Beug, H. 1978. Avian leukemia viruses: Interactions with their target cells in vitro and in vivo. *Biochim. Biophys. Acta* 516: 269–99

Greenberg, M. E., Brackenburg, R., Edelman, G. M. 1984. Alteration of neural cell adhesion molecule (N-CAM) expression after neuronal cell transformation by Rous sarcoma virus. *Proc. Natl. Acad. Sci. USA* 81: 969–74

Hen, R., Dodd, J., Cepko, C., Axel, R. 1987. Construction of a rat olfactory neuronal cell line. *Soc. Neurosci. Abstr.* 13: 1410

Holtzer, H., Biehl, J., Yeoh, G., Meganathan, R., Kaji, A. 1975. Effect of oncogenic virus on muscle differentiation. *Proc. Natl. Acad. Sci. USA* 72: 4051–55

Holtzer, H., Pacifici, M., Croop, J., Boettinger, D., Toyama, Y., Payette, R., Biehl, J., Dlugosz, A., Holtzer, S. 1981. Properties of cell lineages as indicated by effects of ts-RSV and TPA on the generation of cell diversity. *Fortsch. Zool. Bd.* 26: 207–25

Jaenisch, R., Berns, A. 1977. Tumor virus expression during mammalian embryogenesis. In *Concepts in Mammalian Embryogenesis*, ed. M. Sherman, pp. 267–314. Cambridge: MIT Press

Jat, P. S., Cepko, C. L., Mulligan, R. D., Sharp, P. A. 1986. Recombinant retroviruses encoding simian virus 40 large T antigen and polyomavirus large and middle T antigens. *Mol. Cell. Biol.* 6: 1204–17

Karlsson, S., Papayannopoulou, T., Schweiger, S. G., Stamatoyannopoulos, G., Neinhuis, A. 1987. Retroviral-mediated transfer of genomic globin genes leads to regulated production of RNA and protein. *Proc. Natl. Acad. Sci. USA* 84: 2411–15

Keane, R. W., Lipsich, L. A., Brugge, J. S. 1984. Differentiation and transformation of neural plate cells. *Dev. Biol.* 103: 38–52

Land, H., Chen, A. C., Morgenstern, J. P., Parada, L. F., Weinberg, R. A. 1986. Behavior of *myc* and *ras* oncogenes in transformation of rat embryo fibroblasts. *Mol. Cell. Biol.* 6: 1917–25

Land, H., Parada, L. F., Weinberg, R. A. 1983. Tumorigenic conversion of primary embryo fibroblasts requires at least two cooperating oncogenes. *Nature* 304: 596–602

LeDouarin, N., McLaren, A. 1984. *Chimeras in Developmental Biology*. New York: Academic

Lipsich, L., Brugge, J. S., Boettiger, D. 1984.

Expression of the Rous sarcoma virus *src* gene in avian macrophages fails to elicit transformed cell phenotype. *Mol. Cell. Biol.* 4: 1420–24

Magli, M.-C., Dick, J. E., Huszar, D., Bernstein, A., Phillips, R. A. 1987. Modulation of gene expression in multiple hematopoietic cell lineages following retroviral vector gene transfer. *Proc. Natl. Acad. Sci. USA* 84: 789–93

Mallat, M., Mourna Beto, V., Gros, F., Glowinski, J., Prochiantz, A. 1986. Two simian virus 40 (SV40)-transformed cell lines from the mouse striatum and mesencephalon presenting astrocytic characters. II. Interactions with mesencephalic neurons. *Dev. Brain Res.* 26: 23–31

Marks, P. A., Rifkind, R. A. 1978. Erythroleukemic differentiation. *Ann. Rev. Biochem.* 47: 419–48

Mann, R., Mulligan, R. C., Baltimore, D. 1983. Construction of a retrovirus packaging mutant and its use to produce helper-free defective retrovirus. *Cell* 33: 153–59

Miller, A. D., Buttimore, C. 1986. Redesign of retrovirus packaging cell lines to avoid recombination leading to helper virus production. *Mol. Cell. Biol.* 6: 2895–2902

Miller, A. D., Law, M.-F., Vermer, I. M. 1985. Generation of helper-free amphotropic retroviruses than transduce a dominant-acting, methotrexate-resistant dihydrofolate reductase gene. *Mol. Cell. Biol.* 5: 431–37

Mitrani, E., Coffin, J., Boedtker, H., Doty, P. 1987. Rous sarcoma virus is integrated but not expressed in chicken early embryonic cells. *Proc. Natl. Acad. Sci. USA* 84: 2781–84

Moura Neto, V., Mallat, M., Chneiweiss, H., Premont, J., Gros, F., Prochiantz, A. 1986. Two simian virus (SV40)-transformed cell lines from mouse striatum and mesencephalon presenting astrocytic characters. I. Immunological and pharmacological properties. *Dev. Brain Res.* 26: 11–22

Mulligan, R. C. 1983. Construction of highly transmissible mammalian cloning vehicles derived from murine retroviruses. In *Experimental Manipulation of Gene Expression*, pp. 155–73. New York: Academic

Noda, M., Ko, M., Ogura, A., Liu, D-g., Amano, T., Takano, T., Ikawa, Y. 1983. Sarcoma viruses carrying *ras* oncogenes induce differentiation-associated properties in a neuronal cell line. *Nature* 318: 73–75

Pessac, B., Calothy, G. 1974. Transformation of chick embryo neuroretinal cells by Rous sarcoma virus *in vitro*: Induction of cell proliferation. *Science* 185: 709–10

Pessac, B., Girard, A., Romey, G., Crisanti, P., Lorinet, A.-M., Calothy, G. 1983. A neuronal clone derived from a Rous sarcoma virus-transformed quail embryo neuroretina established culture. *Nature* 302: 616–18

Price, J., Turner, D., Cepko, C. 1987. Lineage analysis in the vertebrate nervous system by retrovirus-mediated gene transfer. *Proc. Natl. Acad. Sci. USA* 84: 156–60

Raetz, C. R. H., Wermuth, M. M., McIntyre, T. M., Esko, J. D., Wing, D. C. 1982. Somatic cell cloning in polyester stacks. *Proc. Natl. Acad. Sci. USA* 79: 3223–27

Roberts, B. E., Miller, J. S., Kimelman, D., Cepko, C. L., Lemischka, I. R., Mulligan, R. C. 1985. Individual adenovirus type 5 early region 1A gene products elicit distinct alterations of cellular morphology and gene expression. *J. Virol.* 56: 404–13

Rosenberg, N., Baltimore, B. 1976. A quantitative assay for transformation of bone marrow cells by Abelson murine leukemia virus. *J. Exp. Med.* 143: 1453–63

Ruley, H. E. 1983. Adenovirus early region 1A enables viral and cellular transforming genes to transform primary cells in culture. *Nature* 304: 602–6

Ryder, E., Snyder, E., Deitcher, D., Cepko, C. 1988. Establishment and characterization of neural cells via retrovirus-mediated oncogene transduction. *Neuron.* Submitted

Ryder, E., Snyder, E., Cepko, C. 1987. Establishment of neural cell lines using retrovirus-vector mediated oncogene transfer. *Soc. Neurosci. Abstr.* 13: 700

Sanes, J. R., Rubenstein, J. L. R., Nicolas, J.-F. 1986. Use of a recombinant retrovirus to study post-implantation cell lineage in mouse embryos. *EMBO J.* 5: 3133–42

Seed, B., Aruffo, A. 1987. Molecular cloning of the CD2 antigen, the T-cell erythrocyte receptor, by a rapid immunoselection procedure. *Proc. Natl. Acad. Sci. USA* 84: 3365–69

Soriano, P., Cone, R. D., Mulligan, R. C., Jaenisch, R. 1986. Tissue-specific and ectopic expression of genes introduced into transgenic mice by retroviruses. *Science* 234: 1409–13

Stewart, T. A., Pattengale, P. K., Leder, P. 1984. Spontaneous mammary adenocarcinomas in transgenic mice that carry and express MTV/myc fusion genes. *Cell* 38: 627–37

Tegtmeyer, P. 1975. Function of simian virus 40 gene A in transforming infection. *J. Virol.* 15: 613–18

Teich, N., Weiss, R., Martin, G., Lowy, D. 1977. Virus infection of murine teratocarcinoma stem cell lines. *Cell* 12: 973–82

Trotter, J., Schachner, M., Boulter, C., Wagner, E. 1987. *Abstr. EMBO-EMBL Symp. 1987*, p. 161

Turner, D., Cepko, C. L. 1987. A common progenitor for neurons and glia persists in rat retina late in development. *Nature* 328: 131–36

Watanabe, S., Temin, H. 1983. Construction of a helper cell line for avian reticuloendotheliosis virus cloning vectors. *Mol. Cell. Biol.* 3: 2241–49

Weiss, R., Teich, N., Varmus, H., Coffin, J. 1984, 1985. *RNA Tumor Viruses.* Cold Spring Harbor, NY: Cold Spring Harbor Lab.

Williams, D. A., Orkin, S. H., Mulligan, R. C. 1986. Retrovirus-mediated transfer of human adenosine deaminase gene sequences into cells in culture and into murine hematopoietic cells in vivo. *Proc. Natl. Acad. Sci. USA* 83: 2566–70

Weizacker, F. v., Beug, H., Graf, T. 1986. Temperature-sensitive mutants of MH2 avian leukemia virus that map in the v-*mil* and v-*myc* oncogene, respectively. *EMBO J.* 5: 1521–27

Wyke, J. A. 1973. The selective isolation of temperature-sensitive mutants of Rouse sarcoma virus. *Virology* 52: 587–90

Yoshimura, M., Iwasaki, Y., Kaji, A. 1981. In vitro differentiation of chicken embryo skin cells transformed by Rous sarcoma virus. *J. Cell. Physiol.* 109: 373–85

*Ann. Rev. Neurosci. 1989. 12 : 67–83*

# STRUCTURE OF THE ADRENERGIC AND RELATED RECEPTORS

*Brian F. O'Dowd, Robert J. Lefkowitz, and Marc G. Caron*

Departments of Medicine, Biochemistry, and Cell Biology,
Howard Hughes Medical Institute, Duke University Medical Center,
Durham, North Carolina 27710

## INTRODUCTION

It is now appreciated that a wide variety of receptors for hormones, drugs, and many neutrotransmitters are coupled to guanine nucleotide regulatory proteins, which in turn link them to various biochemical effectors. Among these are receptors for adrenaline and related catecholamines, termed *adrenergic receptors*. Several subtypes of these receptors are known. The $\beta_1$ and $\beta_2$-adrenergic receptors stimulate the membrane-bound enzyme, adenylyl cyclase, thus raising intracellular cyclic AMP levels. $\alpha_2$-Adrenergic receptors inhibit this enzyme, whereas the $\alpha_1$-adrenergic receptor leads to the hydrolysis of polyphosphoinositides, which in turn generates two second messengers, inositoltrisphosphate and diacylglycerol. Because more is known about the molecular and regulatory properties of the adrenergic receptors than any other such receptors, they have become intensively studied models for understanding the nature and function of receptors that are coupled to guanine nucleotide regulatory proteins. The recent cloning of the genes and/or cDNAs for the various adrenergic receptors (Dixon et al 1986, Yarden et al 1986, Kobilka et al 1987a, Frielle et al 1987), as well as for several other related receptors (Kobilka et al 1987b, Bonner et al 1987, Peralta et al 1987, Masu et al 1987), has led to an increase in the understanding of their structure and the structural basis for their functional properties. We review here this recently acquired information about the adrenergic and several other closely related receptors.

All of the adrenergic receptors have been purified essentially to homo-geneity (Benovic et al 1984, Regan et al 1986, Lomasney et al 1986). The success of this endeavor was based on the development of highly specific affinity chromatography matrices and specific affinity and photoaffinity labeling reagents. An SDS polyacrylamide gel electrophoresis pattern of several of the subtypes of adrenergic receptors after their purification is shown in Figure 1. Each receptor consists of a single polypeptide chain of molecular weight varying between 64 and 80,000 daltons. Each is an integral membrane glycoprotein that binds ligands with all the appropriate specificity and stereospecificity characteristics that would be expected for an adrenergic receptor. Moreover, these receptors can be reconstituted in

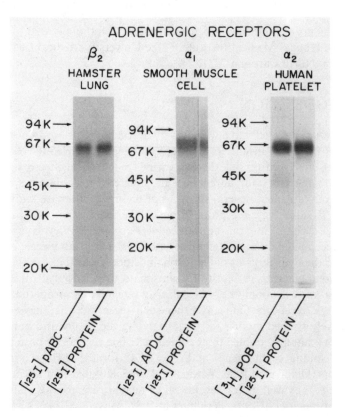

*Figure 1* SDS-polyacrylamide gel electrophoresis profiles of $\beta_2$, $\alpha_1$, and $\alpha_2$-adrenergic recep-tors. The various receptors were purified by affinity chromatography and other procedures, as described (Benovic et al 1984, Regan et al 1986, Lomasney 1986). Purified preparations were visualized by radioiodination of the protein or by affinity and photoaffinity labeling using [³H]phenoxybenzamine or the selective photoaffinity probes [¹²⁵I]pABC and [¹²⁵I]APDQ.

phospholipid vesicles with their appropriate effector guanine nucleotide regulatory proteins (Cerione et al 1986a,b). Thus it can be documented that the single polypeptide chain contains all the functional elements required for ligand binding on the one hand and for activation of coupled guanine nucleotide regulatory proteins on the other.

Based on the purification of these receptors, limited information about protein sequence from several proteolytic fragments was obtained. This information was used to construct oligonucleotide probes that were then used in successful molecular cloning experiments. As a consequence these genes for the various adrenergic receptors and several other members of the family, such as several subtypes of muscarinic cholinergic receptors, have been cloned. The sequences of these genes have revealed that all of these receptors are homologous proteins. Moreover, they all show significant similarities to the visual "receptor" rhodopsin as well (Dixon et al 1986).

# PROTEIN STRUCTURE

## Topography of the Hormone Receptors

The deduced amino acid sequences of the adrenergic receptors and other members of this gene family are outlined in Figure 2. The receptors are each composed of a single polypeptide chain, ranging in size from 477 amino acids (human $\beta_1$) to 413 amino acids (human $\beta_2$) and are clearly similar throughout their sequences, suggesting that their tertiary structures are also similar. The scheme shown in Figure 3 for dividing the sequence into extra cytoplasmic, membranous, and cytoplasmic regions is consistent with experimental observations and also with the topographically better characterized rhodopsin system.

Dohlman et al (1987a) have tested the validity of the hypothesis that the topography of the $\beta_2$-adrenergic receptor ($\beta_2$-AR) includes seven membrane-spanning domains. To do this they reconstituted the hamster $\beta_2$-AR into lipid vesicles and used proteolysis with trypsin and carboxypeptidase Y as well as treatment with endoglycosidase F. By locating key structural landmarks such as a carboxy tail of 7 kD sensitive to carboxypeptidase Y, which contains sites of phosphorylation, a glycosylated amino terminus, an extracellular trypsin sensitive site, and a ligand binding site in the amino terminal domain (see Figure 3), they have demonstrated the existence of five of the proposed seven transmembrane segments.

Other evidence suggests that $\beta_2$-AR contains seven transmembrane regions. The hydropathicity profile of $\beta_2$-AR is very similar to those of rhodopsin and bacteriorhodopsin (Dixon et al 1986), proteins known to have seven membrane spanning regions (Applebury & Hargrave 1986).

RAT m3

BOVINE RHODOPSIN
HUMAN B2AR
HAMSTER B2AR
HUMAN B1AR
TURKEY BAR
HUMAN A2AR
PORCINE m1
PORCINE m2

HUMAN m4
HUMAN G21
BOVINE SUBSTANCE K

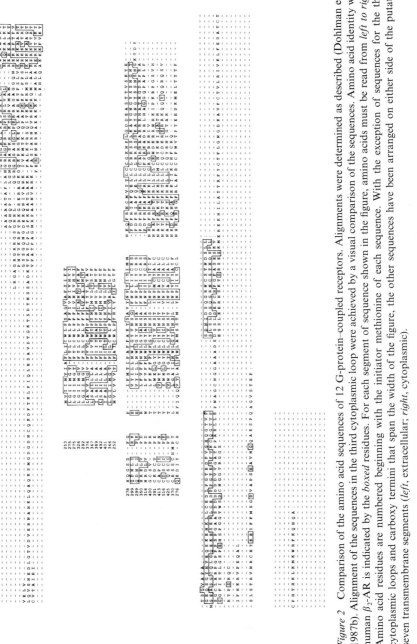

*Figure 2*   Comparison of the amino acid sequences of 12 G-protein-coupled receptors. Alignments were determined as described (Dohlman et al 1987b). Alignment of the sequences in the third cytoplasmic loop were achieved by a visual comparison of the sequences. Amino acid identity with human β₂-AR is indicated by the *boxed* residues. For each segment of sequence shown in the figure, amino acids must be read from *left to right*. Amino acid residues are numbered beginning with the initiator methionine of each sequence. With the exception of sequences for the third cytoplasmic loops and carboxy termini that span the width of the figure, the other sequences have been arranged on either side of the putative seven transmembrane segments (*left*, extracellular; *right*, cytoplasmic).

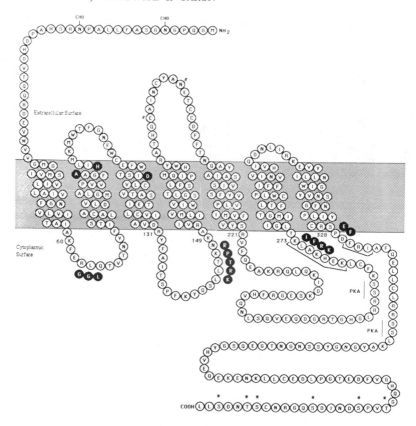

*Figure 3*  A model illustrating the topographical structure of the human $\beta_2$-AR in the membrane. Deletion mutants are indicated by the *brackets*, and substituted amino acids in the cytoplasmic segments of the receptor are *highlighted*. In the first cytoplasmic loop three mutant receptors were prepared: S64 (substitution of the leucine with glycine); S63, 65A (substitution of arginine and glutamine with glycine and leucine); and S63, 65B (substitution of arginine and glutamine with aspartic and glutamic acid). The highlighted amino acids in helix 2 contain the site of [$^{125}$I]-pBABC incorporation (in the hamster $\beta_2$-AR). The asparagine residue highlighted in helix 3 has been proposed as the counterion for the adrenergic ligands (Strader et al 1987a). Also shown are the two sites of PKA phosphorylation, the putative sites of $\beta$ARK phosphorylation (*), and the extracellular trypsin sensitive site (#) found only in the hamster $\beta_2$-AR.

Based on analogy to the structure of rhodopsin and bacteriorhodopsin, the membrane spanning regions of $\beta_2$-AR and the other related hormone binding receptors are believed to be $\alpha$-helical in character. Electron microscopic maps of bacteriorhodopsin have revealed seven closely packed perpendicular helical segments (Henderson & Unwin 1975). Interestingly, three of the helices are tilted, thus suggesting a structural requirement for interlocking of the amino acid side chains from adjacent helices. Results obtained with rhodopsin also predict α-helices and that proline residues, which are located in several of the helices (also found in the hormone binding receptors, Figure 2), will create a 20° kink in these helices (Michel-Villaz et al 1979).

# ADRENERGIC AND MUSCARINIC CHOLINERGIC RECEPTOR PROTEIN SEQUENCES

It appears from a comparison of their coding sequences that segments of the genes for the adrenergic and muscarinic receptors are subject to different selective pressures. The amino termini, the third cytoplasmic loops, and the carboxy termini show less similar sequences than the putative membrane spanning domains, a finding that perhaps indicates less stringent structural requirements for these segments.

## Membrane-Spanning Domains

Of the 49 amino acid differences that exist between the human and hamster $\beta_2$-ARs, only nine of these amino acids are found in the putative transmembrane regions. Thus the most conserved regions in the structures of these receptors are the transmembrane domains. The percentage of amino acids for several related receptors that are identical with those in the human $\beta_2$-AR, both within the membrane-spanning domain (168 residues) and overall for the entire protein, are as follows: human $\beta_1$-AR, 71% and 54%; human $\alpha_2$-AR, 39% and 27%; human G-21, 43% and 30%; porcine m2 receptor, 28% and 22%; and porcine m1 receptor, 33% and 23%.

Thirty-two amino acids in the membrane-spanning domains are conserved in both the adrenergic and muscarinic receptors (Figure 4). Of the residues in the membrane segments conserved between the adrenergic and muscarinic receptors, the majority (78%) are located on the half of the α-helix closest to the cytoplasm (*black circles* in Figure 4). These residues may be conserved between the different classes of receptors because of requirements for interaction with their common cytoplasmic effectors, i.e. G proteins that have highly conserved primary structures. In contrast, the majority of the amino acids that are found exclusively in one or the other type of receptor, either in adrenergic (67%) or muscarinic (64%), are located on the side of the helix closest to the extracellular surface (*white circles* with letters in Figure 4). Thus, among the amino acids in the helices within the cell membrane, the adrenergic and muscarinic receptors show least similarity on the side of the receptor facing the outside of the cell.

Adrenergic extracellular surface

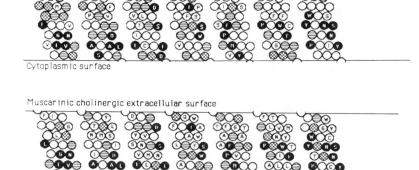

Cytoplasmic surface

Muscarinic cholinergic extracellular surface

Cytoplasmic surface

*Figure 4* Distribution of the conserved and divergent amino acids in the transmembrane regions of the adrenergic receptors (human $\beta_1$ and $\beta_2$, turkey $\beta$ and $\beta_1$, and human $\alpha_2$-AR) and muscarinic cholinergic receptors (porcine m1 and m2, rat m3, and human m4).

Amino acids identical in both the adrenergic and muscarinic receptors are highlighted by the *black circles*. Amino acids exclusive to one or the other type of receptor (either adrenergic or muscarinic) are indicated by letters within *white circles*. Conservative amino acid substitutions in adrenergic and muscarinic are indicated as *striped circles*. Conservative amino acid substitution among all the adrenergic receptors or among the muscarinic receptors are shown as *hatched circles*. Conservative amino acid substitutions were grouped as follows: S, T, P, A, G; M, I, L, V; F, Y, W; N, D, E, Q; R, K, H; C.

This may indicate that the amino acids in this region of the receptors account for their unique binding specificities and that the bound ligands occupy a pocket formed by these regions of the molecule. Two groups have recently confirmed that residues located on the half of the $\alpha$-helices nearest the extracellular surface are involved in ligand binding. Dohlman et al (1988) have reported that the alkylating agent [$^{125}$I]pBABC, a probe that covalently incorporates into the hamster $\beta_2$-AR, binds to either the His-93 or Ser-92 residues (indicated in Figure 3). Site specific mutagenesis of Asp113 → Asn on the third transmembrane helix (indicated in Figure 3) eliminated antagonist binding of the hamster $\beta_2$-AR (Strader et al 1987a). Strader et al suggested that binding of adrenergic and muscarinic ligands involves hydrogen bonding between the carboxylate group of the Asp-113 and the protonated amino group of the ligands.

Interestingly, in the hamster $\beta_2$-AR substitution of the cysteine residue, Cys-106, situated in the extracellular loop joining the second and third transmembrane sequences, caused a dramatic change in agonist binding (Dixon et al 1987). The authors suggested that Cys-106 is involved in a disulphide bond with Cys-184 in the native $\beta_2$-AR and that ligand binding is affected due to a structural destabilization of the receptor in the absence of the disulphide bridge.

Several proline and glycine residues in the membrane $\alpha$-helices are conserved among the adrenergic and muscarinic receptors and the opsins (Figure 2). Proline residues can induce bends in the transmembrane helix and glycine residues can be contained in a bend (Applebury & Hargrave 1986). All the receptors have conserved prolines in five out of seven helices and conserved glycines in three out of the seven helices. It has been suggested that the bent helices might interlock adjacent helices or create a pocket to accommodate a ligand within the transmembrane helices (Applebury & Hargrave 1986). Proline residues in the $\alpha$-helices may also mediate reversible conformational changes in the protein via the *cis-trans* isomerization of the X-Pro peptide bonds (Brandl & Deber 1986).

A positively charged amino acid is apparently required to demarcate the cytoplasmic regions of each of the receptors from the seven hydrophobic membranous regions (e.g. in human $\beta_2$-AR: these residues are $K^{60}$, $R^{131}$, $K^{149}$, $R^{221}$, $K^{273}$, $R^{328}$). The positions are indicated in Figure 3. An analysis by von Heigne (1986) revealed that integral membrane proteins frequently have charged residues in this position and he suggests that they play a role in determining the correct orientation of the protein embedded in the membrane. For rhodopsin, Nathans & Hogness (1984) have suggested that positively charged residues at the cytoplasmic interfaces interact with the negatively charged phospholipid head groups, thus securing the receptor in the membrane.

## Amino Terminus

The lack of significant amino acid similarity and the variations in the lengths of the amino termini among the various receptors suggest that no general functional role exists for the amino terminus. Each of the receptors contains putative glycosylation sites in this region, and $\beta_2$-AR (Stiles et al 1984, Benovic et al 1987a), $\beta_1$-AR (Strasser et al 1984), and $\alpha_2$-AR (Regan 1988) have all been shown to contain oligosaccharides. Dohlman et al (1987a) have established that in the hamster $\beta_2$-AR, of the four putative sites for $N$-linked glycosylation, only those sites in the amino terminal domain ($Asn^6$ and $Asn^{15}$) are utilized and are sensitive to Endo F. The other consensus glycosylation sites are located at the carboxy terminus of the receptor. Also, Benovic et al (1987a) have demonstrated that these oligosaccharides can be complex, hybrid, or high mannose, whereas the $\beta$-AR from turkey erythrocytes contains only complex oligosaccharides (Cervantes et al 1985). These oligosaccharides appear not to take part in processing of the receptors, ligand binding, or biological activating functions (George et al 1986), and may be required to protect the amino terminus from protease attack. Neither bovine rhodopsin nor the hamster $\beta_2$-AR have cleavable $N$-terminal signal sequences. It has been suggested that the first transmembrane region may serve as an uncleaved signal sequence (Singer et al 1987). Blobel and his colleagues have recently reported that five of the seven transmembrane regions of bovine opsin could function as signal sequences (Audigier et al 1987).

## Carboxy Tail and Cytoplasmic Regions

A major function probably associated with the cytoplasmic domains of the adrenergic and related receptors is coupling to effector guanine nucleotide regulatory proteins. In a series of human $\beta_2$-AR mutants, we have explored the function of each of the cytoplasmic structural domains. The mutant receptors were expressed in *Xenopus* oocytes, and the isoproterenol-stimulated adenylyl cyclase activity was determined.

Each of the receptors with mutations in the first loop (Figure 3) led to a marked reduction in the binding of [$^{125}$I]cyanopindolol (probably due to abnormal folding of the protein). Despite the low ligand binding, activation of adenylyl cyclase occurred in each case, suggesting that residues in the first loop are not directly involved in cyclase coupling. The carboxy end of the second cytoplasmic loop in $\alpha_2$-AR, G-21, and the m1 muscarinic receptor contain the homologous sequences (K or R, R, T, P, K, or R). Substitution of this sequence (which is not present in any of the $\beta$-adrenergic receptors) into human $\beta_2$-AR had no effect on either ligand binding or adenylyl cyclase stimulation.

Both the third cytoplasmic loop and the amino terminal region of the carboxy tail of rhodopsin have been directly implicated in rhodopsin-transducin coupling (Takemoto et al 1985, Littman et al 1982, Kuhn & Hargrave 1981). Two mutations in the carboxy terminal region of the third cytoplasmic loop of the human $\beta_2$-AR, one involving a deletion of seven amino acids and the other a substitution of four amino acids (Figure 3), each caused a marked decrease in agonist dependent adenylyl cyclase activation (Figure 5). A mutation involving the deletion of three amino acids and substitution of two amino acids in the amino terminal region of the carboxy tail of $\beta_2$-AR also affected the agonist-dependent adenylyl cyclase activation (Figure 5). Strader and her colleagues (1987b) reported that a deletion either in the amino terminus of the third loop or a large deletion in the carboxy terminal region of the third loop of the hamster $\beta_2$-AR uncoupled the receptor from adenylyl cyclase activity. Thus the data available so far suggest that analogous regions in $\beta_2$-AR and rhodopsin, involving the terminal regions of the third cytoplasmic loop and the amino terminal region of the carboxy tail, are involved in G-protein/transducin interactions.

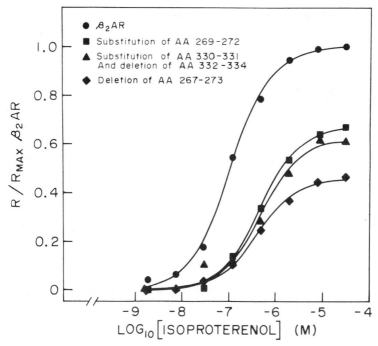

*Figure 5* $\beta_2$-AR- and mutant $\beta_2$-AR–mediated stimulation of adenylyl cyclase in *X. laevis* oocyte membranes by isoproterenol.

The ordinate represents pmol of cAMP formed per hour per mg oocyte membrane protein for the various receptor species relative to the maximum obtained from $\beta_2$-AR controls. Data are representative of two to five experiments with each point determined in duplicate. The levels of expression (i.e. fmol receptor/mg membrane protein) of the various mutant receptor species were within 20% of that obtained with $\beta_2$-AR. Expression level did not influence the maximal extent of stimulation of adenylyl cyclase but did affect the $EC_{50}$ values for isoproterenol.

Light-dependent phosphorylation of ovine rhodopsin results in the phosphorylation by rhodopsin kinase of five out of seven of the serines and threonines contained in the last 15 residues of the carboxy terminal region (Thompson & Findlay 1984). A serine residue on the third loop of rhodopsin may also be phosphorylated by rhodopsin kinase (McDowell et al 1985). All the $\beta$-adrenergic receptors also have concentrations of serine and threonine residues in their carboxy terminal sequences (Figure 2). Recent evidence indicates that the agonist-occupied $\beta_2$-AR may also be phosphorylated by an enzyme analogous to rhodopsin kinase, called $\beta$-adrenergic receptor kinase ($\beta$ARK) (Benovic et al 1986a). $\beta$ARK is also capable of phosphorylating rhodopsin in a light-dependent fashion and extensibly on the same sites as rhodopsin kinase (Benovic et al 1986b). Phosphorylation of rhodopsin by rhodopsin kinase results ultimately in decreased coupling (via a conformational change in the carboxy tail of rhodopsin) to cGMP phosphodiesterase activation (Pellicone et al 1985, Kuhn et al 1982). Phosphorylation of agonist-occupied $\beta_2$-AR by $\beta$ARK leads to uncoupling of the receptor from the guanine nucleotide regulatory protein (Benovic et al 1986b).

The $\alpha_2$-AR can also be phosphorylated in an agonist-dependent fashion by $\beta$ARK (Benovic et al 1987b); however, its carboxy terminus contains no serine or threonine residues. Each of the cytoplasmic loops of $\alpha_2$-AR contains serine or threonine residues, and, by analogy with rhodopsin, the serine and threonine residues in the third loop may serve as substrates for $\beta$ARK. Similarly, the porcine brain and cardiac muscarinic receptors do not have carboxy termini that are serine or threonine rich, but several such regions are found in their third cytoplasmic loops. $\beta$ARK may phosphorylate nonhomologous sequences present in the cytoplasmic tail or third loop of these receptors. The determinants in the coding sequence for recognition by $\beta$ARK, other than the presence of serine or threonine, are not yet clear. The various regions may generate domains that are three-dimensionally similar. Alternatively, the high number of charged residues surrounding the serine and threonine residues may be important. In this context it is interesting that the mouse $\beta_2$-AR specifically interacts with site-directed antibodies raised against rhodopsin sequences corresponding to the serine- and threonine-rich region of the carboxy tail (Weiss et al 1987). These data indicate that this region in rhodopsin exhibits three-dimensional structural similarity with the carboxy tail of the $\beta_2$-AR, even though there is little if any similarity in primary amino acid sequence.

To evaluate the importance of the carboxy terminal serine and threonine residues, truncated receptor mutants, lacking most of these residues, have been expressed in mammalian cell systems (Bouvier et al 1988). Chinese hamster fibroblasts (CHW cells), expressing either the wild type human $\beta_2$-AR or a mutant of the human $\beta_2$-AR truncated after amino acid 365 (T-365), were exposed to the agonist isoproterenol (2 $\mu$M) for 2–180 min. With the wild-type receptor a decrease in the isoproterenol-stimulated adenylyl cyclase activity occurred within the first 2 min and continued up to 180 min. These receptors were also demonstrated to have been phosphorylated two to three fold. However, this rapidly occurring desensitization process at 2 and 10 min was absent with the mutant receptor. Also, no agonist-promoted increase in phosphorylation was observed at any time point (from 5 min to 5 hr). Both Strader et al (1987c) and Kobilka et al (1987c) have reported that $\beta_2$-AR mutants with carboxy-terminus truncations showed desensitization, similar to the wild type receptor, after longer periods of agonist exposure. Thus, the rapid desensitization event, involving phosphorylation of the receptor, apparently requires the participation of the serine and threonine residues in the carboxy tail. Further experiments using the hydrophilic ligand, [$^3$H]CGP 12177, which binds only to surface receptors, indicated that the cells expressing the mutant receptor displayed an amplified agonist-induced "sequestration" of the receptors (Bouvier et al 1988). This may suggest that the carboxy tail of the receptor functions to inhibit sequestration. The truncation mutant, lacking the steric constraint provided by the carboxy tail, would thus display an increase in sequestration.

The presence of related functional elements (e.g. PKA and $\beta$ARK phosphorylation sites) in both the tail and the third cytoplasmic loop of these receptors has suggested that these regions may also have structural features in common. In fact, a comparison of the sequences of the third loop and cytoplasmic tail of the $\beta_2$-AR reveals similarity and suggests the possibility that these regions may have arisen from an internal duplication event, (Figure 6A). The third cytoplasmic loop of the $\alpha_2$-AR is approximately three times longer than the equivalent loop in $\beta_2$-AR and appears to have increased in size by gaining several internal duplications (Figure 6B). Thus, internal duplications, following a complete gene duplication event, may have allowed the $\alpha_2$-AR to diverge in function from the other adrenergic receptors.

## GENE STRUCTURE

The genes encoding the $\beta_2$-adrenergic receptor (Kobilka et al 1987d), $\alpha_2$-adrenergic receptor (Kobilka et al 1987a), each of the muscarinic cholinergic receptors (Bonner et al 1987, Peralta et al 1987), and the putative G-21 protein (Kobilka et al 1987b) lack introns both in the coding and 3′ untranslated regions. The rat and porcine muscarinic receptors contain introns in the 5′ untranslated regions (Bonner et al 1987, Peralta et al 1987). It has been suggested that this family of genes, intronless in the coding regions, may have evolved by a gene-processing event from a common primordial gene. In support of this concept, the $\beta_2$-AR is flanked by the nucleotide sequence, 5′ ATTGTTTG3′, which is closely homologous to the sequences that flank the processed gene phosphoglycerate kinase, 5′ TATGTTT(G)3′ (McCarrey & Thomas 1987). A similar sequence is also found in the 3′ untranslated region of the porcine atrial muscarinic gene (5′ ATGTTT3′).

An examination of the gene structures 5′ to the single expressed sequences of the $\beta_2$-AR and muscarinic receptors has revealed several interesting features. These genes contain ATG codons, which are separated

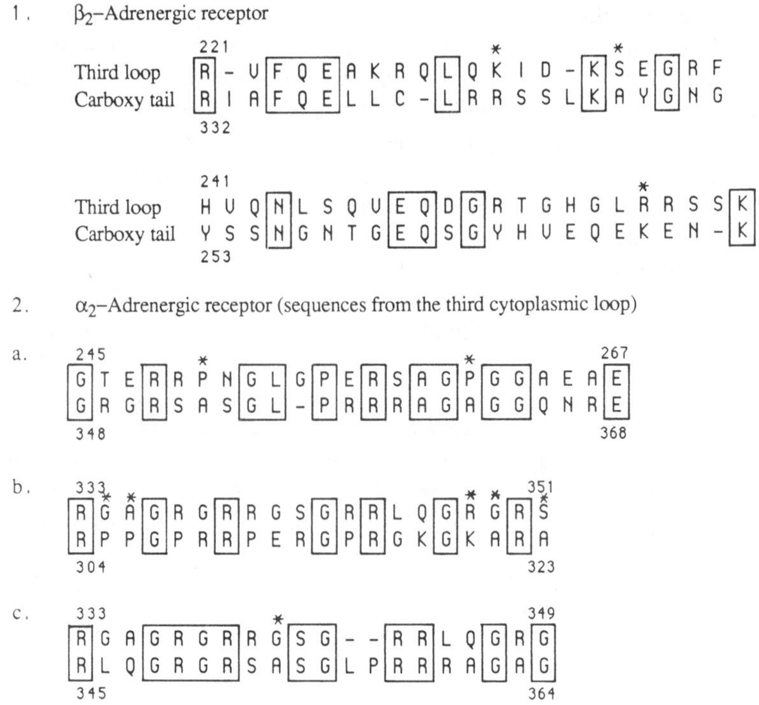

1.   β₂-Adrenergic receptor

Figure 6   Internal repeat amino acid sequences in β₂-AR and α₂-AR. 1. Alignment of amino acids from the third cytoplasmic loop and carboxy terminus of human β₂-AR (see Figure 3). 2. Alignment of amino acid sequences within the third cytoplasmic loop of human α₂-AR. *Boxed* residues indicate identical amino acids; * indicates conserved amino acid substitutions. Gaps (−) were introduced to maximize similarity.

by short open reading frames and termination codons from the initiator codons. Removal of the upstream ATG codons results in a tenfold increase in translation of the β₂-AR mRNA, thus indicating that this region may be involved in control of expression (Kobilka et al 1987c). In addition to the major site of transcription initiation at base −219, situated near a typical "housekeeping" promoter, the β₂-AR gene may also be transcribed under the influence of hormonal signals from a more distal promoter (Kobilka et al 1987d). There is a consensus TATA box at position −1202, a consensus CAAT box at position −1289, and a steroid binding hexamer at position −1438. The observation that addition of hydrocortisone to rabbit TP3 cells, transfected with the human β₂-AR gene, caused a doubling in receptor number also indicates that functional glucocorticoid elements are present in the gene (Emorine et al 1987). However, the genomic clone used in this study did not include the putative steroid binding element at $-1438$, but these workers have identified similar sequences in the 5′ untranslated regions, at base number $-273$ and also in the 3′ untranslated region.

# SUMMARY

The isolation and sequencing of a number of G protein–coupled receptors has now provided extensive primary structure information for this family of homologous proteins. The diverse nature of these receptors suggests that the family of proteins may grow to include receptors for many neurotransmitters and perhaps many peptide hormones. The topography of these receptors, a single polypeptide with seven transmembrane segments, appears to have features well suited for the transmission of signals, via conformational changes, to the interior of the cell. Detailed site-directed mutagenesis studies are now underway in many laboratories to understand the significance of the topography and also the regions of homology evident in the structures of all of these receptors.

Obvious features of interest are the precise residues involved in the coupling of the receptors to the G-proteins and the identification of the residues required for ligand binding in each of the receptors, as well as domains of these receptors involved in the regulation of receptor function. In addition, the availability of molecular probes for this family of proteins will permit the elucidation of mechanisms of regulation at the gene level.

ACKNOWLEDGMENTS

We would like to thank Mark Hnatowich for his contribution to the human $\beta_2$-AR mutagenesis experiments and also for the preparation of Figures 2 and 5. Thanks are also due to Henrik Dohlman for the preparation of Figure 2 and to Mary Holben for typing of this manuscript.

*Literature Cited*

Applebury, M. L., Hargrave, P. A. 1986. Molecular biology of the visual pigments. *Vision Res.* 26: 1881–95

Audigier, Y., Frielander, M., Blobel, G. 1987. Multiple topogenic sequences in bovine opsin. *Proc. Natl. Acad. Sci. USA* 84: 5783–87

Benovic, J. L., Shorr, R. G. L., Caron, M. G., Lefkowitz, R. J. 1984. The mammalian $\beta_2$-adrenergic receptor: Purification and characterization. *Biochemistry* 23: 4510–18

Benovic, J. L., Strasser, R. H., Caron, M. G., Lefkowitz, R. J. 1986a. $\beta$-Adrenergic receptor kinase: Identification of a novel protein kinase that phosphorylates the agonist-occupied from the receptor. *Proc. Natl. Acad. Sci. USA* 83: 2797–2801

Benovic, J. L., Mayor, F. Jr., Somers, R. L., Caron, M. G., Lefkowitz, R. J. 1986b. Light-dependent phosphorylation of rhodopsin by $\beta$-adrenergic receptor kinase. *Nature* 321: 869–72

Benovic, J. L., Staniszewski, C., Cerione, R.

A., Codina, J., Lefkowitz, R. J., Caron, M. G. 1987a. The mammalian $\beta$-adrenergic receptor: Structural and functional characterization of the carbohydrate moiety. *J. Receptor Res.* 7(1–4): 257–81

Benovic, J. L., Regan, J. W., Matsui, H., Mayor, F. M. Jr., Cotecchia, S., Leeb-Lundberg, L. M. F., Caron, M. G., Lefkowitz, R. J. 1987b. Agonist-dependent phosphorylation of the $\alpha_2$-adrenergic receptor by the $\beta$-adrenergic receptor kinase. *J. Biol. Chem.* 262: 17251–53

Bonner, T. I., Buckley, N. J., Young, A. C., Brann, M. R. 1987. Identification of a family of muscarinic acetylcholine receptor genes. *Science* 237: 527–32

Bouvier, M., Hausdorff, W. P., DeBlasi, A., O'Dowd, B. F., Kobilka, B. K., Caron, M. G., Lefkowitz, R. J. 1988. Removal of phosphorylation sites from the $\beta_2$-adrenergic receptor delays onset of agonist-promoted desensitization. *Nature* 333: 370–73

Brandl, C. J., Deber, C. M. 1986. Hypothesis about the function of membrane-buried proline residues in transport proteins. *Proc. Natl. Acad. Sci. USA* 83: 917–21

Cerione, R. A., Benovic, J. L., Codina, J., Birnbaumer, L., Lefkowitz, R. J., Caron, M. G. 1986a. The $\beta$-adrenergic receptor-coupled adenylate cyclase: Reconstitution of the functional interactions of the various purified components. In *The Receptors*, ed. P. M. Conn, 4: 1–36. New York: Academic

Cerione, R. A., Regan, J. W., Nakata, H., Codina, J., Benovic, J. L., Gierschik, P., Somers, R. L., Spiegel, A. M., Birnbaumer, L., Lefkowitz, R. J., Caron, M. G. 1986b. Functional reconstitution of the $\alpha_2$-adrenergic receptor with guanine nucleotide regulatory proteins in phospholipid vesicles. *J. Biol. Chem.* 261: 3901–9

Cervantes-Oliver, P., Durieu-Trautman, O., Delavier-Klutchko, C., Strosberg, A. D. 1985. The oligosaccharide moiety of the $\beta_1$-adrenergic receptor from turkey erythrocytes has a bianterrary, *N*-actyllactosamine-containing structure. *Biochemistry* 24: 3765–70

Dixon, R. A. F., Kobilka, B. K., Strader, D. J., Benovic, J. L., Dohlman, H. G., Frielle, T., Bolanowski, M. A., Bennett, C. D., Rands, E., Diehl, R. F., Mumford, R. A., Slater, E. E., Sigal, I. S., Caron, M. G., Lefkowitz, R. J., Strader, C. D. 1986. Cloning of the gene and cDNA for mammalian $\beta$-adrenergic receptor and homology with rhodopsin. *Nature* 321: 75–79

Dixon, R. A. F., Sigal, I. S., Candelore, M. R., Register, R. B., Scattergood, W., Rands, E., Strader, C. D. 1987. Structural

features required for ligand binding to the $\beta$-adrenergic receptor. *EMBO J.* 6: 3269–75

Dohlman, H. G., Bouvier, M., Benovic, J. L., Caron, M. G., Lefkowitz, R. J. 1987a. The multiple membrane spanning topography of the $\beta_2$-adrenergic receptor. *J. Biol. Chem.* 262: 14282–88

Dohlman, H. G., Caron, M. G., Lefkowitz, R. J. 1987b. A family of receptors coupled to guanine nucleotide regulatory proteins. *Biochemistry* 26: 2657–64

Dohlman, H. G., Caron, M. G., Strader, C. D., Amlaiky, N., Lefkowitz, R. J. 1988. Identification and sequence of a binding site peptide of the $\beta_2$-adrenergic receptor. *Biochemistry* 27: 1813–17

Emorine, L. J., Marullo, S., Delavier-Klutchko, C., Kaveri, S. V., Durieu-Trautmann, O., Strosberg, A. D. 1987. Structure of the gene for human $\beta_2$-adrenergic receptor: Expression and promoter characterization. *Proc. Natl. Acad. Sci. USA* 84: 6995–99

Frielle, T., Collins, S., Daniel, K. W., Caron, M. G., Lefkowitz, R. J., Kobilka, B. K. 1987. Cloning of the cDNA for the human $\beta_1$-adrenergic receptor. *Proc. Natl. Acad. Sci. USA* 84: 7920–24

George, S. T., Ruoho, A. E., Malbon, C. C. 1986. *N*-Glycosylation in expression and function of $\beta$-adrenergic receptors. *J. Biol. Chem.* 261: 16559–64

Henderson, R., Unwin, P. N. T. 1975. Three-dimensional model of purple membrane obtained by electron microscopy. *Nature* 257: 28–32

Kobilka, B. K., Matsui, H., Kobilka, T. S., Yang-Feng, T. L., Francke, U., Caron, M. G., Lefkowitz, R. J., Regan, J. W. 1987a. Cloning, sequencing, and expression of the gene coding for the human platelet $\alpha_2$-adrenergic receptor. *Science* 238: 650–56

Kobilka, B. K., Frielle, T., Collins, T., Yang-Feng, T., Kobilka, T. S., Francke, U., Lefkowitz, R. J., Caron, M. G. 1987b. An intronless gene encoding a potential member of the family of receptors coupled to guanine nucleotide regulatory proteins. *Nature* 329: 75–79

Kobilka, B. K., MacGregor, C., Daniel, K., Kobilka, T. S., Caron, M. G., Lefkowitz, R. J. 1987c. Functional activity and regulation of human $\beta_2$-adrenergic receptors expressed in *Xenopus* oocytes. *J. Biol. Chem.* 262: 15796–15802

Kobilka, B. K., Frielle, T., Dohlman, H. G., Bolanowski, M. A., Dixon, R. A. F., Keller, P., Caron, M. G., Lefkowitz, R. J. 1987d. Delineation of the intronless nature of the genes for the human and hamster $\beta_2$-adrenergic receptor and their putative promoter regions. *J. Biol. Chem.* 262: 7321–27

Kuhn, H., Mommertz, O., Hargrave, P. A. 1982. Light dependent conformational change at rhodopsin's cytoplasmic surface detected by increased susceptibility to proteolysis. *Biochim. Biophys. Acta* 679: 95–100

Littman, B. J., Aton, B., Hartley, J. B. 1982. Functional domains of rhodopsin. *Vision Res.* 22: 1439–42

Lomasney, J. W., Leeb-Lundberg, L. M., Cotecchia, S., Regan, J. W., DeBernardis, J. F., Caron, M. G., Lefkowitz, R. J. 1986. Mammalian $\alpha_1$-adrenergic receptor. *J. Biol. Chem.* 261: 7710–16

Masu, Y., Nakayama, K., Tamaki, H., Harada, Y., Kuno, M., Nakanishi, S. 1987. cDNA cloning of bovine substance-K receptor through oocyte expression system. *Nature* 329: 836–38

McCarrey, J. R., Thomas, K. 1987. Human testis-specific PGK gene lacks introns and possesses characteristics of a processed gene. *Nature* 326: 501–5

McDowell, J. H., Curtis, D. R., Baker, V. A., Hargrave, P. A. 1985. Phosphorylation of rhodopsin: Localization of phosphorylated residues in the helix-V-helix VI connecting loop. *Invest. Ophthalmol. Visual Sci.* 26: 291

Michel-Villaz, M., Saibil, H. R., Chabre, M. 1979. Orientation of rhodopsin $\alpha$-helices in retinal rod outer segment membranes studied by infrared linear dichroism. *Proc. Natl. Acad. Sci. USA* 76: 4405–8

Nathans, J., Hogness, D. S. 1984. Isolation and nucleotide sequence of the gene encoding human rhodopsin. *Proc. Natl. Acad. Sci. USA* 81: 4851–55

Pellicone, C., Nullans, G., Cook, N. J., Virmaux, N. 1985. Light-induced conformational changes in the extradiscal regions of bovine rhodopsin. *Biochem. Biophys. Res. Commun.* 127: 816–21

Peralta, E. G., Winslow, J. W., Peterson, G. L., Smith, D. H., Ashkenazi, A., Ramachandran, J., Schimerlik, M. I., Capon, D. J. 1987. Primary structure and biochemical properties of an $M_2$ muscarinic receptor. *Science* 236: 600–5

Regan, J. W. 1988. The biochemistry of alpha-2 adrenergic receptors. In *The $\alpha_2$-Adrenergic Receptors*, ed. L. E. Limbird, pp. 15–74. Clifton, NJ: Humana

Regan, J. W., Nakata, H., DeMarinis, R. M., Caron, M. G., Lefkowitz, R. J. 1986. Purification and characterization of the human platelet $\alpha_2$-adrenergic receptor. *J. Biol. Chem.* 261: 3894–3900

Singer, S. J., Maher, P. A., Yaffe, M. P. 1987. On the transfer of integral proteins into membranes. *Proc. Natl. Acad. Sci. USA* 84: 1960–64

Stiles, G. L., Benovic, J. L., Caron, M. G., Lefkowitz, R. J. 1984. Mammalian $\beta$-adrenergic receptors: Distinct glycoprotein populations containing high mannose or complex type carbohydrate chains. *J. Biol. Chem.* 259: 8655–63

Strader, C. D., Sigal, I. S., Register, R. B., Candelore, M. R., Rands, E., Dixon, R. A. F. 1987a. Identification of residues required for ligand binding to the $\beta$-adrenergic receptor. *Proc. Natl. Acad. Sci. USA* 84: 4384–88

Strader, C. D., Dixon, R. A. F., Cheung, A. H., Candelore, M. R., Blake, A. D., Sigal, I. S. 1987b. Mutations that uncouple the $\beta$-adrenergic receptor from $G_s$ and increase agonist affinity. *J. Biol. Chem.* 262: 16439–43

Strader, C. D., Sigal, I. S., Blake, A. D., Cheung, A. H., Register, R. B., Rands, E., Zemcik, B. A., Candelore, M. R., Dixon, R. A. F. 1987c. The carboxyl terminus of the hamster $\beta$-adrenergic receptor expressed in mouse L cells is not required for receptor sequestration. *Cell* 49: 855–63

Strasser, R. H., Stiles, G. L., Lefkowitz, R. J. 1984. Differences in the carbohydrate chains of mammalian $\beta_1$ and $\beta_2$-adrenergic receptors. *Circulation* 70: 68 (Abstr.)

Takemoto, D. J., Takemoto, L. J., Hansen, J., Morrison, D. 1985. Regulation of retinal transducin by C-terminal peptides of rhodopsin. *Biochem. J.* 232: 669–72

Thompson, P., Findlay, J. B. C. 1984. Phosphorylation of ovine rhodopsin. *Biochem. J.* 220: 773–80

von Heijne, G. 1986. The distribution of positively charged residue in bacterial inner membrane proteins correlates with the transmembrane topology. *EMBO J.* 5: 3021–27

Weiss, E. R., Hadcock, J. R., Johnson, G. L., Malbon, C. C. 1987. Antipeptide antibodies directed against cytoplasmic rhodopsin sequences recognize the $\beta$-adrenergic receptor. *J. Biol. Chem.* 262: 4319–23

Yarden, Y., Rodriguez, H., Wong, S. K.-F., Brandt, D. R., May, D. C., Burnier, J., Harkins, R. N., Chen, E. Y., Ramachandran, J., Ullrich, A., Ross, E. M. 1986. The avian $\beta$-adrenergic receptor: Primary structure and membrane topology. *Proc. Natl. Acad. Sci. USA* 83: 6795–99

*Ann. Rev. Neurosci. 1989. 12:85–102*

# LONG-TERM DEPRESSION

*Masao Ito*

Department of Physiology, Faculty of Medicine, University of Tokyo, Tokyo 113, Japan

## INTRODUCTION

The term "long-term depression (LTD)" may apply to any form of long lasting depression in synaptic transmission. For example, lowering of transmission efficacy at the dentate area occurs following a low-frequency repetitive stimulation and is called long-term depression (Bramham & Srebro 1987). However, the scope of this article is limited to a special type of LTD that is present in the cerebellar cortex and is now regarded as a memory element for cerebellar motor learning.

Purkinje cells in the cerebellar cortex are supplied two distinct types of excitatory synapses, one from parallel fibers (axons of granule cells) and the other from climbing fibers (axons of inferior olive neurons). Repetition of nearly simultaneous arrival of signals form some parallel fibers and a climbing fiber to a Purkinje cell leads to a long-lasting depression of transmission from the parallel fibers to the Purkinje cell, i.e. LTD. Such synaptic plasticity was suggested to exist around 1970, when dissection of the neuronal network structure of the cerebellum, as summarized by Eccles, Ito & Szentágothai (1967), provoked theoretical considerations of the above-mentioned dual synaptic inputs to Purkinje cells. Whereas Brindley (1964), Marr (1969), and Grossberg (1969) suggested a potentiation of transmission to occur in the parallel fiber-Purkinje cell synapses conjunctively activated with a climbing fiber, Albus (1971) assumed a depression. Early studies aiming at experimental verification of these theoretical possibilities failed for some technical reasons, except for a few studies in which movement-related activity of Purkinje cells provided a support, though indirect, for Albus' view (Gilbert & Thach 1977, Ito 1977). A decade had passed before clear experimental evidence for LTD became available (Ito et al 1982).

LTD constitutes the four major types of synaptic plasticity together

85

0147–006X/89/0301–0085$02.00

with long-term potentiation (LTP) (cf Landfield & Deadwyler 1988), sensitization (cf Byrne 1987), and sprouting of axon terminals (cf Tsukahara 1981). It is interesting to note that LTD is an addition to the ten types of synaptic plasticity that Brindley (1967) postulated to exist in nervous systems, including LTP and sensitization. Recent efforts have been devoted to extend the scope of LTD studies in two directions. Cellular and molecular mechanisms of LTD are investigated with both in vivo and in vitro cerebellar preparations, and the role of LTD as a memory element of the cerebellum is studied with neuronal circuitry dissection, lesion experiments, and recording from Purkinje cells in alert behaving animals. This article summarizes the outcomes of these two lines of recent investigations.

## CHARACTERIZATION OF LONG-TERM DEPRESSION

### Experimental Conditions for Induction and Detection

LTD can be induced by conjunctive stimulation of climbing fibers and parallel fibers, but not by stimulation of climbing fibers or parallel fibers alone. In studies of the cerebellar flocculus of nonanesthetized decerebrate rabbits, climbing fibers were stimulated at their site of origin in the inferior olive, whereas parallel fibers were stimulated indirectly by electrical stimulation of a vestibular nerve that projects mossy fiber afferents to the flocculus (Ito et al 1982). At the exposed cerebellar surface, parallel fibers were stimulated directly through a microelectrode inserted into the cortex (Ito & Kano 1982, Ekerot & Kano 1985). In in vitro slice preparations of guinea pig cerebellum, climbing fibers were stimulated with electrodes placed at the white matter and parallel fibers were stimulated with electrodes at the pial surface. Even though parallel fibers in slices were cut short at 300 $\mu$m length, pial stimulation effectively induced synaptic transmission from parallel fibers to Purkinje cells (Sakurai 1987).

The combined stimulation of climbing fibers and mossy or parallel fibers was repeated at 1–4 Hz, the normal range of discharge in climbing fibers, for 25 s to 10 min. Usually, climbing fiber stimulation was timed so as to precede mossy fiber or parallel fiber stimulation by 10 ms, but the timing of climbing fiber and parallel fiber stimulations has been found to be allowed a relatively wide latitude. Stimulation of parallel fibers during the period between 20 ms prior and 150 ms subsequent to the stimulation of climbing fibers is nearly equally potent in inducing LTD in terms of the probability of occurrence of LTD among trials (Ekerot & Kano 1985, 1988). This contrasts to the critical timing theoretically postulated (Marr 1969), and suggests that LTD depends on some variable factor such as enhanced correlation between the firing of parallel fibers and that of climbing fibers, as assumed in Fujita's (1982a) theory. The idea of critical timing of collision with parallel fiber discharges seems also to be unrealistic in view of the characteristically slow, irregular pattern of climbing fiber discharge (see Ito 1984).

LTD is detected by recording responses of Purkinje cells to stimulation of the parallel fibers or mossy fibers involved in a conjunctive stimulation. In in vivo preparations, extracellular unit spikes are recorded from a Purkinje cell responding to parallel fiber or mossy fiber stimualtion. Only in vitro preparations allow stable intracellular recording from a soma or dendritic shaft of a Purkinje cell. Mass field potentials evoked in the molecular layer of the cerebellar cortex by stimulation of parallel fibers have a component representing postsynaptic excitation in cerebellar cortical neurons. This component, however, was reduced only by 20% at the most, and often the reduction was undetectable, even when prominent LTD occurred in unit spike recording from individual Purkinje cells (Ito & Kano 1982). This probably is because 20 times as many non-Purkinje cells (stellate cells and basket cells) as Purkinje cells are found in the molecular layer of the cerebellar cortex (see Ito 1984). Mass field potentials generated by these non-Purkinje cells would mask the LTD that should happen in the mass field potentials of Purkinje cell origin. Ineffectiveness of mass field potential recording makes studies of LTD relatively difficult, as contrasted to hippocampal LTP which is well represented by mass field potentials (Bliss & Lømo 1973).

## Time Course and Magnitude

LTD consists of an initial phase lasting for about 10 min and a subsequent later phase. Determining the whole time course of LTD is still difficult because of technical difficulties inherent in continuous recording of extracellular unit spikes and intracellular postsynaptic potentials. With extracellular unit spike recording, 1 hr is usually the limit for stable, reliable recording, but in some cases, LTD was followed with reasonable stabiltity for 3 hr (M. Kano and M. Kato, personal communication). Intradendritic recording in slice preparations can hardly be continued stably for more than 1 hr, but it was noted that the depression frequently remained without appreciable recovery at 1 hr. An attempt has been made to follow the time course of LTD by means of a reflex testing (M. Ito and L. Karachot, unpublished). LTD occurring in a mossy fiber-parallel fiber-Purkinje cell pathway was tested by measuring a mossy fiber-evoked Purkinje cell inhibition on a vestibular nerve-evoked excitation of Deiters neurons with extracellular recording of mass field potentials in the nucleus of Deiters. LTD was followed in this way for up to 3 hr without appreciable recovery.

The magnitude of depression at the late phase of LTD is 40% in terms of the firing index of Purkinje cell responses to 2 Hz stimulation of mossy fibers (Ito et al 1982). With direct stimulation of parallel fibers, the depression often amounted to 90% (Ekerot & Kano 1985). With intra-dendritic recording in slice preparations, the parallel fiber-evoked excitatory postsynaptic potentials (EPSPs) were reduced in their peak size by 10–50%, 30% on the average (Sakurai 1987).

*Complications*

It is remarkable that postsynaptic inhibition of Purkinje cell dendrites during conjunctive parallel fiber-climbing fiber stimulation prevents LTD from taking place, for the reason mentioned below (Ekerot & Kano 1985). If a strong stimulation of mossy fibers or parallel fibers is used for conjunctive stimulation with climbing fibers, LTD fails to occur because strong stimulation induces postsynaptic inhibition through the parallel fiber-stellate cell pathway. This may account, at least partly, for the earlier failure in detecting LTD. In an in vitro experiment, this complication was avoided by blocking stellate cell inhibition with picrotoxin (Sakurai 1987). These findings resemble those for LTP in the cerebral neocortex, where abundant postsynaptic inhibition appears to mask the occurrence of LTP (Artola & Singer 1987). No LTD was reported to occur in a cerebellum irrigated with artificial solution (Llinás et al 1981). The reason for this failure is still not clear, but stellate cell inhibition is a factor to be carefully examined as a possible cause.

Do climbing fiber signals act only for induction of LTD or do they play multiple roles? One role might be facilitation of parallel fiber-evoked responses (either excitation or inhibition) in Purkinje cells, which occurs when parallel fiber stimulation is combined with spontaneous complex spikes at 20–50 ms intervals (Ebner & Bloedel 1984). This facilitatory effect, however, is short-lasting, and should be a phenomenon independent of LTD. Another example of multiple climbing fiber roles is a membrane hyperpolarization that follows depolarizing responses of Purkinje cell dendrites to climbing fiber signals (Hounsgaard & Midtgaard 1985, Sakurai 1987). This hyperpolarization summates to a long-lasting membrane hyperpolarization when climbing fibers are stimulated repetitively. It probably underlies the prolonged silence occurring in unit spike activities of Purkinje cells after stimulation of climbing fibers (Rawson & Tilokskulchai 1981). This silence lasts only for a few minutes, however, and hence would not contribute to the major phase of LTD.

The prominent increase of simple spike discharge that follows destruction of the inferior olive (Colin et al 1980) is presumably due to removal of the membrane hyperpolarization built in Purkinje cell dendrites by spontaneous climbing fiber signals. The increase of simple spike discharge diminished gradually in two weeks after destruction of the inferior olive (Benedetti et al 1984, Batini & Billard 1985). Climbing fiber destruction also induced reduction in the inhibitory action of Purkinje cells on Deiters neurons (Ito et al 1979). This reduction, outlasting the increased simple spike discharge (Lopiano & Savio 1986, Karachot et al 1987), would suggest another role of climbing fibers in remote trophic regulation of Purkinje cell axon terminals.

## MOLECULAR MECHANISMS OF LONG-TERM DEPRESSION

Clear evidence is now available for involvement of $Ca^{2+}$ ions and quisqualate-specific glutamate receptors in induction of LTD.

### $Ca^{2+}$ Ions

Climbing fiber impulses evoke a large dendritic spike potential followed by a plateau potential in Purkinje cell dendrites (Ekerot & Oscarsson 1981). These potentials represent inflow of $Ca^{2+}$ ions into Purkinje cell dendrites (Stöckle & ten Bruggencate 1980, Llinás & Sugimori 1980). Postsynaptic inhibition produced in Purkinje cell dendrites through stellate cells effectively depresses this $Ca^{2+}$ inflow (Campbell et al 1983). Involvement of $Ca^{2+}$ ions in induction of LTD was initially suggested by the observation that stellate cell inhibition prevented LTD from occurring (Ekerot & Kano 1985). More direct evidence has recently been provided by injection of EGTA, a $Ca^{2+}$ chelator, into Purkinje cell dendrites in slice preparations, which effectively abolished LTD (Sakurai 1988). This property resembles that of LTP, which is also blocked by injection of EGTA into dendrites of hippocampal pyramidal cells (Lynch et al 1983). A common role of $Ca^{2+}$ ions may be suggested in both the LTD and LTP, even though the direction of the synaptic efficacy change is opposite in them.

### Glutamate Receptors

Accumulating evidence indicates that the neurotransmitter released at parallel fiber-Purkinje cell synapses is L-glutamate (for references, see Ito 1984). A recent immunohistochemical labeling revealed enrichment of glutamate at parallel fiber synapses (Somogyi et al 1986), and a pharmacological study indicated a quisqualate-specific nature of postsynaptic receptors at parallel fiber-Purkinje cell synapses (Kano et al 1988).

Involvement of L-glutamate receptors in induction of LTD was first indicated with iontophoretic application of L-glutamate to a Purkinje cell through a microelectrode. Conjunctive application of L-glutamate with electrical stimulation of climbing fibers at 4 Hz effectively induced long-lasting depression in glutamate sensitivity of that Purkinje cell tested with the same microelectrode (Ito et al 1982). The time course of this depression resembled that of LTD. This effect was not produced by application of L-glutamate or stimulation of climbing fibers alone. It is suggested that the LTD is essentially due to long-lasting lowering of glutamate sensitivity of Purkinje cells.

When L-glutamate was applied iontophoretically to a dendritic region of a Purkinje cell in conjunction with climbing fiber stimulation, synaptic transmission to the Purkinje cell from those parallel fibers passing through that dendritic region exhibited a long-lasting depression equivalent to LTD in both time course and magnitude (Kano & Kato 1987, 1988). This implies that the presence of L-glutamate in synaptic regions, either released from parallel fibers or applied iontophoretically, is necessary for induction of LTD. A similar effect was induced with iontophoretic application of quisqualate to a dendritic region, in place of L-glutamate, but with neither kainate nor aspartate (Kano & Kato 1987, 1988). This implies that quisqualate-specific glutamate receptors that mediate parallel fiber-Purkinje cell transmission are specifically involved in LTD.

Thus, LTD is presumed to be due to a reduction of sensitivity of quisqualate-specific glutamate receptors that occurs under the presence of L-glutamate in a synaptic region. Furthermore, when the excitatory action of iontophoretically applied quisqualate on Purkinje cells was blocked by simultaneous iontophoretic application of a glutamate antagonist, kynurenate, conjunctive stimulation of climbing fibers failed to induce a long-lasting depression of parallel fiber-Purkinje cell transmission (Kano & Kato 1988). This means that for eliciting LTD, L-glutamate or its agonist must react with quisqualate-specific receptors.

Taken together, these results indicate that LTD is due to desensitization of quisqualate-specific glutamate receptors at parallel fiber-Purkinje cell synapses. The glutamate receptor desensitization postulated in Purkinje cells contrasts to the sensitization of glutamate receptors suggested to underlie LTP in the hippocampus (Lynch & Baudry 1984). Testing of glutamate sensitivity of Purkinje cells may be recommended for detecting the postulated desensitization of glutamate receptors underlying LTD. However, selective testing of glutamate sensitivity only in those parallel fiber-Purkinje cell synapses involved in conjunctive stimulation with climbing fibers is technically unrealistic, because iontophoretically applied glutamate for this testing would diffuse to excite also those synapses not involved in the conjunctive stimulation and even extrasynaptic glutamate receptors. Therefore, no attempt has been made to test glutamate sensitivity of parallel fiber-Purkinje cell synapses undergoing LTD.

## Molecular Process that Links $Ca^{2+}$ Inflow to Desensitization of Glutamate Receptors

At present, no evidence suggests a particular molecular process that, following $Ca^{2+}$ inflow, leads to desensitization of glutamate receptors in Purkinje cells. However, several possibilities may be raised on the basis of current knowledge of synaptic transmission.

One possibility is that the increased $Ca^{2+}$ ions may directly facilitate desensitization of L-glutamate receptors. An analogy can be drawn from desensitization of acetylcholine receptors at neuromuscular junctions that is facilitated by $Ca^{2+}$ ions (Miledi 1980). A biochemical binding study also demonstrates that elevated $Ca^{2+}$ concentration facilitates desensitization of acetylcholine receptors (Changeux & Heidman 1987). The time course of the LTD can be simulated with kinetics data of $Ca^{2+}$-facilitated desensitization of acetylcholine receptor molecules. However, no data as yet suggest a facilitatory action of $Ca^{2+}$ ions on glutamate receptor desensitization.

A second possibility is that $Ca^{2+}$ ions facilitate glutamate receptor desensitization via a second messenger process. Since cGMP and cGMP-dependent protein kinase are specifically contained in Purkinje cells (Lohmann et al 1981), a role of cGMP in the LTD may be suggested. $Ca^{2+}$ ions excite guanylate cyclase activity in cerebellar tissues (Ohga & Daly 1977), and this probably accounts for the enhanced cGMP level in the cerebellar cortex following activation of climbing fibers (Biggio & Guidotti 1976). It has also been reported that a high level of cGMP desensitizes glutamate receptors isolated from cerebellar tissues (Sharif & Roberts 1980). Therefore, the possibility has been suggested that climbing fiber signals cause $Ca^{2+}$ inflow into Purkinje cell dendrites, consequently stimulate guanylate cyclase, and thereby enhance cGMP level in Purkinje cell dendrites; glutamate receptors at parallel fiber-Purkinje cell synapses will then be desensitized under conjoint influences of cGMP from the inside and L-glutamate from the outside (Ito 1987). Efforts to reproduce LTD with ionotophoretic application of cGMP analogues, however, has met a difficulty that cGMP analogues by themselves exert an excitatory action on Purkinje cells, probably due to their pharmacological effect (M. Ito and M. Kano, unpublished observation).

A third possibility is that $Ca^{2+}$ ions in collaboration with L-glutamate induce a change in structural configuration of dendritic spines. Since parallel fiber-Purkinje cell synapses are located on the head of spines, the synaptic efficacy would decrease if constriction occurs in the neck of spines, and therefore one may imagine that a long-lasting constriction of spine necks underlies LTD. This contradicts the shortening of spines suggested to underlie LTP in hippocampal pyramidal cells (Crick 1982, Koch & Poggio 1983, Kawato et al 1984). A recent study with rapid-freeze, deep-etch techniques revealed that actin filaments run along the axis of dendritic spines of Purkinje cells (Hirokawa 1988). This finding argues against the occurrence of constriction at spine necks, since these actin filaments would induce shortening, if any, but not constriction of spine necks.

So far, evidence is available only for postsynaptic events involved in LTD, and there is no evidence that LTD also involves a long-lasting presynaptic event such as decreased transmitter release. Presynaptic events have been suggested to contribute to LTP, based on an increased glutamate release from hippocampal tissues (Lynch et al 1985) and an increased quantal number in transmitter release (Yamamoto et al 1987). Because such measurements have not been performed in the cerebellum, argument cannot be made for or against this suggestion. The postsynaptic events described above, however, explain the genesis of LTD satisfactorily.

## ROLES OF LONG-TERM DEPRESSION IN CEREBELLAR FUNCTION

A basic hypothesis so far put forward concerning the learning capabilities of the cerebellum is that a region or cerebellar cortex forms a modifiable side path into a reflex arc or into a command signal pathway for voluntary control through connections of mossy fibers and Purkinje cell axons, and that the signal transfer characteristics of this cerebellar path are modified by climbing fiber signals. Climbing fiber signals represent control errors in the system's performance, and act to depress wiring of the cerebellar network responsible for error generation, through induction of LTD (see Ito 1984).

This hypothesis of cerebellar learning can be critically examined by recording unit spikes from individual Purkinje cells in alert, behaving animals. Activation of Purkinje cells through climbing fibers is represented by generation of *complex* spikes, and activation through parallel fibers is reflected in generation of *simple* spikes, even though simple spike discharge is influenced also by the activities of cortical inhibitory neurons (for example, see Miyashita & Nagao 1984). A number of motor tasks have been introduced for relating Purkinje cell activities to cerebellar function in various animal species.

In an attempt to correlate Purkinje cell activity to behavior, it is crucial to record from Purkinje cells at the very site that is primarily involved in the behavior under examination. According to our current knowledge, the cerebellum is composed of numerous narrow (only a few hundred microns wide) microzones, each receiving climbing fibers from a small separate group of inferior olive neurons (Oscarsson 1976). Each microzone projects Purkinje cell axons to a separate group of nuclear neurons and so is dedicated to a certain bodily function that in general is entirely unrelated to those of two neighboring microzones. Since mossy fiber afferents frequently branch in the cerebellum, they may supply collaterals to neighboring microzones and so induce common simple spike activities in them. Therefore, even if two Purkinje cells exhibit mutually related simple spike activities, this does not necessarily imply their belonging to the same microzone. Similarity in climbing fiber responses in two Purkinje cells should be a more reliable sign for their common microzonal origin, but caution is still needed because even climbing fibers issue collaterals innervating different mirozones (for references, see Ito 1984). The most reliable criterion to identify a microzone in a behaving animal is therefore provided by local cerebellar stimulation that induces a unique effect through Purkinje cell output from the microzone. Complex spike responses of Purkinje cells would next be considered, whereas simple spike responses are the least reliable index for the unity of a microzone. It must be emphasized that insufficient identification of microzones explains most discrepancies of the results reported from different research groups.

Provided that a proper microzone is identified in relation to a certain motor task, two lines of information would be crucial for testing the cerebellar learning hypothesis. First, complex spikes should represent control errors involved in performance of the motor task. Second, sustained correlated activities in complex and simple spikes should lead to a progressive modification in simple spike activity, provided that no disturbance occurs, such as blockade of climbing fiber-induced $Ca^{2+}$ inflow into Purkinje cell dendrites by stellate cell inhibition (see above).

## Vestibulo-ocular Reflex

The involvement of the cerebellar flocculus in adaptive control of the vestibulo-ocular reflex (VOR) has been studied extensively (for references, see Ito 1982, 1984). The cerebellar flocculus is connected to the VOR pathway in two ways, so the flocculus receives vestibular signals as a mossy fiber input and in turn projects Purkinje cell axons to relay cells of the VOR. Floccular Purkinje cells exhibit modulation of simple spike discharges in response to head rotation, and through this modulation, they modify activities of the VOR relay cells, and consequently the VOR. The flocculus also receives visual signals as a climbing fiber input monitoring the performance of the VOR.

Under sustained mismatching of visual and vestibular stimulation, the

VOR undergoes a progressive increase or decrease of its gain that leads to minimization of retinal errors under given stimulating conditions, as first demonstrated in human subjects wearing dove prism goggles (Gonshor & Melvill Jones 1976). The VOR adaptation reproduced in animals is abolished with lesioning of the flocculus and also with interruption of the visual climbing fiber pathway to the flocculus (for references, see Ito 1984). When a rabbit performs the VOR adaptation, floccular Purkinje cells exhibit complex spike activities representing retinal errors (Ghelarducci et al 1975, Miyashita & Nagao 1984, Nagao 1988a), and a progressive change takes place in their simple spike responsiveness to head rotation (Dufossé et al 1978). The direction of the change is consistent with the change of the VOR gain, and hence a causal relationship can be suggested (Ito 1977, 1982).

Taken together, these results strongly support the "Flocculus Hypothesis of the VOR Control" (Ito 1982) that the flocculus is the site of the VOR adaptation and that the LTD plays a key role there. Retinal error signals conveyed by climbing fibers to the flocculus would induce the LTD at those parallel fiber-Purkinje cell synapses transferring vestibular mossy fiber signals at that time, and so would modify transfer characteristics of the flocculus for vestibular mossy fiber signals. The VOR adaptation was reproduced successfully with a computer simulation based on this learning principle (Fujita 1982b).

The VOR adaptation is abolished after deprivation of monoamines from rat brain by means of 6-OHDA (Miyashita & Watanabe 1984). The role of these monoamines in the VOR adaptation is still unclear, but one possibility is that monoamines interfere with the occurrence of LTD. An analogy may be drawn from the fact that noradrenaline enhances the LTP in hippocampal slices (Johnston et al 1988).

An objection against the flocculus hypothesis has been raised based on the failure to detect a change of simple spike modulation correlated to VOR adaptation in Purkinje cells of monkey's flocculus (Miles & Lisberger 1981). This failure, however, is apparently due to ignorance of the microzone structure of the flocculus. When the microzone specifically involved in the horizontal VOR (H-zone) was identified with local stimulation that evoked horizontal eye movement, floccular Purkinje cells in monkeys indeed exhibited adaptive changes similar to those reported in rabbits (Watanabe 1984, 1985).

Another objection has been based on overestimation of the simple spike responsiveness of floccular Purkinje cells to eye velocity. If the eye velocity responsiveness predominates over vestibular responsiveness, the change observed in simple spike responses to head rotation after the VOR adaptation would merely be a reflection of increased or decreased eye velocity due to adaptation, and would not indicate a change of the vestibular mossy fiber responsiveness of floccular Purkinje cells (Miles & Lisberger 1981). However, eye velocity responsiveness actually obtained in systematic surveys through the rabbit's floccular H-zone is generally low so that the reflection of changed eye velocity accounts only for a fraction of the simple spike responsiveness to head rotation actually obtained due to adaptation (Miyashita 1984, Nagao 1988b). Consequently, the change of simple spike responses to head rotation that occurs during VOR adaptation can be regarded as largely representing modification of vestibular mossy fiber responsiveness of floccular H-zone Purkinje cells, as postulated in the flocculus hypothesis.

## Optokinetic Eye Movement Response

The optokinetic eye movement response (OKR) is eye movement following relatively slow movement of the visual environment. Its gain progressively increases under sustained optokinetic stimulation (Collewijn & Grootendorst 1979, Nagao 1988a). This adaptation was abolished by flocculectomy, and floccular H-zone Purkinje cells exhibited changes of simple spike responsiveness to optokinetic stimulation in parallel with the OKR adaptation. It has thus been proposed that the flocculus is the site of adaptation of not only the VOR but also OKR (Nagao 1988a). The observed change of simple spike discharge is correlated to complex spike discharge evoked by the optokinetic stimulation in a manner that suggests the LTD to be its cause. Since the same H-zone Purkinje cells and vestibular relay neurons are involved in both the VOR and OKR, the flocculus appears to adaptively control the sum of these two synergistic reflexes by referring to common retinal error signals. This manner of control has been postulated in terms of the flocculus hypothesis of the coordinated ocular reflex control (Ito 1982).

## Classical Conditioning of the Eye Blink Reflex

Air-puff stimulation of the cornea evokes blinking with the eyelid and nictitating membrane, which can be classically conditioned with tone stimulation. Involvement of the cerebellum in this classic conditioning has recently been demonstrated (Thompson 1986). Auditory stimuli are conveyed to the cerebellum through mossy fiber afferents originating from the pontine nucleus, whereas corneal stimuli are forwarded through climbing fiber afferents. Stimulation of the cornea may imply an error which an eye blink had to prevent from occurring. This error representation by climbing fibers conforms to the general scheme of the cerebellular learning hypothesis.

A role of LTD in this classic conditioning has been suggested under assumptions that tone-eye blink reaction is originally mediated by a cerebellar nucleus, and that it is normally suppressed by Purkinje cells. Repeated combined application of tone and corneal stimuli would induce LTD, which removes driving effects of tone stimuli on Purkinje cells, leading to release of the tone-eye blink reaction from Purkinje cell inhibition (Ito 1984).

Recording from Purkinje cells revealed simple spike activity generally consistent with the above view (Thompson 1986). In the conditioned animals, Thompson and his associates (Donegan et al 1985, Foy & Thompson 1986) found a class of Purkinje cells displaying a decrease in rate of simple spike discharge that preceded and modeled the learned behavioral eyelid response. In the untrained animals, they found Purkinje cells that showed increases in simple spike responses to tone stimuli. Further, the frequency of climbing fiber responses evoked by the corneal airpuff unconditioned stimuli decreased markedly as a result of training. These observations are strongly consistent with the hypothesized roles of LTD and climbing fiber signals in cerebellar learning. In one study, a variety of response patterns have been observed, probably because Purkinje cells were sampled without specifying the microzone responsible for the conditioning (Berthier & Moore 1986). Among Purkinje cells sampled along a track perpendicularly crossing microzones in a hemispheric portion of the lobulus simplex, one cell exhibited inhibitory responses to tone stimuli (record 3 of Berthier & Moore 1986), a finding that conforms to the above view. It is possible that this cell represents the proper microzone for the eye blink conditioning.

Combination of air puff-evoked simple spike responses and spontaneously discharging complex spikes has been reported to fail to produce an LTD-like effect, except for a short-term facilitation of air puff-evoked simple spike responses (Zuo & Bloedel 1987). This observation has been claimed to contradict the cerebellar learning hypothesis, but it is necessary to examine whether any factor prevents LTD from occurring, such as stellate cell inhibition evoked by mossy fiber signals (see above), before drawing this conclusion.

## Locomotion

When walking is disturbed with an obstacle, an animal adaptively modifies patterns of limb movements to avoid the obstacle. In cats, Deiters neurons innervating the forelimb contralateral to the perturbed forelimb exhibited an increased discharge in response to the perturbation, apparently due to a decreased simple spike discharge in those vermal Purkinje cells innervating the Deiters neurons (Matsukawa & Udo 1985). Interlimb coordination during the cat's locomotion may be effected by linked activity of vermal Purkinje cells and Deiters neurons. Perturbation frequently evoked complex spikes in these Purkinje cells, as has also been observed with another type of locomotor perturbation (Andersson & Armstrong 1987) and with a perturbed active forelimb movement (Gellman et al 1985). These complex spikes would represent control errors in locomotion, and the pattern of the simple spike responsiveness to perturbation would be gradually modified during repeated trials by means of the LTD. Recording of these responses during the course of learning will provide a crucial test of this possibility.

Decerebrate locomoting ferrets acquired a conditioned movement of a perturbed forelimb to avoid contacting a bar (Bloedel et al 1987). This conditioning, however, was preserved even after extensive cerebellar lesions, and, therefore, does not represent cerebellar function. Instead, the cerebellum appeared to be critical for the organization of movement required to avoid the bar in successive trials while minimizing modification in the ongoing movement which is in fact a clear instance of learning. Purkinje cell responses related to acquisition of such organization have to be investigated in order to test the validity of the cerebellar learning hypothesis in this particular movement task.

## Posture

Complex spikes in frog cerebellar Purkinje cells appear to be related to errors in postural compensation (Amat 1983). Rats can be operantly conditioned to walk on a rotating rod, and this learning is impaired by lesioning the red nucleus (Kennedy & Humphrey 1987). Since severance of the rubrospinal tract originating from the magnocellular red nucleus neurons was quickly compensated for, the impairment of learning should be acribed to destruction of parvocellular red nucleus neurons projecting to the inferior olive. A peculiar finding is that lesion of the red nucleus no longer impaired the learning after the rat had compensated for severance of the rubrospinal tract. The parvocellular red nucleus neurons appear to play a role in the process of compensating for severance of the rubrospinal tract but not in maintenance of the compensated behavior. An interesting task for the future will be to relate these observations with the unique structure of the rubro-olivo-dentate triangle (see Ito 1984).

## Voluntary Movement

Neuronal organization dedicated to the control of smooth pursuit eye movement has been a subject of recent extensive investigations (see Eckmiller 1987). Activity of complex spikes in monkey cerebellum has been correlated to retinal slip signals in this voluntary movement control (Stone & Lisberger 1986), consistent with cerebellar learning hypothesis.

The monkey's step movements with an arm undergoes adaptative adjustment against changes in a handle load. Purkinje cells exhibited complex spike discharge during arm movement against an altered load, and at the same time spike discharges were reduced (Gilbert & Thach 1977). As soon as the arm movement was readjusted to the altered load, complex spike discharge returned to a control level, while simple spike discharges remained depressed. These observations have been interpreted in accordance with Albus' (1971) plasticity assumption and therefore are in agreement with the cerebellar learning hypothesis.

Purkinje cells in the monkey's cerebellum exhibit complex spike discharges at the onset of visually guided wrist tracking movement. In Mano et al's (1986) records, these complex spike discharges seem to represent control errors in pursuit, i.e. discrepancy between movements of the target visual cue and the cursor trace driven by wrist extension or flexion, in agreement with the cerebellar learning hypothesis. However, these authors considered the complex spikes only in relation with handle velocity, and further misinterpreted the occurrence of complex spike discharges seemingly irrelevant to the degree of training as contradictory to the cerebellar learning hypothesis. It is important to realize that errors in a visually guided tracking task cannot be nullified, no matter how much training is done, because the tracking movement is triggered by an error. Even in a learned state, control errors would remain, probably after their time duration could be minimized. It is crucial to evaluate precisely relationships between complex spike discharges and control errors in the visually guided wrist tracking to find a functional meaning of the complex spike discharges. Mano et al (1986) also claimed that their failure to detect a long-lasting depression of simple spike discharge following a spontaneous complex spike contradicts the cerebellar learning hypothesis. The hypothesis, however, predicts the occurrence of LTD only at those parallel fiber-Purkinje cell synapses repeatedly activated in conjunction with climbing fiber signals at a rate above a certain level. These prerequisite conditions for testing the cerebellar learning hypothesis do not seem to be fulfilled in their experiments.

## SUMMARY

LTD has now been established as a synaptic plasticity specific to the cerebellum. Cellular and molecular mechanisms of LTD have been elucidated to some extent, but still a number of questions are left open. The most crucial question may concern its time course, as to how long the LTD lasts beyond the limit of the present maximum observation time of 3 hr, and whether and how it is eventually transformed to a permanent memory.

Molecular mechanisms underlying LTD should be investigated further in respect to $Ca^{2+}$ binding and storage, protein kinase C, phosphorylation of glutamate receptors, GTP proteins, etc. The ineffectiveness of mass field potentials in representing LTD makes such studies relatively difficult, and a hope for future development may be placed in reproduction of LTD in tissue cultured Purkinje cells or even in isolated glutamate receptors in a simplified form.

The cerebellar neuronal network incorporating LTD as a memory element has been conceived as a simple perceptron-like (Albus 1971) or adaptive filter-like (Fujita 1982a) parallel processing computer. Such a neuronal computer incorporated in a reflex or a more complex movement system would endow the system with subtle capabilities of adaptation and learning. The scheme of the floccular control of the VOR closely resembles that of a self-tuning regulator, a type of adaptive control system. For cerebellar control of voluntary movements, however, another version of the adaptive control system, the model reference control system, seems to be more applicable (Ito 1986). This system continuously readjusts its dynamics by referring to errors derived through comparison of its performance with that of an internal model. It is important to note that a model for an unknown system can be built based on the same principle, by feeding errors derived from their comparison to adjust the model. It may thus be conceived that an internal model is built within the cerebellum in the manner of model reference adaptive control, and that an internal model so formed is utilized for adaptive control of movement. A recent simulation study successfully reproduced learning in formation of an arm trajectory based on these principles of model reference control (Kawato et al 1987).

On the experimental side, however, the complex neural organization for control of locomotion, posture, and voluntary movements still eludes full elucidation. Nevertheless, evidence is accumulating to support the cerebellar learning hypothesis. Some controversies have been raised against the hypothesis, but in view of the often insufficient data on which arguments are based, one may hope that these will be resolved in the future when more data are collected and examined for their consistency with the cerebellar learning hypothesis.

*Literature Cited*

Albus, J. S. 1971. A theory of cerebellar function. *Math. Biosci.* 10: 25–61

Amat, J. 1983. Interaction between signals from vestibular and forelimb receptors in Purkinje cells of the frog vestibulocerebellum. *Brain Res.* 278: 287–90

Andersson, G., Armstrong, D. M. 1987. Complex spikes in Purkinje cells in the lateral vermis (b zone) of the cat cerebellum during locomotion. *J. Physiol. London* 385: 107–34

Artola, A., Singer, W. 1987. Long-term

potentiation and NMDA receptors in rat visual cortex. *Nature* 330: 649–52

Batini, C., Billard, J. M. 1985. Release of cerebellar inhibition by climbing fiber deafferentation. *Exp. Brain Res.* 57: 370–80

Benedetti, F., Montarolo, P. G., Rabacchi, S. 1984. Inferior olive lesion induces longlasting functional modification in the Purkinje cells. *Exp. Brain Res.* 55: 368–71

Berthier, N. E., Moore, J. W. 1986. Cerebellar Purkinje cell activity related to the classically conditioned nictitating membrane response. *Exp. Brain Res.* 63: 341–50

Biggio, G., Guidotti, A. 1976. Climbing fiber activation and 3′,5′-cyclic guanosine monophosphate (cGMP) content in the cortex and deep nuclei of the cerebellum. *Brain Res.* 107: 365–73

Bliss, T. V. P., Lømo, T. 1973. Long-lasting potentiation of synaptic transmission in the dentate area of the anaesthetized rabbit following stimulation of the perforant path. *J. Physiol. (London)* 232: 331–56

Bloedel, J. R., Zuo, C.-C., Ferguson, R., Lou, J.-S. 1987. *Soc. Neurosci.* 17: 69.20 (Abstr.)

Bramham, C. R., Srebro, B. 1987. Induction of long-term depression and potentiation by low- and high-frequency stimulation in the dentate area of the anesthetized rat: Magnitude, time course and EEG. *Brain Res.* 405: 100–7

Brindley, C. S. 1964. The use made by the cerebellum of the information that it receives from sense organs. *IBRO Bull.* 3: 80

Brindley, C. S. 1967. Classification of modifiable synapses and their use in models for conditioning. *Proc. R. Soc. London Ser. B* 168: 361–76

Byrne, J. H. 1987. Cellular analysis of associative learning. *Physiol. Rev.* 67: 329–439

Campbell, N. C., Ékerot, C.-F., Hesslow, G. 1983. Interaction between responses in Purkinje cells evoked by climbing fibre impulses and parallel fibre volleys. *J. Physiol. London* 340: 225–38

Changeux, J.-P., Heidman, T. 1987. Allosteric receptors and molecular models of learning. In *Synaptic Function*, ed. G. M. Edelman, W. E. Gall, W. M. Cowan, pp. 549-601. New York: Wiley. 789 pp.

Colin, F., Manil, J., Desclin, J. C. 1980. The olivocerebellar system. I. Delayed and slow inhibitory effects: An overlooked salient feature of cerebellar climbing fibers. *Brain Res.* 187: 3-27

Collewijn, H., Grootendorst, A. F. 1979. Adaptation of optokinetic and vestibuloocular reflexes to modified visual input in the rabbit. *Progr. Brain Res.* 50: 771–81

Crick, F. 1982. Do dendritic spines twitch? *Trends Neurosci.* 5: 44–46

Donegan, N. H., Foy, N. R., Thompson, R. F. 1985. Neuronal responses of the rabbit cerebellar cortex during performance of the classically conditioned eyelid response. *Neurosci. Abstr.* 11: 835

Dufossé, M., Ito, M., Jastreboff, P. J., Miyashita, Y. 1978. A neuronal correlate in rabbit's cerebellum to adaptive modification of the vestibulo-ocular reflex. *Brain Res.* 150: 611–16

Ebner, T. J., Bloedel, J. R. 1984. Climbing fiber action on the responsiveness of Purkinje cells to parallel fiber inputs. *Brain Res.* 309: 182–86

Eccles, J. C., Ito, M., Szentágothai, J. 1976. *The Cerebellum as a Neuronal Machine.* New York/Heidelberg: Springer-Verlag

Eckmiller, R. 1987. Neural control of pursuit eye movements. *Physiol. Rev.* 67: 797–857

Ekerot, C.-F., Kano, M. 1985. Long-term depression of parallel fibre synapses following stimulation of climbing fibres. *Brain Res.* 342: 357–60

Ekerot, C.-F., Kano, M. 1988. Stimulation parameters influencing climbing fibre-induced long-term depression of parallel fibre synapses. *Neurosci. Res.* 6: In press

Ekerot, C.-F., Oscarsson, O. 1981. Prolonged depolarization elicited in Purkinje cell dendrites by climbing fibre impulses in the cat. *J. Physiol. London* 318: 207–21

Foy, M. R., Thompson, R. F. 1986. Single unit analysis of Purkinje cell discharge in classically conditioned and untrained rabbits. *Neurosci. Abstr.* 12: 518

Fujita, M. 1982a. Adaptive filter model of the cerebellum. *Biol. Cybern.* 45: 195–206

Fujita, M. 1982b. Simulation of adaptive modification of the vestibulo-ocular reflex with an adaptive filter model of the cerebellum. *Biol. Cybern.* 45: 207–14

Gellman, R., Gibson, A. R., Houk, J. C. 1985. Inferior olivary neurons in awake cat: Detection of contact and passive body displacement. *J. Neurophysiol.* 54: 40–60

Ghelarducci, B., Ito, M., Yagi, N. 1975. Impulse discharges from flocculus Purkinje cells of alert rabbits during visual stimulation combined with horizontal head rotation. *Brain Res.* 87: 66–72

Gilbert, P. F. S., Thach, W. T. 1977. Purkinje cell activity during motor learning. *Brain Res.* 128: 309–28

Gonshor, A., Melvill Jones, G. 1976. Extreme vestibulo-ocular adaptation induced by prolonged optical reversal of vision. *J. Physiol. London* 256: 381–414

Grossberg, S. 1969. On learning of spatiotemporal patterns by networks with ordered sensory and motor components.

1. Excitatory components of the cerebellum. *Studies Appl. Math.* 48: 105–32

Hirokawa, H. 1988. The arrangement of actin filaments in the postsynaptic cytoplasm of the cerebellar cortex revealed with the quick freeze, deep etch electronmicroscopy. *Neurosci. Res.* 6: In press

Hounsgaard, J., Midtgaard, J. 1985. *Neurosci. Lett.* 22: S27 (Suppl.)

Ito, M. 1977. Neuronal events in the cerebellar flocculus associated with an adaptive modification of the vestibulo-ocular reflex of the rabbit. In *Control of Gaze by Brain Stem Neurons*, ed. R. Baker, A. Berthoz, pp. 391–98. Amsterdam: Elsevier

Ito, M. 1982. Cerebellar control of the vestibulo-ocular reflex—around the flocculus hypothesis. *Ann. Rev. Neurosci.* 5: 275–96

Ito, M. 1984. *The Cerebellum and Neural Control.* New York: Raven. 580 pp.

Ito, M. 1986. Neural systems controlling movements. *Trends Neurosci.* 9: 515–18

Ito, M. 1987. Long-term depression as a memory process in the cerebellum. *Neurosc. Res.* 3: 531–39

Ito, M., Kano, M. 1982. Long-lasting depression of parallel fiber-Purkinje cell transmission induced by conjunctive stimulation of parallel fibers and climbing fibers in the cerebellar cortex. *Neurosci. Lett.* 33: 253–58

Ito, M., Nisimaru, N., Shibuki, K. 1979. Destruction of inferior olive induces rapid depression in synaptic action of cerebellar Purkinje cells. *Nature* 277: 568–69

Ito, M., Sakurai, M., Tongroach, P. 1982. Climbing fibre induced depression of both mossy fibre responsiveness and glutamate sensitivity of cerebellar Purkinje cells. *J. Physiol. London* 324: 113–34

Johnston, D., Hopkins, W. F., Gray, R. 1988. Noradrenergic enhancement of long-term synaptic potentiation. In *Long-Term Potentiation: From Biophysics to Behavior*, ed. W. Landfield, A. Deadwyler, pp. 355–76. New York: Liss

Kano, M., Kato, M. 1987. Quisqualate receptors are specifically involved in cerebellar synaptic plasticity. *Nature* 325: 276–79

Kano, M., Kato, M. 1988. Mode of induction of long-term depression at parallel fibre-Purkinje cell synapses in rabbit cerebellar cortex. *Neurosci. Res.* 5: 544–56

Kano, M., Kato, M., Chang, H. S. 1988. The glutamate receptor subtype mediating parallel fibre-Purkinje cell transmission in rabbit cerebellar cortex. *Neurosci. Res.* 5: 325–37

Karachot, L., Ito, M., Kanai, Y. 1987. Long-term effects of 3-acetylpyridine-induced destruction of cerebellar climbing fibers on Purkinje cell inhibition of vestibulospinal tract cells of the rat. *Exp. Brain Res.* 66: 229–46

Kawato, M., Kamaguchi, T., Murakami, F., Tsukahara, N. 1984. Quantitative analysis of electrical properties of dendritic spines. *Biol. Cybern.* 50: 447–54

Kawato, M., Furukawa, K., Suzuki, R. 1987. A hierarchical neural-network model for control and learning of voluntary movement. *Biol. Cybern.* 57: 169–85

Kennedy, P. R., Humphrey, D. R. 1987. The compensatory role of the parvocellular division of the red nucleus in operantly conditioned rat. *Neurosci. Res.* 5: 39–62

Koch, C., Poggio, T. 1983. A theoretical analysis of electrical properties of dendritic spines. *Proc. R. Soc. London Ser. B* 218: 455–77

Landfield, P. W., Deadwyler, S. A., eds. 1988. *Long-term Potentiation: From Biophysics to Behavior.* New York: Liss. 548 pp.

Llinás, R., Sugimori, M. 1980. Electrophysiological properties of in vitro Purkinje cell dendrites in mammalian cerebellar slices. *J. Physiol. London* 305: 197–213

Llinás, R., Yarom, Y., Sugimori, M. 1981. Isolated mammalian brain in vitro: New technique for analysis of electrical activity of neuronal circuit function. *Fed. Proc.* 40: 2240–45

Lohmann, S. M., Walter, U., Miller, P. E., Greengard, P., Camilli, P. D. 1981. Immunohistochemical localization of cyclic GMP-dependent protein kinase in mammalian brain. *Proc. Natl. Acad. Sci. USA* 78: 653–57

Lopiano, L., Savio, T. 1986. Inferior olive lesion induces long-term modifications of cerebellar inhibition on Deiters nuclei. *Neurosci. Res.* 4: 51–61

Lynch, G., Baudry, M. 1984. The biochemistry of memory: A new and specific hypothesis. *Science* 224: 1057–63

Lynch, M. A., Errington, M. L., Bliss, T. V. P. 1985. Long-term potentiation of synaptic transmission in the dentate gyrus: Increased release of $^{14}C$ glutamate without increase in receptor binding. *Neurosci. Lett.* 62: 123–29

Lynch, G., Larson, J., Keslo, S., Barrionuevo, G., Schottler, F. 1983. Intracellular injections of EGTA block induction of hippocampal long-term potentiation. *Nature* 305: 719–21

Mano, N., Kanazawa, I., Yamamoto, K. 1986. Complex-spike activity of cerebellar Purkinje cells related to wrist tracking movement in monkey. *J. Neurophysiol.* 56: 137–58

Marr, D. 1969. A theory of cerebellar cortex. *J. Physiol. London* 202: 437–70

Matsukawa, T., Udo, M. 1985. Responses of cerebellar Purkinje cells to mechanical perturbations during locomotion of decerebrate cats. *Neurosci. Res.* 2: 393–98

Miledi, R. 1980. Intracellular calcium and desensitization of acetylcholine receptors. *Proc. R. Soc. London Ser. B* 209: 443–52

Miles, F. A., Lisberger, S. G. 1981. Plasticity in the vestibulo-ocular reflex: A new hypothesis. *Annu. Rev. Neurosci.* 4: 273–99

Miyashita, Y. 1984. Eye velocity responsiveness and its proprioceptive component in the flocculus Purkinje cells of the alert pigmented rabbits. *Exp. Brain Res.* 55: 81–90

Miyashita, Y., Nagao, S. 1984. Contribution of cerebellar intracortical inhibition to Purkinje cell responses during vestibulo-ocular reflex of alert rabbits. *J. Physiol. London* 351: 251:62

Miyashita, Y., Watanabe, E. 1984. Loss of vision-guided adaptation of the vestibulo-ocular reflex after depletion of brain serotonin in the rabbit. *Neurosci. Lett.* 51: 177–82

Nagao, S. 1988a. Behavior of floccular Purkinje cells correlated with adaptation of horizontal optokinetic eye movement response in pigmented rabbits. *Exp. Brain Res.* In press

Nagao, S. 1988b. *Neurosci. Res.* 7: S101. (Suppl.)

Ohga, Y., Daly, J. W. 1977. Calcium ion-elicited accumulations of cyclic GMP in guinea pig cerebellar slices. *Biochim. Biophys. Acta* 498: 61–75

Oscarsson, O. 1976. *Exp. Brain Rex.* 1: 34–42 (Suppl.)

Rawson, J. A., Tilokskulchai, K. 1981. Suppression of simple spike discharges of cerebellar Purkinje cells by impulses in climbing fibre afferents. *Neurosci. Lett.* 25: 125–30

Sakurai, M. 1987. Synaptic modification of

parallel fiber-Purkinje cell transmission in *in vitro* guinea pig cerebellar slices. *J. Physiol. London* 394: 463–80

Sakurai, M. 1988. Depression and potentiation of parallel fiber-Purkinje cell transmission in *in vitro* cerebellar slices. In *Olivo-cerebellar System in Motor Control*, ed. P. Strata. Berlin: Springer-Verlag. In press

Sharif, N. A., Roberts, P. J. 1980. Effect of guanine nucleotides on binding of $L^3H$-glutamate to cerebellar synaptic membranes. *Eur. J. Pharmacol.* 61: 213–14

Somogyi, P., Halasy, K., Somogyi, J., Storm-Mathisen, J., Ottersen, O. P. 1986. Quantification of immunogold labelling reveals enrichment of glutamate in mossy and parallel fibre terminals in cat cerebellum. *Neuroscience* 19: 1045–50

Stöckle, H., ten Bruggencate, G. 1980. Fluctuation of extracellular potassium and calcium in the cerebellar cortex related to climbing activity. *Neuroscience* 5: 893–901

Stone, L. S., Lisberger, S. G. 1986. Detection of tracking errors by visual climbing fiber inputs to monkey cerebellar flocculus during pursuit eye movements. *Neurosci. Lett.* 72: 163–68

Thompson, R. F. 1986. The neurobiology of learning and memory. *Science* 233: 941–47

Tsukahara, N. 1981. Synaptic plasticity in the mammalian central nervous system. *Annu. Rev. Neurosci.* 4: 351–79

Watanabe, E. 1984. Neuronal events correlated with long-term adaptation of the horizontal vestibulo-ocular reflex in the primate flocculus. *Brain Res.* 297: 169–74

Watanabe, E. 1985. Role of the primate flocculus in adaptation of the vestibulo-ocular reflex. *Neurosci. Res.* 3: 20–38

Yamamoto, C., Higashi, M., Sawada, S. 1987. Quantal analysis of potentiating action of phorbol ester on synaptic transmission in the hippocampus. *Neurosci. Res.* 5: 28–38

Zuo, C.-C., Bloedel, J. R. 1987. *Soc. Neurosci.* 13: 69.21 (Abstr.)

*Ann. Rev. Neurosci. 1989. 12 : 103–26*

# NOVEL NEUROTROPHIC FACTORS, RECEPTORS, AND ONCOGENES

*Patricia Ann Walicke*

Department of Neuroscience, University of California at San Diego, La Jolla, California 92093-0608

Theories concerning the regulation of brain growth during development, involutional changes during aging, and responses to pathological insults frequently postulate a principal role for neutrophic factors (NTF). For nearly four decades, the only known NTF has been nerve growth factor (NGF), which, despite its name, addresses a relatively limited spectrum of neuronal types. Detailed models on the role and actions of NTFs have been constructed from studies conducted with NGF and sympathetic or dorsal root ganglia (DRG) neurons. In brief, these propose that NGF is produced by the innervated target tissue, where it is taken up by nerve terminals and retrogradely transported to the neuronal soma to regulate gene transcription. At some developmental stages, NGF is required for neuronal survival. Throughout life, it appears to regulate growth as reflected by somal diameter, axonal arborization, the density of synaptic connections, and the quantity of neurotransmitter produced (Edgar 1985, Thoenen & Edgar 1985, Levi-Montalcini 1987). Relatively recently a role for NGF as an NTF for some CNS neurons, specifically the cholinergic neurons of the septum and basal nucleus, has been recognized. A potential application for NTFs in treatment of CNS disease was suggested by recent studies in which exogeneous NGF was shown to increase survival of septal neurons after lesions of the fornix (Hefti 1986, Williams et al 1986, Kromer 1987) and to reverse the atrophic changes associated with aging in the basal nucleus (Fischer et al 1987). However, only a small minority of the neurons in the brain and spinal cord are responsive to NGF.

The search for novel NTFs has frequently employed neurons in cell culture. Trophic activity has been detcted in many sources, including

103

serum; extracts of brain, cultured astrocytes, or neurons; extracts of other tissues such as heart and skeletal muscle; conditioned medium generated by astrocytes, neurons, and other cell types; fluid collecting in wounds in brain or peripheral nerve (Varon et al 1983, Berg 1984). Not only have these in vitro studies provided provocative insights into cell and tissue interactions regulating neuronal growth, but they have recently led to the identification of several new NTFs.

This review defines NTFs as molecules that either increase neuronal survival or stimulate neurite outgrowth, corresponding to the two classical in vitro bioassays for NGF. Some recent reviews have used separate categories for survival and neurite-promoting factors; however, since many NTFs affect both, the distinction seems rather artificial. Like NGF, NTFs may also enhance expression of other differentiated phenotypic traits, but this is not an essential attribute. Factors that regulate only differentiation, such as to the cholinergic inducing factor (Fukada 1935), are excluded. Although NTFs may be autocrine, they must be active when applied extracellularly, excluding intrinsic growth-related proteins such as GAP 43 (Skene et al 1986). Agents that fulfill obvious metabolic needs, such as glucose or vitamins, are excluded to avoid trivializing the concept of NTF. Although there are numerous interesting partially purified factors, this review focuses only on molecularly defined agents.

The demonstration that an agent promotes neuronal survival or growth in vitro implies that it may be an NTF, but confirmation requires considerably more investigation. Minimal criteria should include (a) demonstration of specific receptors on neurons; (b) availability in the normal neuronal environment; and (c) NTF effects in vivo as well as in vitro. Ideally, positive responses to administration of exogenous NTF and negative responses after interference with endogenous NTF should both be demonstrated. All of these criteria have been fulfilled for NGF, and have been the subject of several recent reviews (Edgar 1985, Thoenen & Edgar 1985, Levi-Montalcini 1987). Considerable progress has been made with several of the novel factors, which may soon be established as NTFs. As these investigations continue, they will profoundly enhance the understanding of molecular regulation of neuronal growth.

## NOVEL NEUROTROPHIC FACTORS

Starting with a tissue extract that promotes trophic responses in an assay system and purifying the active ingredient is an intellectually straightforward, though laborious, process. The identity of the purified factor, however, is often unanticipated. Some investigations have yielded entirely novel proteins, but, not infrequently, the purified NTF is highly homologous or identical to a previously known protein. For convenience, these outcomes are discussed separately.

## Brain-Derived Neurotrophic Factor

Brain-derived neurotrophic factor (BDNF) was initially characterized as a basic 12 kD protein present in brain extracts that increased survival of DRG neurons at concentrations around 0.4 nM (Lindsay et al 1985, Barde et al 1987). Approximately 70% of its sequence has been determined; it may have some limited homology to NGF but not to any other known protein (Y. Barde, personal communication).

The range of neurons responsive to BDNF has been studied extensively in vitro. BDNF supports survival of and stimulates neurite outgrowth from 60–70% of DRG neurons in day E6 chicks, and around 40% in day E12 chicks (Lindsay et al 1985). Many primary sensory neurons are affected, including trigeminal, nodose, vestibular, and petrosal ganglia neurons and mesencephalic trigeminal neurons (Lindsay et al 1985, Davies et al 1986, Barde et al 1987). Recent studies in vivo have shown that repeated injection of BDNF decreased developmental cell death in the DRG and nodose ganglia (Hofer & Barde 1988). These observations have led to the hypothesis that BDNF is a specific regulator of CNS process growth for a variety of peripheral sensory neurons. These neurons are postulated to require two sources of NTFs for survival and growth, one from the periphery (such as NGF) and BDNF from the CNS (Barde et al 1987).

In addition, one population of intrinsic CNS neurons has been reported to respond to BDNF, retinal ganglion cells. In cell cultures from retinas of day E17 rats, BDNF supported survival of and stimulated process growth from 80–90% of ganglion cells, but it only supported about 20% of ganglion cells from postnatal retinas (Johnson et al 1986). Whether this might reflect loss of BDNF receptors during development remains to be determined.

## Purpurin

Purpurin is a 20 kD protein that was purified from chick retinal adherons, which are complexes of proteins and proteoglycans that stimulate neuronal adhesion. It supports the survival of about 50% of day E6 chick retinal neurons in vitro and increases cell-substratum adhesion but does not stimulate neurite outgrowth. The minimal active concentration is around 0.2 nM. The sequence of both purpurin chymotryptic peptides and its full length cDNA clone showed about 50% homology with human serum retinol binding protein. Correct identification was verified by expression of bioactive purpurin in transfected CHO cells (Schubert & LaCorbiere

1985, Schubert et al 1986, Berman et al 1987). Purpurin is related to but distinct from the chick serum retinol binding protein, and although both bind retinol, only purpurin has NTF activity. It is restricted to the retina, and was undetectable on both Western and Northern blots prepared from optic lobe, brain, liver, skeletal, or heart muscle. In vitro it was synthesized only by neural cells, not Müller glial cells or pigmented epithelium (Schubert & LaCorbiere 1985). In situ hybridization with intact retina indicated that synthesis occurs in photoreceptors (Berman et al 1987). Purpurin appears to be an autocrine trophic factor for retinal photoreceptor cells, which are known to require retinol for rhodopsin synthesis. Whether retinol accounts for the effects of purpurin on adhesion and survival is uncertain.

## NOVEL NTFs IDENTIFIED AS PREVIOUSLY KNOWN PROTEINS

In contrast to the studies discussed above, many purification schemes have yielded previously known proteins with newly recognized NTF activities. Generally, the relationship between known biochemical functions and trophic activity is not obvious. Experimental clarification will likely produce insight into the molecular mechanisms underlying such processes as growth cone extension and adhesion to the extracellular matrix.

### Neuroleukin

Neuroleukin (NLK) was originally identified as a factor produced by skeletal muscle that stimulated neurite outgrowth from spinal cord motoneurons. The 63 kD protein was purified to homogeneity from mouse submaxillary gland and a cDNA clone was isolated. Bioactive NLK was produced by transfected COS cells, suggesting correct identification of the gene. At the time that NLK was isolated, it showed limited homology with only one previously known protein, the coat protein of the AIDS virus, HIV-1 (Gurney 1986a). More recently, NLK was recognized to be the glycolytic enzyme, phosphoglucose isomerase (PGI), based on virtual sequence identity and the presence of enzymatic activity in NLK preparations (Chaput et al 1988, Faik et al 1988, Gurney 1988). Whether PGI activity produces the trophic effects is unknown, but antibodies which block NLK activity do not necessarily inhibit enzymatic activity suggesting that they may be separate functions (Gurney 1988). NLK could possibly trigger biological responses by acting as a lectin (Chaput et al 1988). The relationship between the glycolytic and trophic roles of this protein requires further investigation.

NLK has been studied in culture systems employing a variety of neurons, and reported to increase both survival and process formation. The most detailed studies have employed chick DRG neurons, of which about 40% respond to NLK with half-maximal effects at around 10 pM. Survival of neurons from spinal cord, septum, and hippocampus is also increased by NLK (Gurney et al 1986a, Lee et al 1987). Additionally NLK has been reported to be a T cell-derived trophic factor for B lymphocytes (Gurney et al 1986b). Sprouting induced by local injection of botulinum toxin into the soleus muscle was strongly inhibited by administration of specific monoclonal antibodies to NLK, suggesting NTF activity in vivo. However, neither passive nor active immunization led to retraction of normal muscle endplates or spinal cord motoneuron death (Gurney et al 1986a,c).

The observation that NLK contained a 44 amino acid sequence homologous to a conserved region of the HIV-1 gp120 coat protein suggested that NLK might play a role in the pathogenesis of AIDS dementia. In bioassays with DRG neurons in vitro, the HIV-1 coat protein or peptides from the homologous region were potent inhibitors of NLK bioactivity, acting in a manner suggesting competition for a shared neuronal receptor (Lee et al 1987). Further investigation of this bioactive core of NLK and the possible existence of a neuronal receptor might clarify the relationship between the PGI and NTF activities of this protein.

## Apolipoprotein E

Synthesis of several proteins increases in the distal stump of a severed nerve, but the greatest elevation occurs for a 37 kD protein. The purified protein was identified as apolipoprotein E (apoE) based on immunological markers, amino acid composition, and N-amino terminal sequence (Ignatius et al 1986, Snipes et al 1987). ApoE is one of a family of proteins involved in the packaging and transport of cholesterol esters and fatty acids. Lipoproteins containing apoE interact with a specific membrane receptor and are rapidly incorporated into the growth cones of PC12 cells, presumably providing lipids for membrane formation (Ignatius et al 1987). Immunohistochemical studies demonstrate its presence in astrocytes, Schwann cells, and macrophages in injured neural tissue (Boyles et al 1985, Ignatius et al 1986, Snipes et al 1987). Lesser accumulation of apoE in damaged CNS tracts compared to peripheral nerve has been suggested to play a role in poorer CNS recovery; this interesting possibility awaits testing in vivo.

## Glial-derived Nexin

A 43 kD factor stimulating neurite outgrowth from neuroblastoma cells was initially detected in conditioned media from glioma cells. The purified protein is identical to protease nexin I, an irreversible inhibitor of serine proteases, based on reactivity with antisera, receptor competition, amino acid composition, and N-amino terminal sequence (Gloor et al 1986, Rosenblatt et al 1987). It is produced by many cells, including fibroblasts, astrocytes, heart, and skeletal muscle cells (Pittman & Patterson 1987, Rosenblatt et al 1987). Nexin mRNA is present in brain tissue from fetal and adult rats, with an apparent peak in the first postnatal weeks (Gloor et al 1986).

Neurons from both the peripheral and central nervous systems secrete plasminogen activator, particularly from their distal processes and growth cones (Kryostek & Seeds 1984, Pittman 1985). Some interaction between nexin and neuronal proteases likely forms the basis for its neurite promoting action (Monard 1987). So far, only neuroblastoma cells and sympathetic neurons have been shown to respond (Monard 1987, Pittman & Patterson 1987); since both nexin and neuronal proteases are widely distributed (Soreq & Miskin 1981), further studies with other neurons, particularly from the CNS, would be interesting.

## S100b

A factor present in bovine brain extracts was reported to stimulate process elongation from chick telencephalic neurons in serum-free cultures. Amino acid sequence analysis of the purified material demonstrated identity with S100b, a well known immunohistochemical marker for astrocytes. Half-maximal activity required 6.3 nM S100b, but appeared to reside particularly in higher molecular weight fractions containing disulfide-linked S100b oligomers (Kligman & Marshak 1985). These results were supported by an earlier study using a less well-defined S100 protein preparation that reported neurite growth onto S100 coated acrylic spheres in vivo (Hyden & Ronnback 1978).

The biochemistry of S100b has been extensively studied and recently reviewed (Donato 1986). Although it binds calcium, a physiologically significant role in calcium homeostasis is questionable and its true function remains uncertain. S100b is synthesized and secreted principally by astrocytes. High affinity binding sites with a Kd of 7 nM, which have been reported on synaptosomal membranes, could represent neuronal receptors. The available data might suggest a model in which S100b is an NTF made by astrocytes for CNS neurons. Further investigation of its NTF activity both in vitro and in vivo is required.

## Laminin, Heparan Sulfate Proteoglycan, and Fibronectin

Many cells such as cardiac and skeletal myocytes, astrocytes, and Schwannoma cells secrete products that stimulate neurite extension when absorbed on polycationic surfaces (Varon et al 1983, Berg 1984). These neurite-promoting factors appear to be large complexes containing laminin and heparan sulfate proteoglycan (HSPG) (Davis et al 1985, Lander et al 1985). Studies in many laboratories have verified that purified laminin is an excellent substrate for neurite growth from peripheral and CNS neurons (e.g. Manthorpe et al 1983, Edgar et al 1984, Hantaz-Ambroise et al 1987); however, it may not be universally effective. For example, retinal ganglion neurons from day E6 chicks extend neurites on laminin but lose this ability in later development (Cohen et al 1986, Hall et al 1987). Although laminin alone supports few neurons, it can augment neuronal survival when acting in combination with another NTF such as NGF (Edgar 1985).

Laminin is a constituent of basement membranes in many tissues, including peripheral nerve. Although it is absent from the adult CNS except for blood vessels (Liesi 1985), it may be present during critical developmental stages. Axonal outgrowth from retinal ganglion cells may be guided by laminin transiently produced by neuroepithelial cells lining the tract to the optic tectum (Cohen et al 1987). The successful regeneration of sciatic nerve axons along laminin-coated polyester filaments supports its NTF activity in vivo (Yoshii et al 1987). Other studies using basement membrane or extracellular matrix complexes further suggest the possible importance of laminin (Fawcett & Keynes 1986, Davis et al 1987, Madison et al 1987, Müller et al 1987), but ascribing all effects of these complex materials to a single component may be premature.

Laminin is a protein of about $10^6$ daltons containing three disulfhydryl-linked chains arranged to form a cross with three short arms and one long arm (Edgar 1985). Experiments with proteolytic fragments and domain-specific antisera localized the neurite-promoting site to a region near the end of the long arm (Edgar et al 1984, Edgar 1985, Engvall et al 1986). Although nonneuronal cells chiefly bind to another region in the center of laminin, recent evidence suggests that they too can recognize the active domain on the long arm (Aumailley et al 1987).

In addition to laminin, both fibronectin and collagen can serve as substrates for neurite growth (Carbonetto et al 1983, Rogers et al 1985). Two proteolytic fragments of fibronectin retain neurite-promoting activity. One, which chiefly affects peripheral neurons, contains the Arg-Gly-Asp-Ser (RGDS) sequence thought to mediate nonneuronal cell adhesion to fibronectin (Rogers et al 1985, 1987, Ruoslahti & Pierschbacher 1987). The second fragment is more effective for CNS neurons than intact fibronectin, but its active domain has not yet been defined. Since laminin has recently been cloned and sequenced (Graf et al 1987, Pikkarainen et al 1988), its neurite-promoting region will likely soon be determined. It is interesting that the unknown active sites on both laminin and fibronectin are located near heparin-binding domains; possibly they may contain a common domain for neurite outgrowth. If small peptides affecting neurite growth can be produced, they will be extremely useful for both in vitro and in vivo studies.

Several candidate neuronal receptors for these matrix molecules have been advanced. One of these is a large membrane complex of three 140 kD subunits called the cell substratum attachment antigen (CSAT) (Horwitz et al 1985). Antibodies to CSAT block extension of neurites from chick ciliary and retinal neurons on laminin and fibronectin (Tomaselli et al 1986, Cohen et al 1987, Hall et al 1987). CSAT structurally resembles the fibronectin receptors (Ruoslahti & Pierschbacher 1987), and its binding to both proteins is inhibited by the RGDS tetrapeptide (Horwitz et al 1985). Whether it mediates interactions with the RGDS domain alone or also with other neurite-promoting sites is not yet known. The major laminin receptor on nonneuronal cells appears to be a 70 kD protein that recognizes the central portion of laminin. It has been detected on human neuroblastomas (Graf et al 1987), but appears unlikely to be the principal neuronal receptor since it interacts with an inappropriate domain of laminin. Other candidates for potential neuronal laminin receptors include cranin, a 120 kD glycoprotein present in brain membrane preparations (Smalheiser & Schwartz 1987); sulfated glycolipids, specifically galactosylceramide (Roberts et al 1985); and HSPG. It is certainly possible that multiple cell membrane receptors could mediate the adhesive and neurite-promoting effects of laminin and fibronectin.

HSPG copurifies with laminin secreted by many cell types. The only characterized HSPG consists of an 80 kD core protein with many heparan sulfate side chains producing a total molecular weight of about 350 kD (Matthew et al 1985). It does not immediately induce neurite outgrowth (Davis et al 1985, Lander et al 1985) but may after some days (Hantaz-Ambroise et al 1987). Interactions between HSPG and laminin are complex and functionally significant. Although antisera to laminin block neurite growth on the purified protein, they recognize laminin but no longer inhibit neurite growth when it is complexed with HSPG (Davis et al 1985, Edgar 1985, Lander et al 1985). An epitope designated INO (inhibitor of neurite outgrowth) is present in the complex, but not on laminin or HSPG alone (Chiu et al 1986). Unlike laminin antisera, INO antisera block neurite outgrowth on the complex and on peripheral nerve basal lamina both in vitro and in vivo (Sandrock & Matthew 1987a,b). Since both HSPG and INO are present in many tissues (Chiu et al 1986, Eldridge et al 1986, Sandrock & Matthew 1987a), extrapolation from studies with purified laminin in vitro to its role in vivo requires caution. Several other novel NTFs bind to heparin, including apoE, nexin, purpurin, and FGF; HSPG could potentially also affect their activity.

# KNOWN GROWTH FACTORS AS POTENTIAL NTFs

Several of the novel NTFs, such as apoE, laminin, and nexin, are not restricted to neural tissues in distribution or biological activity. Even NGF might have significant extra-neural functions as a modulator in the immune system (Levi-Montalcini 1987). These developments suggest that major signals involved in cellular growth regulation could to some extent be shared among neurons and nonneural cells. In this context, searching for novel NTFs among growth factors previously characterized in other tissues appears warranted. This review discusses epidermal growth factor (EGF), basic and acidic fibroblast growth factors (bFGF; afGF), insulin-like growth factors I and II (IGF-I, IGF-II), and insulin.

Because all of these factors are mitogens for astrocytes (Morrison & de Vellis 1981, Leutz & Schachner 1982, Simpson et al 1982, Lenoir & Honegger 1983, Pettmann et al 1985, Shemer et al 1987) that are well known to support neuronal survival and growth in vitro (Berg 1984), demonstration of direct neuronal stimulation is particularly critical. For afGF, bFGF, and IGF-I, NTF activity has been demonstrated in neuronal cultures lacking any glial cells detectable with standard immunohistochemical cell markers (Aizenman & de Vellis 1987, Unsicker et al 1987, Walicke & Baird 1988a). Responses of the PC12 cell line to EGF (Greenberg et al 1985, Herschman 1986) and the FGFs (Neufeld et al 1987, Schubert et al 1987) and of neuroblastoma cells to IGF-II (Mill et al 1985) are obviously not mediated by glial cells. Together with evidence demonstrating the presence of specific receptors on neurons, the data suggest distinct and direct effects of these trophic factors on both neurons and glia.

## Insulin and the Insulin-like Growth Factors

With the formulation of a serum-free medium for neurons by Bottenstein & Sato in 1978, insulin became the first of the previously known growth factors to be implicated as a potential NTF. Its role, however, is still unresolved, because of difficulty in distinguishing among the activities of insulin, IGF-I, and IGF-II and their respective receptors. The three peptides have extensive sequence homologies ranging between 40% and 60%. The insulin and IGF-I receptors are both tyrosine kinases with significant homologies; the IGF-II receptor appears to be identical to the mannose-6-phosphate receptor. Not surprisingly, insulin, IGF-I, and IGF-II cross-react with each other's receptors, thus complicating interpretation of their effects (see reviews in Nissley & Rechler 1984, Zapf et al 1984, Sasaki et al 1985, Roth 1988).

A peptide with the appropriate antigenic determinants and biological activity for insulin has been extracted from brain tissue (Yalow & Eng 1983, Baskin et al 1987). Immunohistochemical studies have suggested that insulin is widely distributed in neuronal somata (Weyhenmeyer & Fellows 1983, Bernstein et al 1984), but the specificity of staining has been questioned (Baskin et al 1987). There is evidence for insulin synthesis in enriched neuronal cultures (Birch et al 1984, Clarke et al 1986), but attempts to detect proinsulin (Yalow & Eng 1983) or insulin mRNA in adult brain have been less successful (Villa-Komaroff et al 1984, Young 1986). Whether brain insulin is locally synthesized or derived from serum is controversial.

Both IGF-I and IGF-II have been extracted from fetal and adult brain tissue and identified by amino acid analysis and N-amino terminal sequence (Carlsson-Skwirut et al 1986, Sara et al 1986). Immunoreactive IGF-I and IGF-II have been detected in CSF and many regions of the CNS, including neocortex, thalamus, pons, and cerebellum (D'Ercole et al 1984, Haselbacher et al 1985). Fetal cortical cultures produce IGF-I (Birch et al 1987). The mRNAs encoding IGF-I and IGF-II have been detected in fetal and adult brain, and neuronal or glial cell cultures (Brown et al 1986, Lund et al 1986, Irminger et al 1987, Murphy et al 1987, Rotwein et al 1988).

Radioreceptor assays employing membranes from fetal and adult brain have demonstrated the presence of insulin, IGF-I, and IGF-II receptors. The results of these studies are particularly important for interpretation of the NTF bioassays. Each peptide appears to bind to its own receptor with a $K_D$ in the range of 1–7 nM. Insulin acts at the IGF-I receptor with an estimated potency of 0.1–2.0% but has negligible effects on the IGF-II receptor. Both IGF-I and IGF-II have about 1% potency at the insulin receptor and about 20–100% potency at each other's receptors (Sara et al 1982, Goodyer et al 1984, Gammeltoft et al 1985, Van Schravendijk et al 1986, Baskin et al 1987, Shemer et al 1987). Neuronal localization of the insulin and IGF-I receptors is suggested by their presence in enriched cortical cultures (Boyd et al 1985, Burgess et al 1987, Masters et al 1987). Autoradiography demonstrates high levels of insulin binding in neuropil in the intact CNS (Corp et al 1986, Hill et al 1986, Werther et al 1987). Surprisingly, insulin fails to increase glucose uptake in neuronal cultures (Gorus et al 1984, Boyd et al 1985), thus suggesting a unique function for the neuronal insulin receptor.

Further biochemical verification of the identification of the different receptor molecules has employed cross-linking and immunoprecipitation. The brain insulin and IGF-I receptors also have appropriate tyrosine kinase activity (Rees-Jones et al 1984, Gammeltoft et al 1985, Heidenreich et al 1986, Roth et al 1986, Burgess et al 1987, Masters et al 1987, McElduff et al 1987). In cortical membranes, insulin begins to stimulate the IGF-I receptor tyrosine kinase at 10 nM. At a concentration of 100 nM, insulin stimulates as much phosphorylation through activation of the IGF-I receptor as through its own receptor (Lowe & LeRoith 1986, Shemer et al 1987).

The complexity in assessing these peptides as NTFs arises from their ability to cross-react with each other's receptors. Since no study has yet correlated receptor types and NTF responses, interpretation requires comparison of minimal bioactive concentrations with the radioreceptor studies. Quantitative estimates of the concentration of insulin needed for half-maximal effects on survival or neurite outgrowth yielded 0.4 nM for sympathetic (Recio-Pinto et al 1986), 10–50 nM for DRG (Snyder & Kim 1980, Bothwell 1982, Recio-Pinto et al 1986), 160 nM for parasympathetic (Collins & Dawson 1983), 50–175 nM for cortical (Romijn et al 1984, Aizenman & de Vellis 1987), and 350 nM for cerebellar neurons (Huck 1983). An excellent study of IGF-I activity in pure cortical neuronal cultures found half-maximal support of survival at 1.4 nM (Aizenman & de Vellis 1987). IGF-I increased $^3$H uridine incorporation into cortical cultures at 1 nM (Burgess et al 1987). IGF-II increased neurite outgrowth from DRG and sympathetic neurons at 0.1–4 nM (Bothwell 1982, Recio-Pinto et al 1986), and enhanced survival of hippocampal and septal neurons at 0.1 nM (Onifer et al 1987).

The most parsimonious conclusion is that the IGF-I receptor is the mediator of NTF effects. With one exception, the concentrations of either insulin or IGF-II employed were sufficient to stimulate the IGF-I receptor. Insulin receptors may mediate NTF responses in sympathetic neurons, but probably not in the other neurons examined. Since the IGF-II receptor is not stimulated by insulin, its activation may not be required for many NTF responses, but data on its role is really nonexistent. Either IGF-I or IGF-II could act as endogenous ligands mediating NTF responses at the IGF-I receptor. More detailed in vitro studies are required to verify these inferences, and appear likely to be significant given the extensive evidence for the presence of IGFs and their receptors in brain.

## Epidermal Growth Factor

Epidermal growth factor (EGF) is a 53 amino acid peptide originally purified from the male mouse submaxillary gland, which is a mitogen for a variety of epidermal cells (reviewed in Stoschek & King 1986, Gill et al 1987). EGF provokes some of the same responses from PC12 cells as NGF, such as induction of ornithine decarboxylase and tyrosine hydroxylase, but does not cause neuronal differentiation (Greenberg et al 1985, Herschman 1986). Recently, EGF has been reported to enhance survival and process growth of neurons from neonatal rat basal forebrain in enriched cultures containing only about 5% glial cells. Half-maximal response required about 1.0 ng/ml EGF (Morrison et al 1987).

EGF receptors have been detected in radioreceptor assays using membranes from fetal and adult mouse brains (Adamson & Meek 1984, Alm et al 1986). In vitro studies support the presence of EGF receptors on neurons, but at much lower levels than on astrocytes (Leutz & Schachner 1982, Simpson et al 1982). In contrast, an immunohistochemical study of intact brain tissue localized the EGF receptor predominantly to neurons of the basal forebrain nuclei, basal ganglia, and to some extent neocortex, with few EGF receptors present on glial cells (Loy et al 1987). The possibility that other trophic signals present in vivo might affect the relative expression of EGF receptors on neuronal and glial cells has yet to be investigated.

Radioimmunoassays of brain extracts and CSF have yielded conflicting results on whether EGF is present (Roberts et al 1981, Hirata et al 1982, Lakshmanan et al 1986) or absent (Probstmeier & Schachner 1986) from the CNS. In one immunohistochemical study, EGF appeared to be localized to the basal ganglion and basal forebrain (Fallon et al 1984), similar to the distribution of EGF receptors (Loy et al 1987). Since EGF is derived from a 133 kD precursor, some of the variability in results could represent different sensitivities of antibodies to partially processed forms. Yet only low levels of EGF mRNA were detectable in brain polyA selected mRNA (Rall et al 1985). Another protein exists that also activates the EGF receptor, transforming growth factor alpha (Derynck 1986). Some of the inconsistency and controversy in the published studies might suggest participation of either differentially processed forms of EGF or transforming growth factor alpha in some aspects of brain trophic regulation, possibilities that will require considerably more detailed investigation.

## Fibroblast Growth Factor

The two fibroblast growth factors, basic (bFGF) and acidic (aFGF), are peptides of around 17–20 kD that share 53% homology. Both are mitogens for a variety of mesenchymal cells, with bFGF 10–100-fold more potent than aFGF. Both FGFs have a high affinity for heparin and related glycosoaminoglycans. The effect of heparin on FGF actions appears to be complex and possibly variable with heparin concentration and assay system, but in general heparin tends to increase the potency of aFGF and decrease the potency of bFGF. Their effects on endothelial cells are particularly striking, and it appears likely that a number of reported angiogenic factors are either identical or highly homologous to bFGF and aFGF (Baird et al 1986, Gospodarowicz et al 1986, Thomas & Gimenez-Gallego 1986). In studies with PC12 cells, the FGFs fully reproduce the effects of NGF, except that bFGF is about 200-fold more potent. PC12 cells bear a specific receptor used by both bFGF and aFGF, with binding characteristics and molecular weight essentially identical to the mesenchymal receptor (Wagner & D'Amore 1986, Neufeld et al 1987, Schubert et al 1987).

In serum-free cell cultures, both bFGF and aFGF support the survival of neurons from many regions of fetal and neonatal rat brain, including the hippocampus, neocortex, striatum, septum, and thalamus. Variations in the proportions of supported cells suggest that subpopulations of responsive neurons may exist in the cortex and septum, but their phenotypes have not yet been defined (Morrison et al 1986, Walicke et al 1986, Walicke 1988). bFGF also supports cerebellar granule cells (Hatten et al 1988). In addition to these effects on post-mitotic neurons, it is mitogenic for neuroblasts from earlier fetal stages (Gensburger et al 1987). Half-maximal support of neuronal survival has been reported with as little as 1 pM bFGF (Walicke 1988), though other estimates are around 20–70 pM (Morrison et al 1986, Unsicker et al 1987, Hatten et al 1988). As for mesenchymal cells, aFGF is generally less potent (Unsicker et al 1987, Walicke 1988); however, it may have additive effects with bFGF under some conditions (Walicke & Baird 1988b). Retinal ganglion cells have been reported to respond specifically to aFGF rather than bFGF (Lipton et al 1988).

Basic FGF stimulates neurite outgrowth after brief periods of exposure, even on atypical glycosoaminoglycan substrates (Walicke et al 1986), a property that may be related to its affinity for heparin. bFGF is present in extracts of extracellular matrix, where it may be associated with heparan sulfate (Baird & Ling 1987, Vlodavsky et al 1987). Its high affinity for several components of the basal lamina along with its potent biological activity for neurons suggest that it might participate in some neuronal responses to extracellular matrix.

Evidence for neuronal bFGF receptors was obtained in autoradiographic studies showing specific binding to hippocampal neurons in vitro (Walicke & Baird 1988b). Bound bFGF was subsequently internalized and degraded. Radioreceptor assays and cross-linking studies suggest that hippocampal neurons bear a receptor similar to that on mesenchymal cells (P. A. Walicke, unpublished observations). Attempts to demonstrate neuronal receptors in vivo by autoradiography have been less successful; binding has been noted chiefly on vascular basement membranes (Jeanny et al 1987). Immunohistochemical studies with a receptor antibody, when one becomes available, will be useful.

Recent studies have also demonstrated possible NTF effects of the FGFs in vivo as well as in vitro. Inclusion of bFGF inside a polyethylene tube linking the two ends of a severed sciatic nerve increases the rate of axonal regeneration (Cuevas et al 1987, Danielsen et al 1988). Administration of aFGF to the severed stump of the optic nerve augments survival of retinal ganglion cells (Sievers et al 1987). Whether these NTF effects in vivo are caused by direct neuronal stimulation or indirectly through glial or vascular cells is difficult to determine given the pleiotropic trophic effects of the FGFs.

Both bFGF and aFGF have been purified from brain tissue and their identity verified by amino acid analysis (Baird et al 1986, Gospodarowicz et al 1986). Astrocytes have been demonstrated to synthesize bFGF in vitro (Hatten et al 1988). Histochemistry with an antisera recognizing both aFGF and bFGF has demonstrated immunoreactive material in many CNS neurons (Pettman et al 1986). The normal sites of synthesis in situ and the mechanisms governing the distribution of bFGF among potentially responsive cells are questions for further investigation.

## ONCOGENES

Because the oncogenes encode the transforming components of carcinogenic viruses, their normal cellular homologues were anticipated to function principally in mitosis, but several have been related to neuronal growth. When PC12 cells are induced to differentiate into neurons by NGF, there is a rapid 100-fold increase in c-fos expression (Curran & Morgan 1985, Greenberg et al 1985, Kruijer et al 1985, Milbrandt 1986). c-Fos encodes a nuclear protein that presumably regulates the transcription of other genes. Its robust and rapid induction might suggest use as a rapid screen for NTF responses. However, in PC12 cells c-fos is induced by a variety of stimuli including cAMP, elevated potassium, phorbol esters, and the calcium ionophore A23187 (Greenberg et al 1985, Kruijer et al 1985, Milbrandt 1986). In intact brain, epileptic seizures induce c-fos in the dentate gyrus, hippocampus, and cortex (Dragunow & Robertson 1987, Morgan et al 1987). Sensory stimulation increases c-fos in spinal cord neurons (Hunt et al 1987). Although c-fos may help regulate gene transcription in response to NTFs, it may not be specifically related to trophic mechanisms.

A potential role for other oncogenes was suggested by stimulation of neuronal differentiation in PC12 cells injected or transfected by either v-src or members of the ras family (Alema et al 1985, Bar-Sagi & Feramisco 1985, Noda et al 1985, Guerrero et al 1986). Conversely, injections of antibodies to the ras proteins blocked neuronal differentiation in response to NGF (Hagag et al 1986). The ras proteins are a family of G-proteins that are present at high levels in brain tissue and may be localized to neurons (Swanson et al 1986, Tanaka et al 1986). *c-Src* is a tyrosine kinase loosely associated with the cytoplasmic membrane; immunohistochemistry confirms its localization to neurons. Although it is present throughout adulthood, its peak expression occurs during developmental stages characterized by neuronal differentiation and axonal growth (Sorge et al 1984, Brugge et al 1985, Fults et al 1985, Cartwright et al 1987). Whether these proteins might be portions of NTF receptors or serve entirely unrelated functions is open to speculation.

Some oncogenes, like *c-erb-B*, which encodes the EGF receptor, do appear to be growth factor receptors. Not surprisingly, *c-erb-B* is expressed in brain tissue (Gonda et al 1982). A homologous oncogene encoding a membrane-spanning tyrosine kinase has been isolated from neuroblastoma and named *c-neu*. It is expressed in many fetal tissues, including brain (Coussens et al 1985), where it might serve as a receptor for an as yet unknown NTF.

The products of other oncogenes are trophic factors. *c-Sis*, which encodes the B chain of platelet derived growth factor (PDGF), is expressed at high levels in gliomas but has not been detected in normal brain tissue (Heldin et al 1985, Lens et al 1986). PDGF is a mitogen for astrocytes (Besnard et al 1987), but as yet has not been demonstrated to have NTF activity. Newer members of this class include two products with homology to bFGF, *c-hst* and *c-int-2*, but their expression in the CNS has not been demonstrated (Dickson & Peters 1987, Taira et al 1987). Another novel oncogene, *c-int-1*, encodes a secreted 40 kD glycoprotein transiently expressed during CNS development (Shackleford & Varmus 1987, Papkoff et al 1987). In situ hybridization shows a highly distinctive spatial distribution along the margins of the neural tube suggestive of a role in controlling morphogenesis or axonal growth (Wilkinson et al 1987), which might suggest potential NTF activity.

## SUMMARY

The search for novel NTFs is an active field. Many interesting factors have been purified and will likely be sequenced and identified in the near future. Among these are ciliary neuronotrophic factor (CNTF), a 20 kD protein whose effects on parasympathetic, sympathetic, and DRG neurons have been extensively studied in vitro (Barbin et al 1984); an unusual 3 kD protein isolated from lung that supports parasympathetic neurons (Wallace & Johnson 1987); a 16–20 kD seminal vesicle-derived neuronotrophic factor (SVNF) for sensory and sympathetic neurons (Hofmann & Unsicker 1987); a 55 kD serum protein that addresses a variety of CNS neurons (Kaufman & Barrett 1983); and a < 30 kD protein derived from muscle that supports spinal cord motoneurons (Oppenheim et al 1988). Judging from past results, some of these may be identical to known NTFs, others will be homologous to proteins of known function, and some will be completely novel. As other known trophic factors, such as PDGF or the interleukins, are examined more fully in neuronal cultures, it is likely that some may have as yet unrecognized NTF activity. Finally, further examination of oncogene expression likely will also yield new NTFs.

As more NTFs are recognized, they will allow generation of more sophisticated and complete models of nervous system growth. Glial cells have long been recognized to have trophic influences on neurons. The identification of several astrocytic products as potential NTFs—apoE, nexin, laminin, S100b, and bFGF—should allow greater appreciation of the biochemical mechanisms. Some of the new NTFs such as BDNF comply with the models of neuronal growth predicted by NGF, but others depart significantly. If the FGFs and IGFs are substantiated as NTFs, understanding their pleiotropic effects may require viewing neuronal growth as only one facet in the larger question of brain tissue regulation. In some situations these factors might coordinately regulate neuronal and glial growth; for example, matching the sizes of precursor populations generated during development. In other situations competitive interactions might predominate. After injury to the adult CNS, glial cells might use up limited supplies of shared trophic factors, thus decreasing their availability to neurons and impeding regenerative responses.

Most models currently ascribe the failure of axonal regeneration in the CNS to simple absence of NTFs. The reality may be more complex, since sympathetic and DRG neurons cannot be induced to grow on CNS tissue even in the presence of excess NGF (Schwab & Thoenen 1985, Carbonetto et al 1987, Sandrock & Matthew 1987b). Inhibitory growth factors have received considerable attention in general cell biology but have been neglected in neurobiology. Recently, two proteins in CNS myelin that inhibit neurite outgrowth in vitro have been recognized (Caroni & Schwab 1988). Identification of neural growth inhibitors may be even more significant for understanding growth regulation than isolation of additional novel NTFs.

## ACKNOWLEDGMENTS

I would like to thank Drs. Andrew Baird, Darwin Berg, Fred Gage, and Dave Schubert for valuable comments, and Mrs. Ruth Austin for assistance in preparing the manuscript. The preparation of this review and some of the work presented in it was supported by the State of California Alzheimer Disease Program (#86-89621), National Science Foundation grant BNS8705303, and the Pew Neuroscience program. P. W. is a McKnight Foundation Scholar and is supported by a Teacher Investigator Development Award (K07 NS00948-03) from the National Institute of Neurologic and Communicative Disorders and Stroke.

*Literature Cited*

Adamson, E. D., Meek, J. 1984. The ontogeny of epidermal growth factor receptors during mouse development. *Dev. Biol.* 103: 62–70

Aizenman, Y., de Vellis, J. 1987. Brain neurons develop in a serum and glial free environment: Effect of transferrin, insulin, insulin-like growth factor-I and thyroid hormone on neuronal survival, growth and differentiation. *Brain Res.* 406: 32–42

Alema, S., Casalbore, P., Agostini, E., Tato, F. 1985. Differentiation of PC12 phaeochromocytoma cells induced by *v-src* oncogene. *Nature* 316: 557–59

Alm, J., Scott, S., Fisher, D. A. 1986. Epidermal growth factor receptor ontogeny in mice with congenital hypothyroidism. *J. Dev. Physiol.* 8: 377–85

Aumailley, M., Nurcombe, V., Edgar, D., Paulsson, M., Timpl, R. 1987. The cellular interactions of laminin fragments. *J. Biol. Chem.* 262: 11532–38

Baird, A., Esch, F., Mormede, P., Ueno, N., Ling, N., Bohlen, P., Ying, S. Y., Wehrenberg, W. B., Guillemin, R. 1986. Molecular characterization of fibroblast growth factor: Distribution and biological activities in various tissues. *Recent Prog. Horm. Res.* 42: 143–205

Baird, A., Ling, N. 1987. Fibroblast growth factors are present in the extracellular matrix produced by endothelial cells in vitro: Implications for a role of heparinase-like enzymes in the neovascular response. *Biochem. Biophys. Res. Commun.* 142: 428–35

Bar-Sagi, D., Feramisco, J. R. 1985. Microinjection of the *ras* oncogene protein into PC12 cells induces morphological differentiation. *Cell* 42: 841–48

Barbin, G., Manthorpe, M., Varon, S. 1984. Purification of the chick eye ciliary neuronotrophic factor. *J. Neurochem.* 430: 1468–78

Barde, Y. A., Davies, A. M., Johnson, J. E., Lindsay, R. M., Thoenen, H. 1987. Brain derived neurotrophic factor. *Prog. Brain Res.* 71: 185–89

Baskin, D. G., Figlewicz, D. P., Woods, S. C., Porte, D. Jr., Dorsa, D. M. 1987. Insulin in the brain. *Ann. Rev. Physiol.* 49: 335–47

Berg, D. K. 1984. New neuronal growth factors. *Ann. Rev. Neurosci.* 7: 149–70

Berman, P., Gray, P., Chen, E., Keyser, K., Ehrlich, D., Karten, H., LaCorbiere, M., Esch, F., Schubert, D. 1987. Sequence analysis, cellular localization, and expression of a neuroretina adhesion and cell survival molecule. *Cell* 51: 135–42

Bernstein, H., Dorn, A., Reiser, M., Ziegler, M. 1984. Cerebral insulin-like immunoreactivity in rats and mice. *ACTA Histochem.* 74: 33–36

Besnard, F., Perraud, F., Sensenbrenner, M., Labourdette, G. 1987. Platelet-derived growth factor is a mitogen for glial but not neuronal rat brain cells in vitro. *Neurosci. Lett.* 73: 287–92

Birch, N. P., Christie, D. L., Renwick, A. G. C. 1984. Proinsulin-like material in mouse foetal brain cell cultures. *FEBS Lett.* 168: 299–302

Birch, N. P., Christie, D. L., Renwick, A. G. C. 1987. Multiple forms of biologically active insulin-like material from mouse fetal brain cells in culture. *Neuropeptides* 9: 325–31

Bothwell, M. 1982. Insulin and somatomedin MSA promote nerve growth factor-independent neurite formation by cultured chick dorsal root ganglionic sensory neurons. *J. Neurosci. Res.* 8: 225–31

Bottenstein, J. E., Sato, G. H. 1978. Growth of a rat neuroblastoma cell line in serum-free supplemented medium. *Proc. Natl. Acad. Sci. USA* 76: 514–17

Boyd, F. T. Jr., Clarke, D. W., Muther, T. F., Raizada, M. K. 1985. Insulin receptors and insulin modulation of norepinephrine uptake in neuronal cultures from rat brain. *J. Biol. Chem.* 260: 15880–84

Boyles, J. K., Pitas, R. E., Wilson, E., Mahley, R. W., Taylor, J. M. 1985. Apolipoprotein E associated with astrocytic glia of the central nervous system and with nonmyelinating glia of the peripheral nervous system. *J. Clin. Invest.* 76: 1501–13

Brown, A. L., Graham, D. E., Nissley, S. P., Hill, D. J., Strain, A. J., Rechler, M. M. 1986. Developmental regulation of insulin-like growth factor II mRNA in different rat tissues. *J. Biol. Chem.* 261: 13144–50

Brugge, J. S., Cotton, P. C., Queral, A. E., Barrett, J. N., Nonner, D., Keane, R. W. 1985. Neurons express high levels of a structurally, modified, activated form of pp60c-src. *Nature* 316: 554–59

Burgess, S. K., Jacobs, S., Cuatrecasas, P., Sahyoun, N. 1987. Characterization of a neuronal subtype of insulin-like growth factor I receptor. *J. Biol. Chem.* 262: 1618–22

Carbonetto, S., Gruver, M. M., Turner, D. C. 1983. Nerve fiber growth in culture on fibronectin, collagen and glycosoaminoglycan substates. *J. Neurosci.* 3: 2324–35

Carbonetto, S., Evans, D., Cochard, P. 1987. Nerve fiber growth in culture on tissue substrata from central and peripheral nervous systems. *J. Neurosci.* 7: 610–20

Carlsson-Skwirut, C., Jornvall, H., Holmgren, A., Andersson, C., Bergman, T., Lundquist, G., Sjogren, B., Sara, V. R. 1986. Isolation and characterization of variant IGF-I as well as IGF-2 from adult human brain. *FEBS Lett.* 201: 46–50

Caroni, P., Schwab, M. E. 1988. Antibody against myelin-associated inhibitor of neurite growth neutralizes nonpermissive substrate properties of CNS white matter. *Neuron* 1: 85–96

Cartwright, C. A., Simantov, R., Kaplan, P. L., Hunter, T., Eckhart, W. 1987. Alterations in pp60c-src accompany differentiation of neurons from rat embryo striatum. *Mol. Cell. Biol.* 7: 1830–40

Chaput, M., Claes, V., Portetelle, D., Cludts, I., Cravador, A., Burny, A., Gras, H., Tartar, A. 1988. The neurotrophic factor neuroleukin is 90% homologous with phosphohexose isomerase. *Nature* 332: 454–55

Chiu, A. Y., Matthew, W. D., Patterson, P. H. 1986. A monoclonal antibody that blocks the activity of a neurite regeneration-promoting factor: Studies on the binding site and its localization in vivo. *J. Cell Biol.* 102: 1383–98

Clarke, D. W., Mudd, L., Boyd, F. T. Jr., Fields, M., Raizada, M. K. 1986. Insulin is released from rat brain neuronal cells in culture. *J. Neurochem.* 47: 831–36

Cohen, J., Burne, J. F., Winter, J., Bartlett, P. 1986. Retinal ganglion cells lose response to laminin with maturation. *Nature* 322: 465–67

Cohen, J., Burne, J. F., McKinlay, C., Winter, J. 1987. The role of laminin and the laminin/fibronectin receptor complex in the outgrowth of retinal ganglion cell axons. *Dev. Biol.* 122: 407–18

Collins, F., Dawson, A. 1983. An effect of nerve growth factor on parasympathetic neurite outgrowth. *Proc. Natl. Acad. Sci. USA* 80: 2091–94

Corp, E. S., Woods, S. C., Porte, D., Dorsa, D. M., Figlewicz, D. P., Baskin, D. G. 1986. Localization of $^{125}$I-insulin binding sites in the rat hypothalamus by quantitative autoradiography. *Neurosci. Lett.* 70: 17–22

Coussens, L., Yang-Feng, T. L., Liao, Y., Chen, E., Gray, A., McGrath, J., Seeburg, P. H., Libermann, T. A., Schlessinger, J., Francke, U., Levinson, A., Ullrich, A. 1985. Tyrosine kinase receptor with extensive homology to EGF receptor shares chromosomal location with *neu* oncogene. *Science* 230: 1132–39

Cuevas, P., Baird, A., Guillemin, R. 1987. Stimulation of neovascularization and regeneration of the rat sciatic nerve by basic fibroblast growth factor. *J. Cell. Biochem. Suppl.* 11A: 192

Curran, T., Morgan, J. I. 1985. Superinduction of c-fos by nerve growth factor in the presence of peripherally active benzodiazepines. *Science* 229: 1265–68

Danielsen, N., Pettmann, B., Vahlsing, H. L., Manthorpe, M., Varon, S. 1988. Fibroblast growth factor effects on peripheral nerve regeneration in a silicone chamber model. *J. Neurosci. Res.* In press

Davies, A. M., Thoenen, H., Barde, Y. A. 1986. Different factors from the central nervous system and periphery regulate the survival of sensory neurons. *Nature* 319: 497–99

Davis, G. E., Manthorpe, M., Engvall, E., Varon, S. 1985. Isolation and characterization of a rat schwannoma neurite-promoting factor: Evidence that the factor contains laminin. *J. Neurosci.* 5: 2662–71

Davis, G. E., Blaker, S. N., Engvall, E., Varon, S., Manthorpe, M., Gage, F. H. 1987. Human amnion membrane serves as a substratum for growing axons in vitro and in vivo. *Science* 236: 1106–10

D'Ercole, A. J., Stiles, A. D., Underwood, L. E. 1984. Tissue concentrations of somatomedin C: Further evidence for multiple sites of synthesis and paracrine or autocrine mechanisms of action. *Proc. Natl. Acad. Sci. USA* 81: 935–39

Derynck, R. 1986. Transforming growth factor-alpha: Structure and biological activities. *J. Cell. Biochem.* 32: 293–304

Dickson, C., Peters, G. 1987. Potential oncogene product related to growth factors. *Nature* 326: 833

Donato, R. 1986. S-100 proteins. *Cell Calcium* 7: 123–45

Dragunow, M., Robertson, H. A. 1987. Kindling stimulation induces c-fos proteins in granule cells of the rat dentate gyrus. *Nature* 329: 441–42

Edgar, D., Timpl, R., Thoenen, H. 1984. The heparin-binding domain of laminin is responsible for its effects on neurite out-growth and neuronal survival. *EMBO J.* 3: 1463–68

Edgar, D. 1985. Nerve growth factors and molecules of the extracellular matrix in neuronal development. *J. Cell Sci. Suppl.* 3: 107–13

Eldridge, C. F., Sanes, J. R., Chiu, A. Y., Bunge, R. P., Cornbrooks, C. J. 1986. Basal lamina-associated heparan sulphate proteoglycan in the rat PNS: Character-ization and localization using monoclonal antibodies. *J. Neurocytol.* 15: 37–51

Engvall, E., Davis, G. E., Dickerson, K., Ruoslahti, E., Varon, S., Manthorpe, M. 1986. Mapping of domains in human lami-nin using monoclonal antibodies: Local-ization of the neurite-promoting site. *J. Cell Biol.* 103: 2457–65

Faik, P., Walker, J. I., Redmill, A. A., Morgan, M. J. 1988. Mouse glucose-6-phosphate isomerase and neuroleukin have identical 3′ sequences. *Nature* 332: 455–57

Fallon, J. H., Seroogy, K. B., Loughlin, S. E., Morrison, R. S., Bradshaw, R. A., Knauer, D. J., Cunningham, D. D. 1984. Epidermal growth factor immunoreactive material in the central nervous system: Location and development. *Science* 224: 1107–9

Fawcett, J. W., Keynes, R. J. 1986. Muscle basal lamina: A new graft material for peripheral nerve repair. *J. Neurosurg.* 65: 354–63

Fischer, W., Wictorin, K., Bjorklund, A., Williams, L. R., Varon, S., Gage, F. H. 1987. Amelioration of cholinergic neuron atrophy and spatial memory impairment in aged rats by nerve growth factor. *Nature* 329: 65–68

Fukada, K. 1985. Purification and partial characterization of a cholinergic neuronal differentiation factor. *Proc. Natl. Acad. Sci. USA* 82: 8759–99

Fults, D. W., Towle, A. C., Lauder, J. M., Maness, P. F. 1985. pp60*c-src* in the devel-oping cerebellum. *Mol. Cell. Biol.* 5: 27–32

Gammeltoft, S., Haselbacher, G. K., Humbel, R. E., Fehlmann, M., Van Obberghen, E. 1985. Two types of recep-tor for insulin-like growth factors in mam-malian brain. *EMBO J.* 4: 3407–12

Gensburger, C., Labourdette, G., Sensen-brenner, M. 1987. Brain basic fibroblast growth factor stimulates the prolifera-tion of rat neuronal precursor cells in vitro. *FEBS Lett.* 217: 1–5

Gill, G. N., Bertics, P. J., Santon, J. B. 1987. Epidermal growth factor and its receptor. *Mol. Cell. Endocrinol.* 51: 169–86

Gloor, S., Odink, K., Guenther, J., Nick, H.,

Monard, D. 1986. A glia-derived neurite promoting factor with protease inhibitory activity belongs to the protease nexins. *Cell* 5: 687–93

Gonda, T. J., Sheiness, D. K., Bishop, J. M. 1982. Transcripts from the cellular homo-logs of retroviral oncogenes: Distribution among chicken tissues. *Mol. Cell. Biol.* 2: 617–24

Goodyer, C. G., De Stephano, L., Lai, W. H., Guyda, H. J., Posner, B. I. 1984. Characterization of insulin-like growth factor receptors in rat anterior pituitary, hypothalamus, and brain. *Endocrinology* 114: 1187–95

Gorus, F. K., Hooghe-Peters, E. L., Pipe-leers, D. G. 1984. Glucose metabolism in murine fetal cortical brain cells: Lack of insulin effects. *J. Cell. Physiol.* 121: 45–50

Gospodarowicz, D., Neufeld, G., Schwei-gerer, L. 1986. Fibroblast growth factor. *Mol. Cell. Endocrinol.* 46: 187–204

Graf, J., Iwamoto, Y., Sasaki, M., Martin, G. R., Kleinman, H. K., Robey, F. A., Yamada, Y. 1987. Identification of an amino acid sequence in laminin mediating cell attachment, chemotaxis, and receptor binding. *Cell* 48: 989–96

Greenberg, M. E., Greene, L. A., Ziff, E. B. 1985. Nerve growth factor and epidermal growth factor induce rapid transient changes in proto-oncogene transcription in PC12 cells. *J. Biol. Chem.* 260: 14101–10

Guerrero, I., Wong, H., Pellicer, A., Burstein, D. E. 1986. Activated *N-ras* gene induces neuronal differentiation of PC12 rat pheochromocytoma cells. *J. Cell. Physiol.* 129: 71–76

Gurney, M. E. 1988. Letter to the editor. *Nature* 332: 456–57

Gurney, M. E., Heinrich, S. P., Lee, M. R., Yin, H. 1986a. Molecular cloning and expression of neuroleukin, a neurotrophic factor for spinal and sensory neurons. *Science* 234: 566–74

Gurney, M. E., Apatoff, B. R., Spear, G. T., Baumel, M. J., Antel, J. P., Bania, M. B., Reder, A. T. 1986b. Neuroleukin: A lymphokine product of lectin-stimulated T cells. *Science* 234: 574–81

Gurney, M. E., Apatoff, B. R., Heinrich, S. P. 1986c. Suppression of terminal axonal sprouting at the neuromuscular junction by monoclonal antibodies against a muscle-derived antigen of 56,000 daltons. *J. Cell Biol.* 102: 2264–72

Hagag, N., Halegoua, S., Viola, M. 1986. Inhibition of growth factor-induced dif-ferentiation of PC12 cells by microinjec-tion of antibody to *ras* p21. *Nature* 319: 680–82

Hall, D. E., Neugebauer, K. M., Reichardt, L. F. 1987. Embryonic neural retinal cell response to extracellular matrix proteins: Developmental changes and effects of the cell substratum attachment antibody (CSAT). *J. Cell Biol.* 104: 623–34

Hantaz-Ambroise, D., Vigny, M., Koenig, J. 1987. Heparan sulfate proteoglycan and laminin mediate two different types of neurite outgrowth. *J. Neurosci.* 7: 2293–2304

Hatten, M. E., Lynch, M., Rydel, R. E., Sanchez, J., Joseph-Silverstein, J., Moscatelli, D., Rifkin, D. B. 1988. In vitro neurite extension by granule neurons is dependent upon astroglial-derived fibroblast growth factor. *Dev. Biol.* 125: 280–89

Haselbacher, G., Schwab, M. E., Pasi, A., Humbel, R. E. 1985. Insulin-like growth factor II (IGF II) in human brain: Regional distribution of IGF II and of higher molecular mass forms. *Proc. Natl. Acad. Sci. USA* 82: 2153–57

Hefti, F. 1986. Nerve growth factor promotes survival of septal cholinergic neurons after fimbrial transections. *J. Neurosci.* 6: 2155–62

Heidenreich, K. A., Freidenberg, G. R., Figlewicz, D. P., Gilmore, P. R. 1986. Evidence for a subtype of insulin-like growth factor I receptor in brain. *Regul. Peptides* 15: 301–10

Heldin, C. H., Betsholtz, C., Johnsson, A., Nister, M., Ek, B., Ronnstrand, L., Wasteson, A., Westermark, B. 1985. Platelet-derived growth factor: Mechanism of action and relation to oncogenes. *J. Cell Sci. Suppl.* 3: 65–76

Herschman, H. R. 1986. Polypeptide growth factors and the CNS. *Trends Neurosci.* 9: 53–57

Hill, J. M., Lesniak, M. A., Pert, C. B., Roth, J. 1986. Autoradiographic localization of insulin receptors in rat brain: Prominence in olfactory and limbic areas. *Neuroscience* 17: 1127–38

Hirata, Y., Uchihashi, M., Nakajima, H., Fujita, T., Matsukura, S. 1982. Presence of human epidermal growth factor in human cerebrospinal fluid. *J. Clin. Endocrinol. Metab.* 55: 1174–77

Hofer, M. M., Barde, Y. A. 1988. Brain-derived neutrophic factor prevents neuronal death in vivo. *Nature* 331: 261–62

Hofmann, H. D., Unsicker, K. 1987. Characterization and partial purification of a novel neuronotrophic factor from bovine seminal vesicle. *J. Neurochem.* 48: 1425–33

Horwitz, A., Duggan, K., Greggs, R., Decker, C., Buck, C. 1985. The cell substrate attachment (CSAT) antigen has

properties of a receptor for laminin and fibronectin. *J. Cell Biol.* 101: 2134–44

Huck, S. 1983. Serum-free medium for cultures of the postnatal mouse cerebellum: Only insulin is essential. *Brain Res. Bull.* 10: 667–74

Hunt, S. P., Pini, A., Evan, G. 1987. Induction of c-*fos*-like protein in spinal cord neurons following sensory stimulation. *Nature* 328: 632–34

Hyden, H., Ronnback, L. 1978. Effect of S-100 protein immobilized on micro spherules on nerve regeneration. *Brain Res.* 143: 179–80

Ignatius, M. J., Gebicke-Harter, P. J., Skene, J. H. P., Schilling, J. W., Weisgraber, K. H., Mahley, R. W., Shooter, E. M. 1986. Expression of apolipoprotein E during nerve degeneration and regeneration. *Proc. Natl. Acad. Sci. USA* 83: 1125–29

Ignatius, M. J., Shooter, E. M., Pitas, R. E., Mahley, R. W. 1987. Lipoprotein uptake by neuronal growth cones in vitro. *Science* 236: 959–62

Irminger, J., Rosen, K. M., Humbel, R. E., Villa-Komaroff, L. 1987. Tissue-specific expression of insulin-like growth factor II mRNAs with distinct 5' untranslated regions. *Proc. Natl. Acad. Sci. USA* 84: 6330–34

Jeanny, J. C., Fayein, N., Moenner, M., Chevallier, B., Barritault, D., Courtois, Y. 1987. Specific fixation of bovine brain and retinal acidic and basic fibroblast growth factors to mouse embryonic eye basement membranes. *Exp. Cell Res.* 171: 63–75

Johnson, J. E., Barde, Y. A., Schwab, M., Thoenen, H. 1986. Brain-derived neurotrophic factor supports the survival of cultured retinal ganglion cells. *J. Neurosci.* 6: 3031–38

Kaufman, L. N., Barrett, J. N. 1983. Serum factor supporting long-term survival of rat central neurons in culture. *Science* 220: 1394–96

Kligman, D., Marshak, D. R. 1985. Purification and characterization of a neurite extension factor from bovine brain. *Proc. Natl. Acad. Sci. USA* 82: 7136–39

Kromer, L. F. 1987. Nerve growth factor treatment after brain injury prevents neuronal death. *Science* 235: 214–35

Kruijer, W., Schubert, D., Verma, I. M. 1985. Induction of the proto-oncogene *fos* by nerve growth factor. *Proc. Natl. Acad. Sci. USA* 82: 7330–34

Kryostek, A., Seeds, N. W. 1984. Peripheral neurons and Schwann cells secrete plasminogen activator. *J. Cell Biol.* 98: 773–76

Lakshmanan, J., Weichsel, M. E. Jr., Fisher, D. A. 1986. Epidermal growth factor in synaptosomal fractions of mouse cerebral cortex. *J. Neurochem.* 46: 1081–85

Lander, A. D., Fujii, D. K., Reichardt, L. F. 1985. Laminin is associated with the "neurite outgrowth-promoting factors" found in conditioned media. *Proc. Natl. Acad. Sci. USA* 82: 2183–87

Lee, M. R., Ho, D. D., Gurney, M. E. 1987. Functional interaction and partial homology between human immunodeficiency virus and neuroleukin. *Science* 237: 1047–51

Lenoir, D., Honegger, P. 1983. Insulin-like growth factor I (IGF I) stimulates DNA synthesis in fetal rat brain cell cultures. *Dev. Brain Res.* 7: 205–13

Lens, P. F., Altena, B., Nusse, R. 1986. Expression of *c-sis* and platelet-derived growth factor in in vitro-transformed glioma cells from rat brain tissue transplacentally treated with ethylnitrosourea. *Mol. Cell. Biol.* 6: 3537–40

Leutz, A., Schachner, M. 1982. Cell type-specificity of epidermal growth factor (EGF) binding in primary cultures of early postnatal mouse cerebellum. *Neurosci. Lett.* 30: 179–82

Levi-Montalcini, R. 1987. The nerve growth factor 35 years later. *Science* 237: 1154–62

Liesi, P. 1985. Laminin-immunoreactive glia distinguish regenerative adult CNS systems from non-regenerative ones. *EMBO J.* 4: 2505–11

Lindsay, R. M., Thoenen, H., Barde, Y. A. 1985. Placode and neural crest-derived sensory neurons are responsive at early developmental stages to brain-derived neurotrophic factor. *Dev. Biol.* 112: 319–28

Lipton, S. A., Wagner, J. A., Madison, R. D., D'Amore, P. A. 1988. Acidic fibroblast growth factor enhances regeneration of processes by postnatal mammalian retinal ganglion cells in culture. *Proc. Natl. Acad. Sci. USA* 85: 2388–92

Lowe, W. L. Jr., LeRoith, D. 1986. Tyrosine kinase activity of brain insulin and IGF-1 receptors. *Biochem. Biophys. Res. Commun.* 134: 532–38

Loy, R., Springer, J. E., Koh, S. 1987. Localization of cells immunoreactive for epidermal growth factor (EGF) receptor in the adult and developing rat forebrain. *Soc. Neurosci. Abstr.* 13: 575

Lund, P. K., Moats-Staats, B. M., Hynes, M. A., Simmons, J. G., Jansen, M., D'Ercole, A. J., Van Wyk, J. J. 1986. Somatomedin-C/Insulin-like growth factor-I and insulin-like growth factor-II mRNAs in rat fetal and adult tissues. *J. Biol. Chem.* 261: 14539–44

Madison, R. D., DaSilva, C., Dikkes, P., Sidman, R. L., Chiu, T. H. 1987. Peripheral nerve regeneration with entubulation repair: Comparison of biodegradable nerve guides versus polyethylene tubes and the effects of a laminin-containing gel. *Exp. Neurol.* 95: 378–90

Manthorpe, M., Engvall, E., Ruoslahti, E., Longo, F. M., Davis, G. E., Varon, S. 1983. Laminin promotes neuritic regeneration from cultured peripheral and central neurons. *J. Cell Biol.* 97: 1882–90

Masters, B. A., Shemer, J., Judkins, J. H., Clarke, D. W., LeRoith, D., Raizada, M. 1987. Insulin receptors and insulin action in dissociated brain cells. *Brain Res.* 417: 247–56

Matthew, W. D., Greenspan, R. J., Lander, A. D., Reichardt, L. F. 1985. Immunopurification and characterization of a neuronal heparan sulfate proteoglycan. *J. Neurosci.* 5: 1842–50

McElduff, A., Poronnik, P., Baxter, R. C. 1987. The insulin-like growth factor-II (IGF II) receptor from rat brain is of lower apparent molecular weight than the IGF II receptor from rat liver. *Endocrinology* 121: 1306–11

Milbrandt, J. 1986. Nerve growth factor rapidly induces *c-fos* mRNA in PC12 rat pheochromocytoma cells. *Proc. Natl. Acad. Sci. USA* 83: 4789–93

Mill, J. F., Chao, M. V., Ishii, D. N. 1985. Insulin, insulin-like growth factor II, and nerve growth factor effects on tubulin mRNA levels and neurite formation. *Proc. Natl. Acad. Sci. USA* 82: 7126–30

Monard, D. 1987. Role of protease inhibition in cellular migration and neuritic growth. *Biochem. Pharmacol.* 36: 1389–92

Morgan, J. I., Cohen, D. R., Hempstead, J. L., Curran, T. 1987. Mapping patterns of *c-fos* expression in the central nervous system after seizure. *Science* 237: 192–97

Morrison, R. S., de Vellis, J. 1981. Growth of purified astrocytes in a chemically defined medium. *Proc. Natl. Acad. Sci. USA* 78: 7205–9

Morrison, R. S., Sharma, A., de Vellis, J., Bradshaw, R. A. 1986. Basic fibroblast growth factor supports the survival of cerebral cortical neurons in primary culture. *Proc. Natl. Acad. Sci. USA* 83: 7537–41

Morrison, R. S., Kornblum, H. I., Leslie, F. M., Bradshaw, R. A. 1987. Trophic stimulation of cultured neurons from neonatal rat brain by epidermal growth factor. *Science* 238: 72–75

Müller, H., Williams, L. R., Varon, S. 1987. Nerve regeneration chamber: Evaluation of exogenous agents applied by multiple injections. *Brain Res.* 413: 320–26

Murphy, L. J., Bell, G. I., Friesen, H. G.

1987. Tissue distribution of insulin-like growth factor I and II messenger ribonucleic acid in the adult rat. *Endocrinology* 120: 1279–82

Neufeld, G., Gospodarowicz, D., Dodge, L., Fujii, D. K. 1987. Heparin modulation of the neurotrophic effects of acidic and basic fibroblast growth factors and nerve growth factor on PC12 cells. *J. Cell. Physiol.* 131: 131–40

Nissley, S. P., Rechler, M. M. 1984. Somatomedin/insulin-like growth factor tissue receptors. *Clin. Endocrinol. Metab.* 13: 43–67

Noda, M., Ko, M., Ogura, A., Ding-gan, L., Amano, T., Takano, T., Ikawa, Y. 1985. Sarcoma viruses carrying *ras* oncogenes induce differentiation-associated properties in a neuronal cell line. *Nature* 318: 73–75

Onifer, S. M., Farber, S. D., Murphy, S. H., Kaseda, Y., Low, W. C. 1987. Effects of insulin-like growth factor II (IGF-II) on hippocampal and septal neurons maintained in tissue culture. *Soc. Neurosci. Abstr.* 13: 1615

Oppenheim, R. W., Haverkamp, L. J., Prevette, D., McManaman, J. L., Appel, S. H. 1988. Reduction of naturally occurring motoneuron death in vivo by a target-derived neurotrophic factor. *Science* 240: 919–22

Papkoff, J., Brown, A. M. C., Varmus, H. E. 1987. The *int-1* proto-oncogene products are glycoproteins that appear to enter the secretory pathway. *Mol. Cell. Biol.* 7: 3978–84

Pettmann, B., Labourdette, G., Weibel, M., Sensenbrenner, M. 1986. The brain fibroblast growth factor (FGF) is localized in neurons. *Neurosci. Lett.* 68: 175–80

Pettmann, B., Weibel, M., Sensenbrenner, M., Labourdette, G. 1985. Purification of two astroglial growth factors from bovine brain. *FEBS Lett.* 189: 102–8

Pikkarainen, T., Kallunki, T., Tryggvason, K. 1988. Human laminin B2 chain. Comparison of the complete amino acid sequence with the B1 chain reveals variability in sequence homology between different structural domains. *J. Biol. Chem.* 263: 6751–58

Pittman, R. N. 1985. Release of plasminogen activator and a calcium-dependent metalloprotease from cultured sympathetic and sensory neurons. *Dev. Biol.* 110: 91–101

Pittman, R. N., Patterson, P. H. 1987. Characterization of an inhibitor of neuronal plasminogen activator released by heart cells. *J. Neurosci.* 7: 2664–73

Probstmeier, R., Schachner, M. 1986. Epidermal growth factor is not detectable in developing and adult rodent brain by a sensitive double-site enzyme immunoassay. *Neurosci. Lett.* 63: 290–94

Rall, L. B., Scott, J., Bell, G. I., Crawford, R. J., Penschow, J. D., Niall, H. D., Coghlan, J. P. 1985. Mouse prepro-epidermal growth factor synthesis by the kidney and other tissues. *Nature* 313: 228–31

Recio-Pinto, E., Rechler, M. M., Ishii, D. N. 1986. Effects of insulin, insulin-like growth factor-II, and nerve growth factor on neurite formation and survival in cultured sympathetic and sensory neurons. *J. Neurosci.* 6: 1211–19

Rees-Jones, R. W., Hendricks, S. A., Quarum, M., Roth, J. 1984. The insulin-receptor of rat brain is coupled to tyrosine kinase activity. *J. Biol. Chem.* 259: 3470–74

Roberts, A. B., Anzano, M. A., Lamb, L. C., Smith, J. M., Sporn, M. B. 1981. New class of transforming growth factors potentiated by epidermal growth factor: Isolation from non-neoplastic tissues. *Proc. Natl. Acad. Sci. USA* 78: 5339–43

Roberts, D. D., Rao, C. N., Magnani, J. L., Spitalnik, S. L., Liotta, L. A., Ginsberg, V. 1985. Laminin binds specifically to sulfated glycolipids. *Proc. Natl. Acad. Sci. USA* 82: 1306–10

Rogers, S. L., McCarthy, J. B., Palm, S. L., Furcht, L. T., Letourneau, P. C. 1985. Neuron-specific interactions with two neurite-promoting fragments of fibronectin. *J. Neurosci.* 5: 369–78

Rogers, S. L., Letourneau, P. C., Peterson, B. A., Furcht, L. T., McCarthy, J. B. 1987. Selective interaction of peripheral and central nervous system cells with two distinct cell-binding domains of fibronectin. *J. Cell Biol.* 105: 1435–42

Romijn, H. N., van Huizen, F., Wolters, P. S. 1984. Towards an improved serum-free, chemically defined medium for long-term culturing of cerebral cortex issue. *Neurosci. Biobehav. Rev.* 8: 301–34

Rosenblatt, D. E., Cotman, C. W., Nieto-Sampedro, M., Rowe, J. W., Knauer, D. J. 1987. Identification of a protease inhibitor produced by astrocytes that is structurally and functionally homologous to human protease nexin-I. *Brain Res.* 415: 40–48

Roth, R. A. 1988. Structure of the receptor for insulin-like growth factor. II: The puzzle amplified. *Science* 239: 1269–71

Roth, R. A., Morgan, D. O., Beaudoin, J., Sara, V. 1986. Purification and characterization of the human brain insulin receptor. *J. Biol. Chem.* 261: 3753–57

Rotwein, P., Burgess, S. K., Milbrandt, J. D., Krause, J. E. 1988. Differential expression of insulin-like growth factor genes in rat central nervous system. *Proc. Natl. Acad. Sci. USA* 85: 265–69

Ruoslahti, E., Pierschbacher, M. D. 1987. New perspectives in cell adhesion: RGD and integrins. *Science* 238: 491–97

Sandrock, A. W. Jr., Matthew, W. D. 1987a. An in vitro neurite-promoting antigen functions in axonal regeneration in vivo. *Science* 237: 1605–8

Sandrock, A. W. Jr., Matthew, W. D. 1987b. Identification of a peripheral nerve neurite growth-promoting activity by development and use of an in vitro bioassay. *Proc. Natl. Acad. Sci. USA* 84: 6934–38

Sara, V. R., Hall, K., Von Holtz, H., Humbel, R., Sjogren, B., Wetterberg, L. 1982. Evidence for the presence of specific receptors for insulin-like growth factors I (IGF-I) and 2 (IGF-2) and insulin throughout the adult human brain. *Neurosci. Lett.* 34: 39–44

Sara, V. R., Carlsson-Skwirut, C., Andersson, C., Hall, E., Sjogren, B., Holmgren, A., Jornvall, H. 1986. Characterization of somatomedins from human fetal brain: Identification of a variant form of insulin-like growth factor I. *Proc. Natl. Acad. Sci. USA* 83: 4904–7

Sasaki, N., De Vroede, M. A., Rechler, M. M. 1985. Insulin-like growth factor receptors. *J. Cell Sci. Suppl.* 3: 39–51

Schubert, D., LaCorbiere, M. 1985. Isolation of an adhesion-mediating protein from chick neural retina adherons. *J. Cell Biol.* 101: 1071–77

Schubert, D., LaCorbiere, M., Esch, F. 1986. A chick neural retina adhesion and survival molecule is a retinol-binding protein. *J. Cell Biol.* 102: 2295–2301

Schubert, D., Ling, N., Baird, A. 1987. Multiple influences of a heparin-binding growth factor on neuronal development. *J. Cell Biol.* 108: 635–43

Schwab, M. E., Thoenen, H. 1985. Dissociated neurons regenerate into sciatic but not optic nerve explants in culture irrespective of neurotrophic factors. *J. Neurosci.* 5: 2415–23

Shackleford, G. M., Varmus, H. E. 1987. Expression of the proto-oncogene *int-1* is restricted to postmeiotic male germ cells and the neural tube of mid-gestational embryos. *Cell* 50: 89–95

Shemer, J., Raizada, M. K., Masters, B. A., Ota, A., LeRoith, D. 1987. Insulin-like growth factor I receptors in neuronal and glial cells. *J. Biol. Chem.* 262: 7693–99

Sievers, J., Hausmann, B., Unsicker, K., Berry, M. 1987. Fibroblast growth factors promote the survival of adult rate retinal ganglion cells after transection of the optic nerve. *Neurosci. Lett.* 76: 157–62

Simpson, D. L., Morrison, R., de Vellis,

J., Herschman, H. R. 1982. Epidermal growth factor binding and mitogenic activity on purified populations of cells from the central nervous system. *J. Neurosci. Res.* 8: 453–62

Skene, J. H., Jacobson, R. D., Snipes, G. J., McGuire, C. B., Norden, J. J., Freeman, J. A. 1986. A protein induced during nerve growth (GAP-43) is a major component of growth cone membranes. *Science* 233: 783–86

Smalheiser, N. R., Schwartz, N. B. 1987. Cranin: A laminin-binding protein of cell membranes. *Proc. Natl. Acad. Sci. USA* 84: 6457–61

Snipes, G. J., Costello, B., McGuire, C. B., Mayes, B. N., Bock, S. S., Norden, J. J., Freeman, J. A. 1987. Regulation of specific neuronal and nonneuronal proteins during development and following injury in the rat central nervous system. *Prog. Brain Res.* 71: 155–75

Snyder, E. Y., Kim, S. 1980. Insulin: Is it a nerve survival factor. *Brain Res.* 196: 565–71

Soreq, H., Miskin, R. 1981. Plasminogen activator in the rodent brain. *Brain Res.* 216: 361–74

Sorge, L. K., Levy, B. T., Maness, P. F. 1984. pp60c-src is developmentally regulated in the neural retina. *Cell* 36: 249–57

Stoschek, C. M., King, L. E. Jr. 1986. Functional and structural characteristics of EGF and its receptor and their relationship to transforming proteins. *J. Cell. Biochem.* 31: 135–52

Swanson, M. E., Elste, A. M., Greenberg, S. M., Schwartz, J. H., Aldrich, T. H., Furth, M. E. 1986. Abundant expression of *ras* proteins in Aplysia neurons. *J. Cell Biol.* 103: 485–92

Taira, M., Yoshida, T., Miyagawa, K., Sakamoto, H., Terada, M., Sugimura, T. 1987. cDNA sequence of human transforming gene *hst* and identification of the coding sequence required for transforming activity. *Proc. Natl. Acad. Sci. USA* 84: 2980–84

Tanaka, T., Ida, N., Shimoda, H., Waki, C., Slamon, D. J., Cline, M. J. 1986. Organ specific expression of *ras* oncoproteins during growth and development of the rat. *Mol. Cell. Biochem.* 70: 97–104

Thoenen, H., Edgar, D. 1985. Neurotrophic factors. *Science* 229: 238–42

Thomas, K. A., Gimenez-Gallego, G. 1986. Fibroblast growth factors: Broad spectrum mitogens with potent angiogenic activity. *Trends Biochem. Sci.* 11: 81–84

Tomaselli, K. J., Reichardt, L. F., Bixby, J. L. 1986. Distinct molecular interactions mediate neuronal process outgrowth on non-neuronal cell surfaces and extracellular matrices. *J. Cell Biol.* 103: 2659–72

Unsicker, K., Reichert-Preibsch, H., Schmidt, R., Pettmann, B., Labourdette, G., Sensenbrenner, M. 1987. Astroglial and fibroblast growth factors have neurotrophic functions for cultured peripheral and central nervous system neurons. *Proc. Natl. Acad. Sci. USA* 84: 5459–63

Van Schravendijk, C. F. H., Hooghe-Peters, E. L., Van den Brande, J. L., Pipeleers, D. G. 1986. Receptors for insulin-like growth factors and insulin on murine fetal cortical brain cells. *Biochem. Biophys. Res. Commun.* 135: 228–38

Varon, S., Manthorpe, M., Williams, L. R. 1983. Neuronotrophic and neurite promoting factors and their clinical potentials. *Dev. Neurosci.* 6: 73–100

Villa-Komaroff, L., Gonzalez, A., Hou-Yan, S., Wentworth, B., Dobner, P. 1984. Novel insulin-related sequences in fetal brain. *Adv. Exp. Med. Biol.* 181: 65–86

Vlodavsky, I., Folkman, J., Sullivan, R., Fridman, R., Ishai-Michaeli, R., Sasse, J., Klagsbrun, M. 1987. Endothelial cell-derived basic fibroblast growth factor: Synthesis and deposition into subendothelial extracellular matrix. *Proc. Natl. Acad. Sci. USA* 84: 2292–96

Wagner, J. A., D'Amore, P. A. 1986. Neurite outgrowth induced by an endothelial cell mitogen isolated from retina. *J. Cell Biol.* 103: 1363–67

Walicke, P. A., Cowan, W. M., Ueno, N., Baird, A., Guillemin, R. 1986. Fibroblast growth factor promotes survival of dissociated hippocampal neurons and enhances neurite extension. *Proc. Natl. Acad. Sci. USA* 83: 3012–16

Walicke, P. A. 1988. Basic and acidic fibroblast growth factors have trophic effects on neurons from mutiple CNS regions. *J. Neurosci.* 8: 2618–27

Walicke, P. A., Baird, A. 1988a. Neurotrophic effects of basic and acidic fibroblast growth factors are not mediated

through glial cells. *Dev. Brain Res.* 40: 71–79

Walicke, P. A., Baird, A. 1988b. Trophic effects of fibroblast growth factor on neural tissue. *Prog. Brain Res.* In press

Wallace, T. L., Johnson, E. M. Jr. 1987. Partial purification of a parasympathetic neurotrophic factor in pig lung. *Brain Res.* 411: 351–63

Werther, G. A., Hogg, A., Oldfield, B. J., McKinley, M. J., Figdor, R., Allen, A. M., Mendelsohn, F. A. O. 1987. Localization and characterization of insulin receptors in rat brain and pituitary gland using in vitro autoradiography and computerized densitometry. *Endocrinology* 121: 1562–70

Weyhenmeyer, J. A., Fellows, R. E. 1983. Presence of immunoreactive insulin in neurons cultured from fetal rat brain. *Cell. Mol. Neurobiol.* 3: 81–86

Wilkinson, D. G., Bailes, J. A., McMahon, A. P. 1987. Expression of the proto-oncogene *int-1* is restricted to specific neural cells in the developing mouse embryo. *Cell* 50: 79–88

Williams, L. R., Varon, S., Peterson, G. M., Wictorin, K., Fischer, W., Bjorklund, A., Gage, F. H. 1986. Continuous infusion of nerve growth factor prevents basal forebrain neuronal death after fimbria fornix transection. *Proc. Natl. Acad. Sci. USA* 83: 9231–35

Yalow, R. S., Eng, J. 1983. Insulin in the central nervous system. *Adv. Metab. Disord.* 10: 341–54

Yoshii, S., Yamamuro, T., Ito, S., Hayashi, M. 1987. In vivo guidance of regenerating nerve by laminin-coated filaments. *Exp. Neurol.* 96: 469–73

Young, W. S. III. 1986. Periventricular hypothalamic cells in the rat brain contain insulin mRNA. *Neuropeptides* 8: 93–97

Zapf, J., Schmid, Ch., Froesch, E. R. 1984. Biological and immunological properties of insulin-like growth factors (IGF) I and II. *Clin. Endocrinol. Metab.* 13: 3–30

*Ann. Rev. Neurosci. 1989. 12:127–56*

# AXONAL GROWTH-ASSOCIATED PROTEINS

*J. H. Pate Skene*

Department of Neurobiology, Stanford University, Stanford, California 94305-5401

Elongation of axons and active remodeling of their terminal arbors underlies the assembly of neural circuits during development, determines the success or failure of nerve regeneration, and may contribute to some forms of synaptic plasticity in adult brains. For most neurons, elongation of a principal axon is confined to a few days or weeks during development. Remodeling of axon terminal arbors also is most pronounced during transient "critical periods" late in development, although some forms of synaptic remodeling continue throughout life (Lichtman et al 1987). Once past these epochs of axon elongation and dynamic sorting of synaptic terminals, it might be possible to stabilize principal axon branches or their terminal arbors by inactivating some of the molecular processes required for growth and synaptogenesis. Selective inactivation or retention of some growth-related processes in maturing neurons might then define some limits on the mechanisms available for synaptic remodeling in the adult nervous system.

Studies of axon regeneration in vivo and in tissue culture indicate that some aspects of axon growth are indeed repressed in many adult neurons but can be re-induced under some conditions. Although such studies have considered primarily the elongation of primary axons, they also raise the possibility that neuronal processes underlying more subtle aspects of axon growth and synaptogenesis may be down-regulated chronically in mature neurons. At the molecular level, periods of axon outgrowth during development and re-induction of axon growth for regeneration are correlated with large and specific changes in synthesis of a few proteins transported into the growing axons. This suggests the hypothesis that differentiation to a stable mature state can include the selective repression of genes required for axon growth. The list of genes expressed selectively during

127

0147–006X/89/0301–0127$02.00

periods of axon growth is far from complete, but it is already possible to identify a few genes that seem to be tightly correlated with a neuronal "growth state." One of these is a gene encoding an acidic membrane protein designated GAP-43 (also known as B-50, F1, pp46, or p57). Although most neurons selectively reduce GAP-43 expression as they mature, a subset of neurons continue to express high levels of the protein in the adult CNS. Phosphorylation of GAP-43 in adult brain has been correlated with long-term potentiation at some synapses. Biochemical characterization of GAP-43, and emerging evidence of other genes expressed during axon growth, suggest that some "growth-associated" proteins may alter a neuron's responses to extracellular signals by altering intracellular signal transducing systems.

## "GROWTH STATE(S)" AND MATURE NEURONS

When developing neurons are explanted from fetal or neonatal animals, the explanted cells can reinitiate neurite outgrowth in the culture dish within a few hours (Argiro & Johnson 1982, Collins & Lee 1982). In contrast, adult neurons explanted to identical culture conditions do not extend neurites for several days (Agranoff et al 1976, Landreth & Agranoff 1976, Argiro & Johnson 1982, Collins & Lee 1982), thus suggesting that some aspects of axon growth are repressed or impaired in the mature neurons and are reinduced slowly in culture. In some cases, maturation may involve selective loss of individual features of axon growth. Chick retinal neurons, for example, lose much of their ability to extend neurites on a laminin substratum between embryonic days E6 and E12, but retain their capacity for elongating neurites on glial cell surfaces (Cohen et al 1986, Hall et al 1987).

For neurons that regenerate their axons successfully in vivo, a nerve injury made several days prior to explanation substantially reduces the lag before outgrowth of neurites in culture (Agranoff et al 1976, Landreth & Agranoff 1976). Similarly the lag before initiation of axon regeneration in vivo is reduced if the same nerve received an earlier "conditioning" lesion, suggesting that nerve injury induces events in the neuronal cell body that prepare the neuron to extend its axon (reviewed in Grafstein & McQuarrie 1978, McQuarrie 1984). All of these observations are consistent with a model in which some of the molecular processes important for axon growth become repressed or down-regulated in many neurons as they mature, but may be reinduced under some conditions in adult animals. The developmentally regulated events in the cell body might include either fundamental alterations in the machinery for neurite extension or, as illustrated for maturing chick retinal cells, more subtle changes in molecules involved in the recognition and transduction of growth-related signals in the extra-cellular environment.

## MOLECULAR CORRELATES OF AXON GROWTH

The apparent repression of growth-related properties in many neurons during differentiation suggests that some of the genes involved in axon growth might be expressed transiently during development and reinduced during successful axon regeneration. This has prompted a search for genes whose expression is correlated consistently with periods of axon growth. Most of these studies have concentrated on the synthesis of proteins destined for transport into axons and their terminals, the population of neuronal proteins most directly involved in axon functions. In almost all neurons screened in this way, it has been possible to identify one or more axonal proteins whose synthesis is specifically increased an order of magnitude during developmental or regenerative axon growth (Table 1). These proteins are distinguished as rapidly transported or slowly trans-ported proteins, depending on whether they are delivered into the axon within a few hours of synthesis (transport groups I and II or FC; Willard et al 1974, Grafstein & Forman 1980) or move along axons much more slowly as part of a large complex of cytoskeletal and cytoplasmic com-ponents (groups IV and V or SC and SCb; Willard & Hulebak 1977, Hoffman & Lasek 1975, Black & Lasek 1980). The small number of proteins transported at an intermediate velocity (group III of Willard et al 1974) have not been examined in detail for growth-related changes.

When the synthesis of individual proteins is normalized to overall pro-tein synthesis, only a small number of proteins in any one neuronal system show large changes in their relative synthesis during axon growth. Even fewer proteins are consistently expressed at elevated levels during both developmental and regenerative axon growth in many different neurons (Table 1). Among these are tubulin and actin, and a rapidly transported membrane protein designated GAP-43. Periods of axon growth also are characterized in many cases by decreased synthesis or transport of neuro-filament proteins, particularly the largest of the three neurofilament sub-units (Hoffman & Lasek 1980, Shaw & Weber 1982, Willard & Simon 1983, Pachter & Liem 1984, Tetzlaff et al 1987, Kalil & Perdew 1988). It has been suggested that some aspects of axon growth or plasticity may be reduced in maturing neurons by the assembly of an extensive network of neurofilament cross-links mediated by the large neurofilament subunit (Willard et al 1984, Glicksman & Willard 1985).

With the expanding use of subtractive and differential hybridization in screening cDNA libraries, identification of additional "growth-associated"

**Table 1** Large changes in synthesis of individual axonal proteins during period of axon growth[a]

| | GAP-43 | GAP-24 (23–28 K) | Tubulins | Actin | NF decreased transport | Other proteins | References |
|---|---|---|---|---|---|---|---|
| | (Rapidly transported) | | (Slowly transported) | | | | |
| **REGENERATION** | | | | | | | |
| Toad optic n. | ● | ● | ● | ● | | 50 K, 33 K, 42 K | Skene & Willard 1981a |
| Fish optic n. | ● | ● | ● | ● | | 100–140 K | Benowitz et al 1981, 1983 |
| | | | | | | 250 K | Giulian et al 1980 |
| | | | | | | 68–70 K | Heacock & Agranoff 1976, 1982 |
| | | | | | | numerous smaller changes | Perry et al 1987a |
| Frog optic n. | NO (2X)[b] | NO | 2X | | | several smaller changes | Szaro et al 1985 |
| Frog sciatic (DRG) | NO | 2X | | | | | Perry et al 1987b |
| | | | | | | | Perry & Wilson 1981 |
| Rat sciatic n.: | | | | | | | |
| sensory | ● | ● | <2X | <2X | | several smaller changes | Hall 1982 |
| | | | | | | | Redshaw & Bisby 1987 |
| | | | | | | | Basi et al 1987 |
| motor | | | NO | NO | | | Goldstein et al 1987 |
| Rat facial n. | | | ● | ● | | | Hoffman & Lasek 1980 |
| Rat hypoglossal n. | | ● | | | ● ● | | Tetzlaff et al 1987 |
| | | | | | | | Redshaw & Bisby 1984a |
| Rabbit hypoglossal n. | ● ● | no change | | | | | Skene & Willard 1981b |

| | | | | References |
|---|---|---|---|---|
| **DEVELOPMENT** | | | | |
| Rabbit optic n. | • | • | | Skene & Willard 1981b |
| | | | | Willard & Simon 1983 |
| Rat optic n. | • | • | | Freeman et al 1986 |
| | | | | Pachter & Liem 1984 |
| Hamster pyramidal tr. | • | NO | | Kalil & Skene 1986 |
| | | | | Kalil & Perdew 1987 |
| Whole cortex | • | NO | • | Jacobson et al 1986 |
| | | | | Shaw & Weber 1982 |
| | | | | Lewis et al 1985 |
| | | | | Carden et al 1987 |
| **ABORTIVE REGENERATION** | | | | |
| Rabbit optic n. | NO | NO | | Skene & Willard 1981b |
| | | | | Redshaw & Bisby 1984b |
| Rat optic n. | NO | NO | | Freeman et al 1986 |
| | | | | Kalil & Skene 1986 |
| Hamster pyramidal tr. | NO | • | NO | Reh et al 1987 |

[a] All proteins listed are reported to undergo at least a ten-fold change in synthesis correlated with axon growth in at least one system. All entries refer to specific changes in protein synthesis or mRNA abundance. *Solid dots* indicate that synthesis of the protein increases ten-fold or more, compared to uninjured adult neurons. *Filled dots* indicate order-of-magnitude changes. "NO" indicates that synthesis of the indicate protein was assayed and found not to change significantly during axon growth. *Blank spaces* indicate that the protein was not assayed.
[b] See text for discussion.

proteins or genes seems a likely prospect. In the meantime, proteins such as GAP-43 and the cytoskeletal proteins, whose expression is widely correlated with axon growth, provide a starting point for studying how specific changes in gene expression in the cell body are influenced by changing conditions in distal portions of an axon, and how altered expression of those genes might affect various steps in axon growth.

## GAP-43 (A.K.A. B-50/F1/pp46/P-57)

The most extensive correlation with axon growth documented so far is for the synthesis of GAP-43 (Table 1). This assertion rests on evidence, sometimes indirect, that the protein designated "GAP-43" is the same or homologous in each species and neuron type studied. In all cases, "GAP-43" refers to a membrane-bound, rapidly transported (Group I) protein with a very acidic isoelectric point (4.3–4.5) and aberrant migration on SDS gels, such that the protein's *apparent* molecular weight varies from 43–57,000 daltons or greater, depending on the gel concentration used (Jacobson et al 1986, Benowitz et al 1987). Direct electrophoretic comparisons have been used to suggest homologies between the rabbit (Skene & Willard 1981b), rat (Jacobson et al 1986), and hamster (Kalil & Skene 1986) proteins and the definitional toad GAP-43, and between the rat and goldfish proteins (Perrone-Bizzozero et al 1986). Immunological cross-reactivity between the toad and rat proteins has been demonstrated (Jacobson et al 1986). Cloning of rat GAP-43 cDNA (Basi et al 1987, Karns et al 1987, Neve et al 1987) now provides a more definitive basis for identifying GAP-43 in various neurons.

This same protein has been discovered with relentless regularity by investigators working on several aspects of axon growth and synaptic function. A synaptic protein designated B-50 has been suggested as a possible regulator of polyphosphoinositide metabolism (Oestreicher et al 1983, van Dongen et al 1985, van Hoof et al 1988). Other investigators have found that kinase-C-mediated phosphorylation of the same protein is correlated with long-term synaptic potentiation in rat hippocampus (Nelson & Routtenberg 1985, Routtenberg 1986); in that context, the protein was designated F1. A major phosphoprotein of growth cone membranes has been called pp46 (Katz et al 1985, Meiri et al 1986). Finally, an unusual calmodulin-binding protein from bovine and mouse brain, designated p57 (Andreason et al 1983), binds calmodulin preferentially in the absence of calcium; phosphorylation of p57 by kinase C greatly reduces the protein's affinity for calmodulin. Biochemical and immunological evidence that these are all the same protein has been reviewed (Benowitz &

Routtenberg 1987). More recently, direct sequence comparisons (Figure 1) show that mouse and bovine p57 (Wakim et al 1987, Cimler et al 1987) are virtually identical to rat GAP-43 (Basi et al 1987, Karns et al 1987), F1 (Rosenthal et al 1987), and B-50 (Nielander et al 1987).

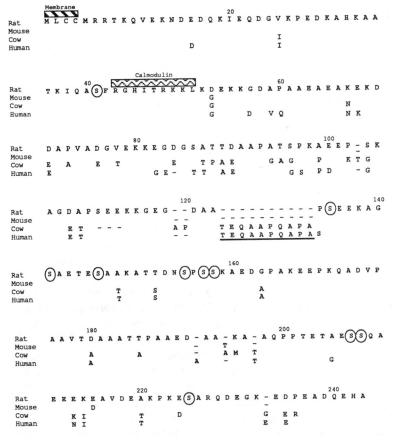

*Figure 1* Amino acid sequence of GAP-43. cDNAs corresponding to rat proteins identified as GAP-43 (Basi et al 1987, Karns et al 1987) or protein F1 (Rosenthal et al 1987) yielded identical sequences. The sequences for mouse and cow were determined for the calmodulin-binding protein p57 (Cimler et al 1987, Wakim et al 1987). Human sequences are predicted from cDNAs isolated with a rat GAP-43 cDNA probe (Ng et al 1988) or with antibodies against purified protein B-50 from rat (Kosik et al 1988). *Shaded bars* indicate the putative membrane binding domain (Basi et al 1987) and calmodulin binding site (Wakim et al 1987, Alexander et al 1988). A 10-amino-acid in the cow and human sequence, not found in rodents, is underlined. *Dashes* in the sequence indicate spaces inserted to optimize alignments. Serine residues conserved in all four sequences are *circled*.

Localization of protein B-50 to presynaptic membranes in adult brain (Gispen et al 1985), and correlation of protein F1 phosphorylation with long-term potentiation (Nelson & Routtenberg 1985, Lovinger et al 1986) has expanded speculation about possible roles of GAP-43 to include physiological or structural plasticity of synaptic terminals, in addition to any role the protein might play in elongation of principal axons.

## Association with Axon Growth

Consistent with a possible role in axon elongation, GAP-43 has been shown to be a prominent protein component of growth cone membranes. A subcellular fraction from developing rat brain highly enriched for growth cones (Pfenninger et al 1983) is also highly enriched for GAP-43 (Skene et al 1986, Meiri et al 1986); GAP-43 is also a prominent substrate for endogenous kinase(s) in these isolated growth cones (Katz et al 1985, deGraan et al 1985, Jacobson et al 1986). GAP-43 is an extremely abundant protein in growth cones, comprising on the order of 1% of the total protein in growth cones membranes (Skene et al 1986). Immunolocalization of GAP-43 in cultured neurons (Meiri et al 1988) confirms that the protein is most heavily concentrated in growth cones. GAP-43 appears to be present throughout the growth cones, including filopodial extensions. Varicosities along the length of neurites also contained high concentrations of GAP-43. Neuron cell bodies show low levels of diffuse staining for GAP-43, in contrast to the punctate staining of neurites and growth cones (Meiri et al 1988).

A growth cone localization for GAP-43 is equally consistent with the protein's involvement in axon elongation or in the formation and active reorganization of synapses once axons reach their target tissues. The time course of GAP-43 expression during axon regeneration is also consistent with either of these possibilities. In regenerating toad retinal ganglion cells, GAP-43 synthesis begins to increase approximately 4 days after axotomy and reaches a maximum during the second week of regeneration (Skene & Willard 1981a); induction follows a similar time course in goldfish (Benowitz et al 1981, Benowitz & Schmidt 1987). Axon elongation commences approximately 4.5 days after injury of goldfish optic nerves (Grafstein & McQuarrie 1978, McQuarrie & Grafstein 1981), and the initial delay for amphibian optic nerves appears to be similar or slightly longer (Agranoff et al 1976). Thus the initial rise in GAP-43 synthesis coincides with, or slightly precedes, initiation of axon outgrowth in fish and amphibian optic nerves, but the protein's synthesis does not peak or plateau until regeneration is well underway. In rat dorsal root ganglia, induction of GAP-43 mRNA begins between 1 and 2 days after sciatic nerve injury (G. S. Basi and J. H. P. Skene, unpublished), again corresponding to the end of the lag period before axon outgrowth (McQuarrie et al 1977, Forman & Berenberg 1978). It is intriguing that the effects of a "conditioning" lesion on subsequent axon regeneration of rat DRG axons are first detectable if the testing lesion is administered 2 days after the conditioning lesion (Forman et al 1980), coincident with the increase in GAP-43 mRNA. Although GAP-43 expression is not fully elevated until axon elongation is underway, induction of GAP-43 does not appear to be a secondary consequence of axon outgrowth. In rat sciatic nerves, application of colchicine at the time of nerve injury or at the end of the lag period (2 days post-crush) should prevent axon outgrowth (e.g. Bamburg et al 1986), but it has no effect on the time course or amplitude of GAP-43 induction (G. S. Basi and J. H. P. Skene, unpublished).

Elevated GAP-43 expression continues throughout the period of axon elongation and synaptogenesis in all developing and regenerating systems examined. In the toad visual system regenerating optic axons begin to reenter their target tissue, the optic tectum, by 3 weeks after nerve crush, but GAP-43 synthesis does not decline until 5–10 weeks after axotomy (Skene & Willard 1981a), during the period when an orderly pattern of retinotectal synapses is reestablished (Freeman 1977). In regenerating goldfish retinal ganglion cells, synthesis of GAP-43 declines slowly beginning 2–3 weeks after a nerve crush, and remains above control levels through 8 weeks, including the period of formation and activity-dependent refinement of the retinotectal projection (Benowitz & Schmidt 1987). The decline of GAP-43 also coincides with synaptogenesis in developing hamster pyramidal tract neurons (Kalil & Skene 1986), and the steady-state amount GAP-43 declines slowly during the "critical period" for plasticity of cat visual cortex (McIntosh et al 1987). The relatively slow decline in GAP-43 synthesis late in axon development or regeneration would therefore permit the protein to play some role in synaptogenesis or in the active sorting out of the terminal arbor.

GAP-43 induction in regenerating systems seems to begin just early enough not to rule out a role in the initial phases of axon outgrowth, and to persist just long enough not to rule out participation in later phases of synaptogenesis and maturation of the axon's terminal arbor. The localization of the protein to growth cones and the distal portions of outgrowing neurites does argue against direct GAP-43 participation in maturation of the axon structure behind the immature axon sprouts, or the slow growth in axon diameter, myelination, and maturation of the axon's electrical properties.

The strong correlation of axon growth and elevated GAP-43 synthesis, and the abundance of GAP-43 in growth cone membranes, suggest that GAP-43 might be essential for some steps in axon growth. However, there have been some reports of successful regeneration with no discernible increase in GAP-43 synthesis or axonal transport (Szaro et al 1985, Hall 1982, Hall et al 1978, Perry et al 1987b). Most of these studies were made before the aberrant SDS gel behavior of GAP-43 was appreciated, and they did not include known GAP-43–containing samples to identify the position of the protein in the gel systems employed. Inspection of the published data of Szaro et al (1985) suggested that their spot 23, which showed a significant induction during regeneration, might correspond to GAP-43 (Jacobson et al 1986). The comparatively modest (two-fold) induction of the GAP-43–like protein during regeneration of *Xenopus* retinotectal axons—compared to more than ten-fold increases in other systems—might reflect the fact that axons were lesioned directly at the tectum and proteins were analyzed after the reestablishment of retinotectal synapses was underway (Szaro et al 1985). Other studies reported no evidence of GAP-43 induction in rat and frog dorsal root ganglia (Hall 1982, Perry et al 1987b) and rat sympathetic ganglia (Hall et al 1978) during axon regeneration. Those investigators employed isoelectric focusing gels with a narrow pH range (4.5–7) in order to gain high resolution, but the very acidic GAP-43 (pI 4.3–4.5) might be difficult to detect reliably on such a system. Indeed, those investigators reported variable detection of a protein somewhat similar to GAP-43 in frog DRG (Perry et al 1987b), and inspection of the data of Hall (1982) suggests that the protein designated 2F1, which showed a highly variable degree of labeling but appeared to increase during regeneration, might be similar to GAP-43. Direct measurement of GAP-43 in one of these systems—rat dorsal root ganglia—shows definitively that GAP-43 mRNA is elevated during axon regeneration (Basi et al 1987; G. S. Basi and J. H. P Skene, unpublished) and development (Karns et al 1987).

## Synaptic Plasticity in Adult Brain

While the correlation of GAP-43 synthesis with axon development and regeneration was being investigated, an independent set of investigations identified the same protein as a potential mediator of synaptic functions in adult brain. Gispen and colleagues described a synaptic membrane protein (B-50) whose phosphorylation was inhibited by an ACTH-derived peptide reported to inhibit performance on a learning task (Zwiers et al 1976). Routtenberg and his colleagues found that phosphorylation of protein F1 was correlated with synaptic long-term potentiation in rat hippocampus (Nelson & Routtenberg 1985, Lovinger et al 1986). In both cases, the phosphorylation was shown to be carried out by the calcium/phospholipid-dependent protein kinase C (Aloyo et al 1983, Nelson & Routtenberg 1985, Akers & Routtenberg 1985).

Long-term potentiation (LTP) occurs at some synapses following a train of high-frequency stimulation. LTP is assayed as an increased post-synaptic response to a given afferent stimulation, and it can be observed for minutes or hours in in vitro preparations and up to several weeks in vivo (reviewed in Teyler & DiScenna 1987, McNaughton & Morris 1987, Smith 1987). Stimulation of LTP can be associative; that is, subthreshold stimulation of one afferent pathway can sum with subthreshold stimulation of a second pathway projecting to the same post-synaptic population to generate LTP. The long duration and associative properties of LTP have made it a popular model for synaptic mechanisms that might be involved in memory (McNaughton & Morris 1987).

Possible roles of protein kinase C in LTP have received increasing attention in recent years. Akers et al (1986) showed that LTP in hippocampus is accompanied by translocation of kinase C from cytosol to membranes. Direct activation of protein kinase C with phorbol esters has been reported to potentiate synaptic transmission in hippocampal slices (Malenka et al 1986), and kinase C activation by phorbol esters or oleic acid has been reported to enhance LTP in intact animals (Routtenberg et al 1986, Linden et al 1986, 1987). Conversely, Lovinger et al (1987) reported that several inhibitors of protein kinase C could inhibit LTP in the intact hippocampus.

In none of these studies could it be determined whether the LTP-related alterations in protein kinase C occurred in the presynaptic terminals or the postsynaptic cells. Although the initial trigger for LTP appears to occur postsynaptically, debate has continued over the proposal that maintenance of LTP occurs through increased presynaptic transmitter release (Teyler & DiScenna 1987, Smith 1987). Preliminary immunolocalization of GAP-43/B-50 by electron microscopy indicates that this protein is exclusively presynaptic in adult brains (Gispen et al 1985, Norden et al 1987). Direct demonstration of kinase-C–mediated phosphorylation of an exclusively presynaptic protein during LTP would show that at least some of the LTP-activated protein kinase C is in presynaptic terminals (Routtenberg 1986). However, assays of GAP-43/F1/B-50 phosphorylation in LTP experiments have been carried out in vitro on homogenized preparations (Akers & Routtenberg 1985, Nelson & Routtenberg 1985, Lovinger et al 1986, Routtenberg 1986); as a result, it has not been shown definitively that phosphorylation of GAP-43 in vitro reflects access of the activated fraction of kinase C to presynaptic GAP-43 in vivo.

GAP-43/F1/B-50 phosphorylation during LTP can be interpreted in several ways. First, kinase C-regulated aspects of LTP might be mediated exclusively through other substrates of the kinase, so that GAP-43 phosphorylation is only coincidental with, and not functionally involved in,

LTP; at least one other kinase C substrate also shows increased phosphorylation during LTP (Nelson et al 1987a). Second, GAP-43 might participate in some steps in activity-stimulated neurotransmitter release. This possibility is not incompatible with a related role in axon growth, since neurotransmitter release and axon elongation have many subcellular mechanisms in common (e.g. vesicle movement, membrane fusion). Finally, if the biochemical actions of GAP-43 pertain exclusively to structural growth of axons, GAP-43/F1 phosphorylated in LTP might participate in structural alterations, such as the growth of additional synaptic terminals during LTP (e.g. Greenough 1984).

Proposals that GAP-43 participates in either axon growth or synaptic plasticity are based so far only on correlative evidence. In the case of developing and regenerating systems, GAP-43 was distinguished from other potential regulators of growth by the consistency with which its synthesis is correlated with growth in many different neuronal systems. GAP-43/F1 phosphorylation during LTP has been investigated primarily in one system, the perforant pathway of rats (Lovinger et al 1986, Routtenberg 1986). Is expression or phosphorylation of GAP-43 generally correlated with synaptic plasticity throughout adult brain?

GAP-43 expression and phosphorylation show a distinctly nonuniform distribution in adult rat and primate brains (Benowitz et al 1988, Neve et al 1987, Kristjansson et al 1982, Oestreicher & Gispen 1986, Rosenthal et al 1987). Both the protein and its mRNA are located predominantly in the prosencephalon, although there are discrete GAP-43-positive nuclei in the medulla and mesencephalon (Benowitz et al 1988). Within neocortex, GAP-43 and is mRNA are both concentrated most heavily in higher order sensory and "associational" cortex, with much lower levels of expression in primary sensory areas and motor cortex (Neve et al 1987, Benowitz et al 1988). Parts of the hippocampus show heavy immunostaining for GAP-43 (Oestreicher & Gispen 1986, Benowitz et al 1988) but low concentrations of GAP-43 mRNA (Neve et al 1987, Rosenthal et al 1987), thus suggesting that the protein is located in synaptic terminals of extrinsic neurons projecting to hippocampus. High concentrations of GAP-43 in adult rat brains also are found in various nuclei of the basal ganglia, thalamus, and hypothalamus (Benowitz et al 1988) and the septal area (Oestreicher et al 1986).

Apparently, GAP-43 is expressed in substantial amounts only in a discrete set of neurons during adult life. Several authors have pointed out that polymodal sensory cortex, the hippocampus, frontal cortex, and many parts of the "limbic system" all show levels of GAP-43 and/or GAP-43 mRNA (Neve et al 1987, Benowitz et al 1988, Nelson et al 1987b), and that these structures are generally thought to be involved in memory and other higher order processing that might well require a good deal of synaptic plasticity. In contrast, primary sensory areas and motor cortex, along with most of the medulla and spinal cord, contain only low concentrations of GAP-43, and these regions often are considered in terms of rather static circuitry. The gradient from low GAP-43 in primary sensory areas to higher GAP-43 expression in polymodal or association cortex has been reported both in rodents and primates (Neve et al 1987, Nelson et al 1987b, Benowitz et al 1988). This sort of generalization leaves one in a quandary over areas such as somatosensory cortex, which displays remarkable functional plasticity in adult life (Kaas et al 1983, Jenkins & Merzenich 1987) but has a relatively low concentration of GAP-43 (Benowitz et al 1988), and many parts of the thalamus and basal ganglia, which contain very high concentrations of GAP-43 but are not usually considered areas of special synaptic plasticity. The immediate challenge is to identify any anatomical, physiological, or functional characteristics shared by the GAP-43–expressing population of adult neurons. Defining these "high GAP-43" populations of neurons should be facilitated by ongoing studies using in situ hybridization to locate cell bodies containing GAP-43 mRNA to complement the immunohistochemical map of GAP-43-containing synaptic terminals (Benowitz et al 1988). The possibility that the subset of neurons expressing high levels of GAP-43 through adult life has a special propensity for some forms of synaptic plasticity (Routtenberg 1986, Jacobson et al 1986, Benowitz et al 1988, Neve et al 1987) remains appealing, but not established.

## Structure and Biochemical Characteristics of GAP-43

Correlative studies have focused attention on GAP-43 as a potential contributor to some forms of axon growth or synaptic plasticity, but the actual effects of GAP-43 on growth cones and synaptic terminals remain unknown. One way to establish the functions of GAP-43 is to build up from its molecular properties a picture of the protein's actions within a neuron. Structural and biochemical studies of GAP-43 reveal a novel protein that appears to interact extensively with several intracellular messenger systems.

The amino acid sequence of GAP-43 is highly conserved among mammals, although cow and human GAP-43 contain a ten–amino acid insert not found in the rodent protein (Figure 1). In Figure 1, the alignment of the bovine and mouse sequences has been altered from the published alignment (Wakim et al 1987) to resemble the alignments of rat and human sequences (Ng et al 1988, Kosik et al 1988). The structural novelty of GAP-43 is underscored by the failure of extensive searches of nucleic acid and amino acid sequence databases to reveal any significant homologies with previously analyzed proteins (Basi et al 1987, Cimler et al 1987, Karns et al 1987, Rosenthal et al 1987). The predicted molecular weight of the unmodified rat protein is 23.6 kD, consistent with earlier indications that the mature protein is much smaller than it appears on SDS gels (Masure et al 1986, Benowitz et al 1988). Secondary structure predictions agree with circular dichroism studies (Masure et al 1986) in showing a large proportion of the protein in random coil, interrupted by 17 proline residues. GAP-43, then, appears to be an elongated (Masure et al 1986) kinked coil, with limited domains of more ordered secondary structure.

MEMBRANE ASSOCIATION    One of the most striking structural features of GAP-43 is its extreme hydrophilicity (Basi et al 1987, Rosenthal et al 1987), somewhat surprising for a protein associated predominantly with membranes. GAP-43 is synthesized as a soluble protein, whose post-translational association with membranes probably is mediated by coval-ent attachment of fatty acid (J. H. P. Skene and I. Virag, unpublished). A short hydrophobic region at the amino terminus of the protein (Figure 1) is the most likely site of fatty acylation and membrane attachment (Basi et al 1987; J. H. P. Skene and I. Virag, unpublished). Although the major fraction of GAP-43 is membrane-bound, there may exist a soluble form of the protein resulting from reversible removal of the bound fatty acid or irreversible cleavage of the putative membrane-binding domain (Basi et al 1987, Rosenthal et al 1987). The extreme hydrophilicity of GAP-43 and the nature of its membrane attachment are consistent with models that envision the protein extending away from the cytoplasmic surfaces of growth cone and synaptic membranes (Skene & Willard 1981c, Gispen et al 1985, Meiri et al 1988), in a position to interact with cytoplasmic or cytoskeletal proteins on one hand, and reversibly attached to the mem-brane on the other.

CALMODULIN BINDING    GAP-43/p57 was identified by Storm and his col-leagues as a unique calmodulin-binding protein, binding calmodulin selec-tively in the *absence* of calcium, and releasing calmodulin at higher calcium concentrations (Andreason et al 1983). This "reversed" calcium depen-dence for calmodulin binding is sufficiently rare that it can be used to purify GAP-43/p57 to homogeneity (Andreason et al 1983, Masure et al 1986). On the basis of its abundance, membrane binding, and its "reversed" pattern of calmodulin binding, GAP-43/p57 has been proposed to act under low calcium conditions to sequester calmodulin in certain regions of neuronal membrane, releasing calmodulin upon influx or mobilization of free calcium (Andreason et al 1983, Cimler et al 1987). However, because the calcium-dependent antagonism of GAP-43–calmodulin binding is strongly affected by ionic strength (Alexander et al 1987), it is not clear whether calcium regulates calmodulin binding to GAP-43 under any or all physiological conditions. An alternative regulator of calmodulin binding to GAP-43 is protein kinase C. Phosphorylation of GAP-43/p57 by protein kinase C strongly inhibits binding of the protein to calmodulin (Alexander et al 1987).

PHOSPHOINOSITIDE METABOLISM    The antagonistic interactions of kinase C and calmodulin with GAP-43 are intriguing, because calmodulin and kinase C participate in separate branches of a bifurcating second-messenger system in many cells (e.g. Berridge 1987). The now-classical version of this second-messenger system begins with the receptor-activated hydrolysis of phosphatidylinositol 4,5-biphosphate ($PIP_2$) to produce diacylglycerol (DAG) and inositol 1,4,5-triphosphate ($IP_3$). DAG, in turn, specifically activates protein kinase C, while $IP_3$ stimulates the release of free $Ca^{2+}$ from intracellular stores (Figure 2).

The precursor for these two second messengers, $PIP_2$ is produced by phosphorylation of phosphatidyl inositol 4-phosphate (PIP). Van Dongen et al (1985) reported that phosphorylated GAP-43/B-50 inhibits the activity of a purified PIP-to-$PIP_2$ kinase; conversely, antibodies against GAP-43/B-50 stimulated production of $PIP_2$ in isolated synaptosomal membranes, while blocking phosphorylation of GAP-43/B-50 (Oestreicher et al 1983). It was therefore suggested that GAP-43/B-50 acts as a feedback inhibitor of kinase C activation and calcium mobilization (Jolles et al 1980, Oestreicher et al 1983, van Dongen et al 1985, van Hoof et al 1988). There is, however, evidence that under some conditions PIP can be hydrolyzed directly to yield DAG plus an inositol biphosphate that does not stimulate calcium mobilization (Berridge 1987). Under those conditions, inhibition of $PIP_2$ production would uncouple kinase C activation from calcium mobilization. In either case, the effect of phosphorylated GAP-43 would be to terminate $IP_3$-dependent calcium mobilization by preventing replacement of hydrolyzed $PIP_2$. It remains to be shown whether the actions of phosphorylated GAP-43 on phosphoinositide phosphorylation in vitro have a substantial effect on the production of $PIP_2$ in vivo.

GAP-43 AND CALCIUM SIGNALING    The possible sequestering of calmodulin by GAP-43 and the potential capability of phosphorylated GAP-43 to terminate calcium mobilization suggest that GAP-43 may play a critical role in modulating intracellular signaling by calcium. The reported interactions of GAP-43 with components of the calcium-calmodulin intracellular messenger system are illustrated in Figure 2. An important unanswered question is whether GAP-43 inhibits the activity of bound calmodulin. If calmodulin bound to GAP-43 is effectively prevented from interacting with other calmodulin-binding effector proteins, the limited

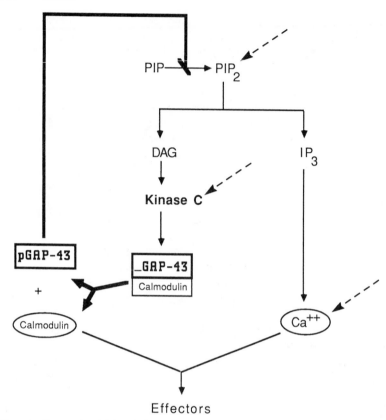

Effectors

*Figure 2*    Proposed activities of GAP-43 in signal transduction pathways. The hydrolysis of phosphatidylinositol 4,5-biphosphate (PIP₂) yields diacylglycerol (DAG), an activator of protein kinase C, and inositol 1,4,5-triphosphate (IP₃), which mobilizes intracellular calcium (reviewed in Berridge 1987). *Bold lines* indicate the binding of calmodulin by GAP-43, and release of calmodulin after phosphorylation of GAP-43 by kinase C (Alexander et al 1987), and the inhibition of phosphatidylinositol 4-phosphate (PIP) phosphorylation by phosphorylated GAP-43 (van Dongen et al 1985). *Dashed lines* indicate potential influences by extracellular signals: receptor-stimulated hydrolysis of PIP₂, influx of extracellular calcium, and alternative activation of protein kinase C discussed in the text.

data available can be assembled into a simplistic model in which GAP-43 acts to sharpen the spatial (Andreason et al 1983, Alexander et al 1987) and temporal resolution of intracellular calcium signals.

In an initial or "resting" state, with low intracellular calcium and a small steady-state pool of PIP₂, a large concentration of GAP-43 would constitute a calmodulin buffer, so that calcium-calmodulin mediated processes in a growth cone or synaptic terminal would have a reduced response to small influxes or mobilization of calcium. Stimulation of $PIP_2$ hydrolysis would cause a rise in free calcium (via $IP_3$) and a parallel activation of protein kinase C. GAP-43 phosphorylation by kinase C would tend to liberate calmodulin from GAP-43, an effect that might be augmented by elevated free calcium. In the presence of GAP-43, then, the cellular response to $PIP_2$ hydrolysis would rise exponentially because GAP-43 makes both the mobilization of calcium and the concentration of free calmodulin responsive to $PIP_2$ breakdown. In the meantime, phosphorylation of GAP-43 would inhibit regeneration of $PIP_2$ from PIP. As a result, mobilization of intracellular calcium would be terminated after hydrolysis of the preexisting pool of $PIP_2$. The effect of GAP-43, therefore, would be to sharpen the temporal resolution of calcium-calmodulin signals resulting from activation of phospholipase C-coupled receptors.

Similar considerations suggest that GAP-43 could act to sharpen the spatial resolution of calcium signals in synaptic terminals and growth cones. Local activation of phospholipase-coupled receptors in one region of a growth cone membrane, for example, would lead to local phosphorylation of GAP-43, particularly if GAP-43 exists in vivo in a complex with protein kinase C (Zwiers et al 1980, Aloyo et al 1983). Unphosphorylated GAP-43 in other regions of the membrane would tend to buffer those regions of the growth cone against calmodulin-dependent responses to mobilized calcium. At the same time, the locally phosphorylated GAP-43 would terminate the $IP_3$-dependent calcium signal, possibly reducing its spread to other parts of the growth cone.

In this model, some effects of GAP-43 depend on the availability of nonclassical activators of protein kinase C. In the classical bifurcating pathway, C kinase is activated only by DAG derived from $PIP_2$. In that case, phosphorylated GAP-43 would be a feedback inhibitor of kinase C. The effect of GAP-43 would be to introduce a (relative) refractory period after phospholipase activation, when neither calcium mobilization nor kinase C activation would occur in response to receptor occupancy, because no $PIP_2$ would be available for hydrolysis. As kinase C was inactivated, recovery of $PIP_2$ could proceed at a rate dependent on the rate of GAP-43 dephosphorylation.

On the other hand, there is evidence that kinase C can be activated independently of $PIP_2$, via hydrolysis of PIP (Berridge 1987), cis-unsaturated fatty acids (Murakami & Routtenberg 1985), or proteolytic cleavage of the kinase itself (Kishimoto 1983, Melloni et al 1986). Activation of kinase C by these mechanisms could keep GAP-43 phosphorylated chronically, inhibiting $IP_3$-dependent calcium mobilization, but providing a large pool of free calmodulin. Under those circumstances, an axon terminal would be highly responsive to an influx of extracellular calcium. Cal-modulin-mediated responses—possibly including neurotransmitter release (Llinas et al 1985, Augustine et al 1987)—would be potentiated so long as kinase C remained activated.

Any account of GAP-43's biological role(s) is almost certain to change substantially with increasing biochemical characterization of this protein. GAP-43 interactions with membranes and possibly with cytoskeletal elements are likely to be important in localizing the protein and may be involved in biological roles beyond any participation in calcium-cal-modulin signaling. Nevertheless, the limited biochemical data already available indicate that one consequence of high concentrations of GAP-43 in a growth cone or synaptic terminal is likely to be a substantial alteration in the way an axon terminal responds to environmental stimuli acting through calcium and calmodulin.

## OTHER GROWTH-ASSOCIATED PROTEINS

GAP-43 is useful in establishing that selective regulation of neuronal gene expression does occur in correlation with periods of axon growth, but as illustrated by Table 1, GAP-43 is likely to be only one example of an emerging class of growth-associated axonal proteins. Identifying these other GAPs is important for understanding the aspects of axon growth subject to this kind of developmental regulation.

A common, though imperfect, correlate of axon outgrowth is elevated synthesis and axonal transport of tubulin. Recently it has been found that only one of several tubulin genes exhibits elevated expression during development and regeneration of axons (Lewis et al 1985, Miller et al 1987a–c). For neurons in which total tubulin synthesis shows little or no specific increase during periods of axon growth, it would be interesting to know whether the high rate of expression of multiple tubulin genes masks growth-associated regulation of one or more tubulin species. In regen-erating nerves, tubulin synthesized in response to an injury does not seem to be required for the initiation of axon outgrowth. The axotomy-induced elevation of tubulin synthesis, when it occurs, begins several days after injury, concurrent with or slightly after the onset of axon elongation (Heacock & Agranoff 1982, Agranoff & Ford-Holevinski 1984, Skene & Willard 1981a). For axons interrupted several millimeters or even several centimeters from their cell bodies, the slowly transported tubulin requires several days more to reach the site of injury. This means that the initial stages of axon regeneration must proceed without benefit of additional tubulin produced in response to axotomy. Nevertheless, an increased supply of a major cytoskeletal protein might contribute to later stages of axon growth. The rate of delivery of slowly transported axon components, including tubulin, has been proposed to set an upper limit on the rate of axon elongation (Hoffman & Lasek 1980, Komiya 1981, Wujek & Lasek 1983, Cancalon 1983). Where the newly synthesized tubulin reaches the distal segments of a growing axon, it may permit an increased velocity of axon extension. For developing neurons extending axons *de novo*, in which axon elongation is initiated at the cell body itself, elevated tubulin synthesis could easily participate in the initial phases of axon outgrowth. Because of its slow delivery along axons, *synthesis* of tubulin does not appear to be a good candidate for the regulation of synaptic remodeling or terminal sprouting far from the cell bodies of maturing or adult neurons. This does not mean, of course, that local alterations in tubulin assembly/disassembly cannot be intimately involved in controlling these modes of axon growth.

Another correlate of axon growth in many systems is elevated synthesis of a 23–28 kD rapidly transported membrane protein (Table 1). Variations in reported molecular weights and isoelectric points make it difficult to determine whether the 23–28 kD growth-associated proteins induced in different neuronal systems are really homologous proteins. Some indirect evidence suggests that the developmentally regulated 23 kD rabbit protein is related to GAP-24 from regenerating toad optic nerves (Skene & Willard 1981b). The difficulty of assessment of homologies among the 23–28 kD proteins in different neurons is aggravated by the number of rapidly transported proteins in this size range, several of which undergo small changes in synthesis during axon growth (e.g. Skene & Willard 1981a,b, Benowitz & Lewis 1983). It is also difficult to evaluate the apparent absence of growth-associated 23–28 kD proteins in several developing and regenerating neurons (Table 1). The rapid degradation of toad GAP-24 (Skene & Willard 1981c) may make the protein difficult to detect in some samples, although efforts to minimize in vivo and artifactual degradation failed to reveal a GAP-24–like protein in developing pyramidal tracts (Kalil & Skene 1986). A more definitive correlation between GAP-24 expression and some forms of axon growth will require specific probes for the protein or its mRNA.

Analysis of metabolically labeled axonally transported proteins has revealed a small number of proteins whose synthesis is widely correlated with periods of axon growth. However, this approach is limited in its ability to detect relatively minor proteins or proteins whose size or charge puts them outside the resolution of the electrophoretic systems used in the analyses. An alternative approach is to investigate the developmental and injury-induced expression of identified genes or proteins known or suspected to participate in cell or axon growth.

The proto-oncogene c-*src* has been found to resemble GAP-43 in its localization and growth-associated expression. Immunodetectable c-*src* and *src*-related tyrosine kinase activity are expressed at elevated levels during neural development (Sorge et al 1984, Maness et al 1986, Lev et al 1984, Simon et al 1985), and are reinduced in regenerating rat sciatic nerves (Le Beau & Walter 1987). Like GAP-43, neural c-*src* is highly enriched in a subcellular fraction containing isolated growth cones (Maness et al 1988). Although the role of c-*src* in growth cones is unknown, expression of the oncogenic viral form of the protein (v-*src*) in the pheochromocytoma cell line PC12 induces neurite outgrowth (Alema et al 1985), thus strongly suggesting a regulatory role for c-*src* in some aspects of axon extension. It remains to be established whether the developmental expression of immunodetectable c-*src* reflects developmentally regulated expression of the c-*src* gene, and whether expression of this gene is widely correlated with axon regeneration a number of different neurons.

Another identified protein whose expression may be correlated with axonal growth is the receptor for apolipoproteins B and E (LDL receptor). Immunostaining reveals extremely high levels of the LDL receptor on the distal portions of regenerating axons in rat sciatic nerves and lesser amounts on remyelinating Schwann cells (Boyles et al 1987). A primary ligand for this receptor, apolipoprotein E (apoE) is synthesized and secreted by nonneuronal cells of the distal nerve stump after injury to mammalian peripheral nerves (Skene & Shooter 1983, Ignatius et al 1986, Snipes et al 1986) and in the developing CNS (Muller et al 1985, Freeman et al 1986). Uptake of lipids via the LDL receptor in neuronal growth cones (Ignatius et al 1987a,b) could contribute to axonal outgrowth. As with c-*src*, it is not yet known whether the growth-associated expression of immunodetectable LDL receptor is mediated by altered synthesis of the protein in neuron cell bodies.

One of the clearest illustrations that maturing neurons alter their responsiveness to environmental cues for axon growth is the developmentally decreasing ability of chick retinal neurons to extend neurites on laminin (Cohen et al 1986, Hall et al 1987). The altered responsiveness of these neurons to laminin occurs between embryonic days 6 and 12, and is accompanied by decreased expression of some protein recognized by the CSAT and JG22 antibodies (Hall et al 1987). These antibodies recognize several members of a family of receptors for extracellular matrix proteins (e.g. Hynes 1987), and the antibodies interfere with neurite outgrowth on a laminin substratum (Tomaselli et al 1986, Bozycko & Horwitz 1986). It will be important to determine whether the developmentally regulated CSAT/JG22 proteins are regulated at the level of protein synthesis, and whether growth-associated expression of these proteins is a common feature of many different developing and regenerating neurons.

Neurite outgrowth on cell surfaces involves axonal receptors separate from the extracellular matrix receptors recognized by the CSAT and JG22 antibodies (Tomaselli et al 1986); these additional receptors apparently include proteins in the N-CAM family of cell-adhesion molecules (Bixby et al 1987). It is therefore particularly interesting that a developmentally regulated growth cone membrane protein, the 5B4 antigen (Wallis et al 1985, Ellis et al 1985), is homologous to neural cell adhesion molecules (NCAMs) (Ellis et al 1987). Immunologically detectable N-CAMs also increase dramatically in rat sciatic nerves during regeneration (Daniloff et al 1986). Again, it remains to be established that the developmental and injury-induced appearance of these axonal proteins reflects altered synthesis rather than local modification, although a transcriptional level for N-CAM regulation has been suggested (Daniloff et al 1986).

These studies based on immunological detection of identified axonal proteins during development and regeneration suggest that the responsiveness of many axons to environmental cues may be altered as the cells mature by changes in cell surface receptors and proteins involved in signal transduction.

# REGULATION OF GROWTH-ASSOCIATED GENE EXPRESSION

Identification of even a few genes whose expression is tightly linked to axon growth supports the suggestion that the decreased propensity for axon growth and remodeling exhibited by many neurons as they mature can be mediated in part by selective repression of neuronal genes involved in growth. Understanding the molecular mechanisms controlling these growth-associated genes should shed light on the kinds of regulatory signals that control a neuron's propensity for axon growth, and may also provide a basis for identifying other neuronal genes able to respond to those same regulatory signals.

Individual growth-associated proteins are not always expressed coordinately; this indicates that multiple signals mediate the retrograde regulation of these genes. In toad optic nerve, for example, synthesis of the protein designated GAP-24 rises sharply 2–4 days after nerve injury, whereas synthesis of GAP-43 and tubulin increases more slowly (Skene & Willard 1981a). In some PNS neurons, the homologue of GAP-24 seems to be expressed throughout adult life, whereas GAP-43 is induced only after nerve injury (Skene & Willard 1981b). Noncoordinate regulation of growth-associated proteins is particularly striking in goldfish retinal ganglion cells whose axons are deprived of their normal synaptic targets by removal of the optic tectum. Tectal removal has no effect (Benowitz et al 1983) or only a small effect (Grafstein et al 1987) on the return of GAP-43 synthesis to pre-injury levels beginning 3–4 weeks after nerve crush. In contrast, synthesis of a 26 K protein, which is induced during regeneration and ordinarily declines 3–4 weeks post-crush, remains elevated if synaptic targets have been removed (Benowitz et al 1983). Tubulin synthesis not only fails to return to prelesion levels, but continues to rise in regenerating neurons deprived of their normal tectal targets (Grafstein et al 1987). Noncoordinate regulation of individual growth-associated proteins indicates that there are likely to be multiple signaling pathways by which events in and around an axon can influence gene expression in neuronal cell bodies. This, in turn, suggests that not all aspects of axon growth must be expressed or repressed coordinately in neurons.

Several kinds of retrogradely transported signals might regulate neuronal gene expression in response to changing conditions in the axon or its synaptic terminals. Signals might originate in synaptic target tissues or from glial cells along the length of axons. Signal molecules might be among those that are transported anterogradely into axons and then return to their own cell bodies of origin by retrograde transport (Bisby 1980, 1984); such molecules could serve as useful signals if they underwent some modification in the distal axon that reflected conditions in the axon. Whatever their source, signals regulating growth-associated gene expression might be negative—chronically repressing GAP expression in uninjured mature neurons, or positive—produced in developing or injured nerves to induce GAP expression. In rat sciatic nerves, removal of the distal nerve stump or blockade of axonal transport with colchicine does not alter the time course and magnitude of GAP-43 mRNA induction in dorsal root ganglia (G. S. Basi and J. H. P. Skene, unpublished), thus indicating that inducers produced distal to the site of nerve injury are not critical for GAP-43 induction. Co-culture of DRG neurons with potential synaptic target cells does reduce GAP-43 synthesis in vitro (Baizer & Fishman 1987), consistent with a target-derived GAP-43 repressor.

Synaptic targets do not seem to play a dominant role in GAP-43 regulation in the CNS. Removal of synaptic target neurons has no effect (Benowitz et al 1983) or has a small effect (Grafstein et al 1987) on the recovery of GAP-43 expression to pre-injury levels after optic nerve regeneraion in goldfish. In the mammalian CNS, injury of axons many millimeters away from their cell bodies does not re-induce GAP-43 expression (Skene & Willard 1981b, Kalil & Skene 1986, Reh et al 1987), despite disconnection of the axons from their synaptic targets. One possibility is that chronic repression of the GAP-43 gene by target-derived signals is replaced or supplemented in some mammalian CNS neurons by inhibitory signals derived from glial cells along the length of the axons. In that case, axotomy far from the cell body, leaving a long segment of the surviving axon in contact with CNS glia, would leave GAP-43 synthesis repressed. Consistent with this explanation, injury of adult rat optic nerves very close to the retina has been reported to induce GAP-43 expression in retinal ganglion cells (Lozano et al 1987). The retinal cells injured close to their cell bodies also showed a far greater propensity to extend axons when provided with favorable environmental conditions in the form of a peripheral nerve graft (Vidal-Sanz et al 1987).

## SUMMARY AND FUTURE PROSPECTS

Axon growth and synaptogenesis involve extensive interactions between an axon and multiple environmental cues, ranging from the availability of glial cell surface molecules extracellular matrix components (Bozycko & Horwitz 1986, Tomaselli et al 1986, Vidal-Sanz et al 1987, Schwab & Caroni 1988) to patterns of neuronal activity (Meyer 1983, Schmidt & Edwards 1983, Fawcett & O'Leary 1985, Stryker & Harris 1986). A neuron's ability to respond to any of these environmental cues, and the nature of its responses to any particular cue, depend on the neuron's expression of relevant receptors, signal transducing proteins, and the structural materials to execute a response—its "preparatory set." The evidence reviewed here indicates that expression of some of the neuronal genes involved in axon growth and/or synaptogenesis decreases sharply as neurons mature, so that extracellular cues that stimulate axon elongation in developing neurons evoke different responses by mature neurons. To borrow a term from neurophysiologists (Evarts et al 1984), the sum of these developmentally regulated genes may constitute a "preparatory set" for axon elongation or synaptic modification. Differential expression of individual growth-associated proteins might create a range of preparatory sets among different neurons. This view of axon growth regulation emphasizes the need to consider not only the "triggering stimuli" that elicit overt axon growth or synaptic remodeling, but also the regulation of internal neuronal events that represent a predisposition for certain forms of growth.

The genes for GAP-43 and a few other axonal proteins are strong candidates for individual elements of a preparatory set for axon growth and/or synaptogenesis. It is not at all clear, however, whether GAP-43 itself is critical for axon elongation or for some other steps in the growth and remodeling of synaptic connections. One important challenge is to identify the full set of genes whose expression confers on developing neurons their predisposition for axon growth and synaptic remodeling, and to determine which genes are involved in individual phases or steps in axonal and synaptic development. A second challenge is to understand the mechanisms by which these genes become repressed during neuronal maturation and conditions under which they can be re-induced in adults.

ACKNOWLEDGMENTS

I am grateful to Eric Shooter, Michael Ignatius, David Schreyer, Michael LaBate, Doug Hess, and Ildiko Virag for their comments on the manuscript, and to Mark Willard, Daniel Storm, Jeannette Norden, Rachel Neve, Patricia Maness, Mark Bisby, Larry Benowitz, Bernice Grafstein, and Aryeh Routtenberg for discussing unpublished work. Collaborative work to establish the identities of GAP-43, B-50, F1, and pp46 benefited from a meeting sponsored by the Neurosciences Research Program and organized by Bernice Grafstein. Research from my laboratory has been supported by National Institutes of Health grants NS20178 and EY07397, and by grants from the Isabelle Niemela Fund and the Weingart Foundation.

*Literature Cited*

Agranoff, B. W., Field, P., Gaze, R. M. 1976. Neurite outgrowth from explanted Xenopus retina: An effect of prior optic nerve section. *Brain Res.* 113: 225–34

Agranoff, B. W., Ford-Holevinski, T. S. 1984. Biochemical aspects of the regenerating goldfish visual system. See Elam & Cancalon 1984, pp. 69–86

Akers, R. F., Routtenberg, A. 1985. Protein kinase C phosphorylates a 47 Mr protein (F1) directly related to synaptic plasticity. *Brain Res.* 334: 147–51

Akers, R. F., Lovinger, D. M., Colley, P. A., Linden, D. J., Routtenberg, A. 1986. Translocation of protein kinase C activity may mediate hippocampal long-term potentiation. *Science* 231: 587–89

Akers, R. F., Routtenberg, A. 1987. Calcium-promoted translocation of protein kinase C to synaptic membranes: Relation to the phosphorylation of an endogenous substrate (protein F1) involved in synaptic plasticity. *J. Neurosci.* 7: 3976–83

Alema, S., Casalbore, P., Agostini, E., Tato, F. 1985. Differentiation of PC12 phaeochromocytoma cells induced by v-src oncogene. *Nature* 316: 557–59

Alexander, K. A., Cimler, B. M., Meier, K. E., Storm, D. R. 1987. Regulation of calmodulin binding to P-57. *J. Biol. Chem.* 262: 6108–13

Alexander, K. A., Wakim, B. T., Doyle, G. S., Walsh, K. A., Storm, D. R. 1988.

Identification and characterization of the calmodulin binding domain of neuromodulin, a neurospecific calmodulin-binding protein. *J. Biol. Chem.* 263: 7544–49

Aloyo, V. J., Zwiers, H., Gispen, W. H. 1983. Phosphorylation of B-50 protein by calcium-activated, phospholipid-dependent protein kinase and B-50 protein kinase. *J. Neurochem.* 41: 649–53

Andreason, T. J., Luetje, C. W., Heidman, W., Storm, D. R. 1983. Purification of a novel calmodulin binding protein from bovine cerebral cortex. *Biochemistry* 22: 4615–18

Argiro, V., Johnson, M. I. 1982. Patterns and kinetics of neurite extension from sympathetic neurons in culture are age dependent. *J. Neurosci.* 2: 503–12

Augustine, G. J., Charlton, M. P., Smith, S. J. 1987. Calcium action in synaptic transmitter release. *Ann. Rev. Neurosci.* 10: 633–93

Baizer, L., Fishman, M. C. 1987. Recognition of specific targets by cultured dorsal root ganglion neurons. *J. Neurosci.* 7: 2305–11

Bamburg, J. R., Bray, D., Chapman, K. 1986. Assembly of microtubules at the tip of growing axons. *Nature* 321: 788–90

Basi, G. S., Jacobson, R. D., Virag, I., Schilling, J., Skene, J. H. P. 1987. Primary structure and transcriptional regulation of

GAP-43, a protein associated with nerve growth. *Cell* 49: 785–91

Benowitz, L. I., Apostolides, P. J., Perrone-Bizzozero, N. I., Finklestein, S. P., Zwiers, H. 1988. Anatomical distribution of the growth-associated proteins GAP-43/B-50 in the adult rat brain. *J. Neurosci.* 8: 339–52

Benowitz, L. I., Lewis, E. R. 1983. Increased transport of 44,000- to 49,000-dalton acidic proteins during regeneration of the goldfish optic nerve: A two-dimensional gel analysis. *J. Neurosci.* 3: 2153–63

Benowitz, L. I., Perrone-Bizzozero, N. I., Finklestein, S. P. 1987. Molecular properties of the growth-associated protein GAP-43 (B-50). *J. Neurochem.* 48: 1640–47

Benowitz, L. I., Routtenberg, A. 1987. A membrane phosphoprotein associated with neural development, axonal regeneration, phospholipid metabolism, and synaptic plasticity. *Trends Neurosci.* 10: 527–31

Benowitz, L. I., Schmidt, J. T. 1987. Activity-dependent sharpening of the regenerating retinotectal projection in goldfish: Relationship to the expression of growth-associated proteins. *Brain Res.* 417: 118–26

Benowitz, L. I., Shashoua, V. E., Yoon, M. G. 1981. Specific changes in rapidly transported proteins during regeneration of the goldfish optic nerve. *J. Neurosci.* 1: 300–7

Benowitz, L. I., Yoon, M. G., Lewis, E. R. 1983. Transported proteins in the regenerating optic nerve: Regulation by interactions with the optic tectum. *Science* 222: 185–88

Berridge, M. J. 1987. Inositol trisphosphate and diacylglycerol: Two interacting second messengers. *Ann. Rev. Biochem.* 56: 159–93

Bisby, M. A. 1980. Retrograde axonal transport. *Adv. Cell. Neurobiol.* 1: 69–117

Bisby, M. A. 1984. Retrograde axonal transport and nerve regeneration. See Elam & Cancalon 1984, pp. 45–67

Bixby, J. L., Pratt, R. S., Lilien, J., Reichardt, L. R. 1987. Neurite outgrowth on muscle cell surfaces involves extracellular matrix receptors as well as $Ca^{2+}$-dependent and -independent cell adhesion molecules. *Proc. Natl. Acad. Sci. USA* 84: 2555–59

Black, M. M., Lasek, R. J. 1980. Slow components of axonal transport: Two cytoskeletal networks. *J. Cell Biol.* 86: 616–23

Boyles, J. K., Hui, D. Y., Weisgraber, K. H., Pitas, R. E., Mahley, R. W. 1987. Expression of apolipoprotein B,E (LDL) receptors and uptake of apolipoprotein E-containing lipoproteins by the regen-erating rat sciatic nerve. *Abstr. Soc. Neurosci.* 13: 294

Bozycko, D., Horwitz, A. F. 1986. The participation of a putative cell surface receptor for laminin and fibronectin in peripheral neurite extension. *J. Neurosci.* 6: 1241–51

Cancalon, P. 1983. Influence of temperature on slow flow in populations of regenerating axons with different elongation velocities. *Dev. Brain Res.* 9: 279–89

Carden, M. J., Trojanowski, J. Q., Schlaepfer, W. W., Lee, V. M.-Y. 1987. Two-stage expression of neurofilament polypeptides during rat neurogenesis with early establishment of adult phosphorylation patterns. *J. Neurosci.* 7: 3489–3504

Cimler, B. M., Giebelhaus, D. H., Wakim, B. T., Storm, D. R., Moon, R. T. 1987. Characterization of murine cDNAs encoding P-57, a neural-specific calmodulin-binding protein. *J. Biol. Chem.* 262: 12158–63

Cohen, J., Burne, J. F., Winter, J., Bartlett, P. 1986. Retinal ganglion cells lose response to laminin with maturation. *Nature* 322: 465–67

Collins, F., Lee, M. R. 1982. A reversible developmental change in the ability of ciliary ganglion neurons to extend neurites in culture. *J. Neurosci.* 2: 424–30

Daniloff, J. K., Levi, G., Grumet, M., Rieger, F., Edelman, G. M. 1986. Altered expression of neuronal cell adhesion molecules induced by nerve injury and repair. *J. Cell Biol.* 103: 929–45

de Graan, P. N. E., van Hoof, C. O. M., Tilly, B. C., Schotman, P., Oestreicher, A. B., Gispen, W. H. 1985. Phosphoprotein B-50 in nerve growth cones from metal rat brain. *Neurosci. Lett.* 61: 235–41

Elam, J. S., Cancalon, P., eds. 1984. *Axonal Transport in Neuronal Growth and Regeneration.* New York: Plenum. 284 pp.

Ellis, L., Wallis, I., Abreu, W., Pfenninger, K. H. 1985. Nerve growth cones isolated from fetal rat brain. IV. Preparation of a membrane subfraction and identification of a membrane glycoprotein expressed on sprouting neurons. *J. Cell Biol.* 101: 1977–89

Ellis, L., Ramos, P., Safaei, R., Kayalar, C. 1987. The 5B4 antigen expressed on sprouting neurons is a member of the NCAM family. *Abst. Soc. Neurosci.* 13: 1223

Evarts, E. V., Shinoda, Y., Wise, S. P. 1984. *Neurophysiological Approaches to Higher Brain Functions.* New York: Wiley. 198 pp.

Fawcett, J. W., O'Leary, D. D. M. 1985. The role of electrical activity in the formation of topographic maps in the nervous system. *Trends Neurosci.* 8: 201–6

Forman, D. S., Berenberg, R. A. 1978. Regeneration of motor axons in the rat sciatic nerve studied by labeling with axonally transported radioactive proteins. *Brain Res.* 156: 213–25

Forman, D. S., McQuarrie, I. G., Labore, F. W., Wood, D. K., Stone, L. S., Braddock, C. H., Fuchs, D. A. 1980. Time course of the conditioning lesion effect on axonal regeneration. *Brain Res.* 182: 180–85

Freeman, J. A. 1977. Possible regulatory functions of acetylcholine receptor in maintenance of retinotectal synapses. *Nature* 269: 218–22

Freeman, J. A., Bock, S., Deaton, M., McGuire, B., Norden, J. J., Snipes, G. J. 1986. Axonal and glial proteins associated with development and response to injury in the rat and goldfish optic nerve. *Exp. Brain Res.* (Suppl.) 13: 34–47

Gispen, W. H., Leunissen, J. L. M., Oestreicher, A. B., Verkleij, A. J., Zwiers, H. 1985. Presynaptic localization of B-50 phosphoprotein: The (ACTH)-sensitive protein kinase substrate involved in rat brain polyphosphoinositide metabolism. *Brain Res.* 328: 381–85

Giulian, D., des Ruisseaux, H., Cowburn, D. 1980. Biosynthesis and intra-axonal transport of proteins during neuronal regeneration. *J. Biol. Chem.* 255: 6494–6501

Glicksman, M. A., Willard, M. 1985. Differential expression of the three neurofilament polypeptides. *Ann. NY Acad. Sci.* 455: 479–91

Goldstein, M. E., Weiss, S., Lazzarini, R., Schneidman, P., Lees, J., Schlaepfer, W. W. 1987. Neurofilament messenger RNA levels decrease following sciatic nerve transection. *Abstr. Soc. Neurosci.* 13: 1707

Grafstein, B., Forman, D. S. 1980. Intracellular transport in neurons. *Physiol. Rev.* 60: 1167–1283

Grafstein, B., Burmeister, D. W., McGuinness, C. M., Perry, G. W., Sparrow, J. R. 1987. Role of fast axonal transport in regeneration of goldfish optic axons. *Prog. Brain Res.* 71: 113–20

Grafstein, B., McQuarrie, I. G. 1978. Role of the nerve cell body in axonal regeneration. In *Neuronal Plasticity,* ed. C. W. Cotman, pp. 155–95. New York: Raven

Greenough, W. T. 1984. Structural correlates of information storage in the mammalian brain: A review and hypothesis. *Trends Neurosci.* 7: 229–33

Hall, M. E. 1982. Changes in synthesis of specific proteins in axotomized dorsal root ganglia. *Exp. Neurol.* 76: 83–93

Hall, D. E., Neugebauer, K. M., Reichardt, L. F. 1987. Embryonic neural retina cell response to extracellular matrix proteins: Developmental changes and effects of the cell substratum attachment antibody (CSAT). *J. Cell Biol.* 104: 623–34

Hall, M. E., Wilson, D. L., Stone, G. C. 1978. Changes in synthesis of specific proteins following axotomy: Detection with two-dimensional gel electrophoresis. *J. Neurobiol.* 9: 353–66

Heacock, A. M., Agranoff, B. W. 1976. Enhanced labeling of a retinal protein during regeneration of optic nerve in goldfish. *Proc. Natl. Acad. Sci. USA* 73: 828–32

Heacock, A. M., Agranoff, B. W. 1982. Protein synthesis and transport in the regenerating goldfish visual system. *Neurochem. Res.* 7: 771–88

Hoffman, P. N., Lasek, R. J. 1975. The slow component of axonal transport. Identification of major structural polypeptides of the axon and their generality among mammalian neurons. *J. Cell Biol.* 66: 351–66

Hoffman, P. N., Lasek, R. J. 1980. Axonal transport of the cytoskeleton in regenerating motor neurons: Constancy and change. *Brain Res.* 202: 317–33

Hynes, R. O. 1987. Integrins: A family of cell surface receptors. *Cell* 48: 549–54

Ignatius, M. J., Gebicke-Harter, P. J., Skene, J. H. P., Schilling, J. W., Weisgraber, K. H., Mahley, R. W., Shooter, E. M. 1986. Expression of apolipoprotein E during nerve degeneration and regeneration. *Proc. Natl. Acad. Sci. USA* 83: 1125–29

Ignatius, M. J., Shooter, E. M., Pitas, R. E., Mahley, R. W. 1987a. Lipoprotein uptake by neuronal growth cones in vitro. *Science* 236: 959–62

Ignatius, M. J., Gebicke-Harter, P. J., Pitas, R. E., Shooter, E. M. 1987b. Apolipoprotein E in nerve injury and repair. *Prog. Brain Res.* 71: 177–84

Jacobson, R. D., Virag, I., Skene, J. H. P. 1986. A protein associated with axon growth, GAP-43, is widely distributed and developmentally regulated in rat CNS. *J. Neurosci.* 6: 1843–55

Jenkins, W. M., Merzenich, M. M. 1987. Reorganization of neocortical representations after brain injury: A neurophysiological model of the bases of recovery from stroke. *Prog. Brain Res.* 71: 249–66

Jolles, J., Zwiers, H., van Dongen, C. J., Schotman, P., Wirtz, K. W. A., Gispen, W. H. 1980. Modulation of brain polyphosphoinositide metabolism by ACTH-sensitive protein phosphorylation. *Nature* 286: 623–25

Kaas, J. H., Merzenich, M. M., Killackey, H. P. 1983. The reorganization of somatosensory cortex following peripheral nerve damage in adult and developing mammals. *Ann. Rev. Neurosci.* 6: 325–56

Kalil, K., Perdew, M. 1988. Expression of two developmentally regulated brain specific proteins is correlated with late outgrowth of the pyramidal tract. *J. Neurosci.* 8: In press

Kalil, K., Skene, J. H. P. 1986. Elevated synthesis of an axonally transported protein correlates with axon outgrowth in normal and injured pyramidal tracts. *J. Neurosci.* 6: 2563–70

Karns, L. R., Ng, S. C., Freeman, J. A., Fishman, M. C. 1987. Cloning of complementary DNA for GAP-43, a neuronal growth-related protein. *Science* 236: 597–600

Katz, F., Ellis, L., Pfenninger, K. H. 1985. Nerve growth cones isolated from fetal rat brain. III. Calcium-dependent protein phosphorylation. *J. Neurosci.* 5: 1402–11

Kishimoto, A., Kajikawa, N., Shiota, M., Nishizuka, Y. 1983. Proteolytic activation of calcium-activated, phospholipid-dependent protein kinase by calcium-dependent neutral protease. *J. Biol. Chem.* 258: 1156–64

Komiya, Y. 1981. Growth, aging and regeneration of axons and slow axonal flow. In *New Approaches to Nerve and Muscle Disorders. Basic and Applied Contributions,* ed. A. D. Kidman, J. K. Tomkins, R. A. Westerman, pp. 173–82. Amsterdam: Elsevier North-Holland

Kosik, K. S., Orecchio, L. D., Bruns, G. A. P., MacDonald, G. P., Cox, D. R., Neve, R. L. 1988. Human GAP-43: Its deduced amino acid sequence and chromosomal localization in mouse and human. *Neuron* 1: 127–32

Kristjansson, G. I., Zwiers, H., Oestreicher, A. B., Gispen, W. H. 1982. Evidence that the synaptic phosphoprotein B-50 is localized exclusively in nerve tissue. *J. Neurochem.* 39: 371–78

Landreth, G. E., Agranoff, B. W. 1976. Explant culture of adult goldfish retina: Effect of prior optic nerve crush. *Brain Res.* 118: 299–303

Le Beau, J. M., Walter, G. 1987. pp60c-src protein-tyrosine kinase is strongly induced in regenerating sciatic nerve. *Abstr. Soc. Neurosci.* 13: 1207

Lev, Z., Leibovitz, N., Segev, O., Shilo, B.-Z. 1984. Expression of the src and abl cellular oncogenes during development of *Drosophila melanogaster. Mol. Cell. Biol.* 4: 982–84

Lewis, S. A., Lee, M. G.-S., Cowan, N. J. 1985. Five mouse tubulin isotypes and their regulated expression during development. *J. Cell Biol.* 101: 852–61

Lichtman, J. W., Magrassi, L., Purves, D. 1987. Visualization of neuromuscular junctions over periods of several months in living mice. *J. Neurosci.* 7: 1215–22

Linden, D. J., Murakami, K., Routtenberg, A. 1986. A newly discovered protein kinase C activator (oleic acid) enhances long-term potentiation in the intact hippocampus. *Brain Res.* 379: 358–63

Linden, D. J., Sheu, F.-S., Murakami, K., Routtenberg, A. 1987. Enhancement of long-term potentiation by cis-unsaturated fatty acid: Relation to protein kinase C and phospholipase A2. *J. Neurosci.* 7: 3783–92

Llinas, R., McGuinness, T. L., Leonard, C. S., Sugimori, M., Greengard, P. 1985. Intraterminal injection of synapsin I or calcium-calmodulin-dependent protein kinase II alters neurotransmitter release at the squid giant synapse. *Proc. Natl. Acad. Sci. USA* 82: 3035–39

Lovinger, D. M., Wong, K. L., Murakami, K., Routtenberg, A. 1987. Protein kinase C inhibitors eliminate hippocampal long-term potentiation. *Brain Res.* 436: 177–83

Lovinger, D. M., Colley, P. A., Akers, R. F., Nelson, R. B., Routtenberg, A. 1986. Direct relation of long-term synaptic potentiation to phosphorylation of membrane protein F1, a substrate for membrane protein kinase C. *Brain Res.* 399: 205–11

Lozano, A. M., Doster, S. K., Aguayo, A. J., Willard, M. B. 1987. Immunoreactivity to GAP-43 in axotomized and regenerating retinal ganglion cells of adult rats. *Abstr. Soc. Neurosci.* 13: 1389

Malenka, R. C., Madison, D. V., Nicoll, R. A. 1986. Potentiation of synaptic transmission in the hippocampus by phorbol esters. *Nature* 321: 175–77

Maness, P. F., Aubry, M., Shores, C. G., Frame, L., Pfenninger, K. H. 1988. The c-src gene product in developing rat brain is enriched in nerve growth cone membranes. *Proc. Natl. Acad. Sci. USA* 85: 5001–5

Maness, P. F., Sorge, L. K., Fults, D. W. 1986. An early developmental phase of pp60c-src expression in the neural ectoderm. *Dev. Biol.* 117: 83–89

Masure, H. R., Alexander, K. A., Wakim, B. T., Storm, D. R. 1986. Physicochemical and hydrodynamic characterization of P-57, a neurospecific calmodulin binding protein. *Biochemistry* 25: 7553–60

McIntosh, H., Meiri, K., Parkinson, D., Willard, M., Daw, N. 1987. Changes in a GAP43-like protein during development of the cat visual cortex. *Abstr. Soc. Neurosci.* 13: 1538

McNaughton, B. L., Morris, R. G. M. 1987.

Hippocampal synaptic enhancement and information storage within a distributed memory system. *Trends Neurosci.* 10: 408–15

McQuarrie, I. G. 1984. Effect of a conditioning lesion on axonal transport during nerve regeneration: The role of slow transport. See Elam and Cancalon 1984, pp. 185–210

McQuarrie, I. G., Grafstein, B. 1981. Effect of a conditioning lesion on optic nerve regeneration in goldfish. *Brain Res.* 216: 253–64

McQuarrie, I. G., Grafstein, B., Gershon, M. D. 1977. Axonal regeneration in the rat sciatic nerve: Effect of a conditioning lesion and of dbcAMP. *Brain Res.* 132: 443–53

Meiri, K. F., Pfenninger, K. H., Willard, M. B. 1986. Growth-associated protein, GAP-43, a polypeptide that is induced when neurons extend axons, is a component of growth cones and corresponds to pp46, a major polypeptide of a subcellular fraction enriched in growth cones. *Proc. Natl. Acad. Sci. USA* 83: 3537–41

Meiri, K. F., Johnson, M. I., Willard, M. 1988. Distribution and phosphorylation of the growth associated protein GAP-43 in regenerating sympathetic neurons in culture. *J. Neurosci.* 8: 2571–81

Melloni, E., Pontremoli, S., Michetti, M., Sacco, O., Sparatore, B., et al. 1986. The involvement of calpain in the activation of protein kinase C in neutrophils stimulated by phorbol myristic acid. *J. Biol. Chem.* 261: 4101–5

Meyer, R. L. 1983. Tetrodotoxin inhibits the formation of refined retinotopography in goldfish. *Dev. Brain Res.* 6: 293–98

Miller, F. D., Naus, C. C. G., Higgins, G. A., Bloom, F. E., Milner, R. J. 1987a. Developmentally regulated rat brain mRNAs: Molecular and anatomical characterization. *J. Neurosci.* 7: 2433–44

Miller, F. D., Naus, C. C. G., Durand, M., Bloom, F. E., Milner, R. J. 1987b. Isotypes of α-tubulin are differentially regulated during neuronal maturation. *J. Cell Biol.* 105: 3065–73

Miller, F. D., Tetzlaff, W., Naus, C. C. G., Bisby, M., Bloom, F. E., Milner, R. J. 1987c. The differential expression of α-tubulin genes during neuronal development is reiterated during neuronal regeneration. *Abstr. Soc. Neurosci.* 13: 1706

Muller, H. W., Gebicke-Harter, P. J., Hangen, D. H., Shooter, E. M. 1985. A specific 37,000-Dalton protein that accumulates in regenerating but not nonregenerating mammalian nerves. *Science* 228: 499–501

Murakami, K., Routtenberg, A. 1985. Direct activation of purified protein kinase C by unsaturated fatty acids (oleate and arachidonate) in the absence of phospholipids and Ca$^{2+}$. *FEBS Lett.* 192: 189–93

Nelson, R. B., Routtenberg, A. 1985. Characterization of protein F1 (47 kDa, 4.5 pI): A kinase C substrate directly related to neural plasticity. *Exp. Neurol.* 89: 213–24

Nelson, R. B., Linden, D. J., Routtenberg, A. 1987a. Two growth cone-enriched C-kinase substrates and two vesicle-associated phosphoproteins are directly correlated with persistence of long-term potentiation: A quantitative analysis of two-dimensional gels. *Abstr. Soc. Neurosci.* 13: 1233

Nelson, R. B., Friedman, D. P., O'Neill, J. B., Mishkin, M., Routtenberg, A. 1987b. Gradients of protein kinase C substrate phosphorylation in primate visual system peak in visual memory storage areas. *Brain Res.* 416: 387–92

Neve, R. L., Perrone-Bizzozero, N. I., Finklestein, S., Zwiers, H., Bird, E., Kurnit, D. M., Benowitz, L. I. 1987. The neuronal growth-associated protein GAP-43 (B-50,F1): Neuronal specificity, developmental regulation and regional distribution of the human and rat mRNAs. *Mol. Brain Res.* 2: 177–83

Ng, S.-C., de la Monte, S. M., Conboy, G. L., Karns, L. R., Fishman, M. C. 1988. Cloning of human GAP-43: Growth association and ischemic resurgence. *Neuron* 1: 133–39

Nielander, H. B., Schrama, L. H., van Rozen, A. J., Kasperaitis, M., Oestreicher, A. B., deGraan, P. N. E., Gispen, P., Schotma, W. H. 1987. Primary structure of the neuron-specific phosphoprotein B-50 is identical to growth-associated protein GAP-43. *Neurosci. Res. Comm.* 1: 163–72

Oestreicher, A. B., Gispen, W. H. 1986. Comparison of the immunocytochemical distribution of the phosphoprotein B-50 in the cerebellum and hippocampus of immature and adult rat brain. *Brain Res.* 375: 267–79

Oestreicher, A. B., Dekker, L. V., Gispen, W. H. 1986. A radioimmunoassay for the phosphoprotein B-50: Distribution in rat brain. *J. Neurochem.* 46: 1366–69

Oestreicher, A. B., van Dongen, C. J., Zwiers, H., Gispen, W. H. 1983. Affinity-purified anti-B-50 protein antibody: Interference with the function of the phosphoprotein B-50 in synaptic plasma membranes. *J. Neurochem.* 41: 331–40

Pachter, J. S., Liem, R. K. H. 1984. The differential appearance of neurofilament triplet polypeptides in the developing rat optic nerve. *Dev. Biol.* 103: 200–10

Perrone-Bizzozero, N. I., Finklestein, S. P., Benowitz, L. I. 1986. Synthesis of a growth-associated protein by embryonic rat cerebrocortical neurons in vitro. *J. Neurosci.* 6: 3721–30

Perry, G. W., Burmeister, D. W., Grafstein, B. 1987a. Fast axonally transported proteins in regenerating goldfish optic axons. *J. Neurosci.* 7: 792–806

Perry, G. W., Krayanek, S. R., Wilson, D. L. 1987b. Effects of a conditioning lesion on bullfrog sciatic nerve regeneration: Analysis of fast axonally transported proteins. *Brain Res.* 423: 1–12

Perry, G. W., Wilson, D. L. 1981. Protein synthesis and axonal transport during nerve regeneration. *J. Neurochem.* 37: 1203–17

Pfenninger, K. H., Ellis, L., Johnson, M. P., Friedman, L. B., Somlo, S. 1983. Nerve growth cones isolated from fetal rat brain: Subcellular fractionation and characterization. *Cell* 35: 573–84

Redshaw, J. D., Bisby, M. A. 1984a. Proteins of fast axonal transport in the regenerating hypoglossal nerve at the rat. *Can. J. Physiol. Pharmacol.* 62: 1387–93

Redshaw, J. D., Bisby, M. A. 1984b. Fast axonal transport in central nervous system and peripheral nervous system axons following axotomy. *J. Neurobiol.* 15: 109–17

Redshaw, J. D., Bisby, M. A. 1987. Proteins of fast axonal transport in regenerating rat sciatic sensory axons: A conditioning lesion does not amplify the characteristic response to axotomy. *Exp. Neurol.* 98: 212–21

Reh, T. A., Redshaw, J. D., Bisby, M. A. 1987. Axons of the pyramidal tract do not increase their transport of growth-associated proteins after axotomy. *Mol. Brain Res.* 2: 1–6

Rosenthal, A., Chan, S. Y., Henzel, W., Haskell, C., Kuang, W.-J., Chen, J. N., Wilcox, A., Ullrich, D. V., Goeddel, E., Routtenberg, A. 1987. Primary structure and mRNA localization of protein F1, a growth-related protein kinase C substrate associated with synaptic plasticity. *EMBO J.* 6: 3641–46

Routtenberg, A. 1986. Synaptic plasticity and protein kinase C. *Prog. Brain Res.* 69: 211–34

Routtenberg, A., Colley, P., Linden, D., Lovinger, D., Murakami, K., Sheu, F.-S. 1986. Phorbol ester promotes growth of synaptic plasticity. *Brain Res.* 378: 374–78

Schmidt, J. T., Edwards, D. L. 1983. Activity sharpens the map during regeneration of the retinotectal projection in goldfish. *Brain Res.* 269: 29–39

Schwab, M. E., Caroni, P. 1988. Oligodendrocytes and CNS myelin are nonpermissive substrates for neurite growth and fibroblast spreading in vitro. *J. Neurosci.* 8: 2381–93

Shaw, G., Weber, K. 1982. Differential expression of neurofilament triplet proteins in brain development. *Nature* 298: 277–79

Simon, M. A., Drees, B., Kornberg, T., Bishop, J. M. 1985. The nucleotide sequence and the tissue-specific expression of *Drosophila c-src. Cell* 42: 831–40

Skene, J. H. P., Jacobson, R. D., Snipes, G. J., McGuire, C. B., Norden, J. J., Freeman, J. A. 1986. A protein induced during nerve growth (GAP-43) is a major component of growth-cone membranes. *Science* 233: 783–86

Skene, J. H. P., Shooter, E. M. 1983. Denervated sheath cells secrete a new protein after nerve injury. *Proc. Natl. Acad. Sci. USA* 80: 4169–73

Skene, J. H. P., Willard, M. 1981a. Changes in axonally transported proteins during axon regeneration in toad retinal ganglion cells. *J. Cell Biol.* 89: 86–95

Skene, J. H. P., Willard, M. 1981b. Axonally transported proteins associated with axon growth in rabbit central and peripheral nervous systems. *J. Cell Biol.* 89: 96–103

Skene, J. H. P., Willard, M. 1981c. Characteristics of growth-associated polypeptides in regenerating toad retinal ganglion cell axons. *J. Neurosci.* 1: 419–26

Smith, S. J. 1987. Progress on LTP at hippocampal synapses: A post-synaptic $Ca^{2+}$ trigger for memory storage? *Trends Neurosci.* 10: 142–44

Snipes, G. J., McGuire, C. B., Norden, J. J., Freeman, J. A. 1986. Nerve injury stimulates the secretion of apolipoprotein E by nonneuronal cells. *Proc. Natl. Acad. Sci. USA* 83: 1130–34

Sorge, L. K., Levy, B. T., Maness, P. F. 1984. pp60c-src is developmentally regulated in the neural retina. *Cell* 36: 249–57

Stryker, M. P., Harris, W. A. 1986. Binocular impulse blockade prevents the formation of ocular dominance columns in cat visual cortex. *J. Neurosci.* 6: 2117–33

Szaro, B. G., Loh, Y. P., Hunt, R. K. 1985. Specific changes in axonally transported proteins during regeneration of the frog (*Xenopus laevis*) optic nerve. *J. Neurosci.* 5: 192–208

Teyler, T. J., DiScenna, P. 1987. Long-term potentiation. *Ann. Rev. Neurosci.* 10: 131–61

Tetzlaff, W., Bisby, M. A., Kreskoski, C. A., Parhad, I. M. 1987. A conditioning lesion does not further stimulate tubulin and actin synthesis but further decreases neurofilament synthesis in the facial nerve of the rat. *Abstr. Soc. Neurosci.* 13: 1706

Tomaselli, K. J., Reichardt, L. F., Bixby, J. L. 1986. Distinct molecular interactions mediate neuronal process outgrowth on extracellular matrices and nonneuronal cell surfaces. *J. Cell Biol.* 103: 2659–72

van Dongen, C. J., Zwiers, H., de Graan, P. N. E., Gispen, W. H. 1985. Modulation of the activity of purified phosphatidyl-inositol 4-phosphate kinase by phosphorylated and dephosphorylated B-50 protein. *Biochem. Biophys. Res. Commun.* 128: 1219–27

van Hoof, C. O. M., deGraan, P. N. E., Oestreicher, A. B., Gispen, W. H. 1988. B-50 phosphorylation and polyphospho-inositide metabolism in nerve growth cone membranes. *J. Neurosci.* 8: 1789–95

Vidal-Sanz, M., Bray, G. M., Villegas-Perez, M. P., Aguayo, A. J. 1987. Axonal regeneration and synapse formation in the superior colliculus by retinal ganglion cells in the adult rat. *J. Neurosci.* 7: 2894–2909

Wakim, B. T., Alexander, K. A., Masure, H. R., Cimler, B. M., Storm, D. R., Walsh, K. A. 1987. Amino acid sequence of p-57, a neurospecific calmodulin-binding protein. *Biochemistry* 26: 7466–70

Wallis, I., Ellis, L., Suh, K., Pfenninger, K. H. 1985. Immunolocalization of a neuronal growth-dependent membrane glycoprotein. *J. Cell Biol.* 101: 1990–98

Willard, M., Hulebak, K. 1977. The intra-axonal transport of polypeptide H: Evidence for a fifth (very slow) group of transported proteins in the retinal ganglion cells of the rabbit. *Brain Res.* 136: 289–306

Willard, M. B., Simon, C. 1983. Modulations of neurofilament axonal transport during the development of rabbit retinal ganglion cells. *Cell* 35: 551–59

Willard, M., Skene, J. H. P., Simon, C., Meiri, K., Glicksman, M. 1984. Regulation of axon growth and cytoskeletal development. See Elam & Cancalon 1984, pp. 171–84

Willard, M., Cowan, W. M., Vagelos, P. R. 1974. The polypeptide composition of intra-axonally transported proteins: Evidence for four transport velocities. *Proc. Natl. Acad. Sci. USA* 71: 2183–87

Wujek, J. R., Lasek, R. J. 1983. Correlation of axonal regeneration and slow component B in two branches of a single axon. *J. Neurosci.* 3: 243–51

Zwiers, H., Schotman, P., Gispen, W. H. 1980. Purification and some characteristics of an ACTH-sensitive protein kinase and its substrate protein in rat brain membranes. *J. Neurochem.* 34: 1689–99

Zwiers, H., Veldhuis, H. D., Schotman, P., Gispen, W. H. 1976. ACTH, cyclic neucleotides, and brain protein phosphorylation in vitro. *Neurochem. Res.* 1: 669–77

*Ann. Rev. Neurosci. 1989. 12:157–83*

# LEARNING ARM KINEMATICS AND DYNAMICS

## Christopher G. Atkeson

Department of Brain and Cognitive Sciences and the
Artificial Intelligence Laboratory, Massachusetts Institute of Technology,
Cambridge, Massachusetts 02139

## INTRODUCTION AND SCOPE OF THE REVIEW

One of the impressive feats of biological motor control systems is their ability to calibrate themselves and improve their performance with experience. Man-made machines are typically calibrated using external measuring devices and references, and often execute the same errors repeatedly. An important step toward understanding biological motor behavior, and also toward building more useful machines, is to understand how motor learning is achieved.

The architecture of the nervous system affects what can be learned and how it might be learned. Hypotheses concerning biological motor control are primarily discussed from the point of view of how well a proposed control scheme might generate desired movements. Few have evaluated hypotheses from a motor learning perspective. This review explores the implications of several proposed motor control schemes for certain issues of learning, such as generalization, learning efficiency, rate of learning, and the ability to interpret and correct performance errors. My own interest in motor learning was sparked by discovering that some control schemes, such as the equilibrium trajectory approach (Berkinblit et al 1986, Hogan et al 1987), do not support efficient movement refinement during practice (C. G. Atkeson, unpublished results).

This review focuses on the information processing issues involved in motor learning. Learning is an information processing problem: A learning algorithm modifies present and future commands by using information from previous experience so as to improve performance. Marr (1982) emphasized the need for a *computational* approach to information pro-

0147–006X/89/0301–0157$02.00

cessing problems in neuroscience. He emphasized separating the different levels of computational theory, algorithm, and implementation. Computational theories describe the information processing problems faced by the brain, whereas algorithms specify how to solve those problems, and implementations specify the mechanisms used to execute algorithms. Recent reviews in this series (Hildreth & Koch 1987; Poggio & Poggio 1984, Robinson 1981, Ullman 1986) and elsewhere (Hildreth & Hollerbach 1985, Hollerbach 1982, Raibert 1986, Saltzman 1979) discuss computational approaches and how these approaches are shaped by and help guide other forms of neuroscience research.

Robotics is an important source of ideas for biologists. The study of robot control and learning contributes to the understanding of biological motor control and learning, and therefore this review also surveys relevant work in robotics. Many proposed methods for robot control and learning are also applicable for biological control, and vice versa. Although the actuators, sensors, structure, and computational architecture of the control systems of mechanical and biological limbs differ, at an information processing level the control problems are similar. The same laws of physics apply, and the mechanical objectives of movement and manipulation are identical. Trying to program a robot to solve a problem that humans routinely solve leads to greater comprehension and insight into what makes that problem difficult. Robots provide a testbed for developing and evaluating hypothesized control schemes and allow exploration of a wide range of possible solutions, some of which may be directly applicable to biological motor control. Specific algorithms and the observed behavior of those algorithms on a machine can guide biological research by suggesting novel experimental tests of hypothesized control schemes. In trying to build a machine that actually works, theories are forced to be specific and concrete, and difficult issues cannot be ignored.

Many control schemes have been criticized for requiring accurate internal models. A reason for renewed interest in motor learning is the recent research showing that it is possible to identify accurate models from movement data. A description of these demonstrations of motor learning in robots is included in this review.

This article emphasizes arm control. Two kinds of motor learning are examined. The first involves building and refining internal models of the motor apparatus and the external world, and using those models to generate the appropriate actuator commands and to guide interpretation and processing of sensory data. The second involves using internal models to transform performance errors into command corrections. The particular tasks used to assess motor learning include moving the hand to a visually observed target and attaining a desired pattern of movement either with or without a load. A recent review of arm motor control psychophysics and neurophysiology is provided by Georgopoulos (1986). Recent reviews of motor learning from a variety of perspectives include Adams (1984, 1987), Annett (1985), Grossberg & Kuperstein (1986), Salmoni et al (1984), Schmidt (1982), and Smyth & Wing (1984).

The study of motor learning may help reveal the forms of representation used by the nervous system. Two alternative representations, structured and tabular, have received a great deal of attention and support different forms of learning. Structured representations can take advantage of prior knowledge of the constraints on a system, whereas tabular representations are general purpose in that they can represent almost any type of transformation. The review focuses on the transformations involved in translating a behavioral objective into muscle activation, and the implications for motor learning of how these transformations are represented. The next section discusses these transformations. In this article the discussion of the issues and the review of the literature have been separated. The third section explores important issues in learning these transformations. The final sections survey how particular proposals have addressed these issues.

## MOTOR CONTROL INVOLVES TRANSFORMATIONS

Motor control may be viewed as a series of transformations from a specified behavioral objective to a plan specifying the desired mechanical output of the motor apparatus and finally to a pattern of activation of the muscles (Hollerbach 1982, Hildreth & Hollerbach 1985). Progress has been made in understanding the planning process (see Hollerbach & Atkeson 1987 for further references). In this section the transformations that may be involved in executing a movement are discussed. It is assumed that a movement plan already exists. Two types of transformations, kinematic and dynamic, are explored.

### Kinematic Transformations

One series of transformations involved in the execution of a movement is kinematic. A limb, as a mechanical device, converts muscle lengths and joint angles to hand positions. This process is referred to as forward kinematics (Figure 1A). In Figure 2A an idealized model of this transformation for movements in a vertical plane is illustrated. The arm is modeled as two rigid links of length $l_1$ and $l_2$. The links rotate about horizontal parallel axes that are fixed with respect to the links. For this idealized model the forward kinematics from joint angles $(\theta_1, \theta_2)$ to hand position $(x, y)$ are given by

$$\begin{pmatrix} x \\ y \end{pmatrix} = \begin{pmatrix} l_1 \cos{(\theta_1)} + l_2 \cos{(\theta_1 + \theta_2)} \\ l_1 \sin{(\theta_1)} + l_2 \sin{(\theta_1 + \theta_2)} \end{pmatrix}. \qquad\qquad 1.$$

The nervous system is faced with the following problem: Given a desired hand position, choose the appropriate muscle lengths and corresponding joint angles to achieve that position. The transformation from desired hand position to the corresponding joint angles and muscle lengths is known as the inverse kinematic transformation (Figure 1A). That we can view a target in space, close our eyes, and move our hand to that target suggests that we have an internal representation or model of the transformation from desired hand positions to joint angles and muscle lengths. Internal models of arm kinematics are also useful in choosing torques at the joints of the arm to achieve a particular force and torque between the hand and some external object. Learning to move accurately may involve building internal models of kinematic transformations.

## Dynamic Transformations

Another set of transformations become important when the pattern of movement is to be controlled. In order to drive the limb along a particular

**A**

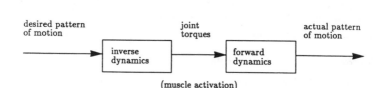

**B**

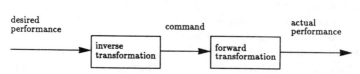

**C**

*Figure 1*   Example motor control transformations.

A

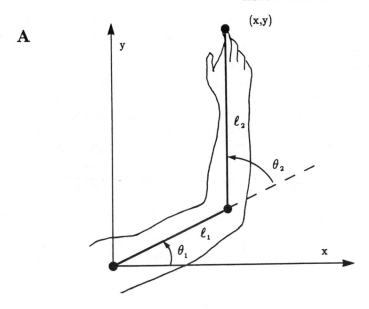

B

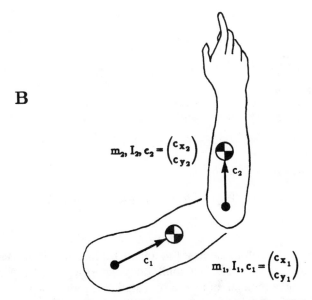

*Figure 2*   Idealized model of two-joint arm: (*A*) Arm kinematics. (*B*) Arm segments and corresponding inertial parameters.

trajectory, the appropriate torques must be applied at the joints. The transformation from joint torques to movement is referred to as forward dynamics (Figure 1*B*). The pattern of muscle activation to achieve these torques may be generated in a variety of ways, including pure feedback (closed loop) control, pure feedforward (open loop) control, or some combination of feedback and feedforward control (Houk & Rymer 1981). Muscle mechanical properties, such as stiffness and viscosity, also contribute to the applied torques (Bizzi et al 1978). For fast movements, such as a baseball pitch, feedback control can play only a small role in controlling the movement because of information transmission delays. Instead, the nervous system must specify the pattern of muscle activation corresponding to the desired pattern of motion. The transformation from a desired pattern of motion to the actuator commands necessary to achieve that motion is referred to as inverse dynamics (Figure 1*B*).

The inverse dynamics transformation from a desired pattern of motion to muscle activation may be broken up into a series of transformations: pattern of motion to joint torques, joint torques to muscle forces, and muscle forces to the necessary muscle activations. Equations 2 and 3 give the joint torques ($\tau_1$ and $\tau_2$) for the idealized two-joint arm model as a function of the desired joint positions ($\theta_1$ and $\theta_2$), joint velocities ($\dot{\theta}_1$ and $\dot{\theta}_2$), and joint accelerations ($\ddot{\theta}_1$ and $\ddot{\theta}_2$):

$$\tau_1 = m_1 c_{x_1} g \cos(\theta_1) - m_1 c_{y_1} g \sin(\theta_1)$$
$$+ I_1 \ddot{\theta}_1 + m_2 [g l_1 \cos(\theta_1) + l_1^2 \ddot{\theta}_1]$$
$$+ m_2 c_{x_2} [g \cos(\theta_1 + \theta_2) + 2 l_1 \ddot{\theta}_1 \cos(\theta_2) + l_1 \ddot{\theta}_2 \cos(\theta_2)$$
$$- 2 l_1 \dot{\theta}_1 \dot{\theta}_2 \sin(\theta_2) - l_1 \dot{\theta}_2^2 \sin(\theta_2)]$$
$$+ m_2 c_{y_2} [-g \sin(\theta_1 + \theta_2) - 2 l_1 \ddot{\theta}_1 \sin(\theta_2) - l_1 \ddot{\theta}_2 \sin(\theta_2)$$
$$- 2 l_1 \dot{\theta}_1 \dot{\theta}_2 \cos(\theta_2) - l_1 \dot{\theta}_2^2 \cos(\theta_2)] + I_2 (\ddot{\theta}_1 + \ddot{\theta}_2) \qquad \text{2.}$$

$$\tau_2 = m_2 c_{x_2} [g \cos(\theta_1 + \theta_2) + l_1 \ddot{\theta}_1 \cos(\theta_2) + l_1 \dot{\theta}_1^2 \sin(\theta_2)]$$
$$+ m_2 c_{y_2} [-g \sin(\theta_1 + \theta_2) - l_1 \ddot{\theta}_1 \sin(\theta_2) + l_1 \dot{\theta}_1^2 \cos(\theta_2)]$$
$$+ I_2 (\ddot{\theta}_1 + \ddot{\theta}_2). \qquad \text{3.}$$

Each segment of the arm has a mass ($m_1$ and $m_2$), a location of the center of mass relative to the proximal joint [the vectors $\mathbf{c}_1 = (c_{x_1}, c_{y_1})$ and $\mathbf{c}_2 = (c_{x_2}, c_{y_2})$], and a moment of inertia for rotations around the joint axis ($I_1$ and $I_2$) (Figure 2*B*). The parameter $g$ is the gravitational constant. The inertial parameters of mass, mass moment (the product of the arm segment mass and its center of mass location), and moment of inertia appear linearly in the inverse dynamics.

Simpler models may be used to approximate the dynamics. Different versions of the equilibrium trajectory approach, for example, either ignore dynamics or use a configuration-independent mass-spring-damper model to approximate the dynamics (Hogan et al 1987, Feldman 1986).

In both engineering and biology, control problems can be posed in the following way: The mechanical apparatus to be controlled transforms its inputs (commands) into some outputs (performance). The control system generates the appropriate commands based on the desired performance of the motor apparatus. To achieve high levels of performance a control system must implement the inverse of the transformation performed by the motor apparatus (Figure 1C). This is true even when using feedback control, and it is especially true in biological systems where signaling delays limit possible feedback gains. One view of motor learning is that its goal is to build an accurate inverse model of the motor apparatus.

## Inverse Transformations Can Be Implemented In Many Ways

Inverse models of the motor apparatus can be represented in many ways. The inverse kinematic transformation is used here as an example. The inverse kinematic transformation for the idealized two-joint arm model can be represented mathematically:

$$\theta_2 = \arccos\left(\frac{x^2 + y^2 - l_1^2 - l_2^2}{2l_1 l_2}\right) \tag{4.}$$

$$\theta_1 = \arctan\left(\frac{y}{x}\right) - \arctan\left(\frac{l_2 \sin(\theta_2)}{l_1 + l_2 \cos(\theta_2)}\right). \tag{5.}$$

To control a two-joint robot arm, the inverse kinematic transformation could be implemented as a digital computer program (Figure 3A), a special purpose analog (or digital) computational circuit that corresponds to the mathematical Expressions 4 and 5 (Figure 3B), or a lookup table (Figure 3C). These are examples of how the same information-processing problem, i.e. computing the inverse kinematic transformation, can be solved by different algorithms and implementations.

The nervous system could also use a variety of mechanisms to implement the inverse kinematic transformation. Neural circuits might exist that correspond to or approximate the circuit of Figure 3B. Many hypothesized brainstem circuits for oculomotor control are of this nature, in that signals are represented by the amount of neural firing, and operations on signals

**A**  Digital computer program

```
theta2 = arccos( (x*x + y*y - l1*l1 - l2*l2)/(2*l1*l2) );
theta1 = arctan( y, x ) - arctan( l2*sin( theta2 ), l1 + l2*cos(theta2) );
```

**B**  Analog computational circuit

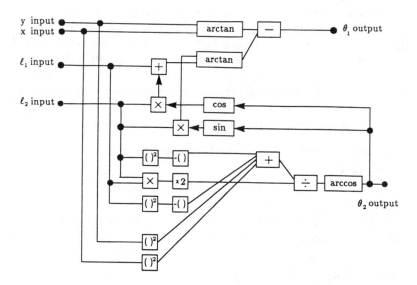

**C**  Lookup table

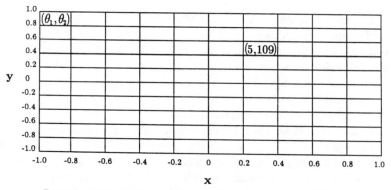

*Figure 3*   Alternative implementations of two-joint inverse kinematics.

are performed by the interaction of signals with operations analogous to addition, multiplication, and integration (Robinson 1981). Other proposed neural representations are similar to the tabular implementation of Figure 3C, in that a signal is represented by activity at a particular location within a neural structure, and operations on signals are performed using patterns of connections or mappings between neural structures (Knudson et al 1987). The superior colliculus is an example of such a tabular representation, in that activity at a particular location in the superior colliculus corresponds to a particular eye movement.

# ISSUES IN REPRESENTING TRANSFORMATIONS

This section discusses the issues and implications involved in using different representations of transformations for motor control and learning. Two extreme alternatives are structured and tabular representations. Models such as the mathematical Expressions 4 and 5, the computer program (Figure 3A), and the computational circuit (Figure 3B) reflect the structure of a transformation, in that only a few parameters are required to describe the transformation and the model is specific to a particular transformation. Such models are referred to as structured models. Tables, such as the example in Figure 3C, are an unstructured method of representing models, in that they can be applied to a wide variety of transformations and require many parameters to be specified. Different forms of representations support different forms of learning. The study of motor learning may help reveal the forms of representation used by the nervous system.

LOCAL GENERALIZATION    The process whereby experience in one activity leads to improved performance in another is referred to as generalization. Typically, transformations such as inverse kinematic and dynamic transformations are continuous, in that similar inputs lead to similar outputs. Therefore, experience of a particular posture or movement should improve the ability to attain similar postures or movements. Tabular representations support this form of local generalization because table entries can be shared between different inputs, and interpolation over several entries can be used to smooth the transformed data.

GLOBAL GENERALIZATION    Representations that capture the structure of a transformation support a more powerful form of generalization, global generalization. In the example of the inverse kinematic circuit shown in Figure 3B, only the lengths of the links and the appropriate calibration of joint angle measurements must be specified for the circuit to represent accurately the entire transformation. The circuit can be calibrated by using only a few sets of inputs and outputs. It is then capable of generalization to all possible inputs and outputs.

Global generalization works as long as the structure of the internal model matches the actual motor apparatus. If the joint axes move relative to the links, due to rolling or sliding of the joint surfaces, or the joint axes are not parallel, then the structure of the idealized two joint kinematic model (Eqs. 4 and 5) is incorrect, and no set of parameters (link lengths, joint angle offsets, etc) will allow the circuit of Figure 3B to predict correctly joint angles for all possible hand positions. The point at which such structural modeling errors degrade the internal model so much that it is no longer useful is difficult to characterize.

LEARNING RATE AND EFFICIENCY    Because structured models incorporate knowledge of the structure of the motor apparatus, they typically are able to improve more quickly and generate better performance than table-based models for a given amount of experience. However, this is true only if the model structure is approximately correct. Learning rate and efficiency are important issues in handling relatively rapid changes in the motor system, as is the case in handling loads and in muscle fatigue.

When a load is lifted, the dynamics of the arm may change dramatically. Structured models can be updated quickly because only a few parameters must be estimated, and knowledge about the load can be generalized to all movements. Table-based models cannot use observed data as efficiently because they do not incorporate knowledge about the structure of the system. Because tables only support local generalization, the effect of the load must be learned again for each new type of movement. Tabular methods can improve their ability to handle loads by having different tables for different loads together with some method for recognizing loads so that the appropriate table is used when that load is present.

WHERE DO MODEL STRUCTURES COME FROM?    How is an appropriate model structure found? Scientists or engineers might use idealized models from physics, such as Newtonian dynamics. Graphs of the inputs versus the outputs of the transformation may suggest appropriate structures. In general, however, an adaptive machine must simply hypothesize possible model structures, estimate the corresponding parameters, and evaluate the model to see how well the model with its assumed structure fits the data.

Identifying a model structure for its arm kinematics and dynamics may not be necessary when designing an adaptive robot. Knowledge about structure may be preprogrammed, with only the parameters adjusted during the lifetime of the machine. Similarly, in biological systems the structure of the motor apparatus is essentially fixed during the lifetime of the organism, and only parametric changes occur.

SPACE REQUIREMENTS OF TABLES    Problems arise in applying straight-forward tabular representations to complex problems, because the size of a table grows exponentially with the number of indices used. In repre-senting the arm inverse dynamics for a six-jointed arm in a table, for example, six joint angles, six joint angular velocities, and six desired joint accelerations would form a set of 18 indices. If each of the 18 dimensions of the table were divided into only 10 parts, there would be $10^{18}$ entries in the table, a number larger than the number of neurons or synaptic con-nections in the brain. More sophisticated table organizations are necessary to reduce the table size.

WHEN ARE DATA PROCESSING DECISIONS MADE?    Another problem with the straightforward tabular representation presented here is that the functions that store and retrieve data must be chosen before any data is collected. These functions play a very important role in smoothing and temporally filtering the data. In general, the amount of smoothing and temporal filtering required depends on the noise present in the data and the density of nearby data points. Better use of the data could result if these data processing decisions could be delayed until the data are actually used. Simons et al (1982) present a scheme for varying the resolution of the table as required by the pattern of input data.

DIFFICULTY OF PARAMETER ESTIMATION    It is not difficult to initialize and maintain a table based on the observed input and output values of a transformation, as long as there is some method for generating the table before enough data are obtained to represent the transformation accurately. On the other hand, a structured model may present a very difficult parameter estimation problem, depending on its mathematical structure. Parameters that appear linearly in a model can be estimated by using simple regressions, but parameters appearing nonlinearly in a model must typically be estimated iteratively. Nonlinear parameter estimation methods also suffer from the problem of finding parameter values that are only locally optimal rather than globally optimal.

CHOICE OF COORDINATE SYSTEMS    It is not clear what coordinate systems are actually used by the brain. For example, instead of a Cartesian co-ordinate system, polar coordinates or some non-orthogonal coordinate system could be used to indicate a hand location in visual space (Soechting & Ross 1984). Tabular representations are typically very flexible in terms of what coordinate systems can be used. The linearization of sensory measurements or muscle commands is not necessary for tabular represen-tations to work. Structures in the brain such as the visual cortex or the superior colliculus, which use location of neural activity to represent where a sensory stimulus is located or where a movement should go, use computational maps rather than sets of coordinates (Knudson et al 1987). Indexing is performed by the pattern of neural connections. The choice of coordinate systems is more important for structured representations. The use of a structured representation based on a particular choice of input and output coordinate systems requires that the input and output coordinates be calibrated according to the coordinate systems used.

Other considerations are the time required for a computation and the accuracy required of the computational elements. Implementations with long series of computations or computational elements may require both high-speed and high-accuracy components. Tabular representations use many computational elements, but they are typically used in parallel so the demands on computational speed and accuracy of the elements may be reduced.

Another problem with both tabular and structured representations is that they are both specialized for a particular transformation direction. A model of the inverse kinematic transformation cannot be used to do the forward kinematic transformation, and vice versa, although the model includes the knowledge necessary to perform this transformation.

The possibility of hybrid or intermediate forms of representation should not be ignored. Computational circuits and tabular representations can be combined by replacing an element of the circuit (the arccos or arctan elements of Figure 3B, for example) with a table of the function that the element implements. Replacing functions with tables is often used in computer programs to speed up computation, as well as in modern electrical circuit design, where complex computational elements are replaced with tabulated values stored in a memory device.

USING SENSORS TO DECOUPLE IDENTIFICATION PROBLEMS    System identification, the building of models, can be simplified by using sensors to decouple different parts of the system. For example, sensing of forces and torques by the hand allows estimation of load parameters independently of the arm dynamics. More generally, covering an arm or any effector system with a sensory barrier such as a tactile sensing skin allows the separate identification of internal system dynamics, external system dynamics, and disturbances. Without a sensory barrier the combined system of arm plus external world may be too complex to identify robustly. Decoupling system identification problems allows different identification procedures to be applied to the different subsystems. In the case of an arm with variable loads, one expects the arm dynamics to vary slowly, whereas load dynamics change rapidly as loads are picked up or put down. Widely different rates and types of system change call for different system identification algorithms to track the changes. Another reason for decoupling arm and load identification is that the arm structure is relatively constant whereas the dynamics of different loads can have quite different structures. Arm identification can be based on a fixed model structure, while a complete load identification system must handle many different types of loads. Similarly, sensing the outputs of the actuators allows internal models of actuators to be identified separately from models of arm dynamics.

## BUILDING INTERNAL MODELS OF TRANSFORMATIONS

This section reviews proposals for how internal models of the motor apparatus and the external world might be built, used, and refined.

### Table-Based Methods

Lookup table-based methods provide a simple yet general approach to representing transformations. Tables also provide an easily modifiable representation appropriate for motor learning. Early work on using tabular or mapping representations to solve motor control problems was inspired by the regular structure of the cerebellum (Grossberg 1969, Marr 1969, Albus 1971, Pellionisz & Llinas 1979, Fujita 1982a,b). Other work attempted to use neuron-like elements or models of neural networks as the components of motor learning systems (Widrow & Smith 1964, Albus 1981, Ersu & Tolle 1984, Barto et al 1983). These schemes store knowledge as weights of the connections between model neurons, and are related to Perceptrons (Rosenblatt 1961, Minsky & Papert 1969) and current work on multilayer "connectionist" networks (Rumelhart et al 1986). The pattern of connections in these proposed networks can be mapped into patterns of table indexing (Michie & Chambers 1968, Albus 1975a,b). In robot control, tabular representations have been proposed as a way to avoid computational delays during real time control (Albus 1975a,b, Raibert 1977, Raibert & Wimberly 1984, Raibert 1986), as underlying mechanisms for adaptive control, and as substrates for motor learning in artificial intelligence (Michie & Chambers 1968, Selfridge et al 1985).

IMPLEMENTING CONTENT-ADDRESSABLE MEMORIES BY USING HASHING   Albus (1975a,b, 1981) proposed using a tabular representation based on Perceptrons (Rosenblatt 1961, Minsky & Papert 1969) to represent internal models of inverse transformations and also to encode trajectories that would achieve particular task goals. Inverse kinematic transformations would be represented by using table lookup to find joint angles corresponding to a desired hand position, as shown in Figure 3C. Solving the inverse dynamics would involve looking up joint torques on the basis of desired joint positions, velocities, and accelerations. A table entry is represented in a distributed fashion by a group of weights, and local generalization and smoothing are performed by sharing weights between neighboring entries. The weights in the table are updated on the basis of observed input and output pairs. The table size is reduced by randomly mapping the originally astronomically large table into a reasonably sized table. This scheme takes advantage of the fact that such a table would be used sparsely, in that only a small number of different types of movements would be made compared to all possible movements that the original table was capable of representing. Albus's approach is equivalent to a computer science technique known as hashing, and is one way to implement a content-addressable memory in software (Kohonen 1980). Miller (1986, Miller et al 1987) has used Albus's scheme to model the inverse kinematic transformation for a robot controlled by a TV camera and to model the inverse dynamics of a robot. El-Zorkany et al (1985) have implemented Albus's scheme for improved trajectory control of a robot.

LOCAL LINEAR MODELS   Pellionisz & Llinas represent a transformation with many local linear models in their tensor-based scheme (Pellionisz & Llinas 1979, 1980, 1982, 1985). A table-like representation is used to store the different local linear models corresponding to different points in the generally nonlinear transformation. Learning is performed by an iterative procedure to find eigenvectors and eigenvalues of the matrix representing the local linear approximation (Pellionisz & Llinas 1985). This approach should work well for transformations that are linear for a given arm configuration, such as the transformation from desired hand velocities to joint velocities (Pellionisz & Llinas 1980) and the transformation from desired joint torques to muscle forces (Gielen & van Zuylen 1986). The approach may work less well for transformations that are nonlinear for all input variables, such as the transformation from desired hand position to joint angles, or the inverse dynamics transformation from desired joint positions, velocities, and accelerations to joint torques. In these cases the approach faces a worse problem of storage size than if the transformation is represented directly in a table. In general, the same number of indices must be used to select the appropriate local transformation matrix as would be needed to select the appropriate output, and the transformation matrices (whose size is the number of inputs multiplied by the number of outputs) must be stored rather than just the outputs. Arbib & Amari (1985) provide a stimulating analysis of the Pellionisz & Llinas approach.

MULTILAYER NEURAL NETWORK APPROACHES   Initial steps are being made toward using multilayer neural network models to address the issues of representing transformations such as inverse kinematic and dynamic transformations (Hinton 1984, Kawato et al 1987, Kuperstein 1987, Jordan 1986). Although successful network learning procedures have been demonstrated, it is not yet clear how well they will scale up to realistically sized problems, how effectively they use data to learn, and how quickly they can learn (Hinton 1986).

BALANCING AN INVERTED PENDULUM    Dynamic balance is an important issue in legged locomotion, and provides a challenging continuous control problem with which to evaluate motor learning schemes. Many studies have dealt with the problem of balancing an inverted pendulum on a movable cart, much as a vertical broomstick can be balanced on the open palm of one's hand by moving the hand (Widrow & Smith 1964, Michie & Chambers 1968, Barto et al 1983, Selfridge et al 1985). A feedback control law was tabulated so that the appropriate movements of the cart are made in response to measured pendulum angle and angular velocity and cart position and velocity. The learning problem in these studies is filling and updating the control law table. Unlike tabulating arm kinematics and dynamics, the appropriate control law cannot be simply observed from measured commands and observed performance. The control law can be specified externally from a teacher (Widrow & Smith 1964) or it may be learned on the basis of control failures occurring when the inverted pendulum falls or the cart moves too far to the left or right (Michie & Chambers 1968, Barto et al 1983, Selfridge et al 1985).

USING TABLES TO CONTROL LOCOMOTION    Raibert (1986, Raibert & Wimberly 1984) used a table to control forward running speed while maintaining balance of a one-legged hopping machine. The table represented the dynamics of the hopper during its stance phase and was used to choose the appropriate foot placement at the beginning of the stance phase to achieve a desired forward velocity, body angle, and body angular rate. The size of the table used was reduced (a) by not using desired performance as indices into the table and looking up the appropriate command directly, but instead by using the table as an aid to searching for the command that would lead to the desired performance, (b) by ignoring some state variables that vary cyclically during the locomotion cycle, and (c) by using polynomials to approximate the contents of the table. The data used in the table were generated by simulating a model of the hopper, but presumably they could also have been generated from actual observations of the dynamics of the hopper. Miura & Shimoyama (1980) have explored the use of tables to control a biped walking machine.

ASSOCIATIVE CONTENT-ADDRESSABLE MEMORIES    Recent developments of massively parallel computers such as the Connection Machine (Hillis 1985) raise the possibility of modeling associative content addressable memories (Kohonen 1980). These types of memories obviate the need for complex indexing schemes to find relevant data points in tables. Current models of the Connection Machine have 64,000 processors and the possibility of simulating many more. Each previously observed data point is assigned to a processor, and the multiple processors search through the data in parallel to find data points close to a requested set of indices. Each processor computes the contribution of its data points to a requested prediction on the basis of distance between the sets of indices. The Connection Machine has a global addition operation to sum the many contributions. It is no longer necessary to allocate storage for all possible combinations of indices, as only data points actually experienced are stored. Having more indices in a data point only adds linearly to storage space size rather than exponentially, since only the size of the data points is increased, not their number. Eventually mechanisms to coalesce and "forget" data points will be needed when the processors or memory is fully utilized. A simple solution to this problem is to store the time the data point was collected as part of the data point and discard data points on the basis of age. A more complex scheme could take into account the density of nearby data points, and combine redundant data points. It is also no longer necessary to organize storage for a particular set of indices. Different sets of indices can be selected by the processors from the same data point, allowing both forward and inverse transformations to be done with the same table.

## Hybrid Approaches

Raibert took advantage of the structure of rigid body dynamics to reduce the size of the table needed to represent the inverse dynamics of a robot arm (Raibert 1977, 1978, Raibert & Horn 1978). The inertial forces are proportional to joint accelerations (note that $\ddot{\theta}_1$ and $\ddot{\theta}_2$ appear linearly in Eqs. 2 and 3), and the Coriolis and centripetal forces are quadratic in joint velocities (joint velocities only appear as the products $\dot{\theta}_1\dot{\theta}_2$, $\dot{\theta}_1^2$, and $\dot{\theta}_2^2$ in Eqs. 2 and 3), and therefore both joint accelerations and joint velocities can be removed as indices of the table at the expense of some additional computation. Local generalization and smoothing in the table was performed by using large table entries and interpolating across neighboring entries, and the table entries were updated by using a regression technique.

## Structured Approaches

KINEMATIC MODEL BUILDING    In robotics, kinematic calibration has been based on estimating parameters of model structures (see Whitney et al 1986, An et al 1988 for examples of such work and further references).

Typically, a model structure is chosen based on the assumptions of rigid links with joint rotation axes fixed relative to the links and linear joint sensors with the correct scale factor. Estimated parameters include link lengths, joint sensor offsets, relative orientations of consecutive axes, compliance in the transmission between the joint position sensor and the rigid link, backlash, and gear transmission errors. The information used to calibrate the robot includes the joint sensor readings and measurements of the positions of various parts of the arm or just the tip of the arm in some external coordinate frame. Vision-based measurement systems have been used as an external measurement source. Often, the parameter estimation procedures are quite complex due to the nonlinear nature of the problem and the number of parameters.

Biological limbs may present a more difficult calibration problem. Joints often have instantaneous axes of rotation that move relative to the adjacent links as the joint surfaces roll or slide on each other (Alexander 1981). Calibrating joint angle or arm configuration measurements based on joint and muscle proprioceptive signals may be a problem (Burgess et al 1982, Mathews 1982). Given the complex model structure required to capture these effects, it may be appropriate to explore table-based kinematic calibration methods (Kuperstein 1987, Jordan 1986).

DYNAMIC MODEL BUILDING: ESTIMATING LOAD PARAMETERS   Methods to estimate the inertial parameters (mass, location of the center of mass, and the moments of inertia) of a rigid body load during general movement have been proposed (An et al 1988, Atkeson et al 1986, Olsen & Bekey 1985, Mukerjee 1984, Mukerjee & Ballard 1985). The rigid body dynamics model structure is linear in the unknown inertial parameters, as in Eqs. 2 and 3, making the parameters straightforward to estimate with regression techniques. Information needed to estimate the inertial parameters include forces and torques exerted by the arm on the load, and arm position, velocity, and acceleration. The estimation equations can be integrated to eliminate the need to measure or calculate arm acceleration. Estimation algorithms have been implemented on robots equipped with a wrist force-torque sensor (Atkeson et al 1986). Good estimates were obtained for load mass and center of mass, and the forces and torques due to the movement of the load could be predicted accurately. Load moments of inertia were more difficult to estimate due to their small effect on the load forces and torques. Because their effect is so small, however, it is not important to estimate these parameters accurately in order to predict the load forces and torques.

DYNAMIC MODEL BUILDING: ESTIMATING ARM PARAMETERS   A method very similar to load inertial parameter estimation can be used to estimate the inertial parameters of the segments of the arm (An et al 1988, Atkeson et al 1986, Mukerjee 1984, Mukerjee & Ballard 1985, Olsen & Bekey 1985, Khosla 1986). Essentially, each segment of the arm is treated as a load on the more proximal segments of the arm. The unknown inertial parameters of these arm links are mass, location of the center of mass, and moments of inertia. Again, since the rigid body dynamics model structure is linear in the unknown inertial parameters, these parameters can be estimated with regression techniques. The information used to estimate the parameters includes the torques at the joints of the arm, and the position, velocity, and acceleration of the arm (the position, velocity, and acceleration of each joint). The estimation equations can be integrated to eliminate the need to estimate arm or joint accelerations. The estimation algorithms have been implemented on a robot, and a good match was obtained between predicted joint torques from the estimated model and actual joint torques (Atkeson et al 1986). The model has also been used for feedforward and computed torque control of the robot (An et al 1987).

## CORRECTING PERFORMANCE ERRORS

Previous sections have dealt with how the motor apparatus could build models of both kinematic and dynamic transformations. Inevitably, the inverse transformations used to control the motor system will not be entirely correct. Errors in the inverse transformation will lead to incorrect commands, and performance errors. This section describes how performance errors can be used to improve future commands.

### A Kinematic Example

Correcting kinematic errors is used as an initial example. Consider a vision system observing our idealized two-joint arm. A visual target is selected and represented in visual coordinates. The inverse kinematic transformation is used to compute the joint angles that will place the hand at the visual target. The arm is moved to those exact joint angles, but due to errors in the inverse kinematic transformation the hand is not exactly at the visual target. The vision system measures this performance error in visual coordinates rather than motor coordinates, and so feedback cannot be used directly to correct the error unless the transformation from visual to motor coordinates is known. The brain must transform the hand position error in visual coordinates (a performance error) into corrections to the joint angles (a command correction).

If a tabular representation of the inverse kinematic transformation is used, then performance errors can be corrected by simply adding the new data point of joint angles and observed hand position to the table. The new data point should slightly modify the nearby mapping so that the predicted joint angles for the desired hand position will be changed in the correct direction. The performance error can be reduced without any explicit error detection or correction (Raibert 1978). There are several possible problems with the above scenario. If the error is too large, compared to the generalization generated by the smoothing done by the table, the original transformation of desired hand position to joint angles may not be modified at all. The same error will be generated again. This problem is particularly acute when tables are initialized and have little or no correct data. The resolution of the table may not be adequate to represent the inverse transformation accurately. In addition, the smoothing and temporal averaging done by the table may reduce the effect of new data and slow down the reduction of the error.

If a structured model is used to represent the inverse kinematic transformation, new parameter estimates can be made to improve the model for arm configurations similar to this one. The approach of reestimating parameters to correct for errors was proposed by Mukerjee (1984) for improving a model of the inverse dynamics transformation during practice of a movement. The use of data from a small range of inputs or states, however, can cause parameter estimation procedures to distort the parameter estimates in order to compensate for structural modeling errors. This can cause predictions for other ranges of data not used in the estimation to deteriorate. Normally, a wide range of inputs and states should be used for parameter estimation of structured models to achieve both a good predictive ability over a corresponding range of outputs and the ability to generalize between quite different inputs.

In order to use feedback directly to correct the error, the measured error (in visual coordinates) must be transformed into the command coordinate system (joint angles). The visual error can be transformed to joint angles by using the same inverse kinematic transformation we used to predict the initial set of joint angles. Although the represented inverse kinematic transformation is inaccurate, it is still the best guess as to how to interpret the performance error.

This problem can be mathematically formulated: Initially the desired hand position $\mathbf{x}_d = (x_d, y_d)$ is transformed into joint angles $\boldsymbol{\theta} = (\theta_1, \theta_2)$ by the inverse kinematic transformation $IK()$ given by Eqs. 4 and 5.

$$\boldsymbol{\theta} = IK(\mathbf{x}_d). \qquad\qquad 6.$$

Unfortunately, when the arm is moved to these joint angles the actual hand position $\mathbf{x} = (x, y)$ as measured by the vision system is different from the desired hand position due to inaccuracies in the inverse kinematic transformation $IK()$. One way to reduce the hand position error is to correct the commanded joint angles by mapping the perceived hand position error into joint coordinates. The hand position error is mapped by using the inverse kinematic transformation to transform the actual and desired hand positions to joint coordinates and then subtract them to estimate the error in joint coordinates:

$$\text{estimated error in joint coordinates} = IK(\mathbf{x}) - IK(\mathbf{x}_d). \qquad 7.$$

This estimated error is then used to correct the commanded joint angles by simply subtracting the error:

$$\theta_{\text{corrected}} = \theta - [IK(\mathbf{x}) - IK(\mathbf{x}_d)]. \qquad 8.$$

Since the inverse kinematic transformation is not entirely accurate, the error will not be reduced completely. The same procedure can be repeated to reduce the error further. The important point is that in order to use feedback to correct errors, perceived errors must be transformed into the appropriate command coordinate system, and the same inverse transformations that were used to generate the initial command can be used to correct that command.

## Learning from Practicing a Movement

The problem of transforming performance errors into command corrections is present when a particular movement is learned by practicing that movement. Performance errors in terms of incorrect positions, velocities, and accelerations of the arm must be used to correct commands that are in terms of joint torques or muscle activations. The dynamic performance errors can be mapped to command corrections by using an internal model of the inverse dynamics of the motor apparatus, much as an internal model of the inverse kinematics was used to correct positioning errors.

In robotics, recent work in a number of laboratories has focused on how to refine feedforward commands for repetitive movements on the basis of previous movement errors. Work on repeated trajectory learning includes that of Arimoto et al (1984, 1985), Casalino & Gambardella (1986), Craig (1984), Furuta & Yamakita (1986), Hara et al (1985), Harokopos (1986), Mita & Kato (1985), Morita (1986), Togai & Yamano (1986), Uchiyama (1978), Wang (1984), and Wang & Horowitz (1985). These papers discuss only linear learning operators and emphasize the stability of the proposed algorithms. Little work has emphasized performance, i.e. the convergence rate of the algorithm. Simulations of several of these algorithms have revealed very slow convergence and large sensitivity to disturbances and sensor and actuator noise (C. G. Atkeson, unpublished results).

In a paper by Atkeson & McIntyre (1986a,b), an explicit model of the nonlinear robot inverse dynamics was used to transform performance errors into command corrections. This technique enabled a robot to improve its ability to follow a desired trajectory by practicing a movement. Figure 4 shows the performance of such a learning scheme when implemented on a three-joint direct drive robot arm. Figure 4*A* shows the initial feedforward torques applied to the robot for each of the three joints.

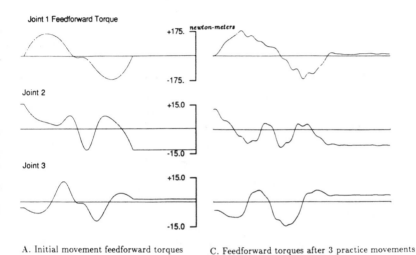

A. Initial movement feedforward torques          C. Feedforward torques after 3 practice movements

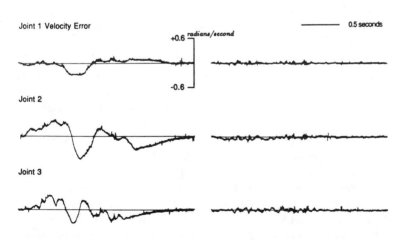

B. Initial movement velocity errors          D. Velocity errors after 3 practice movements

*Figure 4*   Learning from practice in a three-joint robot arm.

These torques were calculated by applying a model of the inverse dynamics of the robot $ID(\ )$ to the desired trajectory joint positions $\theta_d(t)$, joint velocities $\dot{\theta}_d(t)$, and joint accelerations $\ddot{\theta}_d(t)$:

$$\tau_{ff}(t) = ID[\theta_d(t), \dot{\theta}_d(t), \ddot{\theta}_d(t)] \qquad\qquad 9.$$

In addition to the feedforward torques, the robot controller used feedback control to generate torques ($\tau_{fb}$) to compensate for observed trajectory errors. The total commanded joint torques were the sum of the feedforward and feedback torques:

$$\tau(t) = \tau_{ff}(t) + \tau_{fb}(t). \qquad\qquad 10.$$

Figure 4*B* shows the resulting velocity errors for each joint in attempting to follow the desired trajectory.

The model of the robot inverse dynamics was used to transform the trajectory performance errors into corrections to the torque commands to the robot, just as kinematic errors were corrected by using the inverse kinematic model of the arm:

$$\tau_{ff_{corrected}}(t) = \tau(t) - \{ID[\theta(t), \dot{\theta}(t), \ddot{\theta}(t)] - ID[\theta_d(t), \dot{\theta}_d(t)\ddot{\theta}_d(t)]\}. \qquad 11.$$

Figure 4*C* shows the corrected feedforward torques for each of the joints after three practice movements, and Figure 4*D* shows the greatly reduced velocity errors that result. The robot learned to improve its trajectory following performance on this trajectory by practicing a movement and observing and correcting its own errors. One drawback of this approach is that it only produces the appropriate command for a single trajectory. How to modify that command for similar trajectories is currently under investigation.

Motor control schemes that do not include an internal model of the dynamics of the motor apparatus, such as the equilibrium trajectory approach (Berkinblit et al 1986, Hogan et al 1987), do not support efficient movement refinement during practice (C. G. Atkeson, unpublished results). There is a great deal of evidence from research on adaptive human manual control that performance improvements in compensatory tracking tasks are due to improvements of an internal model of the dynamics of the controlled system (Sheridan & Ferrell 1974). Additionally, a source of great frustration to designers of motor psychophysics experiments is their subjects' ability to build internal predictive models of target movement in pursuit tracking tasks (see Rouse 1980 for further references). Recent work in oculomotor control has suggested that an internal model of the motor apparatus is used to maintain accurate performance (Optican & Miles 1985).

# SUMMARY

In this review I have discussed how the form of representation used in internal models of the motor apparatus affects how and what a system can learn. Tabular models and structured models have benefits and drawbacks. Structured models incorporate knowledge of the structure of the controlled motor apparatus. If that knowledge is correct, or close to the actual system structure, the structured models will support global generalization and rapid, efficient learning. Tabular models can play an important role in learning to control systems when either the system structure is not known or only known approximately. Tabular models are general and flexible. Techniques for combining these different representations to attain the benefits of both are currently under investigation.

In the control of multijoint systems such as the human arm, internal models of the motor apparatus are necessary to interpret performance errors. In the study of movements restricted to one joint, the problem of interpreting performance errors is greatly simplified and often overlooked, as performance errors can usually be related to command corrections by a single gain. When multijoint movements of the same motor systems are examined, however, the complex nature of the control and coordination problems faced by the nervous system become evident, as well as the sophistication of the brain's solutions to these problems. Recent progress in the understanding of adaptive control of eye movements provides a good example of this (Berthoz & Melvill-Jones 1985).

Experimental studies of the psychophysics of motor learning can play an important role in bridging the gap between computational theories of how abstract motor systems might learn and physiological exploration of how actual nervous systems implement learning. Quantitative analyses of the patterns of motor learning of biological systems may help distinguish alternative hypotheses about the representations used for motor control and learning. What a system can and cannot learn, the amount of generalization, and the rate of learning give clues as to the underlying performance architecture. It is also important to know the actual performance level of the motor system (Loeb 1983). Different proposed control strategies will be able to attain different performance levels, and the use of simplifying control strategies may be evident in the control and learning performance of motor systems.

ACKNOWLEDGMENTS

I thank the many colleagues who commented on drafts of this article. This article describes research done at the Center for Biological Information

Processing (Whitaker College) and the Artificial Intelligence Laboratory at the Massachusetts Institute of Technology. Support for this research is provided by a grant from the Whitaker Health Sciences Fund. Support for the A. I. Laboratory's research is provided in part by the Office of Naval Research University Research Initiative Program under Office of Naval Research contract N00014-86-K-0685, and the Advanced Research Projects Agency of the Department of Defense under Office of Naval Research contract N00014-85-K-0124.

*Literature Cited*

Adams, J. A. 1984. Learning of movement sequences. *Psychol. Bull.* 96: 3–28

Adams, J. A. 1987. Historical review and appraisal of research on the learning, retention, and transfer of human motor skills. *Psychol. Bull.* 101: 41–74

Albus, J. S. 1971. A theory of cerebellar function. *Math. Biosci.* 10: 25–61

Albus, J. S. 1975a. A new approach of manipulator control: The cerebellar model articulation controller (CMAC). *J. Dynamic Syst. Measure. Control* 97: 220–27

Albus, J. S. 1975b. Data storage in the cerebellar model articulation controller (CMAC). *J. Dynamic Syst. Measure. Control* 97: 228–33

Albus, J. S. 1981. *Brains, Behavior, and Robotics.* Peterborough, NH: Byte

Alexander, R. M. 1981. Mechanics of skeleton and tendons. In *Handbook of Physiology, Section 1: The Nervous System, Volume 2, Motor Control, Part 2,* ed. V. B. Brooks, pp. 17–42. Bethesda, MD: Am. Physiol. Soc.

An, C. H., Atkeson, C. G., Griffiths, J. D., Hollerbach, J. M. 1987. Experimental evaluation of feedforward and computed torque control. *IEEE Conf. on Robot. Automat., Raleigh, NC, March 30–April 3,* pp. 165–68

An, C. H., Atkeson, C. G., Hollerbach, J. M. 1988. *Model-Based Control of a Robot Manipulator.* Cambridge, Mass: MIT Press

Annett, J. 1985. Motor learning: A review. In *Motor Behavior: Programming, Control, and Acquisition,* ed. H. Heuer, U. Kleinbeck, K. H. Schmidt, pp. 189–212. New York: Springer-Verlag

Arbib, M. A., Amari, S. I. 1985. Sensorimotor transformations in the brain (with a critique of the tensor theory of cerebellum). *J. Theoret. Biol.* 112: 123–55

Arimoto, S., Kawamura, S., Miyazaki, F. 1984. Bettering operation of robots by learning. *J. Robot. Syst.* 1: 123–40

Arimoto, S., Kawamura, S., Miyazaki, F., Tamaki, S. 1985. Learning control theory for dynamical systems. *Proc. 24th Conf. on Decision Control, Fort Lauderdale, Fla., Dec. 11–13,* pp. 1375–80

Atkeson, C. G., An, C. H., Hollerbach, J. M. 1986. Estimation of inertial parameters of manipulator loads and links. *Int. J. Robot. Res.* 5: 101–19

Atkeson, C. G., McIntyre, J. 1986a. Robot trajectory learning through practice. *IEEE Conf. on Robot. Automat., San Francisco, Calif., April 7–10,* pp. 1737–42

Atkeson, C. G., McIntyre, J. 1986b. Applications of adaptive feedforward control in robotics. *Proc. 2nd IFAC Worskhop on Adaptive Systems in Control Signal Process., Lund, Sweden, July 1–3,* pp. 137–42

Barto, A. G., Sutton, R. S., Anderson, C. W. 1983. Neuronlike adaptive elements that can solve difficult learning control problems. *IEEE Trans. on Systems, Man, Cybernet.* SMC-13: 834–46

Berkinblit, M. B., Feldman, A. G., Fukson, O. I. 1986. Adaptability of innate motor patterns and motor control mechanisms. *Behav. Brain Sci.* 9: 585–638

Berthoz, A., Melvill Jones, G. 1985. *Adaptive Mechanisms in Gaze Control.* New York, NY: Elsevier

Bizzi, E., Dev, P., Morasso, P., Polit, A. 1978. Effect of load disturbances during centrally initiated movements. *J. Neurophysiol.* 41: 542–56

Burgess, P. R., Wei, J. Y., Clark, F. J., Simon, J. 1982. Signaling of kinesthetic information by peripheral sensory receptors. *Ann. Rev. Neurosci.* 5: 189–218

Casalino, G., Gambardella, L. 1986. Learning of movements in robotic manipulators. See Atkeson & McIntyre 1986a, pp. 572–78

Craig, J. J. 1984. Adaptive control of manipulators through repeated trials. *Proc. Am. Control Conf., San Diego, June 6–8,* pp. 1566–74

El-Zorkany, H. I., Liscano, R., Tondue, B.

1985. A sensor based approach for robot programming. *Proc. SPIE Conf. on Intelligent Robots and Comput. Vision*, pp. 289–97

Ersu, E., Tolle, H. 1984. A new concept for learning control inspired by brain theory. *IFAC 9th World Congr.*, *Budapest, Hungary, July 2–6*, pp. 1039–44

Feldman, A. G. 1986. Once more on the equilibrium-point hypothesis ($\lambda$ model) for motor control. *J. Motor Behav.* 18: 17–54

Fujita, M. 1982a. Adaptive filter model of the cerebellum. *Biol. Cybernet.* 45: 195–206

Fujita, M. 1982b. Simulation of adaptive modification of the vestibulo-ocular reflex with an adaptive filter model of the cerebellum. *Biol. Cybernet.* 45: 207–14

Furuta, K., Yamakita, M. 1986. Iterative generation of optimal input of a manipulator. See Atkeson & McIntyre 1986a, pp. 579–84

Georgopoulos, A. P. 1986. On reaching. *Ann. Rev. Neurosci.* 9: 147–70

Gielen, C. C. A. M., van Zuylen, E. J. 1986. Coordination of arm muscles during flexion and supination: Application of the tensor analysis approach. *Neurosci.* 17: 527–39

Grossberg, S. 1969. On learning of spatio-temporal patterns by networks with ordered sensory and motor components. I: Excitatory components of the cerebellum. *Studies Applied Math.* 48: 105–32

Grossberg, S., Kuperstein, M. 1986. *Neural dynamics of adaptive sensory-motor control*. Amsterdam: North-Holland

Hara, S., Omata, T., Nakano, M. 1985. Synthesis of repetitive control systems and its applcation. See Arimoto et al 1985, pp. 1387–92

Harokopos, E. G. 1986. Optimal learning control of mechanical manipulators in repetitive motions. See Atkeson & McIntyre 1986a, pp. 396–401

Hildreth, E. C., Hollerbach, J. M. 1985. The computational approach to vision and motor control. AI Memo 846, MIT A.I. Lab., Cambridge, Mass.

Hildreth, E. C., Koch, C. 1987. The analysis of visual motion: From computational theory to neuronal mechanisms. *Ann. Rev. Neurosci.* 10: 477–533

Hillis, W. D. 1985. *The Connection Machine.* Cambridge: MIT Press

Hinton, G. E. 1984. Parallel computations for controlling an arm. *J. Motor Behav.* 16: 171–94

Hinton, G. E. 1986. Learning in massively parallel nets. *AAAI-86: 5th Natl. Conf. on Artificial Intelligence, Philadelphia, August 11–15*, p. 1149

Hollerbach, J. M. 1982. Computers, brains, and the control of movement. *Trends Neurosci.* 5: 189–92

Hollerbach, J. M., Atkeson, C. G. 1987. Deducing planning variables from experimental arm trajectories: Pitfalls and possibilities. *Biol. Cybernet.* 56: 279–92

Hogan, N., Bizzi, E., Mussa-Ivaldi, F. A., Flash, T. 1987. Controlling multi-joint motor behavior. *Exercise Sport Sci. Rev.* 15: 153–90

Houk, J. C., Rymer, W. Z. 1981. Neural control of muscle length and tension. See Alexander 1981, pp. 257–324

Jordan, M. I. 1986. Serial order: A parallel, distributed processing approach. Rep. No. 8604, Inst. Cognitive Sci., Univ. Calif., San Diego

Kawato, M., Furukawa, K., Suzuki, R. 1987. A hierarchical neural-network model for control and learning of voluntary movement. *Biol. Cybern.* 57: 169–85

Khosla, P. K. 1986. Estimation of robot dynamics parameters: Theory and application. *2nd Int. IASTED Conf. on Appl. Control and Identification, Los Angeles, Dec.*

Knudsen, E. I., du Lac, S., Esterly, S. D. 1987. Computational maps in the brain. *Ann. Rev. Neurosci.* 10: 41–65

Kohonen, T. 1980. *Content-Addressable Memories.* New York: Springer-Verlag

Kuperstein, M. 1987. Adaptive visual-motor coordination in multijoint robots using parallel architecture. See An et al 1987, pp. 1595–1602

Loeb, G. E. 1983. Finding common ground between robotics and physiology. *Trends Neurosci.* 6: 203–4

Marr, D. 1969. A theory of cerebellar cortex. *J. Physiol. London* 202: 437–70

Marr, D. 1982. *Vision.* San Francisco: Freeman

Matthews, P. B. C. 1982. Where does Sherrington's "muscular sense" originate? Muscles, joints, corollary discharges? *Ann. Rev. Neurosci.* 5: 189–218

Michie, D., Chambers, R. A. 1968. Boxes: An experiment in adaptive control. In *Machine Intelligence 2*, ed. E. Dale, D. Michie, pp. 137–52. New York: Elsevier

Miller, W. T. 1986. A nonlinear learning controller for robotic manipulators. *SPIE 5th Conf. on Intelligent Robots and Computer Vision, Cambridge, Mass., Oct. 28–31*

Miller, W. T., Glanz, F. H., Kraft, L. G. 1987. Application of a general learning algorithm to the control of robotic manipulators. *Int. J. Robot. Res.* 6: 84–98

Minsky, M., Papert, S. 1969. *Perceptrons: An Introduction to Computational Geometry.* Cambridge, MIT Press

Mita, T., Kato, E. 1985. Iterative control

and its application to motion control of robot arm—a direct approach to servo-problems. See Arimoto et al 1985, pp. 1393–98

Miura, H., Shimoyama, I. 1980. Computer control of an unstable mechanism (in Japanese). *J. Fac. Eng. Univ. Tokyo* 17: 12–13

Morita, A. 1986. *A study of learning controllers for robot manipulators with sparse data.* MS thesis, Mechanical Engineering, Cambridge, MIT

Mukerjee, A. 1984. Adaptation in biological sensory-motor systems: A model for robotic control. Vol. 521, *Proc., SPIE Conf. on Intelligent Robots and Computer Vision, SPIE, Cambridge, November*

Mukerjee, A., Ballard, D. H. 1985. Self-calibration in robot manipulators. *Proc. IEEE Conf. Robot. Automat., St. Louis, Mar. 25–28*, pp. 1050–57

Olsen, H. B., Bekey, G. A. 1985. Identification of parameters in models of robots with rotary joints. See Mukerjee & Ballard 1985, pp. 1045–50

Optican, L. M., Miles, F. A. 1985. Visually induced adaptive changes in primate saccadic oculomotor control signals. *J. Neurophysiol.* 54: 940–58

Pellionisz, A., Llinas, R. 1979. Brain modeling by tensor network theory and computer simulation. The cerebellum: Distributed processor for predictive coordination. *Neuroscience* 4: 323–48

Pellionisz, A., Llinas, R. 1980. Tensorial approach to the geometry of brain function: Cerebellar coordination via a metric tensor. *Neuroscience* 5: 1125–36

Pellionisz, A., Llinas, R. 1982. Space-time representation in the brain: The cerebellum as a predictive space-time metric tensor. *Neuroscience* 7: 2949–70

Pellionisz, A., Llinas, R. 1985. Tensor network theory of the metaorganization of functional geometries in the central nervous system. *Neuroscience* 16: 245–73

Poggio, G. F., Poggio, T. 1984. The analysis of stereopsis. *Ann. Rev. Neurosci.* 7: 379–412

Raibert, M. H. 1977. Analytical equations vs. table look-up for manipulation: A unifying concept. *Proc. IEEE Conf. Decision Control, New Orleans, Dec.*, pp. 576–79

Raibert, M. H. 1978. A model for sensorimotor control and learning. *Biol. Cybernet.* 29: 29–36

Raibert, M. H. 1986. *Legged Robots That Balance.* Cambridge: MIT Press

Raibert, M. H., Horn, B. K. P. 1978. Manipulator control using the Configuration Space Method. *Indust. Robot* 5: 69–73

Raibert, M. H., Wimberly, F. C. 1984. Tabu-

lar control of balance in a dynamic legged system. *IEEE Trans. on Systems, Man, Cybernet.* SMC-14: 2: 334–39

Robinson, D. A. 1981. The use of control systems analysis in the neurophysiology of eye movements. *Ann. Rev. Neurosci.* 4: 463–503

Rosenblatt, F. 1961. *Principles of Neurodynamics: Perceptrons and the Theory of Brain Mechanisms.* Washington, DC: Spartan

Rouse, W. B. 1980. *Systems Engineering Models of Human-Machine Interaction.* New York: North-Holland

Rumelhart, D. E., McClelland, J. L., PDP Research Group. 1986. *Parallel Distributed Processing: Explorations in the Microstructure of Cognition*, Vol. 1, *Foundations.* Cambridge: MIT Press

Salmoni, A. W., Schmidt, R. A., Walter, C. B. 1984. Knowledge of results and motor learning: A review and critical reappraisal. *Psychol. Bull.* 95: 355–86

Saltzman, E. 1979. Levels of sensorimotor representation. *J. Math. Psychol.* 20: 91–163

Schmidt, R. A. 1982. *Motor Control and Learning.* Champaign, Ill: Human Kinetics Publ.

Selfridge, O. G., Sutton, R. S., Barto, A. G. 1985. Training and tracking in robotics. *Proc. IJCAI, Los Angeles, Aug. 18–23*, pp. 670–72

Sheridan, T. B., Ferrell, W. R. 1974. *Man-Machine Systems.* Cambridge: MIT Press

Simons, J., Van Brussel, H., De Schotter, J., Verhaert, J. 1982. A self-learning automaton with variable resolution for high precision assembly by industrial robots. *IEEE Trans. on Automatic Control* AC-27: 1109–13

Smyth, M. M., Wing, A. W., eds. 1984. *The Psychology of Human Movement.* New York: Academic

Soechting, J. F., Ross, B. 1984. Psychophysical determination of coordinate representation of human arm orientation. *Neuroscience* 13: 595–604

Togai, M., Yamano, O. 1986. Learning control and its optimality: Analysis and its application to controlling industrial robots. See Atkeson & McIntyre 1986a, pp. 248–53

Uchiyama, M. 1978. Formation of high-speed motion pattern of a mechanical arm by trial. *Trans Soc. Instrum. Control Eng.* (*Jpn.*) 19: 706–12

Ullman, S. 1986. Artificial intelligence and the brain: Computational studies of the visual system. *Ann. Rev. Neurosci.* 9: 1–26

Wang, S. H. 1984. Computed reference error adjustment technique (CREATE) for the control of robot manipulators. *22nd Ann.*

Allerton Conf. on Commun., Control, Computing, Oct.

Wang, S. H., Horowitz, I. 1985. CREATE—a new adaptive technique. Proc. of the 19th Ann. Conf. on Inform. Sci. Systems, March

Widrow, B., Smith, F. W. 1964. Pattern recognizing control systems. In Computer and Information Sciences, ed. J. T. Tou, R. H. Wilcox. Washington, DC: Clever Hume Press

Whitney, D. E., Lozinski, C. A., Rourke, J. M. 1986. Industrial robot forward calibration method and results. J. Dyn. Syst. Meas. Control 108: 1–8

*Ann. Rev. Neurosci. 1989. 12 : 185–204*
*Copyright © 1989 by Annual Reviews Inc. All rights reserved*

# EMERGING PRINCIPLES GOVERNING THE OPERATION OF NEURAL NETWORKS

*Peter A. Getting*

Department of Physiology and Biophysics, The University of Iowa, Iowa City, Iowa 52242

## INTRODUCTION

A basic tenet of neuroscience is that the ability of the brain to produce complex behaviors such as sensory perception or motor control arises from the interconnection of neurons into networks or circuits. Finding out how neural networks are organized and understanding what computational principles underlie their operation remain challenges to modern neuroscience. Advances in anatomical, biochemical, electrophysiological, and computational techniques have provided the tools to begin uncovering concepts underlying neural network function. This review is a summary of insights into the organization and operation of neural circuits acquired through application of these techniques to invertebrate and vertebrate systems. This review is not a survey of the vast variety of known neural networks but rather concentrates on emerging concepts applicable to networks in general. The first section provides a brief historical perspective. The second and third sections summarize two emerging concepts about the operation and modulation of neural circuits. The first concept is that the operation of a neural network depends upon interactions among multiple nonlinear processes at the cellular, synaptic, and network levels. The second concept is that modulation of these underlying processes can alter network operation. The final section suggests potentially profitable research directions for the future.

## HISTORICAL PERSPECTIVE

By the late 1960s, the basic principles of excitability and synaptic transmission were fairly well understood. Ideas about how central neurons and

185

synapses might operate were based upon a rather simple picture provided by the squid axon (Hodgkin & Huxley 1952), the neuromuscular junction (Katz 1966), and the spinal motor neuron (Eccles 1964). Based upon these views, the abilities of a network arose from the interconnection of simple elements into complex networks, thus, from connectivity emerged function. Neural networks were viewed largely as "hard wired" (Bentley & Konishi 1978) and could therefore be defined by their anatomical or monosynaptic connectivity. Activity coming into a network would be operated upon by that network in accordance with the pattern of synaptic connectivity, much like data fed to a digital computer is processed by a preset program. Over the time span it took for a network to process incoming signals and generate an output, circuitry was considered fixed, thus, the operation of a network did not involve the making or breaking of synaptic connections or altering cellular and synaptic properties.

The challenge of uncovering the secrets to brain function lay in the unravelling of neural connectivity. Toward this end, major experimental effort has been expended to identify relevant neurons involved in various behaviors and to characterize their interconnection. Early success with invertebrate systems (Zucker et al 1971) signaled the onslaught of neural network "cracking." These studies were approached with several expectations in mind: First, a knowledge of the connectivity would explain how neural networks operated. Second, it was hoped that for each function (e.g. visual processing or generation of rhythmic motor patterns) we would find only a limited number of ways to implement that function in neural circuitry. Third, it was hoped that circuitry would be conserved, thus, similar functions might be subserved by similar neural networks. After nearly two decades of neural circuit analysis, it is reasonable to ask how well these expectations have fared.

What we found should not have been unexpected but was nonetheless surprising. First, neural networks turned out to be extremely complex and diverse. Even networks underlying simple behaviors in the small neural systems of invertebrates are remarkably complicated (Selverston 1985). Networks that initially appeared relatively straightforward and simple (e.g. crayfish tailflip, Zucker et al 1971; *Tritonia* swimming, Willows 1973) turn out to involve multiple levels of feedforward and feedback pathways imbedded in complicated arrays of connections and cells (Krasne & Wine 1984, Getting 1983c). Second, networks subserving similar functions do not appear to be conserved. This point has become particularly clear in the study of rhythmic motor systems (Getting 1988). In general, these systems produce an alternating pattern of activity between antagonistic motor elements, yet the underlying networks for generating these patterns appear to be unique for each system. Networks with similar connectivity can produce dramatically different motor patterns and, conversely, similar motor patterns can be produced by dramatically different networks (Getting 1988). These observations illustrate a third general finding. Knowledge of connectivity alone is not sufficient to account for the operation and capabilities of neural networks. For example, detailed knowledge of cerebellar circuitry has not provided a clear understanding of its operation (Llinas 1981).

If a knowledge of connectivity is not enough, what does it take to understand how a neural network operates? A debate often arises about what is meant by the term "to understand" how a neural network operates, for understanding can occur at many levels. For example, one can understand how an internal combustion engine works from the principles of thermodynamics without a detailed knowledge of pistons, crankshafts, and fuel injectors. If one's car engine should stall, however, such an understanding will be of little value because the failure was probably not caused by a breakdown in the laws of thermodynamics but a failure in the implementation of these principles. To understand how nervous systems operate, and therefore how they might fail due to disease or injury, requires (a) a knowledge of the overriding principles of neural network organization and function (equivalent to the rules of thermodynamics), and (b) how those principles are implemented by cells and synapses (the "nuts and bolts" of network operation). In an effort to acquire these two levels of understanding, much of the experimental work on neural networks has been guided by a reductionist approach in which a system is successively pared down to its constituent pieces or "building blocks" with the hope of uncovering general principles and how they are implemented.

# THE BUILDING BLOCK BASIS FOR NETWORK OPERATION

A major contribution of the reductionist approach has been the delineation of properties crucial to the operation of neural networks. Although knowledge of connectivity is essential, network operation depends upon the "cooperative interaction" (Selverston et al 1983) among multiple network, synaptic, and cellular properties, many of which are inherently nonlinear. No longer can neural networks be viewed as the interconnection of many like elements by simple excitatory or inhibitory synapses. Neurons not only sum synaptic inputs but are endowed with a diverse set of intrinsic properties that allow them to generate complex activity patterns. Likewise synapses are not just excitatory or inhibitory but possess an equally diverse set of properties. The operation of a neural network must be considered as the parallel action of neurons or classes of neurons, each with potentially different input/output relationships and intrinsic capabilities interconnected by synapses with a host of complex properties. What are the cellular, synaptic, and network properties that constitute the building blocks of network operation, and how do they contribute to network operation? Table 1 summarizes a partial list of cellular, synaptic, and network properties important to neural network operation.

## Cellular Properties

The application of voltage clamp and pharmacological techniques to CNS neurons has uncovered a lengthy and diverse catalog of ionic conductances (Adams et al 1980, Llinas 1984, Jahnsen 1986). These individual conductances can be found mixed and matched in nearly any combination, but more importantly, each combination endows the host neuron with a different set of response properties. For the purposes of this review the natures of the individual conductances are not as important as the properties that they impart, for these response properties will have a direct impact on network operation.

Most CNS neurons fire spikes repetitively when depolarized. The relationship between firing frequency (F) and input (I), generally injected current, is expressed as an F-I plot. Two properties of the F-I plot are important. First is the threshold below which the cell does not fire. Second, above threshold, firing frequency is a monotonically increasing, but not necessarily linear, function of input. On theoretical grounds, nonlinear input/output relationships have been implicated as computationally important characteristics of neurons (McCulloch & Pitts 1943, Hopfield & Tank 1986). A seminal observation is that the F-I plots for different cell types are distinct. For example, the repetitive firing properties of tectal neurons (Lopez-Barnes & Llinas 1988), cerebellar Purkinje neurons (Llinas & Sugimori 1980a,b), hippocampal neurons (Kandel & Spencer 1961),

**Table 1**   Building blocks

| Cellular | Synaptic | Connectivity |
|---|---|---|
| Threshold | Sign | Mutual |
| F-I relationship | Strength | or recurrent inhibition |
| Spike frequency adapt. | Time course | Reciprocal |
| Post-burst hyperpol. | Transmission | or lateral inhibition |
| Delayed excitation | Electrical | Recurrent inhibition |
| Post-inhibitory rebound | Chemical | Recurrent cyclic inhibition |
| Plateau potentials | Release mechanism | Parallel excit./inhib. |
| Bursting | Graded | |
| Endogenous | Spike | |
| Conditional | Multicomponent PSP | |

thalamic neurons (Deschenes et al 1984), neocortical neurons (Connors & Gutnick 1984), brainstem bulbospinal neurons (Dekin & Getting 1987a), olivary neurons (Llinas & Yarom 1981, 1986), and spinal motoneurons (Schwindt & Crill 1984) all differ. Even within a restricted network subserving a single behavior, each interneuron type may display a different F-I relationship (Getting 1983a).

Input/output relationships are not fixed but may undergo a variety of modifications depending upon the recent firing history of the cell. Spike frequency adaptation is a decrease in firing rate during a maintained input. Its effect is to decrease the slope of the F-I plot in time. The mechanism of adaptation appears to be the activation of slow outward potassium currents (Brown & Adams 1979, Meech 1978, Partridge & Stevens 1976, Dekin & Getting 1987b). The degree and time course of adaptation can vary dramatically from one cell type to another and give rise to different temporal firing patterns to the same input (Hume & Getting 1982a). Following a period of activity, many neurons display a transient hyperpolarization and cessation of firing that may last from milliseconds to seconds. This post-burst hyperpolarization appears to be a manifestation of mechanisms similar to those responsible for spike frequency adaptation.

Upon depolarization, many neurons begin firing almost immediately in accord with the charging of their membrane capacitance. Other neurons show a prolonged delay between the onset of depolarization and the beginning of firing. This delay in firing has been termed *delayed excitation* and may range from hundreds of milliseconds in vertebrate neurons (Dekin & Getting 1984, 1987a) to several seconds in molluscan neurons (Byrne 1980a, Getting 1983b). The delay is caused by the activation of a transient potassium current called A-current (Connor & Stevens 1971). This current is ubiquitous (Rogawski 1985) but appears to be expressed as delayed excitation only in cells that are maintained within the proper voltage range for its activation (Dekin & Getting 1987a,b). Delayed excitation provides an intrinsic mechanism for producing long delays and has been implicated in the neural networks controlling inking behavior in *Aplysia* (Byrne 1980b) and swimming in *Tritonia* (Getting 1983b).

Following a hyperpolarization, membrane potential may rebound above resting level to produce a transient depolarization called post-inhibitory rebound (PIR). If sufficiently strong, PIR can lead to a burst of spikes. The mechanism for PIR is not well understood, but it has been implicated in the production of several rhythmic motor patterns (Mulloney & Selverston 1974, Satterlie 1985) and is found in numerous cell types such as tectal neurons (Lopez-Barneo & Llinas 1988), thalamic neurons (Deschenes et al 1984), and brainstem neurons (Johnson & Getting 1987).

Plateau potentials are expressed as two membrane potential states: a resting state and a depolarized state (Russell & Hartline 1978). Small or transient depolarizations can cause transition from the resting state to the depolarized state, where the potential may remain for considerable lengths of time (tens to hundreds of milliseconds) before it either spontaneously reverts or is actively converted by a short hyperpolarizing input back to the resting state. The ability to produce plateau potentials provides a mechanism for translating a transient input into sustained firing, or a prolonged burst. Such mechanisms have been implicated in the generation of rhythmic motor patterns (Dickinson & Nagy 1983, Arshavsky et al 1985), and are seen in the responses of olivary neurons (Llinas & Yarom 1981, 1986), cerebellar Purkinje neurons (Llinas & Sugimori 1980a,b), and spinal motoneurons (Hounsgaard et al 1986).

Many neurons are autoactive and fire continuously in the absence of synaptic input. Pacemaker neurons produce a continuous train of spikes at regular intervals. Other autoactive neurons produce patterned bursts. Endogenous bursting neurons fire bursts of spikes independent of synaptic activation (Alving 1968), while conditional bursters express bursting only upon appropriate synaptic or neurohumoral activation (Anderson & Barker 1981, Miller & Selverston 1982a). The ability to burst arises from particular combinations of intrinsic membrane currents (Adams 1985, Adams & Levitan 1985) and may be expressed in many different forms (Alving 1968, Marder & Eisen 1984, Flamm & Harris-Warrick 1986, Hatton 1984, Dekin et al 1985).

## Synaptic Properties

Synaptic properties also impact directly on network operation. Two properties of obvious importance are the sign (excitation or inhibition) and the strength of synaptic connections. Although determination of the sign of a connection may be relatively straightforward, assessment of synaptic strength is not always clear. Two measures of strength are (a) the amplitude of the post-synaptic potential (PSP), and (b) the degree to which activation of a particular synapse influences the activity of the post-synaptic cell. These two measures may not always covary. For example, an inhibitory synapse with a reversal potential close to rest potential may produce only a small IPSP, but the associated conductance change can have a powerful inhibitory effect by shunting excitatory currents. The relative placement of excitatory and inhibitory synapses on the dendritic structure, therefore, plays an important role in regulating integration (Rall 1981).

Temporal properties of synapses also play an important role in network operation. Individual PSPs may have dramatically different time courses, and thus operate over different time scales. For example, within the network of interneurons that generates the escape swimming motor program of *Tritonia*, the fastest and slowest PSPs differ by a factor of 30 (Getting 1981). Also included under temporal properties are characteristics such as facilitation, depression, and potentiation that modulate the strength of connections in a history dependent manner.

A single synaptic connection may mediate several actions, each with different time courses. For multicomponent synapses, an initial action (either excitation or inhibition) is followed by a second action of either the same or opposite sign (Kehoe 1972). Multicomponent synapses can have a number of interesting integrative properties, including maximal expression *after* the end of a presynaptic burst (Getting 1983a). Complex connections having three and even four different components have been observed (Hume & Getting 1982b).

Mechanisms of synaptic transmission fall into two broad categories— electrical and chemical. Within each category, however, a wide diversity of mechanisms have been described, including both rectifying and non-rectifying electronic synapses, as well as conductance-increase and con-ductance-decrease chemical connections. The mechanism of transmission influences not only the character of the individual synaptic event but also how the PSPs from various sources will interact. For example, non-rectifying electrotonic synapses can have profound effects upon integrative properties, making a network selectively responsive to distributed afferent input (Getting 1974). Conductance-increase chemical PSPs may shunt whereas conductance-decrease PSPs may potentiate other inputs (Rall 1981). The action of each connection, therefore, cannot be considered in isolation but must be integrated with the actions of all other active synapses.

Mechanisms of transmitter release also influence the nature of the infor-mation transmitted at a synapse. Although transmitter release is a con-tinuous function of presynaptic voltage (Graubard 1978), the threshold for detectable release may vary. Some terminals release transmitter only when invaded by a spike, while others release transmitter to graded pre-synaptic voltages. Graded release transmits information about absolute voltage of the presynaptic cell, whereas spike-mediated release transmits information only after the presynaptic signal has been processed into spike-frequency. Different release mechanisms, therefore, have a profound influence upon the nature of the information being conveyed at a particular connection.

## Network Connectivity Patterns

Network connectivity includes patterns of interconnection between neurons within a network. The number of possible pathways between $N$ neurons grows rapidly as $N!/(N-2)!$. Therefore, it is not reasonable to summarize all possible combinations even for relatively small networks. A few patterns, however, are commonly encountered (Figure 1). Mutual or recurrent excitation (Figure 1A) promotes synchrony in firing and is usually found among synergists. Reciprocal inhibition and its cousin lateral inhibition are two forms of mutual inhibition (Figure 1B). Recurrent inhibition occurs when one cell excites a second neuron, which then inhibits the first cell or its synergists (Figure 1C). This pattern of connectivity can serve to regulate excitability (e.g. Renshaw cells) or, under appropriate conditions, can produce patterned output (Friesen & Stent 1978). Recurrent cyclic inhibition is characterized by a ring of neurons interconnected by inhibition (Figure 1D) and can, in theory, produce an oscillatory burst pattern with as many phases as number of cells in the ring (Szekely 1965, Friesen & Stent 1978). In many circuits a single cell may mediate more than one action on its targets. Parallel excitation and inhibition can be mediated by separate pathways (Figure 1E, *left*) or by a single multi-component synapse (Figure 1E, *right*). If the time course of either the excitation or inhibition is longer, then this connectivity scheme can lead to a delayed reversal in the sign of synaptic action (Getting 1983a).

Calling these patterns of connectivity *building blocks* is not meant to imply that all neural networks can be either constructed from, or reduced to, these few connection schemes. For some small systems, relatively com-

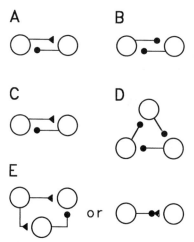

*Figure 1*   Simple patterns of connectivity. *A*. Mutual or recurrent excitation. *B*. Reciprocal or lateral inhibition. *C*. Recurrent inhibition. *D*. Recurrent cyclic inhibition. *E*. Parallel excitation/inhibition. Symbols: *triangles*, excitation; *dots*, inhibition.

plex networks can be simplified in terms of these more restricted connection schemes (Getting & Dekin 1985, Miller & Selverston 1982b), but for larger networks, additional pathways may preclude such reduction. Nor should it be construed that network function, particularly in large networks, can be considered as the simple summation of the action of these components. The properties of these simple schemes may become modified when embedded in larger networks. In addition, large networks may give rise to properties not found in these restricted smaller sets. These simple patterns of connectivity are, however, commonly encountered and appear to form a basis for network function in many diverse systems.

This list of cellular, synaptic, and network properties includes only the more commonly encountered features. It serves, however, to illustrate the vast diversity in properties employed within neural networks. No doubt additional properties will be added as more networks are analyzed. Despite the possible incompleteness of the list, several generalities can be drawn about the use of these properties in neural circuits. First, many of these building blocks are inherently nonlinear, thus the capabilities of neural networks emerge from a complex spatial and temporal interaction of multiple, nonlinear processes at the cellular, synaptic, and network levels. Second similar building block mechanisms have been identified throughout a wide variety of animals, thus suggesting that these constituent building blocks may be conserved. Third, if neural networks acquire their abilities by combining a set of conserved building blocks, then the ability of nervous systems to perform or control diverse behaviors reflects the multitude of ways that these building blocks can be combined. Finally, because of the large number of possible building block mechanisms and ways in which they could be combined, there may be many ways of implementing the same or similar function. For example, many rhythmic motor patterns share features in common, yet the central pattern generator networks underlying the production of these rhythms are disparate (Getting 1988). This disparity suggests that there are numerous ways of combining the building blocks to produce oscillatory, antiphasic patterns, each suited to the particular constraints of the behavior being controlled.

## NETWORKS CAN BE MULTIFUNCTIONAL

If the ability of a neural circuit to perform a function derives from the collective action of the constituent network, synaptic, and cellular building blocks, then altering the properties of building blocks can change the operation of that network. Thus, a single network could subserve several different functions. An important finding in the past decade is that all three classes of building block mechanisms can be controlled by a host of modulatory mechanisms. The implications of these observations for the operation of neural networks are profound. By changing the properties of selected synapses, cells, or pathways, the operation of a network can be dramatically altered. A single network could be multifunctional, participating in or generating more than one behavior. This is not to say that an auditory system can be made into a visual system, but, within the confines of the anatomical substrate, the functional organization of many neural networks appears to be under dynamic control, changing in accordance with the conditions at the moment.

## Definitions of Network Organization and Operation

The idea that the functional organization within a network may be under dynamic control is not new (Sherrington 1906). What is new is an appreciation for the pervasiveness of this principle for neural networks in general (Baldissera et al 1981, Edgerton et al 1976, Getting & Dekin 1985). To provide a framework for discussing the concept of modulation in network operation, the following definitions may prove helpful. One level of network organization is the *anatomical organization*, which is defined by the monosynaptic or anatomical connectivity between neurons. Anatomical organization is specified by the distribution of afferent fibers, the synaptic connectivity within the network, and the projection of efferents. In essence, anatomical organization defines the limits of the network and who talks to whom within the network, but does not give rise to function. The ability of a network to perform a task depends upon what building block mechanisms (network, synaptic, and cellular) are being expressed at a given moment. If these network, synaptic, and cellular mechanisms are under modulatory control, then an anatomical network may be configured into any one of several *modes*, depending upon the particular combination of currently active mechanisms. A *mode* is defined by the distribution and properties of the network, synaptic, and cellular building blocks within the anatomical network. The term *mode* is intended to imply a manner in which a network processes information or generates an output pattern, thus each mode represents the functional organization of the network that gives rise to a function or task. Transitions between modes may occur when afferent or modulatory inputs alter the properties of the constituent building blocks.

To understand how a network operates, the flow of activity within the network must be described quantitatively. One method for quantification is to define *states* of activity within the network. A *state* is defined as the spatial distribution of activity at a given moment in time. If a neuron is considered to either fire an action potential (state 1) or not (state 0), then a network of two neurons has four possible states: both cells firing $(1,1)$, one or the other cell firing (1,0 or 0,1), or neither cell firing (0,0). In this two-cell network, transition between states occurs when either cell starts or terminates an action potential. The temporal sequence of states yields the output pattern of the network. Using the occurrence of individual spikes in every neuron can result in a large number of possible states ($2^N$ where $N$ is the number of neurons in the network), and thus may not be the best criterion to distinguish states. Other criteria such as membrane potential, firing frequency, the onset and termination of bursts, or even the integrated activity across a population may provide better insight into the operation of a particular system. For example, Lennard et al (1980) used a combination of membrane potential and burst times to quantify the temporal sequence of network states underlying swimming in *Tritonia*. This analysis provided important insight into mechanisms for pattern generation in this system (Getting 1983c). When a network is configured into a mode of operation, it expresses a subset of all possible states. The mode, by setting the network, synaptic, and cellular properties, provides an algorithm to produce a temporal sequence of states.

## Modulation of Building Blocks Can Alter Network Operation

Modulation of the network, synaptic, and cellular building blocks can serve to adapt the output pattern to ongoing needs or may dramatically reorganize a network into an entirely new mode mediating a different behavior. The next section deals with mechanisms for modulating each of the three classes of building blocks.

In order to understand how connectivity within a network can be modulated it is necessary to make a distinction between anatomical and functional connectivity. *Anatomical connectivity* refers to the pattern of monosynaptic connections among a group of neurons. *Functional connectivity* refers to the effect of one cell upon another by whatever pathways, monosynaptic or polysynaptic, interconnect the two cells. Anatomical connectivity defines the constraints of a network but functional connectivity determines the activity pattern.

The difference between anatomical and functional connectivity can be illustrated by the network controlling escape swimming in *Tritonia*. Swimming consists of a series of alternating dorsal and ventral flexion movements and is generated by a group of interneurons interconnected by a complex pattern of monosynaptic connections that define the anatomical network (Figure 2A). Within this network, the three dorsal swim interneurons (DSI) excite each other via monosynaptic connections, but this monosynaptic excitation is paralleled by polysynaptic inhibition mediated by the I-cell. From the anatomical connectivity alone, one can not predict

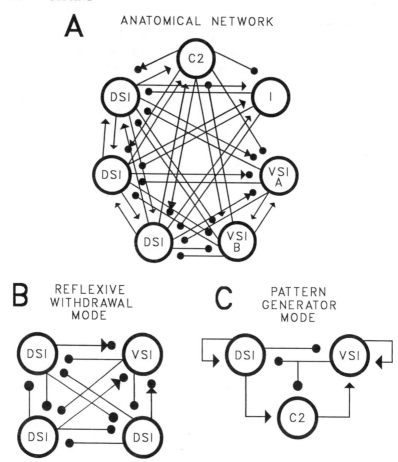

*Figure 2* *A.* Network diagram showing the monosynaptic connectivity between interneurons of the *Tritonia* escape swim system. *B.* Network configuration reflecting the functional connectivity when C2 is silent. In this configuration (mode), the network contributes to reflexive withdrawals. *C.* Network configuration when C2 is active. In this mode the network generates an alternating burst pattern between DSI and VSI which in turn activates moto-neurons for each flexion movement. Pathways with more than one symbol indicate multi-component synapses.

what the effect of driving one DSI would be on another DSI. If the monosynaptic excitation was stronger than the polysynaptic inhibition, then the net effect would be excitatory. In fact, driving one DSI leads to inhibition of another DSI (Getting & Dekin 1985). The polysynaptic inhibitory pathway dominates, thus the functional connectivity between

DSI is inhibitory even though the DSI connect monosynaptically by excitatory synapses.

Functional connectivity reflects the relative strengths of the synaptic connections and the excitability of each neuron along all pathways from one cell to another. Functional connectivity can, therefore, be modulated by alteration in synaptic strengths or in the excitability of neurons along the pathways. The circuit of Figure 2 provides an example of modulation in functional connectivity. When *Tritonia* is not swimming, the C2 neuron is silent and the three DSI functionally inhibit each other via the polysynaptic pathway through the I-cell. During swimming, however, C2 fires in bursts coactive with DSI. Under these conditions, the I-cell is inhibited by C2 and the functional connection among DSI becomes excitatory, mediated by the monosynaptic connections. The functional connection among the DSI can be switched from mutual inhibition to mutual excitation depending upon whether C2 is active.

The consequences of modulation in the functional connectivity within the network of Figure 2A are profound. The anatomical network of Figure 2A can be redrawn to reflect functional connectivity when C2 is silent (Figure 2B) and when C2 is active (Figure 2C). When C2 is silent (Figure 2B), the network is dominated by inhibitory interactions. In this configuration, or mode, each interneuron can be activated independently by afferent input and can contribute to the routing of activity to motor neurons mediating directed reflexive withdrawals (Getting & Dekin 1985). When the animal is stimulated to swim, C2 becomes active, the functional connectivity among DSI becomes excitatory, and the network is reorganized to form the pattern generator circuit shown in Figure 2C. The three DSI have been lumped together because in this configuration they excite each other and act as a single population. In the pattern generator mode, the network produces a sequence of alternating bursts between the DSI and VSI that in turn drives the dorsal and ventral flexion motoneurons. The circuit remains in this mode until the excitation to C2 wanes, the I-cell is disinhibited, and the network reverts to the reflexive withdrawal mode of Figure 2B.

The concept of reordering functional connectivity has been applied to spinal circuitry for some time (Baldissera et al 1981). Many descending pathways to the spinal cord share the same interneurons as peripheral inputs, but the association of these interneurons into functional groups apparently depends upon the task being performed (H. Jankowska, personnal communication). Task-dependent modulation of reflexes has also been observed during different modes of locomotion (Capaday & Stein 1986) and during different phases of the step cycle (Forssberg et al 1975).

Descending influences presumably reorganize the functional connectivity within the spinal cord to fit the task at hand. Reordering of functional connectivity has also been observed in the auditory cortex in response to different sounds (G. L. Gerstein, personal communication). In these cases input may not only *activate* a network but may also *configure* it into an appropriate mode to process that input. Descending commands and afferent inputs should be considered as both instructive and permissive in that they may organize the functional interactions within a network to be appropriate for the task at hand as well as activate the network to perform the task.

Alterations in synaptic properties fall into two general categories: homosynaptic and heterosynaptic. *Homosynaptic* influences depend upon the recent activity of a single synapse, and include such phenomena as facilitation, depression, or potentiation. For *heterosynaptic* modulation, the synaptic efficacy in one pathway can be altered (either increased or decreased) by activity in a second pathway. Heterosynaptic influences can be mediated by direct contact between the two pathways as in presynaptic inhibition or can be mediated by more distant neurohumoral interactions (Shain & Carpenter 1981). The lobster stomatogastric system provides a clear example of application of dopamine selectively altering one IPSP, thus resulting in a phase change in the motor pattern (Eisen & Marder 1984).

Modulation of intrinsic cellular properties also takes many forms (Kaczmarek & Levitan 1987). Since repetitive firing properties reflect the expression of the underlying ionic conductances, modulation of the ionic conductances will alter the input/output relationship of a cell. Many ionic conductances are voltage dependent in the region of resting potential; thus, biasing a neuron more depolarized or hyperpolarized than rest can modulate the expression of these conductances. For example, delayed excitation mediated by A-current can be potentiated by hyperpolarization or nearly abolished by subthreshold depolarization (Getting 1983b, Dekin & Getting 1984). A second mechanism for modulation of intrinsic properties is through receptor-mediated action of neuromodulators. Modulatory substances bind to surface receptors and alter the kinetics of ion channels either directly or by the production of a second messenger (Kaczmarek & Levitan 1987). The effects of modulatory substances can be so profound that cells acquire entirely new properties not seen in the absence of the modulator. The effects of modulators covers the range of intrinsic properties, including increased or decreased excitability, the modulation of spike frequency adaptation, the enhancement of post-inhibitory rebound, the induction of plateau potentials, and the expression of intrinsic bursting.

# PROSPECTS FOR THE FUTURE

The comparative study of neural networks has led to a picture of neural networks as dynamic entities, constrained by their anatomical connectivity but, within these constraints, able to be organized and configured into several operational modes each depending upon the expression and modulation of the constituent cellular, synaptic, and network building blocks. This view has two components: (*a*) neural networks are assembled from a set of cellular, synaptic, and network building blocks, and (*b*) these building blocks can be modulated, thereby altering the operation of the network. These concepts are attractive from several perspectives. First, these organizational principles indicate that a separate neural network is not needed for each behavior or for each modification thereof. A single anatomical set of neurons may perform multiple tasks. Second, these concepts help to reconcile the apparent diversity in neural networks, even those subserving similar functions, since the constituent building block mechanisms appear to be conserved, but not the particular combination. Finally, these concepts provide a framework for approaching neural networks experimentally. In order to understand how a function is implemented one must identify the underlying cellular, synaptic, and network building blocks and how they interact.

These concepts present some formidable problems, however, To understand how a network operates, it will be necessary to analyze the appropriate building blocks while the network is performing the task of interest, otherwise the requisite combination of building block mechanisms may not be operative. This is less of a problem for some invertebrate preparations in which large portions of the nervous system can be isolated while preserving function (Selverston 1985). For vertebrate systems, the situation may be more difficult. Isolation by cell culture or brain slice techniques, although allowing access to and experimental manipulation of individual cells, commonly disrupts the networks sufficiently so that network operation is lost. These preparations can serve an important role, however, in characterizing possible cellular and synaptic mechanisms, but it will be necessary to show how these properties are used within operational networks. For this purpose new methods for gaining access to the relevant building block mechanisms in operational networks will be required. In this regard the advent of a number of in vitro methods for maintaining large portions of the vertebrate CNS hold particular promise (McClellan 1987, O'Donovan 1987, Fulton & Walton 1986, Smith & Feldman 1987, Llinas et al 1981, Richerson & Getting 1987).

Now that some of the building blocks of network operation have been identified, rules for their assembly into neural networks need to be sought.

Further reductionism to the single channel or molecular levels will be important in providing information about mechanisms underlying the building block properties, but these approaches are unlikely to provide insight into principles of network operation. For this purpose we need synthesis, not further reductionism. Do particular combinations of building blocks underlie certain tasks? Are all combinations of building blocks possible or only restricted subsets? What are the processing capabilities of the various building blocks and what do they contribute to network operation? Are there rules governing the assembly of the building blocks into networks?

Answering these questions will require a multidisciplinary approach. In particular, comparative studies will be important to delineate "successful" or useful combinations and may provide insights into rules for network assembly. In this regard, comparative analysis of rhythmic motor pattern generator networks has begun to yield general hypotheses about the assembly and modulation of pattern generator networks (Getting 1988, Harris-Warrick 1988). Ways of manipulating the various building blocks need to be found so that the ways they contribute to network operation can be investigated. In small systems, single cells or small groups of cells can be controlled or deleted (Lennard et al 1980, Miller & Selverston 1979) to allow assessment of their role. In larger systems other methods for altering the constituent building blocks, including possible genetic (Herrup & Sunter 1986, Thomas & Wyman 1984), pharmacologic (Harris-Warrick 1988), and developmental (O'Donovan 1987) manipulations, will be required.

Finally, methods for assessing rules for network assembly and operation must be developed. Biologically realistic computer models should be useful for this purpose. Crude network models based upon the interaction of nonlinear elements are revealing the underpinnings of cognitive function, including content addressable memory and simple pattern recognition (Rumelhart & McClelland 1986, Hopfield & Tank 1986). In terms of the properties of the elements and the complexity of the networks, these models are still barren in comparison with biological systems. What will happen to the emergent properties of these networks when biological reality is incorporated is as yet unclear. The hope is that the capabilities of the networks will increase to approximate that of the CNS, but this remains to be seen. Recent successes such as computer simulation of the establishment and modification of cortical somatosensory maps hold particular promise (Pearson et al 1987). Biologically realistic simulations offer an additional advantage in that individual building blocks or rules governing their assembly must be explicitly stated and therefore can be tested by systematic variation. The ability to manipulate all aspects of model networks should provide a powerful window into the importance of various building blocks and the ways they contribute to overall network function. Computer simulations, however, will be no more useful than the degree of their biological reality. Close interplay between experimentation and simulation must be maintained to ensure the validity of any model of the nervous system.

ACKNOWLEDGMENTS

I am indebted to many colleagues for helpful discussions in the development of the ideas and insights presented. I particularly thank Drs. Michael O'Donovan, Corey Cleland, William Frost, and Andrew McClellan for reading and criticizing the manuscript. I am supported by National Institutes of Health grants NS17328, NS15350, and HL32336.

*Literature Cited*

Adams, W. B. 1985. Slow depolarizing and hyperpolarizing currents which mediate bursting in *Aplysia* neurone R15. *J. Physiol. London* 360: 51–68

Adams, W. B., Levitan, I. B. 1985. Voltage and ion dependences of slow currents which mediate bursting in *Aplysia* neurone R15. *J. Physiol. London* 360: 69–93

Adams, D. J., Smith, S. J., Thompson, S. H. 1980. Ionic currents in molluscan soma. *Ann. Rev. Neurosci.* 3: 141–67

Alving, B. O. 1968. Spontaneous activity in isolated somata of *Aplysia* pacemaker neurons. *J. Gen. Physiol.* 51: 29–45

Anderson, W. W., Barker, D. L. 1981. Synaptic mechanisms that generate network oscillations in the absence of discrete postsynaptic potentials. *J. Exp. Zool.* 216: 187–91

Arshavsky, Y. I., Beloozesova, I. N., Orlovsky, G. N., Panchin, Y. V., Pavlova, G. A. 1985. Control of locomotion in the marine mollusc, *Clione limacina*. IV. Role of type 12 interneurons. *J. Exp. Brain Res.* 58: 285–93

Baldissera, F., Hultborn, H., Illert, M. 1981. Integration in spinal neuronal systems. In *Handbook of Physiology: The Nervous System*, ed. J. M. Brookhart, V. B. Mountcastle, V. B. Brooks, pp. 509–96. Baltimore: William & Wilkins

Bentley, D., Konishi, M. 1978. Neural control of behavior. *Ann. Rev. Neurosci.* 1: 35–59

Brown, D. A., Adams, P. R. 1979. Muscarinic suppression of a novel voltage sensitive $K^+$ current in vertebrate neuron. *Nature* 315: 501–3

Byrne, J. H. 1980a. Analysis of ionic conductance mechanisms in motor cells mediating inking behavior in *Aplysia californica*. *J. Neurophysiol.* 43: 1036–50

Byrne, J. H. 1980b. Quantitative aspects of ionic conductance mechanisms contributing to firing patterns of motor cells mediating inking behavior in *Aplysia californica*. *J. Neurophysiol.* 43: 651–68

Capaday, C., Stein, R. B. 1986. Amplitude modulation of the soleus H-reflex in the human during walking and standing. *J. Neurosci.* 6: 1308–13

Connor, B. W., Gutnick, M. J. 1984. Neocortex: Cellular properties and intrinsic circuitry. In *Brain Slices*, ed. R. Dingledine, pp. 313–42. New York: Plenum

Connor, J. A., Stevens, C. F. 1971. Voltage clamp analysis of a transient outward membrane current in gastropod neural somata. *J. Physiol. London* 213: 21–30

Dekin, M. S., Getting, P. A. 1984. Firing pattern of neurons in the *nucleus tractus solitarius*: Modulation by membrane hyperpolarization. *Brain Res.* 324: 180–84

Dekin, M. S., Getting, P. A. 1987a. In vitro characterization of neurons in the ventral part of the nucleus tractus solitarius. I. Identification of neuronal types and repetitive firing properties. *J. Neurophysiol.* 58: 195–214

Dekin, M. S., Getting, P. A. 1987b. In vitro characterization of neurons in the ventral part of the nucleus tractus solitarius. II. Ionic mechanisms responsible for repetitive firing activity. *J. Neurophysiol.* 58: 215–29

Dekin, M. S., Richerson, G. B., Getting, P. A. 1985. Thyrotropin-releasing hormone induces rhythmic bursting in neurons of the *nucleus tractus solitarius*. *Science* 229: 67–69

Deschenes, M., Paradis, M., Roy, J. P., Steriade, M. 1984. Electrophysiology of neurons of lateral thalamic nuclei in cat: Resting properties and burst discharges. *J. Neurophysiol.* 51: 1196–1219

Dickinson, P. S., Nagy, F. 1983. Control of a central pattern generator by an identified modulatory interneuron in crustacea. II. Induction and modulation of plateau potentials in pyloric neurons. *J. Exp. Biol.* 105: 59–82

Eccles, J. C. 1964. *The Physiology of Synapses*. Berlin: Springer-Verlag

Edgerton, V. R., Frillner, S., Sjostrom, A., Zangger, P. 1976. Central generation of locomotion in vertebrates. In *Neural Control of Locomotion*, ed. R. M. Herman, S. Grillner, P. S. G. Stein, D. G. Stuart, pp. 439–46. New York: Plenum

Eisen, J. S., Marder, E. 1984. A mechanism for production of phase shifts in a pattern generator. *J. Neurophysiol.* 51: 1375–93

Flamm, R. E., Harris-Warrick, R. M. 1986. Aminergic modulation in lobster stomatogastric ganglion. II. Target neurons of dopamine, octopamine and serotonin within the pyloric circuit. *J. Neurophysiol.* 55: 866–81

Forssberg, S., Grillner, S., Rossignol, S. 1975. Phase dependent reflex reversal during walking in chronic spinal cats. *Brain Res.* 85: 103–7

Friesen, W. O., Stent, G. S. 1978. Neural circuits for generating rhythmic movements. *Ann. Rev. Biophys. Bioeng.* 7: 37–61

Fulton, B. P., Walton, K. 1986. Electrophysiological properties of neonatal rat motoneurons studied in vitro. *J. Physiol. London* 370: 651–78

Getting, P. A. 1974. Modification of neuron properties by electronic synapses. I. Input resistance, time constant and integration. *J. Neurophysiol.* 37: 846–57

Getting, P. A. 1981. Mechanisms of pattern generation underlying swimming in *Tritonia*. I. Neuronal network formed by monosynaptic connections. *J. Neurophysiol.* 46: 65–79

Getting, P. A. 1983a. Mechanisms of pattern generation underlying swimming in *Tritonia*. II: Network reconstruction. *J. Neurophysiol.* 49: 1017–34

Getting, P. A. 1983b. Mechanisms of pattern generation underlying swimming in *Tritonia*. III. Intrinsic and synaptic mechanisms for delayed excitation. *J. Neurophysiol.* 49: 1036–50

Getting, P. A. 1983c. Neural control of swimming in *Tritonia*. In *Neural Origin of Rhythmic Movements*, ed. A. Roberts, B. L. Roberts, pp. 89–128. London: Cambridge Univ. Press

Getting, P. A. 1988. Comparative analysis of invertebrate central pattern generators. In *Neural Control of Rhythmic Movements*, ed. A. H. Cohen, S. Rogsignol, S. Grillner, pp. 101–28. New York: John Wiley

Getting, P. A., Dekin, M. S. 1985. Mechanisms of pattern generation underlying swimming in *Tritonia*. IV. Gating of a central pattern generator. 53: 466–80

Graubard, K. 1978. Synaptic transmission without action potentials: Input-output properties of a nonspiking presynaptic neuron. *J. Neurophysiol.* 41: 1014–25

Harris-Warrick, R. M. 1988. Chemical modulation of central pattern generators. See Getting 1988, pp. 285–332

Hatton, G. I. 1984. Hypothalamic neurobiology. In *Brain Slices*, ed. R. Dingledine, pp. 341–74. New York: Plenum

Herrup, K., Sunter, K. 1986. Cell lineage dependent and independent control of Purkinje cell number in the mammalian CNS: Further quantitative studies of *lurcher* chimeric mice. *Dev. Biol.* 117: 417–27

Hodgkin, A. L., Huxley, A. F. 1952. A quantitative description of membrane current and its application to conduction and excitation in nerve. *J. Physiol. London* 117: 500–44

Hopfield, J. J., Tank, D. W. 1986. Neural circuits and collective computation. *Science* 233: 625–33

Hounsgard, J., Hultborn, H., Kiehn, O. 1986. Transmitter-controlled properties of alpha-motoneurones causing long-lasting motor discharge to brief excitatory inputs. *Prog. Brain Res.* 64: 39–49

Hume, R. I., Getting, P. A. 1982a. Motor organization of *Tritonia* swimming. III. Contribution of intrinsic membrane properties to flexion neuron burst formation. *J. Neurophysiol.* 47: 91–102

Hume, R. I., Getting, P. A. 1982b. Motor organization of *Tritonia* swimming. IV. Synaptic drive to flexion neurons from premotor interneurons. *J. Neurophysiol.* 47: 75–90

Jahnsen, H. 1986. Responses of neurons in isolated preparations of the mammalian central nervous system. *Prog. Neurobiol.* 27: 351–72

Johnson, S. M., Getting, P. A. 1987. Repetitive firing properties of neurons within the nucleus ambiguous of adult guinea pigs using the in vitro slice technique. *Neurosci. Abstr.* 13: 825

Kaczmarek, L. K., Levitan, I. B. 1987.

*Neuromodulation.* New York: Oxford Univ. Press

Kandel, E. R., Spencer, W. A. 1961. Electrophysiology of hippocampal neurons. II. After-potentials and repetitive firing. *J. Neurophysiol.* 24: 243–59

Katz, B. 1966. *Nerve, Muscle, and Synapse,* pp. 97–141. New York: McGraw-Hill

Kehoe, J. 1972. Three acetylcholine receptors in *Aplysia* neurons. *J. Physiol. London* 225: 115–46

Krasne, F. B., Wine, J. J. 1984. The production of crayfish tailflip escape responses. In *Neural Mechanisms of Startle Behavior,* ed. R. C. Eaton, pp. 179–211. New York: Plenum

Lennard, P. R., Getting, P. A., Hume, R. I. 1980. Central pattern generator mediating swimming in *Tritonia.* II. Initiation, maintenance and termination. *J. Neurophysiol.* 44: 165–73

Llinas, R. 1981. Electrophysiology of cerebellar networks. See Baldissera et al 1981, pp. 831–76

Llinas, R. 1984. Comparative electrobiology of mammalian central neurons. See Hatten 1984, pp. 7–24

Llinas, R., Sugimori, M. 1980a. Electrophysiological properties of *in vitro* Purkinje cell somata in mammalian cerebellar slices. *J. Physiol. London* 305: 171–95

Llinas, R., Sugimori, M. 1980b. Electrophysiological properties of in vitro Purkinje cell dendrites in mammalian cerebellar slices. *J. Physiol. London* 305: 197–213

Llinas, R., Yarom, Y. 1981. Electrophysiology of mammalian inferior olivary neurons in vitro. Different types of voltage-dependent ionic conductances. *J. Physiol. London* 315: 549–67

Llinas, R., Yarom, Y. 1986. Oscillatory properties of guinea pig olivary neurons and their pharmacological modulation: An in vitro study. *J. Physiol.* 376: 163–82

Llinas, R., Yarom, Y., Sugimori, M. 1981. Isolated mammalian brain in vitro: New techniques for analysis of electrical activity of neural circuit function. *Fed. Proc. Fed. Am. Soc. Exp. Biol.* 40: 2240–45

Lopez-Barneo, J., Llinas, R. 1988. Electrophysiology of mammalian tectal neurons in vitro: I. Transient ionic conductances. *J. Neurophysiol.* 60: 853–68

Marder, E., Eisen, J. S. 1984. Electrically coupled pacemaker neurons respond differently to the same physiological inputs and neurotransmitters. *J. Neurophysiol.* 51: 1362–74

McClellan, A. D. 1987. In vitro CNS preparations: Unique approaches to the study of command and pattern generation systems in motor control. *J. Neurosci. Methods* 21: 251–64

McCulloch, W. S., Pitts, W. H. 1943. A logical calculus of the ideas imminent in nervous activity. *Bull. Math. Biophysics.* 7: 89–93

Meech, R. W. 1978. Calcium-dependent potassium activation in nervous tissues. *Ann. Rev. Biophys. Bioeng.* 7: 1–18

Miller, J. P., Selverston, A. I. 1982b. Mechanisms underlying pattern generation in lobster stomatogastric ganglion as determined by selective inactivation of identified neurons. IV. Network properties of pyloric system. *J. Neurophysiol.* 48: 1416–32

Miller, J. P., Selverston, A. I. 1982a. Mechanisms underlying pattern generation in lobster stomatogastric ganglion as determined by selective inactivation of identified neurons. II. Oscillatory properties of pyloric neurons. *J. Neurophysiol.* 48: 1378–91

Miller, J. P., Selverston, A. I. 1979. Rapid killing of single neurons by irradiation of intracellularly injected dyes. *Science* 206: 702–4

Mulloney, B., Selverston, A. I. 1974. Organization of the stomato-gastric ganglion of spiney lobster. III. Coordination of the two subsets of the gastric system. *J. Comp. Physiol.* 91: 53–78

O'Donovan, M. J. 1987. Developmental approaches to the analysis of vertebrate central pattern generators. *J. Neurosci. Methods* 21: 275–86

Partridge, L. D., Stevens, C. F. 1976. A mechanism for spike frequency adaptation. *J. Physiol. London* 256: 315–32

Pearson, J. C., Finkel, L. H., Edelman, G. M. 1987. Plasticity in the organization of adult cerebral cortical maps: A computer simulation based upon neuronal group selection. *J. Neurosci.* 7: 4209–23

Rall, W. 1981. Functional aspects of neuronal geometry. In *Neurones without Impulses,* ed. A. Roberts, B. M. H. Bush, pp. 223–54. Cambridge, U.K.: Cambridge Univ. Press

Richerson, G. B., Getting, P. A. 1987. Maintenance of complex neural function during perfusion of the mammalian brain. *Brain Res.* 409: 128–32

Rogawski, M. A. 1985. The A-current: How ubiquitous a feature of excitable cells is it. *Trends Neurosci.* 8: 214–19

Rumelhart, D. E., McClelland, J. L. 1986. *Parallel Distributed Processing,* Vols. 1, 2. Cambridge: MIT Press

Russell, D. F., Hartline, D. K. 1978. Bursting neural networks: A reexamination. *Science* 200: 453–56

Satterlie, R. A. 1985. Reciprocal inhibition and post-inhibitory rebound produce reverberation in locomotor pattern generator. *Science* 229: 402–4

Schwindt, P. C., Crill, W. E. 1984. Membrane properties of cat spinal motoneurons. In *Handbook of the Spinal Cord*, ed. R. A. Davidoff, pp. 199–242. New York: Marcel Dekker

Selverston, A. I., ed. 1985. *Model Neural Networks and Behavior*. New York: Plenum

Selverston, A. I., Miller, J. P., Wadepuhl, M. 1983. Cooperative mechanisms for the production of rhythmic movements. In *Neural Origin of Rhythmic Movements*, ed. A. Roberts, B. L. Roberts, pp. 55–88. London: Cambridge Univ. Press

Shain, W., Carpenter, D. O. 1981. Mechanisms of synaptic modulation. *Int. Rev. Neurobiol.* 22: 205–50

Sherrington, C. 1906. *The Integrative Action of the Nervous System*. New Haven: Yale Univ. Press

Smith, J. C., Feldman, J. L. 1987. In vitro brainstem-spinal cord preparations for study of motor systems for mammalian respiration and locomotion. *J. Neurosci. Methods* 21: 321–33

Szekedy, G. 1965. Logical network controlling limb movements in urodela. *Acta Physiol. Acad. Sci. Hung.* 27: 285–89

Thomas, J. B., Wyman, R. J. 1984. Mutations altering synaptic connectivity between identified neurons of Drosophila. *J. Neurosci.* 4: 530–38

Willows, A. O. D. 1973. Interactions between brain cells controlling swimming in a mollusc. In *Neurobiology of Invertebrates*, ed. J. Salanki, pp. 233–47. New York: Plenum

Zucker, R. S., Kennedy, D., Selverston, A. I. 1971. Neuronal circuit mediating escape responses in crayfish. *Science* 173: 645–49

Ann. Rev. Neurosci. 1989. 12:205–25

# THE FUNCTION OF SYNAPTIC TRANSMITTERS IN THE RETINA

Nigel W. Daw, William J. Brunken,[1] and David Parkinson

Department of Cell Biology and Physiology, Washington University
Medical School, St. Louis, Missouri 63110

A large number of different synaptic transmitter types are found within the central nervous system. They include acetylcholine, excitatory amino acids (glutamate and aspartate), inhibitory amino acids ($\gamma$-aminobutyric acid and glycine), several monoamines, and numerous peptides. Clearly one can no longer think of the function of these transmitters simply in terms of excitatory and inhibitory actions, or even direct and modulatory actions. More revealing generalizations are required. To develop such generalizations about neurotransmitter function, one needs to examine the problem of transmitter diversity in a tissue in which all these transmitters are present, the physiology of the cells is well understood, and the anatomical connections of cells containing the transmitters are worked out. The retina is such a tissue (Dowling 1987), and we will argue that generalizations about the function of transmitters can be made from it and successfully extended to other parts of the central nervous system.

Two aspects of the retina make it a particularly appropriate tissue for such a study. First, considerable processing of information takes place in the retina, so that ganglion cells may respond to quite complicated features of the stimulus, such as the direction of movement of the stimulus (Barlow & Levick 1965) or simultaneous contrast between one wavelength of light and another (Daw 1968). Second, the retina can be isolated and bathed or perfused with solutions, while retaining its ability to respond to stimuli.

In general the cell biology and pharmacology of synaptic transmitters in the retina are the same as in other areas of the nervous system, so those

---

[1] Present address: Boston College, Boston, Massachusetts 02167.

aspects of the subject are not discussed in detail here. However, some important observations were first made in the retina and are therefore considered in more detail. For example, a particular type of excitatory amino acid receptor, the APB receptor, was first described in the retina (Slaughter & Miller 1981), as was neurotransmitter-mediated modulation of gap junction permeability (Teranishi et al 1984, Piccolino et al 1984). These results are discussed at the appropriate place. Colocalization of neurotransmitters is not discussed because the evidence at the present time is descriptive rather than functional.

The retina is a tissue that has been investigated in a large variety of species, and the anatomical location of synaptic transmitters shows some variations. Those interested in the details of these variations can consult recent reviews (Ehinger 1982, Brecha 1983, Stell 1985, Massey & Redburn 1987). The focus of our review is to bring out generalizations that can be made about the functions of cells containing a specific transmitter—particularly generalizations that apply independent of where the transmitters are located.

When one considers the function of a cell containing a transmitter—for example, GABA—several conditions need to be met for a definitive series of experiments. First, one needs to know the location and connections of the cell. Second, one needs to identify postsynaptic cells, and characterize the properties of those postsynaptic cells. Third, one needs to observe how the activity of the postsynaptic cells is affected by an antagonist to the synaptic transmitter, or a potentiator of the action of the transmitter. Unfortunately, not very many experiments have been done that meet all these criteria, so much of this review is concerned with situations in which the various bits of information can be put together to synthesize a probable function, even though the definitive experiments may not have been done.

Use of neurotransmitter antagonists is important, because one needs to block actions of a transmitter that is released in response to natural stimulation. This point can be illustrated by considering the role of GABA in directional selectivity. Applying picrotoxin to the retina abolishes the directional selectivity of directionally selective cells. Applying GABA is not expected to produce the reverse effect. Agonists have been useful in working out the cell biology of transmitter action, and in discovering whether a transmitter has an excitatory or inhibitory action, but to get a clear idea of function antagonists clearly have to be used.

A further important point is that multiple receptor types are present for most synaptic transmitters: $D_1$ and $D_2$ for dopamine, multiple nicotinic and at least two muscarinic types ($M_1$ and $M_2$) for acetylcholine, and so on. This point has not always been taken into consideration by physiologists and others working on the function of these transmitters. The coexistence of multiple subtypes of receptors certainly complicates the interpretation of physiological results, particularly where a transmitter acts at two sites, one a postsynaptic site and the other a presynaptic autoreceptor. Thus, agents selective for these subtypes must be employed whenever possible.

## General Organization of the Retina

Photoreceptors in the retina respond to brightness, whereas ganglion cells, whose axons form the fibers of the optic nerve and are the output cells of the retina, respond to a variety of different properties of the stimulus (Table 1): Some respond to objects brighter than the background, others to objects darker; some give a transient response, others sustained; some are color-coded, others not. In certain retinas, some ganglion cells respond best when a particular feature of the stimulus, called a "trigger feature," is presented—for example, its direction of movement or orientation. Clearly, complicated connections are required to provide this variety of responses in the ganglion cells.

The processing takes place in two stages. In the outer plexiform layer of the retina, photoreceptors connect to bipolar cells, and horizontal cells provide interconnections. Whereas photoreceptors respond to brightness and wavelength, bipolar cells respond to contrast. Bipolar cells can be divided into two classes—those that are excited (i.e. depolarized) by a bright spot in a dark surround (ON center), and those that are excited by a dark spot in a bright surround (OFF center) (Werblin & Dowling 1969). In both cases uniform illumination of the whole of the receptive field of the cell gives a reduced response, because illumination of the surround of the receptive field antagonizes illumination of the center. Some bipolar cells respond to color contrast rather than brightness contrast (Kaneko 1973). Some are preferentially driven by cones, and others by rods.

The second stage of processing in the retina takes place in the inner plexiform layer. In this layer, bipolar cells connect to ganglion cells, with amacrine cells providing interconnections. Processing in this layer is very complex. The number of ganglion cell types is much greater than the number of bipolar cell types. In retinas where this number has been carefully studied, such as cat and rabbit, at least 16 distinct ganglion cell types can be recognized by physiological criteria (Cleland & Levick 1974, Stone & Fukuda 1974, Caldwell & Daw 1978a). The number that can be distinguished by Golgi stains is correspondingly large (Kolb et al 1981). Moreover, a large number of amacrine cell types can be distinguished anatomically (Kolb et al 1981), with different transmitters localized to different amacrine cells (Ehinger 1982, Brecha 1983, Massey & Redburn

1987). From the point of view of information processing, amacrine cells and their connections are clearly the most important in the retina.

The concept that at least 20 amacrine cell types provide connections to yield at least 16 ganglion cell types from a limited number of bipolar cell types is one that might daunt the hardiest of scientists. Fortunately, aspects of the organization of the inner plexiform layer make it feasible to tackle this problem. As described by Ramon y Cajal (1892), the inner plexiform layer is stratified. It can be divided into two sublaminae—sublamina a, in which hyperpolarizing bipolar cells are connected to OFF-center ganglion cells, and sublamina b, in which depolarizing bipolar cells are connected to ON-center ganglion cells (Famiglietti & Kolb 1976, Famiglietti et al 1977, Nelson et al 1978).

These sublaminae can be further subdivided into a number of strata—usually five (Ramon y Cajal 1892). Moreover, different transmitters are localized to different strata. Acetylcholine cells ramify primarily in strata 2 and 4 (Baughman & Bader 1977, Masland & Mills 1979). Dopamine cells ramify in stratum 1 in all species and in strata 3 and 5 in some (Ehinger 1982). Indoleamine-accumulating cells ramify in several strata, and are particularly dense in stratum 5 in the rabbit (Ehinger 1982). The processes of peptide-containing neurons also show considerable variety of stratification (Brecha 1983).

In addition to the first and second levels of processing, the inner plexiform layer feeds back to the outer plexiform layer, through interplexiform cells (Boycott et al 1975, Dowling 1987). These cells have their bodies in the same position as amacrine cells, with input and output synapses in the inner plexiform layer and output synapses in the outer plexiform layer.

This general picture of the retina applies primarily to cat and rabbit, and it differs in some respects from that seen in other species. For example, goldfish and macaque have trichromatic rather than dichromatic vision, and ganglion cells in the frog retina have some quite specialized properties. The repertoire of ganglion cells in many retinas may get larger with further study. One has to remember that two types of ganglion cell were recognized in the cat retina in 1953 (Kuffler 1953) compared to at least 16 types recognized now. The assumption of this review, therefore, is that most retinas have a complex organization, and the question to be addressed is whether generalizations about the actions of synaptic transmitters can help us to make sense of the complexities.

## An Excitatory Amino Acid is the Transmitter in the "Straight-through" Pathway in the Retina

Photoreceptors release transmitter continuously in the dark. Since illumination hyperpolarizes them and reduces this output, postsynaptic effects of the photoreceptor transmitter are logically expected to be opposite in sign to the effects produced by light. Excitatory amino acids (EAAs) mimic the action of the photoreceptor transmitter on all second-order neurons. They depolarize horizontal cells and OFF center bipolar cells and hyperpolarize ON bipolar cells (Cervetto & MacNichol 1972, Murakami et al 1972, 1975, Shiells et al 1981, Bloomfield & Dowling 1985), thus strongly supporting their role as the endogenous photoreceptor transmitter.

The EAA receptors that mediate these postsynaptic conductance changes can be separated pharmacologically. Kainate, a selective glutamate agonist in other regions of the central nervous system, mimics the action of the photoreceptor transmitter on all three second-order cell types (Shiells et al 1981, Slaughter & Miller 1983a, Bloomfield & Dowling 1985); however, several other compounds interact selectively with specific cell types. For example, 2-amino-4-phosphonobutyrate (APB) hyperpolarizes the ON bipolar cell, acting like an agonist, but has little effect on horizontal cells or OFF bipolar cells (Slaughter & Miller 1981, Shiells et al 1981, Bloomfield & Dowling 1985). D-O-phosphoserine depolarizes horizontal cells and antagonizes the effect of kainate on this cell type, but has little effect on ON or OFF bipolar cells (Slaughter & Miller 1985b). Finally, cis-2,3-piperidinedicarboxylic acid (PDA) acts as an antagonist on horizontal cells and OFF bipolar cells, but not on ON bipolar cells (Slaughter & Miller 1983c).

It is interesting that the pharmacology of these EAA receptors is different in some respects from those in the spinal cord. Agents that distinguish kainate receptors from quisqualate receptors in the spinal cord, such as D-glutamyl glycine and glutamate diethyl ester (Watkins & Evans 1981), do not have such a specific action in the outer plexiform layer of the retina (Rowe & Ruddock 1982a, Ishida et al 1984, Bloomfield & Dowling 1985, Hals et al 1986). The APB receptor on the ON bipolar cell is particularly interesting, since it is a sign-reversing synapse, involving a transmitter-induced conductance decrease (Slaughter & Miller 1985a, Nawy & Copenhagen 1987).

L-glutamate is the best candidate for the photoreceptor transmitter. Studies of release (Miller & Schwartz 1983), immunocytochemistry (Altschuler et al 1982, Brandon & Lam 1983, Sarthy et al 1986), and uptake do not distinguish between glutamate and aspartate clearly (Lam & Hollyfield 1980, Marc & Lam 1981b, Brandon & Lam 1983, Sarthy et al 1986). However, L-glutamate acts on isolated horizontal and bipolar cells at much lower concentrations than L-aspartate (Lasater & Dowling 1982, Ishida et al 1984, Lasater et al 1984, Attwell et al 1987). The evidence, when put together, therefore suggests that glutamate is the transmitter, acting at three different receptors on horizontal cells, depolarizing bipolar cells, and hyperpolarizing bipolar cells.

EAAs also appear to be involved in the second stage of processing in the retina, as transmitters of bipolar cells and some amacrine cells. EAAs act directly on ganglion cells and many amacrine cells (Slaughter & Miller 1983a, Lukasiewicz & McReynolds 1985, Aizenman et al 1988). PDA blocks ON ganglion cell responses but not ON bipolar cell responses (Slaughter & Miller 1983b), so presumably it acts at the ON bipolar to ON ganglion cell synapse. Use of the nonselective antagonist, kynurenic acid, shows that kainate receptors are involved in the input to a variety of ganglion cells (Coleman et al 1986).

## Are Some "Straight-through" Pathways Sign-Reversing in the Inner Plexiform Layer?

A proposal about retinal connections, known as the push-pull hypothesis, suggests that the center of the receptive field of some ganglion cells may receive tonic input from a sign-reversing bipolar cell as well as sign-conserving input. For ON cells the two influences would both be excitatory in response to light (excitation from a depolarizing bipolar cell, and disinhibition from a hyperpolarizing bipolar cell). For OFF cells the two influences would both be inhibitory in response to light (inhibition and disfacilitation). The evidence comes partly from physiological experiments (Belgum et al 1982) and partly from anatomical observations showing that a variety of bipolar cells synapse onto a single type of ganglion cell (McGuire et al 1984, 1986).

Glycine is a possible transmitter for this role, because some bipolar cells take up glycine (Cohen & Sterling 1986, Pourcho & Goebel 1987). Evidence is strongest in the ON pathway in the cat. Some ON center ganglion cells ramifying in sublamina b are contacted by several types of bipolar cell (Famiglietti 1981, Kolb et al 1981, McGuire et al 1984, 1986). One of these depolarizes in response to light while another hyperpolarizes (Nelson & Kolb 1983). The one that hyperpolarizes in response to light shows strong accumulation of $^3$H-glycine (Pourcho & Goebel 1987) and is therefore a good candidate for a cell with a sign-reversing synapse. Evidence in the OFF pathway is consistent with the hypothesis: Some OFF center ganglion cells increase their discharge rate when APB is iontophoresed nearby (thus implying that they receive inhibitory input via depolarizing bipolar cells), although this input comes at least partly from rod bipolar cells via rod amacrine cells (Müller et al 1988). Whether bipolar cells that ramify in sublamina a (OFF) take up glycine is controversial, and none have been recorded intracellularly and shown to be depolarizing (Cohen & Sterling 1986, Pourcho & Goebel 1987). However, some OFF ganglion cells in necturus certainly receive a disfacilitatory input (Belgum et al 1982, Arkin & Miller 1988).

## Inhibitory Amino Acids Restrict the Range of Stimuli to which Ganglion Cells Respond

Barlow (1969) coined the phrase "trigger feature" to describe the response of a ganglion cell in which one feature, such as directional selectivity, is particularly noticeable. Many cells actually respond to a combination of features: For example, the local edge detectors in the rabbit retina respond best to a small object moving slowly (Barlow et al 1964); the convex edge detectors in the frog retina respond to movement of a small black object (Maturana et al 1960), and the "ON" directional cells in the rabbit retina respond to a bright object moving slowly in a particular direction (Oyster 1968). By comparison, bipolar cells respond to a much broader range of stimuli, a finding that shows that connections within the inner plexiform layer restrict the range of stimuli to which ganglion cells respond.

Restriction brings inhibition to mind, and many years ago Barlow & Levick (1965) suggested lateral inhibition as a mechanism for directional selectively. The two principal inhibitory neurotransmitters are γ-aminobutyric acid (GABA) and glycine. Cells that make lateral connections in the retina are amacrine and horizontal cells. In all vertebrates, GABA and glycine are localized to some amacrine cells, and in lower vertebrates, GABA is localized to some horizontal cells as well (for reviews see Yazulla 1986, Brandon 1985, Mosinger et al 1986).

A number of features of ganglion cell responses are reduced or abolished after perfusing the retina with antagonists to either GABA or glycine. Directional selectivity and orientation selectivity are abolished by the GABA antagonist, picrotoxin (Wyatt & Daw 1976, Caldwell et al 1978, Ariel & Adolph 1985). The specificity of large field units for fast velocities and the specificity of ON directional cells for slow velocities are also reduced by picrotoxin (Caldwell et al 1978). The specificity of local edge detectors for small objects moving slowly is abolished by the glycine antagonist, strychnine (Caldwell et al 1978). In all these cases, lateral inhibition from amacrine cells containing GABA or glycine presumably acts to restrict the range of stimuli to which the ganglion cell responds.

In color processing also, ganglion cells respond to a restricted range of stimuli, compared to photoreceptors and bipolar cells. Goldfish and carp have double opponent ganglion cells that signal red/green contrast but respond little to white light or to uniform red or green illumination (Daw 1968). This is accomplished by having a center mechanism that generates an ON-response to red light and an OFF-response to green light while the surround generates an OFF-response to red light and an ON-response to green light. Ganglion cells with the opposite responses (red OFF-center) are also found.

Connections in the outer plexiform layer produce bipolar cells with an opponent-color response (ON to red light and OFF to green light or vice versa). The H-1 horizontal cells, which hyperpolarize in response to red light, are GABA cells (for review see Lam et al 1980). They feed back onto cones so that GABA hyperpolarizes cones (Murakami et al 1972, Wu & Dowling 1980) and reduces the response to light, particularly in green cones (Murakami et al 1982a, Tachibana & Kaneko 1984). When recordings are made from the H2 horizontal cells, which are opponent-color cells, the sign of the response to red light is inverted by GABA antagonists, thus reflecting the abolition of the feedback (Murakami et al 1982b). All of this conforms to the circuits for color vision proposed by Stell & Lightfoot (1975) from anatomical evidence. The consequence should be that the opponent response in opponent color bipolar cells is abolished by GABA antagonists, although this point has never been tested directly.

Marc (1980) has suggested that the surround of the red ON center double opponent ganglion cell in the goldfish may be formed by input from the Ab amacrine cell. This cell is one of six types that contain glutamic acid decarboxylase (GAD), and can accumulate $^3$H-GABA and/or $^3$H-muscimol (Marc et al 1978, Ball & Brandon 1986, Yazulla et al 1986). The accumulation of GABA is reduced by red light: On the assumption that hyperpolarization of a cell leads to increased accumulation, the Ab amacrine cell should be depolarized by red light. Marc (1980) has also suggested that the surround of the red OFF-center double opponent cell may be mediated by the Aa amacrine cell, a glycine-accumulating cell (Marc & Lam 1981a) that resembles the sustained red-hyperpolarizing cells (Famiglietti et al 1977). The sign of the response and the sign of the output from these two types of amacrine cell are logically consistent with what one expects for cells that mediate the surround of the double opponent ganglion cell, but unfortunately the hypothesis has never been tested directly on ganglion cells. However, it is consistent with the action of GABA on the axon terminals of bipolar cells (Kondo & Toyoda 1983) and is probably consistent with those experiments that have been done so far on ganglion cells (Schellart et al 1984).

The receptive fields of center-surround ganglion cells that are not color-coded are affected much less dramatically than the receptive fields of cells with more complex receptive fields (Kirby 1979, Caldwell & Daw 1978b, Ikeda & Sheardown 1983, Saito 1983, Bolz et al 1985a,b, Priest et al 1985). The heightened response of Y cells for moving stimuli is abolished by GABA antagonists (Caldwell & Daw 1978b, Frishman & Linsenmeier 1982). Spontaneous activity is increased by either GABA or glycine antagonists, reflecting a tonic inhibitory input, with the exception of one report in which bicuculline was iontophoresed onto OFF X cells (Bolz et al 1985a). Most reports also suggest that the light-driven activity increases or decreases along with the spontaneous activity. Some reduction in the surround response has been reported, particularly for experiments in which the whole retina is bathed with an antagonist (Saito 1983). Various correlations between GABA and glycine effects and ON/OFF, X/Y, rod/cone, and area centralis/periphery distinctions have been suggested, but the evidence is contradictory. Since the receptive field of center-surround ganglion cells with sustained responses (X cells) is similar to the receptive field of the bipolar cells that feed into them, one expects little influence from lateral connections in the inner plexiform layer. The transient nature of the response in Y cells might be due to GABA or glycine input, but presently little evidence favors this in mammals, although transient responses of transient ganglion cells in necturus are due to the action of GABA at $GABA_B$ receptors (Maguire et al 1988).

## Acetylcholine Changes the Overall Level of Activity of Ganglion Cells

Two types of acetylcholine-containing neurons can be identified in the retina. One group consists of "true" amacrine cells, with processes that ramify in the middle of sublamina a near stratum 2 of the inner plexiform layer; the other group comprises displaced amacrines, with cell bodies in the ganglion cell layer and processes that ramify in the middle of sublamina b in stratum 4 (Baughman & Bader 1977, Masland & Mills 1979, Famiglietti 1983a). Those that ramify in sublamina a hyperpolarize in response to light, and those that ramify in sublamina b depolarize (Bloomfield & Miller 1986). The two groups form a mirror symmetric matrix with characteristic morphology (Vaney et al 1981), hence the name "starburst amacrines" (Famiglietti 1983a). Their arborization is wide, so that each point in the retina has dendrites from 30–60 cells covering it (Famiglietti 1983, Tauchi & Masland 1984). The same organization has been found in a variety of retinas (for references, see Mariani & Hersh 1988).

Evidence about the function of these cells comes from work on rabbit and cat retinas. The activity of most ganglion cells is increased by acetylcholine and physostigmine (Masland & Ames 1976, Ariel & Daw 1982a, Schmidt et al 1987). There are two main effects: In cells whose response is restricted to particular directions, orientations, and sizes of stimulus, the slope of the response-intensity curve is increased, and in brisk cells with center-surround responses the spontaneous activity is increased, with little effect on the light-driven response (Ariel & Daw 1982a). The opposite effect is obtained in some cells in the rabbit with the nicotinic antagonist mecamylamine: Large reductions are seen in the light-driven response of direction-selective cells and the spontaneous activity of ON sustained cells (Daw & Ariel 1981, Ariel & Daw 1982b). Whether all ganglion cells receive cholinergic input in the normal situation has been questioned, since not all ganglion cells ramify in strata 2 and 4 (Famiglietti 1983b). Some differences between effects on ON and OFF cells and between muscarinic and nicotinic influences have been found, but the functional relevance of these results is not yet fully understood (Masland & Ames 1976, Ariel & Daw 1982a, Cunningham et al 1983, Schmidt et al 1987).

The trigger features of most cells—orientation selectivity, velocity selectivity, motion sensitivity, and size selectivity—are not affected by application of physostigmine (Ariel & Daw 1982a). However, direction selectivity is abolished (Ariel & Daw 1982b, Ariel & Adolph 1985). The exact role that the cholinergic cells play in direction selectivity is a matter of controversy. Anatomical data support the suggestion that the cholinergic amacrine cells are an important source of input to the direction-selective cells in both rabbit and cat retina (Famiglietti 1983b). In addition to direct effects of GABA on ganglion cells (discussed above), GABA also modulates the release of ACh (Massey & Redburn 1982, Cunningham & Neal 1983). Thus lateral inhibition due to GABA is probably exerted on the direction-selective ganglion cells by two routes—one direct and one through the cholinergic amacrines. The direct effect could come partly from GABA released from the cholinergic amacrine cells, where the two transmitters may be colocalized (Vaney & Young 1988). An analysis of the spatial extent of the GABA and ACh inputs to the directional-selective ganglion cell shows that ACh input is located in the center of the receptive field, whereas GABA input comes from a wider area (Ariel & Daw 1982b). Since inhibition from outside the center of the receptive field is needed to explain direction selectivity, the basic mechanism of direction selectivity is almost certainly due to lateral inhibition from cells containing GABA, and ACh is probably there because the ganglion cell needs additional excitatory input to fire.

Interestingly, physostigmine increases the gain of the W cell response-intensity curve. These cells have strong GABA input as suggested by their low spontaneous firing rates and trigger-selection features, both of which are abolished by picrotoxin. Although many W cells are motion sensitive (Masland et al 1984), motion sensitivity is not the crucial factor, because Y cells have motion sensitivity that is abolished by picrotoxin but not by physostigmine (Ariel & Daw 1982a). An alternative generalization is that cells with a strong GABA input for the creation of trigger features need an additional excitatory input as well to fire. This is consistent with the observation that physostigmine and mecamylamine alter the overall levels of activity in ganglion cells without affecting receptive field properties, with the exception of direction selectivity.

## Monoamines Have Different Influences on Rod and Cone Systems in Mammals

The monoamine neurotransmitters include the catecholamines (dopamine, noradrenaline, and adrenaline); the indoleamines (serotonin and melatonin); and histamine. Although some evidence supports all these substances as transmitters in the retina, we focus on dopamine and the indoleamines, because their roles in visual processing have been studied extensively.

Dopamine amacrine cells in the mammalian retina are connected specifically with the rod system. They have processes that ramify in stratum 1 of the IPL (Ehinger 1982) and synapse primarily with other amacrine cells (Dowling & Ehinger 1978). They make extensive connections to the A II amacrine interposed between rod bipolars and ganglion cells (Tork & Stone 1979, Pourcho 1982, Voigt & Wassle 1987). In the primate, dopamine amacrine cells parallel the rods in distribution (Mariani 1984).

Dopamine drugs have an overall modulatory action on the activity of ganglion cells (Thier & Alder 1984, Jensen & Daw 1984). In particular, the $D_1$ antagonist SCH 23390 abolishes the surround of center-surround brisk cells in the rabbit (Jensen & Daw 1986). Dopamine itself reduces the sensitivity of ganglion cell responses, as expected for the transmitter carrying the antagonistic surround input (Jensen & Daw 1986). In addition, the $D_1$ antagonist SCH 23390 turns the center response of sustained ON center cells in the rabbit from sustained excitation to sustained inhibition, a finding that suggests a tonic influence of dopamine on these cells (Jensen & Daw 1986).

Neurochemical experiments show that monoamine neurons have different activities in light and dark. The turnover of dopamine in light is four- to five-fold higher than in darkness (Iuvone et al 1978, Parkinson & Rando 1983). These results argue that dopaminergic neurons are more active in illuminated retinas than in dark-adapted retinas, and may explain why the surround of center-surround cells is absent in the completely dark-adapted state (Barlow et al 1957).

Indoleaminergic amacrine cells are also connected specifically with the rod system. Uptake of serotonin reveals three major classes of indoleamine-accumulating cells that ramify extensively in stratum 5 of the inner plexiform layer (Ehinger 1982, Vaney 1986, Sandell & Masland 1986, Wassle et al 1987), where they contact the terminals of the rod bipolar cells (Ehinger & Holmgren 1979, Holmgren-Taylor 1982). Such cells have been shown to be driven by rod input almost exclusively in both cat and rabbit (Nelson & Kolb 1985, Raviola & Dacheux 1987).

Experiments with indoleamine agonists and antagonists suggest that the endogenous indoleamine potentiates the output of rod bipolar cells (Brunken & Daw 1987). Rod bipolar cells in the mammal synapse in the b sublamina of the inner plexiform layer onto a narrow field bistratified amacrine cell (the A II amacrine), which connects to OFF center ganglion cells through a sign-inverting synapse in sublamina a and to ON center ganglion cells through gap junctions with cone bipolars (Kolb & Famiglietti 1974, Famiglietti & Kolb 1975, Dacheux & Raviola 1986, Müller et al 1988). Consequently, serotonin and its agonists and antagonists tend to have opposite effects on ON and OFF center cells (Thier & Wassle 1984, Brunken & Daw 1987). All the effects of serotonin and its agonists and antagonists on ganglion cells in the rabbit retina are consistent with the hypothesis that the endogenous indoleamine provides positive feedback at the rod bipolar terminal through $5HT_2$ receptors, limited by $5HT_{1A}$ receptors (see below). Consequently the suggested function of the endogenous indoleamine is that it increases the signal/noise ratio in rod pathways.

## Dopamine Has an Effect at a Distance from Its Site of Release

Two humoral effects of dopamine have been seen in some retinas. In lower vertebrates, cones contract in the light and elongate in the dark. Rods move in the opposite direction, and pigment granules in the epithelial cells move to shield the photoreceptors in the light. Exogenous dopamine induces both the contraction of cones and the dispersion of pigment granules through $D_2$ receptors, and light has the same effect (Dearry & Burnside 1986, Pierce & Besharse 1985). Since there are no known processes of dopamine cells ending on photoreceptors, this effect is likely to be due to dopamine diffusing from amacrine or interplexiform cells.

Dopamine also acts at a distance and mimics light in the *Xenopus* retina, where it increases cone input and reduces rod input to horizontal cells in the mesopic state (Witkovsky et al 1988). Melatonin mimics the effect of darkness in that it promotes cone elongation (Pierce & Besharse 1985). The circadian rhythm of melatonin synthesis in *Xenopus* is sustained during in vitro incubations (Besharse & Iuvone 1983). Since melatonin is reported to inhibit the release of dopamine from the rabbit retina (Dubocovich 1983), the melatonin rhythm may modulate dopamine release and hence retinomotor movements (Pierce & Besharse 1985). Interestingly, another monoamine, octopamine, modulates circadian changes in sensitivity in the *Limulus* retina (Kass & Barlow 1984).

Dopamine also has an effect on the gap junctions that connect horizontal cells in the outer plexiform layer. Dopamine closes these gap junctions in fish and turtle, with a decrease in the number of intramembranous particles, so that Lucifer yellow does not diffuse between horizontal cells, and the receptive fields of individual horizontal cells becomes smaller (Negishi & Drujan 1979, Teranishi et al 1984, Piccolino et al 1984, Baldridge et al 1987). In fish this effect is mediated through interplexiform cells which synapse on horizontal and bipolar cells (Dowling & Ehinger 1978). In turtle, however, dopaminergic cells do not project to the outer plexiform layer (Witkovsky et al 1984, 1987), so this effect could also be due to action at a distance. The consequence of this action of dopamine is that the receptive field surround of bipolar cells becomes weaker (Hedden & Dowling 1978).

There are interesting parallels between these effects of dopamine and its influences in the inner plexiform layer of mammals described above. In some cases the monoamine mimics light, as in its effect on photomechanical movements, the density of intramembranous particles between horizontal cells, and the increase in the antagonistic surround that occurs in the light-adapted state. In some cases it appears to mimic dark, as in its effect on the area/response curve measured from fish horizontal cells. In all cases, there is a correlation between the effect of monoamines and light/dark or rod/cone differences, which leads to the conclusion that monoamines play a modulatory role or a role in adaptation in the retina. Neither dopaminergic nor serotonergic drugs affect the trigger features of cells with more complex receptive field properties (Jensen & Daw 1984, Brunken & Daw 1988a).

## Monoamines Have Two Receptor Types, with Opposite Physiological Effects

Each neurotransmitter in the retina has at least two types of receptors associated with it, with the possible exception of glycine and some peptides. This point has been discussed for excitatory amino acids and acetylcholine. For monoamines, different receptors have opposite physiological effects. This first became clear in the rabbit retina, which contains both $D_1$ and $D_2$ receptors for dopamine (Dubocovich & Wiener 1985; see also Gredal et al 1987). $D_1$ antagonists such as SCH 23390 have a variety of effects, and the effects of the $D_2$ agonist, LY 141865, are generally similar, whereas the effects of the $D_2$ antagonist, sulpiride, tend to be opposite (Jensen & Daw 1986). A comparable situation occurs with serotonin, where the 5HT2 antagonists such as ketanserin, methysergide, and LY 52857 tend to have actions similar to those of $5HT_1$ agonists such as 8-OHDPAT and 5-MeODT (Brunken & Daw 1987). Two explanations are possible for this type of result: (a) One of the receptors acts postsynaptically and the other is an autoreceptor, feeding back onto the presynaptic terminal and this is probably the case for $D_1$ and $D_2$ receptors; (b) both receptors are postsynaptic, but one has an excitatory action and the other has an inhibitory action and this is probably the case for $5HT_2$ and $5HT_{1A}$ receptors (Brunken & Daw 1987).

The existence of multiple receptors clearly complicates results obtained by using nonselective agonists, such as the transmitter itself. In both the cat and rabbit retinas, serotonin has some actions opposite those of $5HT_1$-selective agonists (Thier & Wassle 1984, Brunken & Daw 1988b). Presumably this happens because the action of serotonin at $5HT_2$ receptors in this case is more powerful than its effect at $5HT_{1A}$ receptors. The results obtained when dopamine is applied to the rabbit retina are also complicated, and presumably this happens because dopamine is having some effect of both $D_1$ and $D_2$ receptors (Jensen & Daw 1986).

## The Function of Synaptic Transmitters Correlates with Particular Features of Cell Responses

The summary of all this work is that the action of particular neurotransmitters can be correlated with particular features of ganglion cell response, as listed in Table 1. In general, all ganglion cells are affected by GABA, acetylcholine, EAAs, and monoamines in some fashion. The different aspects of their response are affected in different ways by input from horizontal, bipolar, and amacrine cells containing these transmitters.

Since excitatory amino acids carry the message from one level of processing to another, drugs that affect EAA transmission abolish responses altogether or abolish responses in one pathway. The specific glutamate antagonist APB has been particularly useful in abolishing messages in the

**Table 1**    Features of ganglion cell responses

| | |
|---|---|
| Response to bright object (ON) vs. dark object (OFF) | Glutamate, glycine |
| Transient or sustained response | GABA |
| Response in dim light (rod input) vs. bright light (cone input) | Monoamine |
| Spontaneous activity | All |
| Sensitivity (illumination required for a threshold response) | Monoamine |
| Responsiveness (slope of response-intensity curve) | ACh |
| Center-surround balance (response to uniform illumination) | Monoamine |
| Size selectivity (size of center of receptive field) | GABA |
| Color coding | GABA |
| Direction selectivity | GABA |
| Orientation selectivity | GABA |
| Velocity selectivity | GABA |

ON pathway and in analyzing the contributions of this pathway to higher levels of the visual system (Schiller 1982, Horton & Sherk 1984). Excitatory amino acids are found all over the nervous system and frequently carry messages from one level of processing to another (Mayer & Westbrook 1987). Essentially they are the modem lines of the central nervous system, just as acetylcholine cells are the modem lines at the neuromuscular junction.

Where ganglion cells respond to a restricted range of stimuli—giving them direction selectivity, orientation selectivity, selectivity for a particular velocity, selectivity for a particular size of stimulus, sensitivity to movement, or selectivity for color contrast—the response is due to the action of inhibitory amino acids. Inhibitory amino acids produce the direction selectivity and possibly other trigger features of cells in the cat visual cortex, just as they do in the rabbit retina (Sillito 1987). Inhibitory amino acids also affect the size of receptive fields in the somatosensory system (Dykes et al 1984).

Acetylcholine increases the slope of the response intensity curve in a variety of cell types in the retina. In general these cell types are those with strong GABA input, a finding that suggests that acetylcholine cells are there to overcome this input. Acetylcholine increases the response of many cells in the cat visual cortex, just as it does in the rabbit retina (Sillito & Kemp 1983, Sato et al 1987). Interestingly, this occurs even though the receptors in the cortex are primarily muscarinic, whereas the receptors in the retina are primarily nicotinic. More specific functions for acetylcholine may exist, but none has yet been described.

Monoamines affect the overall levels of activity and center-surround balance of ganglion cells. In particular, they mimic the effects of light or dark on a number of different functions within the retina. Monoamine projections from the locus coeruleus and raphe nuclei to other parts of the central nervous system also have an overall modulatory effect on the activity of neurons and are tied into the night/day cycle. The possibility that monoamine terminals in the cortex may release their transmitter in a nontraditional manner was suggested some time ago (Beaudet & Descarries 1984). This is now controversial, and presently the best evidence that dopamine has an action at a distance from its site of release probably comes from the retina. The opposite actions of $D_1$ and $D_2$ receptors, and of $5HT_1$ and $5HT_2$ receptors, have many parallels in other parts of the central nervous system (Clark & White 1987, Gudelsky et al 1986, Andrade & Nicoll 1987).

In many respects, therefore, the generalizations that come from studying the function of neurotransmitters in the retina apply to the visual cortex and other parts of the nervous system. These generalizations have been worked out in the visual system because receptive fields are characterized more easily in this system than in other sensory systems, and in the retina because it can be perfused with solutions while the sensory input is maintained. Generalizations will likely be found to apply to other sensory systems, as the techniques for characterizing receptive fields in those systems are refined.

*Literature Cited*

Aizenman, E., Frosch, M. P., Lipton, S. A. 1988. Responses mediated by excitatory amino acid receptors in solitary retinal ganglion cells from rat. *J. Physiol.* 396: 75–92

Altshuler, R. A., Mosinger, J. L., Harmison, G. G., Parakkal, M. H., Wenthold, R. J. 1982. Aspartate aminotransferase-like immunoreactivity as a marker for aspartate/glutamate in guinea pig photoreceptors. *Nature* 298: 657–59

Andrade, R., Nicoll, R. A. 1987. Pharmacologically distinct actions of serotonin on single pyramidal neurons of the rat hippocampus recorded *in vitro. J. Physiol.* 394: 99–124

Ariel, M., Adolph, A. R. 1985. Neurotransmitter inputs to directionally sensitive turtle retinal ganglion cells. *J. Neurosci.* 54: 1123–43

Ariel, M., Daw, N. W. 1982a. Effects of cholinergic drugs on receptive field properties of rabbit retinal ganglion cells. *J. Physiol.* 324: 135–60

Ariel, M., Daw, N. W. 1982b. Pharmacological analysis of directionally sensitive rabbit retinal ganglion cells. *J. Physiol.* 324: 161–85

Ariel, M., Lasater, E. M., Mangel, S. C., Dowling, J. E. 1984. On the sensitivity of H1 horizontal cells of the carp retina to glutamate, aspartate and their agonists. *Brain Res.* 295: 179–83

Arkin, M. S., Miller, R. F. 1988. Bipolar origin of synaptic inputs to sustained OFF ganglion cells in the mudpuppy. *J. Neurophysiol.* 60: 1122–42

Attwell, D., Mobbs, P., Tessier-Lavigne, M., Wilson, M. 1987. Neurotransmitter-induced currents in retinal bipolar cells of the axolotl, *Ambystoma Mexicanum. J. Physiol.* 387: 125–61

Baldridge, W. H., Ball, A. K., Miller, R. G. 1987. Dopaminergic regulation of horizontal cell gap junction particle density in goldfish retina. *J. Comp. Neurol.* 265: 428–36

Ball, A. K., Brandon, C. 1986. Localization of 3H-GABA, -muscimol, and -glycine in goldfish retinas stained for glutamate decarboxylase. *J. Neurosci.* 6: 1621–27

Barlow, H. B. 1969. Trigger, features, adaptation and economy of impulses. In *Information Processing in the Nervous System*, ed. K. N. Leibovic, pp. 209–30. New York: Springer-Verlag

Barlow, H. B., Fitzhugh, R., Kuffler, S. W. 1957. Change of organization in the receptive fields of the cat's retina during dark adaptation. *J. Physiol.* 137: 338–54

Barlow, H. B., Levick, W. R. 1965. The mechanism of directionally selective units in rabbit's retina. *J. Physiol.* 178: 477–504

Baughman, R. W., Bader, C. R. 1977. Biochemical characterization and cellular localization of the cholinergic system in the chicken retina. *Brain Res.* 138: 469–85

Beaudet, A., Descarries, L. 1984. Fine structure of monoamine axon terminals in cerebral cortex. In *Monoamine Innervation of Cerebral Cortex*, ed. L. Descarries, T. H. Reader, H. Jasper, pp. 77–93. New York: Liss

Belgum, J. H., Dvorak, D. R., McReynolds, J. S. 1982. Sustained synaptic input to ganglion cells of mudpuppy retina. *J. Physiol.* 326: 91–108

Besharse, J. C., Iuvone, P. M. 1983. Circadian clock in the *Xenopus* eye controlling retinal serotonin N-acetyltransferase. *Nature* 305: 133–35

Bloomfield, S. A., Dowling, J. E. 1985. Roles of aspartate and glutamate in synaptic transmission in rabbit retina I. Outer plexiform layer. *J. Neurophysiol.* 53: 699–713

Bloomfield, S. A., Miller, R. F. 1986. A functional organization of ON and OFF pathways in the rabbit retina. *J. Neurosci.* 6: 1–13

Bolz, J., Frumkes, T., Voigt, T., Wassle, H. 1985a. Action and localisation of γ-aminobutyric acid in the cat retina. *J. Physiol.* 362: 369–93

Bolz, J., Thier, P., Voigt, T., Wassle, H. 1985b. Action and localisation of glycine and taurine in the cat retina. *J. Physiol.* 362: 395–412

Boycott, B. B., Dowling, J. E., Fisher, S. K., Kolb, H., Laties, A. M. 1975. Interplexiform cells of the mammalian retina and their comparison with catecholamine-containing retinal cells. *Proc. R. Soc. London Ser. B* 191: 353–68

Brandon, C. 1985. Retinal GABA neurons: Localization in vertebrate species using an antiserum to rabbit brain glutamate decarboxylase. *Brain Res.* 344: 286–95

Brandon, C., Lam, D. M.-K. 1983. L-glutamic acid: A neurotransmitter candidate for cone photoreceptors in human and rat retinas. *Proc. Natl. Acad. Sci. USA* 80: 5117–21

Brecha, N. 1983. Retinal neurotransmitters: Histochemical and biochemical studies. In *Chemical Neuroanatomy*, ed. P. C. Emson, pp. 85–129. New York: Raven

Brunken, W. J., Daw, N. W. 1987. The actions of serotonergic agonists and antagonists on the activity of brisk ganglion cells in the rabbit retina. *J. Neurosci.* 7: 4054–65

Brunken, W. J., Daw, N. W. 1988a. The effects of serotonin agonists and antagonists on the response properties of complex ganglion cells in the rabbit's retina. *Visual Neurosci.* 1: 181–88

Brunken, W. J., Daw, N. W. 1988b. Neuropharmacological analysis of the role of indoleamine-accumulating amacrine cells in the rabbit retina. *Visual Neurosci.* 1: 275–85

Caldwell, J. H., Daw, N. W. 1978a. New properties of rabbit retinal ganglion cells. *J. Physiol.* 276: 257–76

Caldwell, J. H., Daw, N. W. 1978b. Effects of picrotoxin and strychnine on rabbit retinal ganglion cells: Changes in centre surround receptive fields. *J. Physiol.* 276: 299–310

Caldwell, J. H., Daw, N. W., Wyatt, H. J. 1978. Effects of picrotoxin and strychnine on rabbit retinal ganglion cells: Lateral interactions for cells with more complex receptive fields. *J. Physiol.* 276: 277–98

Cervetto, L., MacNichol, E. F. 1972. Inactivation of horizontal cells in turtle retina by glutamate and aspartate. *Science* 178: 767–68

Clark, D., White, F. J. 1987. Review: D1 dopamine receptor—the search for a function: A critical evaluation of the D1/D2 dopamine receptor classification and its functional implications. *Synapse* 1: 347–88

Cleland, B. G., Levick, W. R. 1974. Properties of rarely encountered types of ganglion cells in the cat's retina and an overall classification. *J. Physiol.* 240: 457–92

Cohen, E., Sterling, P. 1986. Accumulation of 3H-glycine by cone bipolar neurons in the cat retina. *J. Comp. Neurol.* 250: 1–7

Coleman, R. A., Massey, S. C., Miller, R. F. 1986. Kynurenic acid distinguishes kainate and glutamate receptors in the vertebrate retina. *Brain Res.* 381: 172–75

Cunningham, J. R., Dawson, C., Neal, M. J. 1983. Evidence for a cholinergic inhibitory feed-back mechanism in the rabbit retina. *J. Physiol.* 340: 455–68

Cunningham, J. R., Neal, M. J. 1983. Effect of γ-aminobutyric acid agonists, glycine, taurine, and neuropeptides on acetylcholine release from the rabbit retina. *J. Physiol.* 336: 563–77

Dacheux, R. F., Raviola, E. 1986. The rod pathway in the rabbit retina: A depolarizing bipolar and amacrine cell. *J. Neurosci.* 6: 331–45

Daw, N. W. 1968. Colour-coded ganglion cells in the goldfish retina: Extension of their receptive fields by means of new stimuli. *J. Physiol.* 197: 567–92

Daw, N. W., Ariel, M. 1981. Effect of synaptic transmitter drugs on receptive fields of rabbit retinal ganglion cells. *Vision Res.* 21: 1643–47

Dearry, A., Burnside, B. 1986. Dopaminergic regulation of cone retinomotor movement in isolated teleost retinas. I. Induction of cone contraction is mediated by D2 receptors. *J. Neurochem.* 46: 1006–21

Dowling, J. E. 1987. *The Retina.* Boston: Harvard Univ. Press

Dowling, J. E., Ehinger, B. 1978. The interplexiform cell system. I. Synapses of the dopaminergic neurons of the goldfish retina. *Proc. R. Soc. London Ser. B* 201: 7–26

Dowling, J. E., Ehinger, B. 1978. Synaptic organization of the dopaminergic neurons in the rabbit retina. *J. Comp. Neurol.* 180: 203–20

Dubocovich, M. L. 1983. Melatonin is a potent modulator of dopamine release in the retina. *Nature* 306: 782–84

Dubocovich, M. L., Wiener, N. 1985. Pharmacological differences between the D-2 autoreceptor and the D-1 dopamine receptor in rabbit retina. *J. Pharmacol. Exp. Ther.* 233: 747–54

Dykes, R. W., Landry, P., Metharate, R., Hicks, T. P. 1984. Functional role of GABA in cat primary somatosensory cortex: Shaping receptive fields of cortical neurons. *J. Neurophysiol.* 52: 1066–93

Ehinger, B. 1982. Neurotransmitter systems in the retina. *Retina* 2: 305–21

Ehinger, B., Holmgren, I. 1979. Electronmicroscopy of the indoleamine-accumulating neurons in the retina of the rabbit. *Cell Tissue Res.* 197: 175–94

Famiglietti, E. V. 1981. Functional architecture of cone bipolar cells in mammalian retina. *Vision Res.* 1554–63

Famiglietti, E. V. 1983a. "Starburst" amacrine cells and cholinergic neurons: Mirror-symmetric ON and OFF amacrine cells of rabbit retina. *Brain Res.* 261: 138–44

Famiglietti, E. V. 1983b. ON and OFF pathways through amacrine cells in mammalian retina: The synaptic connections of "Starburst" amacrine cells. *Vision Res.* 23: 1265–79

Famiglietti, E. V., Kaneko, A., Tachibana, M. 1977. Neuronal architecture of On and Off pathways to ganglion cells in carp retina. *Science* 198: 1267–69

Famiglietti, E. V., Kolb, H. 1975. A bistratified amacrine cell and synaptic circuitry in the inner plexiform layer of the retina. *Brain Res.* 84: 293–300

Famiglietti, E. V., Kolb, H. 1976. Structural basis for ON- and OFF-center responses in retinal ganglion cells. *Science* 194: 193–95

Frishman, L. J., Linsenmeier, R. A. 1982. Effects of picrotoxin and strychnine on non-linear responses of Y-type cat retinal ganglion cells. *J. Physiol.* 324: 347–63

Gredal, O., Parkinson, D., Nielsen, M. 1987. Binding of 3H SCH 23390 to dopamine D-1 receptors in rat retina in vitro. *Eur. J. Pharmacol.* 137: 241–45

Gudelsky, G. A., Koenig, J. I., Meltzer, H. Y. 1986. Thermoregulatory responses to serotonin (5-HT) receptor stimulation in the rat. Evidence for opposing roles of 5-HT$_2$ and 5-HT$_{1A}$ receptors. *Neuropharmacology* 25: 1307–13

Hals, G., Christensen, B. N., O'Dell, T., Christensen, M., Shingai, R. 1986. Voltage-clamp analysis of currents produced by glutamate and some glutamate analogues on horizontal cells isolated from catfish retina. *J. Neurophysiol.* 56: 19–31

Hedden, W. L., Dowling, J. E. 1978. The interplexiform cell system. II. Effects of dopamine on goldfish retinal neurons. *Proc. R. Soc. London Ser. B* 201: 27–55

Holmgren-Taylor, I. 1982. Electron-microscopical observations on the indoleamine-accumulating neurons and their synaptic connections in the retina of the cat. *J. Comp. Neurol.* 208: 144–56

Horton, J. C., Sherk, H. 1984. Receptive field properties in the cat's lateral geniculate nucleus in the absence of ON-center retinal input. *J. Neurosci.* 4: 374–80

Ikeda, H., Sheardown, M. J. 1983. Transmitters mediating inhibition of ganglion cells in the cat retina: Iontophoretic studies in vivo. *Neuroscience* 8: 837–53

Ishida, A. T., Kaneko, A., Tachibana, M. 1984. Responses of solitary retinal horizontal cells from Carassius Auratus to L-

glutamate and related amino acids. *J. Physiol.* 348: 255–70

Iuvone, P. M., Galli, C. L., Garrison-Gund, C., Neff, N. H. 1978. Light stimulates tyrosine hydroxylase activity and dopamine synthesis in retinal amacrine neurons. *Science* 202: 901–2

Jensen, R. J., Daw, N. W. 1984. Effects of dopamine antagonists on receptive fields of brisk cells and directionally selective cells in the rabbit retina. *J. Neurosci.* 4: 2972–85

Jensen, R. J., Daw, N. W. 1986. Effects of dopamine and its agonists and antagonists on the receptive field properties of ganglion cells in the rabbit retina. *Neurosci.* 17: 837–55

Kaneko, A. 1973. Receptive field organization of bipolar and amacrine cells in the goldfish retina. *J. Physiol.* 235: 133–53

Kass, L., Barlow, R. B. 1984. Efferent neurotransmission of circadian rhythm in Limulus lateral eye. I. Octopamine-induced increases in retinal sensitivity. *J. Neurosci.* 4: 908–17

Kirby, A. W. 1979. The effect of strychnine, bicuculline and picrotoxin on X and Y cells in the cat retina. *J. Gen. Physiol.* 74: 71–84

Kolb, H., Famiglietti, E. V. 1974. Rod and cone pathways in the inner plexiform layer of cat retina. *Science* 186: 47–49

Kolb, H., Nelson, R., Mariani, A. 1981. Amacrine cells, bipolar cells and ganglion cells of the cat retina: A Golgi study. *Vision Res.* 21: 1081–14

Kondo, H., Toyoda, J.-I. 1983. GABA and glycine effects on the bipolar cells of the carp retina. *Vision Res.* 23: 1259–64

Kuffler, S. W. 1953. Discharge patterns and functional organization of mammalian retina. *J. Neurophysiol.* 16: 37–68

Lam, D. M.-K., Hollyfield, J. G. 1980. Localization of putative amino acid neurotransmitters in the human retina. *Exp. Eye Res.* 31: 729–32

Lam, D. M.-K., Su, Y. Y. T., Chin, C. A., Brandon, C., Wu, J. Y., Marc, R. E., Lasater, E. M. 1980. GABA-ergic horizontal cells in the teleost retina. *Brain Res. Bull.* 5: 137–40

Lasater, E. M., Dowling, J. E. 1982. Carp horizontal cells in culture respond selectively to L-glutamate and its agonists. *Proc. Natl. Acad. Sci. USA* 79: 936–40

Lasater, E. M., Dowling, J. E., Ripps, H. 1984. Pharmacological properties of isolated horizontal and bipolar cells from the skate retina. *J. Neurosci.* 4: 1966–75

Lukasiewicz, P. D., McReynolds, J. S. 1985. Synaptic transmission at N-methyl-D-aspartate receptors in the proximal retina of the mudpuppy. *J. Physiol.* 367: 99–116

Maguire, G., Lukasiewicz, P., Werblin, F. 1988. Neural interactions underlying the response to change in the tiger salamander retina. *J. Neurosci.* 8: In press

Marc, R. E. 1980. Retinal colour channels and their transmitters. In *Colour Vision Deficiencies*, ed. G. Verriest, 5: 15–29. London: Hilger

Marc, R. E., Lam, D. M.-K. 1981a. Glycinergic pathways in the goldfish retina. *J. Neurosci.* 1: 152–65

Marc, R. E., Lam, D. M.-K. 1981b. Uptake of aspartic and glutamic acid by photoreceptors in goldfish retina. *Proc. Natl. Acad. Sci. USA* 78: 7185–89

Marc, R. E., Stell, W. K., Bok, D., Lam, D. M.-K. 1978. GABA-ergic pathways in the goldfish retina. *J. Comp. Neurol.* 182: 221–45

Mariani, A. P. 1984. Dopamine-containing amacrine cells of rhesus monkey retina parallel rods in spatial distribution. *Brain Res.* 322: 1–7

Mariani, A. P., Hersh, L. B. 1988. Synaptic organisation of cholinergic amacrine cells in the rhesus monkey retina. *J. Comp. Neurol.* 267: 269–80

Masland, R. H., Ames, A. 1976. Responses to acetylcholine of ganglion cells in an isolated mammalian retina. *J. Neurophysiol.* 39: 1220–35

Masland, R. H., Mills, J. W. 1979. Autoradiographic identification of acetylcholine in the rabbit retina. *J. Cell Biol.* 83: 159–78

Masland, R. H., Mills, J. W., Cassidy, C. 1984. The functions of acetylcholine in the rabbit retina. *Proc. R. Soc. London Ser. B* 223: 121–34

Massey, S. C., Redburn, D. A. 1982. A tonic GABA-mediated inhibition of cholinergic amacrine cells in rabbit retina. *J. Neurosci.* 2: 1633–43

Massey, S. C., Redburn, D. A. 1987. Transmitter circuits in the vertebrate retina. *Prog. Neurobiol.* 28: 55–96

Maturana, H. R., Lettvin, J. Y., McCullough, W. S., Pitts, W. H. 1960. Anatomy and physiology of vision in the frog (*Rana Pipiens*). *J. Gen. Physiol.* 16: 129–75

Mayer, M. L., Westbrook, G. L. 1987. The physiology of excitatory amino acids in the vertebrate central nervous system. *Prog. Neurobiol.* 28: 197–276

McGuire, B. A., Stevens, J. K., Sterling, P. 1984. Microcircuitry of bipolar cells in cat retina. *J. Neurosci.* 4: 2920–38

McGuire, B. A., Stevens, J. K., Sterling, P. 1986. Microcircuitry of beta ganglion cells in cat retina. *J. Neurosci.* 6: 907–19

Miller, A. E., Schwartz, E. A. 1983. Evidence for the identification of synaptic transmitters released by photoreceptors of the toad retina. *J. Physiol.* 334: 325–49

Mosinger, J. L., Yazulla, S., Studholme, K. M. 1986. GABA-like immunoreactivity in the vertebrate retina: A species comparison. *Exp. Eye Res.* 42: 631–44

Müller, F., Wässle, H., Voigt, T. 1988. Pharmacological modulation of the rod pathway in the cat retina. *J. Neurophysiol.* 59: 1–15

Murakami, M., Ohtsu, K., Ohtsuka, T. 1972. Effects of chemicals on receptors and horizontal cells in the retina. *J. Physiol.* 227: 899–913

Murakami, M., Ohtsuka, T., Shimazaki, H. 1975. Effects of aspartate and glutamate on the bipolar cells in the carp retina. *Vision Res.* 15: 456–58

Murakami, M., Shimoda, Y., Nakatani, K., Miyachi, E., Watanabe, S. 1982a. GABA-mediated negative feedback from horizontal cells to cones in carp retina. *Jpn. J. Physiol.* 32: 911–26

Murakami, M., Shimoda, Y., Nakatani, K., Miyachi, E., Watanabe, S. 1982b. GABA-mediated negative feedback and color opponency in carp retina. *Jpn. J. Physiol.* 32: 927–35

Nawy, S., Copenhagen, D. R. 1987. Multiple classes of glutamate receptor on depolarising bipolar cells in retina. *Nature* 325: 56–58

Negishi, K., Drujan, B. D. 1979. Reciprocal changes in center and surrounding S potentials of fish retina in response to dopamine. *Neurosci. Res.* 4: 313–18

Nelson, R., Kolb, H. 1983. Synaptic patterns and response properties of bipolar cells in the cat retina. *Vision Res.* 23: 1183–96

Nelson, R., Kolb, H. 1985. A 17: A broad field amacrine cell in the rod system of the cat retina. *J. Neurophysiol.* 54: 592–614

Nelson, R., Famiglietti, E. V., Kolb, H. 1978. Intracellular staining reveals different levels of stratification for On- and Off-center ganglion cells in cat retina. *J. Neurophysiol.* 41: 472–83

Oyster, C. W. 1968. The analysis of image motion by the rabbit retina. *J. Physiol.* 199: 613–35

Parkinson, D., Rando, R. R. 1983. Effect of light on dopamine turnover and metabolism in rabbit retina. *Inv. Ophthalmol. Vis. Sci.* 24: 384–88

Piccolino, M., Neyton, J., Gerschenfeld, H. M. 1984. Decrease of gap junction permeability induced by dopamine and cyclic adenosine 3′:5′ monophosphate in horizontal cells of turtle retina. *J. Neurosci.* 4: 2477–88

Pierce, M. E., Besharse, J. C. 1985. Circadian regulation of retinomotor movements. I. Interaction of melatonin and dopamine in the control of cone length. *J. Gen. Physiol.* 86: 671–89

Pourcho, R. 1982. Dopaminergic amacrine cells in the cat retina. *Brain Res.* 252: 101–9

Pourcho, R. G., Goebel, D. J. 1987. A combined Golgi and autoradiographic study of 3H glycine-accumulating cone bipolar cells in the cat retina. *J. Neurosci.* 7: 1178–88

Priest, T. D., Robbins, J., Ikeda, H. 1985. The action of inhibitory neurotransmitters, γ-aminobutyric acid and glycine may distinguish between the area centralis and the peripheral retina in cats. *Vision Res.* 25: 1761–70

Ramon y Cajal, S. 1892. *La Retine des Vertebres,* tr. S. Thorpe, M. Glickstein, 1972. Springfield, Ill: Thomas

Raviola, E., Dacheux, R. F. 1987. Excitatory dyad synapse in rabbit retina. *Proc. Natl. Acad. Sci. USA* 84: 7324–28

Rowe, J. S., Ruddock, K. H. 1982. Hyperpolarization of retinal horizontal cells by excitatory amino acid neurotransmitter antagonists. *Neurosci. Lett.* 30: 251–56

Saito, H. 1983. Pharmacological and morphological differences between X- and Y-type ganglion cells in the cat's retina. *Vision Res.* 23: 1299–1308

Sandell, J. H., Masland, R. H. 1986. A system of indoleamine-accumulating neurons in the rabbit retina. *J. Neurosci.* 6: 3331–47

Sarthy, P. V., Hendrickson, A. E., Wu, J.-Y. 1986. L-glutamate: A neurotransmitter candidate for cone photoreceptors in the monkey retina. *J. Neurosci.* 6: 637–43

Sato, H., Hata, Y., Masui, H., Tsumoto, T. 1987. A functional role of cholinergic innervation to neurons in the cat visual cortex. *J. Neurophysiol.* 58: 765–80

Schiller, P. H. 1982. Central connections of the retinal ON and OFF pathways. *Nature* 297: 580–83

Schellart, N. A. M., Van Acker, H. F., Spekreijse, H. 1984. Influence of GABA on the spectral and spatial coding of goldfish retinal ganglion cells. *Neurosci. Lett.* 48: 31–36

Schmidt, M., Humphrey, M. F., Wassle, H. 1987. Action and localization of acetylcholine in the cat retina. *J. Neurophysiol.* 58: 997–1015

Shiells, R. A., Falk, G., Naghschineh, S. 1981. Action of glutamate and aspartate analogues on rod horizontal and bipolar cells. *Nature* 294: 592–94

Sillito, A. M. 1987. Synaptic processes and neurotransmitters operating in the central visual system: A systems approach. In *Synaptic Function,* ed. G. Edelman,

W. R. Gall, W. M. Cowan, pp. 329–71. New York: Wiley

Sillito, A. M., Kemp, J. A. 1983. Cholinergic modulation of the functional organisation of the cat visual cortex. *Brain Res.* 289: 143–55

Slaughter, M. M., Miller, R. F. 1981. 2-Amino-4-phosphonobutyric acid: A new pharmacological tool for retina research. *Science* 211: 182–85

Slaughter, M. M., Miller, R. F. 1983a. The role of excitatory amino acid transmitters in the mudpuppy retina: An analysis with kainic acid and N-methyl aspartate. *J. Neurosci.* 3: 1701–11

Slaughter, M. M., Miller, R. F. 1983b. Bipolar cells in the mudpuppy retina use an excitatory amino acid neurotransmitter. *Nature* 303: 537–38

Slaughter, M. M., Miller, R. F. 1983c. An excitatory amino acid antagonist blocks cone input to sign-conserving second-order retinal neurons. *Science* 219: 1230–32

Slaughter, M. M., Miller, R. F. 1985a. Characterization of an extended glutamate receptor of the ON bipolar neuron in the vertebrate retina. *J. Neurosci.* 5: 224–33

Slaughter, M. M., Miller, R. F. 1985b. Identification of a distinct synaptic glutamate receptor on horizontal cells in mudpuppy. *Nature* 314: 96–97

Stell, W. K. 1985. Putative peptide transmitters, amacrine cell diversity and function in the inner plexiform layer. In *Neurocircuitry of the Retina: A Cajal Memorial,* ed. A. Gallego, P. Gouras, pp. 171–87. New York: Elsevier

Stell, W. K., Lightfoot, D. O. 1975. Color-specific interconnections of cones and horizontal cells in the retina of the goldfish. *J. Comp. Neurol.* 159: 473–502

Stone, J., Fukuda, Y. 1974. Properties of cat retinal ganglion cells: A comparison of W-cells with X- and Y-cells. *J. Neurophysiol.* 37: 722–48

Tachibana, M., Kaneko, A. 1984. γ-Aminobutyric acid acts at axon terminals of turtle photoreceptors: Difference in sensitivity among cell types. *Proc. Natl. Acad. Sci. USA* 81: 7961–64

Tauchi, M., Masland, R. H. 1984. The shape and arrangement of the cholinergic neurons in the rabbit retina. *Proc. R. Soc. London Ser. B* 223: 101–19

Teranishi, T., Negishi, K., Kato, S. 1984. Regulatory effect of dopamine on spatial properties of horizontal cells in carp retina. *J. Neurosci.* 4: 1271–80

Thier, P., Adler, V. 1984. Action of iontophoretically applied dopamine on cat retinal ganglion cells. *Brain Res.* 292: 109–21

Thier, P., Wassle, H. 1984. Indoleamine mediated reciprocal modulation of on-centre and off-centre ganglion cell activity in the retina of the cat. *J. Physiol.* 351: 613–30

Tork, I., Stone, J. 1979. Morphology of cate-cholamine-containing amacrine cells in the cat's retina, as seen in retinal whole mounts. *Brain Res.* 169: 261–73

Vaney, D. I. 1986. Morphological identification of serotonin-accumulating neurons in the living retina. *Science* 233: 444–46

Vaney, D. I., Peichl, L., Boycott, B. B. 1981. Matching populations of amacrine cells in the inner nuclear layers of the rabbit retina. *J. Comp. Neurol.* 199: 373–91

Vaney, D. I., Young, H. M. 1988. GABA-like immunoreactivity in cholinergic ama-crine cells of the rabbit retina. *Brain Res.* 438: 369–73

Voigt, T., Wassle, H. 1987. Dopaminergic innervation of A II amacrine cells in mammalian retina. *J. Neurosci.* 7: 4115–28

Wassle, H., Voigt, T., Patel, B. 1987. Morphological and immunocytochemical identification of indoleamine-accumulating neurons in the cat retina. *J. Neurosci.* 7: 1574–85

Wassle, H., Schafer-Trenkler, I., Voigt, T. 1986. Analysis of a glycinergic inhibitory pathway in the cat retina. *J. Neurosci.* 6: 594–604

Watkins, J. C., Evans, R. H. 1981. Excitatory amino acid transmitters. *Ann. Rev. Pharmacol. Toxicol.* 21: 165–204

Werblin, F. S., Dowling, J. E. 1969. Or-ganization of the retina of the mud-puppy, *Necturus maculosis*. II. Intracellular recording. *J. Neurophysiol.* 32: 339–55

Witkovsky, P., Alones, V., Piccolino, M. 1987. Morphological changes induced in turtle retinal neurons by exposure to 6-hydroxydopamine and 5,6-dihydroxy-tryptamine. *J. Neurochem.* 16: 55–67

Witkovsky, P., Eldred, W., Karten, H. J. 1984. Catecholamine- and indoleamine-containing neurons in the turtle retina. *J. Comp. Neurol.* 228: 217–25

Witkovsky, P., Stone, S., Besharse, J. C. 1988. Dopamine modifies the balance of rod and cone inputs to horizontal cells of the *Xenopus* retina. *Brain Res.* 449: 332–36

Wu, S. M., Dowling, J. E. 1980. Effects of GABA and glycine on the distal cells of the cyprinid retina. *Brain Res.* 199: 401–14

Wyatt, H. J., Daw, N. W. 1976. Specific effects of neurotransmitter antagonists on ganglion cells in rabbit retina. *Science* 191: 204–5

Yazulla, S. 1986. GABAergic mechanisms in the retina. In *Progress in Retinal Research*, ed. N. Osborne, J. Chader, 5: 1–52. New York: Pergamon

Yazulla, S., Studholme, K., Wu, J.-Y. 1986. Comparative distribution of 3H-GABA uptake and GAD immunoreactivity in goldfish retinal amacrine cells: A double-label analysis. *J. Comp. Neurol.* 244: 149–62

*Ann. Rev. Neurosci. 1989. 12:227–53*

# FLUORESCENT PROBES OF CELL SIGNALING

*Roger Y. Tsien*

Department of Physiology-Anatomy, University of California, Berkeley, California 94720

## INTRODUCTION

Fluorescence has long been recognized as a powerful tool for probing biological structure and function. Because probe molecules can be very much more fluorescent than the constituents of most biological specimens, the signal for the exogenous fluorophores can be measured continuously and nondestructively with excellent spatial and temporal resolution in living cells (Waggoner 1986). The earliest developed and most straightforward uses of fluorescent groups are simply as positional tags or markers. Examples are immunofluorescence labeling (Nairn 1976), fluorescent analog cytochemistry (Taylor et al 1986a), vital staining of organelles (Pagano & Sleight 1985, Wang & Taylor 1988), assessment of cell morphology or intercellular coupling with microinjected tracers (Stewart 1981), measurement of distances between probes by fluorescence energy transfer (Stryer 1978, Uster & Pagano 1986), and measurement of diffusion coefficients and exchange rates by photobleaching recovery (Elson 1986). The common feature of such applications is that the main role of the fluorescent group is merely to signal its presence and location rather than to sense its environment. The main criteria for such fluorescent tags are simple: wavelengths of excitation and emission, brightness, photostability (Mathies & Stryer 1986), size and charge. Therefore a few fluorophores (e.g. fluoresceins, rhodamines, naphthalimides, phycobiliproteins, nitrobenzoxadiazole) tend to get used over and over, often attached by rather standardized techniques to different macromolecules.

A second type of application of fluorescent probes involves attachment of the fluorophore to a purified macromolecule to sense conformational change of the latter (Yguerabide 1972, Cooke 1982, Lakowicz 1983). The

227

repertoire of tags used here includes some of the above fluorophores, plus various naphthalene derivatives such as dansyl groups. Though such naphthalenes are not very fluorescent and require ultraviolet excitation, they are compact, environmentally sensitive, and sanctified by tradition. The weak fluorescence and short excitation wavelength are less troublesome with isolated macromolecules than with intact cells, so that even endogenous tryptophanes can be monitored (Lakowicz 1983, Kleinfeld 1985).

This review focuses on yet a third domain of fluorescence measurements, the use of indicator molecules to sense membrane potential or concentrations of small molecules or ions in organelles, cells, or tissues. Such indicators need to work at suitable wavelengths, fluoresce brightly, resist bleaching, sense their intended stimulus, reject intefering influences, localize in the correct cellular or tissue compartment, and perturb cell function as little as possible. Because of the stringency of these multiple criteria, progress in this area has so far been critically dependent on custom organic design and synthesis of appropriate fluorophores with built-in sensing ability, rather than attachment of stereotyped fluorophores to various macromolecules as in the first two methodologies. The interdisciplinary requirement for organic chemistry plus cell biology, and the low interest in fluorescence shown by mainstream academic and industrial chemists, have greatly restricted the number of active laboratories in this area. Nevertheless the biological payoff both present and future is enormous.

## Absorbance vs Fluorescence

Most of the types of measurements described in the previous paragraph can in principle be made by absorbance rather than fluorescence. The relative advantages of these two optical readout modes have been much discussed in previous reviews (Grinvald 1985, Cohen & Lesher 1986). In brief, fluorescence becomes increasingly advantageous the smaller the number of probe molecules sampled per detector, since small signals are more easily detected against a background of zero than against the full strength of the transmitted beam. However, fluorescence signals suffer badly if diluted by additional constant background fluorescence, whereas transmittance or absorbance signals are hardly degraded in signal/noise by such contamination. Recent technical developments are shifting the balance of favor toward fluorescence.

BETTER OBJECTIVES    Microscope objectives of low autofluorescence, high UV transmission, and high numerical aperture (NA) have become available, for example the Nikon 40x UV-CF objective, with a NA = 1.3. This lens may be calculated to intercept 39% of the fluorescence isotropically re-emitted in an aqueous medium, compared to the 10% geometrical collection efficiency of a 0.8 NA objective assumed in previous calculations (Cohen & Lesher 1986). Fluorescence benefits strongly from increased NA, whereas absorbance does not, since the incoming and outgoing beam are already relatively well collimated.

BETTER IMAGING DETECTORS    Progress in charge-coupled device (CCD) optosensors has produced cameras with excellent spatial resolution (500–2000 pixels on a side), high quantum efficiency, low dark current when cooled, and good photometric resolution (12–14 bits digitization) (Hiraoka et al 1987). The latter two characteristics provide better matches to fluorescence rather than absorbance measurements. The multiplexed serial output obviates the need for hundreds to thousands of separate amplifiers and connections. The main drawback of these CCDs is the hundreds of milliseconds currently required to read out an entire image at full resolution. For higher speeds one has to sacrifice some spatial resolution or wait for expected further refinement of CCD technology.

RATIO FLUORESCENCE    Fluorescent indicators have been introduced that undergo large spectral shifts when they bind their target cations. The ratios of their spectral amplitudes at two selected wavelengths is sufficient to yield an estimate of the actual free concentration of the cation, providing that the dye is appropriately located in the tissue and that the autofluorescence of the tissue can be neglected or corrected for (Grynkiewicz et al 1985, Tsien & Poenie 1986). In principle the ratio of absorbances at two appropriate wavelengths would give the same information. However, separating the dye-related absorbance from the tissue absorbance and scattering is usually much more difficult than separating dye fluorescence from background (Tsien 1986). Therefore absorbance measurements typically just indicate changes in cation concentration; to establish an absolute scale requires a destructive calibration at the end of each experiment, often not possible or desirable.

CONFOCAL MICROSCOPY    Tremendous interest has developed in scanning confocal microscopy as a means by which relatively thick specimens may be optically sectioned, thus suppressing contributions from out-of-focus planes that would otherwise blur the view of the desired plane (Wilson & Sheppard 1984, White et al 1987). In oversimplified brief terms, the principle of confocal microscopy (Figure 1) is that at any one instant the specimen is illuminated with the image of a pinhole and viewed through an optically conjugate (confocal) pinhole. The illumination rays, which form a converging then diverging conical bundle as they pass through the specimen, cannot help but illuminate parts of the specimen in front and

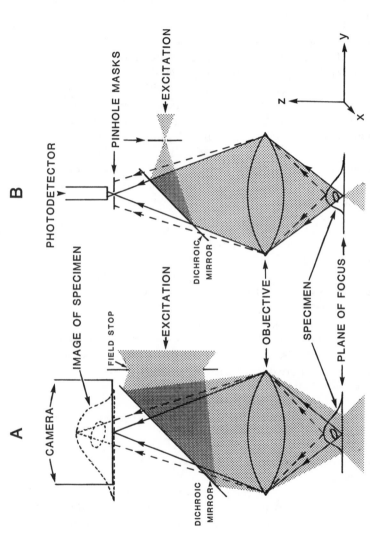

*Figure 1* Comparison of conventional (A) and confocal (B) epifluorescence microscopy, showing how confocal imaging discriminates against blurring from out-of-focus parts of the specimen. Excitation light, shown as *stippling*, enters from the right, reflects off the dichroic mirror, and passes downward through the schematized objective and specimen. In this example the plane of focus is imagined to be the bottom of the cell, where it contacts the substrate. In conventional microscopy (A) the wide-open field stop allows fairly uniform illumination of the specimen; fluorescence emission from the desired plane (*solid lines*) focuses on the camera film or face plate, but emission from out-of-focus portions (*dashed lines*) also reaches the camera, causing an obscuring haze. In confocal microscopy (B) the field stop is narrowed to a pinhole so that intense illumination is tightly focused at one point of the specimen. Other parts receive much less intense excitation or none at all. Emission from the desired zone efficiently passes through another pinhole in front of the photodetector, whereas emission from out-of-focus planes is largely blocked.

back of the desired plane of focus. However, light re-emitted from those conical regions is out of focus when it reaches the plane of the detector pinhole and is therefore almost completely rejected at that pinhole. In order to build up an entire two-dimensional image, the two pinholes must be scanned in $x$ and $y$ over the specimen while maintaining them in optical register. This scanning is greatly simplified if epi-illumination is used, since a single beam deflector between the dichroic mirror and the objective handles both the incident and re-emitted rays. Also, optical access to the specimen is needed only from one side. However, in transmission mode one would require two separate scanners for the in-going and out-going beams, which would be difficult to keep in registration. The alternative of fixing the optics and mechanically scanning the specimen is unattractive for live tissue into which electrodes may be inserted. Transmission mode confocal microscopy is not only technically more difficult but also suffers from some inherent optical inferiorities compared to epifluorescence mode. For example, if nonscattering absorbance were confined to a thin uniform layer extending far enough in the $x$ and $y$ directions to intercept all the rays, even a confocal microscope could not tell at what depth the sheet was located, whereas a sheet of uniform fluorescence is readily localized along the $z$-axis because the fluorescent molecules re-emit isotropically and incoherently. Because confocal scanning microscopy works so well with epifluorescence to image any chosen plane of focus selectively, it promises to become the method of choice for any tissue with significant three-dimensional structure, as almost all neuronal tissues have.

## MEMBRANE POTENTIAL INDICATORS

Dyes sensitive to membrane potential are perhaps the best known and longest used of the fluorescent indicators of dynamic cell function. I do not discuss such potentiometric dyes in detail, since their application has been extensively and expertly reviewed recently by their chief protagonists (Waggoner 1979, Salzberg 1983, Cohen & Lesher 1986, Laris & Hoffman 1986, Salzberg et al 1986, Freedman & Laris 1988, Gross & Loew 1988, Loew 1988a,b, Smith 1988). These indicators are divided into two classes: dyes whose responsiveness to membrane potential depends on translocation across the membrane, vs dyes that sense the electric field without crossing the membrane.

### Permeant Dyes

Permeant dyes have to be lipid-soluble ions in order for the membrane potential to move them across the membrane without endogenous carriers or channels. They give relatively slow responses due to the time required for translocation of the ions from one bulk phase to another. Redistribution affects the overall fluorescence intensity, either because the dye fluoresces more strongly when bound to hydrophobic binding sites in the cell than in the extracellular medium, or because the dye is quenched in the cell by crowding. Because the ions sample the full voltage across the membrane and distribute according to the Nernst equation, the optical responses can be quite large, with several-fold changes in fluorescence for 100 mV change in membrane potential. Also, they tend to work fairly well on a wide variety of cell types, whereas the fast dyes are more idiosyncratic. These "redistributive" dyes are usually dismissed for neurobiological application (Salzberg 1983) because they are too slow to follow individual action potentials, and because dye access to the interior might have extra pharmacological side effects on the cell. But in many current applications, especially in the CNS, the cells are too tightly packed to resolve single cells anyway, so that information on averaged spike activity over tenths of a second would already be of considerable value (Orbach et al 1985, Blasdel & Salama 1986, Grinvald et al 1986, Kauer et al 1987).

Such time resolution is attainable at least with negatively charged dyes (mostly oxonols); the more familiar cationic dyes, cyanines, cross membrane much more slowly due to the positive internal dipole potential of biological membranes (Andersen 1978). The advantage of an oxonol over a non-redistributive dye would be the likelihood of much greater optical sensitivity of the former. As for the worries about pharmacological effects, access to the cytoplasm need not be harmful, since both permeant oxonols and intracellularly injected impermeant dyes have been highly useful without detectable toxicity (Chused et al 1986, Grinvald et al 1987).

## Impermeant Dyes

Fast non-redistributive dyes respond to the membrane potential by a variety of mechanisms, including electrochromism, potential-dependent re-orientation and/or dimerization, etc (Waggoner & Grinvald 1977, Loew et al 1985, Wolf & Waggoner 1986, Loew 1988b). Signal sizes are usually quite small, typically a 0.01% to 1% change in intensity for 100 mV potential change in real neuronal tissue. These amplitudes are difficult to predict in advance: Closely analogous molecular structures often give widely different sensitivities in a given cell type, and a given structure can behave differently even in homologous tissues from different species of the same genus (Ross & Krauthamer 1984). For this reason, the main strategy for finding fast voltage-sensitive indicators has been semirandom screening of nearly 2000 candidates so far (Grinvald 1985). Much of the variability probably arises because most of the voltage-sensing mechanisms are critically dependent on the detailed location, orientation, concentration, or local environment of the probe. Also, the voltage-sensitive component of the signal is usually severely diluted by fluorescence from dye bound to sites other than the plasma membrane of the relevant cell(s). Signals of up to 10–25% change in intensity per 100 mV have been observed in artificial membranes (Fluhler et al 1985) and cultured cells (Grinvald 1985) with clean surfaces; in the latter, the dyes can even reveal the local distribution of transmembrane potential changes due to an externally applied field (Gross et al 1986). If comparable voltage sensitivity could be attained in normal neuronal tissues, the usefulness of this technique would expand enormously, especially because the percentage changes would become large enough to be handled by video equipment (Blasdel & Salama 1986, Kauer 1988) and confocal microscopes. With signal sizes of $<1\%/100$ mV, one is forced to use photodiode arrays with cumbersome arrays of hundreds of individual amplifiers (Grinvald 1985, Cohen & Lesher 1986). However, to achieve the desired sensitivity will probably require new molecular mechanisms, or a way to restrict the fluorescence to that from the plasma membrane, or both.

## ION CONCENTRATION INDICATORS

The design of indicators of ion concentrations poses quite different problems from the design of voltage-sensitive dyes. The ion indicator is usually intended to work in a homogeneous aqueous environment, which is much simpler than the biological membrane with which a potentiometric dye must interact. The spectroscopic effect of binding an ion is much greater than the influence of transmembrane electric fields, simply because the local electrostatic field in the inner coordination sphere of an ion is on the order of $10^8$ V/cm compared to a mere $10^5$–$10^6$ V/cm across a bilayer. Finally, assuming that an ion indicator is loaded into the intracellular volume, one can usually wash away all the extracellular dye, so that the signal comes nearly entirely from the compartment of interest. By contrast it is difficult to prevent a membrane-adherent dye from binding to all sorts of irrelevant membranes and hydrophobic sites other than the membrane whose potential is of interest. Ion indicators face their own problems of binding selectivity, namely discrimination between competing ions such as $Ca^{2+}$ and $Mg^{2+}$, or $Na^+$ and $K^+$. In each pair the minority ion is the one whose concentration changes more dynamically and is of greater need of optical measurement. Fortunately, the desired selectivities have been attainable by using fairly well understood chemical principles (e.g. Vögtle & Weber 1980) to exploit the different sizes and charge densities of the target ions. One particular challenge with an ion indicator is to tune its dissociation constant to near the midpoint of the concentration range to be measured, so that the probe is maximally sensitive and able to respond to fluctuations in either direction. Any sensor whose response can be saturated faces this problem. With "slow," redistributive voltage-sensitive dyes, the sensitivity range can be adjusted by the ratio of extracellular to intracellular space or the strength of hydrophobic binding of the dye to cell constituents so that about half the dye is intracellular at the midpoint of the voltage range of interest. With "fast," non-redistributive dyes the problem is usually less apparent because the voltage sensitivity is too weak to be saturated. Ion-sensitive indicators get tuned in affinity by the laborious process of manipulating the electron affinity or steric properties of substituents (Tsien 1980, Adams et al 1988).

## pH Indicators

BCECF    Cytosolic pH in nearly all cells is tightly regulated to be near 7.0. Variations are limited to a few tenths of a unit at most. Therefore a useful intracellular indicator needs a $pK_a$ very near 7.0. The most popular probe currently is "BCECF," 2′,7′-bis(carboxyethyl)-5(or 6)carboxyfluorescein (Rink et al 1982). BCECF (Figure 2) is a derivative of fluorescein with three extra carboxylate groups to increase hydrophilicity. Two of the carboxylates are attached by short alkyl chains, which serve the additional purpose of raising the $pK_a$ to 6.97–6.99 from the original value of 6.4 in simple fluorescein. BCECF is strongly fluorescent like most fluoresceins, with excitation and emission peaks at 503 nm and 525 nm, respectively. These peaks are strongly pH-dependent in amplitude, being quenched by acidification and enhanced by alkalinification. A valuable additional feature is that at 436–439 nm excitation, the fluorescence is pH-independent (Rink et al 1982, Alpern 1985). The existence of a pH-independent as well as a pH-dependent part of the excitation spectrum is very useful, since the ratio of the latter to the former is a measure of pH that is independent of dye concentration, optical path length, absolute brightness of the overall illumination (assuming the spectral balance is constant), and detector sensitivity. The advantages of ratioing are particularly evident in microscopic imaging of single cells (Bright et al 1987, Paradiso et al 1987).

Though BCECF could be microinjected, it is quite easily introduced into a wide variety of cells ranging from bacteria to mammalian cells by incubating them in a few micromolar of the acetoxymethyl (AM) ester of BCECF, or BCECF/AM. This ester is uncharged, hydrophobic, therefore membrane permeant, yet gradually regenerates BCECF on contact with cytosolic esterases. This convenient trick traps the hydrophilic, relatively impermeant polyanion BCECF in the cells without any microinjection or breaching of the plasma membrane.

BCECF is usually calibrated in situ by subjecting the cells to known

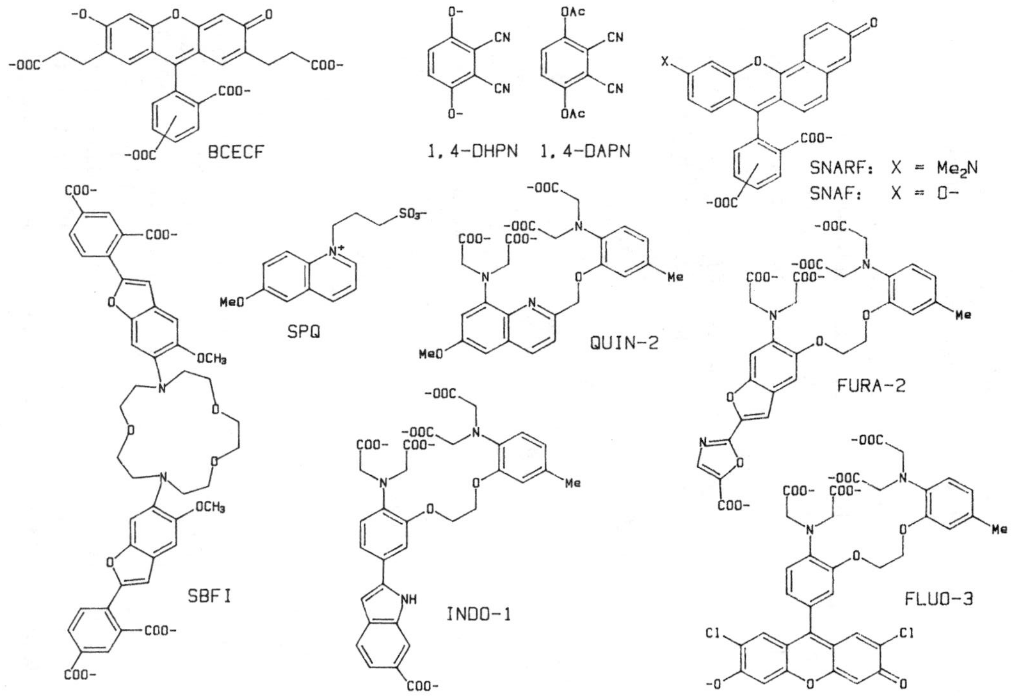

*Figure 2* Structures of fluorescent indicators of ion concentrations. Each is drawn as if ready to bind its target ion. In BCECF, SNARF, and SNAF, the linkage of the carboxylate to the center of the lower benzene ring indicates that the carboxylate could be attached to either of the two adjacent positions, and that the usual preparation consists of a mixture of these two isomers. SNAF-2 actually has one extra chloro atom *ortho* and *para* to the phenol and ring oxygens, respectively. In acetoxymethyl (AM) esters, each –COO⁻ group is replaced by –COOCH₂OAc, where –OAc means –OCOCH₃.

internal pHs by using nigericin, a $K^+/H^+$ ionophore (Thomas et al 1979). When external $K^+$ equals intracellular, nigericin (1–10 $\mu$M) clamps intracellular pH to extracellular pH, which the experimenter sets to various levels while observing the dye fluorescence. In some cells the dye is a few nanometers red-shifted and has about 0.1–0.15 unit higher $pK_a$ than dye in simple aqueous buffers (Rink et al 1982, Paradiso et al 1986), whereas in other cell types such perturbation by cytoplasm was not found (Bright et al 1987).

1,4-DIHYDROXYPHTHALONITRILE ( = 2,3-DICYANOHYDROQUINONE)    Some applications, particularly flow cytometry or scanning confocal microscopy with laser excitation, are much more convenient with a probe that shifts its emission spectrum. Then the ratio of the intensities measured simultaneously at two chosen emission wavelengths can signal the ion concentration. This preserves all the usual advantages of ratioing but eliminates the need to alternate two excitations or multiplex them by frequency modulation (Kurtz 1987). But only the excitation not the emission spectrum of BCECF changes wavelength in response to pH. Currently the only commercially available emission-shifting indicator for intracellular pH is 1,4-dihydroxyphthalonitrile, 1,4-DHPN (Kurtz & Balaban 1985). When excited at 375–407 nm, 1,4-DHPN shifts from prominently blue ($\sim$450 nm) to more greenish ($\sim$480 nm) emission as pH rises from 6 to 10, with a $pK_a$ of 8.0 (Brown & Porter 1977). 1,4-DHPN is readily loaded into cells by hydrolysis of its acetate diester, 1,4-diacetoxyphthalonitrile (1,4-DAPN, Figure 2). The 1,4-DHPN leaks out quite readily, having only 1–2 negative charges, which moreover are partly delocalized. Fortunately the ester precursor is relatively nonfluorescent, so that in many systems the cells may be observed during continuous incubation in the ester to establish a steady state between loading and leakage (Valet et al 1981).

SNA(R)F    Very recently, Molecular Probes, Inc. announced (R. Haugland, personal communication) a promising series of emission-shifting pH indicators with naphthofluorescein chromophores (Figure 2). SNARF-1 excited at 514 nm emits at 588 nm (acid) vs 637 nm (base) with a $pK_a$ of 7.5; SNAF-2 in acid is excited at 490–520 nm and emits at 540 nm, whereas in base the excitation and emission peaks are 586 nm at 630 nm, the $pK_a$ being 7.65–7.7. Biological testing and modification to lower the $pK_a$s are awaited with interest.

## Sodium Indicators

The electrochemical gradient of $Na^+$ across the plasma membrane is the storage battery that powers most action potentials, synaptic depolarizations, and active uptake of nutrients and neurotransmitters. It also plays a major role in the regulation of other ions such as $H^+$ and $Ca^{2+}$. Cells devote a major fraction of their total metabolic energy to pumping intracellular $Na^+$ concentrations, $[Na^+]_i$, to levels much lower than extracellular. Changes of $[Na^+]_i$ are important in mediating the metabolic response to increased cellular activity or injury. Modulation of $[Na^+]_i$ has also been proposed to play an important role in embryonic (Breckinridge & Warner 1982) and NGF-induced neuronal differentiation (Varon & Skaper 1983) and the control of certain $K^+$ currents (Bader et al 1985). Sodium-sensitive indicators are therefore under development in at least two laboratories.

Minta et al (1987) have synthesized a variety of such dyes, of which the current favorite is SBFI, whose structure is shown in Figure 2. It consists of a crown ether of the right size to form an equatorial belt around a $Na^+$ ion, with additional ether oxygens capping both poles. Potassium rejection arises from the size of the crown ether cavity; expansion of the macrocyclic ring has been verified to convert $Na^+$ selectivity to $K^+$ selectivity. Divalent cations are rejected because there are no negative charges lining the cavity. The attached fluorophores are benzofurans rather similar to those in the $Ca^{2+}$ indicator fura-2 (see below), so that the SBFI wavelengths and shift due to $Na^+$-binding are similar to fura-2 and its $Ca^{2+}$ response. SBFI has two identical fluorophores mainly because the organic synthesis was eased by preserving the symmetry around the crown ether ring, though as a side benefit the extinction coefficient is doubled. In the presence of typical vertebrate intracellular $K^+$ levels, the effective dissociation constant for $Na^+$ is 17–18 mM, well suited to monitor $[Na^+]_i$ changes around the typical resting level of 10–20 mM. For example, SBFI has detected a few millimolar increase of $[Na^+]_i$ in single fibroblasts stimulated with mitogens to activate $Na^+/H^+$ exchange. As usual for polycarboxylate dyes, SBFI can be introduced into cells either by microinjection or by hydrolysis of its membrane-permeant acetoxymethyl ester. Calibration is most conveniently performed in intact cells with the pore-forming antibiotic, gramicidin, which rapidly clamps $[Na^+]_i$ and $[K^+]_i$ equal to the extracellular levels of those ions.

Smith et al (1988) have reported a cryptand-based $Na^+$ indicator, FCryp-2, with excellent $Na^+$ affinity (dissociation constant 6 mM) and $K^+$ rejection. When it is excited at 340 nm, the binding of $Na^+$ increases the emission peaking at 395 nm at the expense of that $> 460$ nm. Results inside real cells would be of great interest.

## Chloride Indicators

Chloride ion fluxes are important in several types of inhibitory synapses and in pH regulation by $Cl^-/HCO_3^-$ and related countertransport systems.

Recently, Illsley & Verkman (1987) have shown that 6-methoxy-$N$-(3-sulfopropyl)quinolinium (SPQ) can be used as a $Cl^-$ indicator in vesicles and erythrocyte ghosts. SPQ differs in principle from all the other ion indicators, because in its ground state it does not associate with its target ion. Chloride interacts only with the excited state of SPQ, causing radiationless quenching of the dye fluorescence with no change in the absorbance spectrum. The mechanism for the quenching and the basis for halide selectivity are not really understood, though the effect has long been known in analogous heterocyclic cations such as diprotonated quinine. The Stern-Volmer equation for chloride-dependent quenching is mathematically equivalent to formation of a nonfluorescent $Cl^-$ complex with a dissociation constant of 8.5 mM in free solution. However, inside intact cells of kidney proximal convoluted tubules, the apparent dissociation constant seems to be tenfold greater, 83 mM (Krapf et al 1988a). This drastic weakening of the dye's $Cl^-$ sensitivity has been ascribed to a combination of intracellular anions, viscosity, and dye-binding. Calibration in intact cells therefore must be based on clamping the $[Cl^-]_i$ in situ using ionophores such as tributyltin, a $Cl^-/OH^-$ exchanger, together with nigericin, a $K^+/H^+$ exchanger (Krapf et al 1988a). This calibration procedure, which would have to be done at the end of every experiment, also overcomes the lack of any wavelength shift or ratio capability in the current dye. In the kidney cells, a basal $[Cl^-]_i$ of about 28 mM was determined.

SPQ is surprisingly permeable through membranes, considering that it is a zwitterion with a quaternary nitrogen cation and sulfonate anion. Either group alone is normally sufficient to prevent ready permeation. Perhaps the positive and negative charge nullify each other by ion-pairing. SPQ is loaded into cells and vesicles simply by soaking them in high concentrations of the dye; of course the dye also readily leaks out once the external excess is removed. This is a major current deficiency of SPQ, which may be fixable by additional carboxylate groups protected as acetoxymethyl esters (Krapf et al 1988b).

## Calcium Indicators

More intracellular studies hve been done with indicators for $Ca^{2+}$ than for any other ion. This emphasis reflects the pivotal importance of $Ca^{2+}$ in cellular signal transduction. $Ca^{2+}$ fluctuations play a particularly major role in neurobiology as the key link between membrane depolarization and intracellular biochemical activation (Hille 1984), especially neurotransmitter secretion (Augustine et al 1987) and enzyme activation. Earlier techniques for measuring cytosolic free $Ca^{2+}$, $[Ca^{2+}]_i$ (Blinks et al 1982, Tsien & Rink 1983), such as the luminescent photoprotein aequorin, the absorbance dye arsenazo III, and $Ca^{2+}$-sensitive microelectrodes, all required microinjection or impalements, and were therefore applied mainly to giant cells. More recently, photoproteins have been loaded by various reversible permeabilization procedures (Cobbold & Rink 1987). But the largest expansion in the range of cell types in which $Ca^{2+}$ signals can be quantified has come from the development of new fluorescent indicators that can be loaded by using hydrolyzable esters.

The new dyes are not without their own difficulties and restrictions, but their generic structure is amenable to further optimization along reasonably rational chemical principles. Currently, four fluorescent indicators are in use: quin-2, fura-2, indo-1, and fluo-3. Their structures (Figure 2) share nearly identical binding sites, which are modeled (Tsien 1980) on the well-known $Ca^{2+}$-selective chelator EGTA. This octacoordinate binding site binds $Mg^{2+}$ about five orders of magnitude more weakly than $Ca^{2+}$ because $Mg^{2+}$ is too small to contact more than about half the liganding groups simultaneously. Monovalent cations do not form detectable specific complexes, probably because their charge is inadequate to organize the binding pocket in the face of the electrostatic repulsion of the negative carboxylates. EGTA at pH 7 is normally occupied by two protons, but the incorporation of the aromatic rings in the fluorescent indicators lowers the $pK_a$ of the amine nitrogens to 6.5 or below, thus eliminating almost all the proton interference for pH > 6.8. $Ca^{2+}$ binding diverts the nitrogen lone pair electrons away from the aromatic system, causing large spectral changes that mimic disconnection of the nitrogen substituents. Conversely, the more electron-donating or withdrawing the aromatic nucleus, the higher or lower the $Ca^{2+}$ affinity (Tsien 1980). This principle is exemplified in photochemically reactive chelators, in which photolysis increases or decreases the $Ca^{2+}$ affinity by destroying or creating an electron-withdrawing ketone group *para* to the amine nitrogen (Tsien & Zucker 1986, Gurney et al 1987, Adams et al 1988). Because a wide variety of substituents can be plugged in without changing the geometry of the binding site, the design of this family of tetracarboxylate $Ca^{2+}$ indicators and chelators is quite versatile, rather like a household appliance that accepts a range of attachments for different jobs. Of course the actual organic syntheses are a little more difficult than just fitting pieces together.

Traditionally, fast voltage-sensitive dyes have been the favorites for watching neuronal activity, but $Ca^{2+}$-sensitive dyes may offer complementary information. The latter give larger signals, changing their intensities or ratios many-fold upon cell stimulation instead of a few percent at most, as with non-redistributive voltage dyes. Though $Ca^{2+}$-dyes generally do not resolve single action potentials (but see Schlegel et al 1987), the slower $Ca^{2+}$ signal may itself be as important as the detailed spike pattern, since it is the $[Ca^{2+}]_i$ that controls synaptic activity and in some cases plasticity (Augustine et al 1987, Zucker 1989).

QUIN-2    Quin-2 (reviewed in Rink & Pozzan 1986, Tsien & Pozzan 1989) was the first practical fluorescent indicator of this family, with a simple 6-methoxyquinoline as its fluorophore. The small size of this group means that quin-2 is best excited at fairly short wavelengths (339 nm) and has only a modest extinction coefficient. The brightness of quin-2 fluorescence is not very great, so that relatively high intracellular concentrations, milli-molar to tenths of millimolar, are needed to overcome cellular auto-fluorescence. These levels often buffer fast $Ca^{2+}$ transients. Quin-2 binds $Ca^{2+}$ with a dissociation constant of 60 nM or 115 nM, respectively, in the absence or presence of 1 mM $Mg^{2+}$, both measured under conditions intended to mimic mammalian cytoplasm, pH 7.05, 37°, $\sim 140$ mM ionic strength (Tsien et al 1982). The strong binding (low $K_D$) means that quin-2 is best at measuring submicromolar $[Ca^{2+}]_i$ and saturates near 1–2 $\mu$M. The effect of $Mg^{2+}$ corresponds to a $Mg^{2+}$ dissociation constant on the order of millimolar; quin-2 has poorer $Ca^{2+} : Mg^{2+}$ discrimination (only $10^4 : 1$) than its siblings due to its use of a quinoline nitrogen in place of one ether oxygen. At 339 nm excitation, $Ca^{2+}$-binding increases the fluorescence intensity about six-fold whereas $Mg^{2+}$ has no effect. At longer excitation wavelengths the fluorescence intensity drops off sharply and the proportional effect of $Ca^{2+}$ decreases, whereas the effect of $Mg^{2+}$ increases. Therefore quin-2 does not show a useful $Ca^{2+}$-induced wavelength shift in either excitation or emission spectrum with which to generate a ratio signal (Tsien & Pozzan 1989).

The short excitation wavelengths, modest fluorescence brightness, inade-quacy of ratioing, and poor photostability of quin-2 make it unsuitable for single cell microscopy. Instead, quin-2 has been used mainly in sus-pensions of cells in a cuvet, though occasionally in cell monolayers attached to cover slips inserted diagonally in a cuvet. Its signal is calibrated by lysis of the cells and direct titration of the lysate to known levels of $Ca^{2+}$, or by using ionomycin first to raise $[Ca^{2+}]_i$ to saturating levels then to intro-duce $Mn^{2+}$ into the cells to quench the dye and determine the auto-fluorescence level. Quin-2 does have advantages for some purposes over its more recent relatives. In particular, hydrolysis of quin-2/AM seems easier, reaches higher cytosolic concentrations of chelator and is less often complicated by compartmentation into organellar compartments than is observed with AM esters of higher molecular weight and lesser water solubility. Loading with excess indicator (typically several millimolar intra-cellular concentration) in order to buffer cytosolic $[Ca^{2+}]_i$ is a powerful experimental tool, and quin-2 does it better than any of the other indicators. At a qualitative level, such buffering reveals which cellular responses are truly dependent on elevated $[Ca^{2+}]_i$. At a quantitative level, the dependence of $[Ca^{2+}]_i$ on amount of buffering can reveal the size of the net $Ca^{2+}$ flux into the cytosolic compartment and the amount of the endogenous cellular $Ca^{2+}$ buffering (Tsien & Rink 1983).

FURA-2    Fura-2 (Figure 2) is currently the most popular $Ca^{2+}$ indicator for microscopy of individual cells. Compared to quin-2, the larger fluorophore of fura-2 gives it slightly longer wavelengths of excitation compatible with glass microscope optics, a much larger extinction coefficient, and a higher quantum efficiency, resulting in about 30-fold higher brightness per molecule. $Ca^{2+}$ binding shifts the excitation spectrum about 30 nm to shorter wavelengths, so that the ratio of intensities obtained from 340/380 nm or 350/385 nm excitation pairs is a good measure of $[Ca^{2+}]_i$ unperturbed by variable cell thickness or dye content (Grynkiewicz et al 1985). The green emission from fura-2 peaks at 505–520 nm and does not shift usefully with $Ca^{2+}$-binding. Fura-2 is also very much more resistant to photo-destruction than quin-2. Though fura-2 can be degraded eventually (Becker & Fay 1987), most investigators have found that by attenuating the excitation beam, blocking it whenever measurements are not actually in progress, and using efficient photodetectors, adequate signals can be obtained from single cells for tens of minutes to hours of observations.

Fura-2 binds $Ca^{2+}$ slightly less strongly than quin-2 does. Dissociation constants of 135 nM (no $Mg^{2+}$, 20°, in 100 mM KCl), 224 nM (1 mM $Mg^{2+}$, 37°, 120 mM $K^+$, 20 mM $Na^+$) (Grynkiewicz et al 1985), and 774 nM (no $Mg^{2+}$, 18°, 225 mM $K^+$, 25 mM $Na^+$) (Poenie et al 1985) have been reported, implying that ionic strength, not $Mg^{2+}$, seems to be the most powerful influence on the apparent $K_{DS}$. The $Mg^{2+}$ dissociation constants of 6–10 mM at 37° to 20° represent much better $Ca^{2+}:Mg^{2+}$ discrimination than quin-2 has.

The absolute calibration of fura-2 inside cells is complicated by the fact that dye does seem to have somewhat different spectral characteristics in most cells than in calibration buffers. In cytoplasm the 380–385 nm excitation amplitude is increased 1.1–1.6-fold relative to that at 340–350 nm, thus shifting the ratios downward (Almers & Neher 1985, Tsien et al 1985). This red-shift can be simulated by increasing the viscosity of the calibration medium with agents such as gelatin or sucrose; the amount of viscosity correction to be made can be estimated by fluorescence polarization measurements, or by the ratio of the intensity changes at the two excitation wavelengths when $[Ca^{2+}]_i$ changes but dye content does not (M. Poenie and R. Y. Tsien, manuscript in preparation). The typical net effect of viscosity is to reduce the 350/385 nm ratio by about 15% (Poenie et al

1986). Ideally one would calibrate the dye in situ by clamping the cell to known $[Ca^{2+}]_i$ values with an ionophore such as ionomycin (as in Williams et al 1985 or Chused et al 1987), but such ionophores do not mediate large fluxes of $Ca^{2+}$ when the concentrations are only micromolar or less on both sides of the membrane, so such calibrations are more difficult than pH or $Na^+$ calibrations with nigericin or gramicidin, and are often impossible.

The kinetics of $Ca^{2+}$ binding to fura-2 have been characterized in vitro by stopped-flow (Jackson et al 1987) and temperature jump (Kao & Tsien 1988), giving an association rate constant $k_A$ of $6 \times 10^8$ $M^{-1}$ $s^{-1}$ and a dissociation rate $k_D$ of 84–97 $s^{-1}$ at $20°$ in 0.1 M KCl. The exponential time constant $\tau$ for the response of an indicator of 1 : 1 stoichiometry is given by $\tau = (k_A[Ca^{2+}] + k_D)^{-1}$. Here too there is evidence for perturbation by cytoplasm, in that fura-2 seems to behave in skeletal muscle as though both rate constants were 4–8-fold lower than the in vitro values (Hollingworth & Baylor 1987, Klein et al 1988).

Perhaps the worst problems with fura-2 are that in some tissues AM ester hydrolysis is incomplete (Highsmith et al 1986, Scanlon et al 1987, Oakes et al 1988), the fluorescence becomes compartmentalized into organelles (Almers & Neher 1985, Malgaroli et al 1987, Lukacs et al 1988) or dye is extruded from the cell by anion transport mechanisms (DiVirgilio et al 1987). Many procedures have been empirically developed can ameliorate these problems, but success on every tissue is not guaranteed. Loading is often much improved by mixing the fura-2/AM with amphiphilic dispersing agents before diluting into the incubation medium. Suitable agents include albumin, serum, and Pluronic F-127 (Poenie et al 1986, Barcenas-Ruiz & Wier 1987). Endocytosis and compartmentalization can often be significantly slowed by a reduction in temperature, e.g. from $37°$ to $32°$ during observation (Poenie et al 1986) or to $15°$ just during loading (Malgaroli et al 1987). Anion extrusion can be inhibited by probenecid (DiVirgilio et al 1987) and sulfinpyrazone, which are well known clinically as blockers of uric acid transport. Often, introduction of the dye penta-anion by microinjection (e.g. Cannell et al 1987) or by perfusion with a patch pipet (Almers & Neher 1985) or reversible permeabilization (e.g. Ratan et al 1986) gives "better-behaved" dye. Though these techniques are less convenient than AM ester hydrolysis for loading dye into cells, they are just as applicable to fura-2 than as to traditional $Ca^{2+}$ indicators. Some cells, e.g. sea urchin embryos (M. Poenie, J. Alderton, R. Steinhardt, and R. Y. Tsien, unpublished observations) and plant stamen hair cells (Hepler & Callaham 1987), gradually compartmentalize even injected dye; for such cells, molecular redesign or attachment of fura-2 to a macromolecule such as dextran (R. Haugland, personal communication) may be necessary.

Despite the above litany of cautions, fura-2 has provided much useful new information about the role and regulation of $[Ca^{2+}]_i$. Fortunately, fura-2/AM loading seems less problematical in neurons than in other cell types referenced above. Major compartmentation has not been a problem; for example, Cohan et al (1987, 1988) found microinjected and ester-loaded dye to give essentially identical results. Ester loading seems to give very little fluorescence in glia and other flat cells compared with good amplitudes in cerebellar neurons (Connor et al 1987). $[Ca^{2+}]_i$ can be measured with good signal-to-noise ratio even in neuronal processes as thin as 1 micron (Thayer et al 1987). Signal rise times as fast as 7–10 ms (measured from 10% to 90% of final amplitude) have been observed; dye bleaching and photodynamic damage were insignificant (Lev-Ram & Grinvald 1987). Four major areas of current neurobiological interest are as follows:

*Spatial heterogeneity*    Fura-2 imaging is a powerful tool to study how $[Ca^{2+}]_i$ regulation varies from one part of a neuron to another. Smith et al (1987) observed that tetanization of the squid giant presynaptic terminal caused a wave of $Ca^{2+}$ to spread from the side facing the synaptic cleft, as though the $Ca^{2+}$ channels were co-localized with transmitter release. Hirning et al (1988) found in rat sympathetic neurons that nitrendipine blocked much of the $[Ca^{2+}]_i$ rise due to $K^+$ depolarization, yet hardly affected norepinephrine release. This finding, along with other evidence, points to the dihydropyridine-insensitive "N"-type $Ca^{2+}$ channels rather than the dihydropyridine-sensitive "L"-type channels as controlling nor-epinephrine secretion. Miller (1987) speculated that N channels might be preferentially located in the processes in which the transmitter is stored, whereas L channels might be concentrated in the soma. However, both fura-2 recordings (Thayer et al 1987) and patch clamping (Lipscombe et al 1988a) have failed to confirm such a gross segregation, though the possibility of microheterogeneity in channel distribution remains.

Lipscombe et al (1988a), working on frog sympathetic neurons in culture, did find a difference between cell bodies and processes in their responses to caffeine. This agent, which dumps internal stores of $Ca^{2+}$ without significant effect on $Ca^{2+}$ currents, raised $[Ca^{2+}]_i$ higher and for a longer time in the cell body than in the neurites. These internal stores are surprisingly important even for the response to depolarization; prior depletion of the stores using caffeine greatly weakened the $[Ca^{2+}]_i$ rise evoked by high $K^+$ (Lipscombe et al 1988b). The stores may serve (at least in the cell body) to amplify the effectiveness of $Ca^{2+}$ influx through plasma membrane channels. Evidence for $Ca^{2+}$-induced $Ca^{2+}$-release was found, in that caffeine and $K^+$ depolarization applied together caused $[Ca^{2+}]_i$ oscillations, and high speed imaging of the inward radial spread of $Ca^{2+}$ during stimulated action potentials indicated that $[Ca^{2+}]_i$ continued to climb well after the stimuli were cut off. Yet another role for spatially nonuniform $[Ca^{2+}]_i$ has been suggested by Connor (1986) and Cohan et al (1987) working in cultured rat diencephalon and snail neurons. They found that $[Ca^{2+}]_i$ in actively extending growth cones was moderately elevated compared to the levels in the soma or in stalled neurites. Excessive $[Ca^{2+}]_i$ resulting from action potentials or serotonin application also correlated with cessation of growth, suggesting that growth cone motility and extension require intermediate $[Ca^{2+}]_i$ values.

*Transmitter actions*   Connor et al (1987) have observed a surprising ability of the inhibitory transmitter GABA to induce long-lasting moderate elevations of $[Ca^{2+}]_i$, as well as facilitation by repetitive $K^+$ depolarizations in developing rat cerebellar granule neurons. Connor et al (1987), Murphy et al (1987), and Kudo & Ogura (1986) have examined glutamate-, *N*-methyl-D-aspartate(NMDA)-, and kainate-induced rises in $[Ca^{2+}]_i$ in rodent cerebellar, striatal, and hippocampal cells, respectively. In the latter two cases, voltage-sensitive $Ca^{2+}$ channels, $Na^+/Ca^{2+}$ exchange, and major release of intracellular $Ca^{2+}$ stores could be ruled out, confirming $Ca^{2+}$ influx through receptor-linked channels as the most likely mechanism. Inhibitory agents have also been studied: The ability of somatostatin to block prolactin secretion in pituitary cells can be explained by inhibition of spontaneous $[Ca^{2+}]_i$ spiking (Schlegel et al 1987).

A more complex interaction between excitatory and inhibitory inputs has been demonstrated by Connor et al (1988), who showed that $[Ca^{2+}]_i$ elevations in dendrites of cultured hippocampal CA1 neurons persisted for minutes after a few transient local exposures to glutamate or NMDA in low $Mg^{2+}$ media. Often these responses formed striking $[Ca^{2+}]_i$ gradients along the length of the dendrite. Two or more brief pulses of transmitter spaced a minute or so apart were generally more effective than a single larger application. Pretreatment with sphingosine, an inhibitor of protein kinase C, blocked the longer-term build up of the $[Ca^{2+}]_i$ rises but not the initial transient response. The NMDA antagonist APV or normal $Mg^{2+}$ levels could block the response if given together with NMDA, but were ineffective once the standing gradients were established, whereas GABA was inhibitory at any stage. An obvious interpretation, perhaps relevant to long-term potentiation, is that activation of NMDA channels raises $Ca^{2+}$ and turns on protein kinase C (or another sphingosine-inhibitable enzyme), which opens some other localized $Ca^{2+}$ channels. The latter stay open for a prolonged period but can be shut by GABA even though NMDA antagonists are no longer effective.

*Difficult cell types* Fura-2 has also enabled $Ca^{2+}$ measurement in neuronal preparations not amenable to previous methodologies. For example, Lev-Ram & Grinvald (1987) have shown that myelinated axons in rat optic nerve undergo fast-rising $[Ca^{2+}]_i$ transients during action potential conduction. This finding provides more direct evidence for voltage-sensitive $Ca^{2+}$ channels in myelinated axons, and suggests that $Ca^{2+}$ indicators may be useful for real-time imaging of CNS activity. Imaging of $[Ca^{2+}]_i$ has also been reported in another intact tissue, bullfrog sympathetic ganglia (Nohmi et al 1988). Ratto et al (1988) have loaded fura-2 into intact frog retina, which because of its light sensitivity might seem to be a most unsuitable preparation for experiments with a UV-excited, green-fluorescing dye. But by spreading the excitation and recording over the entire photoreceptor layer, relying on rhodopsin to shield the deeper layers, rod $[Ca^{2+}]_i$ could be seen to drop from 220 nM to 140 nM within 1–2 sec after the onset of nonbleaching illumination. This is the first measurement of $[Ca^{2+}]_i$ in photoreceptors not poisoned by inhibitors of cyclic GMP phosphodiesterase, and provides further direct evidence against a rise in $[Ca^{2+}]_i$ as a step in vertebrate phototransduction.

*$Ca^{2+}$ and secretion* Fura-2 measurements have been instrumental in recent modifications of the classical dogma that elevations in $[Ca^{2+}]_i$ are necessary and sufficient for secretion. Schwartz (1987) has shown that depolarization can release GABA from fish retina horizontal cells even when a $[Ca^{2+}]_i$ rise is prevented. The proposed mechanism is voltage-dependent reversal of a $Na^+/GABA$ co-transporter, not exocytosis of preformed vesicles. But even such exocytosis, at least in nonneuronal cells, may be much less $Ca^{2+}$-dependent than previously thought. Neher (1987) observed that $[Ca^{2+}]_i$ spikes in perfused mast cells were neither necessary nor sufficient for exocytosis, which was precisely measured by capacitance changes. Instead, $Ca^{2+}$ seemed at most to reinforce the effectiveness of more powerful stimulants such as guanosine 3-thiotriphosphate (GTP-$\gamma$-S). Poenie et al (1987) monitored secretion of toxin granules from individual cytolytic T lymphocytes by observing the toxin's "postsynaptic" effect on closely apposed target cells. Though $[Ca^{2+}]_i$ transients did occur in the "presynaptic" T cell, they often peaked well before exocytosis and on the side of the cell remote from the "synaptic cleft." These $[Ca^{2+}]_i$ images, together with experiments on phorbol esters and antibodies against protein kinase C (M. Poenie, A. M. Schmitt-Verhulst, and R. Y. Tsien, manuscript in preparation), suggest that the kinase probably plays a much more important role than $[Ca^{2+}]_i$ in directing exocytosis in this system.

INDO-1    Indo-1 has a rigidized stilbene fluorophore like fura-2 but has the unique property that its emission, not just its excitation spectrum, shifts to shorter wavelengths when the molecule binds $Ca^{2+}$ (Grynkiewicz et al 1985, Tsien 1986). Excitation can therefore be at a single wavelength, typically somewhere between 351 and 365 nm, depending on whether an argon or krypton ion laser or a mercury lamp is used. By measuring the ratio of emissions at 405 nm to that at 485 nm one can estimate the $Ca^{2+}$ concentration; this method has the usual advantages inherent in ratio rather than absolute measurements. The two wavelengths can in principle be separated by a dichroic mirror and measured simultaneously without any chopping. This methodology works well when the measurement is from a single spatial location at any given instant, as in flow cytometry, nonimaging photometry, or laser-scanning or specimen-scanning confocal microscopy, because one can use photomultipliers to read the two emission bands without worries about spatial registration errors. By contrast, it would be quite challenging to bring corresponding pixels into registration over the entire active regions of two separate low-light level TV cameras. Although adaptive algorithms are available to interpolate one image to put its pixels in register with those of another image (Walter & Berns 1986), the long time generally required for the computation negates the main advantage of emission ratioing over excitation ratioing, namely speed. The registration problem could be eased by alternating two filters in front of a single TV camera, but again one might as well chop the excitation. Excitation chopping is easier because maintenance of image quality is not necessary, light transmission efficiency is non-critical, and the chopping apparatus can be located outside the microscopy, where its bulk and possible vibrations can be isolated.

Two other drawbacks of indo-1 compared to fura-2 should be mentioned. The blue and violet wavelengths of indo-1 emission overlap cellular autofluorescence from pyridine nucleotides (Aubin 1979) more severely than the green of fura-2. Also indo-1 bleaches several-fold faster than fura-2 (S. R. Adams and R. Y. Tsien, unpublished observations). On the other hand, complaints about loading and compartmentation seem to crop up more frequently with fura-2 than with indo-1. Some of this difference may be that fura-2 has been tried by more people and on a much greater variety of cells and has been examined more critically by microscopic imaging than indo-1, but some of the difference is probably real. For example, Bush & Jones (1987) have reported that plant cells (barley aleurones) can be loaded with indo-1 by using acidic extracellular pH, whereas fura-2 concentrates in the vacuole. Also, Lee et al (1987) have recorded indo-1 $[Ca^{2+}]_i$ signals from intact (not dissociated) beating mammalian heart, an organ whose pigmentation and pulsation are surely more severe than those of any neurobiological preparation.

FLUO-3   Fluo-3, the newest of our $Ca^{2+}$-indicators (Minta et al 1987), has three main advantages over its predecessors: excitation at visible wavelengths (503–506 nm) rather than near-UV; a very large enhancement in fluorescence intensity, about 40-fold, upon binding $Ca^{2+}$; and a significantly weaker $Ca^{2+}$ affinity, $K_D \sim 400$ nM, permitting measurement to 5–10 $\mu$M $[Ca^{2+}]_i$. These properties are particularly valuable when one monitors the effectiveness of photochemically reactive chelators or caged nucleotides (Gurney & Lester 1987) or caged inositol phosphates (Walker et al 1987) at raising or lowering $[Ca^{2+}]_i$, since those compounds are photolyzed by the same near-UV irradiation that would be used to excite quin-2, fura-2, or indo-1. The visible excitation wavelengths of fluo-3 mean that there is no interference between the actinic and monitoring wavelengths. Those wavelengths are close to the visible output of argon lasers, so that fluo-3 is presently the only $Ca^{2+}$ indicator usable with flow cytometers and laser confocal microscopes that lack UV capability. Another promising domain for fluo-3 will probably be investigations of the interactions of $[Na^+]_i$ and $[Ca^{2+}]_i$, since it should be usable simultaneously with UV-excited SBFI. However binding of $Ca^{2+}$ to fluo-3 causes negligible wavelength shifts in either excitation or emission spectra, so that fluo-3 like quin-2 is limited to intensity changes without wavelength pairs to ratio. Though changes in $[Ca^{2+}]_i$ are readily observed, absolute calibration requires treatment of the cells with ionophores, heavy metals, and/or detergent at the end of every experiment. For these reasons fluo-3 is unlikely to displace ratio indicators like fura-2 and indo-1 from most single-cell imaging applications. In general a single best $Ca^{2+}$ indicator for all applications will probably never be available; every structure is a different compromise between many partially incompatible goals.

## Magnesium Indicators

If the binding sites in $Ca^{2+}$ indicators are reduced in size and coordination number, their $Ca^{2+}$ affinity decreases markedly while their $Mg^{2+}$ affinity is unaffected or increases. The resulting chelators have $Mg^{2+}$ dissociation constants in the millimolar range and are therefore sensitive to physiological $Mg^{2+}$ (Tsien 1980, Levy et al 1988), whereas their $Ca^{2+}$ dissociation constants of tens to hundreds of micromolar are well above the usual $[Ca^{2+}]_i$ values. Efforts to make these chelators fluorescent are in progress; the products are likely to prove quite useful in elucidating the role and control of $Mg^{2+}$, an essential cofactor for many enzymes and an important stabilizer of nuclear, ribosomal, and membrane structure (Grubbs & Maguire 1987, Levy et al 1988).

## PROSPECTS AND CONCLUSIONS

Because of space limitations this review cannot describe the hundreds of biological experiments to which fluorescent indicators have been applied. Nevertheless, I agree with Grinvald (1985) and Cohen (1988) that optical probes are the technique with the greatest long-term promise for monitoring the activity of complex interacting assemblages of neurons as they process information. This opinion is based on the spatial and temporal resolution, nondestructive parallel readout, and varied molecular specificity afforded by optical probes. As discussed above, fluorescence has significant advantages over absorbance in sensitivity and applicability to methods for three-dimensional optical sectioning such as confocal microscopy.

Despite these strong points, currently available fluorescent probes have enough deficiencies that future progress still requires better indicators. The design and production of such molecules exemplifies yet another area in which improved molecular ingenuity and insight would strongly benefit cellular and integrative neuroscience.

### ACKNOWLEDGMENTS

I thank L. B. Cohen, F. DiVirgilio, A. Grinvald, R. Haugland, I. Kurtz, L. Loew, T. Pozzan, T. Rink, B. Salzberg, A. Verkman, and R. Zucker for sending me preprints. The work in my laboratory has been supported by grants from the National Institutes of Health (GM31004 and EY04372), the Searle Scholars Program, and the Cancer Research Coordinating Committee of the University of California.

*Literature Cited*

Adams, S. R., Kao, J. P. Y., Grynkiewicz, G., Minta, A., Tsien, R. Y. 1988. Biologically useful chelators that release $Ca^{2+}$ upon illumination. *J. Am. Chem. Soc.* 110: 3212–20

Almers, W., Neher, E. 1985. The Ca signal from fura-2 loaded mast cells depends strongly on the method of dye-loading. *FEBS Lett.* 192: 13–18

Alpern, R. J. 1985. Mechanism of basolateral membrane $H^+/OH^-/HCO_3^-$ transport in the rat proximal convoluted tubule. *J. Gen. Physiol.* 86: 613–36

Andersen, O. S. 1978. Permeability properties of unmodified lipid bilayer membranes. In *Membrane Transport in Biology*, ed. G. Giebisch, D. C. Tosteson, H. H. Ussing, pp. 369–446. Heidelberg: Springer-Verlag

Aubin, J. E. 1979. Autofluorescence of viable cultured mammalian cells. *J. Histochem. Cytochem.* 27: 36–43

Augustine, G. J., Charlton, M. P., Smith, S. J. 1987. Calcium action in synaptic transmitter release. *Ann. Rev. Neurosci.* 10: 633–93

Bader, C. R., Bernheim, L., Bertrand, D. 1985. Sodium-activated potassium current in cultured avian neurones. *Nature* 317: 540–42

Barcenas-Ruiz, L., Wier, W. G. 1987. Voltage dependence of intracellular $[Ca^{2+}]_i$ transients in guinea pig ventricular myocytes. *Circ. Res.* 61: 148–54

Becker, P. L., Fay, F. S. 1987. Photobleaching of fura-2 and its effect on determination of calcium concentrations. *Am. J. Physiol.* 253: C613–18

Blasdel, G. G., Salama, G. 1986. Voltage-sensitive dyes reveal a modular organization in monkey striate cortex. *Nature* 321: 579–85

Blinks, J. R., Wier, W. G., Hess, P., Prendergast, F. G. 1982. Measurement of $Ca^{2+}$ concentrations in living cells. *Progr. Biophys. Mol. Biol.* 40: 1–114

Breckinridge, L. J., Warner, A. E. 1982. Intracellular sodium and the differentiation of amphibian embryonic neurons. *J. Physiol.* 332: 393–413

Bright, G. R., Fisher, G. W., Rogowska, J., Taylor, D. L. 1987. Fluorescence ratio imaging microscopy: Temporal and spatial measurements of cytoplasmic pH. *J. Cell Biol.* 104: 1019–33

Brown, R. G., Porter, G. 1977. Effect of pH on the emission and absorption characteristics of 2,3-dicyano-*p*-hydroquinone. *J. Chem. Soc. Faraday Trans. I* 73: 1281–85

Bush, D. S., Jones, R. L. 1987. Measurement of cytoplasmic calcium in aleurone protoplasts using indo-1 and fura-2. *Cell Calcium* 8: 455–72

Cannell, M. B., Berlin, J. R., Lederer, W. J. 1987. Effect of membrane potential changes on the calcium transient in single rat cardiac muscle cells. *Science* 238: 1419–23

Chused, T. M., Wilson, H. A., Greenblatt, D., Ishida, Y., Edison, L. J., Tsien, R. Y., Finkelman, F. D. 1987. Flow cytometric analysis of cytosolic free calcium in murine splenic B lymphocytes: Responses to anti-IgM and anti-IgD differ. *Cytometry* 8: 396–404

Chused, T. M., Wilson, H. A., Seligmann, B. E., Tsien, R. Y. 1986. Probes for use in the study of leukocyte physiology by flow cytometry. See Taylor et al 1986, pp. 531–44

Cobbold, P. H., Rink, T. J. 1987. Fluorescence and bioluminescence measurement of cytoplasmic free calcium. *Biochem. J.* 248: 313–28

Cohan, C. S., Connor, J. A., Kater, S. B. 1987. Electrically and chemically mediated increases in intracellular calcium in neuronal growth cones. *J. Neurosci.* 7: 3588–99

Cohen, L. 1988. More light on brains. *Nature* 331: 112–13

Cohen, L. B., Lesher, S. 1986. Optical monitoring of membrane potential: Methods of multisite optical measurement. See De Weer & Salzburg 1986, pp. 71–99

Connor, J. A. 1986. Digital imaging of free calcium changes and of spatial gradients in growing processes in single, mammalian central nervous system cells. *Proc. Natl. Acad. Sci. USA* 83: 6179–83

Connor, J. A., Tseng, H.-Y., Hockberger, P. E. 1987. Depolarization- and transmitter-induced changes in intracellular $Ca^{2+}$ of rat cerebellar granule cells in explant cultures. *J. Neurosci.* 7: 1384–1400

Connor, J. A., Wadman, W. J., Hockberger, P. E., Wong, R. K. S. 1988. Sustained dendritic gradients of $Ca^{2+}$ induced by excitatory amino acids in CA1 hippocampal neurons. *Science* 240: 649–53

Cooke, R. 1982. Fluorescence as a probe of contractile systems. *Methods Enzymol.* 51: 574–93

De Weer, P., Salzberg, B. M., eds. 1986. *Optical Methods in Cell Physiology.* New York: Wiley-Interscience. 480 pp.

DiVirgilio, F., Steinberg, T. H., Swanson, J. A., Silverstein, S. C. 1988. Fura-2 secretion and sequestration in macrophages. *J. Immunol.* 140: 915–20

Elson, E. L. 1986. Membrane dynamics studied by fluorescence correlation spectroscopy and photobleaching recovery. See De Weer & Salzberg 1986, pp. 367–83

Fluhler, E., Burnham, V. G., Loew, L. M. 1985. Spectra, membrane binding, and potentiometric responses of new charge shift probes. *Biochemistry* 24: 5749–55

Freedman, J. C., Laris, P. C. 1988. Optical potentiometric indicators for non-excitable cells. See Loew 1988a, 3: 1–49

Grinvald, A. 1985. Real-time optical mapping of neuronal activity: From single growth cones to the intact mammalian brain. *Ann. Rev. Neurosci.* 8: 263–305

Grinvald, A., Lieke, E., Frostig, R. D., Gilbert, C. D., Wiesel, T. N. 1986. Functional architecture of cortex revealed by optical imaging of intrinsic signals. *Nature* 324: 361–64

Grinvald, A., Salzberg, B. M., Lev-Ram, V., Hildesheim, R. 1987. Optical recording of synaptic potentials from processes of single neurons using intracellular potentiometric dyes. *Biophys. J.* 51: 643–51

Gross, D., Loew, L. M. 1988. Fluorescent indicators of membrane potential: Microspectrofluometry and imaging. In *Quantitative Fluorescence Microscopy: Imaging and Spectroscopy. Methods in Cell Biology,*

Vol. 30, ed. Y.-L. Wang, D. L. Taylor. New York: Academic

Gross, D. M., Loew, L. M., Webb, W. W. 1986. Optical imaging of cell membrane potential changes induced by applied electric fields. *Biophys. J.* 50: 339–48

Grubbs, R. D., Maguire, M. E. 1987. Magnesium as a regulatory cation: Criteria and evaluation. *Magnesium* 6: 113–27

Grynkiewicz, G., Poenie, M., Tsien, R. Y. 1985. A new generation of $Ca^{2+}$ indicators with greatly improved fluorescence properties. *J. Biol. Chem.* 260: 3440–50

Gurney, A. M., Lester, H. A. 1987. Light-flash physiology with synthetic photosensitive compounds. *Physiol. Rev.* 67: 583–617

Gurney, A. M., Tsien, R. Y., Lester, H. A. 1987. Activation of a potassium current by rapid photochemically generated steps of intracellular calcium in rat sympathetic neurones. *Proc. Natl. Acad. Sci. USA* 84: 3496–3500

Hepler, P. K., Callaham, D. A. 1987. Free calcium increases during anaphase in stamen hair cells of *Tradescantia. J. Cell Biol.* 105: 2137–43

Highsmith, S., Bloebaum, P., Snodowne, K. W. 1986. Sarcoplasmic reticulum interacts with the $Ca^{2+}$ indicator precursor fura-2-am. *Bioch. Biophys. Res. Comm.* 138: 1153–62

Hille, B. 1984. *Ionic Channels of Excitable Membranes.* Sunderland, Mass: Sinauer Assoc.

Hiraoka, Y., Sedat, J. W., Agard, D. A. 1987. The use of a charge-coupled device for quantitative optical microscopy of biological structures. *Science* 238: 36–41

Hirning, L. D., Fox, A. P., McCleskey, E. W., Olivera, B. M., Thayer, S. A., et al. 1988. Dominant role of N-type $Ca^{2+}$-channels in evoked release of norepinephrine from sympathetic neurons. *Science* 239: 57–61

Hollingworth, S., Baylor, S. M. 1987. Fura2 signals from intact frog skeletal muscle fibers. *Biophys. J.* 51: 549a

Illsley, N. P., Verkman, A. S. 1987. Membrane chloride transport measured using a chloride-sensitive fluorescent probe. *Biochemistry* 28: 1215–19

Jackson, A. P., Timmerman, M. P., Bagshaw, C. R., Ashley, C. C. 1987. The kinetics of calcium binding to fura-2 and indo-1. *FEBS Lett.* 216: 35–39

Kao, J. P. Y., Tsien, R. Y. 1988. $Ca^{2+}$-binding kinetics of fura-2 and azo-1 from temperature-jump relaxation measurements. *Biophys. J.* 53: 635–39

Kauer, J. S. 1988. Real-time imaging of evoked activity in local circuits of the sala-

mander olfactory bulb. *Nature* 331: 166–68

Kauer, J. S., Senseman, D. M., Cohen, L. B. 1987. Odor-elicited activity monitored simultaneously from 124 regions of the salamander olfactory bulb using a voltage-sensitive dye. *Brain Res.* 418: 255–61

Klein, M. G., Simon, B. J., Szucs, G., Schneider, M. F. 1988. Simultaneous recording of calcium transients in skeletal muscle using high and low affinity calcium indicators. *Biophys. J.* 53: 955–62

Kleinfeld, A. M. 1985. Tryptophan imaging of membrane proteins. *Biochemistry* 24: 1874–82

Krapf, R., Berry, C. A., Verkman, A. S. 1988a. Estimation of intracellular chloride activity in isolated perfused rabbit proximal convoluted tubules using a fluorescent indicator. *Biophys. J.* 53: 955–62

Krapf, R., Illsley, N. P., Tseng, H. C., Verkman, A. S. 1988b. Structure-activity relationships of chloride-sensitive fluorescent indicators for biological application. *Anal. Biochem.* 169: 142–50

Kudo, Y., Ogura, A. 1986. Glutamate-induced increase in intracellular $Ca^{2+}$ concentration in isolated hippocampal neurones. *Br. J. Pharmacol.* 89: 191–98

Kurtz, I. 1987. Apical $Na^+/H^+$ antiporter and glycolysis-dependent $H^+$-ATPase regulate intracellular pH in the rabbit $S_3$ proximal tubule. *J. Clin. Invest.* 80: 928–35

Kurtz, I., Balaban, R. S. 1985. Fluorescence emission spectroscopy of 1,4-dihydroxyphthalonitrile: A method for determining intracellular pH in cultured cells. *Biophys. J.* 48: 499–508

Lakowicz, J. R. 1983. *Principles of Fluorescence Spectroscopy.* New York: Plenum. 496 pp.

Laris, P. C., Hoffman, J. F. 1986. Optical determination of electrical properties of red blood cell and Ehrlich ascites tumor cell membranes with fluorescent dyes. See De Weer & Salzberg 1986, pp. 199–210

Lee, H. C., Smith, N., Mohabir, R., Clusin, W. T. 1987. Cytosolic calcium transients from the beating mammalian heart. *Proc. Natl. Acad. Sci. USA* 84: 7793–97

Lev-Ram, V., Grinvald, A. 1987. Activity-dependent calcium transients in central nervous system myelinated axons revealed by the calcium indicator fura-2. *Biophys. J.* 52: 571–76

Levy, L. A., Murphy, E., Raju, B., London, R. E. 1988. Measurement of cytosolic free magnesium ion concentration by $^{19}F$ NMR. *Biochemistry* 27: 4041–48

Lipscombe, D., Madison, D. V., Poenie, M., Reuter, H., Tsien, R. Y., et al. 1988a. Spatial distribution of calcium channels and cytosolic calcium transients in growth cones and cell bodies of sympathetic neurons. *Proc. Natl. Acad. Sci.* 85: 2398–2402

Lipscombe, D., Madison, D. V., Poenie, M., Reuter, H., Tsien, R. W., et al. 1988b. Changes in calcium distribution in sympathetic neurons resulting from activity of calcium stores and calcium channels. *Neuron* 1: 355–65

Loew, L. M., ed. 1988a. *Spectroscopic Membrane Probes*. Boca Raton: CRC Press

Loew, L. M. 1988b. How to choose a potentiometric membrane probe. See Loew 1988a, 2: 139–51

Loew, L. M., Cohen, L. B., Salzberg, B. M., Obaid, A. L., Bezanilla, F. 1985. Charge-shift probes of membrane potential. Characterization of aminostyrylpyridinium dyes on the squid giant axons. *Biophys. J.* 47: 71–77

Lukács, G. L., Kapus, A., Fonyó, A. 1988. Parallel measurement of oxoglutarate dehydrogenase activity and matrix free $Ca^{2+}$ to fura-2-loaded heart mitochondria. *FEBS Lett.* 229: 219–23

Malgaroli, A., Milani, D., Meldolesi, J., Pozzan, T. 1987. Fura-2 measurement of cytosolic free $Ca^{2+}$ in monolayers and suspensions of various types of animal cells. *J. Cell Biol.* 105: 2145–55

Mathies, R. A., Stryer, L. 1986. Single-molecule fluorescence detection: A feasibility study using phycoerythrin. See Taylor et al 1986, pp. 129–40

Miller, R. J. 1987. Multiple calcium channels and neuronal function. *Science* 235: 46–52

Minta, A., Harootunian, A. T., Kao, J. P. Y., Tsien, R. Y. 1987. New fluorescent indicators for intracellular sodium and calcium. *J. Cell Biol.* 105: 89a

Murphy, S. N., Thayer, S. A., Miller, R. J. 1987. The effects of excitatory amino acids on intracellular calcium in single mouse striatal neurons *in vitro*. *J. Neurosci.* 7: 4145–58

Nairn, R. C. 1976. *Fluorescent Protein Tracing*. Edinburgh: Churchill Livingstone

Neher, E. 1988. The influence of intracellular calcium concentration on degranulation of dialysed mast cells from rat peritoneum. *J. Physiol.* 395: 193–214

Nohmi, M., Kuba, K., Ogura, A., Kudo, Y. 1988. Measurement of intracellular $Ca^{2+}$ in the bullfrog sympathetic ganglion cells using fura-2 fluorescence. *Brain Res.* 438: 175–81

Oakes, S. G., Martin, W. J. II, Lisek, C. A., Powis, G. 1988. Incomplete hydrolysis of the calcium indicator precursor fura-2 pentaacetoxymethyl ester (fura-2 AM) by cells. *Anal. Bioch.* 169: 159–66

Orbach, H. S., Cohen, L. B., Grinvald, A. 1985. Optical mapping of electrical activity in rat somatosensory and visual cortex. *J. Neurosci.* 5: 1886–95

Pagano, R. E., Sleight, R. G. 1985. Defining lipid transport pathways in animal cells. *Science* 229: 1051–57

Paradiso, A. M., Negulescu, P. A., Machen, T. E. 1986. $Na^+$-$H^+$ and $Cl^-$-$OH^-$ ($HCO_3^-$) exchange in gastric glands. *Am. J. Physiol.* 250: G524–34

Paradiso, A. M., Tsien, R. Y., Machen, T. E. 1987. Digital image processing of intracellular pH in gastric oxyntic and chief cells. *Nature* 325: 447–50

Poenie, M., Alderton, J., Steinhardt, R., Tsien, R. Y. 1986. Calcium rises abruptly and briefly throughout the cell at the onset of anaphase. *Science* 233: 886–89

Poenie, M., Alderton, J., Tsien, R. Y., Steinhardt, R. A. 1985. Changes in free calcium levels with stages of the cell division cycle. *Nature* 315: 147–49

Poenie, M., Tsien, R. Y., Schmitt-Verhulst, A.-M. 1987. Sequential activation and lethal hit measured by $[Ca^{2+}]_i$ in individual cytolytic T cells and targets. *EMBO J.* 6: 2223–32

Ratan, R. R., Shelanski, M. L., Maxfield, F. R. 1986. Transition from metaphase to anaphase is accompanied by local changes in cytoplasmic free calcium in $PtK_2$ kidney epithelial cells. *Proc. Natl. Acad. Sci. USA* 83: 5136–40

Ratto, G. M., Payne, R., Owen, W. G., Tsien, R. Y. 1988. The concentration of cytosolic free $Ca^{2+}$ in vertebrate rod outer segments measured with fura-2. *J. Neurosci.* In press

Rink, T. J., Pozzan, T. 1985. Using quin2 in cell suspensions. *Cell Calcium* 6: 133–44

Rink, T. J., Tsien, R. Y., Pozzan, T. 1982. Cytoplasmic pH and free $Mg^{2+}$ in lymphocytes. *J. Cell Biol.* 95: 189–96

Ross, W. N., Krauthamer, V. 1984. Optical measurements of potential changes in axons and processes of neurons of a barnacle ganglion. *J. Neurosci.* 4: 659–72

Salzberg, B. M. 1983. Optical recording of electrical activity in neurons using molecular probes. In *Current Methods in Cellular Neurobiology*, ed. J. L. Barker, J. F. McKelvy, 3: 139–87. New York: Wiley. 320 pp.

Salzberg, B. M., Obaid, A. L., Gainer, H. 1986. Optical studies of excitation and secretion at vertebrate nerve terminals. See De Weer & Salzberg 1986, pp. 133–64

Scanlon, M., Williams, D. A., Fay, F. S. 1987. A $Ca^{2+}$-insensitive form of fura-2 associated with polymorphonuclear leukocytes. *J. Biol. Chem.* 262: 6308–12

Schlegel, W., Winiger, B. P., Mollard, P.,

Vacher, P., Wuarin, F., et al. 1987. Oscillations of cytosolic $Ca^{2+}$ in pituitary cells due to action potentials. *Nature* 329: 719–21

Schwartz, E. A. 1987. Depolarization without calcium can release gamma-aminobutyric acid from a retinal neuron. *Science* 238: 350–55

Smith, G. A., Hesketh, T. R., Metcalfe, J. C. 1988. Design and properties of a fluorescent indicator of intracellular free $Na^+$ concentration. *Biochem. J.* 250: 227–32

Smith, J. C. 1988. Potential-sensitive molecular probes in energy-transducing organelles. See Loew 1988a, 2: 153–97

Smith, S. J., Osses, L. R., Augustine, G. J. 1987. Fura-2 imaging of localized calcium accumulation within squid "giant" presynaptic terminals. *Biophys. J.* 51: 66a

Stewart, W. W. 1981. Lucifer dyes: Highly fluorescent dyes for biological tracing. *Nature* 292: 17–21

Stryer, L. 1978. Fluorescence energy transfer as a spectroscopic ruler. *Ann. Rev. Biochem.* 47: 819–46

Taylor, D. L., Amato, P. A., McNeil, P. L., Luby-Phelps, K., Tanasugarn, L. 1986a. Spatial and temporal dynamics of specific molecules and ions in living cells. See Taylor et al 1986, pp. 347–76

Taylor, D. L., Waggoner, A. S., Murphy, R. F., Lanni, F., Birge, R. R., eds. 1986b. *Applications of Fluorescence in the Biomedical Sciences.* New York: Liss. 639 pp.

Thayer, S. A., Hirning, L. D., Miller, R. J. 1987. Distribution of multiple types of $Ca^{2+}$ channels in rat sympathetic neurons *in vitro. Molec. Pharmacol.* 32: 579–86

Thomas, J. A., Buchsbaum, R. N., Zimniak, A., Racker, E. 1979. Intracellular pH measurements in Ehrlich ascites tumor cells utilizing spectroscopic probes generated in situ. *Biochemistry* 18: 2210–18

Tsien, R. Y. 1980. New calcium indicators and buffers with high selectivity against magnesium and protons: Design, synthesis, and properties of prototype structures. *Biochemistry* 19: 2396–2404

Tsien, R. Y. 1986. New tetracarboxylate chelators for fluorescence measurement and photochemical manipulation of cytosolic free calcium concentrations. See De Weer & Salzberg, pp. 327–45

Tsien, R. Y., Poenie, M. 1986. Fluorescence ratio imaging: A new window into intracellular ionic signaling. *Trends Biochem. Sci.* 11: 450–55

Tsien, R. Y., Pozzan, T. 1989. Measurement of cytosolic free $Ca^{2+}$ with quin2: Practical aspects. *Methods Enzymol.* 172: In press

Tsien, R. Y., Pozzan, T., Rink, T. J. 1982. Calcium homeostasis in intact lympho-cytes: Cytoplasmic free $Ca^{2+}$ monitored with a new, intracellularly trapped fluorescent indicator. *J. Cell Biol.* 94: 325–34

Tsien, R. Y., Rink, T. J. 1983. Measurement of free $Ca^{2+}$ in cytoplasm. See Salzberg 1983, pp. 249–312

Tsien, R. Y., Rink, T. J., Poenie, M. 1985. Measurement of cytosolic free $Ca^{2+}$ in individual small cells using fluorescence microscopy with dual excitation wavelengths. *Cell Calcium* 6: 145–57

Tsien, R. Y., Zucker, R. S. 1986. Control of cytoplasmic calcium with photolabile tetracarboxylate 2-nitrobenzhydrol chelators. *Biophys. J.* 50: 843–53

Uster, P. S., Pagano, R. E. 1986. Resonance energy transfer microscopy: Observations of membrane-bound fluorescent probes in model membranes and in living cells. *J. Cell Biol.* 103: 1221–34

Valet, G., Raffael, A., Moroder, L., Wunsch, E., Ruhenstroth-Bauer, G. 1981. Fast intracellular pH determination in single cells by flow cytometry. *Naturwissenschaften* 68: 265–66

Varon, S., Skaper, S. D. 1983. The $Na^+$, $K^+$ pump may modulate the control of nerve cells by nerve growth factor. *Trends Biochem. Sci.* 8: 22–25

Vögtle, F., Weber, E. 1980. Crown ethers—complexes and selectivity. In *The Chemistry of Ethers, Crown Ethers, Hydroxyl Groups and Their Sulphur Analogues (The Chemistry of Functional Groups. Supplement E, part 1)*, ed. S. Patai, pp. 55–159. New York: Wiley

Waggoner, A. S. 1979. Dye indicators of membrane potential. *Ann. Rev. Biophys. Bioeng.* 8: 47–63

Waggoner, A. S. 1986. Fluorescent probes for analysis of cell structure, function, and health by flow and imaging cytometry. See Taylor, Waggoner, Murphy, Lanni, Birge 1986, pp. 3–28

Waggoner, A. S., Grinvald, A. 1977. Mechanisms of rapid optical changes of potential sensitive dyes. *Ann. N.Y. Acad. Sci.* 303: 217–42

Walker, J. W., Somlyo, A. V., Goldman, Y. E., Somlyo, A. P., Trentham, D. R. 1987. Kinetics of smooth and skeletal muscle activation by laser pulse photolysis of caged inositol 1,4,5-trisphosphate. *Nature* 327: 249–52

Walter, R. J. Jr., Berns, M. W. 1986. Digital image processing and analysis. In *Video Microscopy*, ed. S. Inoue, pp. 327–92. New York: Plenum. 584 pp.

Wang, Y.-L., Taylor, D. L., eds. 1988. *Fluorescent Analogs, Labeling Cells and Basic Microscope System, Methods in Cell Biology*, Vol. 29. New York: Academic

White, J. G., Amos, W. B., Fordham, M.

1987. An evaluation of confocal versus conventional imaging of biological structures by fluorescence light microscopy. *J. Cell Biol.* 105: 41–48

Williams, D. A., Fogarty, K. E., Tsien, R. Y., Fay, F. S. 1985. Calcium gradients in single smooth muscle cells revealed by the digital imaging microscope using fura-2. *Nature* 318: 558–61

Wilson, T., Sheppard, C. J. R. 1984. *Theory and Practice of Scanning Optical Micro-*

*scopy*. London: Academic. 213 pp.

Wolf, B. E., Waggoner, A. S. 1986. Optical studies of the mechanism of membrane potential sensitivity of merocyanine 540. See De Weer & Salzberg 1986, pp. 101–12

Yguerabide, J. 1972. Nanosecond fluorescence spectroscopy of macromolecules. *Methods Enzymol.* 26: 498–578

Zucker, R. S. 1989. Short-term synaptic plasticity. *Ann. Rev. Neurosci.* 12: 13–31

*Ann. Rev. Neurosci. 1989. 12:255–87*

# INVOLVEMENT OF HORMONAL AND NEUROMODULATORY SYSTEMS IN THE REGULATION OF MEMORY STORAGE

*James L. McGaugh*

Center for the Neurobiology of Learning and Memory and Department of Psychobiology, University of California, Irvine, California 92717

## ENDOGENOUS MODULATION OF MEMORY STORAGE

Lasting memory is not formed at the moment that new information is acquired. It is well known that retention can be markedly influenced by treatments administered shortly after learning (McGaugh & Herz 1972). Such findings have generally been interpreted as indicating that the treatments affect retention by altering posttraining neural processes underlying the storage of newly acquired information (McGaugh 1966). The hypothesis that memory traces become consolidated over time was originally proposed by Mueller & Pilzecker (1900). Hebb's (1949) dual-trace hypothesis of memory, which proposed that lasting memory traces are produced by the activity of temporary or short-term memory traces, provided a more explicit view of a possible basis of consolidation. Studies investigating the consolidation of lasting memory traces have focused to a large degree on conditions that disrupt retention. It is clear that retrograde amnesia can be produced by a variety of experimental treatments administered posttraining (McGaugh & Herz 1972). More importantly, susceptibility to retrograde amnesia has been conserved in evolution. Evidence that posttraining treatments alter retention in a retrograde fashion has been obtained in studies with humans, monkeys, cats, rodents, birds, fish, and insects (Agranoff 1980, Bowman et al 1979, Cherkin 1969, Menzel 1983, Weingartner & Parker 1984).

255

It seems reasonable to ask why susceptibility of newly acquired information to posttraining modulating influences is a common feature of memory. One possibility is that lasting memories are best made slowly, perhaps because they require the synthesis of new molecules. The fact that memory storage is impaired by protein synthesis inhibitors argues for such a view (Davis & Squire 1984). Another possibility is that posttraining susceptibility to modulating influences provides an opportunity for endogenous systems to influence the strength of memory traces (Gold & McGaugh 1975, Kety 1972). It would seem to be advantageous as well as economical to be able to regulate memory storage in such a way that the strength of a memory is related to the importance of the experience. This view implies that it should be possible to enhance retention by administration of treatments following training. Extensive evidence supports this implication. Retention can be enhanced by a wide variety of posttraining treatments, including electrical stimulation of the brain (McGaugh & Gold 1976, Kesner 1982), drugs (Dunn 1980, McGaugh 1973, Squire & Davis 1983), and, more importantly for the view emphasized in this paper, hormones and neuromodulators (de Wied 1984, Gold & Zornetzer 1983, McGaugh 1983, McGaugh & Gold 1988).

## INVOLVEMENT OF HORMONES AND NEUROMODULATORS IN REGULATING MEMORY STORAGE

A basic assumption required by this general view is that memory storage processes are modulated by the action of endogenous systems activated by learning experiences. As is summarized below, there is now considerable evidence that retention can be modulated by the administration of hormones and neuromodulators that are normally released by experiences comparable to those used in training, as well as evidence that retention can be altered by treatments that alter the functioning of these systems. In addition, evidence from recent findings argues that hormonal systems influence memory storage through influences on neuromodulatory systems in specific brain regions. It is of particular significance that the effects of such treatments are time-dependent. That is, the treatments are generally most effective in influencing retention when administered immediately following training. Since the animals are not under the direct influence of the treatments during either the original learning or the retention testing, such findings are generally interpreted as indicating that the treatments influence memory by modulating memory storage.

Obviously, treatments acting at the time of learning or retention might directly affect performance for a variety of reasons unrelated to the storage of information. Caution is, of course, required in interpreting the results of experiments using posttraining treatments, since there is evidence that, under some conditions, such treatments may induce state-dependency (e.g. Izquierdo 1984). That is, it may be that the information is stored in a brain state induced by the treatment and is thus less accessible under other brain states. It has also been suggested that posttraining treatments may act as rewards or punishments or act simply to alter the animals' perception of the intensity of the motivational stimuli (e.g. footshock) used in the training task (e.g. Carey 1987, Izquierdo 1986). However, the fact that posttraining drug treatments can enhance latent learning and sensory preconditioning (i.e. learning occurring in the absence of explicit rewards and punishments) argues strongly against these interpretation (Humphrey 1968, Westbrook & McGaugh 1964).

Research in this area has implicated a wide variety of hormonal, neuro-modulatory, and neurotransmitter systems. Recent findings have added to earlier evidence indicating that memory can be influenced—both enhanced and impaired—by posttraining treatments affecting monoaminergic (Gold 1984, Flood & Cherkin 1987, Vanderwolf 1987), cholinergic (Flood & Cherkin 1986), and inhibitory amino acid systems (Castellano & Pavone 1988) as well as peptide systems, including ACTH (de Wied 1974), vaso-pressin (de Wied 1984), and opioid peptides (Castellano 1975, Gallagher & Kapp 1981). In addition, recent findings indicate that memory can be modulated by a posttraining administration of a number of other peptides, including substance P (Schlesinger et al 1986), cholecystokinin (CCK) (Flood et al 1987c), angiotensin II (Yonkov et al 1986), somatostatin (Vecsei et al 1986), and neuropeptide Y (Flood et al 1987b). A central question is whether the memory-modulating effects of treatments affecting hormonal and transmitter systems reflect the operation of processes that are normally involved in memory storage. In view of the known complexity of brain systems it is perhaps not surprising that memory storage is susceptible to treatments affecting such a variety of systems. But the evidence does not require the conclusion that the treatments work through parallel inde-pendent influences. Rather, the recent findings provide increasing evidence of interactions and convergences among the influences of the various memory-modulating treatments.

This review focuses primarily on the involvement of adrenergic, chol-inergic, and opioid peptide systems in memory storage. Earlier research on these systems generally examined one of the systems without reference to other systems. Recent research has clearly emphasized the interactions among these systems as well as the interactions of these systems with other hormonal and neuromodulatory systems in influencing memory storage. Studies of the effects of hormones and transmitters on memory storage must address a number of conceptual and theoretical issues: Are the effects dose and time dependent? Are the effects task specific or are they common to a variety of training tasks and procedures? Can the effects on memory readily be accounted for by alternative interpretations? Are the effects initiated peripherally or centrally? How and where do the hormones act in the brain to modulate the neurobiological processes underlying memory storage? And, ultimately, is there convincing evidence that the hormonal and transmitter systems are involved in the endogenous regulation of memory storage?

## Adrenergic Influences on Memory Storage

A consideration of studies examining adrenergic influences on memory storage will illustrate how recent research has attempted to address the questions raised above. Findings from many laboratories indicate that retention is modulated by posttraining peripheral (ip) administration of the adrenal medullary hormone epinephrine (Gold & van Buskirk 1976, Gold 1984, McGaugh & Gold 1988). Typically, retention is enhanced by low to moderate doses (i.e. 0.5 mg/kg or less) and impaired by higher doses. Thus, an inverted-U dose-response effect is typically obtained with posttraining epinephrine. This form of dose-response effect is typical of many treatments known to enhance retention (Cherkin & Flood 1988). It is well known that stressful stimulation, including the mildly stressful stimulation of the kinds typically used in studies of learning and memory in animals, results in the release of epinephrine from the adrenal medulla. Thus, findings indicating that memory is influenced by posttraining injections of epinephrine are consistent with the view that memory storage is modulated by the release of endogenous epinephrine (Gold & McGaugh 1975).

Gold and his colleagues have obtained extensive evidence in support of this general hypothesis (Gold 1988). They have found, for example, that when rats are trained with a mild footshock punishment on a one-trial inhibitory avoidance task and given posttraining epinephrine (ip), the dose that yields optimal enhancement of memory produces plasma epinephrine levels shortly after training comparable to those found in untreated animals trained with a high footshock that yields good retention (Gold & McCarty 1981, McCarty & Gold 1981). Further, the memory-modulating effect of epinephrine administered posttraining appears to vary with the levels released endogenously: Doses of epinephrine that enhance memory when administered following a mild training footshock impair memory when administered after a high footshock (Gold et al 1977, Izquierdo & Dias 1983). The effect is comparable to that produced by increasing the dose of epinephrine beyond an optimal level when a low footshock is used for training. Other recent findings have shown that the learning and memory impairment produced by adrenal demedullation (Liang & McGaugh 1987) is attenuated by epinephrine administered either shortly before or shortly after training (Borrell et al 1983, Liang et al 1985).

Since most studies of the effect of epinephrine on memory have used one-trial aversive tasks—usually inhibitory (passive) avoidance tasks—it is reasonable to consider the possibility that the aversiveness of the training procedures may make the retention uniquely susceptible to a hormone, such as epinephrine, which is known to be released by aversive stimulation. It might be that posttraining injections of epinephrine might act in some way to amplify the after-effects of noxious stimulation used as punishment (Carey 1987). Available evidence clearly argues against this view. First, posttraining epinephrine does not by itself influence retention of inhibitory avoidance training in control animals that are not given footshock (Liang & McGaugh 1983a). Further, epinephrine effects on memory are not restricted to one-trial inhibitory avoidance tasks in which response latency is used as the measure of retention. Memory-enhancing effects have been found with a multitrial active avoidance task (Liang et al 1985), a one-trial appetitive task, and an aversively motivated discrimination task. In the appetitive task (Sternberg et al 1985a), water-deprived mice were placed in a Y-maze and allowed to explore until they found water at the end of one alley and were then given injections of saline or epinephrine (ip). When retested 24 hours later, the animals given epinephrine entered significantly fewer alleys than did the controls before finding the water spout. In the study using an aversively motivated discrimination task (Introini-Collison & McGaugh 1986), mice were trained to escape from a mild footshock by entering one alley (i.e. the left alley) of a Y-maze and were given immediate posttraining injections (ip) of saline or one of two doses of epinephrine. On a retention test given one day, one week, or one month (i.e. independent groups) later, the position of the correct alley was reversed, and errors made on the discrimination reversal were used as the measure of retention of the original training. As was expected, at each retention interval, the low dose of epinephrine enhanced retention whereas the high dose impaired retention. It is clear from these findings that the memory-modulating effects of posttraining epinephrine are long lasting; and it is important to note that these results, which are based on choice rather than latency as a measure of retention, are highly comparable to those obtained with other training tasks and procedures, including inhibitory avoidance.

These studies have provided answers to several of the questions raised above. First, the effects of posttraining epinephrine are both dose-dependent and time-dependent. Second, although the range of tasks examined is clearly not exhaustive, the effects are not narrowly task-dependent. Third, the effects are not readily accounted for by any alternative interpretation as yet proposed. Obviously, the fact that epinephrine enhances retention in an appetitively motivated task argues against the view that epinephrine affects retention by potentiating the effects of the punishment used in training.

I turn now to the question of whether the effects of epinephrine on memory are initiated peripherally or centrally. Unfortunately, a clear, unequivocal answer cannot as yet be given to this question. It is generally assumed that epinephrine does not readily pass the blood-brain barrier (Weil-Malherbe et al 1959). Certainly, peripheral epinephrine influences the brain (Holdefer & Jensen 1987). But it is not clear whether the effects result from leakage into the brain or from neural activation via peripheral receptors. Conflicting results have been obtained in experiments examining the effects of central injections of epinephrine on memory. Borrell and his colleagues (1985) reported that posttraining intracerebroventricular (icv) injections of epinephrine enhanced retention of an inhibitory avoidance response, whereas de Almeida and his colleagues (1983) found no effects of icv injections of epinephrine administered in a wide range of doses. Borrell has suggested that epinephrine injections, whether peripheral or central, may involve the release of pituitary ACTH. There is, however, clear evidence against this hypothesis. Dexamethasone, which blocks ACTH release, does not block the memory-modulating effects of post-training epinephrine (McGaugh et al 1987).

Another possible clue concerning the locus of action of peripherally administered epinephrine is that the effects of epinephrine on memory in both aversive (Sternberg et al 1986) and appetitive (Sternberg et al 1985a) tasks is blocked by both $\alpha$- and $\beta$-adrenergic antagonists, whereas the memory-modulating effect of clenbuterol, an adrenergic agonist that readily passes the blood-brain barrier, is selectively blocked by centrally, but not peripherally, acting $\beta$-adrenergic antagonists (Introini-Collison & Baratti 1986). Such findings argue that epinephrine effects on memory may be initiated by activation of peripheral adrenergic receptors.

Recent findings of Gold and his colleagues (Gold 1988) suggest that epinephrine influences on memory storage might be mediated, at least in part, by the release of glucose. Posttraining injections (ip) of glucose produce dose- and time-dependent effects on retention of an inhibitory avoidance response like those seen with epinephrine (Gold 1986). Plasma levels of glucose measured shortly after training vary with the footshock intensity used in training. Further, low doses of epinephrine and glucose that are optimal for enhancing retention produce comparable levels of plasma glucose (Hall & Gold 1986). Glucose has also been found to enhance retention in other tasks, including the development of conditioned taste preference and a conditioned emotional response (Messier & White 1984, 1987). Glucose effects on memory are not blocked by adrenergic antagonists (Gold et al 1986). Since glucose readily enters the brain, it may be that glucose affects memory by directly affecting brain glucoreceptors (Oomura et al 1988). The evidence that posttraining icv injections of glucose produce dose- and time-dependent effects on retention (Lee et al 1988) clearly supports the view that the effects seen with peripheral administration of glucose may be due, at least in part, to central influences.

In their studies of the effects of posttraining glucose on retention Messier & White (1987) obtained optimal memory enhancement with a dose (2.0 g/kg), which is considerably higher than those that Gold and his colleagues have found to be optimal (0.1 g/kg or less). With the higher doses, blood glucose levels were not critical for memory enhancement. Although doses of 1.0, 2.0, and 3.0 g/kg produced comparable levels of blood glucose, only the 2.0 g/kg dose affected retention. Further, Messier & White found that memory was also enhanced by fructose, a sugar that has little influence on the brain, as well as glucose analogs that are not readily metabolized. These investigators suggest that in the doses they have used, these sugars may act at peripheral sites involved in activation of a membrane glucose transport mechanism and that in lower doses glucose may affect memory by other mechanisms.

Clearly, additional research is needed to determine whether the memory-modulating effects of epinephrine involve the release of glucose. As is discussed below, extensive evidence suggests that the effects of epinephrine involve the activation of central norepinephrine receptors within the amygdaloid complex. Consequently, it will be of interest to know whether glucose acts independently of such effects in influencing memory storage.

The findings of several studies suggest that rapid forgetting seen in aged rodents (e.g. Gold et al 1981, Sternberg et al 1985b) may be based in part on adrenergic deficits. McCarty (1981) reported that the release of epinephrine induced by footshock declines in aged rats, and Scarpace (1986) found that $\beta$-adrenergic responsiveness decreases with aging. These findings are of interest in view of the evidence indicating that the retention impairment seen in aged rats can be attenuated by posttraining injections of epinephrine (Sternberg et al 1985b). Recently Gold (1988) reported that memory performance in elderly humans is enhanced by glucose. Recent evidence also indicates that memory is impaired in humans by $\beta$-adrenergic antagonists (Solomon et al 1983, Lichter et al 1986, Madden et al 1986). However, in studies with rats and mice, peripherally administered adrenergic antagonists have generally been found not to impair retention (e.g. Sternberg et al 1985a, 1986). This is a puzzling finding in view of the evidence discussed above indicating that retention is readily enhanced by peripherally administerd adrenergic agonists. Further, as is discussed below, there is convincing evidence that retention is impaired by central administration of adrenergic antagonists. The basis of the differential effects of peripherally and centrally administered adrenergic receptor antagonists on retention clearly requires further study.

## Cholinergic Influences on Memory Storage

Extensive evidence indicates that cholinergic systems are involved in memory storage (e.g. Bartus et al 1985). Although perhaps not all investigators would agree with the conclusion that, "if there is a key transmitter for memory . . . the best bet for that transmitter would be acetylcholine" (Davies 1985), the evidence strongly argues that memory storage may be regulated at least partly by cholinergic systems. Whereas evidence of cholinergic involvement in memory has accumulated over the past several decades, interest in cholinergic systems and memory surged dramatically following publication of reports indicating that Alzheimer's disease is accompanied by a decline in cholinergic functioning (Bowen et al 1976, Davies & Maloney 1976, Perry et al 1977). The evidence is summarized in a number of excellent reviews (e.g. Bartus et al 1982, Coyle et al 1983, Bartus et al 1985).

Although recent findings indicate that other brain systems, including noradrenergic, dopaminergic, serotonergic, GABAergic, and somatostatin systems, also undergo decline in Alzheimer's disease (Iversen & Rossor 1984), clearly a particularly marked reduction of cells occurs in the basal cholinergic brain systems, including the nucleus basalis of Meynert and the septum, which are known to provide cholinergic projections to other brain regions including the cortex and hippocampus. Further, extensive evidence also indicates that the functioning of the cholinergic system declines with aging (Decker 1987).

Many recent studies have confirmed the earlier findings of Stratton & Petrinovich (1963) that retention is enhanced by posttraining systemic injections of low doses of the cholinesterase inhibitor, physostigmine. Memory-enhancing effects have been obtained with a wide variety of tasks, including appetitively motivated learning (Stratton & Petrinovich 1963), active avoidance (Flood et al 1981), and inhibitory avoidance (Gower 1987, Kameyama et al 1986) as well as aversively motivated discrimination learning (Introini-Collison & McGaugh 1988). Such findings argue that acetylcholine released by training is involved in the modulation of memory storage. Many studies have also reported that retention is enhanced with posttraining systemic administration of muscarinic cholinergic agonists, including arecoline and oxotremorine (e.g. Baratti et al 1984, Cherkin & Flood 1988, Flood et al 1981, Introini-Collison & McGaugh 1988). When administered posttraining the effects of cholinergic agonists on retention are both dose- and time-dependent.

Gower (1987) reported that in rats, retention on an inhibitory avoidance response is also enhanced by posttraining administration of secoverine, a muscarinic antagonist that blocks autoreceptors mediating the inhibition of acetylcholine release. Further, secoverine, in a dose that was ineffective when administered alone, potentiated the memory-enhancing effect of physostigmine. These results are consistent with those previously reported by Flood and his colleagues in studies using an active avoidance task (Flood et al 1983). These investigators found that the memory-enhancing effects of posttraining icv or ip injections of the cholinergic drugs arecoline, oxotremorine, and edtrophonium chloride (a cholinesterase inhibitor) on memory were markedly potentiated when low doses of two or three of the drugs were administered concurrently. The bases of the remarkable synergistic influences of cholinergic drugs administered in combination have not as yet been addressed.

It is somewhat puzzling that the effects of cholinergic antagonists on memory storage have been less consistent than those seen with agonists and cholinesterase inhibitors. It is clear that injections of cholinergic antagonists, including atropine, scopolamine, and pirenzepine, produce dose-dependent impairment of retention when administered prior to training in a variety of tasks (Aigner & Mishkin 1986, Blozovski & Hennocq 1982, Flood & Cherkin 1986, Caulfield et al 1983, Hagan et al 1986). Such findings might, of course, be due to influences on processes other than memory storage. While many investigators have found that retention in a variety of tasks is impaired by posttraining administration of cholinergic antagonists (e.g. Flood & Cherkin 1986, Baratti et al 1984, Kameyama et al 1986, Intrioini-Collison & McGaugh 1988), other investigators have failed to find significant impairment of retention with posttraining injections (Gower 1987, Hagan et al 1986). Generally, studies finding impairing effects of posttraining injections of cholinergic antagonists have used higher doses than those required for producing impairment with pre-training injections. The bases of these conflicting findings are unclear.

Much of the most recent research concerned with cholinergic influences on memory is motivated by an interest in finding or developing appropriate drug treatments for Alzheimer's disease and other disorders of memory in which cholinergic functioning is impaired. The general assumption underlying this research is that it might be possible to restore cholinergic functioning through the use of cholinergic agonists, cholinesterase inhibitors, or other drugs that may increase the activation of muscarinic cholinergic receptors (Bartus et al 1985). This view has been encouraged by findings of several studies suggesting that physostigmine may produce mild improvement of memory in some Alzheimer's patients (Peters & Levin 1979, Thal et al 1983, Davis et al 1983).

A number of investigators have attempted to develop an animal model of Alzheimer's disease by lesioning cholinergic forebrain nuclei in rats. Although such lesions do not, of course, duplicate the myriad of neurobiological alterations seen in Alzheimer's disease, they do significantly impair learning and retention in a variety of tasks, including inhibitory and active avoidance and appetitively motivated spatial learning (Bartus et al 1985, Haroutunian et al 1985, Miyamoto et al 1985). Further, the magnitude of the impairment is potentiated by scopolamine (Lo Conte et al 1982). More importantly, at least for the assumption that drugs might be effective in the treatment of disorders resulting from cholinergic impairment, several studies have reported finding that the learning deficit induced by basal forebrain lesions can be attenuated by physostigmine administered either before (Murray & Fibiger 1985) or after training (Haroutunian et al 1985). Physostigmine administered to mice before training has also been shown to attenuate memory impairment (assessed in appetitively motivated tasks) produced by pretraining electrical stimulation of the dorsal hippocampus, a treatment that significantly reduces hippocampal choline acetyltransferase (Micheau et al 1985). Tilson and his colleagues (1988) reported that in a spatial memory task, retention impairments induced by colchicine injected into the nucleus basalis were attenuated by pretraining systemic injections of physostigmine, whereas naloxone and vasopressin (each administered in a single dose) were ineffective. Thus, it may be that in animals with impaired functioning of the nucleus basalis, restoration of cholinergic function is required for attentuation of the memory impairment. However, additional studies with wide ranges of doses are needed to determine whether memory-enhancing treatments other than cholinergic drugs can attenuate the memory impairment produced by lesions of cholinergic systems.

Considered together, the findings summarized above provide strong support for the view that memory is influenced by posttraining alterations in cholinergic systems. Thus, the findings are consistent with the view that cholinergic systems activated by training are involved in memory storage. There is little doubt that the effects of cholinergic drugs on memory are centrally mediated, because memory is not affected by cholinergic drugs that do not pass the blood-brain barrier (e.g. Baratti et al 1984, Blozovski & Hennocq 1982, Kameyama et al 1986).

## Opioid Peptide Influences on Memory Storage

Opioid peptides are released both centrally and peripherally by stressful stimulation (Amir et al 1980, Bodnar et al 1980, Izquierdo et al 1980).

Enkephalin is released, together with epinephrine, from the adrenal medulla (Viveros et al 1980), and $\beta$-endorphin is released from the hypothalamus (Carrasco et al 1982) and the pituitary (along with ACTH). The findings of recent studies have provided extensive evidence indicating that retention in rats and mice is influenced by posttraining administration of opioid peptide agonists and antagonists. The memory-modulating effects are both dose- and time-dependent and are obtained with systemic as well as central injections in a variety of tasks employing both appetitive and aversive motivation. Thus, the findings provide strong support for the view that endogenously released opioid peptides are involved in the regulation of memory storage (McGaugh 1983, McGaugh & Gold 1988).

Castellano's initial report (1975) that posttraining injections of opiate receptor agonists impair discrimination learning in mice was followed by a flurry of studies examining the effects of opiate agonists and antagonists on memory in several types of training tasks (McGaugh 1983). Many studies have shown that opiate receptor agonists, including morphine, $\beta$-endorphin, and enkephalin, impair retention when administered posttraining in low doses. The memory-impairing effects of opiate agonists are dose- and time-dependent and are antagonized by opiate antagonists (Castellano et al 1984, Gallagher & Kapp 1978, Gallagher et al 1981, Introini & Baratti 1984, Izquierdo 1979, Messing et al 1979).

Studies of the effects of opiate antagonists on memory provide additional support for the view that opioid systems are involved in regulating memory storage. Numerous studies have shown that retention is enhanced by posttraining administration of opiate antagonists, including naloxone, naltrexone, diprenorphine, levallorphan, nalmenfene, and $\beta$-funaltrexamine (Gallagher 1982, Messing et al 1979, Izquierdo 1979). The memory-enhancing effects of opiate antagonists are dose- and time-dependent and have been obtained in studies with several types of training tasks, including inhibitory avoidance, active avoidance, aversively-motivated Y-maze discrimination, habituation, latent inhibition, and appetitively motivated spatial learning (Gallagher et al 1983, 1987, Introini-Collison & McGaugh 1987, Izquierdo 1979). The finding that posttraining injections of opiate antagonists enhance the acquisition of a spatial memory task in old and young rats (Gallagher et al 1985a) is of interest because aged animals typically perform poorly in spatial tasks (e.g. Barnes 1979). Studies of the effects of opiate antagonists on memory in patients with memory disorders have yielded mixed results (Reisberg et al 1983a,b, Tariot et al 1986).

Recent findings that opiate antagonists block the amnestic effects produced by posttraining electroconvulsive shock (ECS) and direct electrical stimulation of the amygdala and hippocampus suggest that the amnesia produced by such treatments may involve the release of opioid peptides (Carrasco et al 1982, Liang et al 1983, Collier et al 1987). Izquierdo and his colleagues (Izquierdo et al 1984, Perry et al 1983) have reported that novel experiences of various kinds induce the release of brain $\beta$-endorphin (as measured by decreased immunoreactivity) and that the levels remain low for several hours. Recently, McGaugh and Introini-Collison (1987) found that posttraining ECS impaired retention in an inhibitory avoidance response and in a Y-maze discrimination task in control mice, but did not affect the retention of mice given a brief novel exploratory experience one hour prior to training. The findings of other recent experiments have shown that a novel experience given prior to training also blocked the memory-enhancing effects of posttraining naloxone (but did not block memory enhancement produced by post-training epinephrine) (Izquierdo & McGaugh 1985, 1987). These findings are of particular interest in view of evidence indicating that the degree of memory enhancement obtained with posttraining naloxone depends upon the novelty of the training situation (del Cerro & Borrell 1985, Netto et al 1986). These findings suggest that naloxone effects on retention may be due, at least in part, to blocking of the effects of brain $\beta$-endorphin that is released by novel training experiences. That the memory-enhancing effects of posttraining naloxone are due to peripheral influences is unlikely, because retention is not affected by peripheral injections of opiate receptor antagonists that do not pass the blood-brain barrier (Flood et al 1987a, Introini et al 1985, Rush 1986).

The effects of other opioid peptides on memory are somewhat more complex. The amnestic effects of posttraining, peripherally administered Met-enkephalin are blocked by the centrally acting opiate antagonists naloxone and naltrexone and the peripherally acting antagonist MR2263 (Introini et al 1985, Zhang et al 1987). These findings suggest that the effects of enkephalins on memory may be initiated at peripheral opioid receptors but may ultimately involve activation of central opioid systems. Other recent findings suggest that hormones of the adrenal medulla, including enkephalin and epinephrine, may be involved in the modulating effects of opioid peptides on memory. Zhang and her colleagues (1987) found that the memory-modulating effects of peripherally administered Met-enkephalin and naloxone are attentuated in adrenal-denervated rats, and Conte and his colleagues (1986) reported that adrenal demedullation atten-uates the amnestic effects of central (icv) injections of Met-enkephalin administered following active avoidance training. Further, these inves-tigators reported that amnesia could be obtained in adrenal-demedullated rats if epinephrine was administered prior to training. On the basis of these findings they argue that enkephalin acts centrally and that the amnestic actions require an adequate level of circulating epinephrine. These results are of interest in view of the findings (Liang et al 1985) that posttraining electrical stimulation of the amygdala in adrenal demedullated rats impairs retention (in both active and inhibitory avoidance tasks) only in animals administered epinephrine prior to the brain stimulation.

It is clear from the findings of Conte and his colleagues that amnesia can be produced by central injections of Met-enkephalin. Activation of peripheral receptors is not required. However, such findings do not address the role of adrenal enkephalins in the modulation of memory storage. It would be of interest to know whether the dose response effects obtained with central injections are shifted by peripherally-administered Met-enkephalin. Additional research is required to determine whether endogenously released enkephalins act peripherally or centrally (or both) in their influences on memory.

Studies of the effects of $\kappa$-selective agonists on memory have yielded highly conflicting results. Izquierdo and his colleagues (Izquierdo et al 1985) reported that posttraining administration of dynorphin in a wide range of doses in rats did not affect retention in either active or inhibitory avoidance tasks. In mice, dynorphin impaired retention of inhibitory avoidance but did not affect retention in either a discrimination task or a habituation task (Introini 1984, Introini-Collison et al 1987). Jefferys and his colleagues (Jefferys et al 1985) reported that the retention impairment produced by adrenalectomy is attenuated by $\kappa$-selective agonists. Pavone & Castellano (1985) have shown that in mice the effects of $\kappa$-agonists on retention are highly strain dependent. Thus, although there is clear evidence that $\kappa$-agonists can influence memory, the effects are unlike those found with other opioid peptides. As yet no evidence suggests a role for the activation of endogenous $\kappa$-receptors in memory storage.

# INTERACTIONS OF HORMONAL AND NEUROTRANSMITTER SYSTEMS IN MEMORY MODULATION

The findings summarized above clearly indicate that the retention of newly acquired information can be influenced by a variety of treatments affecting neurohormonal and transmitter systems. Much recent research has examined the interactions among these systems in their effects on memory in order to determine whether the various effects might work through common mechanisms. The following section briefly reviews recent findings based on peripheral administration of treatments affecting adrenergic, cholinergic, and opioid peptide systems.

## Adrenergic–opiate Interactions

There is extensive evidence that epinephrine interacts with other neuro-hormone and neuromodulatory systems in influencing memory. As was noted above, for example (Conte et al 1986), epinephrine appears to be required for central injections of enkephalin to induce retrograde amnesia. Borrell and his colleagues (1985) reported that epinephrine is also required for obtaining memory enhancement with posttraining injections of vaso-pressin. Evidence from many recent studies indicates that opioid peptide systems interact with adrenergic (and noradrenergic) systems in their influences on memory. The findings that the memory-impairing effects of high doses of posttraining epinephrine in an inhibitory avoidance task and a Y-maze discrimination task are blocked by naloxone as well as by a novel experience given prior to training (Introini-Collison & McGaugh 1987, Izquierdo & Netto 1985, Izquierdo & Dias 1985) suggest that the impairment may be due to the release of $\beta$-endorphin. The memory enhanc-ing effects of low doses of epinephrine clearly do not involve $\beta$-endorphin, since low doses of epinephrine attenuate the memory-impairing effects of posttraining $\beta$-endorphin (Introini-Collison & McGaugh 1987, Izquierdo & Dias 1985). Further, the effects of low doses of epinephrine and naloxone are additive in their effects on memory. Doses of posttraining epinephrine and naloxone that are ineffective when administered alone enhance reten-tion when administered together (Introini-Collison & McGaugh 1987, Izquierdo & Dias 1985).

The question of the basis (or bases) of the effects of epinephrine and naloxone on memory has been the focus of extensive recent research. Although it remains possible the additive effects of epinephrine and naloxone on memory are based on independent parallel systems, recent evidence strongly suggests that both affect memory through effects result-ing in the activation of central noradrenergic receptors. Much of the recent research on this tissue was stimulated by Izquierdo & Graudenz's (1980) study that reported that the memory-modulating effects of posttraining naloxone are blocked by peripheral injections of the $\beta$-adrenergic receptor antagonist propranolol. They interpreted these findings as suggesting that naloxone acts by blocking the inhibitory effects of opioid peptides on the release of norepinephrine (NE). This interpretation is consistent with extensive evidence from earlier as well as more recent studies indicating that opiate receptors regulate the release of NE in the brain (e.g. Arbilla & Langer 1978, Montel et al 1974, Nakamura et al 1982, Werling et al 1987) and that NE and opioid peptidergic systems interact in many brain regions (Walker et al 1984).

As is indicated below, evidence also suggests that epinephrine affects memory through effects involving the release of NE in the brain (Liang et al 1986). Thus, considered together, these findings suggest that the additive effects of these treatments on memory may result from their influences on the release of NE. Izquierdo and Graudenz's finding that propranolol blocks the memory-modulating effects of naloxone has been confirmed in many experiments (e.g. Introini-Collison & Baratti 1986, McGaugh et al 1988). Further, the findings of other recent experiments have provided additional evidence indicating that naloxone effects on memory involve the activation of NE receptors. Introini-Collison & Baratti (1986) found that naloxone potentiates the memory-enhancing effects of the $\beta$-adrenergic agonist clenbuterol. Because both naloxone and clenbuterol readily pass the blood-brain barrier, they argued that the effects were probably due to activation of central $\beta$-adrenergic receptors. This interpretation is supported by their finding that the memory-enhancing effects of naloxone are not blocked either by sotalol, a peripherally acting $\beta$-adrenergic antagonist, or by the $\alpha$-antagonists phenoxybenzamine or phentolamine. Introini-Collison & Baratti (1986) also found that the effects of naloxone on memory are blocked in animals treated with DSP4, a neurotoxin that produces a relatively specific reduction in brain NE. These findings are consistent with the findings of Gallagher and her colleagues (Gallagher et al 1985b, Fanelli et al 1985) indicating that the memory-modulating effects of naloxone are blocked in animals with 6-OHDA lesions of the dorsal noradrenergic bundle.

The findings of several recent studies suggest that age-related declines in memory may be associated with impaired function of central NE systems. Leslie and her colleagues (Leslie et al 1985) have reported that retention performance of aged mice aged mice on an inhibitory avoidance task is highly correlated with number of cells in the locus coeruleus (LC). These findings are consistent with evidence indicating significant cell loss in LC neurons in aged human brains (e.g. Vijayashankar & Brody 1979). Collier and his colleagues (Collier et al 1985) found that deficits in inhibitory avoidance retention seen in aged rats could be attenuated by icv infusion of NE. Further, Arnsten & Goldman-Rakic (1985) reported that in monkeys, pretraining injections of the $\alpha_{-2}$ agonist clonidine attenuated age-related deficits in performance in a delayed-response task. In view of the evidence that epinephrine affects memory by influencing the release of central NE, the finding that posttraining epinephrine injections attenuate age-related retention impairment (Sternberg et al 1985b) is also consistent with the view that the cognitive decline seen with aging may involve alterations in NE systems.

## Involvement of the Amygdala in Adrenergic and Opiate Influences

Evidence from a number of recent studies strongly suggest that opioid peptide and adrenergic systems may modulate memory through effects involving the amygdaloid complex. It is well known that posttraining electrical stimulation of the amygdala can influence retention (Kesner 1982, McGaugh & Gold 1976). The memory-modulating effects of amygdala stimulation are blocked by lesions of the stria terminalis, a major amygdala pathway, and by administration of naloxone, either systemically or directly into the bed nucleus of the stria terminalis (Liang & McGaugh 1983b, Liang et al 1983). Lesions of the stria terminalis also block the memory-modulating effects of posttrial ip injections of epinephrine (Liang & McGaugh 1983a) as well as naloxone and β-endorphin (McGaugh et al 1986).

The hypothesis that the amygdala is involved in the memory-modulating effects of treatments affecting adrenergic and opiate receptors is supported by extensive evidence indicating that retention can be modulated by post-training intra-amygdala injections of adrenergic and opiate agonists and antagonists. Gallagher and her colleagues (Gallagher et al 1981) have shown that retention is impaired by β-antagonists, including propranolol and alprenolol. The effect is stereospecific and time-dependent. These findings are clearly consistent with the view that memory storage is influenced by the activation of noradrenergic receptors within the amygdala. Kesner and his colleagues reported that posttraining intra-amygdala injections of NE impair retention (Ellis & Kesner 1981, Ellis et al 1983). More recent findings indicate that the effect of intra-amygdala NE depends upon the dose used. In experiments using doses much lower than those used by Kesner and his colleagues, posttraining intra-amygdala injections of NE enhance retention (Liang et al 1986). The effects of NE are dose- and time-dependent and are blocked by concurrent administration of propranolol.

In a series of studies examining the effects of posttraining treatments on olfactory learning in honey bees, Menzel (1984, Menzel & Michelsen 1986) obtained results that are remarkably parallel to those found in research with mammals. Of particular interest in relation to the studies discussed above is Menzel's finding that the retention of newly acquired olfactory information is susceptible to modulating effects of posttraining injections of NE into a specific region of the bee's brain. Further, as has been found in studies with rats, retention is enhanced by low doses and impaired by high doses of NE (Menzel & Michelson 1986).

Other findings provide support for the view that the memory-modulating effects of epinephrine may involve the activation of NE receptors within the amygdala. Posttraining intra-amygdala injections of pro-pranolol also block the memory-enhancing effects of ip injections of epi-nephrine (Liang et al 1986). These results are consistent with evidence that epinephine activates the release of brain NE (Gold & van Buskirk 1978) and that release of NE within the amygdala is highly correlated with plasma catecholamine levels (Dietl 1985). The findings of Tanaka and his colleagues (1982a,b) that naloxone enhances stress-induced NE turnover in the amygdala is consistent with the view that opioid peptides inhibit NE release. Since epinephrine does not readily cross the blood-brain barrier (Weil-Malherbe et al 1959), it is not clear how epinephrine influences the release of brain NE. The effect might be mediated by activation of peri-pheral receptors of neurons that in turn activate the LC. There is, of course, extensive evidence that the LC is activated by arousing and stressful stimulation (Svensson 1987) as well as by administration of epinephrine (Holdefer & Jensen 1987). Since NE systems project to the amygdala in part through the stria terminalis (ST), the findings that the effects of epinephrine (and naloxone) on retention are blocked by ST lesions are consistent with the view that ST lesions may block memory-modulation by attenuating NE release within the amygdala. This suggestion is con-sistent with the findings of Oishi and his colleagues (1979) indicating that stimulation of the LC produced alterations in evoked potentials (induced by olfactory bulb stimulation) recorded from the amygdala and that the effect was selectively blocked by a $\beta$-adrenergic antagonist.

In an extensive series of studies, Gallagher and her colleagues (Gallagher 1982, Gallagher & Kapp 1978, Gallagher et al 1981, 1985b) have found that retention can be modulated by posttraining intra-amygdala adminis-tration of opiate agonists and antagonists. As is found with peripheral injections, retention is impaired by opiate agonists and enhanced by antag-onists, including naloxone, naltrexone, diprenorphine, and levallorphan. The effects are dose-dependent, time-dependent, and stereospecific. Post-training injections of naloxone into the caudate-putamen and cortex dorsal to the amygdaloid complex do not affect retention (Introini-Collison et al 1988). The finding that the memory-enhancing effect of intra-amygdala injections of naloxone is blocked by lesions of the dorsal noradrenergic bundle (Gallagher et al 1985b) argues that the effects involve the release of NE.

Recent experiments have provided additional evidence that naloxone effects on memory, as assessed in several types of tasks, may involve the activation of NE receptors within the amygdala. The memory-enhancing effects of posttraining ip injections of naloxone, in both inhibitory avoid-ance and Y-maze discrimination tasks, are blocked by posttraining intra-amygdala injections of $\beta$-adrenergic antagonists, including propranolol (a

$\beta_{1,2}$-antagonist), atenolol (a $\beta_1$-antagonist), and zinterol (a $\beta_2$-antagonist), in doses that do not affect retention when administered alone (McGaugh et al 1988). The effects of naloxone are blocked neither by propranolol injected into the caudate-putamen or cortex dorsal the amygdala nor by $\alpha$-antagonists (prazosin and yohimbine) administered intra-amygdally. Further, posttraining intra-amygdala injections of propranolol also blocked the memory-enhancing effects of intra-amygdally administered naloxone (Introini-Collison et al 1988).

Considered together, these recent findings provide strong support for the hypothesis that NE receptors within the amygdala are involved in the memory-modulating effects of opiate antagonists and epinephrine, and are consistent with evidence suggesting that opiate-noradrenergic interactions within the amygdaloid complex are involved in appetitively-motivated learning (Oomura et al 1988). Further, they also are consistent with other evidence (Introini-Collison & Baratti 1986, Oishi et al 1979) suggesting that such treatments may affect memory storage through influences selectively involving $\beta$-adrenergic receptors.

## Cholinergic–opiate Interactions

The findings of several recent studies suggest that opioid peptidergic systems may also interact with cholinergic systems in their influences on memory storage. Baratti and his colleagues (Baratti et al 1984) reported that retention in mice is enhanced by posttraining administration of naloxone and oxotremorine injected together ip in low doses that are ineffective when administered alone. Further, they found that the memory-impairing effects of atropine were not blocked by concurrent administration of naloxone and that morphine did not block the memory-facilitating effect of oxotremorine. These findings are consistent with extensive evidence indicating that acetylcholine release is inhibited by opioid agonists and activated by opiate antagonists (e.g. Jhamandas & Sutak 1974, 1983, Jhamandas et al 1975), and suggest that opiate systems may affect memory through modulating influences on cholinergic systems. However, the findings of other recent studies are only partially consistent with this interpretation. Studies by Flood et al (1987a) and Rush (1986) reported that opiate antagonists blocked the effect of scopolamine-induced amnesia in mice. These findings conflict with those of Baratti and his colleagues (1984). However, in agreement with Baratti et al, Flood and his colleagues (1987a) also found that opiate agonists did not block the memory-enhancing effect of a cholinergic agonist. The conflicting results concerning whether the effects of naloxone are blocked by cholinergic antagonists might be due to any of a variety of factors, including the combinations of doses used. Perhaps the effects of opiate and cholinergic antagonists summate and thus cancel each other. Generally, however, the results of these studies suggest that opioid peptidergic systems may modulate retention, at least in part, by regulating the release of acetylcholine. Additional evidence for this conclusion is provided by findings (Bostock et al 1988) indicating that posttraining injections of naloxone into the medial septal region enhanced rats' retention in an appetitively motivated spatial task. Retention was impaired by $\beta$-endorphin. These findings are of interest in view of evidence that administration of opiate agonists and antagonists influence activity in the septo-hippocampal pathway (Moroni et al 1977, Botticelli & Wurtman 1982) and that both enkephalin and $\beta$-endorphin are found in the septum lateralis (Nieuwenhuys 1985).

In a series of recent studies, Blozovski and her colleagues (Blozovski & Hennocq 1982, Blozovski & Dumery 1984, Dumery & Blozovski 1987) found that retention of an inhibitory avoidance response in young rats was impaired by intra-amygdala injections of cholinergic antagonists and that the impairment was antagonized by arecoline. Cholinergic antagonists injected into the hippocampal-entorhinal area also impair retention. Since the drugs were administered prior to training in these experiments, it is not clear whether the drugs' effects were due to influences on posttrial processes. Prado-Alcala and his colleagues (Prado-Alcala et al 1981, 1984) reported that retention of an inhibitory avoidance response is impaired by posttraining injections of atropine into the anterior caudate nucleus. These studies clearly indicate that acquisition and retention can be influenced by altering cholinergic functioning in several brain regions. It is not known whether these effects involve interactions with opiate systems.

## Adrenergic–cholinergic Interactions

Several recent experiments have investigated possible interactions of adrenergic and cholinergic systems in modulating retention. The findings of experiments examining the effects of posttraining systemic administration of epinephrine together with atropine, oxotremorine, and physostigmine on retention of both an inhibitory avoidance task and a Y-maze discrimination task suggest that epinephrine effects on retention may involve cholinergic influences (Introini-Collison & McGaugh 1988). In both tasks atropine blocked the memory-facilitating effects of a low dose of epinephrine, whereas oxotremorine and physostigmine attenuated the memory-impairing effect of a high dose of epinephrine. Further, low doses of epinephrine potentiated the memory-enhancing effects of both oxotremorine and physostigmine. Stone et al (1988) have reported that the retention deficit (in an inhibitory avoidance task) produced by pretraining administration of scopolamine was partially attenuated by posttraining injections of epinephrine and glucose, in doses that have previously been found to enhance memory when administered alone. In contrast to the findings of Flood & Cherkin (1986), Stone and his colleagues (1988) found that the scopolamine-induced retention deficit was not attenuated by posttraining injections of the cholinergic agonist arecoline. Interactions of cholinergic and adrenergic treatments will depend, of course, on the doses used. The finding that scopolamine effects on memory were not attenuated by arecoline is somewhat more puzzling. In another recent study, Decker & Gallagher (1987) have reported that the impairing effects of pretraining injections of scopolamine on performance in a spatial memory task were potentiated by 6-OHDA lesions of the dorsal noradrenergic bundle. Additional studies are needed to clarify the role of posttraining adrenergic-cholinergic interactions in the modulation of memory.

## Other Hormone–neuromodulator Interactions in Memory Storage

Although the focus of the review is on adrenergic, cholinergic and opioid peptide involvement in modulating memory storage, interaction of these systems with treatments affecting other hormonal and transmitter systems warrant brief discussion. Extensive evidence indicates that posttraining injections of the neuropeptide substance P produces dose-dependent and time dependent enhancement of memory in aversive as well as appetitive tasks (Huston & Staubli 1981, Schlesinger et al 1986, Pelleymounter et al 1986). The finding that the memory-enhancing effects of substance P are blocked by naltrexone suggests that the effects may involve interactions with an opioid peptide system (Schlesinger et al 1983). These latter findings are an exception to the general rule, based on findings discussed above, that naloxone potentiates the effects of memory-enhancing treatments. Other findings suggest that substance P effects on memory may involve cholinergic influences. Several studies have reported that retention is enhanced by posttraining injections of substance P into the medial septal nucleus and the nucleus basalis magnocellularis (Staubli & Huston 1980, Nagel & Huston 1988).

Nagel & Huston (1988) also reported that retention is impaired by injections of the GABA agonist muscimol into the basal forebrain. These findings suggest that GABA projections to the basal forebrain may affect memory by inhibiting the release of acetylcholine (Casamenti et al 1986). The findings of other recent studies are consistent with earlier evidence (e.g. Breen & McGaugh 1961, Grecksch & Matthies 1981) that retention is enhanced by posttraining systemic injections of GABA receptor antagonists, including picrotoxin and bicuculline, and impaired by GABA agonists (Castellano & Pavone 1988, Swartzwelder et al 1987, Yonkov et al

1987). The effects appear to be centrally mediated, since retention is not affected by posttraining ip injections of bicuculline methiodide, a peripherally acting GABA receptor antagonist (Brioni & McGaugh 1988).

The effects of vasopressin on memory have been extensively examined in recent experiments (de Wied 1984). Although findings have been conflicting, as well as interpretations of the findings (McGaugh & Gold 1988), there is convincing evidence that posttraining administration of vasopressin enhances retention of training tasks with appetitive as well as aversive motivation (Ettenberg et al 1983a,b). The effects appear to be at least in part centrally mediated, since enhancement of memory is produced by posttraining icv injections of vasopressin and metabolites that lack pressor activity (Burbach et al 1983, de Wied et al 1984). Although there is evidence that vasopressin effects on retention, when administered prior to a retention test, may involve interactions with both noradrenergic and cholinergic systems (Bohus et al 1982, Faiman et al 1988), it is not clear whether these systems are involved in the effects of immediate posttraining injections of vasopressin on retention.

Several recent investigations have reported that retention is influenced by posttraining injections of cholecystokinin-octapeptide (CCK-8). Retention is enhanced by low doses and impaired by high doses and can be obtained with central as well as peripheral injections (Flood et al 1987c, Deupree & Hsiao 1987, Telegdy et al 1985). The effects of peripheral CCK on memory appear to be mediated by the activation of peripheral receptors rather than by direct effects on the brain. Flood and his colleagues (Flood et al 1987c) found that the effects of ip administration of CCK-8 are blocked by vagotomy. Although Telegdy (Telegdy et al 1985) and his colleagues have reported that CCK affects the release of brain catecholamines, there is, as yet, little evidence indicating that the release of catecholamines is critical for the effect of CCK on memory.

## MEMORY STORAGE, MEMORY RETRIEVAL, AND STATE DEPENDENCY

This review emphasizes the effects on subsequent retention of posttraining alterations of hormonal and transmitter systems. The guiding assumption of research in this area is that the alterations affect retention by modulating the storage of newly acquired information. However, many of the treatments considered in this review also affect the performance of learned responses when they are administered prior to retention testing (McGaugh & Gold 1988). Such findings have suggested that hormonal systems may facilitate the retrieval of information (e.g. Riccio & Concannon 1981).

However, when drugs and hormones are administered prior to retention testing it is difficult to distinguish effects of treatments on memory retrieval from other influences on retention performance. This is a particularly complicated problem for interpreting the findings of studies in which response latency is used as the only index of retention.

Evidence also suggests that posttraining treatments affect retention by inducing state-dependency. On the basis of an extensive series of studies, Izquierdo and his colleagues (e.g. Izquierdo 1984, Izquierdo & Netto 1985) have argued, for example, that retention performance is based at least partly on the degree of congruency between endogenous hormonal states after training and at the time of retention testing. For example, they have found that injections of $\beta$-endorphin administered prior to a retention test attenuate the amnesia induced by posttraining injections of $\beta$-endorphin. In animals not given posttraining treatments, retention performance is enhanced by either $\beta$-endorphin or a novel experience (which presumably releases $\beta$-endorphin, as discussed above) prior to the test. Further, if a novel experience is given prior to training (and thus $\beta$-endorphin is not released by the training) administration of $\beta$-endorphin prior to the retention test does not affect retention performance (Izquierdo & McGaugh 1985, 1987). It is not clear from these studies that the memory impairing effect of $\beta$-endorphin (or any other posttraining treatment) is due entirely to the induction of state dependency. Additional experiments with other training tasks commonly employed, including appetitively motivated tasks and discrimination tasks, are needed to address this issue.

That retention enhancement produced by posttraining treatments is based on the induction of state dependency seems highly unlikely. In such experiments, of course, enhanced retention is seen when no treatments are given prior to the retention test. A state-dependency interpretation applied to posttraining treatments that enhance retention would require the ad hoc assumption that the brain state induced following training is congruent with that normally occurring in such animals when no treatment is given prior to the retention test. It is not at all clear why such congruence would be expected, but, if it is assumed that such congruence is the basis of the memory-enhancing effects of posttraining injections of drugs and hormones, administration of the same drugs or hormones prior to testing would be expected to decrease the congruence and, thus, attenuate the enhanced retention. Available evidence clearly argues against this view. For example, picrotoxin administered ip in a range of doses prior to retention testing in an inhibitory avoidance task does not alter the magnitude of the retention enhancement produced by immediate posttraining injections (Castellano & McGaugh 1989).

# ENDOGENOUS MODULATION OF NEURONAL CONNECTIONS

Many recent studies have focused on the possibility that memory-enhancing treatments may act through influences on the development of long-term potentiation (LTP) (Brown et al 1989, this volume). Bloch & Laroche (1984) have shown, for example, that posttrial electrical stimulation of the mesencephalic reticular formation, with parameters that enhance memory, also enhances the development of hippocampal LTP. Gold and his colleagues (Gold et al 1984) have reported that epinephrine enhances the development of LTP in the perforant path-dentate gyrus system. Further, the development of LTP in this brain region is enhanced by direct application of NE and $\beta$-adrenergic agonists and is attenuated by $\beta$-antagonists (Harley & Evans 1988, Hopkins & Johnston 1984, Stanton & Sarvey 1985). However, alteration of NE levels appears not to affect components of hippocampal LTP that persist for a week or longer (Robinson & Racine 1985). It may be that cholinergic inputs into the hippocampus also serve to modulate the development of LTP in this brain region.

Although the hippocampus has been studied in most investigations of LTP, LTP clearly can be obtained in many brain regions (Racine et al 1983, Gerren & Weinberger 1983), including the amygdaloid complex (T. Brown, personal communication). Thus, it might be that treatments affecting the amygdala alter memory by modulating the development of connections within the amygdala. Extensive recent evidence indicates that the amygdaloid complex is involved in learning and retention in a variety of training tasks (LeDoux 1986, Gaffan & Harrison 1987, Hitchcock & Davis 1987, Murray & Mishkin 1985, Gentile et al 1986, Kapp et al 1979, Iwata et al 1986, Saunders et al 1984, Salmon et al 1987). Changes within the amygdala and/or hippocampus probably do not serve as the basis of permanent memory, since lesions of these brain regions do not abolish well-established memories (Milner 1966, Liang et al 1982). However, temporary connections may be produced within the amygdala and hippocampus, and such alterations may in turn serve to influence permanent storage occurring in other brain regions.

# CONCLUDING REMARKS

The findings reviewed here clearly indicate that retention of recently acquired information can be altered by a variety of posttraining treatments affecting hormonal and neuromodulatory systems. The findings provide strong support for the general hypothesis that endogenous hormonal and neuromodulatory systems activated by learning may play an important role in regulating the storage of information. Further, recent research has begun to provide some evidence concerning the brain systems through which hormone and neuromodulatory systems may act to influence memory. It seems clear from these recent findings that an understanding of the neural process underlying lasting memory will require knowledge of how memory storage processes are orchestrated by these modulatory systems.

ACKNOWLEDGMENTS

Preparation of this review was supported by US Public Health Service Grant MH12526 and Contract N00014-87-K-0518 from the Office of Naval Research. I thank Jorge Brioni, Larry Cahill, Nancy Collett, Ines Introini-Collison, Michael Decker, Karl Murray, and Alan Nagahara for their advice and assistance in the preparation of this paper.

*Literature Cited*

Agranoff, B. W. 1980. Biochemical events mediating the formation of short-term and long-term memory. In *Neurobiological Basis of Learning and Memory*, ed. Y. Tsukada, B. W. Agranoff. New York: Wiley

Aigner, T. G., Mishkin, M. 1986. The effects of physostigmine and scopolamine on recognition memory in monkeys. *Behav. Neural Biol.* 45: 81–87

Amir, S., Brown, Z. W., Amit, Z. 1980. The role of endorphin in stress: Evidence and speculation. *Neurosci. Biobehav. Rev.* 4: 77–86

Arbilla, S., Langer, S. Z. 1978. Morphine and β-endorphin inhibit release of noradrenaline from cerebral cortex but not of dopamine from rat striatum. *Nature* 271: 559–61

Arnsten, A. F. T., Goldman-Rakic, P. S. 1985. Alpha2-adrenergic mechanisms in prefrontal cortex associated with cognitive decline in aged nonhuman-primates. *Science* 230: 1273–76

Baratti, C. M., Introini, I. B., Huygens, P. 1984. Possible interaction between central cholinergic muscarinic and opioid peptidergic systems during memory consolidation in mice. *Behav. Neural Biol.* 40: 155–69

Barnes, C. A. 1979. Memory deficits associated with senescence: A behavioral and neurophysiological study in the rat. *J. Comp. Physiol. Psychol.* 93: 74–104

Bartus, R. T., Dean, R. L., Beer, B., Lippa, A. S. 1982. The cholinergic hypothesis of geriatric memory dysfunction. *Science* 217: 408–17

Bartus, R. T., Dean, R. L., Pontecorvo, M. J., Flicker, C. 1985. The cholinergic hypothesis: A historical overview, current perspective, and future directions. See Olton et al 1985, pp. 332–58

Bloch, V., Laroche, S. 1984. Facts and hypotheses related to the search for the engram. See Lynch et al 1984, pp. 249–60

Blozovski, D., Dumery, V. 1984. Implication of amygdaloid muscarinic cholinergic mechanisms in passive avoidance learning in the developing rat. *Behav. Brain Res.* 13: 97–106

Blozovski, D., Hennocq, N. 1982. Effects of antimuscarinic cholinergic drugs injected systemically or into the hippocampoentorhinal area upon passive avoidance learning in young rats. *Psychopharmacology* 76: 351–58

Bodnar, R. J., Kelly, D. D., Brutus, M., Glusman, M. 1980. Stress-induced analgesia: Neural and hormonal determinants. *Neurosci. Biobehav. Rev.* 4: 87–100

Bohus, B., Conti, L., Kovacs, G. L., Versteeg, D. H. G. 1982. Modulation of memory processes by neuropeptides: Interaction with neurotransmitter systems. In *Neuronal Plasticity and Memory Formation*, ed. C. Amjone Marsan, H. Matthies, pp. 75–87. New York: Raven

Borrell, J., del Cerro, S., Guaza, C., Zubiaur, M., de Wied, D. 1985. Interactions between adrenaline and neuropeptides on modulation of memory processes. In *Contemporary Psychology: Biological Processes and Theoretical Issues*, ed. J. L. McGaugh, pp. 17–36. Amsterdam: North-Holland

Borrell, J., de Kloet, E. R., Versteeg, D. H. G., Bohus, B. 1983. Inhibitory avoidance deficit following short-term adrenalectomy in the rat: The role of adrenal catecholamines. *Behav. Neural Biol.* 39: 241–58

Bostock, E., Gallagher, M., King, R. A. 1988. The effects of opioid microinjections into the medial septal area on spatial memory in rats. *Behav. Neurosci.* 102: 643–52

Botticelli, L. J., Wurtman, R. J. 1982. Septo-hippocampal cholinergic neurons are regulated trans-synaptically by endorphin and corticotropin neuropeptides. *J. Neurosci.* 2: 1316–21

Bowen, D. M., Smith, C. B., Whitt, P., Davison, A. N. 1976. Neurotransmitter-related enzymes and indices of hypoxia in senile dementia and other abiotrophies. *Brain* 99: 459–96

Bowman, R. E., Heironiimus, M. P., Harlow, H. F. 1979. Pentylenetetrazol: Posttraining injection facilitates discrimination learning in rhesus monkeys. *Physiol. Psychol.* 7: 265–68

Breen, R. A., McGaugh, J. L. 1961. Facilitation of maze learning with posttrial injections of picrotoxin. *J. Comp. Physiol. Psychol.* 54: 498–501

Brioni, J. D., McGaugh, J. L. 1988. Posttraining administration of GABAergic antagonists enhance retention of aversively motivated tasks. *Psychopharmacology.* In press

Brown, T. H., Ganong, A. H., Kairiss, E. W., Keenan, C. L. 1990. Hebbian synapses: Biophysical mechanisms and algorithms. *Ann. Rev. Neurosci.* 13: Submitted

Burbach, J. P. H., Kovacs, G. L., de Wied, D., van Nispen, J. W., Greven, H. M. 1983. A major metabolite of arginine vasopressin in the brain is a highly potent neuropeptide. *Science* 221: 1310–12

Carey, R. J. 1987. Post-trial hormonal treatment effects: memory modulation or per-

ceptual distortion? *J. Neurosci. Methods* 22: 27–30

Carrasco, M. A., Dias, R. D., Perry, M. L. S., Wofchuk, S. T., Souza, D. O., Izquierdo, I. 1982. Effect of morphine, ACTH, epinephrine, Met-, Leu- and des-Tyr-Met-enkephalin on β-endorphin-like immunoreactivity of rat brain. *Psychoneuroendocrinology* 7: 229–34

Casamenti, F., Deffenu, G., Abbamondi, A. L., Pepeu, G. 1986. Changes in cortical acetylcholine output induced by modulation of the nucleus basalis. *Brain Res. Bull.* 16: 689–95

Castellano, C. 1975. Effects of morphine and heroin on discrmination learning and consolidation in mice. *Psychopharmacology* 42: 235–42

Castellano, C., McGaugh, J. L. 1989. Retention enhancement with posttraining picrotoxin: Lack of state dependence. *Behav. Neural Biol.* In press

Castellano, C., Pavone, F. 1988. Effects of ethanol on passive avoidance behavior in the mouse: involvement of GABAergic mechanisms. *Pharmacol. Biochem. Behav.* 29: 321–24

Castellano, C., Pavone, F., Puglisi-Allegra, S. 1984. Morphine and memory in DBA/2 mice: Effects of stress and of prior experience. *Behav. Brain Res.* 11: 3–10

Caulfield, M. P., Higgins, G. A., Straughan, D. W. 1983. Central administration of the muscarinic receptor subtype—selective antagonist pirenzepine selectively impairs passive avoidance learning in the mouse. *J. Pharmacol. Pharm.* 35: 131–32

Cherkin, A. 1969. Kinetics of memory consolidation: Role of amnesic treatments parameters. *Proc. Natl. Acad. Sci. USA* 63: 1094–1101

Cherkin, A., Flood, J. F. 1988. Behavioral pharmacology of memory. See Woody et al 1988, pp. 343–56

Collier, T. J., Gash, D. M., Bruemmer, V., Sladek, J. R. Jr. 1985. Impaired regulation of arousal in old age and the consequences for learning and memory: Replacement of brain norepinephrine via neuron transplants improves memory performance in aged F344 rats. In *Homeostatic Function and Aging*, ed. B. B. Davis, W. G. Wood, pp. 99–110. New York: Raven

Collier, T. J., Quirk, G. J., Routtenberg, A. 1987. Separable roles of hippocampal granule cells in forgetting and pyramidal cells in remembering spatial information. *Brain Res.* 409: 316–28

Conte, C. O., Rosito, G. B. A., Palmini, A. L. F., Lucion, A. B., de Almeida, A. M. R. 1986. Pre-training adrenaline recovers the amnestic effect of Met-enkephalin in demedullated rats. *Behav. Brain Res.* 21: 163–66

Coyle, J. T., Price, D. L., DeLong, M. R. 1983. Alzheimer's disease: A disorder of cortical cholinergic innervation. *Science* 219: 1184–89

Davies, P. 1985. A critical review of the role of the cholinergic system in human memory and cognition. See Olton et al 1985, pp. 212–17

Davies, P., Maloney, A. J. F. 1976. Selective loss of central cholinergic neurons in Alzheimer's disease. *Lancet* 2: 1403

Davis, H. P., Squire, L. R. 1984. Protein synthesis and memory: A review. *Psychol. Bull.* 96: 518–59

Davis, K. L., Mohs, R. C., Rosen, W. G., Greenwald, B. S., Levy, M. I., et al. 1983. Memory enhancement with oral physostigmine in Alzheimer's disease. *N. Engl. J. Med.* 308: 721

de Almeida, M. A. M. R., Kapczinski, F. P., Izquierdo, I. 1983. Memory modulation by posttraining intraperitoneal but not intracerebronventricular administration of ACTH or epinephrine. *Behav. Neural Biol.* 39: 277–83

Decker, M. W. 1987. The effects of aging on hippocampal and cortical projections of the forebrain cholinergic system. *Brain Res. Rev.* 12: 423–38

Decker, M. W., Gallagher, M. 1987. Scopolamine-disruption of radial arm maze performance: modification by noradrenergic depletion. *Brain Res.* 417: 59–69

del Cerro, S., Borrell, J. 1985. Naloxone influences retention behaviour depending on the degree of novelty inherent to the training situation. *Physiol. Behav.* 35: 667–71

Deupree, D., Hsiao, S. 1987. Cholecystokinin octapeptide, proglumide, and passive avoidance in rats. *Peptides* 8: 25–28

de Wied, D. 1974. Pituitary-adrenal system hormones and behavior. In *The Neurosciences: Third Study Program*, ed. F. O. Schmitt, F. G. Worden, pp. 653–66. Cambridge: MIT Press

de Wied, D. 1984. Neurohypophyseal hormone influences on learning and memory processes. See Lynch et al 1984, pp. 289–312

de Wied, D., Gaffori, O., van Ree, J. M., de Jong, W. 1984. Central target for the behavioural effects of vasopressin neuropeptides. *Nature* 308: 276–78

Dietl, H. 1985. Temporal relationship between noradrenaline release in the central amygdala and plasma noradrenaline secretion in rats and tree shrews. *Neurosci. Lett.* 55: 41–46

Dumery, V., Blozovski, D. 1987. Develop-

ment of amygdaloid cholinergic mediation of passive avoidance learning in the rat. *Exp. Brain Res.* 67: 61–69

Dunn, A. J. 1980. Neurochemistry of learning and memory: An evaluation of recent data. *Ann. Rev. Psychol.* 31: 343–90

Ellis, M. E., Berman, R. F., Kesner, R. P. 1983. Amnesia attenuation specificity: Propranolol reverses norepinephrine but not cycloheximide-induced amnesia. *Pharmacol. Biochem. Behav.* 19: 733–36

Ellis, M. E., Kesner, R. P. 1981. Physostigmine and norepinephrine: Effects of injection into the amygdala on taste association. *Physiol. Behav.* 27: 203–9

Ettenberg, A., Le Moal, M., Koob, G. F., Bloom, F. E. 1983a. Vasopressin potentiation in the performance of a learned appetitive task: Reversal by a pressor antagonist analog of vasopressin. *Pharmacol. Biochem. Behav.* 18: 645–47

Ettenberg, A., van der Kooy, D., Le Moal, M., Koob, G. F., Bloom, F. E. 1983b. Can aversive properties of (peripherally-injected) vasopressin account for its putative role in memory? *Behav. Brain Res.* 7: 331–50

Faiman, C. P., de Erausquin, G. A., Baratti, C. M. 1988. Vasopressin modulates the activity of nicotinic cholinergic mechanisms during memory retrieval in mice. *Behav. Neural Biol.* 50: 112–19

Fanelli, R. J., Rosenberg, R. A., Gallagher, M. 1985. Role of noradrenergic function in the opiate antagonist facilitation of spatial memory. *Behav. Neurosci.* 99: 751–55

Flood, J. F., Cherkin, A. 1986. Scopolamine effects on memory retention in mice: a model of dementia? *Behav. Neural Biol.* 45: 169–84

Flood, J. F., Cherkin, A. 1987. Fluoxetine enhances memory processing in mice. *Psychopharmacology* 93: 36–43

Flood, J. F., Cherkin, A., Morley, J. E. 1987a. Antagonism of endogenous opioids modulates memory processing. *Brain Res.* 422: 218–34

Flood, J. F., Hernandez, E. N., Morley, J. E. 1987b. Modulation of memory processing by neuropeptide Y. *Brain Res.* In press

Flood, J. F., Landry, D. W., Jarvik, M. E. 1981. Cholinergic receptor interactions and their effects on long-term memory processing. *Brain Res.* 215: 177–85

Flood, J. F., Smith, G. E., Cherkin, A. 1983. Memory retention: Potentiation of cholinergic drug combinations in mice. *Neurobiol. Aging* 4: 37–43

Flood, J. F., Smith, G. E., Morley, J. F. 1987c. Modulation of memory processing by cholecystokinin dependence on the vagus nerve. *Science* 236: 832–34

Gaffan, D., Harrison, S. 1987. Amygdalectomy and disconnection in visual learning for auditory secondary reinforcement by monkeys. *J. Neurosci.* 7: 2285–92

Gallagher, M. 1982. Naloxone enhancement of memory processes: Effects of other opiate antagonists. *Behav. Neural Biol.* 35: 375

Gallagher, M., Bostock, E., King, R. 1985a. Effects of opiate antagonists on spatial memory in young and aged rats. *Behav. Neural Biol.* 44: 374–85

Gallagher, M., Kapp, B. S. 1978. Manipulation of opiate activity in the amygdala alters memory processes. *Life Sci.* 23: 1973–78

Gallagher, M., Kapp, B. S. 1981. Influence of amygdala opiate-sensitive mechanisms, fear-motivated responses, and memory processes for aversive experiences. See Martinez et al 1981, pp. 445–61

Gallagher, M., Kapp, B. S., Pascoe, J. P., Rapp, P. R. 1981. A neuropharmacology of amygdaloid systems which contribute to learning and memory. In *The Amygdaloid Complex*, ed. Y. Ben-Ari, pp. 343–54. Amsterdam: Elsevier/North-Holland

Gallagher, M., King, R. A., Young, N. B. 1983. Opiate antagonists improve spatial memory. *Science* 221: 975–76

Gallagher, M., Meagher, M. W., Bostock, E. 1987. Effects of opiate manipulations on latent inhibition in rabbits: sensitivity of the medial septal region to intercranial treatments. *Behav. Neurosci.* 101: 315–24

Gallagher, M., Rapp, P. R., Fanelli, R. J. 1985b. Opiate antagonist facilitation of time-dependent memory processes: Dependence upon intact norepinephrine function. *Brain Res.* 347: 284–90

Gentile, C. G., Jarrell, T. W., Teich, A., McCabe, P. M., Schneiderman, N. 1986. The role of amygdaloid central nucleus in the retention of differential Pavlovian conditioning of bradycardia in rabbits. *Behav. Brain Res.* 20: 263–73

Gerren, R. A., Weinberger, N. M. 1983. Long term potentiation in the magnocellular medial geniculate nucleus of the anesthetized cat. *Brain Res.* 265: 138–42

Gold, P. E. 1984. Memory modulation: Roles of peripheral catecholamines. In *Neuropsychology of Memory*, ed. L. R. Squire, N. Butter, pp. 566–78. New York: Guilford

Gold, P. E. 1986. Glucose modulation of memory storage. *Behav. Neural Biol.* 45: 342–49

Gold, P. E. 1988. Plasma glucose regulation of memory storage processes. See Woody et al 1988, pp. 329–42

Gold, P. E., Delanoy, R. L., Merrin, J. 1984. Modulation of long-term potentiation by peripherally administered amphetamine and epinephrine. *Brain Res.* 305: 103–7

Gold, P. E., McCarty, R. 1981. Plasma catecholamines: Changes after footshock and seizure-producing frontal cortex stimulation. *Behav. Neural Biol.* 31: 247–60

Gold, P. E., McGaugh, J. L. 1975. A single-trace, two process view of memory storage processes. In *Short-Term Memory*, ed. D. Deutsch, J. A. Deutsch, pp. 355–78. New York: Academic

Gold, P. E., McGaugh, J. L., Hankins, L. L., Rose, R. P., Vasquez, B. J. 1981. Age-dependent changes in retention in rats. *Exp. Aging Res.* 8: 53–58

Gold, P. E., van Buskirk, R. 1976. Effects of post-trial hormone injections on memory process. *Horm. Behav.* 7: 509–17

Gold, P. E., van Buskirk, R. 1978. Post-training brain norepinephrine concentrations: Correlation with retention performance of avoidance training with peripheral epinephrine modulation of memory processing. *Behav. Biol.* 23: 509–20

Gold, P. E., van Buskirk, R., Haycock, J. 1977. Effects of posttraining epinephrine injections on retention of avoidance training in mice. *Behav. Biol.* 20: 197–207

Gold, P. E., Vogt, J., Hall, J. L. 1986. Posttraining glucose effects on memory: Behavioral and pharmacological characteristics. *Behav. Neural Biol.* 46: 145–55

Gold, P. E., Zornetzer, S. F. 1983. The mnemon and its juices: Neuromodulation of memory processes. *Behav. Neural Biol.* 38: 151–89

Gower, A. J. 1987. Enhancement by secoverine and physostigmine of retention of passive avoidance response in mice. *Psychopharmacology* 91: 326–29

Grecksch, G., Matthies, H. 1981. Differential effects of intrahippocampally or systemically applied picrotoxin on memory consolidation in rats. *Pharmacol. Biochem. Behav.* 14: 613–16

Hagan, J. J., Tweedie, F., Morris, R. G. M. 1986. Lack of task specificity and absence of posttraining effects of atropine on learning. *Behav. Neurosci.* 100: 483–93

Hall, J. L., Gold, P. E. 1986. The effects of training, epinephrine, and glucose injections on plasma glucose levels in rats. *Behav. Neural Biol.* 46: 156–67

Harley, C. W., Evans, S. 1988. Locus-coeruleus-induced enhancement of the perforant-path evoked potential. See Woody et al 1988, pp. 415–24

Haroutunian, V., Kanof, M. D., Davis, M. D. 1985. Pharmacological alleviation of cholinergic lesion induced memory deficits in rats. *Life Sci.* 37: 945–52

Hebb, D. O. 1949. *The Organization of Behavior*. New York: Wiley

Hitchcock, J. M., Davis, M. 1987. Fear-potentiated startle using an auditory conditioned stimulus: Effect of lesions of the amygdala. *Physiol. Behav.* 39: 403–8

Holdefer, R. N., Jensen, R. A. 1987. The effects of peripheral d-amphetamine, 4-OH amphetamine, and epinephrine on maintained discharge in the locus coeruleus with reference to the modulation of learning and memory by these substances. *Brain Res.* 417: 108–17

Hopkins, W. F., Johnston, D. 1984. Frequency-dependent noradrenergic modulation of long-term potentiation in the hippocampus. *Science* 226: 350–51

Humphrey, G. L. 1968. *Effects of Post-Training Strychnine on Memory of Stage I and Stage II Sensory of Preconditioning in Rats*. England: Univ. Microfilms Limited

Huston, J. P., Staubli, U. 1981. Substance P and its effects on learning and memory. See Martinez et al 1981, pp. 521–40

Introini, I. B. 1984. *Participacion de peptidos opioides endogenos en el proceso de consolidacion del la memoria. Su posible interaccion con otros sistemas neuronales*. PhD thesis, Univ. Buenos Aires

Introini, I. B., Baratti, C. M. 1984. The impairment of retention induced by β-endorphin in mice may be mediated by reduction of central cholinergic activity. *Behav. Neural Biol.* 41: 152–63

Introini, I. B., McGaugh, J. L., Baratti, C. M. 1985. Pharmacological evidence of a central effect of naltrexone, morphine and β-endorphin and a peripheral effect of Met- and Leu-enkephalin on retention of an inhibitory response in mice. *Behav. Neural Biol.* 44: 434–46

Introini-Collison, I. B., Baratti, C. M. 1986. Opioid peptidergic systems modulate the activity of β-adrenergic mechanisms during memory consolidation processes. *Behav. Neural Biol.* 46: 227–41

Introini-Collison, I. B., Cahill, L., Baratti, C. M., McGaugh, J. L. 1987. Dynorphin induces task-specific impairment of memory. *Psychobiology* 15: 171–74

Introini-Collison, I. B., McGaugh, J. L. 1986. Epinephrine modulates long-term retention of an aversively-motivated discrimination task. *Behav. Neural Biol.* 45: 358–65

Introini-Collison, I. B., McGaugh, J. L. 1987. Naloxone and β-endorphin alter the effects of posttraining epinephrine on retention of an inhibitory avoidance response. *Psychopharmacology* 92: 229–35

Introini-Collison, I. B., McGaugh, J. L. 1988. Modulation of memory by post-training epinephrine: Involvement of cholinergic mechanisms. *Psychopharmacology* 94: 379–85

Introini-Collison, I. B., Nagahara, A. H., McGaugh, J. L. 1988. Memory-enhancement with intra-amygdala posttraining naloxone is blocked by concurrent administration of propranolol. *Brain Res.* In press

Iversen, L. L., Rossor, M. N. 1984. Human learning and memory dysfunction: Neurochemical changes in senile dementia. See Lynch et al 1984, pp. 363–67

Iwata, J., LeDoux, J. E., Meeley, M. P., Arneric, S., Reis, D. J. 1986. Intrinsic neurons in amygdaloid field projected to by the medial geniculate body mediate emotional responses conditioned to acoustic stimuli. *Brain Res.* 383: 195–214

Izquierdo, I. 1979. Effect of naloxone and morphine on various forms of memory in the rat: Possible role of endogenous opiate mechanisms in memory consolidation. *Psychopharmacology* 66: 199–203

Izquierdo, I. 1984. Endogenous state dependency: Memory depends on the relation between the neurohumoral and hormonal states present after training and at the time of testing. See Lynch et al 1984, pp. 333–50

Izquierdo, I. 1986. Memory consolidation: not a useful hypothesis in the search for memory-enhancing drugs. *Trends Pharmacol. Sci.* 7: 476–77

Izquierdo, I., de Almeida, M. A. M. R., Emiliano, V. R. 1985. Unlike β-endorphin, dynorphin, 1-13 does not cause retrograde amnesia for shuttle avoidance or inhibitory avoidance learning in rats. *Psychopharmacology* 87: 216–18

Izquierdo, I., Dias, R. D. 1983. Effect of ACTH, epinephrine, β-endorphin, naloxone or and of the combination of naloxone or β-endorphin with ACTH or epinephrine on memory consolidation. *Psychoneuroendocrinology* 8: 81–87

Izquierdo, I., Dias, R. D. 1985. Influence on memory of posttraining or pre-test injections of ACTH, vasopressin, epinephrine, and β-endorphin, and their interaction with naloxone. *Psychoneuroendocrinology* 10: 165–72

Izquierdo, I., Graudenz, M. 1980. Memory facilitation by naloxone is due to release of dopaminergic and β-adrenergic systems from tonic inhibition. *Psychopharmacology* 67: 265–68

Izquierdo, I., McGaugh, J. L. 1985. Effect of a novel experience prior to training or testing on retention of an inhibitory avoidance response in mice: Involvement of an opioid system. *Behav. Neural Biol.* 44: 228–38

Izquierdo, I., McGaugh, J. L. 1987. Effect of novel experiences on retention of inhibitory avoidance behavior in mice: The influence of previous exposure to the same or another experience. *Behav. Neural Biol.* 47: 109–15

Izquierdo, I., Netto, C. A. 1985. The role of β-endorphin in behavioral regulation. In *Memory dysfunctions: An integration of animal and human research from preclinical and clinical perspectives*, ed. D. S. Olton, E. Gamzu, S. Corkin, pp. 162–77. New York: NY Acad. Sci.

Izquierdo, I., Souza, D. O., Carrasco, M. A., Dias, R. D., Perry, M. L., et al. 1980. β-endorphin causes retrograde amnesia and is released from the rat brain by various forms of training and stimulation. *Psychopharmacology* 70: 173–77

Izquierdo, I., Souza, D. O., Dias, R. D., Perry, M. L. S., Carrasco, M., et al. 1984. Effect of various behavioral training and testing procedures on brain β-endorphin-like immunoreactivity and the possible role of β-endorphin in behavioral regulation. *Psychoneuroendocrinology* 9: 381–89

Jefferys, D., Boublik, J., Funder, J. W. 1985. A κ-selective opioidergic pathway is involved in the reversal of a behavioral effect of adrenalectomy. *Eur. J. Pharmacol.* 107: 331–35

Jhamandas, K., Hron, V., Sutak, M. 1975. Comparative effects of opiate agonists methadone, levorphanol and their isomers on the release of cortical acetylcholine in vivo and in vitro. *Can. J. Pharmacol. Physiol.* 53: 540–48

Jhamandas, K., Sutak, M. 1974. Modifications of brain acetylcholine release by morphine and its antagonists in normal and morphine-dependent rats. *Br. J. Pharmacol.* 50: 57–62

Jhamandas, K., Sutak, M. 1983. Stereospecific enhancement of evoked release of brain acetylcholine by narcotic antagonists. *Br. J. Pharmacol.* 78: 433–40

Kameyama, T., Nabeshima, T., Noda, Y. 1986. Cholinergic modulation of memory for step-down type passive avoidance in mice. *Res. Commun. Psychol. Psychiatry Behav.* 2: 193–205

Kapp, B. S., Frysinger, R. C., Gallagher, M., Haselton, J. R. 1979. Amygdala central nucleus lesions: effect on heart rate conditioning in the rabbit. *Physiol. Behav.* 23: 1109–17

Kesner, R. P. 1982. Brain stimulation: Effects on memory. *Behav. Neural Biol.* 36: 315–67

Kety, S. 1972. Brain catecholamines, affective states and memory. In *The Chemistry of Mood, Motivation and Memory*, ed.

J. L. McGaugh, pp. 65–80. New York: Raven

LeDoux, J. E. 1986. Sensory systems and emotion: A model of affective processing. *Integr. Psychiatry* 4: 237–48

Lee, M. K., Graham, S., Gold, P. E. 1988. Memory enhancement with posttraining central glucose injections. *Behav. Neurosci.* 102: 591–95

Leslie, F. M., Loughlin, S. E., Sternberg, D. B., McGaugh, J. L., Young, L. E., et al. 1985. Noradrenergic changes and memory loss in aged mice. *Brain Res.* 359: 292–99

Liang, K. C., Bennett, C., McGaugh, J. L. 1985. Peripheral epinephrine modulates the effects of posttraining amygdala stimulation on memory. *Behav. Brain Res.* 15: 93–100

Liang, K. C., Juler, R., McGaugh, J. L. 1986. Modulating effects of posttraining epinephrine on memory: Involvement of the amygdala noradrenergic system. *Brain Res.* 368: 125–33

Liang, K. C., McGaugh, J. L. 1983a. Lesions of the stria terminalis attenuate the enhancing effect of posttraining epinephrine on retention of an inhibitory avoidance response. *Behav. Brain Res.* 9: 49–58

Liang, K. C., McGaugh, J. L. 1983b. Lesions of the stria terminalis attenuate the amnestic effect of amygdaloid stimulation on avoidance responses. *Brain Res.* 274: 309–18

Liang, K. C., McGaugh, J. L. 1987. Effects of adrenal demedullation and stria terminalis lesions on retention of an inhibitory avoidance response. *Psychobiology* 15: 154–60

Liang, K. C., McGaugh, J. L., Martinez, J. L. Jr., Jensen, R. A., Vasquez, B. J., et al. 1982. Posttraining amygdaloid lesions impair retention of an inhibitory avoidance response. *Behav. Brain Res.* 4: 237–49

Liang, K. C., Messing, R. B., McGaugh, J. L. 1983. Naloxone attenuates amnesia caused by amygdaloid stimulation: The involvement of a central opioid system. *Brain Res.* 271: 41–49

Lichter, I., Richardson, P. J., Wyke, M. A. 1986. Differential effects of atenolol and enalapril on memory during treatment for essential hypertension. *Br. J. Clin. Pharmacol.* 21: 641–45

Lo Conte, G., Bartolini, L., Casamenti, F., Marconcini-Pepeu, I., Pepeu, G. 1982. Lesions of cholinergic forebrain nuclei: Changes in avoidance behavior and scopolamine actions. *Pharmacol. Biochem. Behav.* 17: 933–37

Lynch, G., McGaugh, J. L., Weinberger,

N. M., eds. 1984. *Neurobiology of Learning and Memory.* New York: Guilford

Madden, D. J., Blumenthal, J. A., Ekelund, L.-G., Krantz, D. S., Light, K. C., et al. 1986. Memory performance by mild hypertensives following β-adrenergic blockade. *Psychopharmacology* 89: 20–24

Martinez, J. L. Jr., Jensen, R. A., Messing, R. B., Rigter, H., McGaugh, J. L., eds. 1981. *Endogenous Peptides and Learning and Memory Processes.* New York: Academic

McCarty, R. 1981. Aged rats: Diminished sympathetic-adrenal medullary responses to acute stress. *Behav. Neural Biol.* 33: 204–12

McCarty, R., Gold, P. E. 1981. Plasma catecholamines: Effects of footshock level and hormonal modulators of memory storage. *Horm. Behav.* 15: 168–82

McGaugh, J. L. 1966. Time-dependence processes in memory storage. *Science* 153: 1351–58

McGaugh, J. L. 1973. Drug facilitation of learning and memory. *Ann. Rev. Pharmacol.* 13: 229–41

McGaugh, J. L. 1983. Hormonal influences on memory. *Ann. Rev. Psychol.* 34: 297–323

McGaugh, J. L., Bennett, M. C., Liang, K. C., Juler, R. G., Tam, D. 1987. Retention-enhancing effects of posttraining epinephrine is not blocked by dexamethasone. *Psychobiology* 15: 343–44

McGaugh, J. L., Gold, P. E. 1976. Modulation of memory by electrical stimulation of the brain. In *Neural Mechanism of Learning and Memory,* ed. M. R. Rosenzweig, E. L. Bennett, pp. 549–60. Cambridge: MIT Press

McGaugh, J. L., Gold, P. E. 1988. Hormonal modulation of memory. In *Psychoendocrinology,* ed. R. B. Brush, S. Levine. New York: Academic

McGaugh, J. L., Herz, M. J. 1972. *Memory Consolidation.* San Francisco: Albion Publ.

McGaugh, J. L., Introini-Collison, I. B. 1987. Novel experience prior to training attenuates the amnestic effects of posttraining ECS. *Behav. Neurosci.* 101: 296–99

McGaugh, J. L., Introini-Collison, I. B., Juler, R. G., Izquierdo, I. 1986. Stria terminalis lesions attenuate the effects of posttraining naloxone and β-endorphin on retention. *Behav. Neurosci.* 100: 839–44

McGaugh, J. L., Introini-Collison, I. B., Nagahara, A. H. 1988. Memory-enhancing effects of posttraining naloxone: Involvement of β-noradrenergic influences in the amygdaloid complex. *Brain Res.* 446: 37–49

Menzel, R. 1983. Neurobiology of learning and memory: the honeybee as a model system. *Naturwissenschaften* 70: 504–11

Menzel, R. 1984. Short-term memory in bees. In *Primary Neural Substrates of Learning and Behavioral Change,* ed. D. Alkon, J. Farley, pp. 259–74. New York: Cambridge Univ. Press

Menzel, R., Michelsen, B. 1986. Neuropharmacology of associative learning in bees. In *Learning and Memory: Mechanisms of Information Storage in the Nervous System,* ed. H. Matthies, pp. 273–75. Oxford: Pergamon

Messier, C., White, N. M. 1984. Contingent and non-contingent actions of sucrose and saccharin reinforcers: Effects on taste preference and memory. *Physiol. Behav.* 32: 195–203

Messier, C., White, N. M. 1987. Memory improvement by glucose, fructose, and two glucose analogs: A possible effect on peripheral glucose transport. *Behav. Neural Biol.* 48: 104–27

Messing, R. B., Jensen, R. A., Martinez, J. L. Jr., Spiehler, V. R., Vasquez, B. J., et al. 1979. Naloxone enhancement of memory. *Behav. Neural Biol.* 27: 266–75

Micheau, J., Destrade, C., Jaffard, R. 1985. Physostigmine reverses memory deficits produced by pretraining electrical stimulation of the dorsal hippocampus in mice. *Behav. Brain Res.* 15: 75–81

Milner, B. 1966. Amnesia following operation on the temporal lobes. In *Amnesia,* ed. C. W. M. Whitty, O. L. Zangwill, pp. 109–33. London: Butterworths

Miyamoto, M., Shintani, M., Nagaoka, A., Nagawa, Y. 1985. Lesioning of the rat basal forebrain leads to memory impairments in passive and active avoidance tasks. *Brain Res.* 328: 97–104

Montel, H., Starke, K., Weber, F. 1974. Influence of morphine and naloxone on the release of noradrenaline from rat brain cortex slices. *Naunyn-Schmiedeberg's Arch. Pharmacol.* 283: 357–69

Moroni, F., Cheney, D. L., Costa, E. 1977. Inhibition of acetylcholine turnover in rat hippocampus by intraseptal injections of β-endorphin and morphine. *Naunyn-Schmiedeberg's Arch. Pharmacol.* 299: 149–53

Mueller, G. E., Pilzecker, A. 1900. Experimentelle Beitrage zur Lehre vom Gedachtniss. *Z. Psychol.* 1: 1–288

Murray, C. L., Fibiger, H. C. 1985. Learning and memory deficits after lesions of the nucleus basalis magnocellularis: Reversal by physostigmine. *Neuroscience* 14: 1025–32

Murray, E. A., Mishkin, M. 1985. Amygdalectomy impairs crossmodal association in monkeys. *Science* 228: 604–6

Nagel, J. A., Huston, J. P. 1988. Enhanced inhibitory avoidance learning produced by post-trial injections of substance P into the basal forebrain. *Behav. Neural Biol.* 49: 374–85

Nakamura, S., Tepper, J. M., Young, S. J., Ling, N., Groves, P. M. 1982. Noradrenergic terminal excitability: Effects of opioids. *Neurosci. Lett.* 30: 57–62

Netto, C. A., Dias, R., Izquierdo, I. 1986. Differential effect of posttraining naloxone, β-endorphin, Leu-enkephalin and electroconvulsive shock administration upon memory of an open-field habituation and of a water-finding task. *Psychoneuroendocrinology* 4: 437–46

Nieuwenhuys, R. 1985. *Chemoarchitecture of the Brain.* Berlin: Springer-Verlag

Oishi, R., Watanabe, S., Ohmori, K., Shibata, S., Ueki, S. 1979. Effect of stimulation of locus coeruleus on the evoked potential in the amygdala in rats. *Jpn. J. Pharmacol.* 29: 105–11

Olton, D. S., Gamzu, E., Corkin, S., eds. 1985. *Memory Dysfunctions: An Integration of Animal and Human Research from Preclinical and Clinical Perspectives.* New York: NY Acad. Sci.

Oomura, Y., Nakano, Y., Lenard, L., Nishino, H., Aou, S. 1988. Catecholaminergic and opioid mechanisms in conditioned food intake behavior of the monkey amygdala. See Woody et al 1988, pp. 109–18

Pavone, F., Castellano, C. 1985. Effects of Tifluadom on passive avoidance behavior in DBA/2 mice. *Behav. Brain Res.* 15: 177–81

Pelleymounter, M. A., Fisher, Q., Schlesinger, K., Hall, M., Dearmey, P., et al. 1986. The effect of substance P and its fragments on passive avoidance retention and brain monamine activity. *Behav. Brain Res.* 21: 119–27

Perry, E. K., Perry, R. H., Blessed, G., Tomlinson, B. E. 1977. Necropsy evidence of central cholinergic deficit in senile dementia. *Lancet* 2: 189

Perry, M. L. S., Dias, R. D., Carrasco, M. A., Izquierdo, I. 1983. Effect of stepdown inhibitory avoidance training on β-endorphin like immunoreactivity of rat hypothalamus and plasma. *Braz. J. Med. Biol. Res.* 16: 339–43

Peters, B., Levin, H. S. 1979. Effects of physostigmine and lecithin on memory in Alzheimer's Disease. *Ann. Neurol.* 6: 219–21

Prado-Alcala, R. A., Signoret, L., Figueroa, M. 1981. Time-dependent retention deficits induced by post-training injections of atropine into the caudate nucleus. *Pharmacol. Biochem. Behav.* 15: 633–36

Prado-Alcala, R. A., Signoret-Edward, L., Figueroa, M., Giordano, M., Barrientos, M. A. 1984. Post-trial injection of atropine into the caudate nucleus interferes with long-term but not with short-term retention of passive avoidance. *Behav. Neural Biol.* 42: 81–84

Racine, R. J., Milgram, N. W., Hafner, S. 1983. Long-term potentiation phenomena in the rat limbic forebrain. *Brain Res.* 260: 217–31

Reisberg, B., Ferris, S., Anand, R., Mir, P., Beibel, V., et al. 1983a. Effects of naloxone in senile dementia: a double-blind trial. *New Engl. J. Med.* 308: 721–22

Reisberg, B., Ferris, S., Anand, R., Mir, P., DeLeon, M., et al. 1983b. Naloxone effects on primary degenerative dementia (PDD). *Psychopharmacol. Bull.* 19: 45–47

Riccio, D. C., Concannon, J. T. 1981. ACTH and the reminder phenomena. See Martinez et al 1981, pp. 117–42

Robinson, G. B., Racine, R. J. 1985. Longterm potentiation in the dentate gyrus: effects of noradrenaline depletion in the awake rat. *Brain Res.* 325: 71–78

Rush, D. K. 1986. Reversal of scopolamineinduced amnesia of passive avoidance by pre- and post-training naloxone. *Psychopharmacology* 89: 296–300

Salmon, D. P., Zola-Morgan, S., Squire, L. R. 1987. Retrograde amnesia following combined hippocampus-amygdala lesions in monkeys. *Psychobiology* 15: 37–47

Saunders, R. C., Murray, E. A., Mishkin, M. 1984. Further evidence that the amygdala and hippocampus contribute equally to recognition memory. *Neuropsychologia* 22: 786–96

Scarpace, P. J. 1986. Decreased β-adrenergic responsiveness during senescence. *Fed. Pro.* 45: 51–54

Schlesinger, K., Lipsitz, D. U., Peck, P. L., Pelleymounter, M. A., Stewart, J. M., et al. 1983. Substance P enhancement of passive and active avoidance conditioning in mice. *Pharmacol. Biochem. Behav.* 19: 655–61

Schlesinger, K., Pelleymounter, M. A., van de Kamp, J., Bader, D. L., Stewart, K. M., et al. 1986. Substance P facilitation of memory: effects in an appetitively motivated learning task. *Behav. Neural Biol.* 45: 230–39

Solomon, S., Hotchkiss, E., Saravay, S. M., Bayer, C., Ramsey, P., et al. 1983. Impairment of memory function by antihypertensive medication. *Arch. Gen. Psychiatry* 40: 1109–12

Squire, L. R., Davis, H. P. 1983. The pharmacology of memory: A neurobiological perspective. *Ann. Rev. Pharmacol. Toxicol.* 21: 323–56

Stanton, P. K., Sarvey, J. M. 1985. Depletion of norepinephrine, but not serotonin, reduces long-term potentiation in the dentate gyrus of rat hippocampal slices. *J. Neurosci.* 5: 2169–76

Staubli, U., Huston, J. P. 1980. Facilitation of learning by post-trial injection of substance P into the medial septal nucleus. *Behav. Brain Res.* 1: 245–55

Sternberg, D. B., Isaacs, K., Gold, P. E., McGaugh, J. L. 1985a. Epinephrine facilitation of appetitive learning: Attenuation with adrenergic receptor antagonists. *Behav. Neural Biol.* 44: 447–53

Sternberg, D. B., Martinez, J. L. Jr., Gold, P. E., McGaugh, J. L. 1985b. Age-related memory deficits in rats and mice: Enhancement with peripheral injections of epinephrine. *Behav. Neural Biol.* 44: 213–20

Sternberg, D. B., Korol, D., Novack, G., McGaugh, J. L. 1986. Epinephrine-induced memory facilitation: Attenuation by adrenergic receptor antagonists. *Eur. J. Pharmacol.* 129: 189–93

Stone, W. S., Croul, C. E., Gold, P. E. 1988. Attenuation of scopolamine-induced amnesia in mice. *Psychopharmacology.* In press

Stratton, L. O., Petrinovich, L. F. 1963. Post-trial injections of an anticholinesterase drug and maze learning in two strain of rats. *Psychopharmacologia* 5: 47–54

Svensson, T. H. 1987. Peripheral, autonomic regulation of locus coeruleus noradrenergic neurons in brain: putative implications for psychiatry and psychopharmacology. *Psychopharmacology* 92: 1–7

Swartzwelder, H. S., Tilson, H. A., McLamb, R. L., Wilson, W. A. 1987. Baclofen disrupts passive avoidance retention in rats. *Psychopharmacology* 92: 398–401

Tanaka, M., Kohno, Y., Nakagawa, R., Ida, Y., Ilimori, K., et al. 1982a. Naloxone enhances stress-induced increases in noradrenaline turnover in specific brain regions in rats. *Life Sci.* 30: 1663–69

Tanaka, M., Kohno, Y., Nakagawa, R., Ida, Y., Takeda, S., et al. 1982b. Time-related differences in noradrenaline turnover in rat brain regions by stress. *Pharmacol. Biochem. Behav.* 16: 315–19

Tariot, P. N., Sunderland, T., Weingartner, H., Murphy, D. L., Cohen, M. R., et al. 1986. Naloxone and Alzheimer's disease: cognitive and behavioral effects of a range of doses. *Arch. Gen. Psychiatry* 43: 727–32

Telegdy, G., Kadar, T., Fekete, M. 1985.

Cholecystokinin, learning and memory. In *Brain Plasticity, Learning and Memory*, ed. B. E. Will, P. Schmitt, J. C. Dalrymple-Alford, pp. 303–9. New York: Plenum

Thal, L. J., Fuld, P. A., Masur, D. M., Sharpless, N. D. 1983. Oral physostigmine and lecithin improve memory in Alzheimer's Disease. *Ann. Neurol.* 13: 491–94

Tilson, H. A., McLamb, R. L., Shaw, S., Rogers, B. C., Pediaditakis, P., et al. 1988. Radial-arm maze deficits produced by colchicine administered into the area of the nucleus basalis are ameliorated by cholinergic agents. *Brain Res.* 438: 83–94

Vanderwolf, C. H. 1987. Near-total loss of "learning" and "memory" as a result of combined cholinergic and serotonergic blockade in the rat. *Behav. Brain Res.* 23: 43–57

Vecsei, L., Bollok, I., Penke, B., Telegdy, G. 1986. Somatostatin and (d-trp$^8$, d-cys$^{14}$)-somatostatin delay extinction and reverse electroconvulsive shock induced amnesia in rats. *Psychoneuroendocrinology* 11: 111–15

Vijayashankar, N., Brody, H. 1979. A quantitative study of the pigmental neurons in the nuclei locus coeruleus and subcoeruleus in man as related to aging. *J. Neuropathol. Exp. Neurol.* 38: 490–97

Viveros, O. H., Diliberto, J. R., Hazum, E., Chang, K. J. 1980. Enkephalins as possible adrenomedullary hormones: Storage secretions, and regulation of synthesis. In *Neural Peptides and Neuronal Communication, Advances in Biochemical Psychopharmacology*, ed. E. Costa, M. Trabucchi, pp. 191–204. New York: Raven

Walker, J. M., Khachaturian, H., Watson, S. J. 1984. Some anatomical and physiological interactions among noradrenergic systems and opioid peptides. In *Norepinephrine*, ed. M. G. Ziegler, C. R. Lake, pp. 74–91. Baltimore: Williams & Wilkins

Weil-Malherbe, H., Axelrod, J., Tomchick, R. 1959. Blood-brain barrier for adrenalin. *Science* 129: 1226–28

Weingartner, H., Parker, E. S. 1984. *Memory Consolidation: Psychobiology of Cognition*. Hillsdale: Erlbaum

Werling, L. L., Brown, S. R., Cox, B. M. 1987. Opioid receptor regulation of the release of norepinephrine in brain. *Neuropharmacology* 26: 987–96

Westbrook, W. H., McGaugh, J. L. 1964. Drug facilitation of latent learning. *Psychopharmacology* 5: 440–46

Woody, C. D., Alkon, D. L., McGaugh, J. L., eds. 1988. *Cellular Mechanisms of Conditioning and Behavioral Plasticity*. New York: Plenum

Yonkov, D. I., Georgiev, V. P., Opitz, M. J.

1986. Participation of Angiotensin II in Learning and Memory. II. Interactions of Angiotensin II with Dopaminergic Drugs. *Methods Find. Exp. Clin. Pharmacol.* 8: 203–6

Yonkov, D., Georgiev, V., Kambourova, T., Opitz, M. 1987. Participation of angiotensin II in learning and memory. III. Interactions of angiotensin II with GABA-ergic drugs. *Methods Find. Exp. Clin. Pharmacol.* 9: 205–8

Zhang, S., McGaugh, J. L., Juler, R. G., Introini-Collison, I. B. 1987. Naloxone and [Met5]-enkephalin effects on retention: Attenuation by adrenal denervation. *Eur. J. Pharmacol.* 138: 37–44

*Ann. Rev. Neurosci. 1989. 12:289–327*

# CYCLIC GMP–ACTIVATED CONDUCTANCE OF RETINAL PHOTORECEPTOR CELLS

## K.-W. Yau

Howard Hughes Medical Institute and Department of Neuroscience,
The Johns Hopkins University School of Medicine, Baltimore,
Maryland 21205


## D. A. Baylor

Department of Neurobiology, Sherman Fairchild Center,
Stanford University School of Medicine, Stanford,
California 94305

## Introduction

This review examines the ionic conductance that generates the electrical response to light in rods and cones of the vertebrate retina. In visual physiology this conductance is commonly referred to as the *light-regulated* or *light-sensitive conductance*. The designation *cGMP-activated conductance* is perhaps preferable because it is now known to be controlled by guanosine 3′,5′-cyclic monophosphate (abbreviated here as cGMP); light modulates the conductance solely by changing the intracellular concentration of cGMP. We shall use the different terms for the conductance interchangeably. A review on this channel seems timely in view of the speed of recent progress in understanding its nature and function.

The conductance is interesting for several reasons. First, it has physiological properties that are beautifully suited for its role in visual transduction. Second, it is the first ionic channel found whose gating is directly controlled by a cyclic nucleotide. Third, there are indications that the conductance may belong to a new family of ion channels involved in signal

289

0147–006X/89/0301–0289$02.00

transduction. For instance, a rather similar conductance has recently been identified in olfactory receptors (Nakamura & Gold 1987).

Work on the cGMP-activated conductance is part of a larger arena of investigations on the mechanism of visual transduction. The remarkable recent strides in understanding this process have been documented in a host of reviews (see e.g. Kühn 1984, Koutalos & Ebrey 1986, Lamb 1986, Pugh & Cobbs 1986, Stryer 1986, Liebman et al 1987, Owen 1987). Here we focus on the conductance itself, especially its functional characteristics. The main emphasis is on the cGMP-activated channel of rods, which up to now has been most studied. At the end we briefly take up the cGMP-gated channel of cones, which appears to be similar but not identical to the rod channel.

We begin with brief descriptions of photoreceptor structure and the mechanism of visual transduction.

## Functional Architecture of Photoreceptors

Retinal rods and cones consist of two main parts: an outer segment and an inner segment, which are connected by a modified cilium (Figure 1). Phototransduction takes place within the outer segment. In rods, this structure consists of a cylindrical stack of membranous discs enclosed by the surface membrane. The disc membranes contain the visual pigment rhodopsin, an integral membrane protein present at high density. Cone outer segments have a different membrane topology: The pigment-bearing

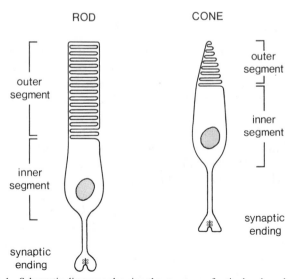

*Figure 1*    Schematic diagrams showing the structure of retinal rods and cones.

membranous discs consist of deep infoldings of the surface membrane (see Rodieck 1973 for details). The inner segment houses the nucleus, endoplasmic reticulum, and mitochondria, and is responsible for energy production and protein synthesis. The base of the inner segment bears a synaptic terminal, where chemical synapses are made on second-order cells.

Characteristic differences in the ionic channels located in the membranes of inner and outer segments suggest that membrane proteins destined for the two locations are sorted with high efficiency. Ions cross the outer segment's surface membrane through the cGMP-gated conductance as well as a $Na^+-Ca^{2+}$ exchange carrier responsible for $Ca^{2+}$ homeostasis (Yau & Nakatani 1984b, 1985a, Schnetkamp 1986, Hodgkin et al 1987, Hodgkin & Nunn 1987, Lagnado et al 1988, Nakatani & Yau 1988a, reviewed by McNaughton 1988; see also Gold & Korenbrot 1980, Yoshikami et al 1980, Schröder & Fain 1984). Light-insensitive channels, which would impair the spread of electrical signals along the outer segment, are present at low density or perhaps are absent altogether. When the cGMP-gated conductance closes in bright light, for instance, the membrane resistance of the rod outer segment approaches that of a pure lipid bilayer (Baylor & Lamb 1982, Baylor & Nunn 1986). Nevertheless, Nöll et al (1982), Attwell & Gray (1984), Kolesnikov et al (1984), and Sather et al (1985) have reported occasionally observing light-insensitive channels in excised or cell-attached patches of plasma membrane from the rod outer segment. Recently Watanabe & Matthews (1988) have also described observing $Ca^{2+}$-activated $K^+$ channels in such patches. It remains to be determined whether light-insensitive channels are a normal constituent of the outer segment or an experimental artifact: In dissociated photoreceptors, the outer and inner segments may fuse, causing a failure of membrane protein segregation (Townes-Anderson et al 1985, Spencer et al 1988).

The membrane of the inner segment and synaptic ending contains several voltage-sensitive conductances that modify the shape of the light-induced voltage change and regulate synaptic transmitter release (see Attwell 1986, Owen 1987 for review). The membrane of the inner segment also contains an ATP-dependent $Na^+/K^+$ pump that is responsible for maintaining the low $Na^+$ and high $K^+$ concentrations inside the cell (see Fain & Lisman 1981). A low density of cGMP-activated channels has also been reported in the membrane of the inner segment (Watanabe & Matthews 1988); again, it is not yet clear whether this reflects a physiological function, imperfect protein targeting, or the experimental artifact described above.

The division of labor between the outer and the inner segments is strikingly underscored by the demonstration that an internally dialyzed outer segment continues to transduce in the absence of the inner segment (Yau & Nakatani 1985b, Hestrin & Korenbrot 1987, Sather & Detwiler 1987, Nakatani & Yau 1988b).

## Mechanism of Visual Transduction

Since the finding two decades ago that light causes the membrane of rods and cones to hyperpolarize (Bortoff 1964, Tomita 1965), it has become clear that the hyperpolarization carries visual information between the outer segment and the synaptic ending. Signals flow as follows. In darkness the light-sensitive conductance is open (Toyoda et al 1969), allowing a steady "dark current" (Hagins et al 1970, Baylor et al 1979a) to enter the surface membrane of the outer segment. The inward current keeps the cell partially depolarized, the membrane potential in darkness being near $-40$ mV. At this voltage transmitter is continuously released from the synaptic terminal (Byzov & Trifonov 1968, Dowling & Ripps 1973). Light closes the channels in the outer segment, hyperpolarizing the membrane and lowering the rate of transmitter release. Curtailment of transmitter release causes characteristic graded responses in second-order cells (Werblin & Dowling 1969, Kaneko 1970). Injection of polarizing currents into single photoreceptors demonstrates that the hyperpolarization is necessary and sufficient to trigger the characteristic light responses of higher-order visual neurons (Baylor & Fettiplace 1977a).

The intracellular signaling process that controls the light-sensitive conductance has aroused intense interest and controversy. Even though the conductance decrease occurs near the site of photon absorption (Hagins et al 1970, Lamb et al 1981, Matthews 1986b), the sizeable effect of a single photon (Baylor & Fuortes 1970, Penn & Hagins 1972, Baylor et al 1979b) as well as the physical separation between the rod disks and surface membrane (Cohen 1970) indicate that a diffusible signal, an internal transmitter, is involved. The two long-standing and sometimes conflicting ideas regarding the identity of the transmitter were the $Ca^{2+}$-hypothesis and the cGMP-hypothesis. The $Ca^{2+}$-hypothesis supposed that during illumination internally liberated $Ca^{2+}$ binds to and blocks the conductance (Hagins 1972). This idea, although attractively simple and supported by circumstantial evidence (see e.g. Miller 1981), was recently ruled out. Internal $Ca^{2+}$ does not directly close the conductance in the intact cell (Yau & Nakatani 1984b, Hodgkin et al 1985), nor does transduction require a rise in internal $Ca^{2+}$ (Korenbrot & Miller 1986, Lamb et al 1986, Nicol et al 1987a). Furthermore, it now appears that the free $Ca^{2+}$ concentration in the rod outer segment decreases during illumination (Yau & Nakatani 1985a, Gold 1986, McNaughton et al 1986, Miller & Korenbrot 1987, Nakatani & Yau 1988a, Ratto et al 1988); according to the

$Ca^{2+}$ hypothesis it should rise. The mechanism that causes the $Ca^{2+}$ concentration to fall in the light is the following. In darkness $Ca^{2+}$ enters the cell (along with $Na^+$) through the light-regulated conductance and leaves the cell via the $Na^+–Ca^{2+}$ exchange. In the light the $Ca^{2+}$ influx is lowered, but the $Ca^{2+}$ efflux continues, thus leading to a net $Ca^{2+}$ efflux and a resulting decrease in the internal free $Ca^{2+}$ (Yau & Nakatani 1985a, Nakatani & Yau 1988a; see also Figure 2 here). Our present picture of the functional effects of the $Ca^{2+}$ changes is described below.

The cGMP-hypothesis supposes that closure of the light-regulated conductance is triggered by a fall in the intracellular concentration of cGMP, a nucleotide known to be present at a relatively high concentration during darkness (Cohen et al 1978, Kilbride & Ebrey 1979, Woodruff & Bownds 1979, Woodruff & Fain 1982). Strong indications for a role of cGMP came initially from biochemical experiments that revealed the existence of a light-activated enzyme cascade (see the next section) that leads to the hydrolysis of cGMP. Suspicions that cGMP was involved in transduction or in light adaptation were strengthened by early electrophysiological experiments that implicated a link between cGMP and the opening of the light-regulated conductance in intact rods (Lipton et al 1977, Miller & Nicol 1979). Nevertheless, the precise role of cGMP remained unsettled (see, for example, Hubbell & Bownds 1979, Miller 1981, Stieve 1986) until recent experiments demonstrated that cGMP opens a cationic conductance in excised patches of plasma membrane from the rod outer segment (Fesenko et al 1985, Nakatani & Yau 1985). This cGMP-gated con-

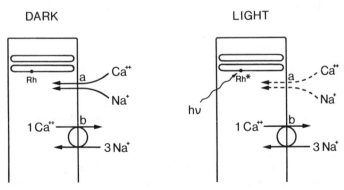

*Figure 2*   $Ca^{2+}$ fluxes across the rod outer segment in darkness and light: *a*, light-regulated conductance; *b*, $Na^+–Ca^{2+}$ exchange carrier; Rh, rhodopsin molecule; Rh*, photo-isomerized rhodopsin molecule; hv, photon. In darkness, there is a balance between $Ca^{2+}$ influx and $Ca^{2+}$ efflux; in the light, the cessation of $Ca^{2+}$ influx (*dashed arrow*) leads to a net $Ca^{2+}$ efflux and hence a decline in internal free $Ca^{2+}$. Reproduced with permission from Yau et al (1988).

ductance had electrical and ionic properties similar to those of the light-sensitive conductance, indicating that they are one and the same (Fesenko et al 1985, 1986, Nakatani & Yau 1985, Zimmerman et al 1985, Matthews 1986a, 1987, Yau et al 1986, Zimmerman & Baylor 1986a, Stern et al 1986, Furman & Tanaka 1987, Matthews & Watanabe 1987). Activation of the conductance by cGMP does not require the presence of other metabolites such as ATP (Fesenko et al 1985). This shows that protein phosphorylation, a common effector mechanism for the actions of cyclic neucleotides, is not involved.

In a truncated rod outer segment preparation, the conductance opened when cGMP was added to the internal solution and was closed by light under conditions that supported the operation of the light-activated enzyme cascade that hydrolyzes cGMP (Yau & Nakatani 1985b, Nakatani & Yau 1988b). These results are expected if the cGMP-activated conductance is the same as the light-regulated conductance. Additional evidence for the role of cGMP in phototransduction has come from experiments that employed 8-bromo-cGMP (MacLeish et al 1984, Zimmerman et al 1985), a hydrolysis-resistant analog of cGMP that opens the channel in excised patches with high efficiency. When introduced into a transducing rod this analog gave a very large inward current that was poorly suppressed by light. These results support the notion that cGMP hydrolysis is required for channel closing and make it unlikely that another messenger can act independently to close the conductance in light.

To summarize, in darkness cGMP acts directly on the conductance to keep it open, while light closes the conductance by removing cGMP. Even though the original $Ca^{2+}$-hypothesis is untenable, $Ca^{2+}$ seems to play an important role in transduction (see below).

## The Light-Triggered cGMP Cascade

We summarize here the present picture of the biochemical cascade that produces light-triggered cGMP hydrolysis in rods. By necessity, the account is brief, and readers are referred to the many excellent reviews available (see e.g. Bitensky et al 1978, Kühn 1984, Chabre 1985, Stryer 1986, Liebman et al 1987).

Upon absorbing a photon, the chromophore of rhodopsin, 11-*cis* retinal, isomerizes and triggers a series of conformational changes in the protein part of the molecule. One transitional intermediate, metarhodopsin II, is enzymatically active. It catalyzes the exchange of GTP for bound GDP on the GTP-binding protein transducin, a peripheral membrane protein of rod discs. Binding of GTP causes the active subunit, $T_\alpha$, of transducin to be liberated from the $T_{\beta\gamma}$ complex. This active subunit in turn removes an inhibitory constraint within the enzyme cGMP phosphodiesterase, another peripheral membrane protein of discs. The activated phospho-diesterase rapidly hydrolyzes the cGMP in the local region of the outer segment. Much amplification is built into the cascade. During the physiological response to light, a single photoisomerized rhodopsin molecule can trigger the liberation of hundreds of $T_\alpha \cdot$ GTPs and hence activate many phosphodiesterase molecules. Each active phosphodiesterase molecule in turn can hydrolyze a large number of molecules of cGMP.

Termination of the response to light involves several events:

1. The isomerized rhodopsin is phosphorylated, a covalent modification that renders it less effective in activating transducin. Phosphorylated rhodopsin also binds another protein, arrestin (also called *48K protein* or S-antigen), which competitively inhibits the binding of transducin to rhodopsin.
2. $T_\alpha$ has GTPase activity and slowly hydrolyzes the bound GTP to GDP. Hydrolysis terminates the ability of $T_\alpha$ to activate the phosphodiesterase and promotes the reunion of $T_\alpha$ with the $T_{\beta\gamma}$ complex.
3. The activated phosphodiesterase may itself be turned off by mechanisms that are still not well understood (see e.g. Deterre et al 1988).
4. Resynthesis of cGMP by the enzyme guanylate cyclase finally restores the cGMP concentration to the dark level.

For several years there were doubts about whether the cGMP cascade could operate fast enough to generate the electrical response to light. It was later found that the activation of transducin as well as the hydrolysis of cGMP in the light can indeed precede the electrical response (Cote et al 1984, Vuong et al 1984, Blazynski & Cohen 1986). Another early problem with the cGMP hypothesis was that the relative reduction in cGMP within the rod outer segment, measured biochemically, was very small even for stimuli that would cause the rods to give a maximal electrical response (Kilbride & Ebrey 1979, Goldberg et al 1983). This problem was somewhat defused by the later finding that the outer segment's free cGMP content, estimated from electrophysiological experiments, represents only a small fraction of the measured total in darkness (Yau & Nakatani 1985b, Nakatani & Yau 1988b). Thus the fractional change in the free cGMP, which gates the conductance, may be much larger than the fractional change in the total cGMP if much of the total is bound. A cellular constituent that may account for the bulk of the cGMP binding in rod outer segments is the cGMP phosphodiesterase, which has been reported to have noncatalytic, high-affinity cGMP binding sites (Yamazaki et al 1980). In addition, there may be other cGMP-binding proteins (Yuen et al 1986, Shinozawa et al 1987). An interesting and important question remains, however: How does the distribution of cGMP between the free and the

bound pools vary in darkness and light? Recently Cote et al (1986) and
Cohen & Blazynski (1989) reported that there is no fixed relation between
changes in the total measured cGMP concentration and the degree of
suppression of the dark current. Perhaps light-dependent dynamic changes
in the ratio of free and bound cGMP are important in transduction.

Figure 3 summarizes key ionic and enzymatic events in phototransduction. In darkness, $Na^+$, $Ca^{2+}$, and $Mg^{2+}$ enter the rod outer
segment through the light-regulated conductance (see below). $Na^+$ is
pumped out by the $Na^+-K^+$ pump at the inner segment, while $Ca^{2+}$ is
pumped out by the $Na^+-Ca^{2+}$ exchange carrier. The exit pathway for
$Mg^{2+}$ is not known. As mentioned above, closure of the cGMP-regulated
conductance in the light together with continued $Ca^{2+}$ efflux through the
exchange causes the free $Ca^{2+}$ concentration to drop in the light.
Conversely, elevation of cGMP levels will increase the influx of $Ca^{2+}$
through the cGMP-regulated conductance, raising the concentration of
$Ca^{2+}$. These changes in $Ca^{2+}$ in response to changes in the cGMP level

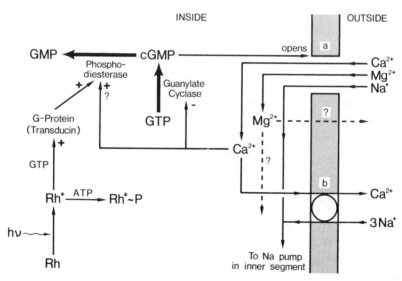

*Figure 3* Scheme of visual transduction in rods. Symbols: hν, photon; Rh, rhodopsin
molecule; Rh*, photoisomerized rhodopsin molecule; Rh* ∼ P, phosphorylated form of
Rh*; *a*, light-sensitive conductance; *b*, $Na^+-Ca^{2+}$ exchange; +, stimulation; −, inhibition.
The possible modulation by $Ca^{2+}$ of the cGMP phosphodiesterase is still poorly understood.
The pathway for $Mg^{2+}$ out of the outer segment is also unknown. Modified from Yau et al
(1986).

point to a negative feedback link between the light-sensitive conductance and the cGMP level. Biochemical measurements have revealed that the concentration of $Ca^{2+}$ strongly affects the level of cGMP in the outer segment: A rise in $Ca^{2+}$ lowers cGMP, and a fall in $Ca^{2+}$ raises cGMP (Cohen et al 1978, Kilbride 1980, Woodruff & Fain 1982). The mechanisms underlying these effects are still not entirely clear, but one site of action appears to be an inhibitory effect of $Ca^{2+}$ on the guanylate cyclase (Krishnan et al 1978, Troyer et al 1978, Fleischman & Denisevich 1979, Lolley & Racz 1982, Pepe et al 1986, Hodgkin & Nunn 1988, Koch & Stryer 1988, Kondo & Miller 1988, Rispoli et al 1988). In addition $Ca^{2+}$ may have an influence on the phosphodiesterase (Robinson et al 1980, Torre et al 1986, Hodgkin 1988). Thus, the negative feedback should operate in the following way. When light triggers the hydrolysis of cGMP, the resulting decline in $Ca^{2+}$ will disinhibit the guanylate cyclase, helping to restore the concentration of cGMP to the dark level. This effect, which will shorten the excitation triggered by a photon, would be assisted if low $Ca^{2+}$ also reduced the duration of the phosphodiesterase activation. In any case, the light-dependent drop in intracellular $Ca^{2+}$ should lower the rod's sensitivity to light, making it a possible mechanism of light adaptation (Yau & Nakatani 1985a). Suggestive results consistent with this notion were obtained by Torre et al (1986) and Korenbrot & Miller (1986). Much stronger supporting evidence, however, is now also available (Matthews et al 1988, Nakatani & Yau 1988c). In darkness the feedback should also reduce spontaneous low-frequency fluctuations in the cGMP level, thus improving the reliability of light detection. Finally, the feedback may serve to protect rods from a harmful or fatal excess cation influx caused by a chance rise in the cGMP level (Baylor 1987, Nakatani & Yau 1988b).

## Ionic Selectivity of the cGMP-Gated Conductance

The ionic selectivity of the light-regulated conductance remained a puzzle until rather recently. Although Sillman et al (1969) rightly concluded nearly two decades ago that the conductance normally passes $Na^+$ ions, initial efforts to detect permeation by other cations had produced negative results (Korenbrot & Cone 1972, Cervetto 1973, Brown & Pinto 1974). This was surprising, for two reasons: (a) no other $Na^+$ channel is known to be completely selective for $Na^+$, and (b) the reversal potential for the light-sensitive current is near zero rather than the $Na^+$ equilibrium potential (Werblin 1975, Bader et al 1979, Bodoia & Detwiler 1985, Baylor & Nunn 1986), suggesting that other cations such as $K^+$ may pass through the conductance as well. To explain this paradox, Bastian & Fain (1982) suggested that external $Na^+$ not only carries current through the conductance but also maintains the conductance in the open state. Thus, upon removing external $Na^+$ in an ionic selectivity experiment, the conductance would inevitably close and the dark current would stop, even if the substituting cation could pass through the open conductance. This idea was supported by the observation that a variety of monovalent cations were indeed permeant when the concentration of external $Ca^{2+}$ was lowered (Yau et al 1981, Bastian & Fain 1982, Woodruff et al 1982, Capovilla et al 1983, Hodgkin et al 1984). More interestingly, permeation by divalent cations was also observed under these conditions (Yau et al 1981, Capovilla et al 1983, Hodgkin et al 1984). Fluxes of ions other than $Na^+$ could also be detected without lowering external $Ca^{2+}$ if a phosphodiesterase inhibitor such as isobutylmethyxanthine (IBMX) was present (Woodruff et al 1982, Capovilla et al 1983, Hodgkin et al 1984). IBMX is well known to increase the light-sensitive conductance in darkness (Lipton et al 1977).

Perhaps the most convincing evidence on the relatively low degree of selectivity of this conductance, however, came from rapid perfusion experiments performed without lowering external $Ca^{2+}$ or adding phosphodiesterase inhibitors (Yau & Nakatani 1984a, Hodgkin et al 1985). In these experiments $Na^+$ was replaced in less than one second by another cation. Removal of $Na^+$ caused the conductance to close gradually over a period of seconds. Before closure occurred, however, a variety of monovalent and divalent cations could be shown to enter the outer segment through the light-sensitive conductance. The initial current observed immediately after another cation was substituted for $Na^+$ (i.e. before the number of open channels had time to change) gave a measure of the permeability for that cation relative to $Na^+$. In this way, for example, the apparent permeability ratios for the common monovalent cations $Li^+$, $K^+$, $Rb^+$, and $Cs^+$ relative to $Na^+$ were found to be around 1.1, 0.7, 0.5, and 0.3, respectively. Rather similar ratios were obtained with low-$Ca^{2+}$ external solution, in which case the currents carried by the other cations were better sustained. Measurements could be made with divalent cations but were more complicated because the divalent cation fluxes saturated rapidly with increasing external concentration (Hodgkin et al 1985). This saturation probably reflects strong interactions between the divalent cations and binding sites within the channel (see below). Owing to this saturation, the permeability of a divalent cation relative to that of $Na^+$ is meaningful only when both concentrations are specified. Under physiological conditions with external $Na^+$, $Ca^{2+}$, and $Mg^{2+}$ concentrations at around 110 mM, 1 mM, and 1.6 mM, respectively, the fractions of inward dark current carried by the three ions were estimated to be around 0.7, 0.15, and 0.05, respectively, giving apparent permeability ratios of about 12.5 ($P_{Ca}/P_{Na}$) and 2.5 ($P_{Mg}/P_{Na}$) (Nakatani & Yau 1988a). Other studies have also indicated that the conductance is more permeable to $Ca^{2+}$ than to $Na^+$ (Capovilla et al 1983, Hodgkin et al 1985, Torre et al 1987). Thus, even though $Na^+$ carries most of the dark current because of its higher extracellular concentration, the conductance is really more selective for $Ca^{2+}$ than $Na^+$. The conductance also passes $Ba^{2+}$, $Sr^{2+}$, $Mn^{2+}$, $Ni^{2+}$, and $Co^{2+}$ to various degrees (Capovilla et al 1983, Yau & Nakatani 1984a, 1985c).

Despite its relatively strong preference for $Ca^{2+}$, the light-regulated conductance behaves quite differently from typical $Ca^{2+}$ channels. For instance, it is not voltage-activated. Furthermore, $Ca^{2+}$ channels have a much higher selectivity for $Ca^{2+}$: They permit the passage of monovalent cations only in the complete absence of divalent cations, and any trace amount of divalent cations will eliminate the monovalent flux (Kostyuk et al 1983, Almers et al 1984, Hess & Tsien 1984). Divalent cations also lower the flux of monovalent cations through the light-sensitive conductance, but the blockage is only partial at physiological concentrations. Thus, as pointed out above, as much as 70% of the normal dark current is carried by $Na^+$ even with 1 mM $Ca^{2+}$ and 1.6 mM $Mg^{2+}$ in the external solution. Removing external $Ca^{2+}$ alone increases the $Na^+$ influx by 2–3-fold (Lamb & Matthews 1988, Nakatani & Yau 1988a), while removing both $Ca^{2+}$ and $Mg^{2+}$ increases the $Na^+$ influx still more (Nakatani & Yau 1988a).

The finding that the channel is nearly as permeable to $K^+$ as to $Na^+$ explains why the reversal potential for the light-sensitive current is near zero voltage. In spite of the relatively large $K^+$ permeability, however, the efflux of $K^+$ from the rod outer segment in darkness is probably rather small. As an example, if the channel were equally permeable to $Na^+$ and $K^+$, and if the channel contained one cation-binding site that accommodated only one cation at a time (see below), the $K^+$ efflux would be only 17% of the $Na^+$ efflux at a membrane potential of $-40$ mV. In reality the fraction will be lower because of the lower relative $K^+$ permeability.

The mechanism by which external $Na^+$ maintains the open state of the conductance is now understood. As mentioned above, internal $Ca^{2+}$ suppresses the level of cGMP in the outer segment (Cohen et al 1978, Kilbride 1980, Woodruff & Fain 1982) and hence can indirectly close the light-sensitive conductance. Furthermore, the $Na^+$–$Ca^{2+}$ exchange that extrudes $Ca^{2+}$ has an absolute requirement for external $Na^+$ (Blaustein & Russell 1975, Bastian & Fain 1982, Hodgkin et al 1984, Yau & Nakatani 1984b). Thus, when $Na^+$ is removed from an external solution containing $Ca^{2+}$, the exchange immediately stops and internal free $Ca^{2+}$ rises, hence lowering cGMP and closing the conductance. If, however, $Na^+$ and $Ca^{2+}$ are removed together (see e.g. Nakatani & Yau 1988a), there is neither $Ca^{2+}$ efflux nor $Ca^{2+}$ influx, and the conductance therefore stays open.

Similarly, in the absence of external $Na^+$ IBMX helps to keep the conductance open by reducing the rate of cGMP hydrolysis.

In retrospect, the ionic selectivity of the conductance has turned out to have more than just academic interest. In particular, the discovery of a $Ca^{2+}$ component in the dark current immediately shed doubt on the $Ca^{2+}$ hypothesis and raised the possibility of a light-induced decline in intracellular $Ca^{2+}$.

## Current-Voltage Relation

Bader et al (1979) pioneered the study of the conductance with voltage clamp techniques. They found that the reversal potential for the light-sensitive current is near zero membrane voltage, and that the current-voltage relation shows an outward rectification that can be described by an exponential function of the form:

$$j(V) \propto \exp[(V - V_r)/V_0] - 1 \qquad \qquad 1.$$

where $j(V)$ is membrane current, $V$ is membrane voltage, $V_r$ is the reversal potential, and the slope constant $V_0$, which specifies the steepness of the rectification, is about 25 mV. Subsequent measurements by others (Bodoia & Detwiler 1985, Baylor & Nunn 1986) are in general agreement with these results, although somewhat steeper relations were obtained. Figure 4 shows the current-voltage relation of the outer segment conductance of a salamander rod in darkness and light.

The original derivation of Eq. 1 was based on a simple model in which the channel is represented as having a single energy barrier at the outer edge of the membrane (see Jack et al 1975). This model does not give slope constants smaller than 25 mV, however, unless the current carriers are multiply charged. More recent studies based on excised-patch recordings (see below) suggest instead that the rectification arises largely from a voltage-dependent block of the channel by divalent cations, with the steepness of rectification dependent on the voltage-dependence of the block.

The flatness of the current-voltage relation at negative voltages (see Figure 4) implies that the rod dark current is rather insensitive to physiological changes in membrane voltage. Such a small dependence of current on voltage allows light-evoked voltage changes to spread efficiently along the outer segment. We return to this point in a following section. Strangely enough, the current-voltage relation of the cone conductance is not flat at negative voltages but instead is inwardly rectifying (Haynes & Yau 1985). This observation is one of several suggesting that the cone and rod conductances are different molecular entities. The functional significance of the form of the cone current-voltage relation remains to be determined.

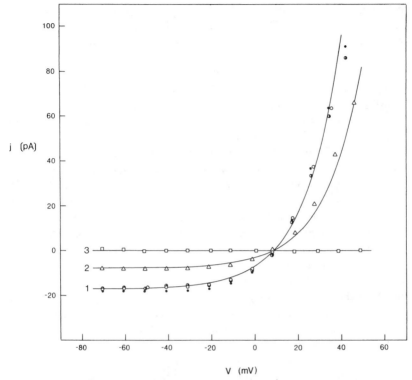

V (mV)

*Figure 4* Current-voltage relations from the outer segment of an isolated salamander rod obtained under voltage clamp. Voltage was clamped with two intracellular electrodes and current collected from the distal half of the outer segment with a suction electrode. Voltage was held at the dark level and current measured at 20 msec after switching to the new level. *Curve 1*, dark; *curve 2*, half-saturating light; *curve 3*, saturating light. Straight line in *curve 3* has a slope of 0.8 pS. Reproduced with permission from Baylor & Nunn (1986).

## Channel Characteristics from Noise Measurements

Measurement of the electrical noise generated by a population of channels can provide information on the properties of individual sites. The first detailed measurements of light-sensitive electrical noise in rods were made by Schwartz (1977) using intracellular voltage recording. At about the same time, Lamb & Simon (1976, 1977) examined voltage noise in cones. From power spectral analysis Schwartz concluded that the unit events constituting the light-sensitive voltage noise in rods had a mean duration of 200–300 msec and an associated conductance change of about 1 pS. Subsequent measurements with cell-attached patch recording and whole-cell voltage clamp, however, revealed a faster component in the light-sensitive noise in rods (Detwiler et al 1982, Bodoia & Detwiler 1985, Gray & Attwell 1985, Zimmerman & Baylor 1985). The relatively slower noise that Schwartz (1977) observed now appears to arise from enzymatic fluctuations in the transduction cascade (Baylor et al 1980, Stryer 1986). The faster noise, which arises directly from the light-regulated conductance, consists of unitary events with a mean duration of about 1 msec and an amplitude of 2–4 fA, corresponding to a unit conductance of about 0.1 pS. Similar estimates for the size and duration of the elementary current were obtained from cGMP-induced current noise in excised patches from the outer segment (Fesenko et al 1985, Zimmerman & Baylor 1986a, Matthews 1986a). The extremely small channel conductance was surprising, considering that typical values for other ion channels are 1–20 pS or more. For a time it seemed possible that this conductance might consist of ligand-activated carrier molecules rather than aqueous pores. This possibility was recently ruled out, however, when excised-patch recordings showed that the conductance assumes much larger values when divalent cations, which block the channel under physiological conditions, are removed (Haynes et al 1986, Zimmerman & Baylor 1986b).

The small effective channel conductance described above is more than a curiosity. Simple statistical calculations (see section on *Relating Conductance to Function*) indicate that it serves the important function of lowering the electrical dark noise resulting from random gating of the channels, thereby allowing single-photon detection to take place.

## Blockage by Divalent Cations

As mentioned above, intracellular $Ca^{2+}$ has little if any direct effect on the light-regulated conductance at physiological concentrations, even though a rise in internal $Ca^{2+}$ concentration can close the conductance indirectly by lowering the concentration of cGMP. In regard to extracellular $Ca^{2+}$, early evidence (Yoshikami & Hagins 1973) indicated that it could inhibit the light-sensitive dark current, an observation later confirmed and extended by others (Yau et al 1981, Bodoia & Detwiler 1985). Most of this effect of external $Ca^{2+}$ now appears to involve changes in internal $Ca^{2+}$, which in turn affect the level of cGMP (Hodgkin et al 1984, Yau & Nakatani 1984a, Hodgkin et al 1985, Lamb & Matthews 1988, Nakatani & Yau 1988a). Nevertheless, external $Ca^{2+}$ also appears to have a direct blocking effect on the conductance. Thus, after the rapid removal of external $Ca^{2+}$ it was possible to identify a fast, 2–3-fold increase in the dark current, which preceded a slow-developing, 10-fold or larger increase (Lamb & Matthews 1988, Nakatani & Yau 1988a). The fast effect is attributed to a direct block of the channel, and the slower effect to a drop in intracellular $Ca^{2+}$, which raises the level of cGMP. By using the same strategy, external $Mg^{2+}$ was also shown to directly block the conductance (Nakatani & Yau 1988a).

The difficulty in studying the block by divalent cations in intact cells is that the intracellular concentrations of divalent cations and cGMP cannot be properly controlled. These problems are avoided with the excised, inside-out membrane patch. Figure 5 shows current-voltage relations of the cGMP-induced current obtained from an experiment in which an excised rod membrane patch was exposed to normal Ringer's solution on its extracellular side and to a pseudointracellular solution containing different concentrations of $Ca^{2+}$ and $Mg^{2+}$ on its cytoplasmic side. At

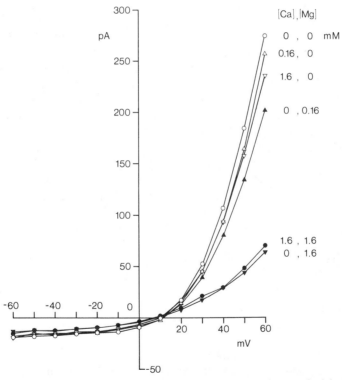

*Figure 5* Effect of divalent cations on the cGMP-activated current in an excised, inside-out membrane patch from the toad rod outer segment. The pipet contained normal Ringer's solution, and the bath contained a succession of pseudointracellular (low $Na^+$-high $K^+$) solutions with the indicated concentrations of $Mg^2$ and $Ca^{2+}$. A saturating concentration of cGMP was present throughout. Adjacent points are joined by straight lines. Reproduced with permission from Yau et al (1986).

negative (i.e. physiological) voltages the cGMP-induced current was quite insensitive to divalent cations on the cytoplasmic side, in good agreement with intact-cell experiments. At positive voltages, however, the divalent cations strongly blocked the current. This indicates that divalent cations on the cytoplasmic side block the conductance in a voltage-dependent manner. The presence of the negative membrane potential explains why internal divalents do not show a blocking effect in measurements on intact cells.

The observation that depolarizing voltages promote channel blockage by divalent cations from the cytoplasmic side, together with the finding that the channel is permeable to these ions, suggest that the divalent cations block after entering the channel. By analogy with the $Ca^{2+}$ channel (Almers & McCleskey 1984, Hess & Tsien 1984), this blockage may result from occupation of one or more binding sites for which cations compete in order to permeate. Prolonged dwell times of the divalent cations in the channel would explain their low flux rates and hence their effective blocking action (Haynes & Yau 1985, Haynes et al 1986). When divalent cations were removed from both sides of a membrane patch so as to remove block, the cGMP-activated current became much larger (Yau et al 1986, Yau & Haynes 1986, Zimmerman & Baylor 1986b). Interestingly, under these conditions the current-voltage relation for the conductance also lost most of its outward rectification (Yau et al 1986, Zimmerman & Baylor 1986b, Matthews 1986a). This observation suggests that the outward rectification of the conductance under physiological conditions is due largely to a voltage-dependent block by divalent cations.

Recently, Zimmerman & Baylor (1988 and in preparation) have been able to account quantitatively for several characteristics of ion permeation through this conductance with a simple model in which the channel is assumed to contain a single cation binding site flanked by an energy barrier on each side. The model assumes that only one ion can bind to the site at a time, and that divalent cations bind much more strongly than monovalents. Because the binding site is located within the channel, the rates of ion movement to and from the site are voltage-dependent.

## Single-Channel Properties

Early attempts to record single-channel currents underlying the light-regulated conductance were unsuccessful (see e.g. Detwiler et al 1982). Instead, as mentioned above, an unusually small single-channel conductance of ca. 0.1 pS was derived from fluctuation analysis (Detwiler et al 1982, Bodoia & Detwiler 1985, Gray & Attwell 1985, Zimmerman & Baylor 1985), and the corresponding single-channel current would be too small to be detectable. Subsequent analysis of cGMP-induced current noise in excised membrane patches also indicated an extremely small channel conductance (Fesenko et al 1985, 1986, Matthews 1986a, Zimmerman & Baylor 1986a). This very small size for the elementary current might have been consistent with conduction by membrane carrier molecules (see Detwiler et al 1982, Fesenko et al 1985, 1986) rather than the water-filled pores typical of other ion channels in nerve cells (Hille 1984). An alternative possibility was that the channel consisted of an aqueous pore whose effective conductance was reduced by blocking by divalent cations. Recent experiments bear out this latter possibility. In the complete absence of divalent cations, single-channel currents induced by cGMP are indeed observed in an excised rod membrane patch (Haynes et al 1986, Zimmerman & Baylor 1986b, Matthews & Watanabe 1987, 1988, Quandt et al 1988). The elementary current amplitudes far exceed carrier transport rates, indicating pore conduction rather than a carrier mechanism.

Figure 6 illustrates some of these single-channel currents; in the traces, upward deflections indicate channel openings. At the low concentrations of cGMP used in this experiment, predominantly one channel was open at a time. Histograms of the single-channel openings indicate more than one conducting state, with a large state at around 25 pS and at least one smaller state with a conductance several times lower (Haynes et al 1986, Zimmerman & Baylor 1986b). These conductances seem roughly constant at different voltages. The physical significance of the multiple conductance levels is still unclear. One possibility is that the different levels correspond to openings of partially and fully liganded channels; another, less likely, possibility is that they represent two different species of channels. Because the channel activity has a characteristic flickery nature, a detailed analysis of the underlying kinetic properties is difficult. Furthermore, the high channel density in excised patches allows single-channel currents to be observed only at very low concentrations of cGMP. The channel openings consist of bursts lasting a few milliseconds, during which the current amplitude fluctuates prominently. One interpretation would be that the burst represents the time that the channel is fully liganded and that the fluctuations result from rapid open-closed transitions in the fully liganded channel. Based on this idea, the short length of the burst indicates that the cGMP is loosely bound to the channel. The functional significance of this feature is described further on. Other measurements on the kinetics of activation are presented in the next section.

When a low concentration of $Ca^{2+}$ or $Mg^{2+}$ is applied to an excised patch, single-channel activity is suppressed. Channel opening events are replaced by bursts of high-frequency flickering that reflect a fast, intermittent block of the current (Figure 7). This phenomenon is consistent with the idea mentioned above that divalent cations intermittently enter the open

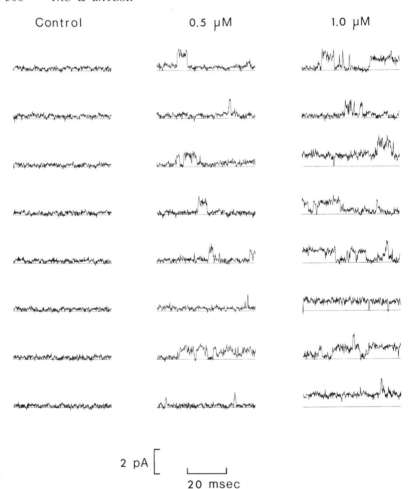

*Figure 6*  Single-channel currents from the cGMP-activated conductance of a salamander rod. cGMP was applied to the cytoplasmic surface of an excised patch by perfusion of the chamber. Recordings were made in symmetrical NaCl solutions containing a total free divalent cation concentration less than 1 $\mu$M. Holding potential +50 mV, outward current across the patch plotted upward. Bandwidth 0.016–2000 Hz. Reproduced with permission from Baylor (1987).

channel and block it from within. A somewhat similar flicker block by divalent cations was previously observed in the $Ca^{2+}$ channel (Lansman et al 1985).

A promising alternative technique for observing channel function is the lipid bilayer method. Here membrane vesicles prepared from rod outer

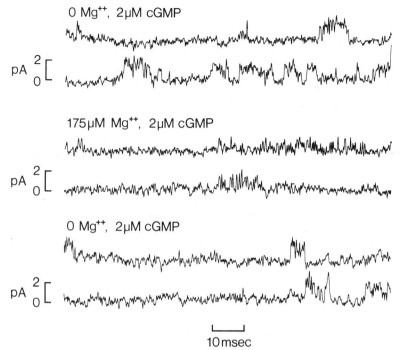

*Figure 7*   Block of open channels by $Mg^{2+}$. An excised, inside-out membrane patch from the salamander rod was held at +60 mV. Both sides of the membrane contained identical NaCl solutions, containing (in mM): 118 NaCl, 0.1 Na EGTA, 0.1 Na EDTA, buffered to pH 7.6 with 5.0 Na HEPES. *Top two* and *bottom two traces* were obtained in the absence of $Mg^{2+}$; *center two traces* had 175 $\mu$M MgCl$_2$ added to the bath (and EDTA and EGTA removed). DC to 5 kHz bandwidth. Reproduced with permission from Haynes et al (1986).

segments (Tanaka et al 1987) or purified preparations of channel protein itself (Cook et al 1987, Hanke et al 1988) are incorporated into an artificial lipid bilayer. Already this method has produced results on the channel that are broadly similar to those described above.

## Density of Channels

The density of channels in the membrane is an important parameter in interpreting transduction. In addition, information on channel density may help in the molecular identification of the channel: A candidate protein should be present in the surface membrane in amounts consistent with known channel density. Estimates of channel density can be made in two ways. The number of open channels in the plasma membrane of the rod outer segment can be inferred from whole-cell channel noise measurements. As mentioned above, the effective single-channel current under physiological conditions (i.e. in the presence of divalent cations) is about 4 fA (Bodoia & Detwiler 1985, Gray & Attwell 1985, Zimmerman & Baylor 1985). Taking the dark current of the salamander rod as 40 pA there would be about 10,000 channels open at any given time. Evidence described below indicates, however, that only a small fraction, perhaps 1–2%, of the channels are open even in darkness. On this basis, the total number of channels in the plasma membrane would be about $5 \times 10^5$. For a surface area of 1000 $\mu m^2$, the average channel density would therefore be about 500 $\mu m^{-2}$. Estimates can also be made from recordings of the single-channel and macroscopic currents from excised patches. This method gave figures of ca. 1000 $\mu m^{-2}$ (Haynes et al 1986) and 400 $\mu m^{-2}$ (Zimmerman & Baylor 1986b). These channel density estimates are subject to two uncertainties, however. First, the membrane area of the excised patches was estimated rather than measured. Second, two single-channel current amplitudes are observed, as described above, thus complicating the interpretation. In different excised patches the number of channels varies widely, possibly indicating spatial variations of channel density in the plasma membrane.

## Mechanism of Activation by cGMP

Even though the conductance is "light-regulated," neither its ability to conduct ions nor its activation by cGMP have been found to be directly affected by light (Yau & Nakatani 1985b, Nakatani & Yau 1988b). The light sensitivity of the channel in transduction therefore appears to result entirely from the light-triggered hydrolysis of its ligand, cGMP.

The dependence of the steady-state macroscopic current on cGMP concentration shows sigmoidicity, indicating that more than one cGMP molecule must bind to open the channel. In the presence of divalent cations, a Hill coefficient ($n$) of 1.7–3.1 was obtained (Fesenko et al 1985, Koch & Kaupp 1985, Zimmerman et al 1985, Puckett & Goldin 1986, Yau & Nakatani 1985b, Yau et al 1986, Nakatani & Yau 1988b). In the absence of divalent cations the value of $n$ was consistently close to 3 (Haynes et al 1986, Zimmerman & Baylor 1986b, Cook et al 1987). The reason for this difference is not clear, though it should be pointed out that the currents at small concentrations of cGMP are the most sensitive indicator of the Hill coefficient and can be measured more accurately with low divalent cation concentrations. Another possibility is that divalent cations modify the cooperativity of activation. What seems certain, nonetheless, is that the channel has more than one binding site for cGMP.

In the steady-state measurements cited above, the cGMP concentration that opened half the channels ($K_{1/2}$) ranged from about 10 to over 50 $\mu$M. The basis for these differences is not clear either. Depolarization appears to promote channel activation by cGMP, leading to a reduction in the value of $K_{1/2}$ (Yau et al 1986, Haynes et al 1986, Karpen et al 1988a).

Single-channel measurements show a burst length of a few milliseconds (see preceding section). Assuming that a burst is terminated when a molecule of cGMP unbinds, the binding of cGMP to the channel would be rather loose. Loose binding is also suggested by the speed of the conductance decrease during the rise of the rod's response to very bright light (Cobbs & Pugh 1987).

Karpen et al (1988a) examined the kinetics of activation over a wide range of cGMP concentrations, using two kinds of rapid perturbation: flash photolysis of "caged" cGMP (Nerbonne et al 1984) and a sudden change in membrane voltage. The results were consistent with a model involving three diffusion-controlled cGMP binding steps followed by rapid, voltage-dependent open-closed transitions in the fully liganded channel. The speed of activation by cGMP (Figure 8) suggested that the

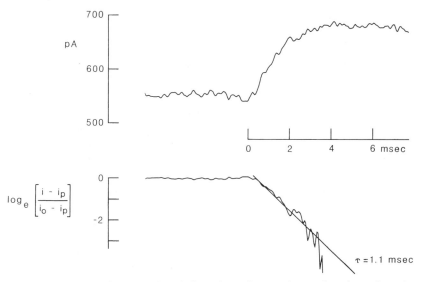

*Figure 8* Response of an excised patch from the surface membrane of a salamander rod outer segment to a sudden jump in cGMP concentration generated by flash photolysis of caged cGMP. The semilog plot below shows that the rise to the final level was exponential, with time constant 1 msec. At time zero a bright 14 nS flash was given from a XeF excimer laser, causing the cGMP concentration to rise in <100 $\mu$sec from 12 $\mu$M to 15 $\mu$M. Low divalent cation concentrations, holding potential +50 mV. Reproduced with permission from Baylor (1988).

cGMP binding sites are located on a single, preformed channel unit in the membrane rather than on several subunits that aggregate after binding cGMP to form a conducting channel.

The channel's requirement for cGMP is very specific, and no other natural nucleotide, including cAMP, has been found to activate it (see e.g. Fesenko et al 1985). Certain analogs of cGMP, such as 8-bromo-cGMP and the fluorescent derivative 8-(5-thioacetamidofluorescein)-cGMP are, however, more effective than cGMP by approximately an order of magnitude (Caretta et al 1985, Koch & Kaupp 1985, Zimmerman et al 1985, Puckett & Goldin 1986, Nakatani & Yau 1988b). Bulky groups at the 8-position may lock cGMP into a syn-like configuration that is favorable for binding to the channel. The phosphorothioate analogs (Sp)-cGMP[S] and (Rp)-cGMP[S] are less effective than cGMP (Zimmerman et al 1985), a finding that indicates the importance of the atoms comprising the cyclic phosphodiester moiety.

Stern et al (1987) recently suggested, from macroscopic current measurements on excised patches, that the conductance can open without cGMP. This suggestion is not supported, however, by observations on the single-channel currents, which indicated that the channel almost never opens without cGMP (Haynes et al 1986, Zimmerman & Baylor 1986b). Ion flux studies on rod membrane preparations have been interpreted as indicating desensitization of the channel's response to cGMP (Koch et al 1987, Puckett & Goldin 1986, Goldin et al 1987). In electrical recordings from excised patches, however, this phenomenon has not been observed (Haynes et al 1986, Zimmerman & Baylor 1986b, Karpen et al 1988a,b). The apparent desensitization observed in membrane vesicle preparations may have arisen from ion depletion in the vesicles. Influx studies with cGMP preincubations of different durations are needed to settle this point. Finally, Koch et al (1987) and Schnetkamp (1987) have recently suggested the existence of two states or species of the conductance with different $K_{1/2}$ values; only one state was blocked by the drug 1-cis diltiazem (see below). The significance of this finding is uncertain.

The conductance in intact cells can increase enormously when a rod is subjected to either low extracellular $Ca^{2+}$ or high intracellular cGMP (Yoshikami & Hagins 1973, Yau et al 1981, MacLeish et al 1984, Matthews et al 1985, Cobbs & Pugh 1985, Zimmerman et al 1985). Similarly, very large current densities can be induced by cGMP in excised patches of rod membrane (see e.g. Yau et al 1986, Zimmerman & Baylor 1986b, Zimmerman et al 1988). These observations indicate that only a very small fraction of the conductance is open in darkness. By comparing the physiological dark current in an intact cell to the fully activated current in the same, subsequently "truncated" cell, it was concluded that the fractional activation in darkness is about 1–2% (Yau & Nakatani 1985b, Nakatani & Yau 1988b). Since light only lowers this fraction, the design seems rather wasteful: Why are 98% of the channels closed even in darkness? In a later section we discuss possible reasons for this "choice." From the measured relation between channel activation and cGMP concentration, and a fractional opening of 1–2%, it can be estimated that the free cGMP concentration in the rod outer segment in darkness is only a few micromolar (Yau & Nakatani 1985b). This is interesting because the total cGMP concentration has been measured biochemically as around 60 $\mu$M (Cohen et al 1978, Kilbride & Ebrey 1979, Woodruff & Bounds 1979, Woodruff & Fain 1982). Thus the free cGMP content in the rod outer segment is only a very small fraction of the total. This may explain the long-standing puzzle of why even bright light causes rather little reduction in the cGMP content of rods. From the above estimate, complete removal of the free cGMP in light might occur with only a few percent reduction in total cGMP.

## Is the Conductance Present in Both Plasma and Disc Membranes?

Attempts to localize the conductance in the rod outer segment have an interesting history. The earliest evidence for a cGMP-activated conductance in rods came from rod "disc" membrane preparations, in which Cavaggioni and co-workers (Caretta et al 1979, Cavaggioni & Sorbi 1981, Caretta & Cavaggioni 1983) first observed cGMP-activated fluxes of $Ca^{2+}$ and other cations. At the time, this conductance was thought to be involved in the regulation of $Ca^{2+}$ movements across disc membranes (see e.g. George & Hagins 1983). Koch & Kaupp (1985) then extended the work of Cavaggioni's group by examining further how the conductance was activated by cGMP. Their findings were similar to those from excised patches of plasma membrane: Activation by cGMP was cooperative and the $K_{1/2}$ values were in the same range. Similar results were obtained by Puckett & Goldin (1986). The agreement between the excised-patch experiments and the flux studies on disc membrane preparations suggested that the same conductance might be present in both the plasma and the disc membranes, though no rational function in transduction could be assigned to the disc conductance. Other groups continued to report the existence of a disc conductance (Matesic & Liebman 1987, Schnetkamp 1987, Schnetkamp & Bounds 1987).

Recently, however, new evidence has suggested that the conductance is not present in the disc membrane after all. Bauer (1988), who measured $Ca^{2+}$ depletion by cGMP and $Na^+$ in rod membrane vesicles, reported that the rod outer segment membrane comprised two fractions: a minor fraction (ca. 6%) that contained both $Na^+$–$Ca^{2+}$ exchange activity and cGMP-activated channels, and a larger fraction (ca. 94%) that contained neither. It is therefore tempting to suppose that the smaller fraction consisted of plasma membrane whereas the larger fraction consisted of disc membrane. Still clearer separation between the two kinds of membranes was achieved by Molday & Molday (1987a,b) using a method that utilized the binding of gold-labeled ricin to the plasma membrane followed by density centrifugation. Only the plasma membrane appeared to contain the 63 kD and 39 kD proteins suggested to be the channel molecule (Cook et al 1986, 1987, Matesic & Liebman 1987). An antibody recognizing the 63 kD candidate bound exclusively to the plasma membrane fraction (N. J. Cook, L. L. Molday, D. Reid, U. B. Kaupp, and R. L. Molday, personal communication). The conductance is likely confined, therefore, to the plasma membrane of the outer segment.

## Pharmacological Blockers for the Conductance

Substances that specifically block this conductance obviously would be useful as pharmacological tools and as markers during channel purification and in subcellular localization studies. So far, two chemicals have been reported to block the conductance. The first is 1-*cis* diltiazem, which blocks the conductance at micromolar concentrations in both rod membrane preparations and excised patches of plasma membrane (Koch & Kaupp 1985, Stern et al 1986). Diltiazem was originally found to block $Ca^{2+}$ channels (see e.g. Lee & Tsien 1983). The *d-cis* isomer of diltiazem is active on the $Ca^{2+}$ channel but is much less effective than the *l-cis* isomer in blocking the cGMP-activated conductance (Koch & Kaupp 1985). Other $Ca^{2+}$ channel blockers, such as verapamil and the dihydropyridines, have no effect (Koch & Kaupp 1985). Although diltiazem blocks the conductance in membrane suspensions or excised patches at concentrations as low as 20 $\mu$M, it fails to block the light-sensitive current in intact cells at concentrations as high as 1 mM (Stern et al 1986). Perhaps the drug fails to reach the conductance in intact cells. Alternatively, removal of the conductance from the intact cell might convert it into a state susceptible to the drug. Recently Koch et al (1987) reported that the cGMP-gated channel in bovine rods exists in two forms, one sensitive to *l-cis* diltiazem and the other not.

The other chemical reported to block the conductance is 3′,4′-dichlorobenzamil, a derivative of the epithelial $Na^+$ channel blocker amiloride (Nicol et al 1987b). The site of action seems to be on the cytoplasmic side of the membrane, and the effective concentration is in the micromolar range. This agent, however, is not very specific and will also block the $Na^+$–$Ca^{2+}$ exchange at concentrations several-fold higher.

In addition, certain divalent cations, such as $Co^{2+}$ and $Cd^{2+}$, have been reported to block this conductance (MacLeish et al 1984, Koch & Kaupp 1985). In these cases, the action is likely to result from binding of these ions within the channel.

## Molecular Identification of the Conductance

A complete understanding of the conductance requires a knowledge of its molecular identification and structure. Information on these points would also facilitate the identification of homologous channel proteins in other types of cell. Although progress has been made in purifying the channel, no consensus has yet been reached on the identity of the protein. Cook et al (1986, 1987) have isolated a protein of 63 kD molecular weight that, when incorporated into liposomes, mediated a cGMP-stimulated $Ca^{2+}$ flux. The activation of this $Ca^{2+}$ flux by cGMP showed a Hill coefficient near 3 and $K_{1/2}$ of about 10 $\mu$M. When the protein was reconstituted into a planar bilayer, single-channel activity elicited by cGMP was observed. At physiological concentrations of monovalent cations and with divalent cations absent the single-channel conductance was 26 pS, quite similar to that of the large state observed in excised patches of plasma membrane. Unlike the native channel in excised patches, however, the purified channel was insensitive to $l$-cis diltiazem.

On the other hand, Matesic & Liebman (1987) reported the purification of a 39 kD protein that, when reconstituted into membrane vesicles, likewise mediated a cGMP-dependent cation flux. Activation by cGMP showed a Hill coefficient of 1.7 and a $K_{1/2}$ near 60 $\mu$M. Interestingly, this channel protein is sensitive to $l$-cis diltiazem, unlike the one purified by Cook et al.

Shinozawa et al (1987) reported a 250 kD protein as a candidate for the channel. Finally, Clack & Stein (1988) recently reported that cGMP-induced channel activity could be observed in membrane patches excised from phospholipid vesicles containing purified bovine opsin, and suggested that rhodopsin or an isoform of rhodopsin may be the channel. This would be most remarkable.

We await definitive identification of the channel protein. The ultimate proof will probably come from cloning of the channel gene and functional expression of the protein in a neutral cell line. An additional puzzle is the relatively small molecular weights of the proteins purified by Cook et al and Matesic & Liebman. Most other channels are all much larger proteins. This feature, together with the reported difference insensitivity to $l$-cis diltiazem between the purified proteins, raises the possibility (Applebury 1987) that what has been isolated may be different subunits of a hetero-polymeric native channel protein, each of which can form a functional cGMP-activated conductance on its own. Alternatively, there may indeed be more than one species of the native conductance, as has been suggested by Koch et al (1987) (see section on the activation of the conductance).

## Relating Conductance Properties to Function

In this section we relate the functional properties of the conductance to its role in phototransduction. Some of the points have been treated previously by Attwell (1986).

IONIC SELECTIVITY    The conductance is permeable to both monovalent and divalent cations. Under normal conditions, the inward dark current through the conductance is carried primarily by $Na^+$. The $Na^+$ component is therefore the main transducer current, suppression of which generates the hyperpolarizing response to light. The $Ca^{2+}$ component provides negative feedback on the concentration of cGMP, the ligand that determines how strongly the conductance is activated. This $Ca^{2+}$ feedback stabilizes the dark current, assists recovery of the flash response, and underlies light adaptation. The $Mg^{2+}$ influx has no known role in phototransduction. Removing $Mg^{2+}$ completely from the extracellular solution had no obvious effect on the sensitivity or the kinetics of the light response (Nakatani & Yau 1988a). The $Mg^{2+}$ influx may simply reflect an inability of the conductance to exclude $Mg^{2+}$ completely while allowing permeation of $Na^+$ and $Ca^{2+}$.

OUTWARD RECTIFICATION    The distance over which voltage changes spread along a leaky cable-like structure varies with the square root of the membrane resistance (see e.g. Jack et al 1975). Since the outer segment has a high specific membrane resistance, it is well designed for efficient longitudinal spread of the small voltage signals that result from absorption of single photons. As shown in Figure 4, the inward current that enters the light-sensitive conductance in darkness is relatively constant at physiological potentials ($-40$ to $-80$ mV). In other words, the slope resistance of the membrane is large even though channels are open, delivering inward current that can be shut off by light. This rectification in the cGMP-gated conductance, together with a low density or absence of leakage channels that would lower the membrane resistance, minimizes attenuation of the hyperpolarization as it propagates passively through the outer segment. Schnapf (1983) calculated that the length constant (distance over which a voltage signal decays to $1/e$ of its initial value) of toad rod outer segments is 474 $\mu m$, compared to a physical length of about 55 $\mu m$.

LOW UNIT CONDUCTANCE    Not only do external $Ca^{2+}$ and $Mg^{2+}$ enter the cell through the channel, but in the process they block the conductance, as described above. Blockage lowers the effective unit conductance from about 25 to only about 0.1 pS; the unitary current at physiological potentials is about 4 fA instead of 1 pA. A dark current of 40 pA therefore flows through 10,000 open channels; 40 open channels would give the same current in the absence of block by divalents. Given a dark current of fixed size, having a large number of small channels open rather than a small number of large channels reduces background noise, a desirable feature for single-photon detection. For a large population of independent channels, each with a very low probability of being open, if the mean number open is $N$ the standard deviation of the statistical fluctuations will be given by $N^{1/2}$. Thus if an average of 10,000 channels are open, and the unitary current is 4 fA, the standard deviation of the current noise will be roughly $100 \times 4$ fA, or 0.4 pA. In comparison, for an average of 40 open channels with unitary currents of 1 pA, the standard deviation of the noise would be 6.3 pA, more than ten times higher. Since photons are encoded by currents of 1 pA or less (Penn & Hagins 1972, Baylor et al 1979b), such considerations are important. Of course, the divalent cation blockage itself introduces noise as well; however, this noise is of relatively high frequency (see e.g. Figure 7) and is mostly removed by low-pass filtering under physiological conditions. Some low-pass filtering is performed within the rod by the passive time constant of its membrane (typically 15–20 msec). Additional low-pass filtering is exerted within the synaptic pathways of the retina, which have unusually slow transfer characteristics (Baylor & Fettiplace 1977b).

The single-photon effect will also be more readily detected if it is highly reproducible, and the small channel conductance helps in this regard as well. The cell is designed so that a single photon blocks about 3% of the open channels, or about 300 in number. The expected variation in this number is about 17, so that the channel component of variability in the photon effect should be small. The somewhat larger variability actually observed (standard deviation being 1/5 the mean response, see Baylor et al 1979b, 1984) may indicate fluctuations in the number of phosphodiesterase molecules activated. For the alternative of a large channel conductance, the mean number of channels blocked would be close to one, and the uncertainty would be correspondingly large. Thus the reliability of signaling a photon's absorption drops precipitously when the pool of open channels goes down.

FAST GATING KINETICS    Because of the low-pass filtering performed by the cell and by synaptic transfer mentioned above, the fast gating kinetics of the channel likewise improve the signal/noise for photon detection (i.e. ratio of response amplitude to standard deviation of dark noise). For a filter that integrates the rod output over a time corresponding to the mean duration or integration time $t_i$ of the photon response, the signal/noise ratio is improved by a factor of $(t_i/t_b)^{1/2}$, where $t_b$ is the channel's effective mean open time (burst duration). For an amphibian rod with $t_i \sim 1$ sec and $t_b \sim 2.5$ msec, this ratio is about 20.

Another advantage of the fast gating kinetics of the conductance is that negligible channel delay is imposed on the generation of the light response. For instance, with a step reduction in the cytoplasmic concentration of cGMP, the ligand will come off and the channel will close within a few milliseconds. This is negligible compared to the rising phase of the electrical response to a dim flash, which takes hundreds of milliseconds in rods of cold-blooded animals (see e.g. Baylor et al 1979a) and apparently reflects the time required for the enzyme cascade to lower the concentration of cGMP. Even in bright light, when phosphodiesterase activation is very large and fast, the rising phase of the electrical response is limited by the membrane time constant of the cell rather than by the channel's gating kinetics (Cobbs & Pugh 1987). Similarly, binding of cGMP to the conductance will not limit the speed of recovery of the dark current (i.e. the falling phase of the light response) after a flash. The flash photolysis and voltage jump experiments described above indicate that the conductance opens within less than 10 msec after a step increase to a cGMP concentration of a few micromolar (Karpen et al 1988a), a range close to the physiological concentration of free cGMP in darkness (see above). In comparison, the decline of the dim flash response takes hundreds of milliseconds (see e.g. Baylor et al 1979a).

LOW FRACTIONAL ACTIVATION    As mentioned above, only 1–2% of the cGMP-activated channels in a rod are normally open in the dark. An alternative to this design would be to have far fewer channels and to keep most of them open by increasing the strength of their cGMP binding or by maintaining a much higher concentration of free cGMP in darkness. Both of these possibilities, however, would cause the channel to operate in an unfavorable part of the dose-response relation, where the cubic relation between fractional channel response and fractional cGMP concentration change no longer applies. Furthermore, since the rate of cGMP binding is already at the diffusion-limited value, stronger binding would require a slower rate of dissociation. This is undesirable for the reasons mentioned in the preceding paragraph. Since there is a basal rate of cGMP turnover in darkness (Goldberg et al 1983), a higher cGMP concentration in darkness would also be expected to increase this steady turnover rate and therefore require larger expenditures of high-energy phosphate compounds (Stryer 1987) because cGMP synthesis consumes GTP.

SIZE OF DARK CURRENT    The statistical factors mentioned above dictate that the light-sensitive current should flow through a large number of channels, but how large should the current be? Whatever the absolute magnitude, efficient signaling requires that the conductance of the outer segment be appropriately matched to that of the inner segment. This may be seen in the following way. Suppose that the rod is represented by a simple equivalent circuit in which the inner segment contains an electromotive force $E$ with lumped internal conductance $G$. The outer segment, which is electrically in parallel with the inner segment, consists of a light-sensitive conductance $G_L$; for simplicity this conductance is assumed to have a reversal potential of 0 mV. The membrane potential of the rod, $V_m$, is given by (see e.g. Baylor & Fuortes 1970):

$$V_m = \frac{EG}{G + G_L}. \qquad \qquad 2.$$

The conductance reduction, $-\Delta G_L$, produced by a photon is a fixed fraction of $G_L$, i.e., $-\Delta G_L = fG_L$. The corresponding voltage change, $\Delta V_m$, is therefore:

$$\Delta V_m = \frac{-EG\Delta G_L}{(G + G_L)^2} = EGf \frac{G_L}{(G + G_L)^2}. \qquad 3.$$

From Eq. 3, $\Delta V_m$ is maximum with $G_L = G$, indicating that a photon will give the largest voltage signal when the dark conductances of the inner and the outer segments are matched and the potential in darkness is half depolarized. The preceding treatment is oversimplified, because the conductances are treated as being ohmic and not voltage-gated. Nevertheless, it is impressive that the design of the cell is close to that suggested by this simple treatment, in that the potential in darkness is usually near $-40$ mV and can reach a value of about $-80$ mV in bright light (see e.g. Torre 1982). An upper limit on the absolute size of the dark current is probably set by the metabolic cost of pumping out the $Na^+$ that enters the cell. A lower limit may be set by the requirements of a match between outer and inner segment conductances and of putting enough channels into the inner segment and synaptic ending in order to produce a synaptic transmitter release that is reliably modulated by the voltage hyperpolarization.

## Cone Conductance

The finding that the light-sensitive conductance in rods is gated by cGMP naturally raised the question of whether the cone conductance is also gated by cGMP. Results from both excised-patch (Haynes & Yau 1985) and whole-cell (Cobbs et al 1985) recordings have demonstrated that it is. Interestingly, the current-voltage relation of the cone conductance is different from that of the rod conductance. In the cone conductance the current increases exponentially with voltage in both the depolarizing and the hyperpolarizing directions (Figure 9). The increase in slope conductance that occurs upon hyperpolarization implies that when the conductance is partially closed by light and the cell hyperpolarizes, there will be an increase in the current through the fraction of the conductance that remains open. This will reduce the size of the light response and thus the cone's sensitivity. This voltage-mediated desensitization, however, is rather weak and will reduce the light sensitivity by less than a factor of two (Haynes & Yau 1985). Overall, cones are about a 100-fold less sensitive

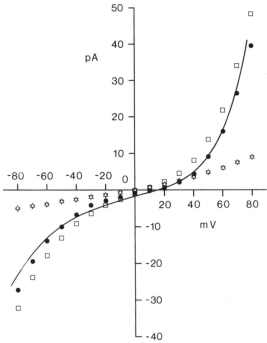

*Figure 9*   Current-voltage relation of the cGMP-induced current measured from an excised, inside-out membrane patch from the outer segment of a catfish cone. $\triangle$, $\triangledown$, relations obtained in control solution before and after application of cGMP; $\square$, relation obtained with a saturating concentration of cGMP; ●, relation for the cGMP-induced current (obtained as difference between the others). *Smooth curve* is a scaling of $j(V) = \exp[(1-\gamma)(V-V_r)/V_0] - \exp[-\gamma(V-V_r)/V_0]$ with $V_r = 15\,mV$, $V_0 = 12.5\,mV$, and $\gamma = 0.35$. The pipet contained normal Ringer's solution, and the bath contained a pseudointracellular (low $Na^+$-high $K^+$) solution with 1.6 mM $MgCl_2$ and 0.1 mM EGTA. Reproduced with permission from Haynes & Yau (1985).

than rods (Schnapf & McBurney 1980, Cobbs et al 1985). It is possible that the peculiar current-voltage relation of the cone conductance has a deeper significance that has not yet been appreciated.

The characteristic current-voltage relation for the cone conductance, as for the rod conductance, arises mostly from a voltage-dependent block by divalent cations, and in the complete absence of divalent cations the relation becomes much more linear (Haynes & Yau 1985). The difference in the rod and cone current-voltage relations may reflect a subtle difference in the profiles of the ion binding site and energy barriers within the channel. Single-channel currents can be recorded from the cone conductance in the absence of divalent cations (Haynes & Yau 1987). Both large and small channel openings are observed, with unit conductances about twice those found for the rod channel. The individual channel openings in a burst are somewhat briefer than those for the rod channel. The bursts, however, last roughly the same time as in the rod channel, thus suggesting a comparable rate for the unbinding of cGMP. The characteristics of activation by cGMP are also broadly similar, with a half-activating cGMP concentration of around 50 $\mu$m and a Hill coefficient of 1.6–3.0 (Haynes & Yau 1985). Depolarization likewise promotes activation of the channel by cGMP (Haynes & Yau 1985). Finally, $l$-$cis$ diltiazem blocks the cone channel at concentrations similar to those that block the rod channel (Haynes & Yau 1988).

Despite the larger total surface area of the outer segment in cones, the density of channels seems to be proportionally lower (L. W. Haynes and K.-W. Yau, in preparation), so that the total number of channels is probably comparable in both kinds of receptors. As in rods, only about 1% of the conductance seems to be open in darkness (Nakatani & Yau 1986). These facts explain why the normal dark currents in rods and cones are similar (see e.g. Schnapf & McBurney 1980, Cobbs et al 1985).

The details of the light-activated cGMP cascade in cones are still largely unknown, though features such as the involvement of a G-protein and a cGMP phosphodiesterase appear to be common (Nakatani & Yau 1986, Lerea et al 1986, Gillespie & Beavo 1988). There also appears to be a negative feedback on light sensitivity that is mediated by changes in internal $Ca^{2+}$ resulting from $Ca^{2+}$ fluxes through the light-regulated conductance and a $Na^+$–$Ca^{2+}$ exchange (Cobbs & Pugh 1986, Yau & Nakatani 1988). In the absence of this $Ca^{2+}$ feedback, cones do not show any sign of light adaptation (Matthews et al 1988, Nakatani & Yau 1988c).

## Conclusion

Exciting advances have been made in our understanding of the light-regulated conductance of retinal photoreceptor cells in the last few years.

The conductance consists of aqueous pores that open after binding more than one molecule of cGMP. Blockage of the pore by divalent cations gives a very low effective unitary conductance under physiological conditions. The speed of gating by cGMP and the small unitary conductance are beautifully suited for the function of these channels in light detection. Several questions still remain. For example, the molecular mechanics of ion transport through the conductance and its gating by cGMP are still not fully understood. The molecular identity of the channel also needs to be settled in order for structure-function correlation to begin. Other important questions are whether the channel is regulated by phosphorylation or other covalent modifications, and how the expression of the channel in the membrane is regulated. Finally, more information about the spatial localization of channels on the outer segment membrane would be useful.

Recently, recordings from excised patches of plasma membrane from olfactory receptors have revealed a conductance that is also directly activated by cyclic nucleotides (Nakamura & Gold 1986). In these cells, unlike rods and cones, cAMP activates the conductance roughly as well as cGMP. Although its significance is presently unknown, this lack of ligand specificity clearly indicates that the conductance in olfactory receptors is different from that in photoreceptors. It seems probable that the three conductances in visual and olfactory receptor cells are only examples of a general class of ion channels that are gated by cyclic nucleotides.

ACKNOWLEDGMENTS

Supported by grants EY 06837 and EY 01543 from the National Eye Institute, United States Public Health Service. We wish to thank Drs. Lawrence W. Haynes, Jeffrey Karpen, and Markus Meister for comments on the manuscript.

*Literature Cited*

Almers, W., McCleskey, E. W. 1984. Non-selective conductance in calcium channels of frog muscle: calcium selectivity in a single-file pore. *J. Physiol. London* 353: 585–608

Almers, W., McCleskey, E. W., Palade, P. T. 1984. A non-selective cation conductance in frog muscle membrane blocked by micromolar external calcium ions. *J. Physiol. London* 353: 565–83

Applebury, M. L. 1987. Biochemical puzzles about the cyclic GMP-dependent channel. *Nature* 326: 546–47

Attwell, D. 1986. Ionic channels and signal processing in the outer retina. *Q. J. Exp. Physiol.* 71: 497–536

Attwell, D., Gray, P. 1984. Patch-clamp recordings from isolated rods of the salamander retina. *J. Physiol. London* 351: 9P

Bader, C. R., MacLeish, P. R., Schwartz, E. A. 1979. A voltage-clamp study of the light response in solitary rods of the tiger salamander. *J. Physiol. London* 296: 1–26

Bastian, B. L., Fain, G. L. 1982. The effects of sodium replacement on the responses of toad rods. *J. Physiol. London* 330: 331–47

Bauer, P. J. 1988. Evidence of two functionally different membrane fractions in bovine retinal rod outer segments. *J. Physiol. London* 401: 309–27

Baylor, D. A. 1987. Photoreceptor signals and vision. *Invest. Ophthalmol. Vis. Sci.* 28: 34–49

Baylor, D. A. 1988. The light-regulated ionic channel of retinal rod cells. See Lam 1988, pp. 31–40

Baylor, D. A., Fettiplace, R. 1977a. Transmission from photoreceptors to ganglion cells in turtle retina. *J. Physiol. London* 271: 391–424

Baylor, D. A., Fettiplace, R. 1977b. Kinetics of synaptic transfer from receptors to ganglion cells in turtle retina. *J. Physiol. London* 271: 425–48

Baylor, D. A., Fuortes, M. G. F. 1970. Electrical responses of single cones in the retina of the turtle. *J. Physiol. London* 207: 77–92

Baylor, D. A., Lamb, T. D. 1982. Local effects of bleaching in retinal rods of the toad. *J. Physiol. London* 328: 49–71

Baylor, D. A., Lamb, T. D., Yau, K.-W. 1979a. The membrane current of single rod outer segments. *J. Physiol. London* 288: 589–611

Baylor, D. A., Lamb, T. D., Yau, K.-W. 1979b. Responses of retinal rods to single photons. *J. Physiol. London* 288: 613–24

Baylor, D. A., Matthews, G., Yau, K.-W. 1980. Two components of electrical dark noise in toad retinal rod outer segments. *J. Physiol. London* 309: 591–621

Baylor, D. A., Nunn, B. J. 1986. Electrical properties of the light-sensitive conductance of salamander rods. *J. Physiol. London* 371: 115–45

Baylor, D. A., Nunn, B. J., Schnapf, J. L. 1984. The photocurrent, noise and spectral sensitivity of rods of the monkey *Macaca fascicularis*. *J. Physiol. London* 357: 575–607

Bitensky, M. W., Wheeler, G. L., Aloni, B., Ventury, S., Matuo, Y. 1978. Light- and GTP-activated photoreceptor phosphodiesterase: regulation by a light-activated GTPase and identification of rhodopsin as the phosphodiesterase binding site. *Adv. Cyclic Nucleotide Res.* 9: 553–72

Blaustein, M. P., Russell, J. M. 1975. Sodium-calcium exchange and calcium-calcium exchange in internally dialyzed squid giant axons. *J. Membr. Biol.* 22: 285–312

Blazynski, C., Cohen, A. I. 1986. Rapid declines in cyclic GMP of rod outer segments of intact frog photoreceptors after illumination. *J. Biol. Chem.* 261: 14142–47

Bodoia, R. D., Detwiler, P. B. 1985. Patch-clamp recordings of the light-sensitive dark noise in retinal rods from the lizard and frog. *J. Physiol. London* 367: 183–216

Bortoff, A. 1964. Localization of slow potential responses in the Necturus retina. *Vision Res.* 4: 627–35

Brown, J. E., Pinto, L. H. 1974. Ionic mechanism for the photoreceptor potential of the retina of *Bufo marinus*. *J. Physiol. London* 236: 575–91

Byzov, A. L., Trifonov, J. A. 1968. The response to electric stimulation of horizontal cells in the carp retina. *Vision Res.* 8: 817–22

Capovilla, M., Caretta, A., Cervetto, L., Torre, V. 1983. Ionic movements through light-sensitive channels of toad rods. *J. Physiol. London* 343: 295–310

Caretta, A., Cavaggioni, A. 1983. Fast ionic flux activated by cyclic GMP in the membrane of cattle rod outer segments. *Eur. J. Biochem.* 132: 1–8

Caretta, A., Cavaggioni, A., Sorbi, R. T. 1979. Cyclic GMP and the permeability of the disks of the frog photoreceptors. *J. Physiol. London* 295: 171–78

Caretta, A., Cavaggioni, A., Sorbi, R. T. 1985. Binding stoichiometry of a flourescent cGMP analogue to membranes of retinal rod outer segments. *Eur. J. Biochem.* 153: 49–53

Cavaggioni, A., Sorbi, R. T. 1981. Cyclic GMP releases calcium from disc membranes of vertebrate photoreceptors. *Proc. Natl. Acad. Sci. USA* 78: 3964–68

Cervetto, L. 1973. Influence of sodium, potassium and chloride ions on the intracellular responses of turtle photoreceptors. *Science* 24: 401–3

Chabre, M. 1985. Trigger and amplification mechanisms in visual phototransduction. *Ann. Rev. Biophys. Biophys. Chem.* 14: 331–60

Clack, J. W., Stein, P. J. 1988. Opsin exhibits cGMP-activated single channel activity. *Invest. Ophthalmol. Vis. Sci.* 29: 386

Cobbs, W. H., Barkdoll, A. E. III, Pugh, E. N. Jr. 1985. Cyclic GMP increases photocurrent and light sensitivity of retinal cones. *Nature* 317: 64–66

Cobbs, W. H., Pugh, E. N. Jr. 1985. Cyclic GMP can increase rod outer segment light-sensitive current 10-fold without delay of excitation. *Nature* 313: 585–87

Cobbs, W. H., Pugh, E. N. Jr. 1986. Two components of outer segment membrane current in salamander rods and cones. *Biophys. J.* 49: 280a

Cobbs, W. H., Pugh, E. N. Jr. 1987. Kinetics and components of the flash photocurrent of isolated retinal rods of the larval salamander, *Ambystoma tigrinum*. *J. Physiol. London* 394: 529–72

Cohen, A. I. 1970. Further studies on the question of the patency of saccules in outer segments of vertebrate photoreceptors. *Vision Res.* 10: 445–51

Cohen, A. I., Blazynski, C. 1989. Light-induced losses and dark recovering rates of guanosine 3′,5′-cyclic monophosphate in rod outer segments of intact amphibian photoreceptors. *J. Gen. Physiol.* In press

Cohen, A. I., Hall, I. A., Ferrendelli, J. A. 1978. Calcium and cyclic nucleotide regulation in incubated mouse retina. *J. Gen. Physiol.* 71: 595–612

Cook, N. J., Hanke, W., Kaupp, U. B. 1987. Identification, purification, and functional reconstitution of the cyclic GMP-dependent channel from rod photoreceptors. *Proc. Natl. Acad. Sci. USA* 84: 585–89

Cook, N. J., Zeilinger, C., Koch, K.-W., Kaupp, U. B. 1986. Solubilization and functional reconstitution of the cGMP-dependent cation channel from bovine rod outer segments. *J. Biol. Chem.* 261: 17033–39

Cote, R. H., Biernbaum, M. S., Nicol, G. D., Bownds, M. D. 1984. Light-induced decreases in cGMP concentration precede changes in membrane permeability in frog rod photoreceptors. *J. Biol. Chem.* 259: 9635–41

Cote, R. H., Nicol, G. D., Burke, S. A., Bownds, M. D. 1986. Changes in cGMP concentration correlate with some, but not all, aspects of the light-regulated conductance of frog rod photoreceptors. *J. Biol. Chem.* 261: 12965–75

Deterre, P., Bigay, J., Forquet, F., Robert, M., Chabre, M. 1988. cGMP phosphodiesterase of retinal rods is regulated by two inhibitory subunits. *Proc. Natl. Acad. Sci. USA* 85: 2424–28

Detwiler, P. B., Conner, J. D. Bodoia, R. D. 1982. Gigaseal patch-clamp recordings from outer segments of intact retinal rods. *Nature* 300: 59–61

Dowling, J., Ripps, H. 1973. Neurotransmission in the distal retina: the effect of magnesium on horizontal cell activity. *Nature* 242: 101–3

Fain, G. L., Lisman, J. E. 1981. Membrane conductances of photoreceptors. *Prog. Biophys. Mol. Biol.* 37: 91–147

Fesenko, E. E., Kolesnikov, S. S., Lyubarsky, A. L. 1985. Induction by cyclic GMP of cationic conductance in plasma membrane of retinal rod outer segment. *Nature* 313: 310–13

Fesenko, E. E., Kolesnikov, S. S., Lyubarsky, A. L. 1986. Direct action of cGMP on the conductance of retinal rod plasma membrane. *Biochim. Biophys. Acta* 856: 661–71

Fleischman, D., Denisevich, M. 1979. Guanylate cyclase of isolated bovine retinal rod axonemes. *Biochemistry* 18: 5060–66

Furman, R. E., Tanaka, J. C. 1987. Ion selectivity of the cGMP-dependent conductance from photoreceptors. *Biophys. J.* 51: 17a

George, J. A., Hagins, W. A. 1983. Control of $Ca^{2+}$ in rod outer segment disks by light and cyclic GMP. *Nature* 303: 344–48

Gillespie, P. G., Beavo, J. A. 1988. Characterization of a bovine cone photoreceptor phosphodiesterase purified by cyclic GMP-sepharose chromatography. *J. Biol. Chem.* 263: 8133–41

Gold, G. H. 1986. Plasma membrane calcium fluxes in intact rods are inconsistent with the calcium hypothesis. *Proc. Natl. Acad. Sci. USA* 83: 1150–54

Gold, G. H., Korenbrot, J. I. 1980. Light-induced calcium release by intact retinal rods. *Proc. Natl. Acad. Sci. USA* 77: 5557–61

Goldberg, N. D., Ames, A. III, Gander, J. E., Walseth, T. F. 1983. Magnitude of increase in retinal cGMP metabolic flux determined by $^{18}O$ incorporation into nucleotide α-phosphoryls corresponds with intensity of photic stimulation. *J. Biol. Chem.* 258: 9213–19

Goldin, S. M., Pearce, B. L., Calhoun, R. D., Burns, P., Vincent, A. 1987. Rapid kinetic measurement of transient cGMP-stimulated movement of $Ca^{++}$ and $Na^+$ across purified disks of rod outer segments. *Invest. Ophthalmol. Vis. Sci. Suppl.* 28: 95

Gray, P., Attwell, D. 1985. Kinetics of light-sensitive channels in vertebrate photoreceptors. *Proc. R. Soc. London Ser. B* 223: 379–88

Hagins, W. A. 1972. The visual process: excitatory mechanisms in the primary receptor cells. *Ann. Rev. Biophys. Bioeng.* 1: 131–58

Hagins, W. A., Penn, R. D., Yoshikami, S. 1970. Dark current and photocurrent in retinal rods. *Biophys. J.* 10: 380–412

Hanke, W., Cook, N. J., Kaupp, U. B. 1988. cGMP-dependent channel protein from photoreceptor membranes: single-channel activity of the purified and reconstituted protein. *Proc. Natl. Acad. Sci. USA* 85: 94–98

Haynes, L. W., Kay, A. R., Yau, K.-W. 1986. Single cyclic GMP-activated channel activity in excised patches of rod outer segment membrane. *Nature* 321: 66–70

Haynes, L. W., Yau, K.-W. 1985. Cyclic GMP-sensitive conductance in outer segment membrane of catfish cones. *Nature* 317: 61–64

Haynes, L. W., Yau, K.-W. 1987. Single cGMP-activated channel activity recorded from excised cone membrane patches. *Biophys. J.* 51: 18a

Haynes, L. W., Yau, K.-W. 1988. L-cis-diltiazem blocks the cGMP-gated channel in cones. *Soc. Neurosci. Abstr.* 14: 160

Hess, P., Tsien, R. W. 1984. Mechanism of ion permeation through calcium channels. *Nature* 309: 453–56

Hestrin, S., Korenbrot, J. I. 1987. Effects of cyclic GMP on the kinetics of the photocurrent in rods and in detached rod outer segments. *J. Gen. Physiol.* 90: 527–51

Hille, B. 1984. *Ionic Channels of Excitable Membranes.* Sunderland, Mass: Sinauer

Hodgkin, A. L. 1988. Modulation of ionic currents in vertebrate photoreceptors. See Lam 1988, pp. 6–30

Hodgkin, A. L., McNaughton, P. A., Nunn, B. J. 1985. The ionic selectivity and calcium dependence of the light-sensitive pathway in toad rods. *J. Physiol. London* 358: 447–68

Hodgkin, A. L., McNaughton, P. A., Nunn, B. J. 1987. Measurement of sodium-calcium exchange in salamander rods. *J. Physiol. London* 391: 347–70

Hodgkin, A. L., McNaughton, P. A., Nunn, B. J., Yau, K.-W. 1984. Effects of ions on retinal rods from *Bufo marinus. J. Physiol. London* 350: 649–80

Hodgkin, A. L., Nunn, B. J. 1987. The effects of ions on sodium-calcium exchange in salamander rods. *J. Physiol. London* 391: 371–98

Hodgkin, A. L., Nunn, B. J. 1988. Control of light-sensitive current in salamander rods. *J. Physiol. London* 403: 439–71

Hubbell, W. L., Bownds, M. D. 1979. Visual transduction in vertebrate photoreceptors. *Ann. Rev. Neurosci.* 2: 17–34

Jack, J. J. B., Noble, D., Tsien, R. W. 1975. *Electrical Current Flow In Excitable Cells.* Oxford: Clarendon

Kaneko, A. 1970. Physiological and morphological identification of horizontal, bipolar and amacrine cells in goldfish retina. *J. Physiol. London* 207: 623–33

Karpen, J. W., Zimmerman, A. L., Stryer, L., Baylor, D. A. 1988a. Gating kinetics of the cyclic GMP-activated channel of retinal rods: flash photolysis and voltage-jump studies. *Proc. Natl. Acad. Sci. USA* 85: 1287–91

Karpen, J. W., Zimmerman, A. L., Stryer, L., Baylor, D. A. 1988b. Molecular mechanics of the cyclic GMP-activated channel of retinal rods. *Cold Spring Harbor Symp. Quant. Biol.* 53: In press

Kilbride, P. 1980. Calcium effects on frog retinal cyclic guanosine 3′:5′-monophosphate levels and their light-initiated rate of decay. *J. Gen. Physiol.* 75: 457–65

Kilbride, P., Ebrey, T. G. 1979. Light-initiated changes of cyclic guanosine monophosphate levels in the frog retina

measured with quick-freezing techniques. *J. Gen. Physiol.* 74: 415–26

Koch, K.-W., Cook, N. J., Kaupp, U. B. 1987. The cGMP-dependent channel of vertebrate rod photoreceptors exists in two forms of different cGMP sensitivity and pharmacological behavior. *J. Biol. Chem.* 262: 14415–21

Koch, K.-W., Kaupp, U. B. 1985. Cyclic GMP directly regulates a cationic conductance in membranes of bovine rods by a cooperative mechanism. *J. Biol. Chem.* 260: 6788–6800

Koch, W.-H., Stryer, L. 1988. Highly cooperative feedback control of retinal rod guanylate cyclase by calcium ions. *Nature* 334: 64–66

Kolesnikov, S. S., Lyubarsky, A. L., Fesenko, E. E. 1984. Single anion channels of frog rod plasma membrane. *Vision Res.* 24: 1295–1300

Kondo, H., Miller, W. H. 1988. Rod light adaptation may be mediated by acceleration of the phosphodiesterase-guanylate cyclase cycle. *Proc. Natl. Acad. Sci. USA* 85: 1322–26

Korenbrot, J. I., Cone, R. A. 1972. Dark ionic flux and the effects of light in isolated rod outer segments. *J. Gen. Physiol.* 60: 20–45

Korenbrot, J. I., Miller, D. L. 1986. Calcium ions act as modulator of intracellular information flow in retinal rod phototransduction. *Neurosci. Res. Suppl.* 4: S11–S34

Kostyuk, J. B., Mironov, S. L., Shuba, Y. M. 1983. Two ion-selecting filters in the calcium channel of the somatic membrane of Mollusc neurons. *J. Membr. Biol.* 76: 83–93

Koutalos, Y., Ebrey, T. G. 1986. Recent progress in vertebrate photoreception. *Photochem. Photobiol.* 44: 809–17

Krishnan, N., Fletcher, R. T., Chader, G. J., Krishna, G. 1978. Characterization of guanylate cyclase of rod outer segments of the bovine retina. *Biochim. Biophys. Acta* 523: 506–15

Kühn, H. 1984. Interactions between photoexcited rhodopsin and light-activated enzymes in rods. *Prog. Retinal Res.* 3: 123–56

Lagnado, L., Cervetto, L., McNaughton, P. A. 1988. Ion transport by the Na:Ca exchange in isolated rod outer segments. *Proc. Natl. Acad. Sci. USA* 85: 4548–52

Lam, D. M. K., ed. 1988. *Proceedings of the Retina Research Foundation Symposium,* Vol. 1. The Woodlands, Texas: Portfolio

Lamb, T. D. 1986. Transduction in vertebrate photoreceptors: the roles of cyclic GMP and calcium. *Trends Neurosci.* 9: 224–28

Lamb, T. D., McNaughton, P. A., Yau, K.-W. 1981. Spatial spread of activation and background desensitization in toad rod outer segments. *J. Physiol. London* 319: 463–86

Lamb, T. D., Matthews, H. R. 1988. External and internal actions on the response of salamander retinal rods to altered external calcium concentration. *J. Physiol. London* 403: 473–94

Lamb, T. D., Matthews, H. R., Torre, V. 1986. Incorporation of calcium buffers into salamander retinal rods: a rejection of the calcium hypothesis of phototransduction. *J. Physiol. London* 372: 315–49

Lamb, T. D., Simon, E. J. 1976. The relation between intercellular coupling and electrical noise in turtle photoreceptors. *J. Physiol. London* 263: 257–86

Lamb, T. D., Simon, E. J. 1977. Analysis of electrical noise in turtle cones. *J. Physiol. London* 272: 435–68

Lansman, J. B., Hess, P., Tsien, R. W. 1985. Direct measurement of entry and exit rates for calcium ions in single calcium channels. *Biophys. J.* 47: 67a

Lee, K., Tsien, R. W. 1983. Mechanism of calcium channel blockade by verapamil, D600, diltiazem and nitrendipine in single dialysed heart cells. *Nature* 302: 790–94

Lerea, C. L., Somers, D. E., Hurley, J. B., Klock, I. B., Milam, A. H. 1986. Identification of specific transducin α subunits in retinal rod and cone photoreceptors. *Science* 234: 77–80

✓Liebman, P. A., Parker, K. R., Dratz, E. A. 1987. The molecular mechanism of visual excitation and its relation to the structure and composition of the rod outer segment. *Ann. Rev. Physiol.* 49: 765–91

Lipton, S. A., Rasmussen, H., Dowling, J. E. 1977. Electrical and adaptive properties of rod photoreceptors in *Bufo marinus*. II. Effects of cyclic nucleotides and prostaglandins. *J. Gen. Physiol.* 70: 771–91

Lolley, R. N., Racz, E. 1982. Calcium modulation of cyclic GMP synthesis in rat visual cells. *Vision Res.* 22: 1481–86

MacLeish, P. R., Schwartz, E. A., Tachibana, M. 1984. Control of the generator current in solitary rods of the *Ambystoma tigrinum* retina. *J. Physiol. London* 348: 645–64

Matesic, D., Liebman, P. A. 1987. cGMP-dependent cation channel of retinal rod outer segments. *Nature* 326: 600–3

Matthews, G. 1986a. Comparison of the light-sensitive and cyclic GMP-sensitive conductances of the rod photoreceptor: noise characteristics. *J. Neurosci.* 6: 2521–26

Matthews, G. 1986b. Spread of the light response along the rod outer segment: an estimate from patch-clamp recordings. *Vision Res.* 26: 535–41

Matthews, G. 1987. Single-channel recordings demonstrate that cGMP opens the light-sensitive ion channel of the rod photoreceptor. *Proc. Natl. Acad. Sci. USA* 84: 299–302

Matthews, G., Watanabe, S.-I. 1987. Properties of ion channels closed by light and opened by guanosine 3′,5′-cyclic monophosphate in toad retinal rods. *J. Physiol. London* 389: 691–716

Matthews, G., Watanabe, S.-I. 1988. Activation of single ion channels from toad retinal rod inner segments by cyclic GMP: Concentration dependence. *J. Physiol.* 403: 389–405

Matthews, H. R., Murphy, R. L. W., Fain, G. L., Lamb, T. D. 1988. Photoreceptor light adaptation is mediated by cytoplasmic calcium concentration. *Nature* 334: 67–69

Matthews, H. R., Torre, V., Lamb, T. D. 1985. Effects on the photoresponse of calcium buffers and cyclic GMP incorporated into the cytoplasm of retinal rods. *Nature* 313: 582–85

McNaughton, P. A. 1988. The sodium-calcium exchange in photoreceptors. In *First International Meeting On Sodium-Calcium Exchange*, ed. T. J. A. Allen. In press

McNaughton, P. A., Cervetto, L., Nunn, B. J. 1986. Measurement of the intracellular free calcium concentration in salamander rods. *Nature* 322: 261–63

Miller, D. L., Korenbrot, J. I. 1987. Kinetics of light-dependent Ca fluxes across the plasma membrane of rod outer segments. A dynamic model of the regeneration of cytoplasmic Ca concentration. *J. Gen. Physiol.* 90: 397–426

Miller, W. H., ed. 1981. *Molecular Mechanisms of Photoreceptor Transduction. Curr. Top. Membr. Transp.* Vol. 15. New York: Academic

Miller, W. H., Nicol, G. D. 1979. Evidence that cyclic GMP regulates membrane potential in rod photoreceptors. *Nature* 280: 64–66

Molday, L. L., Molday, R. S. 1987a. Glycoproteins specific for the retinal rod outer segment plasma membrane. *Biochim. Biophys. Acta* 897: 335–40

Molday, R. S., Molday, L. L. 1987b. Differences in the protein composition of bovine retinal rod outer segment disk and plasma membranes isolated by a ricin-gold-dextran density perturbation method. *J. Cell Biol.* 105: 2589–2601

Nakamura, T., Gold, G. H. 1987. A cyclic nucleotide-gated conductance in olfactory receptor cilia. *Nature* 325: 442–44

Nakatani, K., Yau, K.-W. 1985. cGMP opens the light-sensitive conductance in retinal rods. *Biophys. J.* 47: 356a

Nakatani, K., Yau, K.-W. 1986. Light-suppressible, cyclic GMP-activated current recorded from a dialyzed cone preparation. *Invest. Ophthalmol. Vis. Res. Suppl.* 27: 300

Nakatani, K., Yau, K.-W. 1988a. Calcium and magnesium fluxes across the plasma membrane of the toad rod outer segment. *J. Physiol. London* 395: 695–729

Nakatani, K., Yau, K.-W. 1988b. Guanosine 3′ : 5′-cyclic monophosphate-activated conductance studied in a truncated rod outer segment of the toad. *J. Physiol. London* 395: 731–53

Nakatani, K., Yau, K.-W. 1988c. Calcium and light adaptation in retinal rods and cones. *Nature* 334: 69–71

Nerbonne, J. M., Richard, S., Nargeot, J., Lester, H. A. 1984. New photoactivatable cyclic nucleotides produce intracellular jumps in cyclic AMP and cyclic GMP concentrations. *Nature* 310: 74–76

Nicol, G. D., Kaupp, U. B., Bownds, M. D. 1987a. Transduction persists in rod photoreceptors after depletion of intracellular calcium. *J. Gen. Physiol.* 89: 297–319

Nicol, G. D., Schnetkamp, P. P. M., Saimi, Y., Cragoe, E. J. Jr., Bownds, M. D. 1987b. A derivative of amiloride blocks both the light- and cyclic GMP-regulated conductances in rod photoreceptors. *J. Gen. Physiol.* 90: 651–70

Nöll, G., Neher, E., Baumann, C. 1982. Patch electrode recording of currents from the plasma membrane of rod outer segments of the frog. *Pflügers Arch.* 394: Suppl. R45

Owen, W. G. 1987. Ionic conductances in rod photoreceptors. *Ann. Rev. Physiol.* 49: 743–64

Penn, R. D., Hagins, W. A. 1972. Kinetics of the photocurrent of retinal rods. *Biophys. J.* 12: 1073–94

Pepe, I. M., Panfoli, I., Cugnoli, C. 1986. Guanylate-cyclase in rod outer segments of the toad retina. *FEBS Lett.* 203: 73–76

Puckett, K. L., Goldin, S. M. 1986. Guanosine 3′,5′-cyclic monophosphate stimulates release of actively accumulated calcium in purified disks from rod outer segments of bovine retina. *Biochemistry* 25: 1739–46

Pugh, E. N. Jr., Cobbs, W. H. 1986. Visual transduction in vertebrate rods and cones: A tale of two transmitters, calcium and cyclic GMP. *Vision Res.* 26: 1613–43

Quandt, F. N., Nicol, G. D., Schnetkamp, P. P. M. 1988. Voltage-dependency of gating of cyclic GMP-activated ion channels

in bovine rod outer segments and block by l-cis diltiazem. *Biophys. J.* 53: 390a

Ratto, G. M., Payne, R., Owen, W. G., Tsien, R. Y. 1988. The concentration of cytosolic free calcium in vertebrate rod outer segments measured with Fura-2. *J. Neurosci.* 8(9): 3240–46

Rispoli, G., Sather, W. A., Detwiler, P. B. 1988. Effect of triphosphate nucleotides on the response of detached rod outer segments to low external calcium. *Biophys. J.* 53: 388a

Robinson, P. R., Kawamura, S., Abramson, B., Bownds, M. D. 1980. Control of the cyclic GMP phosphodiesterase of frog photoreceptor membranes. *J. Gen. Physiol.* 76: 631–45

Rodieck, R. W. 1973. *The Vertebrate Retina: Principles of Structure and Function.* San Francisco: Freeman

Sather, W. A., Bodoia, R. D., Detwiler, P. B. 1985. Does the plasma membrane of the rod outer segment contain more than one type of ion channel? *Neurosci. Res. Suppl.* 2: S89–S99

Sather, W. A., Detwiler, P. B. 1987. Intracellular biochemical manipulation of phototransduction in detached rod outer segments. *Proc. Natl. Acad. Sci. USA* 84: 9290–94

Schnapf, J. L. 1983. Dependence of the single photon response on longitudinal position of absorption in toad rod outer segments. *J. Physiol. London* 343: 147–59

Schnapf, J. L., McBurney, R. N. 1980. Light-induced changes in membrane current in cone outer segments of tiger salamander and turtle. *Nature* 287: 239–41

Schnetkamp, P. P. M. 1986. Sodium-calcium exchange in the outer segments of bovine rod photoreceptors. *J. Physiol. London* 373: 25–45

Schnetkamp, P. P. M. 1987. Sodium ions selectively eliminate the fast component of guanosine cyclic 3′,5′-phosphate induced $Ca^{++}$ release from bovine rod outer segment disks. *Biochemistry* 26: 3249–53

Schnetkamp, P. P. M., Bownds, M. D. 1987. $Na^+$- and cGMP-induced $Ca^{2+}$ fluxes in frog rod photoreceptors. *J. Gen. Physiol.* 89: 481–500

Schröder, W. H., Fain, G. L. 1984. Light-dependent release of Ca from rods measured by laser micro mass analysis. *Biophys. J.* 45: 341a

Schwartz, E. A. 1977. Voltage noise observed in rods of the turtle retina. *J. Physiol. London* 272: 217–46

Shinozawa, T., Sokabe, M., Terada, S., Matsusaka, H., Yoshizawa, T. 1987. Detection of cyclic GMP binding protein and ion channel activity in frog rod outer segments. *J. Biochem.* 102: 281–90

Sillman, A. J., Ito, H., Tomita, T. 1969. Studies on the mass receptor potential of the isolated frog retina II. On the basis of the ionic mechanism. *Vision Res.* 9: 1443–51

Spencer, M., Detwiler, P. B., Bunt-Milam, A. H. 1988. Distribution of membrane proteins in mechanically dissociated retinal rods. *Invest. Ophthalmol. Vis. Sci.* 29: 1012–20

Stern, J. H., Kaupp, U. B., MacLeish, P. R. 1986. Control of the light-regulated current in rod photoreceptors by cyclic GMP, calcium and l-cis-diltiazem. *Proc. Natl. Acad. Sci. USA* 83: 1163–67

Stern, J. H., Knutsson, H., MacLeish, P. R. 1987. Divalent cations directly affect the conductance of excised patches of rod photoreceptor membrane. *Science* 236: 1674–78

Stieve, H., ed. 1986. *The Molecular Mechanism of Phototransduction.* Berlin: Dahlem Konferenzen/Springer-Verlag

Stryer, L. 1986. Cyclic GMP cascade of vision. *Ann. Rev. Neurosci.* 9: 87–119

Stryer, L. 1987. Visual transduction: design and recurring motifs. *Chem. Scr.* 27B: 161–71

Tanaka, J. C., Furman, R. E., Cobbs, W. H., Mueller, P. 1987. Incorporation of a retinal rod cGMP-dependent conductance into planar bilayers. *Proc. Natl. Acad. Sci. USA* 84: 724–28

Tomita, T. 1965. Electrophysiological study of the mechanisms subserving color coding in the fish retina. *Cold Spring Harbor Symp. Quant. Biol.* 30: 559–66

Torre, V. 1982. The contribution of the electrogenic sodium-potassium pump to the electrical activity of toad rods. *J. Physiol. London* 333: 315–41

Torre, V., Matthews, H. R., Lamb, T. D. 1986. Role of calcium in regulating the cyclic GMP cascade of phototransduction in retinal rods. *Proc. Natl. Acad. Sci. USA* 83: 7109–13

Torre, V., Rispoli, G., Menini, A., Cervetto, L. 1987. Ionic selectivity, blockage and control of light-sensitive conductance. *Neurosci. Res. Suppl.* 6: S25–S44

Townes-Anderson, E., MacLeish, P. R., Raviola, E. 1985. Rod cells dissociated from mature salamander retina: Ultrastructure and uptake of horseradish peroxidase. *J. Cell Biol.* 100: 175–88

Toyoda, J., Nosaki, H., Tomita, T. 1969. Light-induced resistance changes in single photoreceptors of *Necturus* and *Gekko*. *Vision Res.* 9: 453–63

Troyer, E. W., Hall, I. A., Ferrendelli, J. A. 1978. Guanylate cyclase in CNS: Enzymatic characteristics of soluble and particulate enzymes from mouse cerebellum and retina. *J. Neurochem.* 31: 825–33

Vuong, T. M., Chabre, M., Stryer, L. 1984. Millisecond activation of transducin in the cyclic nucleotide cascade of vision. *Nature* 311: 659–61

Watanabe, S. I., Matthews, G. 1988. Regional distribution of cGMP-activated ion channels in the plasma membrane of the rod photoreceptor. *J. Neurosci.* 8(7): 2334–37

Werblin, F. S. 1975. Regenerative hyperpolarization in rods. *J. Physiol. London* 244: 53–81

Werblin, F. S., Dowling, J. E. 1969. Organization of the retina of the mudpuppy, *Necturus maculosus*. II. Intracellular recording. *J. Neurophysiol.* 32: 339–55

Woodruff, M. L., Bownds, M. D. 1979. Amplitude, kinetics and reversibility of a light-induced decrease in guanosine 3′ : 5′-cyclic monophosphate in frog photoreceptor membranes. *J. Gen. Physiol.* 73: 629–53

Woodruff, M. L., Fain, G. L. 1982. $Ca^{++}$-dependent changes in cyclic GMP levels are not correlated with opening and closing of the light-dependent permeability of toad photoreceptors. *J. Gen. Physiol.* 80: 537–55

Woodruff, M. L., Fain, G. L., Bastian, B. L. 1982. Light-dependent ion influx into toad photoreceptors. *J. Gen. Physiol.* 80: 517–36

Yamazaki, A., Sen, I., Bitensky, M. W., Casnellie, J. E., Greengard, P. 1980. Cyclic GMP-specific, high affinity, non-catalytic binding sites on light-activated phosphodiesterase. *J. Biol. Chem.* 255: 11619–24

Yau, K.-W., Haynes, L. W. 1986. Effect of divalent cations on the macroscopic cGMP-activated current in excised rod membrane patches. *Biophys. J.* 49: 33a

Yau, K.-W., Haynes, L. W., Nakatani, K. 1986. Roles of calcium and cyclic GMP in visual transduction. In *Membrane Control of Cellular Activity*, ed. H. Ch. Lüttgau, pp. 343–66. Stuttgart: Gustav Fischer

Yau, K.-W., Haynes, L. W., Nakatani, K. 1988. Study of the phototransduction mechanism in rods and cones. See Lam 1988, pp. 41–58

Yau, K.-W., McNaughton, P. A., Hodgkin, A. L. 1981. Effects of ions on the light-sensitive current in retinal rods. *Nature* 292: 502–5

Yau, K.-W., Nakatani, K. 1984a. Cation selectivity of light-sensitive conductance in retinal rods. *Nature* 309: 352–54

Yau, K.-W., Nakatani, K. 1984b. Electrogenic Na-Ca exchange in retinal rod outer segment. *Nature* 311: 661–63

Yau, K.-W., Nakatani, K. 1985a. Light-induced reduction of cytoplasmic free calcium in retinal rod outer segment. *Nature* 313: 579–82

Yau, K.-W., Nakatani, K. 1985b. Light-suppressible, cyclic GMP-sensitive conductance in the plasma membrane of a truncated rod outer segment. *Nature* 317: 252–55

Yau, K.-W., Nakatani, K. 1985c. Study of the ionic basis of visual transduction in vertebrate retinal rods. In *Contemporary Sensory Neurobiology*, ed. M. J. Correia, A. A. Perachio, pp. 245–56. New York: Liss

Yau, K.-W., Nakatani, K. 1988. Sodium-dependent calcium efflux at the outer segment of the retinal cone. *Biophys. J.* 53: 473a

Yoshikami, S., George, J. S., Hagins, W. A. 1980. Light-induced calcium fluxes from outer segment layer of vertebrate retinas. *Nature* 286: 395–98

Yoshikami, S., Hagins, W. A. 1973. Control of the dark current in vertebrate rods and cones. In *Biochemistry and Physiology of Visual Pigments*, ed. H. Langer, pp. 245–55. Berlin: Springer-Verlag

Yuen, P. S. T., Walseth, T. F., Panter, S. S., Sundby, S. R., Goldberg, N. D.

1986. Identification of rod outer segment cGMP binding proteins by direct photoaffinity labelling. *Biophys. J.* 49: 278a

Zimmerman, A. L., Baylor, D. A. 1985. Electrical properties of the light-sensitive conductance of salamander retinal rods. *Biophys. J.* 47: 357a

Zimmerman, A. L., Baylor, D. A. 1986a. Gating and conduction in the light-sensitive channels of retinal rods. *Biophys. J.* 49: 408a

Zimmerman, A. L., Baylor, D. A. 1986b. Cyclic GMP-sensitive conductance of retinal rods consists of aqueous pores. *Nature* 321: 70–72

Zimmerman, A. L., Baylor, D. A. 1988. Ionic permeation in the cGMP-activated channel of retinal rods. *Biophys. J.* 53: 472a

Zimmerman, A. L., Karpen, J. W., Baylor, D. A. 1988. Hindered diffusion in excised patches from retinal rod outer segments. *Biophys. J.* 54: 351–55

Zimmerman, A. L., Yamanaka, G., Eckstein, F., Baylor, D. A., Stryer, L. 1985. Interaction of hydrolysis-resistant analogs of cyclic GMP with the phosphodiesterase and light-sensitive channel of retinal rod outer segments. *Proc. Natl. Acad. Sci. USA* 82: 8813–17

*Ann. Rev. Neurosci. 1989. 12:329–53*

# THE CELL BIOLOGY OF VERTEBRATE TASTE RECEPTORS

*Stephen D. Roper*

Department of Anatomy and Neurobiology, Colorado State University, Fort Collins, Colorado 80523 and the Rocky Mountain Taste and Smell Center, University of Colorado Health Sciences Center, Denver, Colorado 80262

## Introduction

The study of taste cells has lagged far behind investigations on the biology of receptor cells in other sensory modalities such as auditory reception, mechanoreception, and photoreception. In large part this is due to the relative inaccessibility of taste cells and their intractibility for detailed studies: In most vertebrate species, taste cells are small cells embedded in a tough, waterproof epithelium. These features render them difficult to impale with intracellular microelectrodes and difficult to preserve with conventional light and electron microscopic fixation protocols.

Nonetheless, the mechanism of taste transduction is important in neurobiology, particularly sensory neurobiology. This is because, in one sense, taste receptor cells represent a model for membrane chemosensitivity in general, whether to sapid stimuli, to hormones, to neurotransmitters, or to changes in the ionic milieu. The cellular mechanisms whereby chemical stimuli are converted into transmembrane voltage changes in taste cells are likely to have much in common with and shed light upon chemosensitivity at other sites in the nervous system. In another sense, taste cells are large, chemosensitive presynaptic terminals: The chemosensitivity is primarily at one end of the cell and the synaptic release sites at the opposite end, but a single intracellular electrode in the cell interior monitors both regions effectively. It may be possible to study aspects of presynaptic function by investigating taste cells. Lastly, taste receptors represent a renewing population of neuroepithelial cells that have a life span of only a few days. Within this short life span, the receptor cells differentiate,

329

0147–006X/89/0301–0329$02.00

acquire many features similar to those of neurons (such as specialized ion channels and membrane receptors), form synapses and communicate with sensory afferents leading to the CNS, lose their synapses, and die. This presents a unique opportunity to study a dynamic population of excitable cells and to investigate aspects of neuronal development such as synaptogenesis or age-related changes in chemosensitivity.

New electrophysiological techniques and the use of experimental species in which taste cells are unusually large and accessible have combined to result in major advances in the cell biology of taste. Within a few short years our understanding of taste reception will likely approach that of auditory reception, mechanoreception, and photoreception.

## A Heuristic Model

This review is centered around a new heuristic model for the vertebrate taste bud. At the outset it should be clarified that this model is a working hypothesis and is speculative. However, lines of evidence gathered throughout the past several decades and from recent years support numerous facets of the model, as becomes clear below. Furthermore, the model provides a conceptual framework to help guide the reader, whether a researcher in the chemical senses or a newcomer to the field of taste.

Vertebrate taste buds are focal collections of differentiated chemosensory receptor cells embedded in a multilaminar, or stratified, squamous epithelium. Each taste bud can be envisioned as a small microprocessor (Figure 1). It consists of an array of approximately 100 chemosensory receptor cells that are interconnected in specific ways by chemical and electrical synapses. A good deal of information processing and sensory integration occurs in the taste bud, itself. For example, the responses of a taste receptor cell can be modified by activity in other taste cells in the taste bud. The integrated output of the taste bud is transmitted to the brain via sensory afferent axons. Each sensory afferent axon ramifies within the taste bud to contact a number of individual taste cells and in a specific manner: The receptor cells that contact a single sensory afferent axon form a "sensory unit," analogous to a motor unit in neuromuscular innervation. The coordinated output of the sensory afferent axons reaches higher centers in the nervous system where the information is decoded into the perception of taste, for example, in the case of humans: sweet, sour, bitter, and salty. Some investigators include a fifth category of taste perception, umami (Kawamura & Kare 1987).

Evidence for this hypothetical model, including an analysis of cell types, cell function, synapses, cell lineage, and chemosensory transduction, are presented below, in that order.

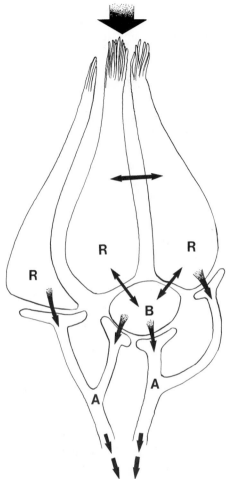

*Figure 1*   Schematic diagram of the vertebrate taste bud, illustrating the proposed integrative mechanisms. *Arrow at top* represents the chemical stimulus. *Small arrows* indicate synaptic interactions among receptor cells (R), basal cells (B), and afferent axons (A). *Small arrows at bottom* represent impulse activity in afferent axons leading to the CNS.

## Taste Cells—a Varied Population

Early light microscopists categorized many different types of cells in taste buds from a variety of species. Most cytologists, beginning with Schwalbe (1867), Loven (1868), Merkel (1880), Ranvier (1888), Hermann (1889), and others, differentiated between chemoreceptor cells proper and a variety of putative nongustatory, supportive cells, such as pier cells, rod cells, roof cells, tegmental cells, and so forth. The distinction of sensory receptor cells from nonsensory cells was based solely on the observation that some cells possessed fine apical "hairs" and tended to be located more centrally within the taste bud. Synaptic relationships with nerve terminals could not, of course, be resolved with the techniques available to these early microscopists, although afferent fibers were seen in abundance throughout the taste bud and came within close apposition to the taste cells. (For descriptions in English of these early findings, cf. Frey 1880, Satterthwaite 1881, Schafer 1885, Bohm et al 1904, Bailey 1906, Ramon y Cajal 1933.)

Closer inspection of taste cells in the last three decades with the electron microscope has revealed that no apical "hairs" exist on any of the taste cells. In rare cases, a taste cell possesses an apical protrusion resembling a stereocilium (Cummings et al 1987, cf. Murray 1986), but the incidence is too low to correspond to the taste "hairs" described by the cytologists. Apical "hairs" were artifacts of preservation and probably represented coagulated microvilli and mucus (de Lorenzo 1958, 1963, Murray & Murray 1960).

Rather extensive catalogs of cell morphologies based upon ultra-structural features of taste cells have been generated to replace the former cytological nomenclature. This has been accompanied by speculation on different functions for these classes of cells. At present, more than four cell "types" have been described, which in many cases do not necessarily correspond to or clarify the distinctions made by the earlier microscopists. The cell types that are currently favored are not, for the most part, well-delineated by unambiguous ultrastructural characteristics. It is often difficult, if not impossible, to categorize many taste cells unequivocally and to draw sharp boundaries among the various categories of cells. Indeed, many investigators interpret the different morphologies as *transitional* forms of but a single cell type, the chemosensory receptor cell.

Furthermore, many electron microscopic studies to date have relied upon random thin sections of taste buds. This has made it difficult to examine the full extent of taste cells and the ultrastructural variations that occur along the length of a single cell. Consequently, some of the discrepancies and controversies in electron microscopic studies may have arisen from investigators describing different cellular regions. Nonethe-less, the essential features of taste cells are remarkably consistent from animal to animal within the vertebrate kingdom,[1] thus suggesting an im-

---

[1] A noteworthy exception to this generalization is the frog, where gustatory chemosensory structures are organized into a rather unique structure, the taste disc (Graziadei & DeHan 1971, During & Andres 1976). The ultrastructure of the component cells is sufficiently different from most other vertebrates, including fish, mammals, and even other amphibia, to exclude this species from the general descriptions of cell types provided in this review.

portant similarity among species regarding chemosensory transduction mechanisms.[2]

Notwithstanding the caveat about distinguishing actual "types" of taste cells (below), the commonly employed descriptions of cell types, as revealed by electron microscopy, are reviewed below. Classes other than these have been described in the literature, but only the principal categories, recognized by most workers in the field, are included here. A summary of this terminology is provided in Table 1.

TYPE I CELLS    Type I (also termed dark cells) are long and narrow and extend from the base of the taste bud to its apex, the taste pore. It is universally accepted that Type I cells are the predominant cells in vertebrate taste buds, comprising 55–75% of the total population (e.g. Murray 1973, Delay et al 1986). The most characteristic feature of this cell, which distinguishes it from the other cell types, is the presence of membrane-bound granules in the apical cytoplasm. These granules may be secreted and contribute to the amorphous substance found in the taste pore (reviewed by Murray 1973, Whitear 1976, Cummings et al 1987). Other ultrastructural features that investigators have ascribed to Type I cells, listed roughly in order of their reliability as diagnostic elements, include electron-dense cytoplasm; deeply-indented, irregularly shaped nuclei; numerous intermediate filaments; abundant rough endoplasmic reticulum; and lysosomes. In *Necturus*, Type I cells possess characteristically long, branched microvilli that reliably set them apart from other taste cells (Cummings et al 1987).

TYPE II CELLS    Light cells, or Type II cells, are also elongated cells that extend from the base of the taste bud to the taste pore. These cells represent a smaller fraction of cells in the taste bud than Type I cells. Investigators have described the following distinctive features of this cell: electron-lucent cytoplasm; large, oval nucleus; presence of smooth endoplasmic reticulum; scant rough endoplasmic reticulum; sparse cytofilaments. In *Necturus*, Type II cells possess regular, short microvilli and, occasionally, a stereocilium (Cummings et al 1987).

TYPE III CELLS    Type III taste cells extend from the base of the taste bud to the taste pore as do Types I and II cells. Morphologically, these cells are similar to the Type II cell, but with the addition of numerous dense-cored vesicles in the cytoplasm, particularly in the basal portion of the

---

[2] Reutter (1978), in a careful and comprehensive study of the ultrastructure of taste organs in the catfish, stresses *distinctions* between taste cells in mammals compared with other vertebrates. He concludes that there are fundamental differences in the structure and function of taste buds in mammalian versus nonmammalian species.

cell. These vesicles perhaps represent monoamine storage sites (*vide infra*). Not all investigators agree upon the separate identity or existence of this cell. In many species, this cell type has not been observed (cf. Table 1). Other investigators claim that this cell is the primary, if not exclusive, chemoreceptor cell on the basis that synaptic contacts with afferent axons

**Table 1**  Descriptions of the ultrastructure of taste cells and their synapses from a variety of vertebrate species, taken from a small sample of the extant literature

| Animal | Synapses | Taste cell type(s) | Reference |
|--------|----------|--------------------|-----------|
| Frog | ●[a] | Taste cells, supporting cells | Uga & Hama (1967) |
|  | ● | Sensory cells, associate cells | Graziadei & DeHan (1971) DeHan & Graziadei (1973) |
| Mudpuppy |  | Light cells, dark cells, basal cells | Farbman & Yonkers (1971) |
|  | ● ● ● | Light cells, dark cells, basal cells | Delay & Roper (1988) |
| Fish | ● ● | Sensory cells, basal cells, supporting cells | Uga & Hama (1967) |
|  | ● ● ● | Light cells, dark cells, basal cells | Reutter (1971, 1978) |
|  | ○ ○ | Light cells, dark cells, intermediate cells, basal cells | Ezeasor (1982) |
|  | ● ● | Receptor cells, basal cells, supporting cells | Toyoshima et al (1984) |
| Turtle | ● ○ | Type 1 (dark) cells, type 2 (light) cells, type 3 (transition) cells, basal cell A, basal cell B | Korte (1980) |
| Mouse | ● | Type I cells, type II cells, type III cells | Takeda et al (1985) |
|  | ● | Dark cells, | Kinnamon et al (1985) |

**Table 1**  *continued*

| Animal | Synapses | Taste cell type(s) | Reference |
|---|---|---|---|
| | ● | light cells, | |
| | ● | intermediate cells | |
| Rat | | Type I (dark) cells, | Farbman (1965) |
| | | type II (light) cells, | |
| | ○ | basal cells, | |
| | | peripheral cells, | |
| | ● | Taste cells | Graziadei (1970) |
| | | Sensory cells, | Chan & Byers (1985) |
| | | supporting cells, | |
| | | basal cells | |
| | | Light cells, | Settembrini (1987) |
| | ● | dark cells, | |
| | | perigemmal cells | |
| Hamster | ○ | Light cells, | Miller & Chaudhry (1976) |
| | | dark cells, | |
| | | basal cells | |
| Rabbit | ● | Gustatory cells, | de Lorenzo (1958) |
| | | supporting cells | |
| | ○ | Receptor cells | de Lorenzo (1963) |
| | | Type I (dark) cells, | Murray et al (1969) |
| | | type II (light) cells, | |
| | ● | type III cells, | |
| | | Type I (dark) cells, | Murray (1973) |
| | | type II (light) cells, | |
| | ● | type III cells, | |
| | | type IV (basal) cells | |
| Monkey | | Gustatory cells | Murray & Murray (1960) |
| | ● | Type III cells | Murray (1973) |
| | ● | Sensory cells, | Zahm & Munger (1983) |
| | ● | basal cells, | |
| | | supporting cells | |
| | | Type I (dark) cells, | Farbman et al (1985) |
| | | type II (light) cells, | |
| | ● | type III cells | |
| Human | ● | Taste cells | Graziadei (1970) |
| | | Type I (light) cells, | Paran et al (1975) |
| | ○ | type II (dark) cells, | |
| | ○ | type III cells | |
| | ● | Taste cells | Arvidson et al (1981) |

[a] ● indicates that synapses were identified on these cells; ○ indicates that the authors observed "synaptic-like" structures but did not identify them as *bona fide* synapses.

are found on the Type III cell in some animals (reviewed by Murray 1986). However, as is discussed below, a number of reports reveal that synaptic contacts are found on all taste cell types, thereby challenging the uniqueness of the Type III cell. Type III cells form a very small proportion (5–15%) of cells in taste buds in which they do occur (Murray 1973, 1986).

BASAL CELLS    Basal cells are flattened or oblate cells that lie at the base of the taste bud on the basal lamina that separates the epithelium from the subjacent lamina propria. Basal cells do not send elongate processes to the taste pore. The most distinctive features of basal cells, other than their position within the taste bud, listed in order of their reliability as a descriptor, are presence of electron-lucent and dense-core vesicles; intermediate filaments associated with the nuclear envelope; and rough endoplasmic reticulum. In some species, the basal cells possess short (1/4 $\mu$ long) cytoplasmic spines that invaginate adjacent cells (Ezeasor 1982, Delay & Roper 1988).

Many investigators state that the basal cell is a *stem cell* or *precursor* to the other cellular elements of the taste bud (Reutter 1978, Delay et al 1986). The close resemblance of basal cells to Merkel cells in some animals (e.g. mudpuppies, fish) has led to the suggestion that the basal cell, like the Merkel cell, is a *mechanoreceptor* (Reutter 1971). An additional function of the basal cell, discussed below, is that of an *interneuron* in the taste bud. The precise role(s) that the basal cell may play in taste reception is not yet established.

The multiplicity of taste cell types underscores the lack of well-defined, unequivocal categories. The most parsimonious explanation is that the ambiguity of cell types represents the presence of transitional forms, i.e. that there is a single, broad spectrum of morphologies reflecting different stages of the lifespan of a single population of taste cells, the sensory chemoreceptor cells. This is described in greater detail below.

## Chemical Synapses in the Taste Bud

Although synaptic transmission undoubtedly occurs between taste cells and gustatory afferent axons, few unambiguous, well-defined synapses in taste buds have been documented to date. Numerous close appositions of nerve terminals and taste cells are observed in electron micrographs of taste buds, but unequivocal synapses are rare. The ultrastructure of those synapses identified is largely ill-defined or primitive. Interestingly, some synaptic contacts on other sensory cells, such as cochlear hair cells and photoreceptors, also lack conventional ultrastructural features (Dunn & Morest 1975, Lasansky 1980).

Large clusters of synaptic vesicles adjacent to prominent pre- and postsynaptic membrane densities rarely occur at sites of contact between taste cells and afferent axons. Furthermore, in taste cell synapses that can be recognized as such, the polarity is often ambiguous. Putative synaptic vesicles are found on both sides of the junctions. The lack of clear polarity with regard to disposition of synaptic vesicles has been confusing from the earliest electron microscopic studies of taste buds (de Lorenzo 1958): "If they (synaptic vesicles) contain chemical transmitters, they are most awkwardly disposed" (de Lorenzo 1963). Identifying synaptic connections reliably in taste buds may be beyond the capacity of conventional electron microscopy and may require sophisticated immunocytochemical techniques instead, such as localizing synapse-related proteins; for example, synapsin (De Camilli et al 1983), N-CAM (Covault & Sanes 1986), protein p38 (synaptophysin) (Navone et al 1986), or other synaptic constituents (Beesley et al 1987; cf. reviews by Kelly & Cotman 1977, Kelly et al 1983, Rash et al 1988).

The somewhat primitive structure of synapses in taste buds might be explained by the relatively short life span of the taste cells (*vide infra*) and the fact that synapses are in a constant state of withdrawal, sprouting, and reinnervation as the taste cell population is renewed (cf. de Lorenzo 1963, Uga & Hama 1967, Graziadei 1970). Indeed, there are surprisingly many similarities between the morphology of synapses in taste buds and newly formed synapses elsewhere (e.g. Kelly & Zacks 1969, Kullberg et al 1977, Favre et al 1986, Maslim & Stone 1986, Nishimura & Rakic 1987, Takahashi et al 1987).

Alternatively, taste cell synapses may lack well-defined ultrastructural features because transmission does not require vesicles. Recent studies by Schwartz (1987) indicate that neurotransmission at some retinal synapses may not involve vesicular release, but instead involve the transmembrane transport of amino acids. Morphologically, there are many parallels between nerve terminals in the taste bud and those at some retinal sites (Lasansky 1980). If an analogy between taste cell synapses and photoreceptor synapses can be made, it might explain the lack of well-defined clusters of presynaptic vesicles and active zones in synapses on taste cells.

Curiously, nerve terminals and synapses in taste buds are frequently closely associated with the nuclei of receptor cells. Nerve terminals often indent deeply into the thin rim of cytoplasms adjacent to the taste cell nucleus, even deforming the nucleus (Murray 1973, Paran et al 1975, Kinnamon et al 1985). This is not a universal finding, and, for example, in *Necturus*, synapses are found principally at the base of the taste bud (Delay & Roper 1988). The significance of the association of nerve endings and synapses with the taste cell nucleus in some species is unknown. However, abundant evidence indicates that there is a strong trophic dependence of taste cells upon their nerve input (reviewed by Oakley 1985): If the nerve supply to the lingual epithelium is severed, taste cells disappear (Guth 1971, but cf. Whitehead et al 1987). Recent studies have revealed that proteins can be transferred from the nerve supply to the taste cells, thus suggesting a potential mechanism for neurotrophic interactions (Chan & Byers 1985). The proximity of nerve ending to taste cell nucleus in some species may be an important factor in this transfer.

Are synapses in taste buds, rudimentary as they might appear, associated with any particular type of taste cell? In rabbits, synapses originally were reported to occur between gustatory afferents and only one category of receptor cells, Type III cells (Murray 1973). This led to the inference that the Type III cell was the chemoreceptor cell and the other categories represented sustentacular cells. However, a number of researchers have observed synapses on taste cells other than the Type III cell. For example, in mouse vallate taste buds, Type I, Type II, and transition cells intermediate between these cells all make synaptic contact with afferent axons (Kinnamon et al 1985, Kinnamon et al 1988). Zahm & Munger (1983) described synapses on "cells resembling basal undifferentiated cells" in taste buds from fetal monkeys. Farbman (1965) reported synaptic-like contacts on basal cells in the rat. Several workers have described basal cell synapses in fish (Uga & Hama 1967, Hirata 1966, Reutter 1978, Ezeasor 1982, Toyoshima et al 1984), turtles (Korte 1980), and mudpuppies (Delay & Roper 1988).[3] In brief, all cell "types" have been observed to possess synaptic connections. Table 1 summarizes a number of these observations.

## Synapses and the Organization of Taste Cells

The branching pattern of afferent nerves and the distribution of synapses in the murine taste bud observed by Kinnamon et al (1988) suggests an organizational pattern among taste cells. A single axon synaptically contacts a group of cells ($\leq 5$), comprised either of Type I or of Type II cells. A single axon does not, however, contact a mixed group of cells, i.e. a group consisting of Type I and Type II cells. The organizational unit of the taste bud may be a small number of taste cells having similar morphology and similar physiological characteristics, which are synaptically connected to a single afferent output. The presence of electrical connections between subsets of taste cells, discussed below, also suggests that taste cells

---

[3] Synapses on basal cells have been more reliably documented in cold-blooded species. Possibly the expected slower rate of taste cell turnover in poikilotherms (Raderman-Little 1979) allows synaptic specializations on basal cells to stabilize and to develop more distinctive ultrastructural features, which are missing in mammalian species.

are organized into small, functional units comprised of only a very few cells.

Synaptic relationships may be more complicated than the simple scheme presented so far, i.e. synapses occurring between taste cells and sensory afferent terminals alone. Synaptic contacts have been found to occur between taste cells and afferent axons and also between taste cells themselves. Reutter (1978) observed that basal cells form synapses with other taste cells in the fish. In a comprehensive survey of the ultrastructure of taste cells in *Necturus*, Delay & Roper (1988), using serial reconstructions of taste buds, reported that Type I and Type II cells formed synapses with basal cells. The presence of synapses between basal cells and other taste cells raises the possibility that basal cells function as interneurons interposed between taste cells and afferent axons (Reutter 1971, 1978, Grover-Johnson & Farbman 1976, Delay & Roper 1988). Verification of this intriguing hypothesis awaits physiological studies and further ultrastructural observations. The limiting factor, as discussed above, may be the ability to resolve synaptic sites in taste buds with electron microscopic and other techniques.

## Is There Efferent Synaptic Control in Taste Buds?

Efferent synaptic input onto taste cells has been the source of speculation and controversy for several years. To date efferent input to taste cells has not been demonstrated unequivocally, although structures that have been interpreted as efferent synapses have been tentatively identified in electron micrographs from fish (Desgranges 1966, Graziadei 1970), mudpuppy (Delay & Roper 1988), rabbit (Fujimoto & Murray 1970, but cf. Murray 1973), mouse (Takeda 1976), fetal monkey (Zahm & Munger 1983), and human (Graziadei 1970). A characteristic feature of these synapses is the presence of clusters of vesicles at active sites in the sensory afferent axon, which suggests a polarity opposite that which is expected for afferent synapses. Often there are subsynaptic cisternae in the taste cell near putative efferent synapses. Subsynaptic cisternae have been identified as characteristic of efferent synapses in other tissues (discussed in Zahm & Munger 1983). However, serial reconstruction of synapses on taste cells in *Necturus* indicate that in most cases clusters of vesicles occur in both elements of synapses in taste buds, thus making the distinction between pre- and postsynaptic structures very difficult, if not impossible (Delay & Roper 1988). This has been mentioned frequently in other electron microscopic studies of taste cell synapses. Conceivably, some synaptic contacts on taste cells are bidirectional (Graziadei 1970). In any event, a convincing case for efferent synapses in taste buds remains to be made.

Although there are several examples in the literature that antidromic impulses in gustatory afferent axons affect taste transduction (e.g. Murayama & Ishiko 1986), physiological evidence for *bona fide* efferent synaptic control of taste cells is scanty but suggestive (Esakov 1961, Brush & Halpern 1970, Morimoto & Sato 1975, Glenn & Erickson 1976, cf. Murayama 1988).

## Neurotransmitters at Taste Cell Synapses

The unequivocal identification of neurotransmitters at synapses in taste buds has not yet been achieved. Morphologically, synapses in taste cells frequently are characterized by dense-core and clear spherical vesicles juxtaposed at putative active zones. The presence of acetylcholinesterase in taste buds has been reported, a finding that suggests that some synapses may be cholinergic (Tsuchiya & Aoki 1967, Murray & Fujimoto 1970, Fujimoto 1973, DeHan & Graziadei 1973). Likewise, fluorescence histochemistry of taste cells suggests the occurrence of monoaminergic synapses (Reutter 1971, Hirata & Nada 1975, DeHan & Graziadei 1973, Nada & Hirata 1975, Takeda & Kitao 1980, Takeda et al 1985). Recent reports indicate that dense-core vesicles in taste cells may contain serotonin, and, as the authors suggest, monoamines may modulate cholinergic transmission at taste cell synapses (Fujimoto et al 1987, but cf. Morimoto & Sato 1977). Putative transmitters and their antagonists have been applied either topically or via lingual arterial perfusion to test cholinergic transmission (Landgren et al 1954, Duncan 1964) and catecholaminergic transmission (Morimoto & Sato 1977, 1983, Nagahama & Kurihara 1985). Norepinephrine and agents affecting catecholaminergic transmission had the greatest influence on taste responses recorded in the glossopharyngeal nerve, thus implicating norepinephrine as a possible transmitter (Nagahama & Kurihara 1985). However, the inaccessibility of the tissue for detailed physiological studies of synaptic transmission has prevented more concrete identification of transmitters. This is clearly a fruitful area for research.

## Other Intercellular Connections: Electrical Synapses

The evidence for electrical synapses between taste cells is now very compelling. Electrophysiological studies by West & Bernard (1978) suggested that taste cells in *Necturus* are electrically coupled. However, the difficulty of achieving microelectrode impalements in these early studies, particularly simultaneous impalement of two adjacent taste cells, prevented an unequivocal acceptance of these data. Nonetheless, the existence of electrical coupling has been confirmed in this and other species by more recent studies that employed fluorescent dye-coupling as an indirect indication of electrical synapses (Teeter 1985, Yang & Roper 1987). In *Necturus*,

Yang & Roper (1987) showed that dye-coupling between taste cells is strong, indicating tight electrical coupling. Furthermore, 20% of the taste cells in *Necturus* are dye-coupled and coupling occurs between discrete groups of two, or at most three cells. Dye-coupled cells have quite similar morphology when examined at the light and electron microscopic level. In fact, dye-coupling exists exclusively between Type I (dark) cells and not, for example, between Type II cells or combinations of Type I and Type II cells (J. Yang, R. J. Delay, and S. D. Roper, in preparation). This suggests that the dye-coupled cells may represent an organizational unit similar to and complementary with the organizational unit described above.

## Cell Lineage in Taste Buds

Taste cells are part of a dynamic, renewing epithelium in the tongue. There is an ongoing turnover of cells in the taste bud, comparable to that in the surrounding stratified squamous lingual epithelium. Beidler & Smallman (1965) measured the lifespan of taste cells to be approximately 10 days in the rat. This is compared with a lifespan of only 4–8 days in the surrounding non-taste lingual epithelium (Bertalanffy 1964). Other workers have confirmed this finding and extended the data to include other species.

Kolmer (1910) had proposed that the vertebrate taste bud was comprised of only a single population of cells, taste receptor cells, that were transformed morphologically as they passed through their lifespan. This process gave rise to transitional forms, discussed above. This interpretation had resolved the confusion about the numbers and types of cells found in taste buds by the early microscopists and provided a unifying hypothesis that gained widespread acceptance earlier in this century (cf. review by Parker 1922). In the intervening years, two camps have arisen. Some investigators hold the hypothesis that a single cell population exists, sensory chemoreceptors. Other workers report different lines of cells within the taste bud (cf. Farbman 1980): Each cell type represents a separate cell line with differing functions such as chemosensory versus sustentacular or secretory. Many current textbooks subscribe to the latter camp.

In an attempt to find evidence supporting either a single cell lineage or multple lineages for taste cells, Delay et al (1986) studied taste cell turnover in the mouse, utilizing tritiated thymidine labeling combined with electron microscopy to distinguish cell types. Prior studies had been limited to light microscopy, which blurs the distinctions between the different categories of cells. Delay et al (1986) confirmed that basal cells represent the precursor or stem cell, since basal cells were first to incorporate radioactive thymidine. Over succeeding days, label was found over the nuclei of Type I cells, followed by intermediate (transition) and then Type II cells. Although the limited number of taste cells that could be analyzed by electron microscopic autoradiography prevented these workers from developing a statistically verifiable, mathematical model for the kinetics of cell turnover, these data strongly support the hypothesis of a single cell type in the vertebrate taste bud. As the basal cell differentiates and progresses through its lifespan, it transforms into a Type I cell, then to an intermediate cell, and lastly to a Type II cell. The presence of transitional cells with morphological features intermediate between Type I and Type II cells supports this interpretation.

This conclusion casts serious doubt on the current view that there are separate taste cell lines and that there are chemosensory and nonsensory, sustentacular cells in the taste bud. Indeed, a critical analysis of the literature leads one to conclude that compelling evidence to support the existence of a sustentacular cell has never been presented. The designation "sustentacular" was based on the absence of apical "taste hairs" and a somewhat more peripheral location of these "hairless" cells within the taste bud. Functional analyses or experimental verification of these inferences have not been made. Furthermore, the occurrence of synapses on all cell types in the taste bud, reviewed above, indicates that no one category of taste cells can be pinpointed as being sustentacular, and not sensory, in function.

It is possible, if not frankly likely, that cells at different stages of their cell cycle may have different membrane properties (reviewed by Spitzer 1983, O'Dowd et al 1988), including different chemosensitivities and different morphologies. However, a critical review of the evidence suggests it is unlikely that separate population(s) of nonsensory cells reside in the taste bud.

## Biophysical Properties of Taste Cells and Chemosensory Transduction

Until recently, a great deal of information about mechanisms of chemosensory transduction in taste cells has been inferred from extracellular recordings from the chorda tympani or glossopharyngeal nerves. Reliable, stable impalements from receptor cells have been difficult to obtain and membrane properties and receptor potentials difficult to measure directly. Since even the simplest description of taste transduction reveals a complex sequence of events that must occur to produce a change in the activity of sensory afferent nerve axons, these indirect measurements were never as satisfactory as intracellular recordings from taste cells. The following sequence indicates the likely chain of events after chemical stimulation and illustrates how far removed is the final change in sensory afferent axon activity from the initial events of chemosensory transduction:

chemical stimulus
↓
ligand-receptor interactions on apical membrane of taste cells
↓
conductance changes in apical membrane
↓
transmembrane current fluxes
↓
intracellular membrane potential change (receptor potential)
↓
alteration in neurotransmitter release at taste cell synapses
↓
diffusion of transmitter across synaptic cleft
↓
transmitter binding to postsynaptic (gustatory axon) membrane
↓
postsynaptic membrane conductance changes
↓
postsynaptic membrane potential change
↓
alteration in impulse activity in sensory afferent axon.

Early attempts to record membrane properties in taste cells suggested that taste cells have low resting potentials ($-20$ to $-40$ mV), low input resistances (17–81 megohms), and that a single cell can respond to a wide variety of taste stimuli (reviewed by Sato 1986). However, these results must be viewed with great caution, since the taste cells in these earlier studies were inevitably damaged by the microelectrode impalements.[4] Recent improvements in intracellular microelectrode recording techniques, in patch recording techniques, and the selection of appropriate animal

---

[4] The quality of the microelectrode impalement of taste cells is vital for interpreting intracellular responses evoked by applied chemical stimuli. Some chemosensory responses are mediated by decreases in membrane conductance (Sato 1986, Kinnamon & Roper 1988b, Tonosaki & Funakoshi 1988), and these decreases are readily shunted by damage-induced leakage at the site of the microelectrode impalement. Similarly, receptor potentials generated by increases in membrane conductance are also reduced by shunting, though to a lesser extent. The low resting input resistances reported in many publications indicates that the taste cells were damaged by the impalement. This may explain the confusing and sometimes contradictory observations about chemosensory receptor potentials, the lack of action potentials, and the wide responsiveness of individual taste cells to applied chemical stimuli: Exogenously-applied ions and charged molecules may leak into the cell through the damaged area around the microelectrode and generate artifactual "receptor potentials," and conductance changes may be masked by damaged-induced shunting.

models have now made it possible to record membrane properties more reliably from taste cells, particularly in amphibians. As a consequence, a good deal of detailed information is rapidly being gathered about the function of individual taste cells. For example, taste cells have been shown to have resting potentials in excess of $-60$ mV and input resistances between 350–1000 megohms (Roper 1983, Kinnamon & Roper 1987a, Avenet & Lindemann 1987b). As newer techniques, particularly patch recording from isolated cells, are applied to other species, including mammals, these data will form a broad base of information from which to make valid generalizations. For the present, this information is lacking and conclusions must be based upon data primarily from amphibian species.

## Ion Channels in Taste Cells

A new finding regarding chemosensory transduction is that taste cells possess a number of ionic channels identical to those found in neurons. Taste cells have now been shown to possess voltage-gated channels for $Na^+$, $Ca^{2+}$, and $K^+$, and also Ca-mediated cation channels (Akabas et al 1987, Kinnamon & Roper 1987a,b, Avenet & Lindemann 1987a, Sugimoto & Teeter 1987, Kinnamon & Roper 1988a). There is also evidence for an amiloride-sensitive (presumably not voltage-gated) Na channel in taste cells in some species (Heck et al 1984, Simon et al 1986, Avenet & Lindemann 1988, but cf. Yoshii et al 1986, Herness 1987) but not in *Necturus* (McPheeters & Roper 1985). Most of these channels, particularly the voltage-gated ones, are unique to the neuroepithelial taste cells and are not found in the surrounding nontaste epithelial cells.

Taste cells, by virtue of the voltage-gated channels, are electrically excitable and generate action potentials in response to electrical and chemical stimulation (Roper 1983, Kashiwayanagi et al 1983, Avenet & Lindemann 1987b, Kinnamon & Roper 1987a, 1988b). The electrical excitability of the taste cells may play a role in chemosensory transduction (Avenet & Lindemann 1987b, Kinnamon & Roper 1988b) as, for example, is the case for coding duration and intensity in mechanosensory receptors (Granit 1955). However, Kinnamon & Roper (1987a) found that even intense and prolonged electrical or chemical stimulation of taste cells in *Necturus* only evoked one or a few impulses at the onset of the stimulation, thus making it difficult to understand how action potentials can faithfully encode sensory stimuli. Furthermore, inactivation of voltage-gated $Na^+$ channels, which inevitably occurs during gradual depolarizations, prevents any but the most rapid (milliseconds) chemical stimulation from eliciting an impulse. It is questionable whether such rapid changes in stimulus concentration occur at the taste cell under physiological conditions. Instead, perhaps the action potential in taste cells serves some function other than direct chemosensory coding, *per se*. Conceivably, impulses may help shape the responses to taste stimuli, for example by providing a rapid transient excitation at the onset of taste stimulation.

The role that voltage-dependent ion channels in taste cells play in taste transduction is just now being appreciated and understood. For example, it is believed that the voltage-dependent $K^+$ channel is critical for taste transduction, particularly of sour and bitter stimuli (Kinnamon & Roper 1988b). The voltage-dependent $K^+$ channels are normally open, even at rest (Kinnamon & Roper 1987a), an unusual characteristic for this type of channel in other excitable tissues. Furthermore, this $K^+$ channel is located predominantly on the apical membrane of the taste cells, precisely where the initial events in chemosensory transduction must take place (McBride & Roper 1988, Kinnamon et al 1988). Blocking the K channels pharmacologically or by hyperpolarizing the cell eliminates intracellular receptor potentials evoked by sour or bitter chemical stimulation in taste cells (Kinnamon & Roper 1988b). Sour and bitter stimuli reduce $K^+$ conductance in taste cells (Sugimoto & Teeter 1987, Akabas et al 1987, Kinnamon & Roper 1988b). Furthermore, $K^+$ channels in taste cells are closed by intracellular cyclic nucleotides (Avenet et al 1988, Tonosaki & Funakoshi 1988). Thus, a hypothesis that emerges for one mechanism for chemosensory transduction is the following (cf. Teeter & Gold 1988):

chemical stimulus

$\downarrow$

activates GTP-binding protein

$\downarrow$

stimulates adenylate cyclase

$\downarrow$

increases intracellular cAMP (or cGMP)

$\downarrow$

closes apical $K^+$ channels

$\downarrow$

depolarizes taste cell

$\downarrow$

opens voltage-dependent $Ca^{2+}$ channels

$\downarrow$

allows influx of $Ca^{2+}$ near synapses

$\downarrow$

causes neurotransmitter release at synapses.

Evidence in support of this mechanism has also come from biochemical studies that indicate that GTP-binding protein is located in taste cell membranes and GTP-dependent cAMP synthesis is increased in taste cells in response to chemosensory stimulation, including amino acids and sugars (Bruch & Kalinoski 1987, Kalinoski et al 1987, Lancet et al 1987). A variation on this is that chemical stimulation may elicit the release of $Ca^{2+}$ from intracellular stores via a second messenger mechanism involving $IP_3$, and the intracellular $Ca^{2+}$ evokes transmitter release (cf. Akabas et al 1987).

Another mechanism for transduction, particularly for simple salts such as KCl or NaCl, is that ions simply pass through the existing ion channels in the apical membrane and diffuse down their electrochemical gradients. For example, for $K^+$ salts:

$$K^+ \text{ stimulation}$$
$$\downarrow$$
$$K^+ \text{ enters cell via apical } K^+ \text{ channels}$$
$$\downarrow$$
$$\text{membrane potential depolarizes}$$
$$\downarrow$$
$$\text{voltage-dependent } Ca^{2+} \text{ channels open}$$
$$\downarrow$$
$$\text{influx of } Ca^{2+} \text{ near synapses}$$
$$\downarrow$$
$$\text{neurotransmitter release.}$$

Similar mechanisms would apply to $Na^+$, $Cl^-$, and $Ca^{2+}$ salts and small charged molecules that can pass through their respective ionic channels; the membrane potential, and hence the status of the voltage-gated $Ca^{2+}$ channels, is determined by the equilibrium potential for the distribution of ionic species across the membrane. The key factors would be (a) whether the appropriate ionic channels are located on the apical membrane and (b) to what extent these channels are open (conducting) at the time of chemostimulation.

The other voltage-gated and $Ca^{2+}$-dependent ion channels presumably also play a role in chemosensory transduction, since they are found in taste cells but not in the surrounding, nontaste epithelial cells, but we still do not understand their role in transduction. Voltage-dependent $Ca^{2+}$ channels, which according to the above explanation might be expected to be situated preferentially near synapses (i.e. on the basolateral membrane), are not exclusively found there: They exist on both basolateral and apical membrane (McBride & Roper 1988). Clearly, this is an area for further study.

An intriguing variation on these schemata for transduction is a synaptic mechanism that has recently been proposed in the retina (Schwartz 1987); namely, release of an amino-acid neurotransmitter is nonvesicular and is independent of $Ca^{2+}$ influx. Instead, transmitter release is controlled by a voltage-regulated transmembrane transport of amino acids. If amino acids act as neurotransmitters at taste cell synapses, it is possible that voltage-sensitive, $Na^+$ cotransport carriers (cf. Villereal & Cook 1978) regulate the efflux of transmitter onto postsynaptic gustatory nerve fibers. Thus, chemical stimulation would not elicit a sudden release of transmitter-laden vesicles, as has been shown exhaustively at other synapses (reviewed by Rash et al 1988). Instead, a change or even a reversal of amino acid transport, caused by receptor potentials, would modulate sensory afferent discharge:

chemical stimulation

↓

produces taste cell membrane potential changes

↓

changes rate or reverses basolateral
amino-acid (neurotransmitter) transport

↓

alters concentration of transmitter at sensory terminals

↓

modulates postsynaptic activity in sensory afferent axons.

If such a mechanism exists in taste cells, this might explain the lack of well-defined, conventional synapses with clusters of synaptic vesicles situated near membrane densities in the taste bud. A precedence for such a mechanism and a corresponding lack of ultrastructurally defined synapses appears to exist in the retina (Lasansky 1980, Schwartz 1987).

A number of other possibilities for chemosensory transduction mechanisms have been proposed, particularly for non-ionic stimulants such as sugars. The most straightforward mechanism is that ligand-receptor binding opens (or closes) ionic channels and leads to a current flow across the apical membrane. On the other hand, more exotic mechanisms, such as the alteration of apical membrane surface charges without alterations of membrane conductance to produce intracellular receptor potentials, have been proposed (Miyake & Kurihara 1983, Kurihara et al 1986). Recent reviews have discussed these and other mechanisms in greater detail (Kurihara et al 1986, Teeter et al 1987, Teeter & Brand 1987; cf. Roper & Atema 1987).

## Summary

New technologies in neurophysiology and ultrastructural research are bringing about rapid advances in our understanding of taste, particularly at the cellular level. The model of chemosensory processing in the taste bud presented here can now be explored in great detail. The synaptic organization of the taste bud indicates a potential for intriguing peripheral integrative mechanisms, including cross-talk between taste cells, summation of chemoreceptor responses by interneurons (basal cells) in the taste bud, and centrifugal control of taste buds via efferent input from the CNS. Figure 2 summarizes these findings. The existence of voltage-gated

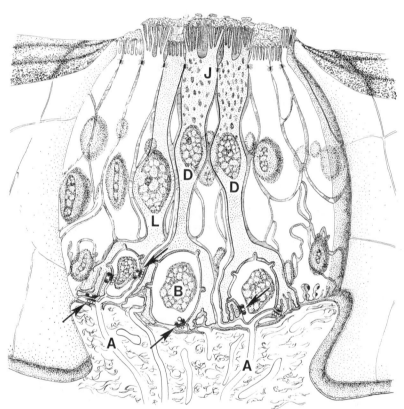

*Figure 2*  Diagram of receptor cells and synaptic connections in a taste bud from *Necturus maculosus*. This drawing illustrates the taste cells and the synapses that they form. A taste bud from this species is shown as a summary figure, since, with only a few differences in cytological details, examples of these connections have been shown to occur in many species. Type I, or dark cells (D), and Type II, or light cells (L), synapse with gustatory axons (A). Synaptic sites (*arrows*) often appear bidirectional, based on ultrastructural observations. Apical junctional complexes (J) may or may not be the sites of electrical connections between adjacent dark cells. Basal cells (B) form synapses with other taste cells and with gustatory axons, suggesting that they may be a form of interneuron in the taste bud.

ionic channels on taste cells and their unequal distribution in apical and basolateral membrane suggests mechanisms for chemosensory transduction: A primary event in the transduction process for many taste stimuli is likely to be the closure of apical potassium channels, thus leading to a depolarizing receptor potential. The closure of these apical potassium channels is probably mediated via cyclic nucleotides or intracellular $Ca^{2+}$.

ACKNOWLEDGMENTS

I would like to express my gratitude to Rona Delay, Stanley B. Kater, Sue C. Kinnamon, and John C. Kinnamon for their many discussions and assistance in composing this review. I would also like to thank Lesley Sealing and the Program in Biomedical Illustration and Communication at Colorado State University for illustrations. The work was supported, in part from National Institute of Health grants NS24107, NS20486, and AG06557.

*Literature Cited*

Akabas, M., Dodd, J., Al-Awqati, Q. 1987. Mechanism of transduction of bitter taste in rat taste bud cells. *Neurosci. Soc. Abstr.* 13: 361

Arvidson, K., Cottler-Fox, M., Friberg, U. 1981. Fine structure of taste buds in the human fungiform papilla. *Scand. J. Dent. Res.* 89: 297–306

Avenet, P., Hofmann, F., Lindemann, B. 1988. Transduction in taste receptor cells requires cAMP-dependent protein kinase. *Nature* 331: 351–54

Avenet, P., Lindemann, B. 1987a. Patch-clamp study of isolated taste receptor cells of the frog. *J. Membr. Biol.* 97: 223–40

Avenet, P., Lindemann, B. 1987b. Action potentials in epithelial taste receptor cells induced by mucosal calcium. *J. Membr. Biol.* 95: 265–69

Avenet, P., Lindemann, B. 1988. Amiloride-blockable sodium currents in isolated taste receptor cells. *J. Membr. Biol.* 105: In press

Bailey, F. R. 1906. *A Textbook of Histology*, pp. 452–53. New York: William Wood. 2nd ed.

Beesley, P. W., Paladino, T., Gravel, C., Hawkes, R. A., Gurd, J. W. 1987. Characterization of gp50, a major glycoprotein present in rat brain synaptic membranes, with a monoclonal antibody. *Brain Res.* 408: 65–78

Beidler, L. M., Smallman, R. L. 1965. Renewal of cells within taste buds. *J. Cell Biol.* 27: 263–72

Bertalanffy, F. D. 1964. Tritiated thymidine versus colchicine in the study of cell population cytodynamics. *Lab. Invest.* 13: 871–86

Böhm, A. A., von Davidoff, M., Huber, G. C. 1904. *Textbook of Histology*, pp. 248–51. Philadelphia: Saunders. 2nd ed.

Bruch, R. C., Kalinoski, D. L. 1987. Interaction of GTP-binding regulatory proteins with chemosensory receptors. *J. Biol. Chem.* 262: 2401–4

Brush, A. D., Halpern, B. 1970. Centrifugal control of gustatory responses. *Physiol. Behav.* 5: 743–46

Chan, K. Y., Byers, M. R. 1985. Anterograde axonal transport and intercellular transfer of WGA-HRP in trigeminal-innervated sensory receptors of rat incisive papilla. *J. Comp. Neurol.* 234: 201–17

Covault, J., Sanes, J. R. 1986. Distribution of N-CAM in synaptic and extrasynaptic portions of developing and adult skeletal muscle. *J. Cell Biol.* 102: 716–30

Cummings, T. A., Delay, R. J., Roper, S. D. 1987. Ultrastructure of apical specializations of taste cells in the mudpuppy, *Necturus maculosus. J. Comp. Neurol.* 261: 604–15

De Camilli, P., Cameron, R., Greengard, P. 1983. Synapsin-1 (protein-1), a nerve terminal-specific phosphoprotein. I. Its general distribution in synapses of the central and peripheral nervous-system demonstrated by immunofluorescence in frozen and plastic sections. *J. Cell Biol.* 96: 1337–54

DeHan, R. S., Graziadei, P. 1973. The innervation of frog's taste organ. "A Histochemical Study". *Life Sci.* 13: 1435–49

Delay, R. J., Kinnamon, J. C., Roper, S. D. 1986. Ultrastructure of mouse vallate taste buds. II. Cell types and cell lineage. *J. Comp. Neurol.* 253: 242–52

Delay, R. J., Roper, S. D. 1988. Ultrastructure of taste cells and synapses in the mudpuppy *Necturus maculosus*. *J. Comp. Neurol.* 277: 268–80

de Lorenzo, A. J. 1958. Electron microscopic observations on the taste buds of the rabbit. *J. Biophys. Biochem. Cytol.* 4: 143–50

de Lorenzo, A. J. 1963. Studies on the ultrastructure and histophysiology of cell membranes, nerve fibers and synaptic junctions in chemoreceptors. In *Olfaction and Taste*, ed. Y. Zotterman, pp. 5–17. New York: Macmillan

Desgranges, J. C. 1966. Sur la double innervation des cellules sensorielles des bourgeons du gout des barbillons du poisson-chat. *C. R. Acad. Sci. Ser. D* 263: 1103–6

Duncan, C. J. 1964. Synaptic transmission at taste buds. *Nature* 203: 875–76

Dunn, R. A., Morest, D. K. 1975. Receptor synapses without synaptic ribbons in the cochlea of the cat. *Proc. Natl. Acad. Sci.* 72: 3599–603

During, M. V., Andres, K. H. 1976. The ultrastructure of taste and touch receptors of the frog's taste organ. *Cell Tissue Res.* 165: 185–98

Esakov, A. I. 1961. The efferent control of receptors (on the example of the chemoreceptors of the tongue). *Bull. Exp. Biol. Med.* 51: 257–62 (Engl. transl.).

Ezeasor, D. N. 1982. Distribution and ultrastructure of taste buds in the oropharyngeal cavity of the rainbow trout, *Salmo gairdneri* Richardson. *J. Fish Biol.* 20: 53–68

Farbman, A. I. 1965. Fine structure of the taste bud. *J. Ultrastruct. Res.* 12: 328–50

Farbman, A. I. 1980. Renewal of taste bud cells in rat circumvallate papillae. *Cell Tissue Kinet.* 13: 349–57

Farbman, A. I., Hellekant, G., Nelson, A. 1985. Structure of taste buds in foliate papillae of the rhesus monkey, *Macaca mulatta*. *Am. J. Anat.* 172: 41–56

Farbman, A. I., Yonkers, J. D. 1971. Fine structure of the taste bud in the mudpuppy, *Necturus maculosus*. *Am. J. Anat.* 131: 353–70

Favre, D., Denemses, D., Sans, A. 1986. Microtubule organization and synapto-genesis in the vestibular sensory cells. *Brain Res.* 390: 137–42

Frey, H. 1880. *The Microscope and Microscopical Technology*, pp. 562–65. New York: William Wood & Co.

Fujimoto, S. 1973. On the Golgi-derived vesicles in the rabbit taste bud cells. *Kurume Med. J.* 20: 133–48

Fujimoto, S., Murray, R. G. 1970. Fine structure of degeneration and regeneration in denervated rabbit vallate taste buds. *Anat. Rec.* 168: 393–413

Fujimoto, S., Ueda, H., Kagawa, H. 1987. Immunocytochemistry on the localization of 5-hydroxytryptamine in monkey and rabbit taste buds. *Acta Anat.* 128: 80–83

Glenn, J. F., Erickson, R. P. 1976. Gastric modulation of gustatory afferent activity. *Physiol. Behav.* 16: 561–68

Granit, R. 1955. Receptors and Sensory Perception. Haven: Yale Univ. Press. 366 pp.

Graziadei, P. P. C. 1970. The ultrastructure of taste buds in mammals. In *2nd Symposium on Oral Sensation and Perception*, ed. J. F. Bosma, pp. 5–35. Springfield: Thomas

Graziadei, P. P. C., DeHan, R. S. 1971. The ultrastructure of frog's taste organs. *Acta Anat.* 80: 563–603

Grover-Johnson, N., Farbman, A. I. 1976. Fine structure of taste buds in the barbel of the catfish, *Ictalurus punctatus*. *Cell Tissue Res.* 169: 395–403

Guth, L. 1971. Degeneration and regeneration of taste buds. In *Handbook of Sensory Physiology*, Vol. 4, *Chemical Senses*, ed. L. M. Beidler, pp. 63–74. Berlin: Springer-Verlag

Heck, G. L., Mierson, S., DeSimone, J. A. 1984. Salt taste transduction occurs through an amiloride-sensitive sodium transport pathway. *Science* 223: 403–5

Hermann, F. 1888. Studien über den feineren Bau des Geschmacksorgans. Sitzb. Math. Phys. K. *Akad. Wiss. Muenchen* 18: 277–318

Herness, M. S. 1987. Are apical membrane ion channels involved in frog taste transduction? See Roper & Atema 1987, pp. 362–65

Hirata, K., Nada, O. 1975. A monoamine in the gustatory cell of the frog's taste organ. *Cell Tissue Res.* 159: 101–8

Hirata, Y. 1966. Fine structure of the terminal buds on the barbels of some fishes. *Arch. Histol. Jpn.* 26: 507–23

Kalinoski, D. L., LaMorte, V., Brand, J. G. 1987. Characterization of a taste-stimulus sensitive adenylate cyclase from the gustatory epithelium of the channel catfish, *Ictalurus punctatus*. *Soc. Neurosci. Abstr.* 13: 1405

Kashiwayanagi, M., Miyake, M., Kurihara, K. 1983. Voltage-dependent $Ca^{2+}$ channel and $Na^+$ channel in frog taste cells. *Am. J. Physiol.* 244: C82–C88

Kawamura, Y., Kare, M. R. 1987. *Umami: a Basic Taste.* New York: Dekker. 450 pp.

Kelly, A. M., Zacks, S. I. 1969. The fine structure of motor endplate morphogenesis. *J. Cell Biol.* 42: 154–69

Kelly, P. T., Cotman, C. W. 1977. Identification of glycoproteins and proteins at synapses in the central nervous system. *J. Biol. Chem.* 252: 786–93

Kelly, R. B., Miljanich, G., Pfeffer, S. 1983. Presynaptic mechanisms of neuromuscular transmission. In *Myasthenis Gravis*, ed. E. X. Albuquerque, A. T. Eldefrawi, pp. 43–104. New York: Chapman & Hall

Kinnamon, S. C., Dionne, V. E., Beam, K. G. 1988. Sour taste-modulated $K^+$ channels are restricted to the apical membrane of mudpuppy taste cells. *Proc. Natl. Acad. Sci. USA.* In press

Kinnamon, S. C., Roper, S. D. 1987a. Passive and active membrane properties of mudpuppy taste receptor cells. *J. Physiol.* 383: 601–14

Kinnamon, S. C., Roper, S. D. 1987b. Voltage-dependent ionic currents in dissociated mudpuppy taste cells. See Roper & Atema 1987, pp. 413–16

Kinnamon, S. C., Roper, S. D. 1988a. Membrane properties of isolated mudpuppy taste cells. *J. Gen. Physiol.* 91: 351–71

Kinnamon, S. C., Roper, S. D. 1988b. Evidence for a role of voltage-sensitive apical $K^+$ channels in sour and salt taste transduction. *Chem. Senses* 13: 115–21

Kinnamon, J. C., Sherman, T. A., Roper, S. D. 1988. Ultrastructure of mouse vallate taste buds. III. Patterns of synaptic connectivity. *J. Comp. Neurol.* 270: 1–10

Kinnamon, J. C., Taylor, B. J., Delay, R. J., Roper, S. D. 1985. Ultrastructure of mouse vallate taste buds. I. Taste cells and their associated synapses. *J. Comp. Neurol.* 235: 48–60

Kolmer, W. 1910. Über Strukturen im Epithel des Sinnesorgane. *Anat. Anz.* 36: 281–99

Korte, G. E. 1980. Ultrastructure of the tastebuds of the red-eared turtle, *chrysemys scripta elegans. J. Morphol.* 163: 231–52

Kullberg, R. W., Lentz, T. L., Cohen, M. W. 1977. Development of the myotomal neuromuscular junction in *Xenopus laevis*: An electrophysiological and fine-structural study. *Dev. Biol.* 60: 101–29

Kurihara, K., Yoshii, K., Kashiwayanagi, M. 1986. Transduction mechanisms in chemoreception. *Comp. Biochem. Physiol. A* 85: 1–22

Lancet, D., Striem, B. J., Pace, U., Zehari, U., Naim, M. 1987. Adenylate cyclase and GTP binding protein in rat sweet taste transduction. *Soc. Neurosci. Abstr.* 13: 361

Landgren, S., Liljestrand, G., Zotterman, Y. 1954. Chemical transmission in taste fiber endings. *Acta Physiol. Scand.* 30: 105–14

Lasansky, A. 1980. Lateral contacts and interactions of horizontal cell dendrites in the retina of the larval tiger salamander. *J. Physiol.* 301: 59–68

Loven, C. 1868. Beiträge zur Kenntnis vom Bau der Geschmackswarzchen der Zunge. *Arch. Mikrosk. Anat.* 4: 96–109

Maslim, J., Stone, J. 1986. Synaptogenesis in the retina of the cat. *Brain Res.* 373: 35–48

McBride, D. W., Roper, S. D. 1988. Distribution of Na, K, Ca and Ca-mediated channels in taste cells of the mudpuppy. *Chem. Senses.* 13(4): In press

McPheeters, M., Roper, S. D. 1985. Amiloride does not block taste transduction in the mudpuppy, *Necturus maculosus. Chem. Senses* 10: 341–52

Merkel, F. 1880. Ueber die Endigungen der sensiblen Nerven in der Haut der Wirbelthiere. *Rostok.* 214 pp.

Miller, R. L., Chaudhry, A. P. 1976. Comparative ultrastructure of vallate, foliate and fungiform taste buds of golden Syrian hamster. *Acta Anat.* 95: 75–92

Miyake, M., Kurihara, K. 1983. Resting potential of the mouse neuroblastoma cells. II. Significant contribution of the surface potential to the resting potential of the cells under physiological condition. *Biochim. Biophys. Acta* 762: 256–64

Morimoto, K., Sato, M. 1975. Inhibition of taste nerve response by antidromic stimuli of the glossopharyngeal nerve. *J. Physiol. Soc. Jpn.* 37: 110

Morimoto, K., Sato, M. 1977. Is serotonin a chemical transmitter in the frog taste organ? *Life Sci.* 21: 1685–96

Morimoto, K., Sato, M. 1983. Role of monoamines in afferent synaptic transmission in frog taste organ. *Jpn. J. Physiol.* 32: 855–71

Murayama, N. 1988. Interaction among different sensory units within a single fungiform papilla in the frog tongue. *J. Gen. Physiol.* 91: 685–702

Murayama, N., Ishiko, N. 1986. Selective depressant action of antidromic impulses on gustatory nerve signals. *J. Gen. Physiol.* 88: 219–36

Murray, R. G. 1973. The ultrastructure of taste buds. In *The Ultrastructure of Sensory Organs*, ed. I. Friedmann, pp. 1–81. New York: Elsevier

Murray, R. G. 1986. The mammalian taste

bud type III cell: A critical analysis. *J. Ultrastruct. Mol. Struct. Res.* 95: 175–88

Murray, R. G., Fujimoto, S. 1970. Demonstration of cholinesterase in rabbit foliate taste buds. *Proc. Septieme Congres Int. Microscopie Electronique, Grenoble*, ed. P. Favard, pp. 757–58. Paris: Soc. Francaise de Microsc. Electronique

Murray, R. G., Murray, A. 1960. Fine structure of taste buds of rhesus and cynomalgus monkeys. *Anat. Rec.* 138: 211–33

Murray, R. G., Murray, A., Fujimoto, S. 1969. Fine structure of gustatory cells in rabbit taste buds. *J. Ultrastruct. Res.* 27: 444–61

Nada, O., Hirata, K. 1975. Occurence of cell type containing a specific monoamine in taste bud of rabbit's foliate papilla. *Histochemistry* 43: 237–40

Nagahama, S., Kurihara, K. 1985. Norepinephrine as a possible transmitter involved in synaptic transmission in frog taste organs and Ca-dependence of its release. *J. Gen. Physiol.* 85: 431–42

Navone, F., Jahn, R., DiGioia, G., Stukenbrok, H., Greengard, P., DeCamilli, P. 1986. Protein-p38: An integral membrane protein specific for small vesicles of neurons and neuroendocrine cells. *J. Cell Biol.* 103: 2511–27

Nishimura, Y., Rakic, P. 1987. Development of the Rhesus monkey retina: II. A three-dimensional analysis of the sequences of synaptic combinations in the inner plexiform layer. *J. Comp. Neurol.* 262: 290–313

Oakley, B. 1985. Trophic competence in mammalian gustation. In *Taste, Olfaction, and the Central Nervous System*, ed. D. Pfaff, pp. 92–103. New York: Rockefeller Univ. Press

O'Dowd, D. K., Ribera, A. B., Spitzer, N. C. 1988. Development of voltage-dependent calcium, sodium and potassium currents in Xenopus spinal neurons. *J. Neurosci.* 8: 792–805

Paran, N., Mattern, C. F. T., Henkin, R. I. 1975. Ultrastructure of taste bud of human fungiform papilla. *Cell Tissue Res.* 161: 1–10

Parker, G. H. 1922. *Smell, Taste and Allied Senses in the Vertebrates*, Chap. VI, pp. 110–31. Philadelphia: Lippincott Co.

Raderman-Little, R. 1979. Effect of temperature on the turnover of taste bud cells in catfish. *Cell Tissue Kinet.* 12: 269–80

Ramon y Cajal, S. 1933. *Histology*, revised by J. F. Tello-Muñoz, transl. by M. Fernán-Núñez, pp. 350–53. Baltimore: William & Wood. 10th ed.

Ranvier, L. 1888. *Traite technique d'Histologie*. Paris. 1109 pp.

Rash, J. E., Walrond, J. P., Morita, M. 1988. Structural and functional correlates of synaptic transmission in the vertebrate neuromuscular junction. *J. Electron Microsc. Tech.* 10: 153–85

Reutter, K. 1971. Die Geschmacksknospen des Zwergwelses *Amiurus nebulosus* (Leseur). Morphologische und histochemische Untersuchungen. *Z. Zellforsch. Mikrosk. Anat.* 120: 280–308

Reutter, K. 1978. Taste organ in the bullhead (*Teleostei*). *Adv. Anat. Embryol. Cell Biol.* 55: 1–98

Roper, S. 1983. Regenerative impulses in taste cells. *Science* 220: 1311–12

Roper, S. D., Atema, J., eds. 1987. *Olfaction and Taste IX. Ann. NY Acad. Sci.* 510. 747 pp.

Sato, T. 1986. Receptor potential in rat taste cells. *Prog. Sensory Physiol.* 6: 1–37

Satterthwaite, T. E. 1881. *A Manual of Histology*, pp. 380–85. New York: William Wood & Co.

Schäfer, E. A. 1885. *The Essentials of Histology*, pp. 132–35. Philadelphia: Lea Brothers & Co.

Schwalbe, G. A. 1867. Das Epithel der Papillae vallatae. *Arch Mikrosk. Anat.* 3: 504–8

Schwartz, E. A. 1987. Depolarization without calcium can release gamma-aminobutyric acid from a retinal neuron. *Science* 238: 350–55

Settembrini, B. P. 1987. Papilla palatina, nasopalatine duct and taste buds of young and adult rats. *Acta Anat.* 128: 250–55

Simon, S. A., Robb, R., Garvin, J. L. 1986. Epithelial responses to rabbit tongue and their involvement in taste transduction. *Am. J. Physiol.* 251: R598–608

Spitzer, N. C. 1983. The development of neuronal membrane properties *in vivo* and in culture. In *Developing and Regenerating Vertebrate Nervous Systems*, ed. P. W. Coates, R. R. Markwald, A. D. Kenny, pp. 41–59. New York: Liss

Sugimoto, K., Teeter, J. H. 1987. Voltage-dependent and chemically-modulated ionic currents in isolated taste receptor cells of the tiger salamander. *Soc. Neurosci. Abstr.* 13: 1404

Takahashi, T., Nakajima, Y., Hirosawa, K., Nakajima, S., Onodera, K. 1987. Structure and physiology of developing neuromuscular synapses in culture. *J. Neurosci.* 7: 473–81

Takeda, M. 1976. An electron microscopic study on the innervation in the taste buds of the mouse circumvallate papillae. *Arch. Histol. Jpn.* 39: 257–69

Takeda, M., Kitao, K. 1980. Effect of monoamines on the taste buds in the mouse. *Cell Tissue Res.* 210: 71–78

Takeda, M., Suzuki, Y., Shishido, Y. 1985.

Effects of colchicine on the ultrastructure of mouse taste buds. *Cell Tissue Res.* 242: 409–16

Teeter, J. H. 1985. Dye-coupling in catfish taste buds. *Proc. 19th Japanese Symp. on Taste and Smell*, ed. S. Kimura, A. Miyoshi, I. Shimida, pp. 29–33. Hozumicho, Gifu: Asahi Univ.

Teeter, J., Funakoshi, M., Kurihara, K., Roper, S., Sato, T., Tonosaki, K. 1987. Generation of the taste cell potential. *Chem. Senses* 12: 217–34

Teeter, J. H., Brand, J. G. 1987. Peripheral mechanisms of gustation: physiology and biochemistry. In *Neurobiology of Taste and Smell*, ed. T. E. Finger, W. L. Silver, pp. 299–329. New York: Wiley

Teeter, J. H., Gold, G. H. 1988. A taste of things to come. *Nature* 331: 298–99

Tonosaki, K., Funakoshi, M. 1988. Cyclic nucleotides may mediate taste transduction. *Nature* 331: 354–56

Toyoshima, K., Nada, O., Shimamura, A. 1984. Fine structure of monoamine-containing basal cells in the taste buds on the barbels of three species of teleosts. *Cell Tissue Res.* 235: 479–84

Tsuchiya, S., Aoki, T. 1967. Cholinesterase activities in the gustatory region of the rat tongue and their inhibition by bitter-tasting substances. *Tohoku J. Exp. Med.* 91: 41–52

Uga, S., Hama, K. 1967. Electron microscopic studies on the synaptic region of the taste organ of carps and frogs. *J. Electron Microsc.* 16: 269–76

Villereal, M. L., Cook, J. S. 1978. Regulation of active amino acid transport by growth-related changes in membrane potential in a human fibroblast. *J. Biol. Chem.* 253: 8257–62

West, C. H. K., Bernard, R. A. 1978. Intracellular characteristics and responses of taste bud and lingual cells of the mudpuppy. *J. Gen. Physiol.* 72: 305–26

Whitear, M. 1976. Apical secretion from taste bud and other epithelial cells in amphibians. *Cell Tissue Res.* 172: 389–404

Whitehead, M. C., Frank, M. E., Hettinger, T. P., Hou, L. T., Nah, H. D. 1987. Persistence of taste buds in denervated fungiform papillae. *Brain Res.* 405: 192–95

Yang, J., Roper, S. D. 1987. Dye-coupling in taste buds in the mudpuppy, *Necturus maculosus*. *J. Neurosci.* 7: 3561–65

Zahm, D. S., Munger, B. L. 1983. Fetal development of primate chemosensory corpuscles. I. Synaptic relationships in late gestation. *J. Comp. Neurol.* 213: 146–62

*Ann. Rev. Neurosci. 1989. 12:355–75*
*Copyright © 1989 by Annual Reviews Inc. All rights reserved*

# STARTLE, CATEGORICAL RESPONSE, AND ATTENTION IN ACOUSTIC BEHAVIOR OF INSECTS

*Ronald R. Hoy*

Section of Neurobiology and Behavior, Cornell University, Ithaca, New York 14853-2702

## INTRODUCTION

There are three categories of auditory behavior that are often considered to be the province of psychophysicists and neurobiologists who study higher mammals, including humans, for which behavioral analogs might be expected to be found in insects. These behaviors are (*a*) categorical perception, (*b*) acoustic startle, and (*c*) selective attention. *Categorical perception* can be defined as the behavioral segmentation of a stimulus that varies continuously along some physical parameter. For example, in the case of sound, sharp behavioral (or perceptual) boundaries that define distinctly different behavioral acts can be drawn along the dimension of spectral frequency. *Acoustic startle* can be defined as a short-latency, characteristic sequence of muscle contractions, elicited by the sudden onset of a sound, especially one that is loud, unexpected, and potentially threatening. *Selective attention* has been defined as "that process by which an organism chooses to deal more effectively with one type of sensory stimulus at the expense of other stimuli" (Robinson & Petersen 1986). I present evidence that with appropriate amendments of definition to take into account surplus meaning (cognitive implications) in the original concepts as applied to humans, categorical perception, acoustic startle, and selective attention can be studied in such creatures of modest behavioral means as crickets, moths, praying mantises, and green lacewings.

355

0147–006X/89/0301–0355$02.00

The sense of hearing is well-developed in only two major groups of terrestrial animals: vertebrates and insects. In both groups, hearing subserves two major behavioral "jobs" critical to an animal's survival. First, many vertebrates and insects employ acoustic signals in the service of social communication—the business of finding and choosing mates and warning off potential rivals for mates or territorial preserves. Second, the sense of hearing is used to detect potential predators or prey; hearing a "dinner bell" is, evolutionarily speaking, very old. Some insects have evolved ears to serve both functions, others one or the other. Acoustic signals, whether they emanate from members of the same species in the context of social behavior or from another species in the context of a predator-prey interaction can be described in terms of physical properties such as amplitude ("loudness"), frequency ("pitch"), and temporal properties ("rhythm"). Although human hearing is obviously extraordinarily sensitive and has a broad frequency range, insectan ears can exceed human capabilities: Whereas the upper range of adult human hearing is about 15–20 kHz (depending on age), several insect species hear to 100 kHz and beyond. Moreover, the high rate of pulse repetition of some insectan signals as well as the hunting signals of some vertebrate predators suggests that the temporal acuity of insectan ears is very high (Michelsen & Nocke 1974).

To keep this review within length limits, I have chosen to restrict coverage severely, and I make no claim to have thoroughly reviewed all of the literature on insect audition. For example, directionality in insect hearing is not touched upon at all, and, taxonomically speaking, I have concentrated more on crickets and moths than other groups. However, this limitation is alleviated by the availability of several recent books (Sales & Pye 1974, Michelsen & Nocke 1974, Zhantiev 1981, Huber & Markl 1983, Kalmring & Elsner 1985, Huber et al 1989) and review articles (Michelsen & Larsen 1985, Elsner & Popov 1978) that cover insect audition comprehensively.

## STARTLE BEHAVIOR OCCURS IN MANY ANIMALS

Who has not experienced the "behavioral rush" that is elicited from hearing the cracking of a twig, however softly, as one walks along an unfamiliar street, on a dark, moonless night? That same sound might not even enter the consciousness in the same setting, at high noon. There can be no question that auditory stimuli have always been highly negotiable currency in the economics of natural selection. The detection of a predator or a prey that "gives itself away" by sound certainly carries survival value, and it is in this evolutionary context that the acoustic startle response (sometimes called "reflex") is found to be common among animals that have evolved a sense of hearing.

The concept of the acoustic startle response has its origins in the study of mammalian reflexes (Forbes & Sherrington 1914), and, as Bullock (1984) points out, early workers considered startle reflexes to be phylogenetically restricted to mammals (Landis & Hunt 1939). However, when comparative physiologists and behaviorists turned their attention to the simple behavioral acts of invertebrate animals, it became clear that startle responses (which could be triggered by nonacoustic as well as acoustic stimuli) are ubiquitous in the animal kingdom. Indeed, a small cottage industry within comparative neurobiology was spawned by the finding that numerous examples of startle (or "escape") behavior in invertebrates are associated with the occurrence of so-called "giant" neurons in the CNS (Bullock 1984). The taxonomically widespread occurrence of startle behaviors in the animal kingdom and the convergence of neural systems to subserve these rapid, simple acts was the subject of a recent volume (Eaton 1984). It seems highly likely that startle responses evolved because of their high survival value, such as in the context of predator/prey relationships; thus it is not at all surprising that many, if not most mammals, which are among the most auditory of animals, demonstrate a well-developed acoustic startle response. Similarly, it should be expected that those insects that have a sense of hearing might also have an acoustic startle response that would be elicited in an evolutionarily adaptive context, such as a predator/prey interaction. I review the evidence that many insects that can hear also hear their predators, specifically, that nocturnally active, flying insects are vulnerable to the predations of insectivorous bats that use ultrasonic biosonar signals to detect and locate their prey on the wing. I review some of an extensive and growing literature that documents the ability of insects to hear ultrasound and employ it to initiate an acoustic startle response that rivals its better-known counterpart in mammals.

The reason that the acoustic startle response is so intensively studied in mammals is because of the premium (adaptively speaking) placed on the speed and stereotypy of the evoked act. Thus, the neural circuitry underlying the acoustic startle response is presumed to be simple (a minimum of neuronal elements between input and output) so as to minimize behavioral latency, which can be impressively short in mammals—under 10 ms in some cases (Davis 1984). It is presumed that significant insights into the neural processing operations that control behavior can be gleaned from the study of "simplified" behaviors such as the mammalian acoustic startle response (Davis 1984). For the same reasons, the startle or escape behaviors of invertebrates have been studied (Bennett 1984, Bullock 1984), and the study of startle responses in insects is presumed to be simple whether escape behavior is initiated by acoustic stimuli (Roeder 1967, Nolen & Hoy 1984) or nonacoustic stimuli (Pearson & O'Shea 1984, Wine & Krasne 1972, Wyman et al 1984).

## Acoustic Escape Behavior in Moths

The first well-established case of an insectan auditory escape behavior (in moths) stems from the work of K. D. Roeder and his colleagues (Roeder 1967), which stands even today as a paradigm of neuroethological research. Roeder combined natural field observations, lab behavior, and sensory-motor neurophysiological studies to probe this acoustic escape response in moths. Roeder et al showed that free-flying moths altered their flight course abruptly when stimulated with high intensity ultrasound (in the range 20–80 kHz), generally steering away from a directional ultrasound source. Roeder & Treat (1957, 1961) showed that the tympanal hearing organ ("ear") of some species of moths (including members of the noctuids, arctiids, and geometrids) is a simple scolopophorous organ consisting of two auditory receptor cells, A-1 and A-2, and a receptor of presumed proprioceptive function, B. The auditory receptors in all species examined are sensitively tuned to ultrasound in a range that matches the spectral frequencies found in the biosonar signals of insectivorous bats (Roeder 1967, Simmons et al 1979, Fenton & Fullard 1983), and this is the sensory basis for the escape behavior of moths in response to bats. The escape behavior is certainly adaptive; moths are eaten by bats that hunt them down using echolocation (Griffin 1958), and moths that can hear bat biosonar signals evade bats by steering away from them on the wing (Roeder 1967). Roeder then made some investigations of auditory inter-neurons in the moth CNS (Roeder 1966, 1969a,b); however, these studies were made before the widespread application of intracellular dyes that have enabled physiologists to identify neurons by unique structure/ function criteria (Stretton & Kravitz 1968, Stewart 1978).

Recently, dye injection techniques have been applied to the moth CNS that have resulted in the identification of several classes of auditory inter-neurons (Boyan & Fullard 1986). These workers report seven different interneurons in the thoracic ganglia of noctuid moth *Heliothis virescens*. These cells are excited by bat-like ultrasonic stimuli and are either tonic or phasic/tonic in their response; thus far no inhibitory responses have been reported. Most cells appear to be driven by the lower threshold A-1 receptor, but a few are also driven by the higher threshold A-2 cell. Moreover, Boyan & Fullard constructed a hypothetical neural circuit to account for the negative phonotactic response of moths to directional ultrasound based on acoustic information processed in the thorax alone, without influences from the brain. They point out, however, that in their experiments the moth is not actually flying, and it is known from work in other insects that the performance of specific neurons in acoustic behavior critically depends on behavioral context (Nolen & Hoy 1984).

However one interprets Boyan & Fullard's findings, there is no doubt that significant auditory processing occurs in the brains of moths. Typically, Roeder (1969a,b) was far ahead of his time when he made single unit recordings in the brains of several moth species while stimulating them with simulated and recorded bat biosonar signals. Since his work preceded the invention of modern single-cell dye technologies, none of the auditory units he studied qualifies as an "identified neuron," which is now the operating standard in invertebrate neurobiology when the objective is to describe a neutral system that underlies a given behavior (Selverston 1985). Nonetheless, his findings warrant a brief review. He discovered brain neurons that responded tonically as well as phasically in response to repetitive stimuli; some tonic units were excited while the stimulus was maintained at low amplitude, but were suppressed upon elevating the sound level (Roeder 1969a). An intriguing finding was that some auditory units showed spontaneous "lapses" in their acoustic responsiveness, lasting from seconds to an hour, and such lapses appeared unrelated to obvious stimulus conditions (Roeder 1969b). Are these lapses related to changes in behavioral context, or perhaps to "attention?" Future investigations with modern techniques applied to the brain will likely yield significant insights into the acoustically driven escape behavior of the moth.

Before leaving this discussion of moths, it is worthwhile recalling that the evasive behavior consists of two strategies (Roeder 1967). The first occurs when a tympanate moth is stimulated by relatively low intensity, directional sources of ultrasound (as would be emitted by an echolocating bat at distances from the moth beyond the detection threshold of the bat). In this case, the moth simply steers its flapping flight directly away from the ultrasound source. The second strategy occurs when the moth is stimulated by very high intensity ultrasound (as would be emitted by a bat so close that detection of the moth is nearly certain). In this "last chance" escape mode, the moth responds with unpredictable movements, including power dives, rolls, or cessation of wing flapping—a dead drop. Roeder predicted that the first escape mode could be initiated by activation of the low threshold A-1 pathway, whereas the second could be initiated in conjunction with the higher threshold A-2 pathway. Although definitive tests of Roeder's simple conjecture remain to be made, Surlykke (1984) has employed the comparative method to illuminate this question. Surlykke studied a species of notodontid moth that has an ear consisting of but a single auditory receptor. Her behavioral observations revealed that these moths displayed both strategies, directional steering and "last chance," of ultrasound evasive behavior. Yet her physiological tests showed that the single auditory receptor had the response characteristic of an A-1 type; this was confirmed anatomically, as well. Thus, for some moth species it is not necessary to have A-2 type receptors in order to initiate "last chance" escape behaviors; however, it still remains a possibility in moths whose ears contain two auditory receptors.

The auditory response of moths to ultrasound has been classified as an escape behavior, and not an acoustic startle response. However, many of the properties of this behavior fulfill the description of an acoustic startle response. The response latencies tend to be very short, on the order of tens of milliseconds, the escape movements are stereotyped in their expression, and the movements may be graded with sound amplitude (Roeder 1967). When the response of moths to ultrasound is compared with those of other orders of insects (below), the case for calling these escape behaviors acoustic startle responses becomes apparent.

## Green Lacewings

Another nocturnal, flying insect, the green lacewing (genus *Chrysopidae*), is also subject to bat predation. This particular bat-bug story has been investigated by Miller and his colleagues. Like moths, this insect has well-developed scolopophorous hearing organs. Its "ears" are borne within the radial wing veins of each forewing (Miller 1970). This tympanal organ contains two chordotonal organs, comprising about 25 scolopales (Miller 1970). The lacewing ear has a best auditory response in the range between 30 and 60 kHz (Miller 1971). Typically, lacewing ears are about 20 dB less sensitive than the average moth ear, and at best frequencies have thresholds of about 55–60 dB SPL (Miller 1975). The exact number of ultrasound-responsive units in the ear at best frequency is not yet known. Presumably, equipped with a pair of such ears, night-flying lacewings detect the presence of insectivorous bats, and these insects engage in a variety of evasive maneuvers to avoid capture, as shown in both laboratory and natural field encounters that have been observed between insect and bat (Miller & Olesen 1979). These behaviors include a passive nosedive (in which flying lacewings cease to beat their wings and simply drop) when the bat is distant. When the bat is close, indicated by high amplitude levels of ultrasound, the lacewings perform a combined nosedive and wingflip (an active wingbeat interpolated at an unpredictable point during is passive nosedive); clearly, this maneuver resembles the moth's "last chance" behavior. It is interesting to note that these two behaviors appear to be keyed to the pulse repetition rate of the ultrasound stimulus as well as to sound level (Miller & Olesen 1979). At low pulse repetition rates (10–50 Hz), which would be typical of distant bats engaged in the "search" phase of echolocation (Griffin 1958), simple nosedives are more likely, whereas at high stimulus repetition rates (100–200 Hz), typical of bats that have detected their prey and are closing in for capture (Griffin 1958), drop-flips are more likely. Madsen & Miller (1987) also report that acoustic responses can be recorded as synaptic and spiking potentials in the motor neurons of the flight muscles of green lacewings.

## Crickets

Only in moths and green lacewings has it been possible to manipulate experimentally the behavioral interaction between insect and bat under natural conditions in the out-of-doors. However, many insects are active and fly at night. Any insect that takes to the wing after sunset is at risk from bats. Thus, is is not surprising to find reports that some crickets (*Acheta domesticus*) that fly at night are in fact captured and eaten by bats on the wing (Cranbrook & Barrett 1965) or that field crickets have been observed to drop out of the air to the ground in the vicinity of hunting bats (Popov & Shuvalov 1977). Bats alone among mammals have taken to the air, mostly at night, to earn their living, and their species diversity, sheer numbers, and adaptive radiations are a testament to their evolutionary success (Fenton & Fullard 1983). Thus, there is little wonder that some nocturnally active, flying insects have evolved countermeasures against bat predation. Apparently, moths and green lacewings have evolved ears in response to bat predation. It should be no surprise given that insects that had already evolved a sense of hearing (in the context of social communication in crickets) that they might have also extended their auditory sensitivities to encompass the ultrasonic frequencies of bat biosonar. This is especially true for nocturnally active, flying insects.

Following the lead of Roeder, it has been shown in the laboratory that tethered, flying crickets respond to bat-like ultrasound signals played from directionally placed loudspeakers by steering their flight away from the loudspeaker (Popov & Shuvalov 1977, Moiseff et al 1978). In the past decade, an intensive study has been made on the auditory responses of the Australian field cricket, *Teleogryllus oceanicus*, to ultrasound (Moiseff et al 1978, Moiseff & Hoy 1983, Nolen & Hoy 1984, 1986a,b, 1987, Hoy & Nolen 1987). *T. oceanicus* performs stable, persistent flight when suspended in the air by a tether and placed in a windstream. Moreover, these crickets respond readily to acoustic signals by making steering responses either toward (positive phonotaxis) or away from (negative phonotaxis) the loudspeaker emitting the signal. They steer toward cricket songs and away from ultrasonic pulses. This dichotomous behavior has an obvious teleological interpretation: These crickets are attracted to the social calls of other crickets, and they attempt to fly away from (escape) the ultrasonic signals of predacious bats (Hoy et al 1982). The former behavior is discussed at length below; I consider the latter, negative phototactic response because this behavior, like its counterpart in moths and green lacewings, is an acoustic startle response.

BEHAVIORAL ASPECTS OF ACOUSTIC STARTLE IN CRICKETS    Moiseff et al (1978) found that tethered, flying *T. oceanicus* make steering movements (phonotaxis) involving nearly every moveable part of the body when stimulated by sound. In particular, the abdomen is used like a rudder to steer the body through space, such that the abdomen flexes at the thoracic-abdominal joint, in effect turning the abdomen into a convenient "pointer," indicating the direction of the turn. Of course, asymmetrical wingbeats also comprise part of the directional steering response (May & Brodfuehrer 1987), and lateralized movements of the legs, antennae, mouthparts, and head round out the "body English" that accompanies the act of directional steering in a tethered, flying cricket.

The negative phonotactic steering response fulfills several of the criteria of an acoustic startle response: (*a*) it occurs with short latency (25–35 msec, Nolen & Hoy 1986a), (*b*) it is highly stereotyped, involving nearly all moveable body parts (Moiseff et al 1978), (*c*) it tends to have an all-or-none characteristic, although full expression of the behavior is evoked only by very loud stimulus intensities (above 80 dB SPL), (*d*) the response can be evoked by even a single short pulse of ultrasound (5 msec, 30 kHz, 60 dB, Nolen & Hoy 1986a), and (*e*) the response is directed away from the sound source (Moiseff et al 1978, Nolen & Hoy 1986a). Thus, in *Teleogryllus oceanicus*, it is clear that negative phonotaxis to ultrasound is a directional-evasive response that is analogous to other well-known escape behaviors in animals as varied as crayfish (Wine & Krasne 1972), cockroaches (Roeder 1967, Camhi & Tom 1978, Ritzmann 1984), *Drosophila* (Wyman et al 1984), earthworms (Drewes 1984), and teleost fishes (Eaton & Hackett 1984). In all of these examples cited, one can make the case that the escape behaviors are also startle behaviors (Aljure et al 1980, Eaton 1984). It seems quite evident that ultrasound-initiated negative phonotaxis in *T. oceanicus* is an acoustic startle response.

THE NEURAL BASIS OF THE ACOUSTIC STARTLE RESPONSE IN CRICKETS    The cricket ear is also a scolopophorous hearing organ, consisting of about 70 receptors in the tibial segment of foreleg (Schwabe 1906, Michel 1974). Because of the wide range of frequencies (3 kHz to 100 kHz) that must be encoded by the cricket auditory system, only some (as yet unknown number) of the receptors are tuned to ultrasound. However, their anatomical representation is tonotopic, just as in the mammalian cochlea (Zhantiev

& Korsunovskaya 1978, Oldfield et al 1986). Thus, the linear array of auditory receptors in the hearing organ is arranged such that low frequency tuned receptors are found at the proximal end of the tibia and high frequencies are encoded by those at the distal end, with intermediate frequencies progressing from lower to higher in an orderly proximal to distal sequence.

The central processing of ultrasound and its ultimate translation to steering movements in the motor system have only begun to be elucidated. A key question is how directional steering movements are initiated once appropriate stimulation enters the CNS. Are there specific "bat detectors" in the central sensory pathways? Are there "command neurons" that control specific steering motor acts? An approach to these questions was undertaken in *Teleogryllus oceanicus* (Moiseff & Hoy 1983, Nolen & Hoy 1984). It was found that a bilateral pair of identical auditory interneurons (int-1) in the prothoracic ganglion is strongly and tonically excited by ultrasound over a range that overlaps with negative phonotactic steering behavior in tethered, flying crickets (Moiseff & Hoy 1983). This parallelism between neural activity in an identified interneuron and steering behavior in the intact animal raised the question as to whether they were causally related. In brief, direct electrical stimulation of int-1 in the absence of acoustic stimulation caused the animal to perform negative phonotactic steering movements and thus showed that activity in int-1 is sufficient to initiate steering. Moreover, an acoustically triggered act of negative phonotaxis can be nullified by preceding the acoustic stimulus with a pulse of hyperpolarizing current injected intracellularly into int-1 that is adequate to suppress its neural activity. This demonstrated that activity in int-1 is necessary to initiate steering, even in the presence of an otherwise suprathreshold ultrasound stimulus. These experiments (Nolen & Hoy 1984) were designed to test the causal role of int-1 in steering behavior. Incidentally, these tests fulfill the criterion set forth by Kupferman & Weiss (1978) to define a command neuron. Since int-1 is probably a second-order auditory neuron, and given its key role in initiating steering, it can also be considered a "bat detector," but whether int-1 should be considered a bat detector or a command neuron is not important and is largely a matter of emphasis; what matters is that a relatively complex act can be initiated and controlled by activity in a single neuron (Hoy & Nolen 1987). Perhaps the most interesting aspect of these experiments was the finding that int-1 exerted its initiator/command role only when the cricket was flying, as judged by activation of the central flight pattern generator; a cricket that is not in flight ("standing," grooming, etc) cannot be set into steering movements no matter how strongly int-1 is activated (Nolen & Hoy 1984). Thus, the flight central pattern generator appears to gate int-1 into the steering circuitry; this finding is similar to those of Steeves & Pearson (1982) and Ritzmann et al (1980) in showing the importance of behavioral context in shaping the behavioral outcome of afferent stimulation of a given interneuron.

The motor effects (activation of circuitry downstream from int-1) in crickets have yet to be worked out, but preliminary studies (Pollack & Hoy 1981a) show that it is possible to record neural correlates of steering from flight muscles and abdominal steering muscles, as well as their motor neurons. Moreover, whereas decapitated crickets perform apparently normal flight behavior, they no longer steer in response to sound stimuli. Thus, while Boyan & Fullard (1986) posit that flight phonotaxis in moths in response to ultrasound might occur without influences from the brain, behavioral tests in the cricket indicate otherwise; although these initial studies are promising, there remains much work yet to be done before we can claim to understand the neural system for acoustic escape behavior in either species.

## The Acoustic Startle Response in Praying Mantises

Moths, green lacewings, and crickets are by common knowledge creatures of the night; however, such is not the perception of the praying mantis. These are among the largest insects, and their large eyes appear to suggest a mostly diurnal existence; indeed, their visually guided predatory strike behavior is a classic example of a sensorimotor feedback loop (Mittelstaedt 1957). Thus, it was all the more surprising to discover that these creatures, which are also nocturnally active, have well-developed hearing organs that are tuned to ultrasonic frequencies. The range of ultrasound sensitivity in the mantis overlaps that of moths, green lacewings, and crickets, and suggests that possibly these insects have also evolved ears to detect insectivorous bats (Yager & Hoy 1986). The mantis ear (in *Mantis religiosa*) is unique in the animal kingdom. The ear is located between the metathoracic legs, in the midline plane, rendering the mantis an auditory "cyclops." The hearing organ is a conventional scolopophorous type (Yager & Hoy 1987). In the CNS, several auditory neurons have been identified, and one ultrasound-sensitive cell appears to bear striking morphological similarities to the G-neuron in locusts, first identified by Rehbein (1976) on the basis of acoustic sensitivity. Finally, the evidence that ultrasound initiates an acoustic startle response in the praying mantis comes from the observation that tethered, flying mantises (*Creobroter gemmatus*) make short-latency steering responses to pulses of ultrasound, such as found in the biosonar cries of bats (Yager & Hoy 1985). The mantis's phonotactic response consists of a pronounced abdominal dorsi-flexion and extension of the raptorial forelegs; these movements would cause a deviation in the flight course of a free-flying animal. The short latency, stereotypy, and frequency dependence (only ultrasound elicits this response) suggests that this behavior is an acoustic startle response, and that it possibly evolved as a bat-evading behavior.

In summary, the case has been presented that ultrasound is an aversive stimulus in four orders of insects, all of which share a nocturnal habit and take to the wing at some stage in their lives. All flying, nocturnally active insects are potential prey to insectivorous bats. Since most acoustic startle responses appear to serve a warning/alerting function, these acoustic escape behaviors in insects fit right into that classification. Although much work needs to be done to establish the neural bases of the acoustic startle response in insects, it is encouraging that some aspects of the neural circuitry appear to be simple (in number—as few as one auditory receptor in the ear of some moths, and one auditory interneuron to initiate escape in crickets), leading to a hope that the neural system subserving insectan acoustic startle response can be fully understood.

## DOES CATEGORICAL PERCEPTION OCCUR IN INSECTS?

In the introduction to this review, I refer to categorical perception as a process of "segmenting" a stimulus (say, sound frequency) that would otherwise vary continuously into discrete bands that elicit discretely different behaviors. In the mammalian, psychophysical sense, categorical perception studies are best known in speech perception (Liberman et al 1957). Since it is obvious that crickets have neither speech nor perception in the conventional senses of those terms, and that the term "perception" carries surplus, anthropomorphic meaning when applied to insects, I use *categorical response* in this article instead. Despite the obvious differences, there are similarities between categorical perception and categorical response nonetheless if we consider the biological context in which the behavior occurs. In both, the problem is how it is that even when an animal is shown to be sensitive to sound over a broad and continuous range of frequencies, it responds differentially or discriminates only to certain discrete frequencies as though those key frequencies contained "labels." In the most general sense, the question deals with how an animal assigns meaning to particular combinations of stimuli, literally dividing its stimulus world into different categories that command attention. This is a critical problem in the study of neuroethology (Camhi 1984) as well as in cognitive psychology (Marler 1978).

In the section on acoustic startle, I pointed out that moths and crickets are both sensitive to a wide range of ultrasound, and moreover that ultrasound elicits escape behaviors, as though it carries the label "predatory bat." The following section shows that crickets extend their range of hearing to lower frequencies, where social communication among conspecifics occurs. Following the metaphor about labeled stimuli, these lower frequencies might indicate "potential mate—conspecific male," "potential rival," or "heterospecific male." In some of these situations, crickets do not escape but approach the sound source. I present evidence that such decisions are made categorically, and that it is possible to predict the insect's response from the physical description of the acoustic signal (spectral frequency, temporal pattern, intensity). Thus, I try to establish parallelisms between categorical perception as it occurs in mammals and categorical response as found in insects, without arguing for similarity of underlying mechanisms. However, given that categorical perception is considered to be the province of "higher" animals, it is worth inquiring how such a ubiquitous biological task as labeling the sensory world is done in other animals.

## Social Communication in Crickets

It has long been known that female field crickets find potential mates by hearing their calling songs (Regen 1913). These calling songs are species-specific and sung only by adult, sexually mature males singing on their territories (Alexander 1961). Females of the same species detect species-specific features of the male's calling song and walk directly to its source over a distance of tens of meters; this behavior is called positive phonotaxis (Alexander 1961, Hill et al 1972). As pointed out above, flying crickets also perform positive phonotaxis to calling songs. Species-specificity of the calling song derives from two sources. First and most important, the temporal pattern of sound pulses varies markedly from species to species (Alexander 1962), and second, the carrier frequency is also species-specific, ranging within a narrow band from 3 to 6 kHz (Alexander 1961, 1962, Hill 1974). In the laboratory, female field crickets (*G. campestris*) walking on a servo-controlled tracking device have been shown to be highly discriminating in their photontactic response to electronic models of calling song: Only those models that closely match the pulse rhythm of the natural call are attractive to walking females (Weber et al 1981). Similarly in other lab studies, tethered flying female crickets (*Gryllus bimaculatus* and *Teleogryllus oceanicus*) steer directionally toward calling songs (Popov & Shuvalov 1977, Moiseff et al 1978). Mole crickets (*Scapteriscus spp.*) have been shown to be attracted on the wing to conspecific calling songs that are played from a tape recorder under natural conditions in the field (Ulagaraj & Walker 1973). Thus, it is clear that the 3–100 kHz range of auditory sensitivity of the field cricket is divided into at least two categories of sound sources: lower frequencies (3 to 6 kHz) are "recognized" as crickets, and elicit positive phonotaxis, whereas ultrasonic frequencies (20 kHz and above) are "recognized" as bats, and elicit negative phonotaxis. This dichotomous response to acoustic frequency has been shown to operate in the lab with tethered, flying *T. oceanicus,* and certainly argues for an elementary form of categorical response (Hoy et al 1982).

It is possible that CR is even further refined in field crickets. Male field crickets sing at least two other kinds of "songs," besides the calling song, that accompany social acts. Once a calling male has attracted a conspecific female and the pair is within physical contact distance, the male sings a "courtship song" that is thought to facilitate copulation, and if one male cricket is in physical contact with another (rival) male, fighting often breaks out over territory, and such encounters are accompanied by the production of an "aggresive song" by one or both contestants (Alexander 1961). Both aggression and courtship songs are quite different in their temporal structure from the stereotyped calling songs. Interestingly, the courtship songs of the genus *Gryllus* differ markedly from the calling song in spectral frequency: the dominant frequency of calling song is the same as the carrier frequency (5 kHz), but the dominant frequency of the courtship song is often the third harmonic (15 kHz) (Nocke 1972). Unfortunately, few investigations have been made that define the behavioral role of either courtship or aggression songs, but we might expect that further behavioral "segmentation" of the frequency band in cricket hearing to occur. Preliminary studies indicate that the ultrasonic region might be broken up such that 15 kHz would "label" potential mates, whereas higher frequencies (up to 100 kHz) would label predatory bats (Murray & Hoy 1988). However, the behavioral context in which these frequencies are heard must also be considered, as is discussed below.

## The Neural Basis of Song "Recognition" in Crickets

The species-specificity of the calling song in field crickets and the relatively unambiguous behavioral responses made to it by conspecific females, have drawn the attention of physiologists seeking to define the neural systems that underlie song recognition. Since it is the pattern of amplitude modulation of the call (the call rhythm or temporal pattern) that confers species-specificity, the search for neural correlates of call recognition has focused on finding single units that respond selectively to the temporal pattern of the calling song. The strategy has been to work on interneurons, beginning with second-order cells in the prothoracic ganglion and then to work up the hierarchy to cells in the brain. In field crickets, between 6 and 10 interneurons have been identified in the prothoracic ganglion that respond to one feature or another of cricket song (Popov & Shuvalov 1977, Popov et al 1978, Casaday & Hoy 1977, Wohlers & Huber 1978, 1982, Stout et al 1985). Interestingly, none of these neurons appears to respond differentially to the temporal pattern of calling song, although several, notably AN-1 (Wohlers & Huber 1982), have an afferent-like response that relays a precise "copy" of the temporal pattern of song to the brain. The failure to find "song detectors" in the prothoracic ganglion has taken the search to the brain, and with very promising results. The remarkable findings of Schildberger (1984, 1985) in *G. campestris* are clearly significant. Schildberger discovered several classes of interneurons in the brain, among them local circuit neurons called the BNC-1 and BNC-2 cells. They are presumably at least third-order auditory cells, and appear to get input from the AN-1 cells. It is currently believed on the basis of recordings to date that the BNC-1 cells act as temporal low-pass filters, that some members of the BNC-2 class act as temporal high-pass filters, and that others of the BNC-2 class are temporal band-pass filters. In particular, it turns out that the BNC-2 band-pass type appear to respond selectively to pulse repetition rates in model calling song stimuli near that of naturally occurring conspecific calling songs. The model presented by Schildberger for crickets, and interestingly, a remarkably similar model presented by Rose & Capranica for the detection of amplitude-modulated mating calls in frogs (1984), represent the best efforts of neuroethology to demonstrate that "song detectors" sensitive to specific patterns of amplitude modulation in acoustic signals in the brain occur at the level of single neurons. Naturally, an ever-present caveat accompanies such claims: What has been shown so far is correlative evidence. The data demonstrate that neurons such as BNC-2 are differentially responsive to the temporal pattern of a stimulus song, a pattern that matches that which occurs in the natural species song. However, it is not shown that "recognition," whatever that is, occurs in BNC-2 cells. Perhaps the BNC-2 cells drive other postsynaptic cells that in turn drive the motor act of phonotaxis. It would be desirable to put the BNC-2 cells to the test of necessity and sufficiency (Kupfermann & Weiss 1978) for phonotaxis. However, this test might not be applicable to the BNC cells because it appears that at least several different cells make up each class. Thus, unlike the situation that was outlined above for negative phonotaxis in flying crickets, in which a particular interneuron, int-1, was shown to be both necessary and sufficient to initiate steering, the process of initiating positive phonotaxis is likely to be complicated by larger numbers of cells at critical points in the neuronal hierarchy.

In summary, the case for categorical response in crickets can be made from neurophysiological as well as behavioral evidence. The act of mating-call recognition appears to be mediated by a separate neural pathway to the brain (with AN-1 as the front-end interneuron), and involves several classes of higher order brain interneurons (BNC-1, BNC-2). In the case of mating-call recognition, not only must the spectral frequency of the call be correct (3 to 6 kHz, depending on species), but so must the temporal pattern of the sound pulses. At the other end of the frequency spectrum, a separate interneuron, int-1, mediates ultrasound processing in the brain. Not only is int-1 the front end neuron in the circuit, but it is a causal link in initiating the response. It is tempting to speculate that whereas there is a partitioning of stimulus information early in the phonotactic steering networks (low frequency afferents, AN-1, BNC cells in positive phonotaxis, and high frequency afferents, int-1 in negative phonotaxis), these pathways will converge upon a common motor system for the actual act of steering.

## Other Possible Examples of Categorical Response

The long-horned grasshoppers, the tettigoniids, are related to crickets. These insects also produce songs, but unlike the tone-like calls of crickets, tettigoniid calls are noisy and contain ultrasonic harmonics (DuMortier 1963). In fact, some tropical tettigoniidae are inaudible to the human ear because their calls contain only ultrasonic frequencies in the range 30–70 kHz (Suga 1968), which overlap those frequencies found in bat echolocation calls (Griffin 1958). The fact that the calls of these insects serve as species-specific mating calls requires that females of the species be able to discriminate among ultrasound sources. Just as in crickets, one source of species-specific information in the calls of tettigoniidae occurs in the temporal pattern of the sound pulses (DuMortier 1963). However, these long-horned grasshoppers are also nocturnal and capable of flight, thus raising the question of bat predation. Indeed, some species of tethered, flying *Neoconocephalus* respond to pulsed ultrasound (30 kHz) by making abdominal flexion movements directed away from the loudspeaker (R. R. Hoy, unpublished); this act is reminiscent of the rudder-like steering response observed in crickets (Moiseff et al 1978). In fact, tettigoniids do not even have to fly in order to be preyed upon by bats. Recent work by Belwood & Morris (1987) suggests that gleaning bats detect the songs of singing insects and use the mating call as a beacon to locate their prey. Thus, it appears that tettigoniids, too, have a serious bat predation problem. Since in many species of these insects, ultrasonic frequencies occur in their social communication signals as well as in the echolocation signals of their predators, some sort of "labeling" process must take place, because the insects must make categorically different behavioral responses to mating calls and predatory echolocation signals.

# SELECTIVE ATTENTION

The third and final behavior that can be studied in crickets as well as mammals is selective attention. The basic problem for humans is of commonplace familiarity and is known as the "cocktail party phenomenon" (Cherry 1966). In such surroundings, one can be distracted by many simultaneous conversations within easy earshot, yet it is easy to "tune into" a single conversation selectively, particularly if cued by the sudden mention of one's name, for example.

The cricket-equivalent of the "cocktail party phenomenon" is also commonplace to these semisocial insects. Singing crickets, sometimes even several different species, can crowd themselves to such an extent that the nearest-neighbor distances between calling males competing for the attention of a conspecific female is on the order of a meter or less, and any female walking into a den and din of simultaneously calling males must sort out the call of a male of her own species (Campbell & Clarke 1971). Laboratory tests demonstrate that female crickets accurately discriminate the calls of their own species from that of others (Hill et al 1972, Popov & Shuvalov 1977, Pollack & Hoy 1981b, Weber et al 1981). Hill et al (1972) reported that in natural areas where sympatric species of field crickets congregate, there is no evidence of hybridization among them, and these authors inferred that song discrimination is sufficient to maintain the separation of species.

Although it may seem a stretch to analogize female song choice in crickets to selective listening in a crowd of humans, there are parallels, and recently Pollack (1986, 1988) has presented evidence for selective attention in *T. oceanicus*, as well as a neural mechanism to account for it. Pollack showed that an identified auditory interneuron, the omega neuron, selectively encodes the pulse pattern of an ipsilateral calling song when a contralateral calling song is presented simultaneously. This is true even though the contralateral stimulus was shown to be unambiguously encoded by the same omega neuron when the contralateral stimulus was presented by itself. Thus, although both acoustic stimuli were demonstrated to be "heard" by the neuron, only the ipsilateral one is attended to when the stimuli occur simultaneously. Pollack (1988) interprets his findings as the equivalent in crickets of selective attention. It is interesting to note that Cherry's demonstration of the "cocktail party" effect requires the presence of binaural sound cues (Cherry 1953, 1966). Pollack (1988) demonstrates that the physiological basis for the selective response to the ipsilateral sound stimulus results from an apparent intensity difference between ipsilateral and contralateral stimuli, arising from the directionality of the auditory system. The more intense (ipsilateral) stimulus results in an intensity-dependent inhibition of an identified neuron so that the less intense, contralateral stimulus is rendered subthreshold.

At present, the identity of the inhibitory neural system is unknown. Nonetheless, Pollack has presented an attractive model system for the study of selective attention. Although sound intensity appears to be a dominant factor in commanding the attention of crickets, it is clearly not necessarily the case in SA in humans, because one can focus in on a specific conversation occurring in the midst of several simultaneous conversations even though the one of interest is more "softly" spoken.

As an aside, another related auditory phenomenon that is the subject of studies in mammalian psychoacoustics, echo suppression, can also be examined in crickets, and preliminary evidence (G. B. Casaday and R. R. Hoy, unpublished) suggests that this phenomenon can also be demonstrated in the phonotactic behavior of tethered, flying crickets.

## CONCLUDING REMARKS

Behaviors that have survival value are likely to be found in all animals facing similar problems. In many animals a sense of hearing is the keystone of survival acts, such as detecting predators or recognizing and attracting mates. This is no less true for insects than mammals, for whom the sense of hearing also serves these roles. Yet, in mammals it is held that certain kinds of auditory phenomena require higher order, cognitive processing. Certainly, selective attention and categorical perception are examples. There are analogs to both of these examples in the acoustic behavior of crickets. The evolutionary pressure for being able to assign unambiguous meaning along a stimulus dimension (frequency), or focusing attention in a noisy environment, is common to all animals that use hearing for making their way in the world. Thus, with appropriate "pruning" of their definitions, both categorical perception (here, categorical response) and selective attention can be studied in crickets. The case for common kinds of evolutionary pressures (predators) in the case of the acoustic startle response is much clearer, and the parallels in these behaviors are apparent. The acoustic startle response appears to initiate an escape response in a variety of animals (Eaton 1984).

Why bother to point out these similarities in behavior? Besides the intrinsic satisfaction to be derived from comparative studies that indicate behavioral convergence in the face of similar evolutionary selection pressures, the comparative neurobiologist might be heartened to know that in the insect nervous system it is possible to understand the underlying neural system in the acoustic startle response, categorical response, and selective attention in terms of identified neurons, operating in networks of considerably less complexity those of mammals. No claims are made that the operation of neural systems in insects will necessarily explain the mechanisms that underlie the acoustic startle response, categorical perception, and selective attention of mammals. It will be rewarding enough to know how complex auditory behavior works in even so modest a creature as a cricket or a moth.

ACKNOWLEDGMENTS

I am fortunate to be a member of a lively, stimulating, international community of bioacousticians who study insects, and I acknowledge my debt to them. I thank Jerry Pollack, Bob Wyttenbach, and Margaret Nelson for commenting on this manuscript. The research from my laboratory has been generously supported by the NINCDS, including a Jacob Javits Neuroscience Investigator Award; I am greatly indebted to them.

*Literature Cited*

Alexander, R. D. 1961. Aggressiveness, territoriality, and sexual behavior in field crickets (Orthoptera : Gryllidae). *Behavior* 17: 130–223

Alexander, R. D. 1962. Evolutionary change in cricket acoustical communication. *Evolution* 16: 443–67

Aljure, E., Day, J. W., Bennett, M. V. L. 1980. Postsynaptic depression of Mauthner cell-mediated startle reflex, a possible contributor to habituation. *Brain Res.* 188: 261–68

Belwood, J. J., Morris, G. K. 1987. Bat predation and its influence on calling behavior in neotropical katydids. *Science* 238: 64–67

Bennett, M. V. L. 1984. Escapism: some startling revelations. See Eaton 1984, pp. 353–63

Boyan, G. S., Fullard, J. H. 1986. Interneurones responding to sound in the tobacco budworm moth *Heliothis virescens* (Noctuidae): morphological and physiological characteristics. *J. Comp. Physiol. A* 158: 391–404

Bullock, T. H. 1984. Comparative neuroethology of startle, rapid escape, and giant-fiber mediated responses. See Eaton 1984, pp. 1–11

Camhi, J., Tom, W. 1978. The escape behavior of the cockroach *Periplaneta americana*. I. Turning response to wind puffs. *J. Comp. Physiol.* 128: 193–201

Camhi, J. 1984. *Neuroethology*. Sunderland: Sinauer

Campbell, D. J., Clarke, D. J. 1971. Nearest neighbour tests of significance for non-randomness in the spatial distribution of singing crickets (*Teleogryllus commodus* Walker). *Anim. Behav.* 19: 750–56

Casaday, G. C., Hoy, R. R. 1977. Auditory interneurons in the cricket *Teleogryllus oceanicus*: Physiological and anatomical properties. *J. Comp. Physiol.* 121: 1–13

Cherry, C. 1953. Some experiments on the recognition of speech, with one and two ears. *J. Acoust. Soc. Am.* 25: 975–79

Cherry, C. 1966. *On Human Communication*. Cambridge: MIT Press

Cranbrook, the Earl of, Barrett, H. F. 1965. Observations on noctule bats (*Nyctalus noctula*) captured while feeding. *Proc. Zool. Soc. London* 144: 1–24

Davis, M. 1984. The mammalian startle response. See Eaton 1984, pp. 287–342

Drewes, C. D. 1984. Escape reflexes in earthworms and other annelids. See Eaton 1984, pp. 43–86

DuMortier, B. 1963. The physical characteristics of sound emissions in arthropoda. In *Acoustic Behavior of Animals*, ed. R. G. Busnel, pp. 346–73. New York: Elsevier

Eaton, R. C., ed. 1984. *Neural Mechanisms of Startle Behavior*. New York: Plenum

Eaton, R. C., Hackett, J. T. 1984. The role of the Mauthner cell in Fast-starts involving escape in teleost fishes. See Eaton 1984, pp. 213–62

Elsner, N., Popov, A. V. 1978. Neuroethology of acoustic communication. *Adv. Ins. Physiol.* 13: 229–355

Fenton, M. B., Fullard, J. H. 1983. Moth

hearing and the feeding strategies of bats. *Am. Sci.* 69: 266–75

Forbes, A., Sherrington, C. S. 1914. Acoustic startle reflexes in the decerebrate cat. *Am. J. Psychiatr.* 35: 367–76

Griffin, D. R. 1958. *Listening in the Dark.* New Haven: Yale Univ. Press

Hill, J. G. 1974. Carrier frequency as a factor in phonotactic behavior of female crickets (*Teleogryllus commodus*). *J. Comp. Physiol.* 93: 7–18

Hill, K. G., Loftus-Hills, J. J., Gartside, D. F. 1972. *Aust. J. Zool.* 20: 153–63

Hoy, R. R., Pollack, G. S., Moiseff, A. 1982. Species-recognition in the field cricket, *Teleogryllus oceanicus*: behavioral and neural mechanisms. *Am. Zool.* 22: 597–607

Hoy, R. R., Nolen, T. G. 1987. The role of behavioral context in decision making by an identified interneuron in the cricket. In *Higher Brain Functions*, ed. S. P. Wise. New York: Wiley

Huber, R., Markl, H. 1983. *Neuroethology and Behavioral Physiology.* Heidelberg: Springer-Verlag

Huber, F., Moore, T. E., Loher, W. 1989. *Cricket Behavior and Neurobiology.* Ithaca: Cornell Univ. Press. In press

Kalmring, K., Elsner, N. 1985. *Acoustic and Vibrational Communication in Insects.* Berlin/Hamburg: Verlag Paul Parey

Kupfermann, I., Weiss, K. R. 1978. The command neuron concept. *Behav. Brain Sci.* 1: 3–39

Landis, C., Hunt, W. A. 1939. *The Startle Pattern.* New York: Farrar & Rinehart

Liberman, A. M., Harris, K. S., Hoffman, H. S., Griffith, B. C. 1957. The discrimination of speech sounds within and across phoneme boundaries. *J. Exp. Psychol.* 54: 358–68

Madsen, B. M., Miller, L. A. 1987. Auditory input to motor neurons of the dorsal longitudinal flight muscles in a noctuid moth (*Barathra brassicae* L). *J. Comp. Physiol.* 160: 23–31

Marler, P. 1978. Perception and innate knowledge. In *The Nature of Life*, ed. W. H. Heidcamp, pp. 111–39. Baltimore: University Park Press

May, M. L., Brodfuehrer, P. D. 1987. Changes in wing parameters in Teleogryllus oceanicus due to ultrasound stimulation. *Soc. Neurosci. Abstr.* 13: 398

Michel, K. 1974. Das tympanalorgan von *Gryllus bimaculatus* DeGeer (Saltatoria, Gryllidae). *Z. Morphol. Tiere* 77: 295–315

Michelsen, A., Nocke, H. 1974. Biophysical aspects of sound communication in insects. *Adv. Ins. Physiol.* 10: 247–96

Michelsen, A., Larsen, O. N. 1985. Hearing and sound. In *Comprehensive Insect Physiology Biochemistry and Pharmacology*, ed.

G. A. Kerkut, L. I. Gilbert, pp. 495–555. New York: Pergamon

Miller, L. A. 1970. Structure of the green lacewing tympanal organ (*Chrysopa carnea*). *J. Morphol.* 131: 359–82

Miller, L. A. 1971. Physiological responses of green lacewings (*Chrysopa*, Neuroptera) to ultrasound. *J. Insect Physiol.* 17: 491–506

Miller, L. A. 1975. The behavior of flying green lacewings, *Chrysopa carnea*, in the presence of ultrasound. *J. Insect Physiol.* 21: 205–19

Miller, L. A., Olesen, J. 1979. Avoidance behavior in green lacewings. *J. Comp. Physiol.* 131: 113–20

Mittelstaedt, H. 1957. Prey capture in mantids. In *Recent Advances in Invertebrate Physiology*, ed. B. T. Scheer, pp. 41–71. Eugene: Univ. Oreg. Publ.

Moiseff, A., Pollack, G. S., Hoy, R. R. 1978. Steering responses of flying crickets to sound and ultrasound: mate attraction and predator avoidance. *Proc. Natl. Acad. Sci. USA* 75: 4052–56

Moiseff, A., Hoy, R. R. 1983. Sensitivity to ultrasound in an identified auditory interneuron in the cricket: a possible neural link to phonotactic behavior. *J. Comp. Physiol.* 152: 155–67

Murray, J. J., Hoy, R. R. 1988. Courtship success in the cricket *Gryllus bimaculatus*: Differential contribution of carrier frequency and harmonics in courtship song. *Soc. Neurosci. Abstr.* 14(p. 1): 310

Nocke, H. 1972. Physiological aspects of sound communication in crickets (*Gryllus campestris* L.). *Z. Vgl. Physiol.* 80: 141–62

Nolen, T. G., Hoy, R. R. 1984. Initiation of behavior by single neurons: the role of behavioral context. *Science* 226: 992–94

Nolen, T. G., Hoy, R. R. 1986a. Phonotaxis in flying crickets. I. Attraction to the calling song and avoidance of bat-like ultrasound are discrete behaviors. *J. Comp. Physiol.* 159: 423–39

Nolen, T. G., Hoy, R. R. 1986b. Phonotaxis in flying crickets. II. Physiological mechanisms of two-tone suppression of the high frequency avoidance steering behavior by the calling song. *J. Comp. Physiol.* 159: 441–56

Nolen, T. G., Hoy, R. R. 1987. Postsynaptic inhibition mediates high-frequency selectivity in the cricket *Teleogryllus oceanicus*: implications for flight phonotaxis behavior. *J. Neurosci.* 7: 2081–96

Oldfield, B. P., Kleindienst, H. U., Huber, F. 1986. Physiology and tonotopic organization of auditory receptors in the cricket *Gryllus bimaculatus* DeGeer. *J. Comp. Physiol.* 159: 457–64

Pearson, K. G., O'Shea, J. 1984. The escape behavior of the locust: the jump and its initiation by visual stimuli. See Eaton 1984, pp. 163–77

Pollack, G. S., Hoy, R. R. 1981a. Phonotaxis is flying crickets: neural correlates. *J. Insect Physiol.* 27: 41–45

Pollack, G. S. 1986. Discrimination of calling song models by the cricket, *Teleogryllus oceanicus*: the influence of sound direction on neural encoding of the stimulus temporal pattern and on phonotactic behavior. *J. Comp. Physiol.* 158: 549–61

Pollack, G. S. 1988. Selective attention in an insect auditory neuron. *J. Neurosci.* 8: 2635–39

Pollack, G. S., Hoy, R. R. 1981b. Phonotaxis to individual rhythmic components of a complex cricket calling song. *J. Comp. Physiol.* 144: 367–73

Pollack, G. S., Huber, F., Weber, T. 1984. Frequency and temporal pattern-dependent phonotaxis of crickets (*Teleogryllus oceanicus*) during tethered flight and compensated walking. *J. Comp. Physiol.* 154: 13–26

Popov, A. V., Shuvalov, V. F. 1977. Phonotactic behavior of crickets. *J. Comp. Physiol.* 119: 111–26

Popov, A. V., Markovitch, A. M., Andjan, A. S. 1978. Auditory neurons in the prothoracic ganglion of the cricket, *Gryllus bimaculatus* DeGeer. *J. Comp. Physiol.* 126: 183–92

Regen, J. 1913. Über die Anlockung des Weibchens ven *Gryllus campestris* L. dürch telephonisch übertragene Stridulations-laute des Mannchens. *Plügers Arch.* 155: 193–200

Rehbein, H. G. 1976. Auditory neurons in the ventral cord of the locust: morphological and functional properties. *J. Comp. Physiol.* 110: 233–50

Ritzmann, R. E., Tobias, M. L., Fourtner, C. R. 1980. Flight activity initiated by giant interneurons of the cockroach: evidence for bifunctional trigger interneurons. *Science* 210: 443–45

Ritzmann, R. E. 1984. The cockroach escape response. See Eaton 1984, pp. 93–128

Robinson, D. L., Petersen, S. E. 1986. A neurobiology of attention. In *Mind and Brain*, ed. J. E. LeDoux, W. Hirst, pp. 142–70. New York: Cambridge Univ. Press

Roeder, K. D. 1966. Interneurons of the thoracic nerve cord activated by tympanic nerve fibers in noctuid moths. *J. Insect Physiol.* 12: 1227–44

Roeder, K. D. 1967. *Nerve Cells and Insect Behavior*. Cambridge: Harvard Univ. Press

Roeder, K. D. 1969a. Acoustic interneurons in the brain of noctuid moths. *J. Insect Physiol.* 15: 825–38

Roeder, K. D. 1969b. Brain interneurons in noctuid moths: differential suppression by high sound intensities. *J. Insect Physiol.* 15: 1713–18

Roeder, K. D., Treat, A. E. 1957. Ultrasonic reception by the tympanic organs of noctuid moths. *J. Exp. Zool.* 134: 127–58

Roeder, K. D., Treat, A. E. 1961. The detection and evasion of bats by moths. *Am. Sci.* June: 135–48

Rose, G. J., Capranica, R. R. 1984. Processing amplitude-modulated sounds by the auditory midbrain of two species of toads: matched temporal filters. *J. Comp. Physiol.* 154: 211–19

Sales, G., Pye, D. 1974. *Ultrasonic Communication By Animals*. London: Chapman & Hall

Schildberger, K. 1984. Temporal selectivity of identified auditory neurons in the cricket brain. *J. Comp. Physiol.* 155: 171–85

Schildberger, K. 1985. Recognition of temporal patterns by identified auditory neurons in the cricket brain. See Kalmring & Elsner 1985, pp. 41–50

Schwabe, J. 1906. Beitrage zur morphologie und histologie der tympanalen sinnesapparet det orthopteren. *Zoologica* 50: 1–154

Selverston, A. I. 1985. *Model Neural Networks and Behavior*. New York: Plenum

Simmons, J. A., Fenton, M. B., O'Farrell, M. J. 1979. Echolocation and pursuit of prey by bats. *Science* 203: 16–21

Steeves, J. D., Pearson, K. G. 1982. Proprioceptive gating of inhibitory pathways to hindleg flexor motoneurons in the locust. *J. Comp. Physiol.* 146: 507–15

Stewart, W. W. 1978. Functional connections between cells as revealed by dye-coupling with a highly fluorescent naththalidimide tracer. *Cell* 14: 741–59

Stout, J. F., Atkins, G., Burghardt, F. 1985. The characterization and possible importance for phonotaxis of "L" shaped ascending acoustic interneurons in the cricket *Acheta domesticus*. See Kalmring & Elsner 1985, pp. 89–100

Stretton, A. O. W., Kravitz, E. A. 1968. Neuronal geometry: determination with a technique of intracellular dye injection. *Science* 162: 132–34

Suga, N. 1968. Ultrasonic production and its reception in some neotropical tettigoniidae. *J. Insect Physiol.* 12: 1039–50

Surlykke, A. 1984. Hearing in notodontid moths. A hearing organ with only a single auditory neurone. *J. Exp. Biol.* 113: 323–35

Ulagaraj, S. M., Walker, T. J. 1973. Phonotaxis of crickets in flight: attraction of male and female crickets to male calling songs. *Science* 182: 1278–79

Weber, T., Thorson, J., Huber, F. 1981. Auditory behavior of the cricket. I. Dynamics of compensated walking and discrimination paradigms on the Kramer treadmill. *J. Comp. Physiol.* 141: 215–32

Wine, J. J., Krasne, F. B. 1972. The organization of escape behavior in the crayfish. *J. Exp. Biol.* 56: 1–18

Wohlers, D. W., Huber, F. 1978. Intracellular recording and staining of cricket auditory interneurons (*Gryllus campestris* L., *Gryllus bimaculatus* DeGeer). *J. Comp. Physiol.* 127: 11–28

Wohlers, D. W., Huber, F. 1982. Processing of sound signals by six types of neurons in the prothoracic ganglion of the cricket, *Gryllus campestris* L. *J. Comp. Physiol.* 146: 161–73

Wyman, R. J., Thomas, J. B., Salkoff, L., King, D. G. 1984. The *Drosophila* giant fiber system. See Eaton 1984, pp. 133–60

Yager, D. D., Hoy, R. R. 1985. Neuroethology of audition in the praying mantis, Creobroter gemmatus. *Soc. Neurosci. Abstr.* 11: 202

Yager, D. D., Hoy, R. R. 1986. The cyclopean ear: a new sense for the praying mantis. *Science* 231: 727–29

Yager, D. D., Hoy, R. R. 1987. The midline metathoracic ear of the praying mantis, *Mantis religiosa*. *Cell Tissue Res.* 250: 531–41

Zhantiev, R. D. 1981. *Bioacoustics of Insects.* Moscow: Moscow Univ. Press (In Russian)

Zhantiev, R. D., Korsunovskaya, O. S. 1978. Morphological organization of tympanal organs in *Tettigonia cantans* (Orthoptera, Tettigoniidae). *Zool. J.* 57: 1012–16 (In Russian)

*Ann. Rev. Neurosci. 1989. 12:377–403*

# VISUAL AND EYE MOVEMENT FUNCTIONS OF THE POSTERIOR PARIETAL CORTEX

*Richard A. Andersen*

Department of Brain and Cognitive Sciences, Massachusetts Institute of Technology, Cambridge, Massachusetts 02139

## INTRODUCTION

Lesions of the posterior parietal area in humans produce interesting spatial-perceptual and spatial-behavioral deficits. Among the more important deficits observed are loss of spatial memories, problems representing spatial relations in models or drawings, disturbances in the spatial distribution of attention, and the inability to localize visual targets. Posterior parietal lesions in nonhuman primates also produce visual spatial deficits not unlike those found in humans. Mountcastle and his colleagues were the first to explore this area, using single cell recording techniques in behaving monkeys over 13 years ago. Subsequent work by Mountcastle, Lynch and colleagues, Hyvarinen and colleagues, Robinson, Goldberg & Stanton, and Sakata and colleagues during the period of the late 1970s and early 1980s provided an informational and conceptual foundation for exploration of this fascinating area of the brain. Four new directions of research that are presently being explored from this foundation are reviewed in this article.

1. The anatomical and functional organization of the inferior parietal lobule is presently being investigated with neuroanatomical tracing and single cell recording techniques. This area is now known to be comprised of at least four separate cortical fields.
2. Neural mechanisms for spatial constancy are being explored. In area 7a information about eye position is found to be integrated with visual inputs to produce representations of visual space that are head-centered

377

0147–006X/89/0301–0377$02.00

(the meaning of a head-centered coordinate system is explained on p. 13).

3. The role of the posterior parietal cortex, and the pathways projecting into this region, in processing information about motion in the visual world is under investigation. Visual areas within the posterior parietal cortex may play a role in extracting higher level motion information including the perception of structure-from-motion.

4. A previously unexplored area within the intraparietal sulcus has been found whose cells hold a representation in memory of planned eye movements. Special experimental protocols have shown that these cells code the direction and amplitude of intended movements in motor coordinates and suggest that this area plays a role in motor planning.

## ANATOMICAL AND FUNCTIONAL ORGANIZATION

The posterior parietal cortex comprises the caudal aspect of the parietal lobe (see Figure 1). This cortical area consists of the superior and inferior parietal lobules. Brodmann (1905) designated the superior parietal lobule area 5 and the inferior parietal lobule area 7. Area 5 contains exclusively somatosensory association cortex. Area 7 has been further subdivided into two areas based on cytoarchitectural criteria: a caudalmedial area designated 7a by Vogt & Vogt (1919) or PG by von Bonin & Bailey (1947) and a more lateral-rostral area designated 7b (Vogt & Vogt 1919) or PF (von Bonin & Bailey 1947). The inferior parietal lobule includes not only the cortex on the gyral surface but also the cortex in the lateral bank of the intraparietal sulcus, the cortex in the anterior bank of the caudal third of the superior temporal sulcus, and even extends to include a small section of cortex on the medial wall of the cerebral hemisphere.

Description of precise homologies for areas in the posterior parietal cortices of monkeys and humans is difficult. Brodmann asserted that the superior parietal lobule of man was cytoarchitecturally comparable to the inferior parietal lobule of the monkey. If this were true then there would be no homologous area in the monkey to Brodmann's areas 39 and 40, which comprise the inferior parietal lobule of humans. Von Bonin & Bailey disagreed with Brodmann's scheme in the monkey and suggested that their areas PG and PF that comprise the inferior parietal lobule in monkey were equivalent to the inferior parietal lobule in man. The fact that lesions of the inferior parietal lobule produce similar visual disorders in monkey and man whereas lesions of the superior parietal lobule generally result in somatosensory disorders in the two species argues for von Bonin & Bailey's view.

Based on functional differences and differing cortico-cortical connections and subcortical connections, the inferior parietal lobule has been found to contain at least four distinct subdivisions: areas 7b, LIP, MST, and 7a. A flattened reconstruction of the cortex of the inferior parietal lobule and adjacent prelunate gyrus are pictured in Figure 1 with the various cortical areas of the inferior parietal lobule indicated in the figure.

## Area 7b

A majority of the cells in area 7b respond to somatosensory stimuli (Hyvarinen & Shelepin 1979, Robinson & Burton 1980a,b, Hyvarinen 1981, Andersen et al 1985c). Robinson & Burton (1980a,b) reported a crude

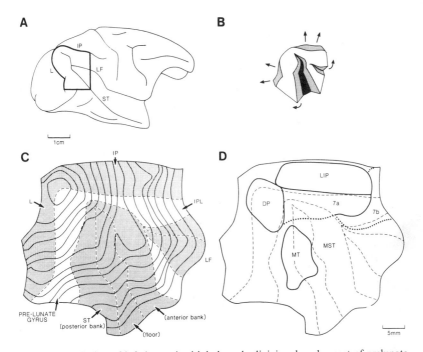

*Figure 1* Parcellation of inferior parietal lobule and adjoining dorsal aspect of prelunate gyrus based on physiological, connectional, myeloarchitectural, and cytoarchitectural criteria. Cortical areas are represented on flattened reconstructions of cortex. (*A*) Lateral view of monkey hemisphere. *Darker lines* outline flattened area. (*B*) Same cortex isolated from the rest of the brain. *Stippled areas*, cortex buried in sulci; *blackened area*, floor of superior temporal sulcus (ST); *arrows*, movement of local cortical regions resulting from mechanical flattening. (*C*) Completely flattened representation of same frontal sections through this area. (*D*) Locations of several cortical areas. *Dotted lines*, borders of cortical fields not precisely determinable; DP, dorsal prelunate area; IP, intraparietal sulcus; IPL, inferior parietal lobule; L, lunate sulcus; LF, lateral fissure; LIP, lateral intraparietal area; MST, medial superior temporal area; MT, middle temporal area. From Andersen (1987).

topographic arrangement of the body represented in area 7b, although the receptive fields of the cells were often very large and as a result obscured the topographic order. Cells have been studied in this area that are responsive to reaching and hand manipulation (Mountcastle et al 1975, Andersen et al 1985c).

A minority of the cells in this region (10%) have also been reported to be responsive to visual or visual and somatosensory stimuli (Robinson & Burton 1980a,b, Hyvarinen 1981). As would be expected, area 7b has been found to possess cortico-cortical connections primarily with other areas involved in somatosensory processing. These other somatosensory areas include the insular cortex, and area 5 (Andersen 1987 for review). Area 7b also has been found to receive its primary thalamic input from the oral subdivision of the pulvinar (Asanuma et al 1985). This thalamic nucleus connects to other somatosensory areas such as area 5.

## Area LIP

The lateral intraparietal area (LIP) is located in the lateral bank of the intraparietal sulcus. It appears to play a role in saccadic eye movements. Shubutani and colleagues (1984) have reported evoking saccades with electrical stimulation to the area at lower thresholds than to other regions they have examined in the posterior parietal cortex. However, the currents they used in area LIP were still rather high compared to those required to elicit eye movements from the frontal eye fields or superior colliculus. Andersen et al (1985a) have found many more saccade-related neurons in this area than in area 7a. Area LIP demonstrates a much stronger projection than that arising from area 7a to the frontal eye fields and superior colliculus (Barbas & Mesulam 1981, Asanuma et al 1985, Lynch et al 1985), two structures involved in the generation of saccades. LIP is the recipient of inputs from several extrastriate cortical areas, including the middle temporal area (MT), a cortical field implicated in visual motion processing (Ungerleider & Desimone 1986, Andersen 1987).

## Area MST

Recent experiments suggest that the medial superior temporal area (MST) is specialized for motion analysis and smooth pursuit eye movements. Sakata and colleagues (1983) and Wurtz & Newsome (1985) have found that most cells responding during smooth pursuit are located in this brain area. Sakata et al (1985) and Saito et al (1985) have found that many cells in this area are sensitive to relative motion, responding to such parameters as rotation and size change. Lesions to MST produce deficits in smooth pursuit eye movements; the speed required to initiate smooth pursuit in order to match target speed is underestimated and the maintenance of pursuit is subsequently defective for tracking toward the side of the lesion (Dursteler et al 1986). Area MST has been shown to receive direct projections from several extrastriate visual areas, including area MT. It projects to area 7a and LIP (Maunsell & Van Essen 1983, Seltzer & Pandya 1984, Colby et al 1985, Siegel et al 1985, Andersen 1987).

## Area 7a

Area 7a appears to play a role in spatial analysis through the integration of eye position and retinotopic visual information. A majority of the cells examined in this area have visual receptive fields (Hyvarinen 1981, Motter & Mountcastle 1981, Andersen et al 1987). Many of these visual cells also carry eye position signals and some display saccade-related activity (Andersen et al 1987). For many neurons visual excitability varies as a function of the position of the eyes in the orbit, that is, the angle of gaze. This gating of visual signals by eye position produces a tuning for locations in what is known as head-centered space (Andersen et al 1985b); these notions are described in more detail in the section on Spatial Constancy.

Area 7a has been found to have more extensive connections with high-order areas in the frontal and temporal lobes and the cingulate gyrus than do the other areas in the posterior parietal cortex. While 7a also projects strongly to the prefrontal cortex in and around the principal sulcus (area 46 of Walker), unlike area LIP, it is only weakly connected to the frontal eye fields (Barbas & Mesulam 1981, Andersen 1987). Area 7a has very strong interconnections with the entire cingulate gyrus; the most dense connections are to area LC in the posterior half of the gyrus (Pandya et al 1981, Andersen 1987). In contrast area 7b is connected primarily, if not exclusively, to area LA in the anterior aspect of the cingulate gyrus (Andersen 1987). Area 7a additionally demonstrates the most extensive connections of all the posterior parietal areas with the cortex that is buried in the superior temporal sulcus, including area MST (Andersen 1987).

Differences in connections define at least two additional visual areas: the ventral intraparietal area (VIP) located in the fundus of the intraparietal sulcus (Maunsell & Van Essen 1983) and the dorsal prelunate gyrus which lies above area V4 (Asanuma et al 1985, Andersen 1987). The functional properties of the neurons in these areas have not yet been explored.

The above observations indicate that the monkey inferior parietal lobule can be subdivided into a larger somatosensory area essentially coextensive with PF, and several visual areas within PG.

## Visual Pathways into the Inferior Parietal Lobule

Visual inputs into the inferior parietal lobule arrive from extrastriate cortex rather than directly from the primary visual cortex (V1) itself. A second possible source of input lies in the pathway from the retinorecipient areas of the superior colliculus and pretectum through the pulvinar into the inferior parietal lobule (IPL). Visual inputs from this pathway are likely to be of minor significance, however. The retinorecipient (superficial) layers of the superior colliculus project to the inferior pulvinar and to the ventral aspect of the lateral pulvinar (Benevento & Fallon 1975, Benevento & Standage 1983, Harting et al 1980, Trojanowski & Jacobson 1975) but these areas of the thalamus do not project to the IPL (Asanuma et al 1985, Yeterian & Pandya 1985). Area 7a receives its pulvinar input from the medial pulvinar (Asanuma et al 1985, Yeterian & Pandya 1985), and it is only the deep, oculomotor layers (not the superficial visual layers) of the superior colliculus that project to the medial pulvinar. Also the medial pulvinar does not receive descending corticothalamic projections from other visual cortices (Bisiach & Luzzatti 1978, Benevento & Standage 1983, Harting et al 1980). Thus, with the exception of a minor projection from the pretectum (Benevento & Standage 1983), no obvious visual inputs enter the medial pulvinar that could then be relayed up to area 7a. Areas DP and LIP receive their principal thalamic inputs from the visual non-retinotopic dorsal aspect of the lateral pulvinar (Asanuma et al 1985). The lateral pulvinar receives inputs only from the oculomotor part of the superior colliculus and from the pretectum, and its cells are weakly driven by visual stimuli (Benevento et al 1977, Benevento & Standage 1983, Harting et al 1980).

The flow of visual processing can be presumed by determining the laminar distributions of the sources and terminations of corticocortical projections in visual cortex (Rockland & Pandya 1979, Maunsell & Van Essen 1983). Early in the visual pathway, feedforward projections originate from cell bodies located in the supragranular layers and end in terminals in layer IV and lower layer III. Feedback projections originate in the supragranular and infragranular layers and end most densely in layers I and VI. The hierarchical progression of visual processing can be traced from area V1 at the base of the hierarchy to area 7a at the top if the following modification is considered for the projections into the inferior parietal lobule: feedforward projections originate in both superficial and deep cortical layers but still end predominantly in layer IV and lower layer III.

The routes of visual input into the inferior parietal lobule (*dashed square*) are shown in Figure 2, arranged in a hierarchical structure determined by the laminar distribution of the sources and terminals of the connections. Each line represents reciporcal corticocortical connections between fields. It can be seen that multiple visual pathways project into the inferior parietal lobule and that area 7a represents the pinnacle of the hierarchy.

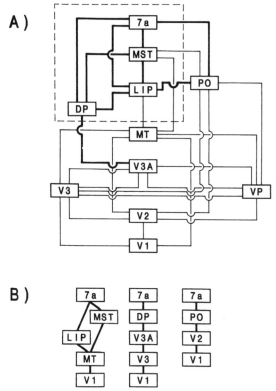

*Figure 2*  (*A*) Hierarchy of visual pathways from area V1 to the inferior parietal cortex determined by laminar patterns of sources and terminations of projections. *Dashed box,* cortical areas of inferior parietal lobule and dorsal aspect of prelunate gyrus. (*B*) Three of the shortest pathways for visual information to travel from area V1 to area 7a. From Andersen (1987).

Figure 2*b* demonstrates the shortest routes from area V1 to area 7a; each of these paths must pass through two or three extrastiate visual areas prior to arriving at area 7a. Of particular importance to motion processing is the pathway that begins in area V1 and passes through areas MT and MST to area 7a. This pathway and its role in motion processing are discussed in the next section.

## MOTION ANALYSIS

A substantial body of evidence suggests that visual motion analysis is treated by specific brain regions. There are several accounts of brain lesions in humans that produce deficits in motion perception without deficits in other forms of vision. Recording experiments by a number of investigators have pointed to areas in the dorsal aspect of extrastriate cortex that may be specialized for motion analysis. Taken together, these various recording experiments delineate a pathway that originates in primary visual cortex and terminates in posterior parietal cortex. The anatomical determination of this presumed motion processing pathway is an important first step in understanding the neural mechanisms that account for motion perception.

At the beginning of the pathway, direction-selective cells in layers 4b and 6 of primary visual cortex project to the middle temporal area (MT) (V5 of Zeki). Results of recording experiments in MT indicate that nearly all of the cells in this area are direction selective compared to only about 20% in area V1. A functional architecture for direction selectivity exists in MT similar to the functional architecture described for orientation in V1 (Albright et al 1984). The neurons of area MT also exhibit interesting responses that may account for certain features of motion perception. For example, some cells are selective for the global direction of motion of patterns rather than for the motion of those components of the pattern that occur orthogonally to the preferred orientation of the neurons (Movshon et al 1985). Other cells are speed invariant over a wide range of spatio-temporal frequencies (Movshon 1985). MT cells also exhibit opponent center/surround organizations for direction selectivity; strong inhibition results when motion in the surround is in the same direction as motion in the center (Allman et al 1985, Tanaka et al 1986). These surround mechanisms are quite large and often include the entire visual field. Some have speculated that these cells play a role in processing motion parallax (important for extracting depth from motion cues) or in distinguishing the external movement of objects in the world from motion generated by the eyes moving in their orbits (Allman et al 1985). While these data have led to the suggestion that area MT is specialized for processing motion, it had not been shown until recently that damage restricted to area MT disrupts motion perception.

Area MT then projects to the immediately adjacent medial superior temporal (MST) area, which is located in the posterior parietal cortex within the anterior bank of the superior temporal sulcus. Recording experiments suggest that MST contains cells that are selective for rotation and expansion of velocity fields (Saito et al 1985, Sakata et al 1985, 1986). Area MST in turn projects to area 7a in the posterior parietal cortex. Cells in this region demonstrate an opponent direction organization with respect to the fixation point, a feature that has been suggested to be important to the analysis of visual flow fields during locomotion (Motter & Mountcastle 1981).

There is evidence for a functional hierarchy for visual motion analysis within the dorsal extrastriate cortex. Area MT processes more complex aspects of motion than do the direction-selective cells of V1, and in turn areas MST and 7a appear to analyze still more complex aspects of motion, such as aspects of structure-from-motion.

## Motion Psychophysics

In order to understand the functional processes that occur along this pathway, Andersen and colleagues, Wurtz and colleagues, and Newsome and colleagues employed a strategy whereby the ability of monkeys to perceive various aspects of motion is first determined psychophysically. Lesions are then made at different locations in the motion pathway, and the monkeys are retested to determine the contribution of the different cortical areas to movement perception. These psychophysical and lesion experiments are described in the next two sections.

Until recently, no psychophysical experiments had been performed to measure the ability of monkeys to see motion. Andersen and colleagues (Golomb et al 1985, Siegel & Andersen 1988, Andersen & Siegel 1988) undertook these psychophysical studies and chose to examine two types of relative motion perception in monkeys—the ability to detect shear motion and the ability to detect two- and three-dimensional structure in velocity fields. The spatio-temporal integration characteristics of structure-from-motion perception were also examined. The results of these experiments suggest that the brain forms neural representations of surfaces by using structure-from-motion information.

In our first psychophysical investigations, Golomb et al (1985) studied the detection of shearing motion with psychophysical tasks in monkeys and humans. Shearing motion is a class of relative motion in which the change in direction of motion occurs along the axis orthogonal to the direction of motion. Because this form of motion accounts for depth perception from motion parallax, it is of great physiological importance and also provides a cue for the recognition of foreground versus background using motion discontinuities. Recording experiments in area MT of monkey (Allman et al 1985), the pigeon optic tectum (Frost & Nakayama 1985), and area V2 of monkey (Orban et al 1986) have identified neurons with receptive fields that have opponent-direction center-surround organization; these cells would be maximumly active for a shear stimulus. Due to its importance, a large body of psychophysical data has already been collected in humans with respect to shear motion detection abilities (Nakayama & Tyler 1981, Nakayama et al 1984, Nakayama 1981, Rogers & Graham 1979, 1982), making comparison of our results in the monkey and human with those of previous work possible.

The shear motion stimulus consisted of a standing transverse wave of sinusoidly varying spatial and temporal frequency. Thresholds for detection of shear were measured for various combinations of spatial frequency, temporal frequency, and amplitude. For all these combinations of parameters, humans and monkeys showed similar thresholds. Figure 3 shows an example of the stimulus and the psychophysical results.

A second type of motion perception we have examined is the ability of monkeys and humans to detect structure in velocity fields, i.e. structure from motion. We have recently developed a novel set of stimuli that

A.

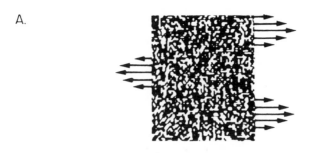

B.

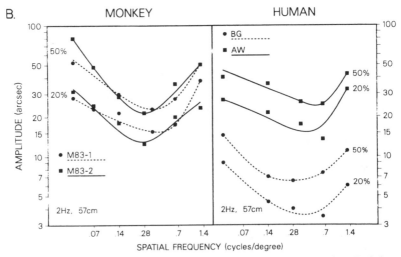

*Figure 3* A random dot display that undergoes a horizontal shearing motion. Each *horizontal row of dots* moves as a rigid unit with a velocity that is a sinusoidal function of the vertical position of the row. The *arrows* represent the instantaneous velocity vectors. Note that the direction of motion reverses at the zero crossings of the sinusoidal velocity function. From Golomb et al (1985).

test the ability of subjects to detect structure in velocity fields without introducing confounding positional cues. The three types of structures have been examined using these stimuli with both monkey and human subjects (Siegel & Andersen 1988, Andersen & Siegel 1988). Two types of two-dimensional structures (expansion and rotation) and a three-dimensional structure [a revolving hollow cylinder (Figure 4a)] have been developed for these tasks. The subjects are required to perform a reaction-time task in which they must detect the transition from an unstructured velocity field to a partial or completely structured velocity field. The structured and unstructured displays contain the same vectors; however, in the unstructured case the vectors have been randomly shuffled, thus destroying structure.

Figure 4b shows psychometric functions for the 3-D cylinder detection task for a monkey and a human in which performance is plotted as a function of the percentage of structure in the display. It can be seen that the psychometric functions for the two species are quite similar. Such equivalent performance was seen in monkeys and humans for both the 2-D stimuli detection tasks as well.

Having established that out subjects were able to detect three-dimensional structure from the velocity fields in our psychophysical paradigm, we then examined how information about structure-from-motion might be integrated over space and time by the visual system. The important theoretical groundwork for this problem was laid by Ullman (1979) when he proved in his structure-from-motion theorem that "given three distinct orthographic views of four non-coplanar points in a rigid configuration, the structure and motion compatible with the three views are uniquely determined." Ullman (1979) realized that his theorem described an algebraic limit and that the brain may require more points and frames, since it is a "noisy" system, and it might also use a less than optimum algorithm. In this set of experiments the question of spatial integration was investigated by varying the density of the points, whereas the temporal integration problem was investigated by varying the point life times in the 3-D cylinder task.

The performance of a monkey and a human subject for different point lives is illustrated in Figure 4c with the standard 128 points present in the display; it can be seen that below 100 msec the human's performance was poor and below 50 msec the monkey's was poor. Using a fixed point life of 532 msec and studying the effect of the number of points present in the display, the human and monkey subjects both showed poor performance below 32 points. Interestingly there was space-time trading such that equivalent performance could be achieved with longer lifetimes and fewer points or vice versa. These experiments demonstrate that motion infor-

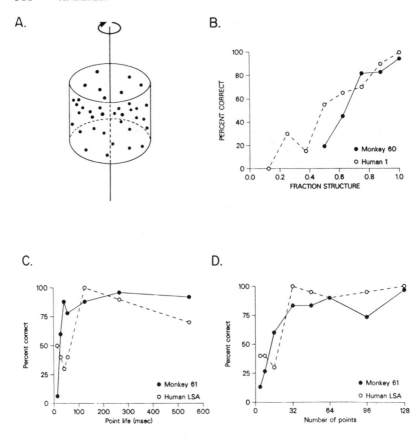

*Figure 4* (*A*) The 3-D structure-from-motion stimulus. *Dots* are projected onto the surface of a revolving hollow cylinder. (*B*) The percentage of trials in which the subject released the key within the requisite time window is plotted as a function of the fraction of structure. The human and monkey have similar psychometric functions that have a decrease in detection when the fraction of structure is approximately less than 0.65. Statistical analysis using the $\chi^2$ for independent samples showed no difference between the human and monkey subjects ($P < 0.05$). In this experiment, 128 points were viewed with a point lifetime of 532 msec; the cylinder revolved at 35 deg/sec; the display was refreshed at 35 Hz. The ability of the monkey and human subjects to perform the task depended on both the point lifetime (*C*) and number of points (*D*) of the display. The significant difference in minimal point lifetime required to perform the task successfully between the human and primate subject is likely due to the additional training given the monkey subjects. It can be seen that decreasing the number of points makes it more difficult for the subjects to perform the task. In figures (*C*) and (*D*), the fraction of structure was 0.875; the refresh rate was 70 Hz; the cylinder revolved at 35 deg/sec. From Siegel & Andersen (1988).

mation is integrated in both time and space to form neural representations of three-dimensional surfaces.

These space-time integration experiments provide information on the neural mechanisms responsible for computing structure-from-motion. An algorithm proposed by Ullman (1984) required the brain to track continuously the location of individual points in absolute coordinates to determine whether their trajectories conformed to the movement of a rigid object. Thus the brain, by the algorithm, would be solving algebraic equations using the points in the image. We know from the above experiments that the brain does not require continuously visible points to compute structure from motion, indicating that at least under these conditions the brain uses a different algorithm. The reaction time data indicate that the 3-D cylinder structure-from-motion computation required several hundreds of milliseconds. Therefore each point would have to be tracked over an extended period of time. The time integration experiments showed good performance when points were visible for only 100 msec and were then replotted at random new positions, making any continuous, point-bound computation impossible. These results indicate that a 3-D surface can be computed by using a large number of intermittent sampling points across that surface.

## Area MT Plays an Important Role in the Perception of Motion

Results from recent lesioning experiments establish area MT as part of a motion processing system responsible for the perception of motion. Equally interesting is the observation that after at least small area MT lesions the perception of motion recovers in a matter of days.

Newsome et al (1985) first tested the effect of MT lesions on smooth pursuit eye movements, a behavior requiring motion analysis. They discovered that small ibotenic acid lesions placed at retinotopically identified loci in area MT resulted in defects of smooth pursuit. During the early, so-called "open loop" stage of tracking the animals underestimated the target speed. The deficit was specific for the retinal locus of the lesion, thus indicating a sensory rather than a motor defect. Since the animals could make saccades accurately to the retinotopic locus of the lesion, the deficit was specific for motion perception and was not due to a general blindness.

These important experiments did not directly establish that the monkeys were unable to see motion, but rather that they could not use this information to make smooth pursuit eye movements. Two recent studies have shown directly that the perception of motion is in fact disrupted with area MT lesions. In the first of these two experiments Siegel & Andersen trained monkeys to perform motion psychophysical tasks and obtained prelesion thresholds for the detection of shear motion and structure-from-motion (Andersen & Siegel 1986, 1988, Siegel & Andersen 1986). Small ibotenic acid lesions were then made to area MT; these produced an eight-fold increase in shear motion thresholds. As expected, the increase in thresholds was found only for those areas in the visual field that corresponded to the retinotopic locus of the lesion in area MT (Figure 5). The deficit was specific for motion, since contrast sensitivity thresholds were also tested and found not to be affected at the same locations in the visual field as the motion deficit. Since the animals were required only to determine whether or not they saw motion, and not any of the parameters of the motion such as stimulus direction, it was interpreted that the animals had a deficit in the detection of motion. Thus restricted lesions to area MT produced motion scotomas in the visual field. Figure 5 illustrates the most interesting finding that the deficits were transient, recovering in four to five days and demonstrating a similar time course to the recovery as the tracking deficit recorded in the experiments of Newsome et al (1985). The effects of MT lesions on structure-from-motion thresholds were also tested in one hemisphere. The task the monkey performed was the 3-D revolving hollow cylinder task detailed in the preceding section of this article. As with the psychophysical studies, the amount of structure in the stimulus was varied to generate psychometric functions. The animal could not do this structure-from-motion task after the area MT lesion even when the velocity field was completely structured. Moreover, the structure-from-motion deficit remained for 23 days (at which time the experiment was terminated), long after the motion detection thresholds had returned to normal.

In a second set of experiments, Newsome & Pare (1986, 1988) tested the ability of monkeys to determine the direction of correlated motion imbedded in noise. They then made ibotenic acid lesions to area MT and found large increases in the measured motion thresholds. These lesions did not affect contrast thresholds. They also found the motion perception deficit to be transient; thresholds recovered to normal values within a few days.

The above experiments indicate that area MT is a component of a motion processing system important for perceiving motion. The thresholds recorded for the shear detection tasks after MT lesions were so large that it is possible that the animals were using positional cues to solve the task at very large amplitudes of shear. The transient nature of the deficit is a common feature of all the experiments, and is seen for pursuit eye movements, motion detection, and direction detection. The recovery may result either from a reorganization of area MT or from recruitment of some parallel pathway which normally does not play a primary role in motion

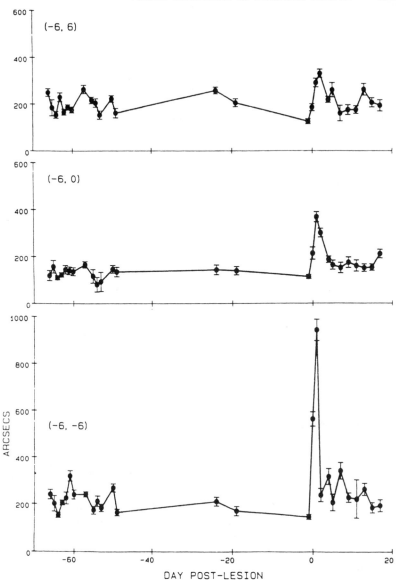

*Figure 5* The effect of ibotenic acid lesions in area MT on shear motion thresholds. Fifty percent hit rates are plotted from psychometric curves in which spatial and temporal frequency is held constant and amplitude is varied. In this case the lesion was placed in the lower visual field representation of area MT in the contralateral (*right*) hemisphere. Note that the greatest effect is focused at the lower field location; thresholds increase after one day to over 900 arc sec. Also note that the thresholds recover rapidly in the matter of a few days. From Andersen & Siegel (1988).

perception. These two possibilities could be tested by completely destroying area MT. If the motion thresholds do not return, the recovery is a result of the reorganization of area MT; if they do still recover, parallel pathways are involved in the recovery.

An interesting question regarding recovery is whether training is required for it to occur. The monkeys perform thousands of trials on the motion tasks before the thresholds recover. If the animals were to be placed in the dark immediately after MT lesions and brought out a week later, would the thresholds have recovered spontaneously or would they still be elevated, indicating that retraining has played a significant role?

The single observation of a more permanent deficit in structure-from-motion perception, if it is reproduced in subsequent experiments, will indicate that (a) the structure-from-motion computation is performed in area MT and no other pathway can assume this task, or alternatively (b) some other area such as area MST requires preprocessing for the structure-from-motion computation that only area MT can provide.

## SPATIAL CONSTANCY

Because the projection from the eyes to the primary visual cortex is retinotopic, an "image" of whatever the eyes are looking at appears over the visual cortex, and this "image" changes as the eyes view different parts of the world. Such retinotopic mapping is widespread throughout the visual system. Two reasons to believe that the brain contains non-retinotopic representations of visual space, however, are as follows:

1. When we move our eyes, our perceived visual world is stationary. This result suggests that eye position information is used to compensate for movements of the visual image focused on the retinas. Although it could be argued that purely sensory signals are used to stabilize the visual world (i.e. that a complete translation of the retinal image indicates an eye movement and not movement of the world), other evidence indicates that this explanation is unlikely. Moving the eyes passively, which generates the same retinal stimulation as willed eye movements, produces the impression that the world has moved.
2. Motor movements such as reaching are made accurately to visual targets without visual feedback during the movement. This observation indicates that the motor system uses representations of the visual stimulus mapped in body-centered rather than retinal coordinates.

A most likely area of the brain to find non-retinotopic representations of visual space is the posterior parietal cortex. Lesions in this region in humans produce a syndrome called visual disorientation in which patients cannot reach accurately to visual targets and have difficulty navigating around seen obstacles. These patients are not blind but appear to be unable to associate the locations of what they see with their body position.

Recent recording experiments have begun to establish how visual space is represented in area 7a of the posterior parietal cortex (Andersen et al 1985b). Visual receptive fields were first mapped in these experiments with monkeys fixating at different eye positions. The heads were fixed simplifying the coordinate space to a head-centered frame. Two possible results under these conditions would be either that the retinal receptive fields move with the eyes so that visual space is mapped in retinal coordinates, or that the receptive fields do not move with the eyes but remain constant for a location in head-centered space. Neither representation was found. Instead the receptive fields were found to move with the eyes (i.e. they were retinotopic), but the responsiveness of these receptive fields to retinotopically identical stimuli varied as a function of the eye position (Figure 6).

The interaction of eye position and retinal position was found to be multiplicative. The activity of area 7a neurons to visual stimuli was described as a gain that was a function of eye position, multiplied by the response profile of the retinal receptive field. This interaction produces a tuning for locations of targets in head-centered space but in a fashion dependent on eye position. To illustrate this result in a simple example, consider a cell that has a receptive field 10 deg to the left and is only responsive when the animal looks 10 deg to the right from straight ahead. Such a cell will respond only for stimuli located straight ahead in head-centered space (spatial tuning) but only if the animal is looking 10 deg to the right (eye-position dependence). Cells have never been found that code the location of a target in space in an eye-position-independent fashion; it can be shown, however, that sufficient information for such a coding is present in the population response of area 7a neurons. How such a distributed code is interpreted by the brain is an open question. One possible solution would be the presence of a topographic organization for spatial tuning in the cortex; however, so far recording experiments have not revealed such a map and the present data suggest that if such a map exists it is likely to be either crude or highly fractured.

Recently a parallel network model was created by Zipser & Andersen (1988, Andersen & Zipser 1988) that learns spatial location by combining eye position and retinal position inputs. The model is a three layer network trained for spatial location using a back-propagation learning algorithm. This mathematical model generates receptive field properties in its middle layer units that are similar to those found for actual posterior parietal neurons. Figure 7 illustrates the model. The first, input layer consists of

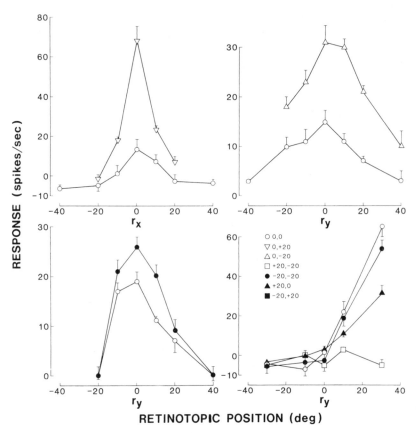

*Figure 6*   Mean response rates for different eye positions plotted in retinal coordinates along horizontal ($r_x$) or vertical ($r_y$) axes passing through the centers of the receptive fields of four neurons; each graph shows data for one neuron. Each point represents the mean response (+ or − standard error) to eight repetitions of the stimulus presented at the same retinal location. A randomized block design was used to present stimuli to different retinal locations in the receptive field of each cell. The reported response at each retinal location is equal to the activity during the presentation of the stimulus minus the background activity determined before the stimulus presentation. From Andersen et al (1985).

64 units in an 8 by 8 array that samples at equidistant points from a continuous two-dimensional retinal field and 4 sets of eye position inputs that code horizontal and vertical eye positions similar to the eye position cells found in area 7a. The intermediate layer has 9 units (and in some simulations up to 36 units) that receive inputs from all the input units and in turn project to either of two representations in the output layer that code location in head-centered space for any pair of arbitrary retinal and

eye position inputs. After training is complete the middle layer units have receptive fields that remain retinotopic, but their activity becomes modulated by eye position in a manner similar to the recording data from area 7a neurons. The visual receptive fields of the model cells appear very similar to area 7a fields, as they are large and occasionally complex but smoothly varying in shape.

These interesting results show that the trained parallel network and the posterior parietal cortex compute coordinate transformations in the same way. The network model was used to program an algorithm for solving the problem using a parallel architecture; however, that the network arrives at the same solution as area 7a does not mean that the solution was achieved by the same learning algorithm in the brain; i.e. back propagation. Moreover, in its strictest sense, back propagation could never be used by the brain, since it requires information to pass rapidly backwards through synapses. It will be important to investigate other parallel networks that have more realistic structures and mechanisms analogous to those found in the brain to determine how they might produce this form of distributed coding.

The model results suggest that the posterior parietal cortex may learn to associate visual inputs with eye position. This neural coding may be an example of a distributed associative memory. The spatial code in the model is by definition non-topographic, and yet it can be interpreted by the brain because it was learned and thus became an inherent property of the synaptic structure of the network. This learning mechanism suggests that area 7a need not require a topography for spatial tuning in order to read out eye-position-independent spatial locations, although it also does not rule against one. Finally, the similarity of the retinal receptive fields of the network units and those of the cells suggests that parietal neurons may have access to the entire retina (as a result of the multistage divergence of the cortico-cortical projections from V1 to area 7a) and that the receptive fields that eventually develop for area 7a neurons are a result of competition during the learning process.

# ROLE OF PARIETAL LOBE IN VISUAL-MOTOR INTEGRATION

The pioneering neurophysiological studies by Mountcastle and colleagues showed that the cells of the posterior parietal cortex have activity correlated with the motor and oculomotor behaviors of the animals (Mountcastle et al 1975, Lynch et al 1977). They proposed a command hypothesis for the function of the posterior parietal cortex that stated that this area synthesized sensory and motivational information from several cortical areas and issued commands of a general nature for motor behaviors. In a subsequent study by Robinson et al (1978) it was found that many of these same neurons responded to sensory stimuli. They argued that the motor-related activity seen by Mountcastle and colleagues was a result of sensory stimulation, either from the targets for movement or as a result of sensory

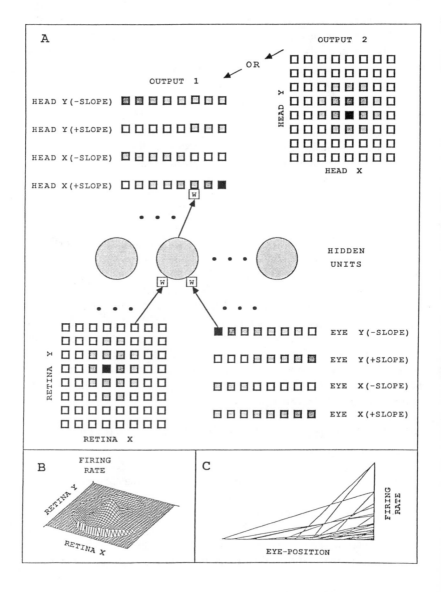

stimulation due to the movement. They also found that the responses to sensory stimuli were often enhanced if those stimuli were behaviorally relevant, and they proposed that the parietal lobe was important for attentional rather than motor command functions.

A number of investigators have subsequently designed experiments to separate sensory from motor-related responses, which were generally linked in the earlier studies (Sakata et al 1980, 1983, Andersen et al 1987, Wurtz & Newsome 1985, Bioulac & Lamarre 1979, Seal et al 1982, Seal & Commenges 1985). Neurons generally have both sensory and movement-related responses. Cells responding to reaching behavior also have somatosensory inputs, and cells responding to smooth pursuit, saccades, or fix-

---

*Figure 7* (*A*) Back propagation network used to model area 7a. The visual input consists of 64 units with Gaussian receptive fields with 1/e widths of 15 deg. The center of each receptive field occupies a position in an 8 by 8 array with 10 deg spacings. The *shading* represents the level of activity for a single spot stimulus. The darker shading represents higher rates of activity. The units have been arrayed topographically for illustrative purposes only; this pattern is not an aspect of the model, since each hidden unit receives input from every one of the 64 retinal input units. The eye position input consists of 4 sets of 8 units each. Two sets code horizontal position (one for negative slope and one for positive slope), and two sets code vertical position. *Shading* represents the level of activity. The intercepts have been ordered for illustrative purposes only and do not represent information available to the hidden layer. Each eye position cell projects to every unit in the hidden layer. Two output representations were used; the Gaussian output format is shown on the *right* and the monotonic format on the *left*. The Gaussian format units have Gaussian-shaded receptive fields plotted in head-centered coordinates. They have 1/e widths of 15 deg and are centered on an 8 by 8 array in head coordinate space with 10 deg spacings. The monotonic format units have firing rates that are a linear function of position of the stimulus in head-centered coordinates. There are four sets of 8 units with two sets of opposite slope for vertical position and two sets for horizontal position in head-centered coordinates. Again, *shading* represents the degree of activity, and the topographic ordering is for illustrative purposes only. The *small boxes* with w's indicate the location of the synapses whose weights are trained by back propagation. Each hidden unit projects to every cell in the output layer.

The output activity of the hidden and output layer units is calculated by the logistic function: output $= 1/(1-e^{-net})$ where net $=$ (weighted sum of inputs) $+$ threshold. The *arrow* for the connections represents the direction of activity propagation; error was propagated back in the opposite direction. The back propagation procedure guarantees that the synaptic weight changes will always move the network toward lower error by implementing a gradient descent in error in the multi-dimensional synaptic weight space.

(*B*) Area 7a visual neuron receptive field with a single peak near the fovea. Visual cells that had no eye position related activity or modulation of their responses by eye position were used to model the retinal input to the network.

(*C*) A composite of 30 area 7a eye position units whose firing rates are plotted as a function of horizontal or vertical eye deviation. The slopes and intercepts are experimental values for eye position neurons. From Zipser & Andersen (1988).

ations also respond to visual stimuli. From these findings has emerged the concept that the parietal lobe should not be viewed as primarily a sensory or primarily a motor structure (Andersen 1987). Rather it apparently occupies a location somewhere between these two points, integrating sensory information to be used for the formulation of motor behaviors. Put quite simply, the area is involved in sensory-motor integration.

As mentioned above, a functional segregation appears to exist within the posterior parietal cortex (Hyvarinen 1981, Andersen et al 1985a,c); area MST contains cells responding to smooth pursuit, area 7b contains the reach cells, area LIP is involved in saccades, and area 7a contains many of the space-tuned and fixation neurons.

## Area LIP and its Role in Motor-planning

An area has recently been discovered within the lateral bank of the intraparietal sulcus (the lateral intraparietal area—LIP) that appears to play a role in programming saccadic eye movements. This area is similar to the frontal eye fields in many respects, having many cells that fire before saccades and code eye movements in motor coordinates (Gnadt & Andersen 1986, 1988, Andersen & Gnadt 1988).

Gnadt & Andersen have recently found that many of the LIP neurons have memory-linked activity (Gnadt & Andersen 1988, Andersen & Gnadt 1988). The memory-related responses are present in saccade tasks in which the animal must remember, in total darkness, the spatial location of a briefly flashed target for up to 1.6 sec before making an eye movement to that location (Figure 8). The cells become active 50 to 100 msec after the onset of the flash and remain active in the absence of any visual stimuli until the saccade is made. Thus, these neurons appear to be acting as latches, maintaining a steady rate of firing during the entire period that the saccade is withheld.

This interesting response property can be interpreted in three possible ways: (a) that it represents the memory of the retinotopic locus of the stimulus, (b) that it codes the memory of the spatial location of the stimulus, or (c) that it represents the memory of the movement that the animal intends to make. These three alternatives have been experimentally tested by asking in what coordinate frame the memory-linked response is represented. Using special tasks that separate sensory from motor coordinates, Gnadt & Andersen found that these cells encode the intended amplitude and direction of movements in motor coordinates (Gnadt & Andersen 1988, Andersen & Gnadt 1988). Thus the cells hold in register the intent to make movements of a particular metric. This observation suggests that area LIP has activity related to aspects of motor planning.

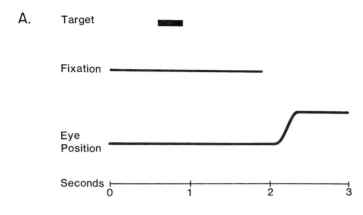

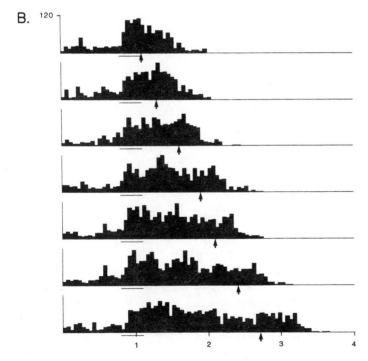

*Figure 8*   (*A*) Diagram of the sequence of stimulus events in the memory saccade task. (*B*) Activity histograms of an intended movement cell during saccades to a remembered target location in the cell's motor field. Trials are grouped according to increasing response delay times from top to bottom. The *horizontal bar* below each histogram indicates the stimulus presentation time. The *arrow* indicates when the fixation light was extinguished commanding the saccade. From Gnadt & Andersen (1988).

# CONCLUSION

This review has covered new findings on the role of the posterior parietal lobe in coordinate transformations for spatial perception and behavior, motion perception, and the processing of saccadic eye movements. The experiments have indicated the distributed nature of the non-retinotopic representation of space in the parietal cortex. The Zipser-Andersen model has been valuable in demonstrating that this spatial representation could be learned. The learning of the associations between visual space and body position is a reasonable hypothesis, since a system for spatial representation would need constant recalibration during growth. Experiments to determine the plasticity of this representation would be most interesting. For instance, distortions of space that result from wearing prisms can be rapidly compensated for by learning (prismatic adaptation). Does this learning result in the cells in area 7a changing their spatial tuning characteristics? Would the Zipser-Andersen model make predictions about changes that may turn up in the recording data as a result of such learning?

Recent experiments indicate that the posterior parietal cortex is at the pinnacle of a presumed hierarchy in motion processing. Area MT feeds motion information into the parietal cortex and has been shown to be part of a motion processing pathway important for perceiving motion. The fast recovery of motion thresholds following MT lesions suggests several experiments on the mechanisms of cortical compensation and reorganization after cerebral lesions. The motion psychophysics/MT lesion paradigm is well suited for asking these questions, since the pathways involved are well understood and the experiments can be rigorously controlled.

Area LIP contains neurons that reflect the intention of the animals to make motor movements. The response properties of these neurons may provide a useful handle for dissecting how motor plans are formulated for sequences of movements and how spatial transformations are recomputed following each movement of a sequence. These problems can be addressed by observing the dynamic changes in the representation of intended movements in area LIP as animals perform sequences of saccades that are programmed by memorizing sequences of flashed sensory targets.

The results of single cell recording and lesion experiments in the posterior parietal cortex have been instrumental in establishing its role in spatial functions. These same techniques will be able to address, among other issues, how the brain recovers visual-spatial functions after injury, how associations between visual space and body position are learned and stored, and how sequential motor activities are planned and executed. The posterior parietal cortex should prove to be an interesting place for physiologists and anatomists to work in the next few years.

ACKNOWLEDGMENTS

I wish to thank C. Andersen for editorial assistance and D. Duffy for typing the manuscript. This work was supported by National Institutes of Health grants EY05522 and EY07492, the Sloan Foundation and the Whitaker Health Sciences Foundation.

*Literature Cited*

Albright, R. D., Desimone, R., Gross, C. G. 1984. Columnar organization of directionally selective cells in visual area MT of the macaque. *J. Neurophysiol.* 51: 16–31

Allman, J., Miezen, F., McGuinness, E. 1985. Stimulus specific responses from beyond the classical receptive field: Neurophysiological mechanism for local-global comparisons in visual neurons. *Ann. Rev. Neurosci.* 8: 407–30

Andersen, R. A. 1987. The role of the inferior parietal lobule in spatial perception and visual-motor integration. In *The Handbook of Physiology, Section 1: The Nervous System Volume V, Higher Functions of the Brain Part 2*, ed. F. Plum, V. B. Mountcastle, S. R. Geiger, pp. 483–518. Bethesda: Am. Physiol. Soc.

Andersen, R. A., Asanuma, C., Cowan, W. M. 1985a. Callosal and prefrontal associational projecting cell populations in area 7a of the macaque monkey: A study using retrogradely transported fluorescent dyes. *J. Comp. Neurol.* 232: 443–55

Andersen, R. A., Essick, G. K., Siegel, R. M. 1985b. The encoding of spatial location by posterior parietal neurons. *Science* 230: 456–58

Andersen, R. A., Siegel, R. M., Essick, G. K., Asanuma, C. 1985c. Subdivision of the inferior parietal lobule and dorsal prelunate gyrus of macaque by connectional and functional criteria. *Invest. Ophthalmol. Visual Sci.* 23: 266 (Abstr.)

Andersen, R. A., Essick, G. K., Siegel, R. M. 1987. Neurons of area 7 activated by both visual stimuli and oculomotor behavior. *Exp. Brain Res.* 67: 316–22

Andersen, R. A., Gnadt, J. W. 1988. Role of posterior parietal cortex in saccadic eye movements. *Reviews in Oculomotor Research*, Vol. 3, ed. R. Wurtz, M. Goldberg. Amsterdam: Elsevier. In press

Andersen, R. A., Siegel, R. M. 1986. Two- and three-dimensional structure from motion sensitivity in monkeys and humans. *Soc. Neurosci. Abstr.* 12: 1183

Andersen, R. A., Siegel, R. M. 1988. Motion processing in primate cortex. In *Signal and*

*Sense: Local and Global Order in Perceptual Maps*, ed. G. M. Edelman, W. E. Gall, W. M. Cowan. New York: Wiley. In press

Andersen, R. A., Zipser, D. 1988. The role of the posterior parietal cortex in coordinate transformations for visual-motor integration. *Can. J. Physiol. Pharmacol.* 66(4): 488–501

Asanuma, C., Andersen, R. A., Cowan, W. M. 1985. The thalamic relations of the caudal inferior parietal lobule and the lateral prefrontal cortex in monkeys: Divergent cortical projections from cell clusters in the medial pulvinar nucleus. *J. Comp. Neurol.* 241: 357–81

Baleydier, C., Mauguiere, R. 1980. The duality of the cingulate gyrus in monkey. Neuroanatomical study and functional hypothesis. *Brain* 103: 525–54

Barbas, H., Mesulam, M. M. 1981. Organization of afferent input to subdivisions of area 8 in the rhesus monkey. *J. Comp. Neurol.* 200: 407–31

Benevento, L. A., Fallon, J. H. 1975. The ascending projections of the superior colliculus in the rhesus monkey (*Macaca mulatta*). *J. Comp. Neurol.* 160: 339–62

Benevento, L. A., Rezak, M., Santos-Anderson, R. 1977. An autoradiographic study of the projections of the pretectum in the rhesus monkey (*Macaca mulatta*): Evidence of sensorimotor links to the thalamus and oculomotor nuclei. *Brain Res.* 127: 197–218

Benevento, L. A., Standage, G. P. 1983. The organization of projections of the retinorecipient and non-retinorecipient nuclei of the pretectal complex and layers of the superior colliculus to the lateral pulvinar and medial pulvinar in the macaque monkey. *J. Comp. Neurol.* 217: 307–36

Bioulac, B., Lamarre, Y. 1979. Activity of postcentral cortical neurons of the monkey during conditioned movements of a deafferented limb. *Brain Res.* 172: 427–37

Bisiach, E., Luzzatti, C. 1978. Unilateral neglect of representational space. *Cortex* 14: 129–33

Brodmann, K. 1905. Beitrage zür histologischen localisation der grosshirnrinde, dritte mitteilung: Die rinderfelder der niederen affen. *J. Psychol. Neurol.* 4: 177–226

Colby, C. L., Olson, C. R. 1985. Visual topography of cortical projections to monkey superior colliculus. *Soc. Neurosci. Abstr.* 11: 1244

Dursteler, M. R., Wurtz, R. H., Yamasaki, D. S. 1986. Pursuit and OKN deficits following ibotenic acid lesions in the medial superior temporal area (MST) of monkey. *Soc. Neurosci. Abstr.* 12: 1182

Frost, B. J., Nakayama, K. 1985. Single visual neurons code opposing motion independent of direction. *Science* 13(220): 744–45

Gnadt, J. W., Andersen, R. A. 1986. Spatial, memory, and motor-planning properties of saccade-related activity in the lateral intraparietal area (LIP) of macaque. *Soc. Neurosci. Abstr.* 13: 454

Gnadt, J. W., Andersen, R. A. 1988. Memory related motor planning activity in posterior parietal cortex of macaque. *Exp. Brain Res.* 70: 216–20

Goldman-Rakic, P. S. 1988. Topography of cognition: Parallel distributed networks in primate association cortex. *Ann. Rev. Neurosci.* 11: 137–56

Golomb, B., Andersen, R. A., Nakayama, K., MacLeod, D. I. A., Wong, A. 1985. Visual thresholds for shearing motion in monkey and man. *Vision Res.* 25: 813–20

Harting, J. K., Huerta, M. F., Frankfurter, A. J., Strominger, N. L., Royce, F. J. 1980. Ascending pathways from the monkey superior colliculus: An autoradiographic analysis. *J. Comp. Neurol.* 192: 853–82

Hyvarinen, J. 1981. Regional distribution of functions in parietal association area 7 of the monkey. *Brain Res.* 206: 287–303

Hyvarinen, J., Shelepin, Y. 1979. Distribution of visual and somatic functions in the parietal associative area 7 of the monkey. *Brain Res.* 169: 561–64

Lynch, J. C., Graybiel, A. M., Lobeck, L. J. 1985. The differential projection of two cytoarchitectural subregions of the inferior parietal lobule of macaque upon the deep layers of the superior colliculus. *J. Comp. Neurol.* 235: 241–54

Lynch, J. C., Mountcastle, V. B., Talbot, W. H., Yin, T. C. T. 1977. Parietal lobe mechanisms for directed visual attention. *J. Neurophysiol.* 40: 362–89

Maunsell, J. H. R., Van Essen, D. C. 1983. The connections of the middle temporal visual area (MT) and their relationship to a cortical hierarchy in the macaque monkey. *J. Neurosci.* 3: 2563–86

Motter, B. C., Mountcastle, V. B. 1981. The functional properties of the light-sensitive neurons of the posterior parietal cortex studied in waking monkeys: Foveal sparing and opponent vector organization. *J. Neurosci.* 1: 3–26

Mountcastle, V. B., Lynch, J. C., Georgopoulos, A., Sakata, H., Acuna, C. 1975. Posterior parietal association cortex of the monkey: Command function for operations within extrapersonal space. *J. Neurophysiol.* 38: 871–908

Movshon, J. A. 1985. Processing of motion information by neurons in the striate extrastriate visual cortex of the macaque. *Invest. Ophthalmol. Vis. Sci.* 26S: 133

Movshon, J. A., Adelson, E. H., Gizzi, M. S., Newsome, W. T. 1985. The analysis of moving visual patterns. *Exp. Brain Res.* 11: 117–51 (Suppl.)

Nakayama, K. 1981. Differential motion hyperacuity under conditions of common image motion. *Vision Res.* 21: 1475–82

Nakayama, K. Silverman, G., MacLeod, D. I. A., Muligan, J. 1985. Sensitivity to shearing and compressive motion in random dots. *Perception* 14(2): 225–38

Nakayama, K., Tyler, C. W. 1981. Psychophysical isolation of movement sensitivity by removal of familiar position cues. *Vision Res.* 21: 427–33

Newsome, W. T., Pare, E. B. 1986. MT lesions impair discrimination of direction in a stochastic motion display. *Neurosci. Abstr.* 12: 1183

Newsome, W. T., Pare, E. B. 1988. A selective impairment of motion perception following lesions of the middle temporal visual area (MT). *J. Neurosci.* In press

Newsome, W. T., Wurtz, R. H., Dursteler, M. R., Mikami, A. 1985. The middle temporal visual area of the macaque monkey. Deficts in visual motion processing following ibotenic acid lesions in MT. *J. Neurosci.* 5: 825–40

Orban, G. A., Spileers, W., Gulyas, B., Bishop, P. O. 1986. Motion in depth selectivity of cortical cells revisited. *Soc. Neurosci. Abstr.* 12: 584

Pandya, D. A., Van Hoesen, G. W., Mesulam, M. M. 1981. Efferent connections of the cingulate gyrus in the rhesus monkey. *Exp. Brain Res.* 42: 319–30

Robinson, C. J., Burton, H. 1980a. Organization of somatosensory receptive fields in cortical areas 7b, retroinsular postauditory and granular insula of *M. fascicularis. J. Comp. Neurol.* 192: 69–92

Robinson, C. J., Burton, H. 1980b. Somatic submodality distribution within the second somatosensory (SII), 7b, retroinsular postauditory, and granular insular cortical areas of *M. fascicularis. J. Comp. Neurol.* 192: 93–108

Robinson, D. L., Goldberg, M. E., Stanton, G. B. 1978. Parietal association cortex in the primate: Sensory mechanisms and behavioral modulations. *J. Neurophysiol.* 41: 910–32

Rockland, K. S., Pandya, D. N. 1979. Laminar origins and terminations of cortical connections of the occipital lobe in the rhesus monkey. *Brain Res.* 179: 3–20

Rogers, B., Graham, M. 1979. Motion parallax as an independent cue for depth perception. *Perception* 8: 125–34

Rogers, B., Graham, M. 1982. Similarities between motion parallax and stereopsis in human depth perception. *Vision Res.* 22: 261–70

Saito, H., Yukio, M., Tanaka, K., Hikosaka, K., Fukada, Y., Iwai, E. 1985. Integration of direction signals of image motion in the superior temporal sulcus of the macaque monkey. *J. Neurosci.* 6: 145–57

Sakata, H., Shibutani, H., Kawano, K. 1980. Spatial properties of visual fixation neurons in posterior parietal association cortex of the monkey. *J. Neurophysiol.* 43: 1654–72

Sakata, H., Shibutani, H., Kawano, K. 1983. Functional properties of visual tracking neurons in posterior parietal association cortex of the monkey. *J. Neurophysiol.* 49: 1364–80

Sakata, H., Shibutani, H., Kawano, K., Harrington, T. 1985. Neural mechanisms of space vision in the parietal association cortex of the monkey. *Vision Res.* 25: 453–64

Sakata, H., Yukio, M., Tanaka, K., Hikosaka, K., Fukada, Y., Iwai, E. 1986. Integration of direction signals of image motion in the superior temporal sulcus of the macaque monkey. *J. Neurosci.* 6: 145–57

Seal, J., Commenges, D. 1985. A quantitative analysis of stimulus and movement-related responses in the posterior parietal cortex of the monkey. *Exp. Brain Res.* 58: 144–53

Seal, J., Gross, C., Bioulac, B. 1982. Activity of neurons in area 5 during a simple arm movement in monkeys before and after deafferentation of trained limb. *Brain Res.* 250: 229–43

Seltzer, B., Pandya, D. N. 1984. Further observations on parieto-temporal connections in the rhesus monkey. *Exp. Brain Res.* 5: 301–12

Shubutani, H., Sakata, H., Hyvarinen, J. 1984. Saccade and blinking evoked by microstimulation of the posterior parietal association cortex of the monkey. *Exp. Brain Res.* 55: 1–8

Siegel, R. M., Andersen, R. A. 1986. Motion perceptual deficits following ibotenic acid lesions of the middle temporal area (MT) in the behaving rhesus monkey. *Soc. Neurosci. Abstr.* 12: 1183

Siegel, R. M., Andersen, R. A. 1988. Perception of three-dimensional structure from two-dimensional visual motion in monkey and man. *Nature* 331(6153): 259–61

Siegel, R. M., Andersen, R. A., Essick, G. K., Asanuma, C. 1985. The functional and anatomical subdivision of the inferior parietal lobule. *Soc. Neurosci. Abstr.* 11: 1012

Tanaka, K., Hikosaka, K., Saito, H., Yuki, M., Fukada, Y., Iwai, E. 1986. Analysis of local and wide-field movements in the superior temporal visual areas of the macaque monkey. *J. Neurosci.* 6: 134–44

Trojanowski, J. Q., Jacobson, S. 1975. Peroxidase labeled subcortical afferents to pulvinar in rhesus monkey. *Brain Res.* 97: 144–50

Ullman, S. 1979. *The Interpretation of Visual Motion.* Cambridge, Mass: MIT Press

Ullman, S. 1984. Maximizing rigidity: The incremental recovery of 3-D structure from rigid and nonrigid motion. *Perception* 13: 255–74

Ungerleider, L. G., Desimone, R. 1986. Cortical connections of visual area MT in the macaque. *J. Comp. Neurol.* 248: 190–222

Vogt, C., Vogt, O. 1919. Allgemeine Ergebnisse unserer hirnforschung. *J. Psychol. Neurol.* 25: 279–462

von Bonin, G., Bailey, P. 1947. *The Neocortex of Macaca Mulatta.* Urbana: Univ. Illinois Press

Wurtz, R. H., Newsome W. T. 1985. Divergent signals encoded by neurons in extrastriate areas MT and MST during smooth pursuit eye movements. *Soc. Neurosci. Abstr.* 11: 1246

Yeterian, E. H., Pandya, D. N. 1985. Corticothalamic connections of the posterior parietal cortex in the rhesus monkey. *J. Comp. Neurol.* 237: 408–26

Zipser, D., Andersen, R. A. 1988. A back propagation programmed network that simulates response properties of a subset of posterior parietal neurons. *Nature* 331: 679–84

*Ann. Rev. Neurosci. 1989. 12:405–14*

# SPIDER TOXINS:
# Recent Applications
# in Neurobiology

*Hunter Jackson and Thomas N. Parks*

Department of Anatomy, University of Utah School of Medicine,
Salt Lake City, Utah 84132, and Natural Product Sciences, Inc.,
Salt Lake City, Utah 84108

## Introduction

During their long evolutionary history, spiders have developed remarkably diverse morphological and behavioral specializations for predation (Foelix 1981). All but a few of the 30,000 known spider species employ venoms to paralyze or kill their prey. Spider venoms are complex mixtures of enzymes, nucleic acids, amino acids, inorganic salts, monoamines, and proteinaceous and nonproteinaceous toxins. Although there have been few comprehensive biochemical characterizations of spider venoms, most appear to be about 25% protein by weight; some contain free glutamate at concentrations of more than 400 mM and nonproteinaceous toxins at more than 20 mM. SDS gel electrophoresis indicates considerable heterogeneity in venom peptides within a single venom and great heterogeneity among different taxa. Ecotypic variation in venom composition occurs in at least two species and gender differences are apparent in others. Many toxic components of spider venoms are heat- and pH-stable, whereas others appear comparatively labile (Geren & Odell 1984, Early & Michaelis 1987, Adams et al 1987; H. Jackson and T. N. Parks, unpublished observations).

The single greatest impediment to research on spider venoms has been the difficulty in obtaining adequate amounts of material. Thus, most studies have been carried out on venom-gland extracts from relatively large spiders, which often have larger glands than smaller species. This limitation has resulted in recent investigations being confined almost exclusively to members of only four of the 60 known spider families, with a large majority of studies being carried out on three genera of orb-weaving spiders (family *Araneidae*). The recent availability of venoms obtained by repeated milking

0147–006X/89/0301–0405$02.00

of living spiders (and therefore uncontaminated by digestive secretions, hemolymph, or venom gland debris) has removed a major obstacle to further progress in this field.

It has long been known that the venoms of a few spiders [e.g. the black widows (*Latrodectus* spp.) and Sydney funnel web spiders (*Atrax* spp.)] contain toxins that affect the vertebrate nervous system, and, until recently, research on spider venoms has focused almost exclusively on such medically important species (Duchen & Gomez 1984, Geren & Odell 1984, Vest 1987). During the last few years, however, three trends have led to broader investigation of spider venoms. First, work on other arthropods, notably scorpions (Watt & Simard 1984) and wasps (Piek 1986), has shown that arthropod venoms contain toxins with unique and interesting actions on excitable tissues in diverse species. The fact that particular scorpion toxins bind to sodium channels in both insects and mammals, for example, has made it seem more likely that study of arthropod venoms will yield useful tools for investigation of both invertebrate and vertebrate nervous systems. Second, it was discovered (Kawai et al 1982b) that the venoms of spider taxa without any known medical importance can also contain neurotoxins with effects on the vertebrate CNS. Finally, the importance of excitatory amino acid (EAA) neurotransmitters and calcium channels to both neuromuscular transmission in arthropods (Huddart 1985, Piek 1985) and the pathophysiology of several important neurological disorders (e.g. Greenamyre 1986, Greenberg 1987) has raised hopes that spider venoms might provide lead structures for synthesis of new EAA receptor antagonists or calcium channel antagonists for use in agriculture and medicine.

The present review concentrates on recent studies involving the use of spider venoms to affect EAA neurotransmission and calcium channel function. Space constraints preclude detailed discussion of evidence that some spider venoms also contain toxins that affect axonal conduction, in at least some cases by acting at sodium channels (Cruz-Hofling et al 1985, Jackson et al 1985, Adams 1988). Work on venoms of medically important spiders has been viewed comprehensively by Tu (1984).

## Effects of Araneid Venoms on Excitatory Amino Acid Neurotransmission

The discovery that venoms from certain orb-weaving spiders contain potent neurotoxins that block neuromuscular transmission in marine (Kawai et al 1982a) and terrestrial (Usmanov et al 1983, Tashmukhamedov et al 1983) arthropods has stimulated considerable interest in the physiological actions and chemical structures of these toxins (Jackson & Usherwood 1988). Research has concentrated on three orb-weaver genera (*Nephila*, *Argiope* and *Araneus*), as discussed below.

VENOMS FROM THE GENUS *NEPHILA*    Kawai and colleagues have shown that a low molecular weight toxin (Joro spider toxin, JSTX) from *Nephila clavata* venom blocks the excitatory postsynaptic potential (EPSP) and the potential elicited by iontophoretically applied glutamate at EAA synapses in the lobster neuromuscular junction (Kawai et al 1982a, Abe et al 1983, Miwa et al 1987); venom from *Araneus ventricosus* had similar effects (Kawai et al 1983a). In studies on the giant synapse of the squid stellate ganglion, JSTX blocked EPSP's and glutamate-induced depolarization without affecting the antidromic response (Kawai et al 1983b). Similarly, in catfish lateral line organ this toxin abolished afferent impulses elicited by focal application of glutamate (Nagai et al 1984). Using guinea pig hippocampal slice preparations, this group showed that JSTX blocks the responses of $CA_1$ pyramidal neurons either to stimulation of the Schaffer collateral fibers or to direct application of glutamate, without affecting antidromically elicited responses (Saito et al 1985). The action of JSTX is described as being irreversible or "practically irreversible," with an apparent $IC_{50}$ of $10^{-7}$ M.

In a recent paper (Miwa et al 1987), this group notes that in a neuromuscular preparation from spiny lobster, the toxin produces an irreversible, voltage-independent blockade at very low concentrations. The decay phase of the excitatory postsynaptic current exhibits a single exponential in the presence of toxin, leading Miwa et al (1987) to the conclusion that JSTX does not resemble a typical ion channel blocker and probably "binds to the normal attachment site of glutamate in the receptor molecule" or, possibly, to "the external site of the ionic channel to prevent entry of ions." Either possibility is consistent with the finding that [125]I-labeled JSTX binds to synaptic sites on the surface of lobster leg muscle sarcolemma (Shimazaki et al 1987).

Shiells & Falk (1987) report that JSTX hyperpolarizes horizontal cells in the dogfish retina, blocks their responses to light, and antagonizes glutamate-induced depolarization of these cells in a slowly reversible manner. In contrast, JSTX acts like an agonist at "on-bipolar" cells, closing the same ion channels as glutamate. Thus, JSTX appears capable of exerting specific effects at two classes of EAA receptor-channel complex in vertebrates.

The structures of JSTX and a related toxin from the New Guinea spider *Nephila maculata* (NSTX) have been determined (Aramaki et al 1987a,b). Together with the low molecular weight toxins discovered in the venom of various species of the genus *Argiope* (see below), NSTX and JSTX appear to represent a family of closely related toxins in the venoms of orb-weaving (araneid) spiders. The basic structure of these molecules comprises an amino acid (arginine) linked via a peptide bond to a spermine-like polyamine, which, in turn, is connected to the α-carboxyl group of an asparagine whose α-amino group is coupled with a phenolic moiety (e.g. 2,4-dihydroxyphenylacetic acid) or an indolic one (e.g. 4-hydroxyindole-3-acetic acid). Since the araneid toxins can be synthesized (e.g. Hashimoto et al 1987), it has been possible to begin structure-activity studies. This work has shown that 2,4-dihydroxyphenylacetic acid (2,4-DHPA) alone can block glutamate binding to rat brain synaptic membranes (Pan-Hou & Suda 1987) and that 2,4-dihydroxyphenylacetylasparagine is more effective in this regard than 2,4-DHPA and nearly as effective as whole JSTX (Pan-Hou et al 1987). A spermine-containing synthetic analogue of JSTX is ten times more potent than the cadaverine-containing analogue in suppressing lobster neuromuscular transmission; the blocking effects of both analogues are reversible, whereas JSTX is apparently irreversible (Shudo et al 1987).

VENOMS FROM THE GENERA *ARGIOPE* AND *ARANEUS*    Usmanov et al (1983) reported that a venom-gland extract from *Argiope lobata* blocks neuromuscular transmission in both frog and locust. This venom reversibly blocks frog muscle responses to iontophoretically applied acetylcholine and, at lower doses, irreversibly blocks the responses of locust muscle to applied glutamate. Subsequently, this group reported that it had isolated the active venom fraction by gel chromatography and used the toxin to purify "glutamate receptors" from crab muscle by affinity chromatography. The purified receptors were inserted into artificial membranes, and a pharmacologically specific glutamate-induced ion conductance across the membrane was described (Tashmukhamedov et al 1985). The findings of this study highlight a major potential application of these toxins but remain to be confirmed by other laboratories.

Work on the active component of *Argiope lobata* venom yielded a toxin called argiopine (Grishin et al 1986), a 636 dalton molecule closely related to JSTX and NSTX and identical to the argiotoxin$_{636}$ described subsequently by other workers (see below). Magazanik et al (1987) report that argiopine acts as a postsynaptic open-channel blocker in frog and blowfly larval muscles, and estimate $K_D$s of $10^{-5}$ M and $10^{-7}$ M, respectively. Given the time-dependence of araneid toxin effects, however, accurate comparisons of relative potencies across preparations may be quite difficult to obtain (Jackson & Usherwood 1988).

Usherwood and colleagues have studied the effects of low molecular weight toxins from the venoms of *Argiope trifasciata*, *Argiope florida*, and *Araneus gemma* on neuromuscular transmission in the locust retractor unguis preparation. These "argiotoxins" block neurally evoked muscle twitch, the junctional potential evoked by glutamate iontophoresis, and the voltage-clamped EPSP in a slowly reversible fashion (Usherwood et al

1984, Bateman et al 1985, Usherwood & Duce 1985). Single channel studies of locust muscle EAA receptors suggest that argiotoxins are extremely potent noncompetitive antagonists of EAA receptor-channel complexes whose principal effect is a use-dependent blockade of open ion channels (Kerry et al 1988), although some effect on closed channels is probable (Jackson & Usherwood 1988). The apparent combination in araneid venoms of a high concentration of free glutamate (Early & Michaelis 1987) to open cation channels with a potent open-channel antagonist to block them provides a sophisticated and powerful mechanism for immobilizing insect prey. This synergism may be responsible for the physiological effects of whole *Araneus* venom on chick spinal cord neurons reported by Vyklicky et al (1986). Budd et al (1988) investigated the structures of the active toxins in *A. trifasciata*, *A. florida*, and *A. gemma*. These authors report finding two families of argiotoxins: The first includes the prominent 636-dalton toxin (with 2,4-dihydroxyphenylacetic acid as the chromophore) and its methylene analogues; the second family of toxins includes a prominent 659-dalton toxin with an indolic chromophore.

Adams et al (1987) describe the complete chemical characterization of the argiotoxins and, by using synthesized toxins, derive comparative potency information with in vivo insect paralysis assays and measurement of stimulus-evoked muscle EPSPs. Argiotoxins of molecular weights 636, 659, and 673 daltons are described by these authors. The 636- and 659-dalton toxins (which are present in whole venom at concentrations of about 20 mM each) have an amino-terminal arginine attached by an amide linkage to a polyamine (1,13-diamino-4,8-diazatridecane), whereas the 673-dalton toxin has dimethylated spermine as the polyamine. In each case, a second amide linkage is formed to the carboxyl group of asparagine, and the amino group of asparagine is coupled through an amide linkage to one of two aromatic acids. The 659- and 673-dalton toxins contain 4-hydroxyindole-3-acetic acid, whereas the 636-dalton toxin has 2,4-dihydroxyphenylacetic acid. In house-fly paralysis assays, the synthetic 659-dalton toxin has an $ED_{50}$ of 0.9 pmol/mg, whereas the synthetic 636-dalton toxin's $ED_{50}$ is 3.3 pmol/mg. In *in vitro* assays, however, the potencies of the two toxin families are similar.

Jackson et al (1985, 1986b) report that *Argiope aurantia* venom blocks synaptic transmission in the chick cochlear nucleus that is known to be mediated by non-$N$-methyl-D-aspartate receptors. This blockade is produced primarily by the 659-dalton argiotoxin, with some contribution from the 636-dalton molecule. In contrast to the effects of the argiotoxins on insect muscle, we found the suppression of responses in the chick CNS to be use-*independent*, suggesting a mechanism of action in vertebrates other than the open-channel blockade described by Usherwood and colleagues. Pagnozzi et al (1988) report that JSTX and the 636- and 659-dalton argiotoxins are functional antagonists of both NMDA and non-NMDA glutamate receptors in rat brain. These authors also find that the polyamine toxins greatly increase glycine binding to rat brain and conclude that the toxins' mechanism of action is unlike that of either standard competitive or noncompetitive antagonists. We have also isolated a toxin of 5–10 kDa from the venom of the funnel-web spider *Hololena curta* (family *Agelenidae*) that irreversibly blocks synaptic transmission in the chick cochlear nucleus preparation (Jackson et al 1985). This toxin, which is about 1000 times more potent an antagonist in this preparation than *cis*-2,3-piperidine dicarboxylic acid, a standard EAA antagonist, also appears to block responses to directly applied EAA receptor agonists in the chick cochlear nucleus. Thus, it may represent a class of relatively large and irreversible EAA antagonists.

## Calcium Channel Antagonists

The essential role of calcium in mediating a large range of cellular functions, including muscle contraction in insects, would seem to make calcium channels a worthy target for spider toxins. In the course of surveying venoms for effects on EAA transmission, we found indications of such a toxin in the venom of the *Agelenopsis aperta* spider (Jackson et al 1986a). This venom irreversibly blocks synaptic transmission in the chick cochlear nucleus without significantly affecting depolarization caused by direct application of EAA agonists. The toxin's effects are, however, countered by increasing extracellular calcium. The toxin responsible for this effect, tentatively called AG1, is contained in a fraction of about 4500 $M_r$.

AG1 has recently been shown to have potent and selective effects in guinea pig cerebellar Purkinje cells, blocking dendritic calcium spikes at a concentration of about 20 nM with no effect on rapid or low-threshold, noninactivating sodium conductances (Sugimori & Llinas 1987). It was also found that synaptic potentials in Purkinje cells elicited by climbing fiber stimulation were not blocked by AG1, a finding that suggests a selectivity of the toxin for calcium channels mediating dendritic spikes as opposed to those involved in transmitter release. Such selectivity is also apparent in the lack of effect on synaptically evoked twitch, miniature endplate potential (MEPP) frequency, or MEPP amplitude in mouse diaphragm (H. Jackson, M. R. Urnes, and T. N. Parks, unpublished observations).

At least one other calcium antagonist toxin active in the vertebrate CNS is also contained in *Agelenopsis aperta* venom. This smaller toxin (AG2), in contrast to AG1, reversibly blocks the slow inward calcium current in heart muscle (Jackson et al 1988) and produces *reversible* suppression of chick cochlear nucleus synaptic transmission. The different actions of AG1 and AG2 also have been demonstrated in competitive binding studies (Kerr et al 1987, 1988). With rat brain membrane preparations, AG2 was shown to decrease [$^3$H]-nitrendipine binding, whereas AG1 increased binding of this dihydropyridine. Conversely, when tested against [$^{125}$I]-omega-conotoxin, AG1 blocked binding whereas AG2 enhanced binding of the labeled toxin. This result was quite unexpected in light of the fact that omega-conotoxin (from the marine snail *Conus geographus*) is thought to affect N-type calcium channels preferentially whereas dihydropyridines are associated with L-type channels (Miller 1987). Scatchard analysis of nitrendipine binding with and without AG1 indicated that enhancement of nitrendipine binding by AG1 results from an increase in $B_{max}$ with no significant change in $K_D$, thus suggesting that AG1 increases the number of sites that bind nitrendipine. The tentative conclusion from this rather complicated experiment is that these spider toxins may reveal an ability of calcium channel subtypes, otherwise thought to be discrete anatomical and functional entities, to assume certain "characteristic" features of other subtypes in the presence of appropriate modifiers. The corollary, of course, is that calcium channel properties may be mutable under various physiological, as well as pharmacological, circumstances.

In addition to the use of spider toxins in research as probes for receptors and ion channels, their unusual properties make them potentially applicable to a variety of nervous system disorders. The best illustration of this to date is provided by AG2 toxin, which has been found to be a very effective anticonvulsant under some experimental conditions (Jackson & Parks 1987, 1988). We have found that AG2, administered to rats either intravenously or intraventricularly, provides dose-dependent protection against the chemical convulsants kainic acid, picrotoxin, and bicuculline. It is significant that the toxin is apparently equally effective against all three of these convulsants, which differ both in their chemical properties and mechanisms of action. This finding supports suggestions of a fundamental involvement of calcium channels in epileptiform activity (e.g. Prince & Connors 1986).

The intravenous dose of AG2 that we found to be effective was only about 2 $\mu$M/kg. At this dose, subjects appeared mildly sedated for about an hour but retained complete postural control and responsiveness to external stimuli. Administration of AG2 alone at about 6 $\mu$M/kg produced only a small increase in the magnitude of that response but no loss of consciousness or deaths. It is apparent, therefore, that at least under these conditions, the dose of toxin required to suppress seizures did not have the potentially serious side-effects that one might expect of a compound acting on CNS calcium channels.

As work in this field progresses, it is becoming apparent that calcium channel antagonists may be a fairly common component of spider venoms. Studies by Jan and colleagues (Bowers et al 1987, Branton et al 1987), for example, have shown that the venoms of the *Hololena curta* and *Plectreurys tristes* spiders contain peptides that produce long-lasting presynaptic blockade of neuromuscular transmission in *Drosophila* larvae. Although the physiological effects of the calcium-antagonist toxins from the two species are similar, they are apparently structurally distinct as judged by chromatographic, immunologic, and competitive binding criteria. It is also interesting to note that the *Hololena* toxin had no effect on frog neuromuscular preparations (Bowers et al 1987) or chick cochlear nucleus; *Plectreurys* venom also did not suppress transmission in the chick cochlear nucleus (Jackson et al 1985). These results probably reflect species-specific properties of calcium channels and highlight spider venoms as possible sources of insect-specific toxins for agricultural applications.

*Literature Cited*

Abe, T., Kawai, N., Miwa, A. 1983. Effects of a spider toxin on the glutaminergic synapse of lobster muscle. *J. Physiol. London* 339: 243–52

Adams, M. E. 1988. Spider venom toxins affecting multiple targets at the insect neuromuscular junction. *Abstr. 3rd Int. Symp. Insect Neurobiol. Pesticide Action Indust.*, p. 6. London: Soc. Chem.

Adams, M. E., Carney, R. L., Enderlin, F. E., Fu, E. T., Jarema, M. A., et al. 1987. Structures and biological activities of three synaptic antagonists from orb weaver spider venom. *Biochem. Biophys. Res. Commun.* 148: 678–83

Aramaki, Y., Yasuhara, T., Higashijima, T., Miwa, A., Kawai, N., Nakajima, T. 1987a. Chemical characterization of spider toxin, NSTX. *Biomed. Res.* 8: 167–83

Aramaki, Y., Yasuhara T., Shimazaki, K., Kawai, N., Nakajima, T. 1987b. Chemical structure of Joro spider toxin (JSTX). *Biomed. Res.* 8: 241–45

Bateman, A., Boden, P., Dell, A., Duce, I. R., Quicke, D. L. J., Usherwood, P. N. R. 1985. Postsynaptic block of a glutamatergic synapse by low molecular weight fractions of spider venom. *Brain Res.* 339: 237–44

Bowers, C. W., Phillips, H. S., Lee, P., Jan, Y. H., Jan, L. Y. 1987. Identification and purification of an irreversible presynaptic neurotoxin from the venom of the spider *Hololena curta*. *Proc. Nat. Acad. Sci. USA* 84: 3506–10

Branton, W. D., Kolton, L., Jan, Y. N., Jan, L. Y. 1987. Neurotoxins from *Plectreurys* spider venom are potent presynaptic blockers in *Drosophila*. *J. Neurosci.* 7: 4195–4200

Budd, T., Clinton, P., Dell, A., Duce, I. R., Johnson, S. J., et al. 1988. Isolation and characterisation of glutamate receptor antagonists from venoms of orb-web spiders. *Brain Res.* 448: 30–39

Cruz-Hofling, M. A., Love, S., Brook, G., Duchen, L. W. 1985. Effects of *Phoneutria nigriventer* spider venom on mouse peripheral nerve. *Q. J. Exp. Physiol.* 70: 623–40

Duchen, L. W., Gomez, S. 1984. Pharmacology of spider venoms. See Tu 1984, pp. 483–512

Early, S. L., Michaelis, E. K. 1987. Presence of proteins and glutamate as major constituents of the venom of the spider *Araneus gemma*. *Toxicon* 25: 433–42

Foelix, R. 1981. *The Biology of Spiders*. Cambridge: Harvard Univ. Press

Geren, C. R., Odell, G. V. 1984. The biochemistry of spider venoms. See Tu 1984, pp. 441–81

Greenamyre, J. T. 1986. The role of glutamate in neurotransmission and in neurologic disease. *Arch. Neurol.* 43: 1058–64

Greenberg, D. A. 1987. Calcium channels and calcium channel antagonists. *Ann. Neurol.* 21: 317–30

Grishin, E. V., Volkova, T. M., Arseniev, A. S., Reshetova, O. S., Onoprienko,

V. V., et al. 1986. Structure-functional characterization of argiopine—an ion channel blocker from the venom of the spider *Argiope lobata. Bioorg. Khim.* 12: 1121–24 (in Russian)

Hashimoto, Y., Endo, Y., Shudo, K., Aramaki, Y., Kawai, N., Nakajima, T. 1987. Synthesis of spider toxin (JSTX-3) and its analogs. *Tetrahed. Lett.* 28: 3511–14

Huddart, H. 1985. Visceral muscle. In *Comprehensive Insect Physiology, Biochemistry and Pharmacology*, Vol. 11, *Pharmacology*, ed. G. A. Kerkut, L. I. Gilbert, pp. 131–194. Oxford: Pergamon

Jackson, H., Bkaily, G., Urnes, M., Benabderrazik, M., Gray, W. R., Parks, T. N., Sperelakis, N. 1988. Specific $Ca^{2+}$ blockade in heart muscle by a toxin (AG2) from the *Agelenopsis aperta* spider. Submitted for publication

Jackson, H., Parks, T. N. 1987. Suppression of chemically-induced behavioral seizures in rats by a novel spider toxin. *Soc. Neurosci. Abstr.* 13: 1078

Jackson, H., Parks, T. N. 1988. Anticonvulsant action of a toxin isolated from *Agelenopsis aperta* spider venom. Submitted for publication

Jackson, H., Urnes, M. R., Gray, W. R., Parks, T. N. 1985. Spider venoms block synaptic transmission mediated by non-$N$-methyl-$D$-aspartate receptors in the avian cochlear nucleus. *Soc. Neurosci. Abstr.* 11: 107

Jackson, H., Urnes, M. R., Gray, W. R., Parks, T. N. 1986a. Presynaptic blockade of transmission by a potent, long-lasting toxin from *Agelenopsis aperta* spiders. *Soc. Neurosci. Abstr.* 12: 730

Jackson, H., Urnes, M. R., Gray, W. R., Parks, T. N. 1986b. Effects of spider venoms on transmission mediated by non-$N$-methyl-$D$-aspartate receptors in the avian cochlear nucleus. In *Excitatory Amino Acid Transmission*, ed. D. Lodge et al., pp. 51–54. New York: Liss

Jackson, H., Usherwood, P. N. R. 1988. Spider toxins as tools for dissecting elements of excitatory amino acid transmission. *Trends Neurosci.* 11: 278–83

Kawai, N., Miwa, A., Abe, T. 1982a. Spider venom contains specific receptor blocker of glutaminergic synapses. *Brain Res.* 247: 169–71

Kawai, N., Miwa, A., Abe, T. 1982b. Effect of a spider toxin on glutaminergic synapses in mammalian brain. *Biomed. Res.* 3: 353–55

Kawai, N., Miwa, A., Abe, T. 1983a. Specific antagonism of the glutamate receptor by an extract from the venom of the spider *Araneus ventricosus. Toxicon* 21: 438–40

Kawai, N., Yamagishi, S., Saito, M., Furuya, K. 1983b. Blockade of synaptic transmission in the squid giant synapse by a spider toxin (JSTX). *Brain Res.* 278: 346–49

Kerr, L. M., Filloux, F., Wamsley, J. K., Parks, T. N., Jackson, H. 1987. Effects of spider toxins on L and N CNS calcium channels: Inhibition and enhancement of binding. *Soc. Neurosci. Abstr.* 13: 102

Kerr, L. M., Wamsley, J. K., Parks, T. N., Jackson, H. 1988. Effects of *Agelenopsis aperta* spider toxins on L and N CNS calcium channels: Inhibition and enhancement of binding. Submitted for publication

Kerry, C. J., Ramsey, R. L., Sansom, M. S. P., Usherwood, P. N. R. 1988. Single channel studies on non-competitive antagonism of a quisqualate-sensitive glutamate receptor by argiotoxin$_{636}$—a fraction isolated from orb-weaver spider venom. *Brain Res.* In press

Magazanik, L. G., Antonov, S. M., Federova, I. M., Volkova, T. M., Grishin, E. V. 1987. Argiopin—A naturally occurring blocker of glutamate-sensitive synaptic channels. In *Receptors and Ion Channels*, ed. Y. A. Ovchinnikov, F. Hucho, pp. 305–12. New York: DeGruyter

Miller, R. J. 1987. Multiple calcium channels and neuronal function. *Science* 235: 46–52

Miwa, A., Kawai, N., Saito, M., Pan-Hou, H., Yoshioka, M. 1987. Effect of a spider toxin (JSTX) on excitatory postsynaptic current at neuromuscular synapse of spiny lobster. *J. Neurophysiol.* 58: 319–26

Nagai, T., Obara, S., Kawai, N. 1984. Differential blocking effects of a spider toxin on synaptic and glutamate responses in the afferent synapse of the acoustico-lateralis receptors of *Plotosus. Brain Res.* 300: 183–87

Pagnozzi, M. J., Saccamano, N. A., Gnllak, M. F., Volkmann, R. A., Mena, E. E. 1988. Polyamine spider venom components as excitatory amino acid antagonists in the rat CNS. *Soc. Neurosci. Abstr.* 14: In press

Pan-Hou, H., Suda, Y. 1987. Molecular action mechanism of spider toxin on glutamate receptor: Role of 2,4-dihydroxyphenylacetic acid in toxin molecule. *Brain Res.* 418: 198–200

Pan-Hou, H., Suda, Y., Sumi, M., Yoshioka, M., Kawai, N. 1987. Inhibitory effect of 2,4-dihydroxyphenylacetylasparagine, a common moiety of spider toxin, on glutamate binding to rat brain synaptic membranes. *Neurosci. Lett.* 81: 199–203

Piek, T. 1985. Neurotransmission and neuromodulation of skeletal muscle. See Huddart 1985, pp. 55–118

Piek, T. 1986. *Venoms of the Hymenoptera.* New York: Academic

Prince, D. A., Connors, B. W. 1986. Mechanisms of interictal epileptogenesis. *Advances in Neurology,* Vol. 44, *Basic Mechanisms of the Epilepsies,* ed. A. V. Delgado-Escueta et al. pp. 275–99. New York: Raven

Saito, M., Kawai, N., Miwa, A., Pan-Hou, H., Yoshioka, M. 1985. Spider toxin (JSTX) blocks glutamate synapse in hippocampal pyramidal neurons. *Brain Res.* 346: 397–99

Shiells, R. A., Falk, G. 1987. Joro spider venom: Glutamate agonist and antagonist on the rod retina of the dogfish. *Neurosci. Lett.* 77: 221–25

Shimazaki, K., Hagiwara, K., Hirata, Y., Nakajima, T., Kawai, N. 1987. An autoradiographic study of binding of iodinated spider toxin to lobster muscle. *Neurosci. Lett.* 84: 173–77

Shudo, K., Endo, Y., Hashimoto, Y., Aramaki, Y., Nakajima, T., Kawai, N. 1987. Newly synthesized analogues of the spider toxin block the crustacean glutamate receptor. *Neurosci. Res.* 5: 82–85

Sugimori, M., Llinas, R. 1987. Spider venom blockade of dendritic calcium spiking in Purkinje cells studied *in vitro. Soc. Neurosci. Abstr.* 13: 228

Tashmukhamedov, B. A., Usmanov, P. B., Kazakov, I., Kalikulov, D., Yukelson, L. Y., Atakuziev, B. U. 1983. Effects of different spider venoms on artificial and biological membranes. in *Toxins as Tools in Neurochemistry,* ed. Y. A. Ovchinni-

kov, F. Hucho, pp. 331–23. New York: DeGruyter

Tashmukhamedov, B. A., Makhumudova, E. M., Usmanov, P. B., Kazakov, I. 1985. Reconstitution in bilayer lipid membranes of the crab *Potamon transcapsicum* spider venom sensitive glutamate receptors. *Gen. Physiol. Biophys.* 4: 625–30

Tu, A. T. 1984. *Insect Poisons, Allergens, and Other Invertebrate Venoms. Handbook of Natural Toxins,* Vol. 2. New York: Dekker

Usherwood, P. N. R., Duce, I. R. 1985. Antagonism of glutamate receptor channel complexes by spider venom polypeptides. *NeuroToxicol.* 6: 239–50

Usherwood, P. N. R., Duce, I. R., Boden, P. 1984. Slowly-reversible block of glutamate receptor-channels by venoms of the spiders, *Argiope trifasciata* and *Araneus gemma. J. Physiol. Paris* 79: 241–45

Usmanov, P. B., Kalikulov, D., Shadyeva, N., Tashmukhamedov, B. A. 1983. Action of the venom of the spider *Argiope lobata* on the glutametergic and cholinergic synapses. *Dokl. Akad. Nauk SSSR* 273: 1017–18. (In Russian)

Vest, D. K. 1987. Necrotic arachnidism in the northwest United States and its probable relationship to *Tegenaria agrestis* (Walckenaer) spiders. *Toxicon* 25: 175–84

Vyklicky, L. Jr., Krusek, J., Vyklicky, L., Vyskocil, F. 1986. Spider venom of *Araneus* opens and desensitizes glutamate channels in chick spinal cord neurons. *Neurosci. Lett.* 68: 227–31

Watt, D. D., Simard, J. M. 1984. Neurotoxic proteins in scorpion venom. *J. Toxicol. Toxin Rev.* 3: 181–221

*Ann. Rev. Neurosci. 1989. 12:415–61*

# ACUTE REGULATION OF TYROSINE HYDROXYLASE BY NERVE ACTIVITY AND BY NEUROTRANSMITTERS VIA PHOSPHORYLATION

*Richard E. Zigmond,*[1] *Michael A. Schwarzschild, and Ann R. Rittenhouse*

Department of Biological Chemistry and Molecular Pharmacology, Harvard Medical School, Boston, Massachusetts 02115

## INTRODUCTION

Although stimulation of adrenergic nerves or adrenal chromaffin cells causes the release of catecholamines, in many cases it does not cause a decrease in the catecholamine content of these cells. One of the first reports of this phenomenon was made by T. R. Elliot (1912) in a study designed to establish whether the release of catecholamines by the adrenal medulla was controlled by the splanchnic nerves. Although Elliot obtained considerable data supporting this view, he also found that prolonged electrical stimulation of the splanchnic nerves did not produce a pronounced depletion of the content of adrenal catecholamines. In fact, Elliot wrote, "so slight is the change in the residual adrenalin caused by faradisation of the splanchnic nerves, that it would never have sufficed to convince me of the existence of the splanchnic control." Holland & Schumann (1956) later extended these findings by measuring directly the amount of catecholamines released during splanchnic nerve stimulation and the content of catecholamines in the adrenal gland before and after the stimulation. They observed that the output of catecholamines into the blood stream

[1] Current address: Center for Neurosciences, Case Western Reserve University, School of Medicine, 2119 Abington Road, Cleveland, Ohio 44106.

415

0147–006X/89/0301–0415$02.00

always exceeded the decrease in adrenal catecholamine content. From these data, they hypothesized that nerve stimulation increases the rate of catecholamine biosynthesis in the adrenal gland, thus compensating for the increase in catecholamine release. A similar conclusion was reached by Hökfelt & McLean (1950).

Since these early studies, a great deal has been learned about the biochemical pathway involved in the biosynthesis of catecholamines from the circulating amino acid tyrosine and about the mechanisms by which this pathway is regulated. As originally suggested by Blaschko (1939) and by Holtz (1939), hydroxylation of tyrosine is followed by decarboxylation, $\beta$-hydroxylation, and finally $N$-methylation to form epinephrine. The mammalian enzymes that catalyze these four reactions—tyrosine hydroxylase (TH),[2] dopa decarboxylase, dopamine $\beta$-hydroxylase (DBH), and phenylethanolamine $N$-methyltransferase (PNMT)—have now been well characterized (Usdin et al 1981) and messenger RNAs for three of them have been cloned and sequenced [i.e. rat and human TH (see below), human DBH (Lamouroux et al 1987), and bovine PNMT (Baetge et al 1986)]. Which catecholamines (i.e. dopamine, norepinephrine, or epinephrine) are synthesized and released by a particular cell depends simply on which of these four enzymes are expressed in the cell. Thus, dopaminergic neurons contain TH and dopa decarboxylase but not DBH or PNMT, whereas noradrenergic neurons are missing only PNMT (Hökfelt et al 1984). Early studies by Levitt et al (1965) indicated that the first step in catecholamine biosynthesis, the hydroxylation of tyrosine, is the rate-limiting step, and it is this step that has been found to be regulated by nerve activity and by neurotransmitters.

Two mechanisms for the regulation of tyrosine hydroxylation have now been studied in some detail. This review focuses on one of these mechanisms, the one that involves a rapid increase in the activity of TH in response to stimulation of adrenal medullary cells or catecholaminergic neurons. As discussed below, this mechanism involves covalent modification of TH via phosyphorylation. The second mechanism involves an increase in TH that occurs only after a delay that lasts from 12–48 hr following a stimulus, and that has been shown to be caused by an increase in the amount of TH enzyme protein. (For reviews see Costa et al 1974,

[2] Abbreviations used: BH$_4$, tetrahydrobiopterin; cAMP, adenosine 3′,5′-cyclic monophosphate; cGMP, guanosine 3′,5′-cyclic monophosphate; DBH, dopamine $\beta$-hydroxylase; DMPP, dimethylphenylpiperazinium; NGF, nerve growth factor; OAG, 1-oleoyl-2-acetylglycerol; PDBu, 4$\beta$-phorbol 12,13-dibutyrate; PHI, peptide histidine isoleucine amide; PI, phosphoinositide; PMA, 4$\beta$-phorbol-12$\beta$-myristate-13$\alpha$-acetate; PNMT, phenylethanolamine $N$-methyltransferase; SCG, superior cervical ganglion; TH, tyrosine hydroxylase; VIP, vasoactive intestinal peptide.

Thoenen et al 1979, Zigmond 1980). This delayed increase in TH activity is blocked by inhibitors of RNA and protein synthesis (Mueller et al 1969, Zigmond & Mackay 1974) and involves an increase in the amount of TH mRNA (e.g. Mallet et al 1983, Black et al 1985, Stachowiak et al 1985, Tank et al 1985, Faucon Biguet et al 1986, Berod et al 1987).

Interest in these mechanisms of regulation of catecholamine biosynthesis has increased with the findings that both acute and long-term increases in TH activity in sympathetic neurons and/or adrenal chromaffin cells occur when animals are exposed to a variety of conditions. These include stressful situations such as cold (Thoenen 1970, Fluharty et al 1985a), immobilization (Weinhold & Rethy 1969, Kvetnansky et al 1971), and hypoglycemia (Fluharty et al 1983, 1985b); treatment with hypotensive agents, such as phenoxybenzamine (Dairman et al 1968, Thoenen et al 1969), reserpine (Thoenen et al 1979), and nitroprusside (Takimoto & Weiner 1981a); and a number of other conditions, including electroconvulsive shock (Masserano & Weiner 1979a, Masserano et al 1981) and partial sympathectomy, caused by 6-hydroxydopamine (Fluharty et al 1985a). In addition, under less extreme experimental conditions, TH activity has been found to increase acutely at night in the pineal gland (see Table 1) and during the day in the retina (Iuvone et al 1978) and to be higher in the adrenal gland in spontaneously hypertensive rats than in normotensive control animals (Nagatsu et al 1971). In most of these cases, it has been hypothesized that an increase in the firing of sympathetic nerves innervating these tissues or of the dopaminergic amacrine cells in the retina occurs and that this change in firing rate triggers the increases in TH activity.

The literature on the acute regulation of TH is too extensive to be reviewed exhaustively here, and we have, therefore, chosen to focus this review on the sympathetic nervous system. (See Masserano & Weiner 1983, Goldstein & Greene 1987, for reviews on TH regulation in the central nervous system.) Studies on the central nervous system are cited, however, when they are particularly relevant either historically or conceptually.

## BASIC PROPERTIES OF TYROSINE HYDROXYLASE

TH is a mixed function oxidase or monooxygenase that catalyzes the hydroxylation of tyrosine to form dopa by using molecular oxygen and the cofactor tetrahydrobiopterin ($BH_4$) (Kaufman 1977, 1985). $BH_4$ is oxidized to $BH_2$ during the course of this reaction, and the reaction to regenerate the reduced cofactor is catalyzed by a second enzyme, dihydropteridine reductase (Kaufman 1977). TH exists as a homotetramer, with each subunit having a molecular weight of approximately 60,000 as judged by gel electrophoresis (Vulliet & Weiner 1981). cDNAs for TH mRNA prepared from rat and human pheochromocytoma have been cloned (Lamouroux et al 1982, Lewis et al 1983, Kobayashi et al 1987) and sequenced (Grima et al 1985, Grima et al 1987, Kaneda et al 1987). Although it has been reported that multiple transcripts are synthesized from a single TH gene in human pheochromocytoma (Grima et al 1987), only a single mRNA for TH has been found in the rat brain and adrenal gland (C. Boni and J. Mallet, in preparation). Based on the sequence of the rat cDNA, the predicted molecular weight of the TH monomer is 55,903 (Grima et al 1985). Interestingly, the deduced amino acid sequence of the enzyme shows considerable sequence homology to two other monooxygenases that also use $BH_4$ as a cofactor and molecular oxygen as a substrate—phenylalanine hydroxylase (Kwok et al 1985, Dahl & Mercer 1986) and trytophan hydroxylase (Grenett et al 1987).

It has been suggested by a number of authors that the concentration of $BH_4$ in adrenergic cells is subsaturating for TH. Measurements of $BH_4$ in the adrenal medulla have yielded values of about 10 $\mu$M (Lloyd & Weiner 1971, Abou Donia & Viveros 1981). As discussed below, the $K_m$ values of TH for cofactor vary depending on the state of activation of the enzyme, the subcellular fraction in which enzyme activity is examined, the type of pterin cofactor used, and the pH at which the enzyme is assayed, although conflicting results have been obtained concerning the effect of pH on the $K_m$ (cf Markey et al 1980, Pollock et al 1981). Nevertheless, when TH activity was measured at near physiological pH (6.8) in the supernatant fraction of the unstimulated rat adrenal gland, the $K_m$ for the natural cofactor, $BH_4$, was found to be 2 mM (Lovenberg et al 1981). This value is well above the reported concentration of $BH_4$, thus raising the possibility that changes in the concentration of the cofactor or in the affinity of TH for the cofactor would significantly affect the rate of tyrosine hydroxylation in situ.

TH activity is inhibited by high concentrations of catechols (e.g. dopa, dopamine, norepinephrine, and epinephrine) (Nagatsu et al 1964). It has been found that TH activity in intact tissues decreases when catecholamine levels are increased in response to inhibition of catecholamine breakdown by a monoamine oxidase inhibitor (Spector et al 1967, Weiner et al 1972). However, it is not certain whether TH is exposed to high concentrations of catecholamines in situ under nonpharmacological conditions (Perlman 1981). In addition, the $K_i$ for the inhibition of TH by catechols typically increases when the enzyme is activated (see below), reducing the extent of any feedback inhibition that may occur. The $K_i$ of the nonactivated form of TH also varies with pH (Hegestrand et al 1979, Lazar et al 1982); however, this does not seem to be the case for the activated form of the enzyme (Lazar et al 1982). Interestingly, the inhibition of TH by catechols is competitive with pterin cofactor (Udenfriend et al 1965, Nagatsu et al 1972). Thus, in assessing the extent of feedback inhibition of the enzyme in vivo, one must take into account the concentration of both cytoplasmic catechols and $BH_4$ as well as the $K_i$ and $K_m$ for catechols and $BH_4$ respectively.

Although conflicting results have been obtained concerning the subcellular localization of TH (Nagatsu 1973), a number of studies have suggested that the enzyme exists both in a cytoplasmic and in a membrane-bound form (e.g. Kuczenski & Mandell 1972). In the striatum, the membrane-bound form has a lower $K_m$ value for both cofactor and tyrosine and a lower $K_i$ value for catechols than does the soluble form (Kuczenski & Mandell 1972, Kuczenski 1973b). The functional significance of a membrane-bound form of TH is unclear, although it has been shown that newly synthesized norepinephrine is released preferentially over norepinephrine that has been stored by the cells (Kopin et al 1968), perhaps reflecting an association of TH with synaptic vesicles (Stephens et al 1981, Morita et al 1987) or with the plasma membrane in the vicinity of transmitter release sites. The subcellular localization of TH with respect to the localization of catecholamines is also relevant to the question of whether end-product inhibition is likely to be an important mechanism in the regulation of TH activity in vivo.

In most of the studies that are discussed below, one of two types of assays has been performed: either (a) the rate of tyrosine hydroxylation has been measured in intact tissues or cells, usually in short-term organ or cell culture, or (b) the activity of TH has been measured in homogenates. In the case of the former, the measure is referred to below as the rate of tyrosine hydroxylation or the rate of dopa synthesis in situ. In the case of the latter, the measure is referred to as TH activity in vitro and has been used to assess the extent of TH activation. Due to the relatively high concentrations of catecholamines in homogenates of adrenal chromaffin cells (e.g. Meligeni et al 1982), pheochromocytoma cells (e.g. Chalfie & Perlman 1977), and corpora striata (e.g. Ames et al 1978), care is often taken to extract the catecholamines from the homogenates, prior to assaying TH, to avoid complications caused by changes in the extent of feedback inhibition of the enzyme. Two other factors that are important in studies on enzyme activation are the concentration of pterin cofactor used and the pH at which the assay is performed. TH is generally assayed with one of two synthetic cofactors (i.e. 6,7-dimethyltetrahydropterin or 6-methyltetrahydropterin), both of which have a $K_m$ that is 4–5 times higher than that of the natural cofactor tetrahydrobioptern (Nagatsu et al 1972). Although the pH optimum reported for TH varies for different tissues (e.g. Hegstrand et al 1979, Acheson et al 1981), a value of approximately 6 is most commonly observed (Nagatsu et al 1964, Shiman et al 1971). Interestingly, increases in enzyme activity produced by certain stimuli [e.g. adenosine 3′,5′-cyclic monophosphate (cAMP)] are more prominent when the enzyme is assayed at a more physiological pH (i.e. around pH 7) (Lloyd & Kaufman 1975, Goldstein et al 1976). As discussed below, in many cases TH activation is caused by a decrease in $K_m$ of the enzyme for cofactor, and, thus, in those cases, increases in TH activity can most easily be seen if assayed at a low cofactor concentration.

## ACUTE REGULATION OF TH

The first evidence that an increase in the rate of catecholamine biosynthesis produced by nerve stimulation resulted from a change in the first step in the biosynthetic pathway, i.e. the rate of tyrosine hydroxylation, came from the studies by Alousi & Weiner (Alousi & Weiner 1966, Weiner & Rabadjija 1968a) on the hypogastric nerve–vas deferens preparation and by Gordon et al (1966a) on the sympathetic innervation of the heart. Both groups observed that when sympathetic nerve fibers were stimulated, the synthesis of radiolabeled catecholamines from radiolabeled tyrosine was increased rapidly, whereas the synthesis from radiolabeled dopa was unchanged. The change in catecholamine synthesis in the vas deferens was shown to occur in the absence of any change in incorporation of $^3$H-tyrosine into protein and to be insensitive to puromycin, an inhibitor of protein synthesis. These results indicated that nerve stimulation produced a rather selective metabolic change and that this change did not require new protein synthesis.

The initial hypothesis proposed to explain these increases in tyrosine hydroxylation in situ was that they were caused by a decrease in end-product inhibition of TH that resulted from the increased release of norepinephrine by the sympathetic neurons themselves (Alousi & Weiner 1966, Gordon et al 1966a). However, Morgenroth et al (1974) subsequently reported that a direct "activation" of TH was involved. These workers found that when TH was assayed in supernatant fractions prepared from vasa deferentia following hypogastric nerve stimulation, conditions under which endogenous catecholamines are diluted considerably, a decrease in the $K_m$ for the synthetic cofactor dimethyltetrahydropterin and an increase in the $K_i$ for norepinephrine were seen. Activation of TH occurs quite rapidly after nerve stimulation. For example, TH activity measured in vitro at a subsaturating cofactor concentration was significantly elevated after only 1 min of stimulation of the hypogastric nerve (25 Hz for 10 out of every 20 sec) (Weiner et al 1978). Similar changes in the kinetic properties of TH have since been found in many other neural systems, under a variety of conditions that produce an acute increase in tyrosine hydroxylation (see below). Although Morgenroth et al (1974) also reported a decrease in the $K_m$ for tyrosine, this result has been obtained in few other studies. In addition, the physiological significance of such a change is unclear, since it is generally believed that the concentration of tyrosine in adrenergic cells is near saturation for TH (e.g. Nagatsu et al 1964, Gordon et al 1966b, Perlman 1981; but see Wurtman et al 1974, Milner & Wurtman 1986). Finally, in certain studies in which TH activation was found, including that of Morgenroth et al (1974), an increase in $V_{max}$ of the enzyme was observed (e.g. Joh et al 1978, Greene et al 1984, Pollock et al 1981).

# MECHANISMS OF ACUTE REGULATION

## First Messengers

As is discussed in the following sections, a number of transmitters have been found to produce an acute stimulation of dopa synthesis in situ and TH activity in vitro in adrenergic cells (Table 1). In addition, in several preparations, direct cell depolarization has been found to produce similar effects. In subsequent sections of this review, the second messenger systems involved in these effects are discussed.

SYNAPTIC ACTIVITY    In their studies, Alousi & Weiner (1966) and Morgenroth et al (1974) stimulated preganglionic sympathetic nerve fibers in the hypogastric nerve and measured subsequent changes in catecholamine synthesis in postganglionic adrenergic neurons in the vas deferens. Similar studies have been done more recently in the superior cervical ganglion (SCG) (Steinberg & Keller 1978, Ip et al 1983). In the latter preparation, stimulation of the preganglionic cervical sympathetic trunk at 10 Hz for 15 min led to a four-fold stimulation in the rate of dopa synthesis in postsynaptic adrenergic cells (Ip et al 1983). To examine the neurotransmitter(s) involved in this transsynaptic biochemical effect, the preganglionic nerve was stimulated in the presence of a nicotinic (hexamethonium, 3 mM) and a muscarinic (atropine, 6 $\mu$M) antagonist. These agents together reduced the increase in dopa synthesis seen with preganglionic nerve stimulation by about 50%, a finding that suggests that acetylcholine is responsible for at least part of the transsynaptic signal (Ip et al 1983). However, increasing the concentration of these cholinergic antagonists ten-fold produced no further reduction, thus suggesting that the residual stimulation is mediated by a noncholinergic preganglionic neurotransmitter (Ip et al 1983). Later studies, discussed below, have raised the possibility that a neuropeptide of the secretin-glucagon family is involved.

Interestingly, the pharmacology of the transsynaptic effect varied with the pattern of preganglionic nerve stimulation. Thus, when the preganglionic nerve was stimulated at 10 Hz for one out of every 6 sec for 30 min, rather than at 10 Hz continuously, the increase in dopa synthesis was entirely insensitive to cholinergic antagonists (Ip & Zigmond 1984b). However, when the nerve was stimulated at 10 Hz for 5 min or at 1.67 Hz for 30 min, both conditions that involve an equal number of pulses as occurred during the discontinuous pattern of stimulation, the increase in dopa synthesis was blocked by about 50% by cholinergic antagonists. It has been hypothesized that these differences in pharmacology are caused by a longer half-life in the synaptic cleft of the noncholinergic transmitter than exists for acetylcholine, allowing the former to be effective during either continuous or discontinuous stimulation (Ip & Zigmond 1984b).

CHOLINERGIC AGONISTS    In addition to investigations of the effects of synaptic stimulation, several laboratories have investigated the acute effects of cholinergic agonists on catecholamine synthesis in postganglionic sympathetic neurons and chromaffin cells. The first report, a study on neonatal SCG in vitro, demonstrated that the mixed cholinergic agonist carbachol increased tyrosine hydroxylation. The authors concluded from experiments with selective cholinergic antagonists that carbachol acted by stimulating muscarinic receptors (Lloyd et al 1979, Ikeno et al 1981). This finding was of particular interest because previous studies had suggested that the long-term stimulation of TH activity produced by cholinergic stimulation of the adult SCG was mediated by nicotinic, but not by muscarinic, receptors (Otten & Thoenen 1976, Chalazonitis et al 1980, Chalazonitis & Zigmond 1980). In a subsequent study on the adult SCG, it was found that the acute stimulation of tyrosine hydroxylation by carbachol could only be blocked entirely when antagonists for both types of cholinergic receptors were included (Ip et al 1982a). Also, hexamethonium by itself produced a larger decrease in the effect of carbachol than did atropine. The effects of carbachol could be seen following a 15 min incubation and were fully reversible within 15 min of removing the agonist (Ip et al 1982a). When the effects of selective nicotinic (i.e. dimethylphenylpiperazinium, DMPP) and muscarinic (i.e. bethanechol or muscarine) agonists were examined, both types of agonist were found to produce increases in tyrosine hydroxylation in situ and increases in TH activity in vitro, with the nicotinic agonist producing the largest maximal effects (Ip et al 1982a, Horwitz & Perlman 1984b).

**Table 1** Signals that acutely increase the activity and/or phosphorylation of TH

| Signal | TH activity[a] | TH phosphorylation[b] |
|---|---|---|
| **Environmental signals** | | |
| Day/night | Pineal gland[1,2,3,4] | |
| Physiological stress (cold, insulin) | Adrenal gland[5,6,7], heart[6] | |
| Non-physiological stress (decapitation, electroshock, hypotensive agents, immobilization, 6-OH-DA) | Adrenal gland[8,9,10,11,12,13,14,6], heart[8,6], blood vessels[15,16,17], celiac and superior mesenteric ganglia[17] | |
| Transsynaptic nerve stimulation | SCG[18,19,20,21,22,23], vas deferens[24,25,26,27,28] | SCG[29,30] |
| **First messengers** | | |
| Cholinergic | Chromaffin cells[31,32], PC12[33], SCG[34,35,36,19d,37d] | Chromaffin cells[31,38,39,40,41,42,32], PC12[33] |
| Nicotinic | Chromaffin cells[42], SCG[36,43,44,45,46] | Chromaffin cells[48,39,41,49,42], PC12[33], SCG[29,50,51,30] |
| Muscarinic | SCG[36,43,52,44,46] | SCG[50,51,30] |
| Peptidergic | | |
| Secretin family | | |
| VIP | Chromaffin cells[53,54,55,56], PC12[53,54,57], SCG[58d,44,59d,45,60], iris[61], pineal gland[61], salivary gland[61] | Chromaffin cells[55], SCG[50,51,30] |
| Secretin | PC12[57,54], SCG[58d,59,45,60], iris[61], pineal gland[61], salivary gland[61] | |
| Others (rat GRF, PHI, glucagon) | PC12[57,54], SCG[59,45,60] | |
| Bradykinin | SCG[62] | |
| Adrenergic (isoproterenol) | | SCG[50] |
| Purinergic (adenosine and adenosine analogues) | Pheo[63], PC12[64] | PC12[33] |

**Table 1** · *continued*

| Signal | TH activity[a] | TH phosphorylation[b] |
|---|---|---|
| Prostaglandin E$_1$ | SCG[62] | |
| Trophic factors | | |
| NGF | Chromaffin cells[65], PC12[66,67] | PC12[68,69,67,70,71,72] |
| EGF | PC12[67] | PC12[68,67] |
| Ionic changes[c] | | |
| Direct electrical stimulation (nonsynaptic) | SCG[23d], Heart[73], salivary gland[74,21,22], iris[22], pineal gland[22] | |
| High K$^+$ | Chromaffin cells[31,42], pheo[75,76], PC12[77,78,67,79,80], SCG[19d,22d,23d], salivary gland[61], iris[22] | Chromaffin cells[40,42], PC12[78,69,67,33,80,79,30] |
| Veratridine | Chromaffin cells[31], SCG[37d] | SCG[50,51,30] |
| Second messengers | | |
| cAMP (cAMP, cAMP analogues, forskolin, cholera toxin) | Adrenal gland[81,82,83,11,84,85,12,14], chromaffin cells[86,87,42,65,88,89,32], pheo[90,76,91], PC12[77,92,64,67,80,54,79,89,93], SCG[94,35,19,37,43,47,45d,46], nictitating membrane[94], salivary gland[61], iris[61], vas deferens[27,28] | Adrenal gland[84,14], chromaffin cells[86,87,38,95,39,96,42,55,32], pheo[91], PC12[68,67,70,79,80,72,93], SCG[50,51,30] |
| Ca$^{2+}$ (Ca$^{2+}$; ionomycin, A23187) | Adrenal gland[98], chromaffin cells[31,88], PC12[80,93], SCG[47] | Chromaffin cells[39,96], PC12[33,80,93], SCG[47] |
| Diacylglycerol (diacylglycerol analogues, phorbol esters) | Chromaffin cells[49,88], PC12[67,93], SCG[46d] | Chromaffin cells[41,49], PC12[67,70,72,93], SCG[46,30,72] |
| cGMP (cGMP, cGMP analogue, nitroprusside) | Chromaffin cells[89], PC12[89], vas deferens[27] | |

Protein kinases

| | | |
|---|---|---|
| cAMP-dependent kinase | Adrenal gland[82,10,101,106,14,6,99], pheo[91,100,80,97], PC12[92], heart[6] | Adrenal gland[101,14,99], chromaffin cells[96], PC12[92,104,96], pheo[91,14,100,102,103,80,97], SCG[105] |
| Ca$^{2+}$/calmodulin-dependent kinase | Adrenal gland[106,98] | Adrenal gland[106], pheo[100,102,103,80], SCG[105] |
| Ca$^{2+}$/phospholipid-dependent kinase | PC12[104] | Pheo[102], PC12[104,108], SCG[105] |
| cGMP-dependent kinase | Pheo[97] | Pheo[97] |
| TH-associated kinase(s) | | Pheo[103,107] |
| NGF-activated kinase | PC12[108] | PC12[108] |

[a] References listed under "TH Activity" demonstrate a rapid increase in either (a) dopa production, (b) catecholamine synthesis from radiolabeled tyrosine but not from radiolabeled dopa, or (c) synthesis of dopamine and dihydroxyphenylacetic acid in the presence of a dopamine β-hydroxylase inhibitor.

[b] References listed under "TH Phosphorylation" demonstrate a rapid increase in $^{32}$P incorporation into TH.

[c] References listed under "ionic changes" include only those which demonstrate a stimulation likely to occur via a nonsynaptic mechanism.

[d] Reference numbers followed by a "d" indicate that the signal examined was shown to produce a stimulation in both normal and decentralized ganglia.

References: 1 (McGeer & McGeer 1966), 2 (Miller et al 1981), 3 (Craft et al 1984), 4 (Abrau et al 1987), 5 (Fluharty et al 1983), 6 (Fluharty et al 1985a), 7 (Fluharty et al 1985b), 8 (Dairman et al 1968), 9 (Kvetnansky et al 1971), 10 (Masserano & Weiner 1979a), 11 (Masserano & Weiner 1979b), 12 (Masserano & Weiner 1981), 13 (Masserano et al 1981), 14 (Tank et al 1984), 15 (Takimoto & Weiner 1979), 16 (Takimoto & Weiner 1981b), 17 (Takimoto & Weiner 1981a), 18 (Steinberg & Keller 1978), 19 (Ip et al 1983), 20 (Ip & Zigmond 1984b), 21 (Anden et al 1986), 22 (Rittenhouse & Zigmond 1987), 23 (Rittenhouse et al 1988), 24 (Weiner & Rabadjija 1968a), 25 (Weiner & Rabadjija 1968b), 26 (Morgenroth et al 1974), 27 (Weiner et al 1976), 28 (Weiner et al 1978), 29 (Cahill & Perlman 1984b), 30 (Cahill & Perlman 1987b), 31 (Haycock et al 1982b), 32 (Waymire et al 1988), 33 (Nose et al 1985), 34 (Lloyd et al 1979), 35 (Ikeno et al 1981), 36 (Ip et al 1982a), 37 (Horwitz & Perlman 1984b), 38 (Haycock et al 1982a), 39 (Niggli et al 1984), 40 (Cote et al 1986), 41 (Michener et al 1986), 42 (Pocotte et al 1986), 43 (Horwitz & Perlman 1984), 44 (Ip & Zigmond 1984a), 45 (Ip et al 1985), 46 (Wang et al 1986), 47 (Cahill et al 1985), 48 (Frye & Holz 1983), 49 (Pocotte & Holz 1986), 50 (Cahill & Perlman 1984a), 51 (Cahill & Perlman 1986), 52 (Horwitz et al 1984), 53 (Tischler et al 1985), 54 (Tischler et al 1986), 55 (Craviso et al 1987), 56 (Houchi et al 1987), 57 (Roskoski 1987), 58 (Ip et al 1982b), 59 (Ip et al 1982b), 60 (Schwarzschild et al 1985), 61 (Schwarzschild & Zigmond 1989), 62 (Bucciarelli de Simone & Rubio 1980), 63 (Erny et al 1981), 64 (Erny & Wagner 1984), 65 (Acheson & Thoenen 1987), 66 (Greene et al 1984), 67 (McTigue et al 1985), 68 (Halegoua & Patrick 1980), 69 (Lee et al 1985), 70 (Cremins et al 1986), 71 (Greene et al 1986), 72 (Cahill & Perlman 1987a), 73 (Gordon et al 1966a), 74 (Sedvall & Kopin 1967), 75 (Chalfie & Perlman 1977), 76 (Vaccaro et al 1980), 77 (Greene & Rein 1978), 78 (Yanagihara et al 1984), 79 (Yanagihara et al 1986), 80 (Tachikawa et al 1986), 81 (Lovenberg et al 1975), 82 (Hoeldtke & Kaufman 1977), 83 (Ames et al 1978), 84 (Yamauchi & Fujisawa 1979b), 85 (Yamauchi & Fujisawa 1980), 86 (Waymire et al 1979), 87 (Meligeni et al 1982), 88 (Yanagihara et al 1987), 89 (Roskoski & Roskoski 1987), 90 (Chalfie et al 1979), 91 (Vulliet et al 1980), 92 (Markey et al 1980), 93 (Tachikawa et al 1987), 94 (Rubio 1977), 95 (Treiman et al 1983), 96 (Haycock et al 1985), 97 (Roskoski et al 1987), 98 (Yamauchi et al 1981), 99 (Kiuchi et al 1987), 100 (Vulliet et al 1984), 101 (Yamauchi & Fujisawa 1979a), 102 (Vulliet et al 1985), 103 (Campbell et al 1986), 104 (Albert et al 1984), 105 (Cahill & Perlman 1985), 106 (Yamauchi & Fujisawa 1981), 107 (Pigeon et al 1987), 108 (Rowland et al 1987).

Abbreviations: EGF, epidermal growth factor; GRF, growth hormone-releasing factor; 6-OH-DA, 6-hydroxydopamine; NGF, nerve growth factor; pheo, pheochromocytoma cells.

Cholinergic stimulation of catecholamine biosynthesis has also been seen in bovine chromaffin cells. In this preparation, the nicotinic agonist DMPP, but not the muscarinic agonist methacholine (0.3 mM), was found to increase dopa synthesis in situ (Pocotte et al 1986). Carbachol also increased TH activity in PC12 cells (Nose et al 1985), though the type of cholinergic receptor(s) involved has not been reported.

PEPTIDERGIC AGONISTS    The implication that a noncholinergic pre-ganglionic neurotransmitter is involved in the transsynaptic regulation of dopa synthesis in the SCG led to the screening of a variety of possible transmitter candidates. A number of biogenic amines and purinergic compounds were examined, including norepinephrine, dopamine, serotonin, histamine, and 2-chloroadenosine, but none of these compounds was found to affect the rate of dopa synthesis (Ip et al 1985). However, of 24 peptides examined, a subgroup of peptides of the secretin-glucagon family was found to increase the rate of tyrosine hydroxylation, namely secretin, vasoactive intestinal peptide (VIP), peptide histidine isoleucine amide (PHI), rat growth hormone-releasing factor, and helodermin H38 (Ip et al 1982b, 1984, 1985, Schwarzschild et al 1985; M. A. Schwarzschild and R. E. Zigmond, unpublished observations). Of these peptides, secretin was the most potent, having an $EC_{50}$ of about 5 nM and a maximal effect at 100 nM (Ip et al 1982b). Five other members of this peptide family—glucagon, gastric inhibitory peptide, human growth hormone-releasing factor, and glucagon-like peptides I and II—produced no effects (Ip et al 1982b, 1985; M. A. Schwarzschild and R. E. Zigmond, unpublished observations). Of the peptides that increased dopa accumulation in situ, those that were tested also increased TH activity in vitro (Schwarzschild et al 1985). As with the effects of cholinergic agonists, the effect of VIP on dopa synthesis occurred within minutes and was rapidly and fully reversible (Ip et al 1982b). In addition to increasing TH activity in adrenergic cell bodies/dendrites in the SCG, VIP increases dopa synthesis in situ and/or TH activity in vitro in adrenergic nerve terminals in the iris, salivary gland, and pineal gland (Schwarzschild & Zigmond 1989). VIP has also been found to increase tyrosine hydroxylation in rat and bovine adrenal chromaffin cells (Tischler et al 1985, Craviso et al 1987) and in PC12 cells (Tischler et al 1985). Secretin, interestingly, had no effect on dopa synthesis in bovine chromaffin cells (Craviso et al 1987) but was effective in the iris, salivary gland, and pineal gland (Schwarzschild & Zigmond 1989).

Immunohistochemical studies have demonstrated that fibers containing VIP-like (Hökfelt et al 1977, Sasek et al 1987, Sasek & Zigmond 1989) and PHI-like (Sasek & Zigmond 1989) immunoreactivities are present in the rat SCG. These immunoreactivities are also present in the preganglionic cervical sympathetic trunk, and they build up on the spinal cord side of a ligature of that trunk (Sasek et al 1987). Also, VIP- and PHI-like immunoreactive neurons have been found in the intermediolateral cell column in the rat thoracic spinal cord, the area where the cell bodies of preganglionic sympathetic neurons are located (Sasek et al 1987). Finally, a proportion of these peptide-containing neurons was directly shown—by a double-labeling experiment using a fluorescent dye that is retrogradely transported by nerve fibers—to send preganglionic fibers to the superior cervical ganglion (Baldwin et al 1988). All of these data raise the possibility that VIP, PHI, or a chemically related peptide serves as a preganglionic neurotransmitter in the rat SCG. VIP-like immunoractivity has also been found to be present in the adrenal gland (Hökfelt et al 1981, Hozwarth 1984) and in parasympathetic nerve endings in certain autonomic end organs, such as the submaxillary gland (Lundberg et al 1980, Johansson & Lundberg 1981). The latter data, together with the finding that VIP increases dopa synthesis in the submaxillary gland, raise the possibility of an interaction between the parasympathetic and sympathetic nervous systems at the level of the autonomic end organ, leading to the regulation of catecholamine synthesis in sympathetic nerve endings (Schwarzschild & Zigmond 1989).

Bradykinin has been reported to increase dopa synthesis in the cat SCG at a concentration of 1 nM (Bucciarelli de Simone & Rubio 1980), though 10 $\mu$M bradykinin had no effect on dopa synthesis in the rat SCG (Ip et al 1985). Angiotensin-II-amide increased the synthesis of [14]C-catecholamines from [14]C-tyrosine but not from [14]C-dopa in the guinea pig atrium maintained in vitro (Boadle-Biber et al 1972). Angiotensin also increased catecholamine synthesis in the guinea pig portal vein and the rat—though not the guinea pig—vas deferens. Several characteristics of these effects of angiotensin raise doubts as to whether they are mechanistically similar to the acute effects of nerve stimulation and neuropeptides discussed above. First, there is a delay of 30–60 min after addition of angiotensin before an increase in catecholamine biosynthesis is seen (Boadle-Biber et al 1972). Second, this effect can be blocked by addition of the protein synthesis inhibitor puromycin (Roth & Hughes 1972). Third, angiotensin also increased the synthesis of [14]C-labeled proteins from [14]C-tyrosine or [14]C-leucine.

In addition to the peptides discussed above that increase TH activity, two peptides have been found to have interesting modulatory effects, in that they block the increases produced by other agonists. One of these interactions is that of substance P with the nicotinic agonist DMPP in the rat SCG. Substance P by itself produced no change in the rate of dopa synthesis, but the peptide blocked completely the increase in dopa synthesis produced by the nicotinic agonist (Ip & Zigmond 1984a). Substance P did not affect the stimulation of dopa synthesis produced by VIP or by the muscarinic agonist bethanechol. Also, in the SCG, a C-terminal 26 amino acid fragment of atrial natriuretic factor ($Arg^{101}$-$Tyr^{126}$) has been shown to block partially the stimulation by carbachol of $^3$H-dopamine synthesis from $^3$H-tyrosine in the presence of a DBH inhibitor (Debinski et al 1987b). Interestingly, both substance P–like and atrial natriuretic peptide–like immunoreactivities have been found in the rat SCG (Gamse et al 1981, Debinski et al 1987a).

PURINERGIC AGONISTS  Erny et al (1981) found that adenosine and 2-chloroadenosine increased the rate of synthesis of dopa in pheochromocytoma cells. Addition of the enzyme adenosine deaminase inhibited this effect of adenosine and, interestingly, also inhibited by about 50% the basal rate of dopa synthesis. The latter finding raises the possibility that the chronic release of adenosine by these cells participates in the regulation of their basal rate of catecholamine synthesis. It is interesting to note that ATP is stored in chromaffin granules together with catecholamines and can be converted to adenosine in the extracellular space (Stevens et al 1972). However, 2-chloroadenosine did not stimulate dopa synthesis in guinea pig or bovine chromaffin cells (Erny 1983) or in rat SCG (Ip et al 1985), a finding that raises doubts as to whether adenosine represents a physiologically important mediator involved in TH regulation.

OTHER AGONISTS  Nerve growth factor (NGF) produces a rapid activation of TH in PC12 (Greene et al 1984) and calf adrenal chromaffin (Acheson & Thoenen 1987) cells. In the case of PC12 cells, this trophic factor was shown to increase the $V_{max}$ of the enzyme without changing its $K_m$ for cofactor. Epidermal growth factor has been found to increase dopa synthesis in PC12 cells (McTigue et al 1985). Interestingly, in addition to stimulating catechol synthesis acutely, both of these growth factors have been found to produce long-term increases in TH activity, presumably by increasing TH levels (Thoenen et al 1979, Goodman et al 1980, Lewis & Chikaraishi 1987). NGF is known to be present in and released by a number of sympathetic end organs, and sympathetic nerve terminals contain receptors for NGF (Harper & Thoenen 1981). Prostaglandin $E_1$, at about 30 nM, was found to increase tyrosine hydroxylation in the cat (Bucciarelli de Simone & Rubio 1980), though not in the rat (Ip et al 1985), SCG.

DIRECT (NONSYNAPTIC) ELECTRICAL STIMULATION  Because acetylcholine, acting via nicotinic receptors, leads to depolarization of postganglionic neurons in the SCG, producing an increase in their firing frequency, the effect of increasing the action potential frequency of these neurons directly, via antidromic stimulation, was examined. Electrical stimulation of the postganglionic internal and external carotid nerves (10 Hz for 15 min), under conditions in which all known synaptic stimulation of the ganglion cells had been eliminated, increased dopa synthesis two-fold and also increased TH activity in vitro (Rittenhouse et al 1988). In addition to increasing TH activity in sympathetic ganglia, pre- or postganglionic stimulation also has been shown to increase dopa synthesis or TH activity acutely in sympathetic nerve terminals in the heart (Gordon et al 1966a), submaxillary gland (Sedvall & Kopin 1967), iris (Rittenhouse & Zigmond 1987), and pineal gland (Rittenhouse & Zigmond 1987). It is interesting to note that antidromic stimulation of the SCG, unlike orthodromic stimulation of the ganglion, does not produce a long-term increase in TH (Chalazonitis & Zigmond 1980).

DEPOLARIZATION    In their study on the vas deferens, Morgenroth et al (1974) observed that in addition to hypogastric nerve stimulation, depolarization by an elevated concentration of $K^+$ also increased TH activity. Exposure to a high concentration of $K^+$ (usually around 50 mM) has since been observed to increase dopa synthesis and/or TH activity in pheochromocytoma (Chalfie & Perlman 1977), PC12 (Greene & Rein 1978), and bovine chromaffin (Haycock et al 1982b) cells, in slices of adrenal medulla (Togari et al 1982), and in the SCG (Ip et al 1983), iris (Rittenhouse & Zigmond 1987), and submaxillary gland (Schwarzschild & Zigmond 1989). Similarly, depolarization of adrenal chromaffin cells (Haycock et al 1982b) and SCG (Horwitz & Perlman 1984a) with veratridine also increases TH activity. In the studies on the effects of $K^+$ and veratridine on the SCG, prior decentralization of the ganglion diminished, but did not abolish, the increase in tyrosine hydroxylation (Ip et al 1983, Horwitz & Perlman 1984a). These data suggest that part of the effect of these depolarizing agents in the intact ganglion is mediated by release of transmitter(s) from the preganglionic nerve terminals, while part of their effect is mediated by a direct depolarization of the postganglionic neurons.

## Second Messengers and Kinases

As discussed below, the acute activation of TH involves covalent modification of the enzyme by protein phosphyorylation. TH was first shown to be phosphorylated in situ by Letendre et al (1977a,b), who maintained SCG in organ cultures in the presence of $^{32}$P-potassium phosphate, and isolated labeled TH by immunoprecipitation with an antiserum to the enzyme. Since this report, a great deal of information has been obtained concerning the roles of specific second messenger systems and specific protein kinases in the acute regulation of TH. All four of the best-characterized second messenger systems—cAMP, $Ca^{2+}$, diacylglycerol, and guanosine 3′,5′-cyclic monophosphate (cGMP)—and their respective protein kinases have been implicated in this regulation. Below, data on each of these systems are reviewed in terms of the following three questions: (*a*) what is the effect of directly increasing each of these second messengers on catecholamine biosynthesis, (*b*) what is the effect of each of the protein kinases on the phosphorylation and activation of TH, and (*c*) which of the first messengers that alter TH activity are known to increase (or decrease) the levels of specific second messengers in peripheral adrenergic cells.

cAMP AND cAMP-DEPENDENT PROTEIN KINASE

*Effects of cAMP, cholera toxin, and forskolin on dopa synthesis and on TH activity* In the late 1960s and early 1970s, Greengard and co-workers reported that in both sympathetic ganglia and the corpus striatum, dopamine agonists elevate cAMP levels (Greengard 1978). Since dopaminergic agonists had been found to inhibit dopamine biosynthesis in slices of the striatum, Goldstein et al (1973) examined the effects of dibutyryl cAMP on this process and, to their surprise, found that this cyclic nucleotide stimulated, rather than inhibited, dopamine biosynthesis. In later studies, both in striatal slices (Anagnoste et al 1974) and in striatal synaptosomes (Harris et al 1974), analogues of cAMP were found to increase the synthesis of radiolabeled dopamine from radiolabeled tyrosine, but not from radiolabeled dopa. Kinetic studies on TH activity in the cytosol fractions prepared from tissue exposed to cAMP analogues showed a decrease in the $K_m$ of the enzyme for pterin cofactor with no change in $V_{max}$ (Harris et al 1974).

cAMP analogues have since been shown to stimulate tyrosine hydroxylation in situ and/or TH activity in vitro in normal adrenal chromaffin (Meligeni et al 1982), pheochromocytoma (Chalfie et al 1979), and PC12 (Greene & Rein 1978) cells as well as in vas deferens (Weiner et al 1978), SCG (Ikeno et al 1981, Ip et al 1985), submaxillary gland (Schwarzschild & Zigmond 1989), and nictitating membrane (Rubio 1977). In addition, cholera toxin or forskolin, agents that increase cAMP levels, have been found to increase dopa synthesis in situ in pheochromocytoma (Chalfie et al 1979) and PC12 (Erny & Wagner 1984) cells, SCG (Ip et al 1985), iris (Schwarzschild & Zigmond 1989), and salivary gland (Schwarzschild & Zigmond 1989), and to cause activation of TH in PC12 cells (Tachikawa et al 1987).

The first indication that cAMP acted in these systems via protein phosphorylation was the finding that when supernatant fractions prepared from rat brain were incubated under "cAMP-dependent phosphorylating conditions" (i.e. incubation with cAMP, theophylline, ATP, $Mg^{2+}$, NaF, and EGTA), TH activity was increased (Lovenberg et al 1975, Morgenroth et al 1975). Kinetic studies indicated that this effect was due to a decrease in $K_m$ of TH for cofactor, with no change in $V_{max}$ (Lovenberg et al 1975). Pollock et al (1981) have reported, however, that under certain conditions an increase in $V_{max}$ is also seen. No activation was seen if ATP was omitted from the incubation mixture as would be expected if phosphorylation was involved. Further evidence for this hypothesis came from the demonstration that TH activation under these conditions can be blocked by the addition of a specific inhibitor of cAMP-dependent protein kinase (Morgenroth et al 1975, Andrews et al 1983). TH activation was also found under these phosphorylating conditions with supernatant fractions from rat adrenal gland (Lovenberg et al 1975). Subsequently, other workers reported similar findings in bovine adrenal gland (Yamauchi & Fujisawa 1979b) and in pheochromocytoma (Vulliet et al 1980) and PC12 (Roskoski & Roskoski 1987) cells.

*Effects of cAMP-dependent protein kinase on the phosphorylation and activation of TH*    Early studies attempting to determine whether TH itself was phosphorylated when it was incubated under such "phosphorylating" conditions indicated that it was not (Lloyd & Kaufman 1975, Lovenberg et al 1975), thus raising the possibility that a second protein might be phosphorylated and might subsequently "activate" TH. However, in 1978, Joh et al (1978) and Edelman et al (1978) reported that TH purified from the corpus striatum was both phosphorylated and activated when exposed to cAMP and a purified cAMP-dependent protein kinase. The phosphorylation of the enzyme was detected by subjecting the incubation mixture to gel electrophoresis and examining autoradiographically the band corresponding to the molecular weight of TH, i.e. approximately 60,000. Markey et al (1980) and Vulliet et al (1980) made similar observations when TH, purified from PC12 and pheochromocytoma cells, respectively, was exposed to the catalytic subunit of cAMP-dependent protein kinase. Further evidence that TH itself is phosphorylated under such conditions was obtained by introducing an immunoprecipitation step with an antiserum raised against TH, prior to gel electrophoresis (Yamauchi & Fujisawa 1979b).

Analysis of the kinetics of the phosphorylated enzyme gave conflicting results in different studies, and the authors came to significantly different conclusions concerning the activity of the nonphosphorylated form of TH. Joh et al (1978) found that phosphorylation of TH produced an increase in $V_{max}$ with no change in $K_m$ for cofactor. They hypothesized that the non-phosphorylated form of the enzyme was completely inactive. Vulliet et al (1980) found that under control conditions TH activity did not exhibit linear kinetics with respect to cofactor; however, after phosphorylation, the kinetics became linear. They hypothesized that TH exists in two forms under control conditions, a high $K_m$ form (which represents nonphosphorylated enzyme) and a low $K_m$ form (which represents phos-phorylated enzyme). They proposed that when TH was completely phos-phorylated, all of the enzyme molecules were present in the low $K_m$ form. When their control data were fitted to a model based on two $K_m$s, the data indicated that under control conditions about 30% of the enzyme was already in the low $K_m$ form. Interestingly, they found that after incubating TH with a saturating amount of the catalytic subunit of the cAMP-dependent protein kinase, 0.7 moles of $^{32}P$ were incorporated per mole of TH subunit. It should be pointed out, however, that even with this interpretation of the data, the actual activity of the nonphosphorylated form of the enzyme in situ may, in fact, be negligible, depending on the $BH_4$ concentration and the actual $K_m$ of this form of the enzyme in intact cells. [See discussion by Miller & Lovenberg (1985).] In a study similar to that of Vulliet et al (1980), Markey et al (1980) observed an increase in the $K_i$ of the phosphorylated enzyme for catechols, in addition to a decrease in its $K_m$ for cofactor.

The first demonstration that a stimulus that leads to enzyme activation can indeed cause increased phosphorylation of TH in situ was made by Meligeni et al (1982). These workers found that when bovine chromaffin cells were exposed to 8-bromo-cAMP and $^{32}P_i$, catecholamine synthesis in situ, TH activity in vitro, and TH phosphorylation were all increased. TH phosphorylation has also been observed to occur in situ in PC12 cells in response to dibutyryl-cAMP (Halegoua & Patrick 1980) and in SCG in response to 8-bromo-cAMP and forskolin (Cahill & Perlman 1984a, 1987b).

*First messengers that increase cAMP levels*    Many neurotransmitters and putative neurotransmitters have been found to act via specific receptors to increase cAMP levels in their effector cells (e.g. Greengard 1978). In most cases, this effect is caused by an activation of adenylate cyclase. Since cAMP analogues lead to the activation of TH in the sympathetic nervous system, it is of interest to determine which of the first messengers that activate TH also elevate cAMP levels in adrenergic cells (Table 2).

Volle & Patterson (1982) reported that VIP (0.1–5.0 $\mu M$) increased cAMP levels in the rat SCG three- to eight-fold, with an $EC_{50}$ of approximately 1 $\mu M$. In subsequent studies, Zigmond and co-workers confirmed this finding and found that a number of related peptides (i.e. secretin, PHI, rat growth hormone-releasing factor, and helospectin) also increased cAMP levels in this ganglion (Ip et al 1985, Schwarzschild et al 1985; M. A. Schwarzschild and R. E. Zigmond unpublished observations). Of these peptides, secretin was the most potent, with an $EC_{50}$ of about 3 nM. Many other peptides tested, including other members of the secretin-glucagon family (i.e. glucagon, gastric inhibitory peptide, and human growth hormone-releasing factor) produced no change in cAMP levels (Ip et al 1985). These data, together with the results described above concerning peptidergic stimulation of TH (see PEPTIDERGIC AGONISTS) and concerning the effects of cAMP-dependent protein kinase on TH activity (see CAMP-DEPENDENT PROTEIN KINASE), are consistent with the hypothesis that certain peptides of the secretin-glucagon family increase TH activity in the SCG by increasing cAMP levels (Figure 1). VIP also increases cAMP levels (and, as noted above, TH activity) in bovine chromaffin cells (Craviso et al 1987, Wilson 1988, but see Houchi et al 1987) and in PC12 cells (Tischler et al 1985). On the other hand, it should be noted that VIP has also been found to increase phosphoinositide (PI) turnover in the SCG with an $EC_{50}$ of approximately 1 $\mu$M, and it has been suggested that VIP increases dopa synthesis via this mechanism (Audigier et al 1986, Durroux et al 1987). Although this remains a possibility, secretin, PHI, and rat growth hormone-releasing factor do not affect PI turnover in the SCG (Audigier et al 1986).

Other agents that have been shown to activate adenylate cyclase or increase cAMP levels in adrenergic cells include NGF in PC12 cells (Schubert & Whitlock 1977) and in SCG (Nikodijevic et al 1975) as well as adenosine analogues in pheochromocytoma (Erny et al 1981) and PC12 (Guroff et al 1981, Erny & Wagner 1984) cells. With both types of agonists, evidence suggests that the increase in TH activity and/or phosphorylation is mediated, at least in part, via an increase in cAMP (See *First messengers that increase PI turnover* and MULTIPLE SITES OF TH PHOSPHORYLATION.) For example, the spectrum of proteins—including TH—that are phosphorylated by NGF in PC12 cells is identical to that produced by cholera toxin, dibutyryl cAMP, or cAMP plus theophylline (Halegoua & Patrick 1980). In the case of the adenosine analogue phenylisopropyladenosine, a mutant PC12 clone was isolated in which the analogue neither stimulated cAMP levels nor increased the rate of dopa synthesis (Erny & Wagner 1984). In these same cells, cholera toxin and dibutyryl cAMP did increase dopa synthesis, indicating that the defect in responsiveness to phenylisopropyladenosine occurred at, or prior to, the stimulation of adenylate cyclase by the agonist.

The nicotinic agonist DMPP increases cAMP levels in bovine chromaffin

**Table 2**  Associations between second messenger systems and first messengers known to stimulate tyrosine hydroxylase activity

| First messenger | Second messenger systems | | | |
|---|---|---|---|---|
| | cAMP[a] | Calcium[b] | PI Metabolism[c] | cGMP[d] |
| **Cholinergic[e]** | | | | |
| Nicotinic | Adrenal gland[1]<br>Chromaffin cells[2,21]<br>PC12[3] | Chromaffin cells[4,5,6,7,8,9,2*,10]<br>PC12[11]<br>SCG[12,13*,14*] | | |
| Muscarinic | | Chromaffin cells[8,15,9]<br>PC12[16,26,17,18,19] | Adrenal gland[20]<br>Chromaffin cells[22,23,24,25]<br>PC12[16,17,18,27,19]<br>SCG[28,29,30,31,32,33,34,35,36] | Chromaffin cells[37,38,8]<br>SCG[39,12,40,41,42] |
| **Peptidergic** | | | | |
| Secretin family | | | | |
| VIP | Chromaffin cells[43,21]<br>PC12[44,45,46]<br>SCG[47,41,48,49,14,35] | | SCG[35,51] | |
| Secretin and others | Pheochromocytoma[52]<br>PC12[46], SCG[49,14,53] | | | |
| Bradykinin[f] | | | PC12[54], SCG[33] | |

| NGF | PC12[56,57,60] | PC12[57] | PC12[59,61] |
| | SCG[58] | | SCG[62] |
| Adenosine | Pheochromocytoma[63] | | |
| | PC12[26,50,55] | | |

[a] References under "cAMP" demonstrate that the first messenger indicated acutely increases either the cAMP content, the formation of cAMP from a radiolabeled precursor, or the adenylate cyclase activity in the cells or tissue indicated.

[b] References under "Calcium" demonstrate that the first messenger indicated either increases $^{45}Ca^{2+}$ flux, raises intracellular $Ca^{2+}$ concentration, or (when noted by an asterisk,*) stimulates TH activity in a $Ca^{2+}$-dependent manner in the cells or tissue indicated.

[c] References under "PI Metabolism" (phosphoinositide metabolism) demonstrate that the first messenger indicated increases the accumulation of radiolabeled phosphoinositides, inositol phosphates, or diacylglycerol following incubation of the indicated cells or tissue with a radiolabeled precursor.

[d] References under "cGMP" demonstrate that the first messenger indicated increases the cGMP content of the cells or tissue indicated.

[e] Under "cholinergic," only those references in which selective nicotinic or muscarinic agonists or antagonists were used are cited.

[f] See First messengers that increase PI turnover for further discussion of bradykinin.

References: 1 (Jaanus & Rubin 1974), 2 (Pocotte et al 1986), 3 (Baizer & Weiner 1985), 4 (Holz et al 1982), 5 (Kilpatrick et al 1982), 6 (Frye & Holz 1983), 7 (Knight & Kesteven 1983), 8 (Ohsako & Deguchi 1983), 9 (Kao & Schneider 1986), 10 (Sasakawa et al 1986), 11 (Stallcup 1979), 12 (Volle et al 1981), 13 (Horwitz & Perlman 1984b), 14 (Ip et al 1985), 15 (Kao & Schneider 1985), 16 (Vicentini et al 1985), 17 (Pozzan et al 1986), 18 (Vicentini et al 1986), 19 (Rabe et al 1987), 20 (Mohd. Adnan & Hawthorne 1981), 21 (Wilson 1988), 22 (Fisher et al 1981), 23 (Forsberg et al 1986), 24 (Swilem et al 1986), 25 (Eberhard & Holz 1987), 26 (Guroff et al 1981), 27 (Horwitz 1987), 28 (Lapetina et al 1976), 29 (Pickard et al 1977), 30 (Bone et al 1984), 31 (Horwitz et al 1984), 32 (Patterson & Volle 1984), 33 (Bone & Michell 1985), 34 (Horwitz et al 1985), 35 (Audigier et al 1986), 36 (Horwitz et al 1986), 37 (Yanagihara et al 1979), 38 (Derome et al 1981), 39 (Kebabian et al 1975), 40 (Briggs et al 1982), 41 (Volle et al 1982), 42 (Vente et al 1987), 43 (Craviso et al 1987), 44 (Tischler et al 1985), 45 (Tischler et al 1986), 46 (Roskoski 1987), 47 (Volle & Patterson 1982), 48 (Volle & Patterson 1983), 49 (Dvorkin et al 1984), 50 (Erny & Wagner 1984), 51 (Durroux et al 1987), 52 (Levey et al 1975), 53 (Schwarzschild et al 1985), 54 (van Calker et al 1987), 55 (Nose et al 1985), 56 (Schubert & Whitlock 1977), 57 (Schubert et al 1978), 58 (Nikodijevic et al 1975), 59 (Traynor et al 1982), 60 (Traynor & Schubert 1984), 61 (Traynor 1984), 62 (Lakshmanan 1978), 63 (Erny et al 1981).

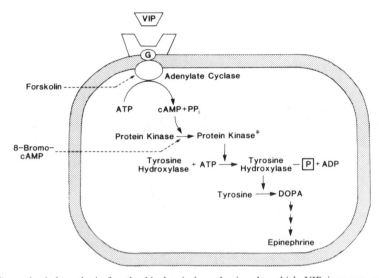

*Figure 1* A hypothesis for the biochemical mechanism by which VIP increases cate-cholamine biosynthesis in adrenal chromaffin cells, based on data from Craviso et al (1987). VIP increases cAMP levels, presumably by stimulating adenylate cyclase activity via acti-vation of a G protein. It is assumed that cAMP in these cells, as in other eukaryotic cells, binds to the regulatory subunit of a cAMP-dependent protein kinase, causing this subunit to dissociate from and thereby activate the catalytic subunit of the enzyme. (The activated enzyme is denoted in the figure by the *asterisk*.) TH is phosphorylated in response to VIP, causing TH activation, which produces an increase in the enzyme's ability to catalyze the conversion of tyrosine to dopa in situ. Further evidence that VIP acts via a cAMP-dependent protein kinase is provided by the findings that forskolin and 8-bromo-cAMP mimic the effects of VIP on the activity and phosphorylation of TH in situ. Analysis of the phos-phorylation sites in TH by chromatography of the tryptic peptides indicates that all three of these agents increase phosphorylation of the same phosphopeptides.

cells (Pocotte et al 1986), but not in bovine (Kebabian et al 1975) or rat (Volle et al 1982, Ip et al 1985) SCG. Since nicotinic agonists in other systems act by increasing permeability to $Na^+$ and $Ca^{2+}$ ions, rather than by stimulating adenylate cyclase, one possibility is that their effect on cAMP levels in bovine chromaffin cells is indirect, via the stimulation of secretion of some other agonist (e.g. norepinephrine or VIP) that acts back on the cells via autoreceptors. [See introduction of Baizer & Weiner (1985) for further discussion.] It has been suggested that the muscarinic stimu-lation of cAMP in the SCG is also mediated indirectly, via the stimulation of dopamine release (Kalix et al 1974).

$CA^{2+}$ AND $CA^{2+}$/CALMODULIN-DEPENDENT PROTEIN KINASE

*Effects of $Ca^{2+}$ and $Ca^{2+}$ ionophores on dopa synthesis and TH acti-vity*  Catecholamine release, like the release of other neurotransmitters, is evoked by the influx of $Ca^{2+}$ resulting from cell depolarization. Since the acute regulation of catecholamine biosynthesis is thought to be, at least in certain instances, an adaptive response to stimuli that cause increased catecholamine release, it is of particular interest that $Ca^{2+}$ influx also triggers TH activation. An early indication, though indirect, that $Ca^{2+}$ influx might be involved in the regulation of catecholamine biosynthesis was the finding that a number of depolarizing stimuli increased tyrosine hydroxylation in adrenergic tissues only if $Ca^{2+}$ was present in the incubation medium (see below). A more direct demonstration of this relationship came from studies with $Ca^{2+}$ ionophores, such as ionomycin and A23187. Thus, for example, incubation of SCG with ionomycin leads to an increase in dopa synthesis in situ and in TH activity in vitro (Cahill et al 1985). Ionomycin produces no such effect if $Ca^{2+}$ is removed from the incubation medium. Similarly, A23187, in the presence of $Ca^{2+}$, increases the rate of tyrosine hydroxylation in bovine chromaffin cells (Haycock et al 1982b) and causes TH activation in PC12 cells (Tachikawa et al 1986). The role of $Ca^{2+}$ has also been examined in digitonin-permeabilized chromaffin cells (Yanagihara et al 1987). Increasing the concentration of $Ca^{2+}$ in the medium from 0.03 to 1.0 $\mu$M increased TH activity about 2.5-fold. The $EC_{50}$ for this effect of $Ca^{2+}$ was about 0.2 $\mu$M. It should be pointed out that in the case of the SCG, and even in the case of cultured cells, it has not been possible to rule out the possibility that the effect of increased $Ca^{2+}$ influx on TH activity is indirect, via the release of a transmitter that acts back on the cell via an autoreceptor.

Evidence that these $Ca^{2+}$-dependent effects are mediated by a $Ca^{2+}$/calmodulin-dependent protein kinase first came from the observation of Yamauchi & Fujisawa (1980) that incubation of cytosol fractions prepared from brainstem tissue under conditions that facilitate this kinase activity (i.e. addition of $Ca^{2+}$, calmodulin, ATP, and $Mg^{2+}$) led to TH activation. When any of these components was left out, no stimulation of enzyme activity was seen. The increase in TH activity represented a decrease in $K_m$ for cofactor (Yamauchi & Fujisawa 1980).

*Effects of $Ca^{2+}$ ionophores and $Ca^{2+}$/calmodulin-dependent protein kinase on TH phosphorylation and activation*    In parallel with their stimulation of TH activity, $Ca^{2+}$ ionophores also stimulate the phosphorylation of the enzyme. Thus, ionomycin and A23187 have been shown to increase phosphorylation of TH in the SCG (Cahill et al 1985) and in PC12 cells (Nose et al 1985), respectively. Early studies on cytosol fractions from rat brain, analogous to those described above for a cAMP-dependent protein kinase, indicated that both TH phosphorylation and activation can be mediated by a $Ca^{2+}$/calmodulin-dependent protein kinase (Yamauchi &

Fujisawa 1980). However, subsequent studies indicated that $Ca^{2+}$-dependent activation of TH was somewhat more complex. Yamauchi & Fujisawa reported in 1981 that incubation of TH, purified from the adrenal medulla, with $Ca^{2+}$, calmodulin, and a purified $Ca^{2+}$/calmodulin-dependent protein kinase led to the phosphorylation of TH but not to its activation. Other workers have found similarly that phosphorylation of TH purified from PC12 cells by a $Ca^{2+}$/calmodulin-dependent protein kinase does not lead to activation of TH (Albert et al 1984, Vulliet et al 1984, Tachikawa et al 1986). Subsequently, it was shown that if an "activator" protein purified from brain was added to the incubation, both phosphorylation and activation of TH were seen (Yamauchi & Fujisawa 1981). In these experiments, phosphorylation and activation could be studied sequentially. It was found that the activator protein was equally effective whether it was added at the same time as the kinase or after the phosphorylation reaction had been stopped by addition of a $Ca^{2+}$ chelator (Yamauchi & Fujisawa 1981). The activator protein had no apparent effect on the extent of TH phosphorylation (Yamauchi & Fujisawa 1981).

An activator protein with these properties has been purified from cerebral cortex ($M_r = 70,000$) (Yamauchi et al 1981). Interestingly, this protein has an activating effect on tryptophan hydroxylase similar to that on TH, namely, that it activates both $BH_4$-dependent monooxygenases after they have been phosphorylated by a $Ca^{2+}$/calmodulin-dependent protein kinase (Yamauchi et al 1981). On the other hand, the activator protein does not stimulate the activity of phosphorylated TH if the phosphorylation has been produced by a cAMP-dependent protein kinase (Yamauchi & Fujisawa 1981). It will be of interest to determine whether the concentration of this activator protein or its accessibility to TH is regulated in adrenergic tissues.

*First messengers that increase* $Ca^{2+}$ *influx*    Certain "first messengers" that have been found to increase dopa synthesis, TH activity, and TH phosphorylation in adrenergic tissues do so only if $Ca^{2+}$ is present in the incubation medium (Table 2). Such stimuli include cholinergic agonists (Haycock et al 1982b), selective nicotinic agonists (Horwitz & Perlman 1984b), depolarizing agents such as $K^+$ and veratridine (Haycock et al 1982b, Ip et al 1983), and antidromic stimulation (Rittenhouse et al 1988). In certain instances, the stimuli have been shown directly to increase $Ca^{2+}$ influx (Table 2). Thus, for example, Douglas & Poisner (1961) demonstrated an increase in $^{45}Ca^{2+}$ uptake in the cat adrenal medulla in vivo in response to acetylcholine. In PC12 cells, carbachol, 50 mM $K^+$, and veratridine all increased $^{45}Ca^{2+}$ influx, and this effect of carbachol was blocked by *d*-tubocurarine but not by atropine (Schubert et al 1978).

Similarly, in bovine chromaffin cells, the nicotinic agonists DMPP and nicotine, and 56 mM $K^+$ increased the uptake of $^{45}Ca^{2+}$, whereas the muscarinic agonists muscarine and methacholine produced no such effect (Holz et al 1982).

With respect to studies on $^{45}Ca^{2+}$ influx in the intact SCG, it is important to distinguish between uptake into presynaptic nerve terminals and into postsynaptic cell bodies and dendrites. Both Blaustein (1971) and Volle et al (1981) reported that the increase in $^{45}Ca^{2+}$ uptake in the SCG produced by preganglionic nerve stimulation was partially blocked by a nicotinic antagonist. This result raises the possibility that a significant portion of the uptake was into postganglionic neurons. In the cases of both DMPP and high $K^+$, it has been shown that $^{45}Ca^{2+}$ uptake occurs to a similar extent both in intact and in decentralized ganglia, thus indicating again that a major part of the uptake stimulated by these agents occurs in postganglionic neurons (Blaustein 1976, Volle et al 1981).

The interpretation that nicotinic agonists and depolarizing agents increase catecholamine biosynthesis by increasing $Ca^{2+}$ influx is strengthened by the finding, in a number of instances, that the effects of these stimuli on TH can be blocked by antagonists of voltage-sensitive $Ca^{2+}$ channels. Thus, the effect of acetylcholine on the phosphorylation of TH in bovine chromaffin cells is blocked by $Mn^{2+}$, an inorganic $Ca^{2+}$ channel antagonist (Haycock et al 1982b), whereas the similar effect of carbachol on PC12 cells is inhibited by the organic $Ca^{2+}$ channel antagonist nitrendipine (Nose et al 1985). Recently, it has been found that omega-conotoxin, another $Ca^{2+}$ channel antagonist, inhibits the activation of TH in the iris produced by depolarization with an elevated concentration of $K^+$ (Rittenhouse & Zigmond 1989). Also, as discussed above, the potential importance of $Ca^{2+}$ influx in the regulation of the phosphorylation and activation of TH is indicated by the effects of $Ca^{2+}$ ionophores on these processes. Finally, it is interesting to note that substance P, which antagonizes the effects of nictotinic agonists on dopa synthesis, has been shown to inhibit the ability of nicotinic agonists to increase cation permeability in adrenal chromaffin (Clapham & Neher 1984) and PC12 (Stallcup & Patrick 1980) cells and in sympathetic neurons (Role 1984).

DIACYLGLYCEROL AND $CA^{2+}$/PHOSPHOLIPID-DEPENDENT PROTEIN KINASE
*Effects of OAG and phorbol esters on dopa synthesis and TH activity*    The discovery by Nishizuka (1984) that certain phorbol esters activate a $Ca^{2+}$/phospholipid-dependent protein kinase (protein kinase C), mimicking the effects normally produced by the PI metabolite diacylglycerol, has provided a tool for investigating the potential physiological roles for this kinase. Incubation of PC12 cells with one of these active compounds,

$4\beta$-phorbol-$12\beta$-myristate-$13\alpha$-acetate (PMA; also referred to as 12-O-tetra-decanoylphorbol-13-acetate or TPA), increased the rate of dopa synthesis in situ approximately two-fold over a 30 min period (McTigue et al 1985). Similarly, $4\beta$-phorbol 12,13-dibutyrate (PDBu) in the SCG (Wang et al 1986) and both PMA and PDBu in bovine chromaffin cells rapidly increased dopa synthesis (Pocotte & Holz 1986). However, $4\beta$-phorbol and $4\alpha$-phorbol-12,13-didecanoate, both compounds that do not activate protein kinase C, had no effect on dopa synthesis in the SCG or in bovine chromaffin cells, respectively (Wang et al 1986, Pocotte & Holz 1986). In addition to the phorbol esters, the compound 1-oleoyl-2-acetyl-glycerol (OAG), a synthetic diacylglycerol, has been found to activate protein kinase C, and this compound has been found also to activate TH in PC12 cells (Tachikawa et al 1987).

*Effects of $Ca^{2+}$/phospholipid-dependent protein kinase on TH phosphorylation and activation*    Incubation of TH in the presence of protein kinase C together with $Ca^{2+}$, phosphatidylserine, and diolein (like OAG, a synthetic diacylglycerol) led to phosphorylation and activation of TH (Albert et al 1984). The alteration in this enzyme activity produced a four-fold decrease in the $K_m$ for cofactor and a ten-fold increase in $K_i$ for feedback inhibition by dopamine. Additional evidence that protein kinase C may play an important role in the regulation of the state of phosphorylation of TH comes from in situ experiments with phorbol esters. As discussed in the previous section, PMA and PDBu increase dopa synthesis in adrenal chromaffin and PC12 cells and in SCG. Under these conditions, the phorbol esters also led to parallel increases in phosphorylation of TH (McTigue et al 1985, Pocotte & Holz 1986, Wang et al 1986, Tachikawa et al 1987). On the other hand, $4\beta$-phorbol and $4\beta$-phorbol-12-monomyristate had no effect on the phosphorylation of TH (Cremins et al 1986). These two compounds are also ineffective in stimulating protein kinase C, and the former compound is known to be ineffective in increasing dopa synthesis (see previous paragraph). Finally, 1,2-dioctanoyl-glycerol (also a synthetic analogue of diacylglycerol) stimulates TH phosphorylation in PC12 cells (Cremins et al 1986).

*First messengers that increase PI turnover*    The stimulation of PI turnover in the SCG by acetylcholine was one of the first instances in which a neurotransmitter was shown to stimulate this second-messenger system (Hokin 1965, 1966). Subsequent studies in the SCG and adrenal gland and in adrenal chromaffin and PC12 cells demonstrated that this effect can be blocked by muscarinic antagonists and can be mimicked by muscarinic agonists (Table 2). Despite their inability to stimulate $Ca^{2+}$ influx in adrenal chromaffin cells, described above, muscarinic agonists raise the concentration of cytosolic free $Ca^{2+}$ in these cells independent of extracellular $Ca^{2+}$ (Kao & Schneider 1985). These data suggest that muscarinic stimulation releases $Ca^{2+}$ from an intracellular source, as do other agonists that stimulate PI hydrolysis (Berridge 1987). PI metabolism in adrenergic tissues is also stimulated by NGF (Table 2). Thus, considering the evidence cited above for its action on TH via cAMP, NGF may regulate TH activity through multiple second-messenger systems.

Recently, it has been reported that the nicotinic agonist DMPP can increase PI turnover in bovine chromaffin cells; however, data with the $Ca^{2+}$ channel blocker D600 suggests that this effect may be secondary to an influx of $Ca^{2+}$ (Eberhard & Holz 1987). Although bradykinin stimulates PI turnover in the rat SCG and in PC12 cells (Table 2), it is not clear that this is the mechanism by which the peptide increases dopa synthesis. As discussed above, bradykinin has not been found to stimulate dopa synthesis in the rat SCG, although it has in the cat SCG. It should be noted also that the increase in dopa synthesis produced by bradykinin is blocked by indomethacin, a finding that raises the possibility that an alteration in prostaglandin synthesis is involved (Bucciarelli de Simone & Rubio 1980). Finally, as noted above, although VIP has been shown to increase PI turnover in the SCG (Audigier et al 1986, Durroux et al 1987), it is the only member of the secretin-glucagon family of peptides tested that produces such an effect, thus raising a question as to the importance of this second messenger effect in the stimulation of dopa synthesis produced by these peptides. (See *First messengers that increase cAMP levels*, above).

cGMP AND cGMP-DEPENDENT PROTEIN KINASE

*Effects of cGMP analogues, sodium nitroprusside, and cGMP-dependent protein kinase on dopa synthesis, TH activity, and TH phosphorylation*    The possible role of cGMP as a second messenger in the regulation of catecholamine metabolism has been examined in relatively few studies (Ikeno et al 1981). Ip et al (1985) found no effect of cGMP analogues on dopa synthesis in the SCG, and Meligeni et al (1982) found no effect in adrenal chromaffin cells. The only reported stimulation of either dopa synthesis in situ or TH activity in vitro produced by a cGMP analogue (8-bromo-cGMP) is that of Roskoski & Roskoski (1987) in PC12 cells. These authors also demonstrated that sodium nitroprusside, an activator of guanylate cyclase, increased cGMP levels, TH activity in vitro, and dopa synthesis in situ in these cells. (In the same study, it was reported that nitroprusside also activated TH in bovine chromaffin cells and striatal synaptosomes.) When cell free extracts from PC12 cells were incubated under cGMP-dependent phosphorylating conditions (i.e. incubation with cGMP, theo-phylline, ATP, $Mg^{2+}$, and EGTA), TH was activated as it had been in earlier studies when extracts were incubated under cAMP-dependent phosphorylating conditions. Evidence that cGMP and cAMP activate different kinases in PC12 cells was provided by the finding that the heat stable inhibitor protein of cAMP-dependent protein kinase only blocked the activation of TH that occurred when cAMP was added, and had no effect on the activation in response to cGMP (Roskoski & Roskoski, 1987).

Only a single study has been reported in which the effects on TH of a cGMP-dependent protein kinase have been examined. Incubation of puri-fied TH with this kinase led to activation and phosphorylation of TH. The activation was expressed as a five-fold decrease in $K_m$ for cofactor with no change in $V_{max}$ (Roskoski et al 1987).

*First messengers that increase cGMP*    Unlike for cAMP, few first mess-engers have been identified that use cGMP as a second messenger. [Cur-rently atrial natriuretic factor is the only example of an agonist that stimulates membrane-bound guanylate cyclase (Murad et al 1987).] Although muscarinic agonists have been shown to increase cGMP levels both in adrenal chromaffin cells and in the SCG (Table 2; but see Frey & McIsaac 1981), analogues of cGMP (at millimolar concentrations) do not increase dopa synthesis in either tissue, as noted above.

APPARENT INCONSISTENCIES BETWEEN THE EFFECTS OF FIRST AND SECOND MESSENGERS ON DOPA SYNTHESIS    In a few instances, studies on the effects of first and second messengers have produced seemingly inconsistent results. For example, in the SCG, isoproterenol (10 $\mu$M) increases cAMP levels about ten-fold (Volle & Patterson 1982), but it produced no effect on the rate of dopa synthesis (Ip et al 1985), even though ana-logues of cAMP produce such an increase (e.g. Ip et al 1983). Similarly, vasopressin produces a large increase in PI turnover in the SCG (Bone et al 1984) but did not increase dopa synthesis in this tissue (Ip et al 1985, Horwitz et al 1986), even though other stimulators of PI turnover do produce such an increase (Wang et al 1986). In both of these instances, it is possible that the apparent discrepancies arise from the cellular hetero-geneity of the SCG. Thus, isoproterenol has been reported to increase cAMP levels predominantly in satellite cells in the SCG (Ariano et al 1982). Similarly, the increase in incorporation of [3]H-inositol into phospholipids in response to vasopressin has been reported to be primarily over neuropil (Horwitz et al 1986). Muscarine, on the other hand, an agonist that increases TH activity in the same tissue, increased labeling in the cytoplasm of postganglionic neurons (Horwitz et al 1986).

# MULTIPLE SITES OF TH PHOSPHORYLATION

As discussed above, TH is a substrate for at least four different protein kinases. Therefore, several questions arise, such as whether a single amino acid is phosphorylated by all of the kinases or whether there are multiple sites of phosphorylation on TH, and, if the latter is true, whether these sites are phosphorylated differentially by different kinases. All of these kinases are known to phosphorylate serine or threonine residues; however, differences have been found in the preferences of specific kinases for different amino acid sequences in the vicinity of these phosphorylated residues. In all the studies on phosphorylation of TH, only serine residues have been found to be phosphorylated (McTigue et al 1985, Yanagihara et al 1986, Pigeon et al 1987, Waymire et al 1988). If one examines the amino acid sequence of rat TH deduced from its cDNA, there are three serines preceded at positions $-3$ and $-4$ by either arginine or lysine residues and at position $-2$ by a neutral amino acid (Ser$^{40,153,172}$, i.e. the serine residues at positions 40, 153, and 172 in the TH molecule). These sequences are characteristic of substrates for cAMP- and cGMP-dependent protein kinases (Krebs & Beavo 1979, Glass & Krebs 1979). In addition, there are four serine residues preceded at position $-4$ by an arginine residue (Ser$^{19,269,319,457}$) (Grima et al 1985), sequences found in substrates for a number of protein kinases (O'Brian et al 1984, Pearson et al 1985, Davis et al 1986, House et al 1987).

Early studies with enzymatic cleavage of TH led to the concept that the enzyme has regulatory and catalytic domains (Hoeldtke & Kaufman 1977). Thus, partial proteolytic digestion with trypsin results in a fragment [probably at the C-terminal part of the TH monomer (Grima et al 1985)] that has a molecular weight of approximately 34,000 (i.e. slightly more than half the size of the subunit). This fragment retains enzyme activity and, in fact, shows increased activity over the native enzyme (Musacchio et al 1971, Kuczenski 1973a, Vigny & Henry 1981). The fragment, however, cannot be further activated by exposing it to phosphorylating conditions (Hoeldtke & Kaufman 1977). Such data suggest that the C-terminal portion of the enzyme is the catalytic domain and that, of the consensus sequences for phosphorylation in the molecule, those in the N-terminal portion play the most important role in enzyme activation (Figure 2). In light of these ideas, it is interesting to note the finding that it is the C-terminal portion of TH that has the most sequence homology with the two other BH$_4$-dependent monooxygenases, phenylalanine and tryptophan hydroxylase. Thus, when the deduced sequences of rat TH and rat phenylalanine hydroxylase are compared, little sequence homology exists for the first 140 amino acids (Dahl & Mercer 1986). On the other hand,

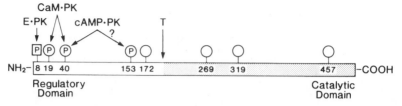

*Figure 2* Schematic diagram of the primary structure of rat TH (after Campbell et al 1986). The *shading* indicates the area of maximum sequence homology between TH and the other two BH$_4$-requiring monooxygenases, phenylalanine and tryptophan hydroxylases (Ledley et al 1985). T indicates the hypothesized site of tryptic cleavage that yields a catalytically active COOH-terminal fragment that cannot be further activated by phosphorylation (see Grima et al 1985, Ledley et al 1985). The square containing a P refers to Ser[8], a site found to be phosphorylated by an endogenous protein kinase (E · PK) that copurifies with TH (Campbell et al 1986). This serine residue is not preceded by a consensus sequence for a known protein kinase. The circles containing a P refer to serine residues 19, 40, and 153, sites that have been demonstrated to be phosphorylated by a cAMP-dependent (cAMP · PK) and/or a Ca$^{2+}$/calmodulin-dependent (CaM · PK) protein kinase. It should be noted that these serine residues are all on the NH$_2$-terminal fragment of the enzyme produced by trypsin, consistent with the hypothesis that this portion of the enzyme constitutes its regulatory domain. These serines are preceded by an arginine or a lysine residue(s) either four or three and four residues on their amino-terminus side. These sequences are typical of those found in substrates of cAMP · PK and CaM · PK. The *open circles* indicate serines 172, 269, 319, and 457, residues that are also preceded by such basic amino acids. It is not known, however, whether these serine residues are ever phosphorylated in situ. The *question mark* refers to the fact that phosphorylation of serine[153] was only seen in a single experiment in which TH had been purified in the absence of protease inhibitors, thus raising the possibility that this residue is not normally accessible to phosphorylation by a cAMP-dependent protein kinase (Campbell et al 1986).

considerable homology is found in the central and C-terminal regions, i.e. over 73% for residues 140–353. Similarly, comparison of the deduced amino acid sequence of rat TH with that of rabbit or rat tryptophan hydroxylase reveals the most homology in the central and C-terminal regions (Grenett et al 1987, Darmon et al 1988). These results are consistent with the findings that all three enzymes not only use the same cofactor, but also use molecular oxygen and an aromatic amino acid as substrates (Kaufman & Fisher 1974).

The first direct evidence both that multiple sites of phosphorylation exist on the enzyme and that specific stimuli alter phosphorylation at these sites differentially came from the study of Haycock et al (1982a). Bovine chromaffin cells were exposed briefly to either acetylcholine or 8-bromo-cAMP in the presence of $^{32}$P$_i$. TH, identified by molecular weight following SDS gel electrophoresis, was then exposed to trypsin, and the resulting peptide fragments were separated by electrophoresis and chromatography.

Cells incubated under control conditions yielded two radiolabeled phosphopeptides, and the incorporation of $^{32}$P into both of these peptides was stimulated by acetylcholine. 8-bromo-cAMP, on the other hand, only stimulated the phosphorylation of one of the peptides.

Since this first report, multiple sites of phosphorylation of TH have been demonstrated in a variety of adrenergic cell types, including pheochromocytoma (Vulliet et al 1984), PC12 cells (Lee et al 1985, McTigue et al 1985), and SCG (Cahill & Perlman 1984a). However, the number of different phosphorylation sites reported in such studies varies from two to seven (cf Yanagihara et al 1986, Waymire et al 1988). One difficulty encountered in trying to make comparisons among the phosphopeptides observed in different studies is that often different proteolysis conditions and different separation procedures have been used. It should be noted that even within a single study, a number of problems arise in trying to determine the exact number of such sites. For example, there may be more than one phosphorylation site on a particular tryptic peptide. Another difficulty arises because some of the tryptic peptides separated in these studies may have overlapping sequences due to incomplete enzymatic digestion. This problem, which is present in many studies using trypsin, is particularly acute in studies on phosphorylated peptides. Certain of the recognition sequences for specific protein kinases include pairs of basic residues, and basic residues are the sites of cleavage by trypsin. Thus, if tryptic hydrolysis of a protein containing pairs of basic residues is incomplete, one can generate tryptic peptides, which differ by one basic residue at either their N- or C-terminus or both.

Studies with purified protein kinases and purified TH have suggested that all four kinases examined stimulate the phosphorylation of the same tryptic peptide, and that, in addition, the $Ca^{2+}$/calmodulin-dependent protein kinase phosphorylates a second site. Thus, Albert et al (1984) found that incubation of TH with protein kinase C or cAMP-dependent protein kinase led to the phosphorylation of a single chymotryptic peptide. Similar data were obtained by Vulliet and colleagues, who showed that these kinases and a $Ca^{2+}$/calmodulin-dependent protein kinase led to the phosphorylation of the same tryptic peptide. In addition, they found that the latter kinase led to the phosphorylation of a second peptide (Vulliet et al 1984, 1985, Campbell et al 1986). Roskoski et al (1987) reported that a cGMP-dependent protein kinase stimulated the phosphorylation of the same tryptic peptide as did the cAMP-dependent protein kinase.

A complete analysis of the sites on TH where phosphorylation is stimulated by specific stimuli requires sequencing of the phosphopeptides produced under each condition. Thus far such data are only available for the phosphorylation of purified rat TH resulting from addition of cAMP-dependent or $Ca^{2+}$/calmodulin-dependent protein kinase (Campbell et al 1986). cAMP-dependent protein kinase increased phosphorylation reproducibly only at $Ser^{40}$ (Figure 2). Interestingly, in view of the discussion above, two phosphopeptides were isolated in this experiment that differed only in whether their C-terminus was $Arg^{46}$ or $Arg^{46}$ $Lys^{47}$. $Ca^{2+}$/calmodulin-dependent protein kinase also increased the phosphorylation of $Ser^{40}$, but had a larger effect at $Ser^{19}$. Again $Ser^{19}$ was found in two phosphopeptides, one of which had an N-terminal $Arg^{16}$ $Ala^{17}$ and another that had an N-terminal $Ala^{17}$. In addition to $Ser^{19}$ and $Ser^{40}$, Campbell et al (1986) found that $Ser^8$, which is not preceded by a consensus sequence for a known kinase, was phosphorylated by a kinase that copurified with TH. Although $Ser^{153}$ phosphorylation was catalyzed by a cAMP-dependent protein kinase, this reaction only occurred in an experiment in which TH was purified in the absence of protease inhibitors, thus suggesting that the site may only be revealed after partial proteolysis of the enzyme (Campbell et al 1986).

A number of studies have been done in which specific second messenger systems have been stimulated in situ, and the effects on phosphorylation of TH have been examined by analysis of tryptic fragments. For example, Niggli et al (1984) examined bovine chromaffin cells that had been made "leaky" by brief exposure to high-voltage electric fields. The cells were then exposed to medium containing $[\gamma\text{-}^{32}P]ATP$, cAMP, and a low concentration of $Ca^{2+}$ (80 nM). When TH was subsequently isolated by immunoprecipitation and subjected to partial proteolysis, with a protease from *Streptomyces griseus*, predominantly one phosphopeptide was seen. When the cells were exposed to a higher concentration of $Ca^{2+}$ (10 $\mu$M), a phosphopeptide with the same electrophoretic mobility was seen together with a second phosphopeptide. It is impossible to determine with certainty which serine residues were phosphorylated in this study (and in the studies cited below) given the information available. However, the data of Niggli et al resemble those of Campbell et al in that the latter predict that if a high concentration of $Ca^{2+}$ leads to the activation of a $Ca^{2+}$/calmodulin-dependent protein kinase, then a second site should be phosphorylated in addition to the site phosphorylated in response to the cAMP-dependent protein kinase.

Data obtained on other studies, however, cannot be easily fit into a two-site model. When bovine chromaffin cells were exposed to $^{32}P_i$ and cell proteins subsequently separated by two-dimensional gel electrophoresis, three spots were seen at a molecular weight of about 56,000 daltons, the spots varying in isoelectric point from 6.37 to 6.15 (Pocotte et al 1986). All three spots were identified as TH on Western blots, using an antiserum raised against TH, as was a fourth nonphosphorylated spot that had a pI of 6.5. Similar data were obtained in studies on PC12 cells (Nose et al 1985) and on cultured embryonic dopaminergic neurons (Kapatos 1987). It has been hypothesized that the four molecular species represent TH with 0, 1, 2, and 3 phosphate groups (Pocotte et al 1986, Kapatos 1987). Further, McTigue et al (1985) examined the phosphorylation of TH in PC12 cells in situ that had been exposed to high $K^+$ (in the presence of $Ca^{2+}$ in the extracellular medium), cholera toxin, dibutyryl cAMP, or PMA. Analysis of the tryptic peptides obtained from TH indicated that phosphorylation was increased in a single peptide (labeled T3 by the authors) under all four conditions. High $K^+$ also led to increased phosphorylation of a second peptide (T2), whereas cholera toxin and dibutyryl cAMP increased phosphorylation of a third peptide (T1). Cahill & Perlman (1987b), examining the effects of veratridine, forskolin, 8-bromo-cAMP, and PDBu, and Tachikawa et al (1987), examining the effects of ionomycin, forskolin, PMA, and OAG, also came to the conclusion that three or more sites on TH can be phosphorylated in situ.

The effects of certain first messengers on the phosphorylation of specific sites on TH have also been examined. In the SCG (Cahill & Perlman 1987b) and in bovine chromaffin cells (Craviso et al 1987), VIP stimulates the incorporation of phosphate into a single tryptic peptide, which is the same peptide whose phosphorylation is enhanced by forskolin and 8-bromo-cAMP. Thus, the hypothesis that VIP stimulates dopa synthesis via stimulation of cAMP levels and activation of a cAMP-dependent protein kinase (see cAMP AND cAMP-DEPENDENT PROTEIN KINASE) is further supported at this level of analysis (Figure 1). In the SCG, the nicotinic agonist DMPP has been reported to increase phosphorylation of four tryptic peptides, whose phosphorylation is also stimulated by veratridine (Cahill & Perlman 1987b).

In the study by McTigue et al (1985) cited above, NGF increased phosphorylation of peptides T1 and T3 in PC12 cells, peptides that are also phosphorylated in response to cholera toxin and dibutyryl cAMP. The effects of NGF were also examined in a mutant PC12 cell line that is deficient in cAMP-dependent protein kinase (Cremins et al 1986). Under these conditions, phosphorylation was seen in peptide T3 but not in T1, thus suggesting that phosphorylation of T1 in wild type PC12 cells occurs via a cAMP-dependent protein kinase. Based on other data, the authors suggest that NGF stimulation of phosphorylation of T3 is mediated via protein kinase C. The involvement of these two second messenger systems in the phosphorylation of TH produced by NGF is compatible with the data discussed above concerning the effects of NGF on cAMP (see *First messengers that increase cAMP levels*) and on PI turnover (see *First messengers that increase PI turnover*). However, other kinases may be present that are activated by NGF and phosphorylate TH in these cells. Rowland et al (1987) have isolated a novel NGF-activated kinase from PC12 cells that phosphorylates a serine residue(s) of purified TH.

The functional significance of phosphorylation at each of the relevant serine residues on TH is not known at present. It is interesting to note, however, that the time courses (Waymire et al 1988) as well as the extents (McTigue et al 1985) of phosphorylation of different sites can differ in response to different stimuli. Also, little is known at present about the phosphatase(s) that dephosphorylate TH, and their effectiveness at different serine residues (Nelson & Kaufman 1987). However, particularly with the advent of new tools in molecular biology, such as site-directed mutagenesis, the effects of phosphorylation at specific serine residues will undoubtedly be examined further.

## OVERVIEW

In the 1940s and 1950s neurophysiologists turned their attention to studying synaptic integration, i.e. the way multiple synaptic signals—some excitatory and others inhibitory—summed to produce a postsynaptic potential. Recent studies on the acute regulation of TH have revealed that a number of neurotransmitters and neuromodulators can influence the activity of this enzyme in postsynaptic targets, both in adrenergic cell bodies and/or dendrites and in adrenergic terminals. At present it appears that all of these changes can be mediated by the phosphorylation of TH via the activation of specific protein kinases. Since TH has multiple sites of phosphorylation and since different kinases appear to phosphorylate preferentially different sites, a series of mechanisms exist that may lead to a quite subtle regulation of the activity of this enzyme and, perhaps, of other aspects of its behavior (e.g. its half-life). During the next several years, the mechanisms and the significance of the multiple pathways leading to the acute regulation of TH will continue to be an area of active investigation.

ACKNOWLEDGMENTS

We thank Drs. David Glass, Simon Halegoua, Robert Perlman, and P. Richard Vulliet for helpful comments on our manuscript. The work cited from our laboratory was supported by United States Public Health Service grant NS12651. R. E. Z. was supported by a Research Scientist Award (MH00162), M. A. S. and A. R. R. by training grants GM07306 and NS07009, respectively.

*Literature Cited*

Abou Donia, M. M., Viveros, O. H. 1981. Tetrahydrobiopterin increases in adrenal medulla and cortex: a factor in the regulation of tyrosine hydroxylase. *Proc. Natl. Acad. Sci. USA* 78: 2703–6

Abreu, P., Santana, C., Hernandez, G., Calzadilla, C. H., Alonso, R. 1987. Day-night rhythm of rat pineal tryrosine hydroxylase activity as determined by HPLC with amperometric detection. *J. Neurochem.* 48: 665–68

Acheson, A., Thoenen, H. 1987. Both short- and long-term effects of nerve growth factor on tyrosine hydroxylase in calf adrenal chromaffin cells are blocked by S-adenosylhomocysteine hydrolase inhibitors. *J. Neurochem.* 48: 1416–24

Acheson, A. L., Kapatos, G., Zigmond, M. J. 1981. The effects of phosphorylating conditions on tyrosine hydroxylase activity are influenced by assay conditions and brain region. *Life Sci.* 28: 1407–20

Albert, K. A., Helmer-Matyjek, E., Nairn, A. C., Muller, T. H., Haycock, J. W., Greene, L. A., Goldstein, M., Greengard, P. 1984. Calcium/phospholipid-dependent protein kinase (protein kinase C) phosphorylates and activates tyrosine hydroxylase. *Proc. Natl. Acad. Sci. USA* 81: 7713–17

Alousi, A., Weiner, N. 1966. The regulation of norepinephrine synthesis in sympathetic nerves: Effect of nerve stimulation, cocaine, and catecholamine-releasing agents. *Proc. Natl Acad. Sci. USA* 56: 1491–96

Ames, M. M., Lerner, P., Lovenberg, W. 1978. Tyrosine hydroxylase. Activation by protein phosphorylation and end product inhibition. *J. Biol. Chem.* 253: 27–31

Anagnoste, B., Shirron, C., Friedman, E., Goldstein, M. 1974. Effect of dibutyryl cyclic adenosine monophosphate on $^{14}$C-dopamine biosynthesis in rat brain striatal slices. *J. Pharmacol. Exp. Ther.* 191: 370–76

Andén, N.-E., Grabowska-Andén, M., Klaesson, L. 1986. Stimulation of the synthesis of catecholamines in a sympathetic ganglion via cholinergic and non-cholinergic mechanisms. *Naunyn Schmiedeberg's Arch. Pharmacol.* 333: 17–22

Andrews, D. W., Langan, T. A., Weiner, N. 1983. Evidence for the involvement of a cyclic AMP-independent protein kinase in the activation of soluble tyrosine hydroxylase from rat striatum. *Proc. Natl. Acad. Sci. USA* 80: 2097–2101

Ariano, M. A., Briggs, C. A., McAfee, D. A. 1982. Cellular localization of cyclic

nucleotide changes in rat superior cervical ganglion. *Cell. Mol. Neurobiol.* 2: 143–56

Audigier, S., Barberis, C., Jard, S. 1986. Vasoactive intestinal polypeptide increases inositol phospholipid breakdown in the rat superior cervical ganglion. *Brain Res.* 376: 363–67

Baetge, E. E., Suh, Y. H., Joh, T. H. 1986. Complete nucleotide and deduced amino acid sequence of bovine phenylethanolamine N-methyltransferase: Partial amino acid homology with rat tyrosine hydroxylase. *Proc. Natl. Acad. Sci. USA* 83: 5454–58

Baldwin, C., Sasek, C. A., Zigmond, R. E. 1988. Localization of vasoactive intestinal peptide (VIP)- and peptide histidine isoleucine amide (PHI)-like immunoreactivities (IR) in preganglionic neurons that project to the rat superior cervical ganglion (SCG). *Soc. Neurosci. Abstr.* 14: 355 (Abstr.)

Baizer, L., Weiner, N. 1985. Nerve growth factor treatment enhances nicotine-stimulated dopamine release and increases in cyclic adenosine 3′:5′-monophosphate levels in PC12 cell cultures. *J. Neurosci.* 5: 1176–79

Berod, A., Faucon Biguet, N., Dumas, S., Bloch, B., Mallet, J. 1987. Modulation of tyrosine hydroxylase gene expression in the central nervous system visualized by in situ hybridization. *Proc. Natl. Acad. Sci. USA* 84: 1699–1703

Berridge, M. J. 1987. Inositol trisphosphate and diacylglycerol: Two interacting second messengers. *Annu. Rev. Biochem.* 56: 159–93

Black, I. B., Chikaraishi, D. M., Lewis, E. J. 1985. Trans-synaptic increase in RNA coding for tyrosine hydroxylase in a rat sympathetic ganglion. *Brain Res.* 339: 151–53

Blaschko, H. 1939. The specific action of l-DOPA decarboxylase. *J. Physiol. London* 96: 50–51P

Blaustein, M. P. 1971. Preganglionic stimulation increases calcium uptake by sympathetic ganglia. *Science* 172: 391–93

Blaustein, M. P. 1976. Barbiturates block calcium uptake by stimulated and potassium-depolarized rat sympathetic ganglia. *J. Pharmacol. Exp. Ther.* 196: 80–86

Boadle-Biber, M. C., Hughes, J., Roth, R. H. 1972. Acceleration of catecholamine biosynthesis in sympathetically innervated tissues by angiotensin-II-amide. *Br. J. Pharmacol.* 46: 289–99

Bone, E. A., Fretten, P., Palmer, S., Kirk, C. J., Michell, R. H. 1984. Rapid accumulation of inositol phosphates in isolated rat superior cervical sympathetic ganglia exposed to $V_1$-vasopressin and muscarinic cholinergic stimuli. *Biochem. J.* 221: 803–11

Bone, E. A., Michell, R. H. 1985. Accumulation of inositol phosphates in sympathetic ganglia. Effects of depolarization and of amine and peptide neurotransmitters. *Biochem. J.* 227: 263–69

Briggs, C. A., Whiting, G. J., Ariano, M. A., McAfee, D. A. 1982. Cyclic nucleotide metabolism in the sympathetic ganglion. *Cell. Mol. Neurobiol.* 2: 129–41

Bucciarelli de Simone, A., Rubio, M. C. 1980. Differential effects of prostaglandins $E_1$ and $E_2$ on tyrosine hydroxylase activity of the superior cervical ganglion of the cat. *Gen. Pharmacol.* 11: 539–41

Cahill, A. L., Horwitz, J., Perlman, R. L. 1985. Low-$Na^+$ medium increases the activity and the phosphorylation of tyrosine hydroxylase in the superior cervical ganglion of the rat. *J. Neurochem.* 44: 680–85

Cahill, A. L., Perlman, R. L. 1984a. Phosphorylation of tyrosine hydroxylase in the superior cervical ganglion. *Biochim. Biophys. Acta* 805: 217–26

Cahill, A. L., Perlman, R. L. 1984b. Electrical stimulation increases phosphorylation of tyrosine hydroxylase in superior cervical ganglion of rat. *Proc. Natl. Acad. Sci. USA* 81: 7243–47

Cahill, A. L., Perlman, R. L. 1985. Protein kinases in the superior cervical ganglion. *Soc. Neurosci. Abstr.* 11: 814 (Abstr.)

Cahill, A. L., Perlman, R. L. 1986. Nicotinic and muscarinic agonists, phorbol esters, and agents which raise cyclic AMP levels phosphorylate distinct groups of proteins in the superior cervical ganglion. *Neurochem. Res.* 11: 327–38

Cahill, A. L., Perlman, R. L. 1987a. The phosphorylation of tyrosine hydroxylase in PC12 cells after down regulation of protein kinase C. *Soc. Neurosci. Abstr.* 13: 806 (Abstr.)

Cahill, A. L., Perlman, R. L. 1987b. Preganglionic stimulation increases the phosphorylation of tyrosine hydroxylase in the superior cervical ganglion by both cAMP-dependent and calcium-dependent protein kinases. *Biochim. Biophys. Acta* 930: 454–62

Campbell, D. G., Hardie, D. G., Vulliet, P. R. 1986. Identification of four phosphorylation sites in the N-terminal region of tyrosine hydroxylase. *J. Biol. Chem.* 261: 10489–92

Chalazonitis, A., Rice, P. J., Zigmond, R. E. 1980. Increased ganglionic tyrosine hy-

droxylase and dopamine-$\beta$-hydroxylase activities following preganglionic nerve stimulation: Role of nicotinic receptors. *J. Pharmacol. Exp. Ther.* 213: 139–43

Chalazonitis, A., Zigmond, R. E. 1980. Effects of synaptic and antidromic stimulation on tyrosine hydroxylase activity in the rat superior cervical ganglion. *J. Physiol. London* 300: 525–38

Chalfie, M., Perlman, R. L. 1977. Regulation of catecholamine biosynthesis in a transplantable rat pheochromocytoma. *J. Pharmacol. Exp. Ther.* 200: 588–97

Chalfie, M., Settipani, L., Perlman, R. L. 1979. The role of cyclic adenosine 3′: 5′-monophosphate in the regulation of tyrosine 3-monooxygenase activity. *Mol. Pharmacol.* 15: 263–70

Clapham, D. E., Neher, E. 1984. Substance P reduces acetylcholine-induced currents in isolated bovine chromaffin cells. *J. Physiol. London* 347: 255–77

Costa, E., Guidotti, A., Hanbauer, I. 1974. Do cyclic nucleotides promote the transsynaptic induction of tyrosine hydroxylase? *Life Sci.* 14: 1169–88

Cote, A., Doucet, J.-P., Trifaro, J.-M. 1986. Phosphorylation and dephosphorylation of chromaffin cell proteins in response to stimulation. *Neuroscience* 19: 629–45

Craft, C. M., Morgan, W. W., Reiter, R. J. 1984. 24-Hour changes in catecholamine synthesis in rat and hamster pineal glands. *Neuroendocrinology* 38: 193–98

Craviso, G. L., Waymire, J. C., Lickteig, K., Baldwin, C., Zigmond, R. E. 1987. Characterization of the VIP-induced phosphorylation and activation of tyrosine hydroxylase in bovine adrenal chromaffin cells. *Soc. Neurosci. Abstr.* 13: 1644 (Abstr.)

Cremins, J., Wagner, J. A., Halegoua, S. 1986. Nerve growth factor action is mediated by cyclic AMP- and $Ca^{+2}$/phospholipid-dependent protein kinases. *J. Cell Biol* 103: 887–93

Dahl, H.-H. M., Mercer, J. F. B. 1986. Isolation and sequence of a cDNA clone which contains the complete coding region of rat phenylalanine hydroxylase. Structural homology with tyrosine hydroxylase, glucocorticoid regulation, and use of alternate polyadenylation sites. *J. Biol. Chem.* 261: 4148–53

Dairman, W., Gordon, R., Spector, S., Sjoerdsma, A., Udenfriend, S. 1968. Increased synthesis of catecholamines in the intact rat following administration of α-adrenergic blocking agents. *Mol. Pharmacol.* 4: 457–64

Darmon, M. C., Guibert, B., Leviel, V., Ehret, M., Maitre, M., Mallet, J. 1988.

Sequence of two mRNAs encoding active rat tryptophan hydroxylase. *J. Neurochem.* 51: 312–16

Davis, R. J., Johnson, G. L., Kelleher, D. J., Anderson, J. K., Mole, J. E., Czech, M. P. 1986. Identification of serine 24 as the unique site on the transferrin receptor phosphorylated by protein kinase C. *J. Biol. Chem.* 261: 9034–41

Debinski, W., Gutdowska, J., Kuchel, O., Racz, K., Buu, N. T., Cantin, M., Genest, J. 1987a. Presence of an atrial natriuretic factor-like peptide in the rat superior cervical ganglia. *Neuroendocrinology* 46: 236–40

Debinski, W., Kuchel, O., Buu, N. T., Cantin, M., Genest, J. 1987b. Atrial natriuretic factor partially inhibits the stimulated catecholamine synthesis in superior cervical ganglia of the rat. *Neurosci. Lett.* 77: 92–96

Derome, G., Tseng, R., Mercier, P., Lemaire, I., Lemaire, S. 1981. Possible muscarinic regulation of catecholamine secretion mediated by cyclic GMP in isolated bovine adrenal chromaffin cells. *Biochem. Pharmacol.* 30: 855–60

deVente, J. D., Garssen, J., Tilders, F. J. H., Steinbusch, H. W. H., Schipper, J. 1987. Single cell quantitative immunocytochemistry of cyclic GMP in the superior cervical ganglion of the rat. *Brain Res.* 411: 120–28

Douglas, W. W., Poisner, A. M. 1961. Stimulation of uptake of calcium-45 in the adrenal gland by acetylcholine. *Nature* 192: 1299

Durroux, T., Barberis, C., Jard, S. 1987. Vasoactive intestinal polypeptide and carbachol act synergistically to induce the hydrolysis of inositol containing phospholipids in the rat superior cervical ganglion. *Neurosci. Lett.* 75: 211–15

Dvorkin, B., Makman, M. H., Kessler, J. A. 1984. Adenylate cyclase of cultures of rat sympathetic neurons and striatum: Influence of VIP, secretin and dopamine. *Soc. Neurosci. Abstr.* 10: 762 (Abstr.)

Eberhard, D. A., Holz, R. W. 1987. Cholinergic stimulation of inositol phosphate formation in bovine adrenal chromaffin cells: Distinct nicotinic and muscarinic mechanisms. *J. Neurochem.* 49: 1634–43

Edelman, A. M. Raese, J. D., Lazar, M. A., Barchas, J. D. 1978. In vitro phosphorylation of a purified preparation of bovine corpus striatal tyrosine hydroxylase. *Commun. Psychopharmacol.* 2: 461–65

Elliot, T. R. 1912. The control of the suprarenal glands by the splanchnic nerves. *J. Physiol. London* 44: 374–409

Erny, R. E. 1983. Regulation of adenylate cyclase and catecholamine synthesis by adenosine in pheochromocytoma cells. PhD thesis. Harvard Univ., Cambridge, Mass.

Erny, R. E. Berezo, M. W., Perlman, R. L. 1981. Activation of tyrosine 3-monooxygenase in pheochromocytoma cells by adenosine. *J. Biol. Chem.* 256: 1335–39

Erny, R., Wagner, J. A. 1984. Adenosine-dependent activation of tyrosine hydroxylase is defective in adenosine kinase-deficient PC12 cells. *Proc. Natl. Acad. Sci. USA* 81: 4974–78

Faucon Biguet, N., Buda, M., Lamouroux, A., Samolyk, D., Mallet, J. 1986. Time course of the changes of TH mRNA in rat brain and adrenal medulla after a single injection of reserpine. *EMBO J.* 5: 287–91

Fisher, S. K., Holz, R. W., Agranoff, B. W. 1981. Muscarinic receptors in chromaffin cell cultures mediate enhanced phospholipid labeling but not catecholamine secretion. *J. Neurochem.* 37: 491–97

Fluharty, S. J., Snyder, G. L. Stricker, E. M., Zigmond, M. J. 1983. Short- and long-term changes in adrenal tyrosine hydroxylase activity during insulin-induced hypoglycemia and cold stress. *Brain Res.* 267: 384–87

Fluharty, S. J., Rabow, L. E., Zigmond, M. J., Stricker, E. M. 1985a. Tyrosine hydroxylase activity in the sympathoadrenal system under basal and stressful conditions: Effect of 6-hydroxydopamine. *J. Pharmacol. Exp. Ther.* 235: 354–60

Fluharty, S. J., Snyder, G. L., Zigmond, M. J., Stricker, E. M. 1985b. Tyrosine hydroxylase activity and catecholamine biosynthesis in the adrenal medulla of rats during stress. *J. Pharmacol. Exp. Ther.* 233: 32–38

Forsberg, E. J., Rojas, E., Pollard, H. B. 1986. Muscarinic receptor enhancement of nicotine-induced catecholamine secretion may be mediated by phosphoinositide metabolism in bovine adrenal chromaffin cells. *J. Biol. Chem.* 261: 4915–20

Frey, E. A., McIsaac, R. J. 1981. A comparison of cyclic guanosine 3':5'-monophosphate and muscarinic excitatory responses in the superior cervical ganglion of the rat. *J. Pharmacol. Exp. Ther.* 218: 115–21

Frye, R. A., Holz, R. W. 1983. Phospholipase $A_2$ inhibitors block catecholamine secretion and calcium uptake in cultured bovine adrenal medullary cells. *Mol. Pharmacol.* 23: 547–50

Gamse, R., Wax, A., Zigmond, R. E., Leeman, S. E. 1981. Immunoreactive substance P in sympathetic ganglia: Distribution and sensitivity towards capsaicin. *Neuroscience* 6: 437–41

Glass, D. B., Krebs, E. G. 1979. Comparison of the substrate specificity of adenosine 3′ : 5′-monophosphate- and guanosine 3′ : 5′-monophosphate-dependent protein kinases. Kinetic studies using synthetic peptides corresponding to phosphorylation sites in histone H2B. *J. Biol. Chem.* 254: 9728–38

Goldstein, M., Anagnoste, B., Shirron, C. 1973. The effect of trivastal, haloperidol and dibutyryl cyclic AMP on ($^{14}$C)dopamine synthesis in rat striatum. *J. Pharm. Pharmacol.* 25: 348–51

Goldstein, M., Bronaugh, R. L., Ebstein, B., Roberge, C. 1976. Stimulation of tyrosine hydroxylase activity by cyclic AMP in synaptosomes and in soluble striatal enzyme preparations. *Brain Res.* 109: 563–74

Goldstein, M., Greene, L. A. 1987. Activation of tyrosine hydroxylase by phosphorylation. In *Psychopharmacology: The Third Generation of Progress*, ed. H. Y. Meltzer, pp. 75–80. New York: Raven

Goodman, R., Slater, E., Herschman, H. R. 1980. Epidermal growth factor induces tyrosine hydroxylase in a clonal pheochromocytoma cell line, PC-G2. *J. Cell Biol.* 84: 495–500

Gordon, R., Reid, J. V. O., Sjoerdsma, A., Udenfriend, S. 1966a. Increased synthesis of norepinephrine in the rat heart on electrical stimulation of the stellate ganglia. *Mol. Pharmacol.* 2: 610–13

Gordon, R., Spector, S., Sjoerdsma, A., Udenfriend, S. 1966b. Increased synthesis of norepinephrine and epinephrine in the intact rat during exercise and exposure to cold. *J. Pharmacol. Exp. Ther.* 153: 440–47

Greene, L. A., Rein, G. 1978. Short-term regulation of catecholamine biosynthesis in a nerve growth factor responsive clonal line of rat pheochromocytoma cells. *J. Neurochem.* 30: 549–55

Greene, L. A., Seeley, P. J., Rukenstein, A., DiPiazza, M., Howard, A. 1984. Rapid activation of tyrosine hydroxylase in response to nerve growth factor. *J. Neurochem.* 42: 1728–34

Greene, L. A., Drexler, S. A., Connolly, J. L., Rukenstein, A., Green, S. H. 1986. Selective inhibition of responses to nerve growth factor and of microtubule-associated protein phosphorylation by activators of adenylate cyclase. *J. Cell Biol.* 103: 1967–78

Greengard, P. 1978. *Cyclic Nucleotides, Phosphorylated Proteins, and Neuronal Function.* New York: Raven

Grenett, H. E., Ledley, F. D., Reed, L. L.,

Woo, S. L. C. 1987. Full-length cDNA for rabbit tryptophan hydroxylase: Functional domains and evolution of aromatic amino acid hydroxylases. *Proc. Natl. Acad. Sci. USA* 84: 5530–34

Grima, B., Lamouroux, A., Blanot, F., Faucon Biguet, N., Mallet, J. 1985. Complete coding sequence of rat tyrosine hydroxylase mRNA. *Proc. Natl. Acad. Sci. USA* 82: 617–21

Grima, B., Lamouroux, A., Boni, C., Julien, J. F., Javoy-Agid, F., Mallet, J. 1987. A single human gene encoding multiple tyrosine hydroxylases with different predicted functional characteristics. *Nature* 326: 707–11

Guroff, G., Dickens, G., End, D., Londos, C. 1981. The action of adenosine analogs on PC12 cells. *J. Neurochem.* 37: 1431–39

Halegoua, S., Patrick, J. 1980. Nerve growth factor mediates phosphorylation of specific proteins. *Cell* 22: 571–81

Harper, G. P., Thoenen, H. 1981. Target cells, biological effects, and mechanism of action of nerve growth factor and its antibodies. *Ann. Rev. Pharmacol. Toxicol.* 21: 205–29

Harris, J. E., Morgenroth, V. H. III, Roth, R. H., Baldessarini, R. J. 1974. Regulation of catecholamine synthesis in the rat brain in vitro by cyclic AMP. *Nature* 252: 156–58

Haycock, J. W., Bennett, W. F., George R. J., Waymire, J. C. 1982a. Multiple site phosphorylation of tyrosine hydroxylase. Differential regulation in situ by 8-bromo-cAMP and acetylcholine. *J. Biol. Chem.* 257: 13699–13703

Haycock, J. W., Meligeni, J. A., Bennett, W. F., Waymire, J. C. 1982b. Phosphorylation and activation of tyrosine hydroxylase mediate the acetylcholine-induced increase in catecholamine biosynthesis in adrenal chromaffin cells. *J. Biol. Chem.* 257: 12641–48

Haycock, J. W., George R. J., Waymire, J. C. 1985. In situ phosphorylation of tyrosine hydroxylase in chromaffin cells: Localization to soluble compartments. *Neurochem. Int.* 7: 301–8

Hegstrand, L. R., Simon, J. R., Roth, R. H. 1979. Tyrosine hydroxylase:—Examination of conditions influencing activity in pheochromocytoma, adrenal medulla and striatum. *Biochem. Pharmacol.* 28: 519–23

Hoeldtke, R., Kaufman, S. 1977. Bovine adrenal tyrosine hydroxylase: Purification and properties. *J. Biol. Chem.* 252: 3160–69

Hökfelt, B., McLean, J. 1950. The adrenaline and noradrenaline content of suprarenal glands of the rabbit under normal conditions and after various forms of stimulation. *Acta Physiol. Scand.* 21: 258–70

Hökfelt, T., Elfvin, L.-G., Schultzberg, M., Fuxe, K., Said, S. I., Mutt, V., Goldstein, M. 1977. Immunohistochemical evidence of vasoactive intestinal polypeptide-containing neurons and nerve fibers in sympathetic ganglia. *Neuroscience* 2: 885–96

Hökfelt, T., Lundberg, J. M., Schultzberg, M., Fahrenkrug, J. 1981. Immunohistochemical evidence for a local VIP-ergic neuron system in the adrenal gland of the rat. *Acta Physiol. Scand.* 113: 575–76

Hökfelt, T., Johansson, O., Goldstein, M. 1984. Central catecholamine neurons as revealed by immunohistochemistry with special reference to adrenaline neurons. In *Handbook of Chemical Neuroanatomy*, Vol. 2: *Classical Transmitters in the CNS*, ed. A. Bjorklund, T. Hökfelt, pp. 157–276. Amsterdam: Elsevier

Hokin, L. E. 1965. Autoradiographic localization of the acetylcholine-stimulated synthesis of phosphatidylinositol in the superior cervical ganglion. *Proc. Natl. Acad. Sci. USA* 53: 1369–76

Hokin, L. E. 1966. Effects of acetylcholine on the incorporation of $^{32}P$ into various phospholipids in slices of normal and denervated superior cervical ganglia of the cat. *J. Neurochem.* 13: 179–84

Holland, W. C., Schumann, H. J. 1956. Formation of catechol amines during splanchnic stimulation of the adrenal gland of the cat. *Br. J. Pharmacol.* 11: 449–53

Holtz, P. 1939. Dopadecarboxylase. *Naturwissenschaften* 27: 724–25

Holz, R. W., Senter, R. A., Frye, R. A. 1982. Relationship between $Ca^{2+}$ uptake and catecholamine secretion in primary dissociated cultures of adrenal medulla. *J. Neurochem.* 39: 635–46

Holzwarth, M. A. 1984. The distribution of vasoactive intestinal peptide in the rat adrenal cortex and medulla. *J. Auton. Nerv. Syst.* 11: 269–83

Horwitz, J. 1987. Muscarine increases the diacylglycerol content of PC12 cells. *Soc. Neurosci. Abstr.* 13: 486 (Abstr.)

Horwitz, J., Perlman, R. L. 1984a. Stimulation of DOPA synthesis in the superior cervical ganglion by veratridine. *J. Neurochem.* 42: 384–89

Horwitz, J., Perlman, R. L. 1984b. Activation of tyrosine hydroxylase in the superior cervical ganglion by nicotinic and muscarinic agonists. *J. Neurochem.* 43: 546–52

Horwitz, J., Anderson, C. H., Perlman, R. L. 1986. Comparison of the effects of muscarine and vasopressin on inositol phospholipid metabolism in the superior cervical ganglion of the rat. *J. Pharmacol. Exp. Ther.* 237: 312–17

Horwitz, J., Tsymbalov, S., Perlman, R. L. 1984. Muscarine increases tyrosine 3-monooxygenase activity and phospholipid metabolism in the superior cervical ganglion of the rat. *J. Pharmacol. Exp. Ther.* 229: 577–82

Horwitz, J., Tsymbalov, S., Perlman, R. L. 1985. Muscarine stimulates the hydrolysis of inositol-containing phospholipids in the superior cervical ganglion. *J. Pharmacol. Exp. Ther.* 233: 235–41

Houchi, H., Oka, M., Misbahuddin, M., Morita, K., Nakanishi, A. 1987. Stimulation by vasoactive intestinal polypeptide of catecholamine synthesis in isolated bovine adrenal chromaffin cells. Possible involvement of protein kinase C. *Biochem. Pharmacol.* 36: 1551–54

House, C., Wettenhall, R. E. H., Kemp, B. E. 1987. The influence of basic residues on the substrate specificity of protein kinase C. *J. Biol. Chem.* 262: 772–77

Ikeno, T., Dickens, G., Lloyd, T., Guroff, G. 1981. The receptor-mediated activation of tyrosine hydroxylation in the superior cervical ganglion of the rat. *J. Neurochem.* 36: 1632–40

Ip, N. Y., Perlman, R. L., Zigmond, R. E. 1982a. Both nicotinic and muscarinic agonists acutely increase tyrosine 3-monooxygenase activity in the superior cervical ganglion. *J. Pharmacol. Exp. Ther.* 223: 280–83

Ip, N. Y., Ho, C. K., Zigmond, R. E. 1982b. Secretin and vasoactive intestinal peptide acutely increase tyrosine 3-monooxygenase in the rat superior cervical ganglion. *Proc. Natl. Acad. Sci. USA* 79: 7566–69

Ip, N. Y., Perlman, R. L., Zigmond, R. E. 1983. Acute transsynaptic regulation of tyrosine 3-monooxygenase activity in the rat superior cervical ganglion: Evidence for both cholinergic and noncholinergic mechanisms. *Proc. Natl. Acad. Sci. USA* 80: 2081–85

Ip, N. Y., Zigmond, R. E. 1984a. Substance P inhibits the acute stimulation of ganglionic tyrosine hydroxylase activity by a nicotinic agonist. *Neuroscience* 13: 217–20

Ip, N. Y., Zigmond, R. E. 1984b. Pattern of presynaptic nerve activity can determine the type of neurotransmitter regulating a postsynaptic event. *Nature* 311: 472–74

Ip, N. Y., Baldwin, C., Zigmond, R. E. 1984. Acute stimulation of ganglionic tyrosine hydroxylase activity by secretin, VIP and PHI. *Peptides* 5: 309–12

Ip, N. Y., Baldwin, C., Zigmond, R. E. 1985. Regulation of the concentration of adenosine 3′,5′-cyclic monophosphate and the activity of tyrosine hydroxylase in the rat superior cervical ganglion by three neuropeptides of the secretin family. *J. Neurosci.* 5: 1947–54

Iuvone, P. M., Galli, C. L., Neff, N. H. 1978. Retinal tyrosine hydroxylase: Comparison of short-term and long-term stimulation by light. *Mol. Pharmacol.* 14: 1212–19

Jaanus, S. D., Rubin, R. P. 1974. Analysis of the role of cyclic adenosine 3',5'-monophosphate in catecholamine release. *J. Physiol. London* 237: 465–76

Joh, T. H., Park, D. H., Reis, D. J. 1978. Direct phosphorylation of brain tyrosine hydroxylase by cyclic AMP-dependent protein kinase: Mechanism of enzyme activation. *Proc. Natl. Acad. Sci. USA* 75: 4744–48

Johansson, O., Lundberg, J. M. 1981. Ultrastructural localization of VIP-like immunoreactivity in large dense-core vesicles of 'cholinergic-type' nerve terminals in cat exocrine glands. *Neuroscience* 6: 847–62

Kalix, P., McAfee, D. A., Schorderet, M., Greengard, P. 1974. Pharmacological analysis of synaptically mediated increase in cyclic adenosine monophosphate in rabbit superior cervical ganglion. *J. Pharmacol. Exp. Ther.* 188: 676–87

Kaneda, N., Kobayashi, K., Ichinose, H., Kishi, F., Nakazawa, A., Kurosawa, Y., Fujita, K., Nagatsu, T. 1987. Isolation of a novel cDNA clone for human tyrosine hydroxylase: Alternative RNA splicing produces four kinds of mRNA from a single gene. *Biochem. Biophys. Res. Commun.* 146: 971–75

Kao, L.-S., Schneider, A. S. 1985. Muscarinic receptors on bovine chromaffin cells mediate a rise in cytosolic calcium that is independent of extracellular calcium. *J. Biol. Chem.* 260: 2019–22

Kao, L.-S., Schneider, A. S. 1986. Calcium mobilization and catecholamine secretion in adrenal chromaffin cells. A Quin-2 fluorescence study. *J. Biol. Chem.* 261: 4881–88

Kapatos, G. 1987. Multiple states of phosphorylation of tyrosine hydroxylase in cultured dopamine neurons. *Soc. Neurosci. Abstr.* 13: 1180 (Abstr.)

Kaufman, S. 1977. Mixed function oxygenases—General considerations. In *Structure and Function of Monoamine Enzymes*, ed. E. Usdin, N. Weiner, M. B. H. Youdim, pp. 3–22. New York: Dekker

Kaufman, S. 1985. Regulatory properties of phenylalanine, tyrosine and tryptophan hydroxylases. *Biochem. Soc. Trans.* 13: 433–36

Kaufman, S., Fisher, D. B. 1974. Pterin-requiring aromatic amino acid hydroxylases. In *Molecular Mechanisms of Oxygen Activation*, ed. O. Hayaishi, pp. 285–369. New York: Academic

Kebabian, J. W., Steiner, A. L., Greengard, P. 1975. Muscarinic cholinergic regulation of cyclic guanosine 3,5-monophosphate in autonomic ganglia: Possible role in synaptic transmission. *J. Pharmacol. Exp. Ther.* 193: 474–88

Kilpatrick, D. L., Slepetis, R. J., Corcoran, J. J., Kirshner, N. 1982. Calcium uptake and catecholamine secretion by cultured bovine adrenal medulla cells. *J. Neurochem.* 38: 427–35

Kiuchi, K., Kiuchi, K., Togari, A., Nagatsu, T. 1987. Effect of spermine on tyrosine hydroxylase activity before and after phosphorylation by cyclic AMP-dependent protein kinase. *Biochem. Biophys. Res. Commun.* 148: 1460–67

Knight, D. E., Kesteven, N. T. 1983. Evoked transient intracellular free $Ca^{2+}$ changes and secretion in isolated bovine adrenal medullary cells. *Proc. R. Soc. London Ser. B* 218: 177–99

Kobayashi, K., Kaneda, N., Ichinose, H., Kishi, F., Nakazawa, A., Kurosawa, Y., Fujita, K., Nagatsu, T. 1987. Isolation of full-length cDNA clone encoding human tyrosine hydroxylase type-3. *Nucleic Acids Res.* 15: 6733

Kopin, I. J., Breese, G. R., Krauss, K. R., Weise, V. K. 1968. Selective release of newly synthesized norepinephrine from the cat spleen during sympathetic nerve stimulation. *J. Pharmacol. Exp. Ther.* 161: 271–78

Krebs, E. G., Beavo, J. A. 1979. Phosphorylation-dephosphorylation of enzymes. *Ann. Rev. Biochem.* 48: 923–59

Kuczenski, R. 1973a. Rat brain tyrosine hydroxylase. Activation by limited tryptic proteolysis. *J. Biol. Chem.* 248: 2261–65

Kuczenski, R. 1973b. Striatal tyrosine hydroxylases with high and low affinity for tyrosine: Implications for the multiplepool concept of catecholamines. *Life Sci.* 13: 247–55

Kuczenski, R. T., Mandell, A. J. 1972. Regulatory properties of soluble and particulate rat brain tyrosine hydroxylase. *J. Biol. Chem.* 247: 3114–22

Kvetňanský, R., Weise, V. K., Gewirtz, G. P., Kopin, I. J. 1971. Synthesis of adrenal catecholamines in rats during and after immobilization stress. *Endocrinology* 89: 46–49

Kwok, S. C. M., Ledley, F. D., DiLella, A. G., Robson, K. J., Woo, S. L. 1985. Nucleotide sequence of a full-length complementary DNA clone and amino acid sequence of human phenylalanine hydroxylase. *Biochemistry* 24: 556–61

Lakshmanan, J. 1978. Nerve growth factor induced turnover of phosphatidylinositol in rat superior cervical ganglia. *Biochem. Biophys. Res. Commun.* 82: 767–75

Lamouroux, A., Faucon Biguet, N., Samolyk, D., Privat, A., Salomon, J. C., Pujol, J. F., Mallet, J. 1982. Identification of cDNA clones coding for rat tyrosine hydroxylase antigen. *Proc. Natl. Acad. Sci. USA* 79: 3881–85

Lamouroux, A., Vigny, A., Faucon Biguet, N., Darmon, M. C., Franck, R., Henry, J.-P., Mallet, J. 1987. The primary structure of human dopamine β-hydroxylase: Insights into the relationship between the soluble and membrane-bound forms of the enzyme. *EMBO. J.* 6: 3931–37

Lapetina, E. G., Brown, W. E., Michell, R. H. 1976. Muscarinic cholinergic stimulation of phosphatidylinositol turnover in isolated rat superior cervical sympathetic ganglia. *J. Neurochem.* 26: 649–51

Lazar, M. A., Lockfeld, A. J., Truscott, R. J. W., Barchas, J. D. 1982. Tyrosine hydroxylase from bovine striatum: Catalytic properties of the phosphorylated and nonphosphorylated forms of the purified enzyme. *J. Neurochem.* 39: 409–22

Ledley, F. D., DiLella, A. G., Kwok, S. C. M., Woo, S. L. C. 1985. Homology between phenylalanine and tyrosine hydroxylases reveals common structural and functional domains. *Biochemistry* 24: 3389–94

Lee, K. Y., Seeley, P. J., Müller, T. H., Helmer-Matyjek, E., Sabban, E., Goldstein, M., Greene, L. A. 1985. Regulation of tyrosine hydroxylase phosphorylation in PC12 pheochromocytoma cells by elevated K⁺ and nerve growth factor. Evidence for different mechanisms of action. *Mol. Pharmacol.* 28: 220–28

Letendre, C. H., MacDonnell, P. C., Guroff, G. 1977a. The biosynthesis of phosphorylated tyrosine hydroxylase by organ cultures of rat medulla and superior cervical ganglia. *Biochem. Biophys. Res. Commun.* 74: 891–97

Letendre, C. H., MacDonnell, P. C., Guroff, G. 1977b. The biosynthesis of phosphorylated tyrosine hydroxylase by organ cultures of rat adrenal medulla and superior cervical ganglia: A correction. *Biochem. Biophys. Res. Commun.* 76: 615–17

Levey, G. S., Weiss, S. R., Ruiz, E. 1975. Characterization of the glucagon receptor in a pheochromocytoma. *J. Clin. Endocrinol. Metab.* 40: 720–23

Levitt, M., Spector, S., Sjoerdsma, A., Udenfriend, S. 1965. Elucidation of the rate-limiting step in norepinephrine biosynthesis in the perfused guinea-pig heart. *J. Pharmacol. Exp. Ther.* 148: 1–7

Lewis, E. J., Chikaraishi, D. M. 1987. Regulated expression of the tyrosine hydroxylase gene by epidermal growth factor. *Mol. Cell. Biol.* 7: 3332–36

Lewis, E. J., Tank, A. W., Weiner, N., Chikaraishi, D. M. 1983. Regulation of tyrosine hydroxylase mRNA by glucocorticoid and cyclic AMP in a rat pheochromocytoma cell line. Isolation of a cDNA clone for tyrosine hydroxylase mRNA. *J. Biol. Chem.* 258: 14632–37

Lloyd, T., Kaufman, S. 1975. Evidence for the lack of direct phosphorylation of bovine caudate tyrosine hydroxylase following activation by exposure to enzymatic phosphorylating conditions. *Biochem. Biophys. Res. Commun.* 66: 907–13

Lloyd, T., Montgomery, P., Guroff, G. 1979. The receptor-mediated, short-term activation of tyrosine hydroxylase in organ cultures of rat superior cervical ganglia. *Biochem. Biophys. Res. Commun.* 86: 1058–62

Lloyd, T., Weiner, N. 1971. Isolation and characterization of a tyrosine hydroxylase cofactor from bovine adrenal medulla. *Mol. Pharmacol.* 7: 569–80

Lovenberg, W., Bruckwick, E. A., Hanbauer, I. 1975. ATP, cyclic AMP, and magnesium increase the affinity of rat striatal tyrosine hydroxylase for its cofactor. *Proc. Natl. Acad. Sci. USA* 72: 2955–58

Lovenberg, W., Levine, R., Miller, L. 1981. The hydroxylase cofactor and catecholamine synthesis. See Usdin et al 1981, pp. 225–30

Lundberg, J. M., Änggärd, A., Fahrenkrug, J., Hökfelt, T., Mutt, V. 1980. Vasoactive intestinal polypeptide in cholinergic neurons of exocrine glands: Functional significance of coexisting transmitters for vasodilation and secretion. *Proc. Natl. Acad. Sci. USA* 77: 1651–55

Mallet, J., Faucon Biguet, N., Buda, M., Lamouroux, A., Samolyk, D. 1983. Detection and regulation of the tyrosine hydroxylase mRNA levels in rat adrenal medulla and brain tissues. *Cold Spring Harbor Symp. Quant. Biol.* 48: 305–8

Markey, K. A., Kondo, S., Shenkman, L., Goldstein, M. 1980. Purification and characterization of tyrosine hydroxylase from a clonal pheochromocytoma cell line. *Mol. Pharmacol.* 17: 79–85

Masserano, J., Weiner, N. 1979a. Similarities between the in vivo activation of adrenal tyrosine hydroxylase and the in vitro activation of the enzyme by an adenosine 3',5'-monophosphate dependent protein phosphorylating system. In *Catecholamines: Basic and Clinical Frontiers*, ed. E. Usdin, I. Kopin, J. Barchas, 1: 100–2. New York: Pergamon

Masserano, J. M., Weiner, N. 1979b. The rapid activation of adrenal tyrosine hydroxylase by decapitation and its relationship to a cyclic AMP-dependent phosphorylating mechanism. *Mol. Pharmacol.* 16: 513–28

Masserano, J. M., Weiner, N. 1981. The rapid activation of tyrosine hydroxylase by the subcutaneous injection of formaldehyde. *Life Sci.* 29: 2025–29

Masserano, J. M., Takimoto, G. S., Weiner, N. 1981. Electroconvulsive shock increases tyrosine hydroxylase activity in the brain and adrenal gland of the rat. *Science* 214: 662–65

Masserano, J. M., Weiner, N. 1983. Tyrosine hydroxylase regulation in the central nervous system. *Mol. Cell. Biochem.* 53: 129–52

McGeer, E. G., McGeer, P. L. 1966. Circadian rhythm in pineal tyrosine hydroxylase. *Science* 153: 73–74

McTigue, M., Cremins, J., Halegoua, S. 1985. Nerve growth factor and other agents mediate phosphorylation and activation of tyrosine hydroxylase. A convergence of multiple kinase activities. *J. Biol. Chem.* 260: 9047–56

Meligeni, J. A., Haycock, J. W., Bennett, W. F., Waymire, J. C. 1982. Phosphorylation and activation of tyrosine hydroxylase mediate the cAMP-induced increase in catecholamine biosynthesis in adrenal chromaffin cells. *J. Biol. Chem.* 257: 12632–40

Michener, M. L., Dawson, W. B., Creutz, C. E. 1986. Phosphorylation of a chromaffin granule-binding protein in stimulated chromaffin cells. *J. Biol. Chem.* 261: 6548–55

Miller, L. P., Pradhan, S., Mikus, M., Lovenberg, W. 1981. Analysis of fluctuations in the activity of rat pineal tyrosine hydroxylase. *Neurochem. Int.* 3: 263–67

Miller, L. P., Lovenberg, W. 1985. The use of the natural cofactor (6R)-L-erythrotetrahydrobiopterin in the analysis of nonphosphorylated and phosphorylated rat striatal tyrosine hydroxylase at pH 7.0. *Neurochem. Int.* 7: 689–97

Milner, J. D., Wurtman, R. J. 1986. Catecholamine synthesis: Physiological coupling to precursor supply. *Biochem. Pharmacol.* 35: 875–81

Mohd. Adnan, N. A., Hawthorne, J. N. 1981. Phosphatidylinositol labelling in

response to activation of muscarinic receptors in bovine adrenal medulla. *J. Neurochem.* 36: 1858–60

Morgenroth, V. H. III, Boadle-Biber, M., Roth, R. H. 1974. Tyrosine hydroxylase: Activation by nerve stimulation. *Proc. Natl. Acad. Sci. USA* 71: 4283–87

Morgenroth, V. H. III, Hegstrand, L. R., Roth, R. H., Greengard, P. 1975. Evidence for involvement of protein kinase in the activation by adenosine 3':5'-monophosphate of brain tyrosine 3-monooxygenase. *J. Biol. Chem.* 250: 1946–48

Morita, K., Teraoka, K., Oka, M. 1987. Interaction of cytoplasmic tyrosine hydroxylase with chromaffin granule. In vitro studies on association of soluble enzyme with granule membranes and alteration in enzyme activity. *J. Biol Chem.* 262: 5654–58

Mueller, R. A., Thoenen, H., Axelrod, J. 1969. Inhibition of trans-synaptically increased tyrosine hydroxylase activity by cycloheximide and actinomycin D. *Mol. Pharmacol.* 5: 463–69

Murad, F., Leitman, D. C., Bennett, B. M., Molina, C., Waldman, S. A. 1987. Regulation of guanylate cyclase by atrial natriuretic factor and the role of cyclic GMP in vasodilation. *Am. J. Med. Sci.* 294: 139–43

Musacchio, J. M., Wurzburger, R. J., D'Angelo, G. L. 1971. Different molecular forms of bovine adrenal tyrosine hydroxylase. *Mol. Pharmacol.* 7: 136–46

Nagatsu, T. 1973. *Biochemistry of Catecholamines.* Baltimore: University Park Press

Nagatsu, T., Levitt, M., Udenfriend, S. 1964. Tyrosine hydroxylase: The initial step in norepinephrine biosynthesis. *J. Biol. Chem.* 239: 2910–17

Nagatsu, I., Nagatsu, T., Mizutani, K., Umezawa, H., Matsuzaki, M., Takeuchi, T. 1971. Adrenal tyrosine hydroxylase and dopamine $\beta$-hydroxylase in spontaneously hypertensive rats. *Nature* 230: 381–82

Nagatsu, T., Mizutani, K., Nagatsu, I., Matsuura, S., Sugimoto, T. 1972. Pteridines as cofactor or inhibitor of tyrosine hydroxylase. *Biochem. Pharmacol.* 21: 1945–53

Nelson, T. J., Kaufman, S. 1987. Activation of rat caudate tyrosine hydroxylase phosphatase by tetrahydropterins. *J. Biol. Chem.* 262: 16470–75

Niggli, V., Knight, D. E., Baker, P. F., Vigny, A., Henry, J. P. 1984. Tyrosine hydroxylase in "leaky" adrenal medullary cells: Evidence for in situ phosphorylation by separate $Ca^{2+}$ and cyclic AMP-dependent systems. *J. Neurochem.* 43: 646–58

Nikodijevic, B., Nikodijevic, O., Yu, M.-Y., Pollard, H. Guroff, G. 1975. The effect of nerve growth factor on cyclic AMP levels in superior cervical ganglia of the rat. *Proc. Natl. Acad. Sci. USA* 72: 4769–71

Nishizuka, Y. 1984. Turnover of inositol phospholipids and signal transduction. *Science* 225: 1365–70

Nose, P. S., Griffith, L. C., Schulman, H. 1985. $Ca^{2+}$-dependent phosphorylation of tyrosine hydroxylase in PC12 cells. *J. Cell Biol.* 101: 1182–90

O'Brian, C. A., Lawrence, D. S., Kaiser, E. T., Weinstein, I. B. 1984. Protein kinase C phosphorylates the synthetic peptide Arg-Arg-Lys-Ala-Ser-Gly-Pro-Pro-Val in the presence of phospholipid plus either $Ca^{2+}$ or a phorbol ester tumor promoter. *Biochem. Biophys. Res. Commun.* 124: 296–302

Ohsako, S., Deguchi, T. 1983. Phosphatidic acid mimics the muscarinic action of acetylcholine in cultured bovine chromaffin cells. *FEBS Lett.* 152: 62–66

Otten, U., Thoenen, H. 1976. Mechanisms of tyrosine hydroxylase and dopamine $\beta$-hydroxylase induction in organ cultures of rat sympathetic ganglia by potassium depolarization and cholinomimetics. *Naunyn Schmiedeberg's Arch. Pharmacol.* 292: 153–59

Patterson, B. A., Volle, R. L. 1984. Muscarinic receptors and [³H]inositol incorporation in a rat sympathetic ganglion. *J. Auton. Nerv. Syst.* 10: 69–72

Pearson, R. B., Woodgett, J. R., Cohen, P., Kemp, B. E. 1985. Substrate specificity of a multifunctional calmodulin-dependent protein kinase. *J. Biol. Chem.* 260: 14471–76

Perlman, R. L. 1981. The control of tyrosine hydroxylase activity: Enzyme activation versus feedback inhibition. See Usdin et al 1981, pp. 45–53.

Pickard, M. R., Hawthorne, J. N., Hayashi, E., Yamada, S. 1977. Effects of surugatoxin and other nicotinic and muscarinic antagonists on phosphatidylinositol metabolism in active sympathetic ganglia. *Biochem. Pharmacol.* 26: 448–50

Pigeon, D., Ferrera, P., Gros, F., Thibault, J. 1987. Rat pheochromocytoma tyrosine hydroxylase is phosphorylated on serine 40 by an associated protein kinase. *J. Biol. Chem.* 262: 6155–58

Pocotte, S. L., Holz, R. W. 1986. Effects of phorbol ester on tyrosine hydroxylase phosphorylation and activation in cultured bovine adrenal chromaffin cells. *J. Biol. Chem.* 261: 1873–77

Pocotte, S. L., Holz, R. W., Ueda, T. 1986. Cholinergic receptor-mediated phosphorylation and activation of tyrosine hydroxylase in cultured bovine adrenal chromaffin cells. *J. Neurochem.* 46: 610–22

Pollock, R. J., Kapatos, G., Kaufman, S. 1981. Effect of cyclic AMP-dependent protein phosphorylating conditions on the pH-dependent activity of tyrosine hydroxylase from beef and rat striata. *J. Neurochem.* 37: 855–60

Pozzan, T., Di Virgilio, F., Vicentini, L. M., Meldolesi, J. 1986. Activation of muscarinic receptors in PC12 cells. Stimulation of $Ca^{2+}$ influx and redistribution. *Biochem. J.* 234: 547–53

Rabe, C. S., Delorme, E., Weight, F. F. 1987. Muscarine-stimulated neurotransmitter release from PC12 cells. *J. Pharmacol. Exp. Ther.* 243: 534–41

Rittenhouse, A. R., Schwarzschild, M. A., Zigmond, R. E. 1988. Both synaptic and antidromic stimulation of neurons in the rat superior cervical ganglion acutely increase tyrosine hydroxylase activity. *Neuroscience* 25: 207–15

Rittenhouse, A. R., Zigmond, R. E. 1987. Antagonists of voltage-sensitive $Ca^{++}$ channels block the acute increase in tyrosine hydroxylase activity produced by $K^+$ depolarization in both the superior cervical ganglion and the iris. *Soc. Neurosci. Abstr.* 13: 1338 (Abstr.)

Rittenhouse, A. R., Zigmond, R. E. 1989. Blockers of voltage sensitive $Ca^{++}$ channels inhibit $K^+$ stimulation of tyrosine hydroxylase activity in sympathetic neurons. In *Calcium Channels: Structure and Function. Ann. NY Acad. Sci.* In press (Abstr.)

Role, L. W. 1984. Substance P modulation of acetylcholine-induced currents in embryonic chicken sympathetic and ciliary ganglion neurons. *Proc. Natl. Acad. Sci. USA* 81: 2924–28

Roskoski, R. Jr. 1987. Regulation of tyrosine hydroxylase activity in rat PC12 cells by neuropeptides of the secretin family. *Soc. Neurosci. Abstr.* 13: 1179 (Abstr.)

Roskoski, R. Jr., Roskoski, L. M. 1987. Activation of tyrosine hydroxylase in PC12 cells by the cyclic GMP and cyclic AMP second messenger systems. *J. Neurochem.* 48: 236–42

Roskoski, R. Jr., Vulliet, P. R., Glass, D. B. 1987. Phosphorylation of tyrosine hydroxylase by cyclic GMP-dependent protein kinase. *J. Neurochem.* 48: 840–45

Roth, R. H., Hughes, J. 1972. Acceleration of protein synthesis by angiotensin-correlation with angiotensin's effect on catecholamine biosynthesis. *Biochem. Pharmacol.* 21: 3182–87

Rowland, E. A., Müller, T. H., Goldstein, M., Greene, L. A. 1987. Cell-free detection and characterization of a novel nerve growth factor-activated protein kinase in PC12 cells. *J. Biol. Chem.* 262: 7504–13

Rubio, M. C. 1977. Effects of db cAMP on tyrosine hydroxylase activity of ganglia and nerve endings. *Naunyn Schmiedeberg's Arch. Pharamacol.* 299: 69–75

Sasakawa, N., Ishii, K., Kato, R. 1986. Nicotinic receptor-mediated intracellular calcium release in cultured bovine adrenal chromaffin cells. *Neurosci. Lett.* 63: 275–79

Sasek, C. A., Landis, S., Zigmond, R. E. 1987. The distribution and possible cellular origins of vasoactive intestinal peptide (VIP) and peptide histidine isoleucine amide (PHI) in the rat superior cervical ganglion (SCG). *Soc. Neurosci. Abstr.* 13: 1337 (Abstr.)

Sasek, C. A., Zigmond, R. E. 1989. Localization of vasoactive intestinal peptide- and peptide histidine isoleucine amide-like immunoreactivities in the rat superior cervical ganglion and its nerve trunks. *J. Comp. Neurol.* In press

Schubert, D., Whitlock, C. 1977. Alteration of cellular adhesion by nerve growth factor. *Proc. Natl. Acad. Sci. USA* 74: 4055–4058

Schubert, D., LaCorbiere, M., Whitlock, C., Stallcup, W. 1978. Alterations in the surface properties of cells responsive to nerve growth factor. *Nature* 273: 718–23

Schwarzschild, M. A., Rivier, J., Vale, W. W., Ip, N. Y., Zigmond, R. E. 1985. Activation of tyrosine hydroxylase in the rat superior cervical ganglion by peptides of the glucagon-secretin family. *Soc. Neurosci. Abstr.* 11: 1203 (Abstr.)

Schwarzschild, M. A., Zigmond, R. E. 1989. Secretin and vasoactive intestinal peptide activate tyrosine hydroxylase in sympathetic nerve endings. *J. Neurosci.* 9(1): In press

Sedvall, G. C., Kopin, I. J. 1967. Acceleration of norepinephrine synthesis in the rat submaxillary gland in vivo during sympathetic nerve stimulation. *Life Sci.* 6: 45–51

Shiman, R., Akino, M., Kaufman, S. 1971. Solubilization and partial purification of tyrosine hydroxylase from bovine adrenal medulla. *J. Biol. Chem.* 246: 1330–40

Spector, S., Gordon, R., Sjoersma, A., Udenfriend, S. 1967. End-product inhibition of tyrosine hydroxylase as a possible mechanism for regulation of norepinephrine synthesis. *Mol. Pharmacol.* 3: 549–55

Stachowiak, M., Sebbane, R., Stricker, E. M., Zigmond, M. J., Kaplan, B. B. 1985. Effect of chronic cold exposure on tyrosine hydroxylase mRNA in rat adrenal gland. *Brain Res.* 359: 356–59

Stallcup, W. B. 1979. Sodium and calcium fluxes in a clonal nerve cell line. *J. Physiol. London* 286: 525–40

Stallcup, W. B., Patrick, J. 1980. Substance P enhances cholinergic receptor desensitization in a clonal nerve cell line. *Proc. Natl. Acad. Sci. USA* 77: 634–38

Steinberg, M. I., Keller, C. E. 1978. Enhanced catecholamine synthesis in isolated rat superior cervical ganglia caused by nerve stimulation: Dissociation between ganglionic transmission and catecholamine synthesis. *J. Pharmacol. Exp. Ther.* 204: 384–99

Stephens, J. K., Masserano, J. M., Vulliet, P. R., Weiner, N., Nakane, P. K. 1981. Immunocytochemical localization of tyrosine hydroxylase in rat adrenal medulla by the peroxidase labeled antibody method: Effects of enzyme activation on ultrastructural distribution of the enzyme. *Brain Res.* 209: 339–54

Stevens, P., Robinson, R. L., Van Dyke, K., Stitzel, R. 1972. Studies on the synthesis and release of adenosine triphosphate-8-$^3$H in the isolated perfused cat adrenal gland. *J. Pharmacol. Exp. Ther.* 181: 463–71

Swilem, A.-M. F., Yagisawa, H., Hawthorne, J. N. 1986. Muscarinic release of inositol trisphosphate without mobilization of calcium in bovine adrenal chromaffin cells. *J. Physiol. Paris* 81: 246–51

Tachikawa, E., Tank, A. W., Yanagihara, N., Mosimann, W., Weiner, N. 1986. Phosphorylation of tyrosine hydroxylase on at least three sites in rat pheochromocytoma PC12 cells treated with 56 mM $K^+$: Determination of the sites on tyrosine hydroxylase phosphorylated by cyclic AMP-dependent and calcium/calmodulin-dependent protein kinases. *Mol. Pharmacol.* 30: 476–85

Tachikawa, E., Tank, A. W., Weiner, D. H., Mosimann, W. F., Yanagihara, N., Weiner, N. 1987. Tyrosine hydroxylase is activated and phosphorylated on different sites in rat pheochromocytoma PC12 cells treated with phorbol ester and forskolin. *J. Neurochem.* 48: 1366–76

Takimoto, G. S., Weiner, N. 1979. Properties of tyrosine hydroxylase in rabbit portal vein: Comparison of pentobarbital anaesthesia with stunning and decapitation. *Proc. West. Pharmacol. Soc.* 22: 169–73

Takimoto, G. S., Weiner, N. 1981a. Effects of nitroprusside-induced hypotension on sympathetic nerve activity, tyrosine hydroxylase activity and [$^3$H]norepinephrine uptake in rabbit vascular tissues. *J. Pharmacol. Exp. Ther.* 219: 420–27

Takimoto, G. S., Weiner, N. 1981b. Evidence for the transsynaptic activation of tyrosine hydroxylase in rabbit portal vein mediated through a cyclic AMP-independent mechanism. *J. Pharmacol. Exp. Ther.* 219: 97–106

Tank, A. W., Meligeni, J., Weiner, N. 1984. Cyclic AMP-dependent protein kinase is not involved in the in vivo activation of tyrosine hydroxylase in the adrenal gland after decapitation. *J. Biol. Chem.* 259: 9269–76

Tank, A. W., Lewis, E. J., Chikaraishi, D. M., Weiner, N. 1985. Elevation of RNA coding for tyrosine hydroxylase in rat adrenal gland by reserpine treatment and exposure to cold. *J. Neurochem.* 45: 1030–33

Thoenen, H., Mueller, R. A., Axelrod, J. 1969. Increased tyrosine hydroxylase activity after drug-induced alteration of sympathetic transmission. *Nature* 221: 1264

Thoenen, H. 1970. Induction of tyrosine hydroxylase in peripheral and central adrenergic neurones by cold-exposure of rats. *Nature* 228; 861–62

Thoenen, H., Otten, U., Schwab, M. 1979. Orthograde and retrograde signals for the regulation of neuronal gene expression: The peripheral sympathetic nervous system as a model. In *The Neurosciences: Fourth Study Program*, 911–928, eds. F. O. Schmitt, F. G. Worden. Cambridge, MA: The MIT Press

Tischler, A. S., Perlman, R. L, Costopoulos, D., Horwitz, J. 1985. Vasoactive intestinal peptide increases tyrosine hydroxylase activity in normal and neoplastic rat chromaffin cell cultures. *Neurosci. Lett.* 61: 141–46

Tischler, A. S., Perlman, R. L., Jumblatt, J. E., Costopoulos, D., Horwitz, J. 1986. Vasoactive intestinal peptide increases tyrosine hydroxylase activity in cultures of normal and neoplastic chromaffin cells. In *Neural and Endocrine Peptides and Receptors*, ed. T. W. Moody, pp. 449–60. New York: Plenum

Togari, A., Kato, T., Nagatsu, T. 1982. Studies on the regulatory mechanism of the tyrosine hydroxylase system in adrenal slices by using high-performance liquid chromatography with electrochemical detection. *Biochem. Pharmacol.* 31: 1729–34

Traynor, A. E., Schubert, D., Allen, W. R. 1982. Alterations of lipid metabolism in response to nerve growth factor. *J. Neurochem.* 39: 1677–83

Traynor, A. E. 1984. The relationship between neurite extension and phospholipid metabolism in PC12 cells. *Brain Res.* 316: 205–10

Traynor, A. E., Schubert, D. 1984. Phospholipases elevate cyclic AMP levels and promote neurite extension in a clonal nerve cell line. *Brain Res.* 316: 197–204

Treiman, M., Weber, W., Gratzl, M. 1983. 3′,5′-Cyclic adenosine monophosphate- and $Ca^{2+}$-calmodulin-dependent endogenous protein phosphorylation activity in membranes of the bovine chromaffin secretory vesicles: Identification of two phosphorylated components as tyrosine hydroxylase and protein kinase regulatory subunit type II. *J. Neurochem.* 40: 661–69

Udenfriend, S., Zaltzman-Nirenberg, P., Nagatsu, T. 1965. Inhibitors of purified beef adrenal tyrosine hydroxylase. *Biochem. Pharmacol.* 14: 837–45

Usdin, E., Weiner, N., Youdim, M. B. H. 1981. *Function and Regulation of Monoamine Enzymes—Basic and Clinical Aspects*. London: Macmillan

Vaccaro, K. K., Liang, B. T., Perelle, B. A., Perlman, R. L. 1980. Tyrosine 3-monooxygenase regulates catecholamine synthesis in pheochromocytoma cells. *J. Biol. Chem.* 255: 6539–41

van Calker, D., Assmann, K., Greil, W. 1987. Stimulation by bradykinin, angiotensin II, and carbachol of the accumulation of inositol phosphates in PC-12 pheochromocytoma cells: Differential effects of lithium ions on inositol mono- and polyphosphates. *J. Neurochem.* 49: 1379–85

Vicentini, L. M., Ambrosini, A., Di Virgilio, F., Pozzan, T., Meldolesi, J. 1985a. Muscarinic receptor-induced phosphoinositide hydrolysis at resting cytosolic $Ca^{2+}$ concentration in PC12 cells. *J. Cell Biol.* 100: 1330–33

Vicentini, L. M., Ambrosini, A., Di Virgilio, F., Meldolesi, J., Pozzan, T. 1986. Activation of muscarinic receptors in PC12 cells. Correlation between cytosolic $Ca^{2+}$ rise and phosphoinositide hydrolysis. *Biochem. J.* 234: 555–62

Vigny, A., Henry, J.-P. 1981. Bovine adrenal tyrosine hydroxylase: Comparative study of native and proteolyzed enzyme, and their interaction with anions. *J. Neurochem.* 36: 483–89

Volle, R. L., Patterson, B. 1983. cAMP in guinea-pig superior cervical ganglia during preganglionic nerve stimulation. *Experientia* 39: 1345–46

Volle, R. L., Patterson, B. A. 1982. Regulation of cyclic AMP accumulation in a rat sympathetic ganglion: Effects of vasoactive intestinal polypeptide. *J. Neurochem.* 39: 1195–97

Volle, R. L., Quenzer, L. F., Patterson, B. A., Alkadhi, K. A., Henderson, E. G. 1981.

Cyclic guanosine 3′:5′-monophosphate accumulation and $^{45}Ca$-uptake by rat superior cervical ganglia during preganglionic stimulation. *J. Pharmacol. Exp. Ther.* 219: 338–43

Volle, R. L., Quenzer, L. F., Patterson, B. A. 1982. The regulation of cyclic nucleotides in a sympathetic ganglion. *J. Auton. Nerv. Syst.* 6: 65–72

Vulliet, P. R., Langan, T. A., Weiner, N. 1980. Tyrosine hydroxylase: A substrate of cyclic AMP-dependent protein kinase. *Proc. Natl. Acad. Sci. USA* 77: 92–96

Vulliet, P. R., Weiner, N. 1981. A schematic model for the allosteric activation of tyrosine hydroxylase. See Usdin et al 1981, pp. 15–24

Vulliet, P. R., Woodgett, J. R., Cohen, P. 1984. Phosphorylation of tyrosine hydroxylase by calmodulin-dependent multiprotein kinase. *J. Biol. Chem.* 259: 13680–83

Vulliet, P. R., Woodgett, J. R., Ferrari, S., Hardie, D. G. 1985. Characterization of the sites phosphorylated on tyrosine hydroxylase by $Ca^{2+}$ and phospholipid-dependent protein kinase, calmodulin-dependent multiprotein kinase and cyclic AMP-dependent protein kinase. *FEBS Lett.* 182: 335–39

Wang, M., Cahill, A. L., Perlman, R. L. 1986. Phorbol 12,13-dibutyrate increases tyrosine hydroxylase activity in the superior cervical ganglion of the rat. *J. Neurochem.* 46: 388–93

Waymire, J. C., Haycock, J. W., Meligeni, J. A., Browning, M. D. 1979. Activation and phosphorylation of tyrosine hydroxylase by cAMP. In *Catecholamines: Basic and Clinical Frontiers*, ed. E. Usdin, I. Kopin, J. Barchas, 1: 40–45. New York: Pergamon

Waymire, J. C., Johnston, J. P., Hummer-Lickteig, K., Lloyd, A., Vigny, A., Craviso, G. L. 1988. Phosphorylation of bovine adrenal chromaffin cell tyrosine hydroxylase: Temporal correlation of acetylcholine's effect on site phosphorylation, enzyme activation and catecholamine synthesis. *J. Biol. Chem.* 263: 12439–47

Weiner, N., Cloutier, G., Bjur, R., Pfeffer, R. I. 1972. Modification of norepinephrine synthesis in intact tissue by drugs and during short-term adrenergic nerve stimulation. *Pharmacol. Rev.* 24: 203–21

Weiner, N., Lee, G. F.-L., Barnes, E. 1976. Further studies on the enhancement of norepinephrine synthesis during nerve stimulation. In *Catecholamines and Stress*, ed. E. Usdin, R. Kvetňanský, I. Kopin, pp. 343–51. Oxford: Pergamon

Weiner, N., Lee, F.-L., Dreyer, E., Barnes, E. 1978. The activation of tyrosine hydroxylase in noradrenergic neurons during acute nerve stimulation. *Life Sci.* 22: 1197–1215

Weiner, N., Rabadjija, M. 1968a. The effect of nerve stimulation on the synthesis and metabolism of norepinephrine in the isolated guinea-pig hypogastric nerve-vas deferens preparation. *J. Pharmacol. Exp. Ther.* 160: 61–71

Weiner, N., Rabadjija, M. 1968b. The regulation of norepinephrine synthesis. Effect of puromycin on the accelerated synthesis of norepinephrine associated with nerve stimulation. *J. Pharmacol. Exp. Ther.* 164: 103–14

Weinhold, P. A., Rethy, V. B. 1969. Comparison of halogenated phenylalanine analogues as inhibitors as particle-bound and soluble tyrosine hydroxylase. *Biochem. Pharmacol.* 18: 677–80

Wilson, S. P. 1988. Vasoactive intestinal peptide elevates cyclic AMP levels and potentiates secretion in bovine adrenal chromaffin cells. *Neuropeptides* 11: 17–21

Wurtman, R. J., Larin, F., Mostafapour, S., Fernstrom, J. D. 1974. Brain catechol synthesis: Control by brain tyrosine concentration. *Science* 185: 183–84

Yamauchi, T., Fujisawa, H. 1979a. In vitro phosphorylation of bovine adrenal tyrosine hydroxylase by adenosine 3′:5′-monophosphate-dependent protein kinase. *J. Biol. Chem.* 254: 503–7

Yamauchi, T., Fujisawa, H. 1979b. Regulation of bovine adrenal tyrosine 3-monooxygenase by phosphorylation-dephosphorylation reaction, catalyzed by adenosine 3′:5′-monophosphate-dependent protein kinase and phosphoprotein phosphatase. *J. Biol. Chem.* 254: 6408–13

Yamauchi, T., Fujisawa, H. 1980. Involvement of calmodulin in the $Ca^{2+}$-dependent activation of rat brainstem tyrosine 3-monooxygenase. *Biochem. Int.* 1: 98–104

Yamauchi, T., Fujisawa, H. 1981. Tyrosine 3-monooxygenase is phosphorylated by $Ca^{2+}$-calmodulin-dependent protein kinase, followed by activation by activator protein. *Biochem. Biophys. Res. Commun.* 100: 807–13

Yamauchi, T., Nakata, H., Fujisawa, H. 1981. A new activator protein that activates tryptophan 5-monooxygenase and tyrosine 3-monooxygenase in the presence of $Ca^{2+}$-calmodulin-dependent protein kinase. Purification and characterization. *J. Biol. Chem.* 256: 5404–9

Yanagihara, N., Isosaki, M., Ohuchi, T., Oka, M. 1979. Muscarinic receptor-mediated increase in cyclic GMP level in isolated bovine adrenal medullary cells. *FEBS Lett.* 105: 296–98

Yanagihara, N., Tank, A. W., Weiner, N. 1984. Relationship between activation and phosphorylation of tyrosine hydroxylase by 56 mM K$^+$ in PC12 cells in culture. *Mol. Pharmacol.* 26: 141–47

Yanagihara, N., Tank, A. W., Langan, T. A., Weiner, N. 1986. Enhanced phosphorylation of tyrosine hydroxylase at more than one site is induced by 56 mM K$^+$ in rat pheochromocytoma PC12 cells in culture. *J. Neurochem.* 46: 562–68

Yanagihara, N., Uezono, Y., Koda, Y.,

Wada, A., Izumi, F. 1987. Activation of tyrosine hydroxylase by micromolar concentrations of calcium in digitonin-permeabilized adrenal medullary cells. *Biochem. Biophys. Res. Commun.* 146: 530–36

Zigmond, R. E. 1980. The long-term regulation of ganglionic tyrosine hydroxylase by preganglionic nerve activity. *Fed. Proc.* 39: 3003–8

Zigmond, R. E., Mackay, A. V. P. 1974. Dissociation of stimulatory and synthetic phases in the induction of tyrosine hydroxylase. *Nature* 247: 112–13

*Ann. Rev. Neurosci. 1989. 12:463–90*

# BIOCHEMISTRY OF ALTERED BRAIN PROTEINS IN ALZHEIMER'S DISEASE

*Dennis J. Selkoe*

Department of Neurology (Neuroscience), Harvard Medical School, and Center for Neurologic Diseases, Brigham and Women's Hospital, Boston, Massachusetts 02115

In contrast to other unsolved diseases of the human brain, Alzheimer's disease (AD) presents features that make it particularly amenable to studies of pathogenesis at the molecular level. The progressive dysfunction of limbic and association cortices in patients with AD is accompanied by the formation of unusual intraneuronal and extracellular proteinaceous filaments. These striking structural changes of neurons, their processes, and adjacent cerebral microvessels provide a cytopathological signature of the disease, a feature often lacking in other neurodegenerative disorders. The filamentous lesions have served as a starting point for much of the recent progress in deciphering macromolecular alterations in AD. The density and distribution of these lesions correlate to a considerable but not complete degree with several phenotypic features of AD, including degree of intellectual impairment, extent of neuronal loss in hippocampus and certain other brain regions, and decline of neurotransmitter markers, particularly cortical choline acetyltransferase activity.

Alzheimer's disease also shares certain phenotypic characteristics with a common chromosomal abnormality, Trisomy 21 (Down's syndrome). The fact that the characteristic cerebral lesions of AD accumulate in highly similar fashion in the brains of virtually all Down's syndrome subjects at an early age has provided valuable clues to the pathogenesis of AD. Another advantage, not yet fully exploited, that accrues from the study of Alzheimer's disease is the fact that similar structural changes in the cortical neuropil and microvessels occur spontaneously and progressively during

463

0147–006X/89/0301–0463$02.00

aging in normal primates and certain other mammals. This phenomenon should enable future studies of the genesis of certain AD-type brain lesions to be conducted in a more dynamic fashion than is possible in postmortem human tissue. Unfortunately, inexpensive laboratory rodents have not yet been shown to develop the characteristic lesions of Alzheimer's disease.

An additional observation about the neuropathology of AD further heightens interest in molecular approaches. Virtually all very old humans develop lesions in small numbers and restricted topographical distribution that are qualitatively similar, if not identical, to those occurring abundantly in AD brain. This fact, coupled with the age-linked development of amyloid-bearing neuritic plaques in lower mammals, suggests that knowledge about the molecular pathogenesis of AD will shed light on certain mechanisms of brain aging in general.

In the following pages, emerging information about macromolecular abnormalities in AD brain is reviewed. An attempt is made to synthesize numerous, sometimes disparate observations into a model that describes some aspects of the pathogenesis of the disease. To interpret recent protein chemical and molecular genetic findings in the proper context, one must first review the characteristics of the profound structural alterations that occur in the brain tissue of Alzheimer victims.

## Morphological Changes Accompanying Brain Aging and Alzheimer's Disease

Shrinkage and loss of certain neurons, alterations in the fine structure of innumerable neuronal processes, and abnormal cytoplasmic fibers in neuronal cell bodies represent the principal neuropathological changes that define the diagnosis of AD. Of these changes, the proliferation of spherical clusters of dystrophic neurites surrounding extracellular amyloid filaments (neuritic or senile plaques) is the most specific feature of the disease. As mentioned above, all of the lesions of AD occur to a limited extent in the brains of aged, nondemented humans, so that no alteration is entirely specific to AD. However, large numbers of neuritic plaques in the cerebral neocortex have been found in only two neuropathological disorders: AD and Trisomy 21. In contrast, neurofibrillary tangles similar if not identical to those in AD occur in a wide range of etiologically diverse human brain diseases.

SENILE PLAQUES AND AMYLOID ANGIOPATHY    The senile plaque is a complex alteration of the neuropil that occurs not in one form but as a spectrum of cellular and extracellular elements whose relative contributions to any one plaque evolve over time. Nothing is known about the time required for the genesis of the lesion that the light microscopist refers to as a "classical" senile plaque, nor is it clear which molecular changes serve as the nidus for plaque formation. Senile plaques of varying morphologies usually contain some abnormally dilated axonal terminals and dendrites that may have excess numbers of mitochondria and lysosomes (Terry et al 1964, Kidd 1964, Gonatas et al 1967). Some but not all of these dystrophic neurites contain abnormal straight or helical filaments; these closely resemble the straight filaments and paired helical filaments comprising the neurofibrillary tangles of neuronal cell bodies (see below). In many senile plaques, the density of astrocytes and microglial cells is higher than that of the surrounding neuropil.

Many, and probably all, plaques contain extracellular 6–10 nm filaments having the ultrastructural and tinctorial properties of amyloid. *Amyloid* is an ancient and unfortunate histopathological term, coined by Virchow in the last century. It can be defined as clumps of proteinaceous filaments lying in the extracellular space of a tissue and having certain tinctorial properties, e.g. binding of the dye, Congo red, and the fluorochrome, thioflavin S. As knowledge about the molecular structure of various tissue amyloids has advanced, other features have been added to this definition, notably, the presence of considerable $\beta$-pleated sheet conformation in the protein subunits comprising the filaments (Glenner 1980). Amyloid is a pathological term; filaments of this sort are not found in healthy mammalian tissues. However, it has become clear that the abnormal filaments are composed of fragments of either normal gene products (as is currently believed to be the case in AD) or isoforms of normal proteins that have discrete single mutations. Whereas an amyloid subunit is presumed not to occur normally as a filamentous polymer, it can be derived from a structurally normal cellular protein.

Given the growing interest in the molecular genetics and chemistry of amyloid proteins in AD, it is perhaps surprising that two principal questions remain unanswered: (*a*) are amyloid deposits an obligatory constituent of all neuritic plaques in AD and aged brains? and (*b*) does the amyloid lead directly or indirectly to injury of surrounding neural elements, or is it simply an inert by-product of local cellular events? Whereas nothing is known about the second question, there is some published information about the first. Miyakawa and colleagues (1982) examined numerous senile plaques in six AD brains by serial ultrastructural sectioning and discovered that every plaque examined contained amyloid filaments. Moreover, these authors reported that each plaque was associated with a capillary that, at one point along its extent, bore an amyloid deposit just outside the basement membrane of the endothelial cell. Since it has long been known that many if not all AD brains contain amyloid-bearing microvessels in cortex and meninges (see e.g. Scholz 1938, Divry 1952, Mandybur 1975, Vanley

et al 1981, Glenner et al 1981, Joachim et al 1988b), it has been hypothesized that senile plaques may originate from highly focal pericapillary
amyloid deposits. Miyakawa's data are consistent with this postulate but
do not prove it. The presence of an amyloid-laden blood vessel rather than
a typical amyloid core at the center of some senile plaques has long been
observed by the neuropathologist. The occurrence of amyloid fibrils closely
similar to those of plaques in the media of small meningeal arteries (i.e. in
the extracellular space between smooth muscle cells) (Mandybur 1975,
Vanley et al 1981, Glenner et al 1981) places additional emphasis on the
role of vascular amyloid deposition in AD. Models proposing a neuronal
origin for the amyloid in AD (Masters et al 1985a, Masters & Beyreuther
1987) require a special explanation for this extracerebral location of the
$\beta$-amyloid protein.

In aged lower mammals, particularly primates and dogs, $\beta$-amyloid-
bearing neuritic plaques are also accompanied by cortical and meningeal
$\beta$-amyloid angiopathy (Selkoe et al 1987). Indeed, in a group of nine dogs
aged 12 years or more, we found that all animals showed microvascular
amyloid deposits whereas only five of nine had demonstrable senile plaques
in cerebral cortex. These results would suggest that $\beta$-amyloid deposition
(particularly in the perivascular space) can precede or occur independently
of the formation of neuritic plaques. Certain observations in humans may
be consistent with such a sequence. In recent studies, Motte & Williams
(1987) examined the brains of Down's syndrome patients who died between
ages 25 and 59. They found that the youngest Down's syndrome brains in
their series displayed cortical amyloid deposits unassociated with surrounding neuritic or glial abnormality, as judged by conventional histological stains, immunocytochemistry, Golgi impregnation, and electron
microscopy. They concluded that focal deposition of amyloid appeared
to be the earliest detectable event in plaque morphogenesis. Since the
distribution and morphology of senile plaques and neurofibrillary tangles
in Down's syndrome is usually indistinguishable from that in AD, these
results should be applicable to the latter disorder. Indeed, several laboratories have very recently reported the occurrence of so-called "pre-
amyloid" deposits in the cortical neuropil of Down's syndrome, AD, or
normal aged brains (see e.g. Mann & Isiri 1988, Tagliavini et al 1988,
Yamaguchi et al 1988). These diffuse, noncompacted deposits are detectable with antibodies to the $\beta$-amyloid protein but are unrecognized by
throflavin S or Congo red stains. No neuritic or glial alterations have yet
been detected by light and electron microscopy within such diffuse deposits,
which can be found in cerebellar as well as cerebral cortex of AD and
Down's patients. This important observation heightens interest in multi-
focal deposition of $\beta$-amyloid protein in a nonfibrillar form as an early,

perhaps seminal, event in plaque formation. Following this deposition, only some regions of the central nervous system (e.g. cerebral cortex but not cerebellum) may be able to mount a regenerative/degenerative neuritic response to the $\beta$-amyloid and its associated molecules. It should be noted that the triplication of the $\beta$-amyloid gene locus (and all other chromosome 21 genes) in patients with Trisomy 21 but not those with AD (see below) means that the mechanism of plaque formation and the precise role of $\beta$-amyloid protein may differ to some degree in the two disorders.

The discovery of a human brain disease in which amyloid deposits composed of the $\beta$-protein are largely restricted to cerebral blood vessels may provide important clues to the role of $\beta$-amyloid in AD. van Duinen and colleagues (1987) studied members of four Dutch kindreds with an autosomal dominant disease manifested as a brain-restricted micro-vascular amyloidosis that results in multiple, often fatal hemorrhages beginning in middle age. Immunocytochemistry with antibodies to a synthetic $\beta$-amyloid peptide showed that in addition to marked amyloidosis of meningeal and cortical arterioles, diffuse plaque-like lesions were detected in cerebral cortex in some cases. These plaques lacked the com-pacted cores that are labeled by thioflavin S or Congo red but showed silver-positive surrounding neuritic alteration. No neurofibrillary tangles were found. As the Dutch patients were not intellectually impaired at the time of their fatal intracerebral hemorrhages, one may speculate whether more cortical amyloid plaques and eventual cortical dysfunction would have ensued with longer survival. Subsequent isolation and sequencing of the vascular amyloid subunit protein in this Dutch disease showed that it had the same sequence through amino acid 39 as the amyloid in AD and Down's syndrome brains (Prelli et al 1988). The reasons for the preferential vascular localization of the $\beta$-protein deposits in the former disorder are of great interest vis-à-vis AD. Diffuse, noncompacted amyloid plaques that closely resemble those in the Dutch disease are often noted in AD and aged human cortex as well as in that of lower mammals.

PAIRED HELICAL FILAMENTS AND RELATED FIBERS OF THE NEUROFIBRILLARY TANGLE    The term, neurofibrillary tangle, is an appelation of the light microscopist. It refers to a non–membrane-bound mass of thickened fibers detectable by several different histochemical stains in the perinuclear cyto-plasm of selected neurons. At the ultrastructural level, neurofibrillary tangles in AD and numerous other human brain diseases often contain fibers that appear to be tightly adherent pairs of helically wound $\sim 10$ nm filaments, referred to as paired helical filaments (PHF) (Kidd 1963, 1964, Terry 1963, Terry et al 1964, Wisniewski et al 1976). The maximum width of a PHF is $\sim 20$–24 nm, including the fuzzy, granular outer coat that characterizes PHF in situ. Its helical periodicity is approximately 160 nm (half-period $\simeq$ 80 nm). PHF having these particular dimensions have so far only been described in human neurons, although helically-wound filaments resembling PHF but with shorter periodicities have occasionally been found in the neurons of lower mammals (monkey, dog, rat) and even invertebrates (whip spider).

Neurofibrillary tangles may also contain straight, unpaired filaments varying in diameter from 10 to 20 nm, often around 15 nm (Shibayama & Kitoh 1978, Yagashita et al 1981, Metuzals et al 1981, Rudelli et al 1982, Selkoe 1984, Yoshimura 1984). These straight filaments may coexist with PHF in the same tangle. In some cases of clinically and light microscopically typical AD, we have observed that all neurofibrillary tangles examined by electron microscopy are composed of arrays of straight $\sim 15$ nm filaments. Such tangles are thus similar to those characteristic of the distinct degenerative disorder, progressive supranuclear palsy (PSP). That the straight and helical filaments occurring in AD, PSP, Pick's disease, and several other human neurofibrillary disorders are antigenically related has been shown by light and electron microscopic immunolabeling studies with antibodies to isolated AD tangles or to microtubule-associated or neurofilament proteins (see e.g. Probst et al 1983a, Dickson et al 1985, Rasool & Selkoe 1985, Perry et al 1985, 1986, Joachim et al 1987a). It should be noted that dystrophic neurites in AD cortex, whether found within or outside of senile plaques, can also contain straight or helical filaments, or both. These and other observations indicate that the fibrous cytopathology of neurons in AD is more complex and heterogeneous than generally stated. Molecular models for the reorganization of the neuronal cytoskeleton in AD need to take this structural diversity into account.

Although few protein chemical analyses of isolated PHF have been undertaken in diseases other than AD, available ultrastructural and immunochemical data suggest that the neurofibrillary tangles in such etiologically diverse disorders as AD, dementia pugilistica, PSP, Guam Parkinson-dementia (PD) complex, and Hallervorden-Spatz variants (Eidelberg et al 1987) are highly similar. If so, then tangle formation may be viewed as a somewhat nonspecific response of certain neurons to a variety of cellular insults. Such a conclusion in no way lessens the importance of fully understanding the molecular alterations that precede fibrous degeneration of neurons; if anything, it raises interest in this common process. The abundance of neurofibrillary tangles found in AD but not in normal aged neocortex indicates that the occurrence of these inclusions is a sign of significant cellular dysfunction. PHF-bearing neuronal perikarya appear rather intact when examined ultrastructurally in well-preserved, biopsied

AD cortex. Thus, accumulation of PHF and related straight filaments is likely to precede by some time the terminal degeneration of the neuron. Conversely, it cannot be assumed that all neurons that die prematurely in AD brain have passed through a "neurofibrillary stage." The sparing of some human neuronal populations from tangle-formation and the absence of similar structural changes in the neurons of aged or diseased lower mammals indicate that PHF formation, while not unique to AD, is nonetheless a process with specific cellular and molecular determinants.

ALTERATIONS OF AXONS AND DENDRITES IN ALZHEIMER'S DISEASE CORTEX
The notion of the neuropathology of AD as "plaque and tangle disease" is a misleading simplification. A more widespread disruption of the normal cortical architecture is actually observed. This point, while already apparent from earlier histological and ultrastructural analyses (see e.g. Scheibel et al 1976, Buell & Coleman 1979) has been dramatically emphasized by the use of antibody labeling techniques. For example, the monoclonal antibody, Alz 50, sensitively detects a marked alteration of neuritic anatomy in certain areas of AD cerebral cortex (Wolozin et al 1986, Wolozin & Davies 1987). Previously described polyclonal or monoclonal antibodies produced against PHF (see e.g. Ihara et al 1983, Yen et al 1985) or against the major antigenic constituent of PHF, tau (see e.g. Nukina & Ihara 1986, Kosik et al 1986, Grundke-Iqbal et al 1986, Wood et al 1986), also demonstrated this extensive change of fine cortical neurites. Thus, PHF-related cytoskeletal changes are not restricted to tangles and plaque neurites. The cerebral cortex of aged humans without AD shows nothing approaching the degree of neuritic pathology observed in most AD cases. Therefore, the proliferation of such altered cortical neurites, sometimes referred to as curly or kinky fibers (see e.g. Kosik et al 1986, Braak et al 1986, Kowall & Kosik 1987), may actually be the most specific histological change found in AD versus normal cortex. Given the profound degree of neuropil alteration observed in the association cortex of most AD brains, this lesion may well represent a closer morphological substrate of cortical dysfunction in AD patients than do plaques and tangles themselves.

It is not known whether dystrophic neurites that are widely distributed in the cortical neuropil arise from similar pathogenetic mechanisms as those that are concentrated within neuritic plaques. The remarkably spherical form of most senile plaques and the inclusion of neurites of diverse transmitter content within one plaque (see e.g. Struble et al 1982, 1987, Walker et al 1988) suggest the presence of a point-source of trophic and/or toxic substance(s) that affect surrounding neurites and glia in a centrifugal fashion (Selkoe 1986). How the more randomly distributed curly fibers arise vis-à-vis this hypothetical model of neuritic plaque formation is unclear. Ihara (1988) has provided morphological evidence that the curly fibers represent an abortive regenerative response of local cortical neurons to the disease cascade, i.e. a form of dendritic sprouting. Circumstantial evidence that an enhanced trophic response could be occurring in AD cerebral cortex has been put forward (see e.g. Probst et al 1983b, Geddes et al 1985, Uchida et al 1988). Detailed studies of local neuronal sprouting and its role in cortical pathology are likely to emerge as an important theme in AD research in the near future.

## Biochemistry of Paired Helical Filaments

It is now well-recognized that two related problems have hampered the rigorous biochemical characterization of PHF: their insolubility in sodium dodecyl sulfate (SDS), reducing agents, and many other denaturants (Selkoe et al 1982, 1983), and the consequent difficulty in purifying PHF to homogeneity. A method for the full purification of PHF from human brain tissue has not been achieved, although methods that take advantage of their insolubility to produce considerable enrichment of the filaments are available (Ihara et al 1983, Iqbal et al 1984, Selkoe & Abraham 1986, Wischik et al 1988a,b). Investigators have, therefore, been forced to rely on immunochemical approaches to delineate to some extent the protein(s) contributing to PHF.

Early immunocytochemical studies (Grundke-Iqbal et al 1979, Yen et al 1981) showed that proteins present in crude microtubule-enriched fractions prepared by repetitive cycles of temperature-dependent tubulin assembly shared antigens with PHF. Although it became apparent that tubulin itself was not the responsible antigen, the specific proteins in the microtubule fractions that crossreacted with PHF were not identified until recently. In the meantime, several groups showed that polyclonal or monoclonal antibodies to normal mammalian neurofilament proteins, particularly the middle and high molecular weight members of the neurofilament triplet, recognized neurofibrillary tangles, both in situ and, to a lesser extent, following extraction in SDS (Gambetti et al 1980, Ihara et al 1981, Anderton et al 1982, Rasool et al 1984, Rasool & Selkoe 1984, Perry et al 1985, Sternberger et al 1985, Cork et al 1986). In contrast, polyclonal (Ihara et al 1983, Brion et al 1985a) and monoclonal (Yen et al 1985) antibodies produced directly to isolated PHF recognized neither normal neuronal structures in brain sections nor neurofilaments and other proteins in immunoblots of human brain homogenates. This seeming paradox was resolved by the discovery in several laboratories that brain fractions highly enriched in the microtubule-associated phosphoprotein, tau, react strongly with all polyclonal and some monoclonal antibodies to PHF (Brion et al 1985b, Nukina & Ihara 1986, Kosik et al 1986, Grundke-Iqbal et al 1986, Wood et al 1986). Conversely, polyclonal and monoclonal antibodies raised to normal tau purified from brain tissue of various mammals specifically label neurofibrillary tangles and altered neurites in AD and aged human brain. It appears that the level of tau in homogenates of adult postmortem brain is low and its antigenic preservation in normal neuronal structures of routinely fixed autopsied brain is poor. Thus, the reactivity of PHF antibodies with tau was only demonstrable when enriched fractions of tau from fresh rodent or fetal human brain were immunoblotted. It has been further clarified that the monoclonal or polyclonal antibodies to neurofilament proteins that had previously been shown to label PHF crossreact with phosphorylated epitopes on tau (Nukina et al 1987, Ksiesak-Reding et al 1987). Following dephosphorylation of neurofilaments or tau, these particular neurofilament antibodies recognize neither protein, nor do they recognize dephosphorylated neurofibrillary tangles. Other neurofilament monoclonal antibodies that do not recognize tangles show no reaction with tau (Nukina et al 1987). Currently available evidence thus suggests that the neurofilament immunoreactivity sometimes observed in neurofibrillary tangles is largely due to the sharing of phosphorylated epitopes between tau and the higher molecular weight neurofilament polypeptides. Future evidence implicating neurofilament components in PHF will require the labeling of the isolated fibers by antibodies that recognize only neurofilaments and not tau.

How does the growing body of data implicating tau as the major antigenic constituent of PHF relate to information about other PHF-related proteins? Davies and colleagues have produced and characterized a monoclonal antibody, Alz 50, that stains tangles and many dystrophic neurites in AD brain and that recognizes an antigen of $\sim 65$–$70$ kDa at much higher levels in AD than normal cortex (Wolozin et al 1986, Wolozin & Davies 1987). The immunocytochemical pattern obtained with Alz 50 in AD cortex is similar but not identical to that obtained with various polyclonal or monoclonal antibodies to tau and to PHF. Also, Alz 50 recognizes tangles occurring in disorders other than AD (e.g. Guam PD complex, PSP, Kuf's diseases, normal aging), as well as the filamentous Pick bodies characteristic of neurons in Pick's disease (Love et al 1988). This cross-reaction is a property of Alz 50 shared with tau and PHF antibodies. Recently, it has been shown that Alz 50 recognizes both the phosphorylated and dephosphorylated forms of normal rodent and human tau in a fashion highly similar to various tau antibodies (Nukina et al 1988). Moreover, Alz 50 precipitates proteins in the 50–70 kDa MW range from AD but not normal cortex; these proteins are in turn labeled by various tau antibodies, including Alz 50 (Nukina et al 1988). This new information suggests, but does not yet prove, that the antigens recognized by Alz 50 at much higher levels in AD than normal brain are posttranslationally altered forms of tau protein. The nature of any post-translational modifications of tau (e.g. altered phosphorylation) that may occur during neurofibrillary degeneration in AD and other tangle-forming diseases remains unknown.

Two recent lines of evidence have further strengthened the argument that tau is an integral component of the PHF. Kosik and colleagues (1988) have created a tau cDNA sub-library by subjecting cDNAs encoding human tau to nuclease digestion and packaging in a λgt11 expression vector. This sub-library, containing cloned cDNA fragments from throughout the tau sequence, has been screened with various antibodies known specifically to label neurofibrillary tangles, and the resultant immunoreactive tau clones have been sequenced. Such epitope mapping reveals that various PHF-reactive antibodies recognize epitopes of tau distributed throughout most of the molecule (Kosik et al 1988), including the three highly homologous, repeated sequences found in the carboxyl portion that are believed to represent the microtubule-binding domain of tau (Lee et al 1988). When the tau cDNA sub-library was probed with Alz 50, no reactive clones were identified; this finding suggests that Alz 50 is directed against a conformational or posttranslational modification of tau that is not produced by the bacterial expression system. Subsequent epitope mapping on cyanogen bromide-generated fragments of purified bovine tau protein has identified Alz 50–reactive peptides contained in the carboxyl-half of the molecule (Kosik et al 1988).

The second line of evidence implicating tau has come from the first direct protein chemical analyses of PHF fractions (Ihara 1987, Wischik et al 1988a,b). One laboratory has used endopeptidase digestion of partially purified PHF fractions with careful comparison of the released peptides to those obtained from identically treated control brain fractions (Ihara 1987, Mori et al 1987). At least three peptide fragments containing amino acid sequences of human tau were identified only in the PHF-derived digests (Ihara 1987). These tau fragments were again localized to the region around the three homologous, repeating sequences. Simultaneously performed analyses of other PHF-derived proteolytic fragments provided the first evidence for the tight association of the small protein, ubiquitin, with the PHF (Mori et al 1987). Subsequent immunocytochemical studies confirmed the presence of ubiquitin epitopes in neurofibrillary tangles (Perry et al 1987). Another laboratory has approached the problem of PHF composition by treating highly enriched fractions of the fibers with pronase, a step that removes the tau-immunoreactive, fuzzy outer coat of PHF (Wischik et al 1988b). The subsequent pronase-resistant "core" retains the characteristic helical morphology of the PHF (Wischik et al 1988a,b, Selkoe et al 1983). This PHF core was further subjected to sonication in concentrated formic acid, releasing two principal proteins of 9.5 and 12 kDa (Wischik et al 1988b). Sequencing of these peptides showed that they were derived from the conserved region of tau containing the three repeating segments (Wischik et al 1988b, Goedert et al 1988). Thus, these studies localized tau protein to both the pronase-susceptible outer coat and a part of the pronase-resistant core of the PHF. However, the analyses showed that a considerable portion of the core (perhaps more than 75%) remained insoluble following all of the extraction procedures employed; the identity of the protein comprising this major portion remains unknown (Wischik et al 1988b).

The structural studies of Ihara and colleagues and Wischik and co-workers just summarized are entirely consistent with earlier immuno-chemical analyses of PHF. Extensive extraction of isolated PHF in SDS removes their fuzzy, granular coating (Selkoe et al 1983, Yen & Kress 1983) and reduces but does not abolish their reactivity with some tau antibodies (Kosik et al 1986, Ihara et al 1986). Indeed, highly SDS-extracted, "nonfuzzy" PHF, when used as an immunogen, give rise to antisera that recognize normal tau proteins (Ihara et al 1983). Moreover, Nukina and colleagues previously showed that proteinase K digestion of isolated PHF reproducibly releases three peptide fragments ($\sim$ 10, 26, and 36 kDa) that are reactive with both tau- and PHF-antibodies (Nukina & Ihara 1985, 1986, Ihara et al 1986). Extracellular tangles ("ghost tangles"), which appear to represent the insoluble, proteinase-resistant residue of once-intraneuronal tangles, are still immunoreactive with certain tau or PHF antibodies. Taken together, the various available data about PHF composition suggest that the tau molecule is present within the filaments. The question of whether other proteins may also be incorporated awaits further analyses. The precise functions of tau in normal neurons are unclear, although in vitro experiments indicate that tau promotes the polymerization of tubulin into microtubules and stabilizes these fibrous organelles. Whether this function of tau or others are perturbed in AD neurons undergoing neurofibrillary degeneration will be difficult to estab-lish. The role of ubiquitin in the neurofibrillary tangle and the protein to which it is conjugated also remain unknown. It is already clear that this apparent involvement of a ubiquitin-mediated process is not specific to AD neuronal degeneration; the pathological fibers in the intra-neuronal Lewy bodies of Parkinson's disease (which share epitopes with neurofilaments but not tau) also contain ubiquitin (Love et al 1988, Kuzuhara et al 1988).

The findings indicating the involvement of the tau protein in the PHF structure itself may also be relevant to so-called "pre-tangle" abnormalities of pyramidal neurons in AD hippocampus, neocortex, and certain subcortical nuclei. It is known that some neurons in these pathologically involved regions display a diffuse, granular cytoplasmic reaction with antibodies to tau (see e.g. Joachim et al 1987b), including the monoclonal antibody, Alz 50 (Wolozin et al 1986, Wolozin & Davies 1987). Such neurons can be shown not to contain thioflavin- or Congo red–positive neurofibrillary tangles. Since neuronal cell bodies with this type of tau immunoreaction are not found in normal human or rodent brain, it has been assumed that they may represent a stage of cytoskeletal derangement that predates formation of insoluble filaments (i.e. PHF and straight filaments). The sequence of molecular alterations of tau that could be responsible for such progressive changes will be difficult to establish in the absence of an experimental animal that undergoes a similar process.

## How Does Cerebral Amyloidosis Play a Role in Cortical Degeneration in Alzheimer's Disease?

For many decades, investigators have debated the role of amyloid deposition in the pathogenesis of neuritic plaques. It has often been argued that amyloid filaments may accumulate secondarily within clusters of degenerating neurites and glial cells. Indeed, one recent and widely discussed hypothesis states that the low molecular weight subunit peptide of AD-type amyloid filaments [designated the $\beta$-amyloid protein ($\beta$AP) or A4] accumulates first inside neuronal cell bodies and plaque neurites, where it forms the core protein of the PHF. Later, it is argued, the amyloid subunit is exported from the neuron into the extracellular space, where it first forms amyloid plaque cores and eventually reaches local blood vessels to form microvascular deposits (Masters et al 1985a, Masters & Beyreuther 1987).

In contrast, other studies, some of which were summarized above, provide evidence that extracellular deposition of the $\beta$AP in the neuropil and/or in blood vessels occurs very early in the course of cortical degeneration. It has been noted, for example, that young adults with Down's syndrome who die prematurely of other causes show amyloid-containing plaques in cortex before the development of neurofibrillary tangles (Wisniewski et al 1988). Similarly, a significant minority of patients with otherwise typical AD show abundant amyloid-bearing neuritic plaques but essentially no tangle-bearing neuronal perikarya in the cerebral neocortex (Terry et al 1987). It should also be recalled that aged humans, lower primates, and certain other mammals normally develop $\beta$-amyloid bearing plaques and blood vessels in neocortex in the absence of any neurofibrillary tangles.

Such considerations, taken in the context of the remarkable similarity of AD neuropathology to that in Trisomy 21, have resulted in the hypothesis that in AD patients, alterations of an unknown gene on chromosome 21 may lead, through a complex cascade, to altered biosynthetic or proteolytic processing of the $\beta$AP precursor, causing an acceleration of its normal age-related deposition in brain microvessels and neuropil, particularly in limbic and association cortices (see e.g. Glenner 1983, Selkoe 1986). This progressive cerebral amyloidosis may exert, via mechanisms that are direct or indirect (e.g. via $\beta$AP-associated macromolecules), a trophic or toxic effect on local neurites, leading gradually to axonal and dendritic dystrophy and perikaryal cytopathology (including neurofibrillary tangles) in a retrograde or transsynaptic fashion (Selkoe 1986). The usefulness of this model in elucidating the pathogenesis of AD may be better judged after reviewing the observations to date about the $\beta$AP precursor and its gene.

## Biochemistry of the $\beta$-Amyloid Protein and Its Precursor

The first protein chemical studies of amyloid in AD and Down's syndrome brains were conducted by Glenner & Wong (1984a,b). Prior to this work, numerous immunocytochemical analyses had produced conflicting findings regarding the cross-reactivity of senile plaque amyloid with a variety of normal proteins, including immunoglobulins, prealbumin, complement, and neurofilament polypeptides (see, e.g. Ishii et al 1975, Powers et al 1981, Shirahama et al 1982, Eikelenboom & Stam 1982). Glenner and Wong used a long-established procedure for isolating systemic amyloid fibrils to obtain enriched fractions of meningovascular amyloid from AD and Down's syndrome brains. Upon solubilization in guanidine hydrochloride, a reagent conventionally used to dissolve systemic amyloids, they isolated and sequenced the 24 amino-terminal residues of a $\sim$4 kDa, hitherto unknown protein, which they designated the $\beta$-protein ($\beta$AP). Shortly thereafter, the development of methods for the purification of amyloid plaque cores from AD cortex revealed that the major protein isolated from these structures was also a 4–5 kDa peptide having a composition essentially identical to the $\beta$AP from meningovascular amyloid (Masters et al 1985b, Selkoe et al 1986). In contrast to the solubility of the latter amyloid in guanidine hydrochloride, amyloid plaque cores remained insoluble in this reagent. Only use of the stronger chaotropic salt, guanidine thiocyanate, or concentrated formic acid could quantitatively solubilize amyloid cores (Masters et al 1985b, Selkoe et al 1986). A further distinction between these two loci of amyloid deposition was observed upon attempted sequencing of the core-derived $\beta$AP. Whereas one group obtained heterogenous sequences from the plaque core $\beta$AP having four different aminotermini (Masters et al 1985b), three other laboratories were unable to derive any sequence from amyloid cores (Selkoe et al 1986, Gorevic et al

1986, Bobin et al 1987). Loading of up to two nanomoles of purified core βAP on a gas-phase microsequencer produced no detectable sequence, thus suggesting that the βAP found in the compacted amyloid of plaque cores had a buried or blocked amino-terminus (Selkoe et al 1986). This result, coupled with the insolubility of plaque amyloid but not meningo-vascular amyloid in guanidine hydrochloride, suggests that the βAP found in compacted cortical deposits may be further modified from that found in the vessel wall. The report by Masters et al (1985a) that PHF-rich fractions of AD cortex yield sequence data consistent with the sequence of Glenner & Wong (1984a,b) but having even greater N-terminal hetero-geneity (six or seven termini) must be viewed with caution in view of the widely accepted evidence that all PHF fractions are contaminated to a greater or lesser extent with plaque and microvascular amyloid filaments (Masters et al 1985a, Selkoe et al 1986, Gorevic et al 1986, Wischik et al 1988b). Since vascular-derived βAP can be readily sequenced from the amino-terminus (Glenner & Wong 1984a,b, Joachim et al 1988a), this material may be the source of βAP sequences obtained from PHF-enriched cortical fractions. The use of mixed fibril fractions from postmortem AD cortex makes rigorous sequence interpretation hazardous.

Another area of current ambiguity in protein chemical analyses of βAP is the precise identity of the carboxyl terminus of the amyloidogenic frag-ment. Kang and colleagues (1987), when describing the nucleotide sequence of the first full-length cDNA for the β-amyloid protein precursor (βAPP), also reported the remainder of the protein sequence of cortically derived βAP. Previously, Wong et al (1985) had extended the amino-terminal sequence of meningovascular βAP to position 30. Kang and colleagues reported that cortically derived βAP ends at either alanine (position 42) or threonine (position 43). However, Joachim et al (1988a) used cyanogen bromide cleavage of βAP purified from meningovascular amyloid to derive a C-terminal peptide going from methionine (position 35) to valine (position 40); no sequence was obtained beyond residue 40. Prelli and co-workers (1988) sequenced the βAP isolated from the vascular amyloid deposits of the Dutch form of congophilic angiopathy; they obtained the valine at position 39 as the C-terminal residue. Miller and colleagues (1988) found that the meningovascular βAP sequence ends at residue 40 (valine); that the core-derived βAP could not be sequenced due to a blocked amino-terminus, as reported earlier (Selkoe et al 1986); and that its amino acid composition was consistent with a 39- or, at most, 40-residue chain. Further analyses should reveal the extent of carboxyl-terminal heterogeneity among different βAP isolates and any consistent differences in the chain lengths of the meningovascular- and core-derived peptides. This issue is not trivial, since determination of the proteolytic cleavage sites within the precursor that allow release of the amyloidogenic fragment is necessary to understand the local processing of the $\beta$APP in AD cortex.

Following the isolation and amino-terminal sequencing of the $\beta$AP, the production of antibodies to the native, tissue-derived $\beta$AP (Selkoe et al 1986, Joachim et al 1988a) or to synthetic peptides prepared from its sequence (Masters et al 1985a, Wong et al 1985, Allsop et al 1986, Selkoe et al 1986, Gorevic et al 1986) reproducibly demonstrated the cross-reaction of microvascular and plaque $\beta$-amyloid deposits. However, virtually all studies to date have found no reaction of $\beta$AP antibodies with neurofibrillary tangles, senile plaque neurites, or other PHF-containing neuronal structures, even after the use of antigen-exposing techniques such as pretreatment of brain sections with formic acid (Kitamoto et al 1987). Only one study has reported that a subset of tangle-bearing neurons in neocortex show reaction with an antiserum to a synthetic $\beta$1-11 peptide (Masters et al 1985a, Masters & Beyreuther 1987). This finding was not confirmed by Wong et al (1985) using an antiserum to a synthetic $\beta$1-10 peptide. Neither the extracellular ("ghost") tangles found in some regions of AD cortex nor tangles isolated from brain by extensive detergent extraction have shown reaction with $\beta$AP antibodies. Such results are consistent with recent ultrastructural and protein chemical analyses of purified PHF, which have provided no evidence that the $\beta$AP is the core protein of these filaments (Wischik et al 1988a,b).

The separate occurrence of either neurofibrillary tangles or $\beta$-amyloid deposits in certain human and animal pathological conditions is also not supportive of a common protein origin of PHF and amyloid filaments. Guiroy and colleagues (1987) have reported electrophoretic, amino acid compositional and protein sequencing data on partially purified PHF fractions from the brains of patients with Guam PD Complex. Their results led them to conclude that the PHF comprising neurofibrillary tangles in this disorder, which are ultrastructurally indistinguishable from PHF in AD, are composed of the A4 (or $\beta$AP) protein. However, the actual data they published are open to alternative explanations. Guiroy et al chose to study PHF from this disease because Guam PD brains have been classically described as having only neurofibrillary tangles and no senile plaques or amyloid deposits. However, the electron micrographs shown by Guiroy et al demonstrate that their fractions contained nonpaired, 5–15 nm filaments of unknown origin that were not PHF; these resemble amyloid filaments. The most recent neuropatholoical examinations of Guam PD brains apparently reveal variable numbers of amyloid-bearing senile plaques in at least some cases (D. Perl, personal communication). Whether such lesions represent the plaques occurring "normally" with age in these patients or are actually part of the specific Guam PD histopathology is not yet clear. In any event, small numbers of formative amyloid plaques could provide the source of the 5–15 nm filaments observed by Guiroy et al in their PHF fractions. The amino acid composition of their Guam PD PHF fractions shows several residues (including five moles percent proline) that are clearly incompatible with the known A4 ($\beta$AP) composition; the latter peptide contains no proline. Their sequencing of seven peptide chains simultaneously in the Guam PD PHF fractions to provide the A4 sequence is difficult to interpret, given the complexity of unambiguously sequencing mixtures of just two or three peptides. These various considerations, taken together with the ultrastructural and immunochemical distinctions between PHF and amyloid filaments, indicate that the protein chemical identity of these two fibers stated by Masters et al (1985a) and Guiroy et al (1987) is not yet proven.

The issue just discussed is distinct from the question of the cellular origin of the particular $\beta$APP molecules that are cleaved into amyloidogenic fragments in AD and aged brain tissue. Since it is clear from in situ hybridization analyses that many if not all neurons contain abundant $\beta$APP message (see below), a neuronal origin for the extracellular $\beta$-amyloid deposits in AD is by no means excluded.

Knowledge of the partial protein sequence for the $\beta$-amyloid subunit enabled several further research strategies to go forward. The construction of correct or mutant synthetic peptides to the first 28 amino acids of $\beta$AP resulted in the in vitro production of fibrils that display some of the properties of native AD amyloid filaments (Castano et al 1986, Kirschner et al 1987), including extensive cross-$\beta$-pleated sheet protein conformation (Kirschner et al 1986, 1987). Whereas a correct $\beta$1-28 synthetic peptide readily comes out of solution to form abundant $\sim$8 nm filaments that even aggregate into masses resembling portions of cores, a homologous peptide with an alanine-for-lysine substitution at position 16 produces much smaller numbers of mutant filaments that have a different ultra-structure and an an altered x-ray diffraction pattern (Kirschner et al 1987). Further studies of this sort that examine full-length synthetic peptides (39–42 residues long) may produce filaments more closely resembling native $\beta$AP. Such approaches should allow the establishment of the specific molecular sequences and buffer conditions required for fibrillogenesis, thus providing a manipulable in vitro model for AD-type amyloid filament formation.

Another critical area of progress that emerged from the initial protein chemical analysis of the $\beta$-amyloid subunit was the cloning of cDNAs encoding its normal precursor protein. The ramifications of this work are reviewed in the following section.

## Molecular Biology of the β-Amyloid Protein Precursor Molecule

Using oligonucleotides to regions of the β-protein sequence, four laboratories initially reported the isolation, sequencing, and chromosomal assignment of cDNAs from human brain encoding the βAPP (Kang et al 1987, Goldgaber et al 1987, Robakis et al 1987, Tanzi et al 1987b). The isolation of an apparently full-length cDNA (Kang et al 1987) demonstrated that the βAP represented a ~40-residue fragment within a 695-amino acid polypeptide whose predicted sequence showed features of a glycosylated cell-surface protein with a single membrane-spanning domain. The βAP was found toward the carboxyl-terminal region of βAPP, with a third of its sequence inside the putative membrane-spanning region and the remainder situated just outside the membrane. The localization of the gene encoding βAPP to the long arm of human chromosome 21, in the region 21q11.2–21q21, immediately provided one mechanism for the invariant development of βAP deposits in Down's syndrome brain on the basis of increased gene dosage. One laboratory shortly reported that a duplication of a subsegment of chromosome 21 containing the βAPP gene occurred in three sporadic cases of AD (Delabar et al 1987), but subsequent analyses of a much larger number of patients failed to find evidence of βAPP microduplication in AD (Podlisny et al 1987, St. George-Hyslop et al 1987b, Tanzi et al 1987a). Recent studies have further localized the βAPP gene to the region 21q21.15–21q21.2 (Korenberg et al 1988), centromeric to the so-called obligatory Down's syndrome region responsible for that phenotype in patients who have translocations rather than complete trisomy.

Southern analyses of genomic DNA from numerous mammalian species demonstrated that the gene for βAPP was highly conserved in evolution, and Northern analyses showed that its mRNAs were widely distributed in tissues and cell lines (see e.g. Tanzi et al 1987b, Goldgaber et al 1987). Sequencing of cloned cDNAs for the βAPP of the mouse (Yamada et al 1987) and rat (Shivers et al 1988) reveals remarkable homology (97%) to the human homologue. Importantly, these two rodents rarely if ever develop βAP-bearing plaques and vessels in the brain with age, a finding that suggests that the limited sequence differences from the human could turn out to be relevant to the age-related processing of the protein. As might be expected from the strong evolutionary conservation and wide tissue distribution of this common gene product, alternative transcripts for the βAPP have been identified. A 751-amino-acid mRNA encodes a polypeptide containing an additional 56-residue sequence inserted at amino acid 289 of the initially cloned 695-residue precursor (Ponte et al

1988, Tanzi et al 1988, Kitaguchi et al 1988). This insert shows ~50% homology to certain members of the Kunitz family of serine protease inhibitors. Actual protease inhibitory activity of the polypeptide expressed by this mRNA has been demonstrated in transfected mammalian cells (Kitaguchi et al 1988). Another alternative transcript (770 amino acids long) containing a 75-residue insert at the same locus that has 19 additional amino acids beyond the Kunitz-like domain has also been documented (Kitaguchi et al 1988).

The knowledge that there are at least three different mRNAs transcribed from the $\beta$APP gene has led to a reanalysis of initial in situ hybridization studies. Distinct oligonucleotide probes that, under ideal conditions, should specifically hybridize to just one of the three transcripts have been designed. Early data from several laboratories using such probes suggest that the insert-containing mRNAs occur widely in neural and nonneural tissues, whereas the original 695-residue transcript may have a more restricted distribution primarily to nervous tissue. Well controlled, quantitative in situ hybridization and Northern analyses on numerous animal and human brains will be necessary before clear, reproducible patterns of mRNA levels and distribution will emerge. The issue of the over- or underexpesssion of a particular transcript in AD brain versus non-AD diseased brain versus normal aged brain may be difficult to resolve, given the variability in neuronal loss, lesion density, and agonal and postmortem artifacts in autopsied human tissues. Initial reports of in situ hybridization studies indicate that many neurons (and perhaps other cells) in the brain contain $\beta$APP message (Bahmanyar et al 1987, Cohen et al 1988, Higgins et al 1988). Although the presence of a protease inhibitor-like domain within the $\beta$APP is intriguing, its relevance to regional alterations of precursor processing in AD brain is unclear as yet.

Separate evidence for the direct presence of a serine protease inhibitor in AD amyloid deposits has been obtained by using antisera to isolated cortical amyloid filaments to screen cDNA libraries and isolate clones coding for $\alpha_1$-antichymotrypsin (Abraham et al 1988). This approach was undertaken because certain antisera raised to intact AD amyloid filaments reacted weakly with purified or synthetic $\beta$AP and yet strongly labeled vascular and plaque amyloid deposits in AD brain. To identify the responsible antigen, several nonneural tissues were examined immunocytochemically, and liver showed strong and specific reactivity. cDNAs for $\alpha_1$-antichymotrypsin were subsequently isolated from a liver expression library with the amyloid antiserum. Confirmation of the presence of the inhibitor in the amyloid of AD and aged normal brain was provided by epitope mapping and by light and electron microscopic immunocytochemistry with several antibodies to $\alpha_1$-antichymotrypsin (Abraham et al

1988). This molecule is also tightly associated with the βAP deposits occurring in aged primates (C. R. Abraham et al, unpublished data). The role of this protease inhibitor in β-amyloidosis and the question of whether the $\alpha_1$-antichymotrypsin found in AD lesions originates from the serum, where it is abundantly present, or is synthesized locally in brain, where its mRNA has been demonstrated (Abraham et al 1988), are the subjects of further investigation.

## Molecular Genetics of Familial Alzheimer's Disease

It has long been known that certain families show a vertical pattern of inheritance of AD in multiple generations, consistent with an autosomal dominant mutation. While new protein chemical and molecular biological information about amyloid in AD was emerging, genetic linkage analyses focusing on various regions of chromosome 21 identified an anonymous DNA marker (referred to as D21S1/S11) that showed allelic segregation with the AD phenotype in four pedigrees with early onset, autosomal dominant AD (St. George-Hyslop et al 1987a). Within months of this important observation, further analysis demonstrated that there were recombination (cross-over) events between this familial AD (FAD) marker and the βAPP locus on chromosome 21, thus indicating that the latter gene was distinct from the site of the FAD defect (van Broeckhoven et al 1987, Tanzi et al 1987c). Further linkage studies of numerous FAD kindreds are being pursued in order to localize more precisely the region of chromosome 21 containing the FAD mutation. It is clear from recent work that the putative FAD locus is centromeric to the βAPP gene by greater than eight million bases and may actually be very close to the centromere itself. Nonetheless, the βAPP gene and the FAD defect (in the particular pedigrees analyzed to date) occur within a segment of chromosome 21 representing less than 1% of the human genome. This fact, plus the occurrence of β-amyloid lesions typical of AD in all cases of Trisomy 21, raises the possibility that the FAD gene may have some specific interaction with the βAPP locus. Alternatively, the two genes could be localized to chromosome 21 merely by coincidence, and the role of the FAD mutation in ultimately producing accelerated βAP deposition could involve a long and complex cascade of intermediate steps.

The further localization of the FAD genetic defect and the elucidation of its structural abnormality in the disease may prove exceedingly difficult, given the apparent dearth of new, multigenerational kindreds able to provide informative DNA samples. The study of smaller families presents significant theoretical limitations in linkage analysis, depending on how far from the actual mutation a particular marker resides. The occurrence of nonallelic heterogeneity among families with FAD can only be convincingly demonstrated by finding definite linkage to a chromosomal locus distinct from that now known in additional informative pedigrees.

## Further Approaches to Amyloidogenesis in Alzheimer's Disease

If the increase in βAP-containing filaments in AD brain compared to aged normal brain is not explained by a mutation or duplication of the βAPP gene, then transcriptional, translational, or posttranslational changes are implicated. One may postulate, for example, that in normal humans and some lower mammals, there is a gradual alteration during aging of the degradative processing of the βAPP that is specific for certain brain regions, resulting in cleaved fragments that are amyloidogenic. The familial AD genetic defect(s) may somehow accelerate this age-related failure of processing. Based on this hypothesis, we have begun to study the degradation of βAPP in amyloid-prone brain regions versus brain regions and nonneural tissues that remain amyloid free. As an initial step, we have recently identified the native precursor protein in normal tissues and cultured cells of several species (Selkoe et al 1988).

Antibodies to the predicted carboxyl-terminus of βAPP specifically detected a group of membrane associated proteins of 110–135 kDa in brain and nonneural human tissues. These proteins are relatively uniformly distributed throughout human brain, abundant in fetal brain, and detected in numerous nonneural tissues known to contain βAPP mRNAs. Similar-sized proteins are found in rat, cat, cow, and monkey brains and in cultured human HL-60 and HeLa cells. The precise banding patterns in the 110–135 kDa range are heterogeneous among various human and animal tissues and cell lines. Confirmation that the immunodetected tissue proteins are forms of the βAPP was obtained when mouse and human cultured cells transfected with a full length βAPP cDNA showed selectively augmented expression of 110–135 kDa proteins and specific immunocytochemical staining. Unexpectedly, the C-terminal antibodies used in this work label amyloid-containing senile plaques in AD brain (Selkoe et al 1988).

These results indicate that the highly conserved βAP precursor molecule occurs in mammalian tissues as a heterogeneous group of membrane-associated proteins of ~120 kDa. The properties of the 110–135 kDa protein we found are consistent with current postulates about βAPP based on its cDNA-derived primary sequence and reported mRNA distribution. Part of the heterogeneity we observed may be due to the alternative mRNAs for βAPP as well as variation in the glycosylated side chains of its extracellular domain. The latter mechanism could account for the larger size of the immunodetected tissue forms than that of the predicted

polypeptide backbone ($\sim$ 80,000 kDa). The presence of the non-amyloido-
genic carboxyl-terminus with senile plaques suggests that proteolytic
processing of the normal precursor into insoluble filaments occurs
locally in cortical regions that develop $\beta$-amyloid deposits with age and
in AD.

The detection of these forms of the native $\beta$APP in neural and nonneural
tissues can now be related to evidence recently provided by Schubert and
colleagues (1988) suggesting that the $\beta$APP is a heparan sulfate proteoglycan core protein. These investigators reported that a heparan sulfate
proteoglycan core protein shed into the culture medium of rat PC-12 cells
showed sequence homology and antigenic cross-reaction with the $\beta$APP.
Their observation, if confirmed by further studies, indicates that the normal precursor molecule is, indeed, localized to the cell surface, where it
may function in cell-cell interaction rather than as a receptor protein. How
this putative functional role relates, if at all, to the altered processing of
the molecule that occurs in selected regions of aged and AD brains is not
clear. The fact that some cell-surface heparan sulfates have been shown to
bind cellular growth factors (see e.g. Roberts et al 1988) could turn out to
have relevance for the altered neuritic growth that may be occurring in the
vicinity of amyloid plaques in AD cortex.

## Toward a Molecular Model of Alzheimer's Disease Pathogenesis

The extensive experimental observations reviewed in this chapter have
a common initiating thread: the striking structural abnormalities that
accompany progressive cerebral dysfunction in Alzheimer patients.
Despite doubts expressed over the years about the utility of analyzing the
fibrous deposits in AD brain, it is increasingly apparent that molecular
studies of these lesions can provide important insights into the pathogenesis of cortical degeneration in AD. Perhaps the most striking example
of this fact to date is the localization of the genetic defect causing at least
some cases of familial AD to chromosome 21. This chromosome was only
chosen as the site of attack for initial linkage studies because all patients
with Trisomy 21 develop fibrous lesions indistinguishable from those in
AD brain. Perhaps it will not turn out to be coincidental that the gene
encoding the precursor of the amyloid deposits invariably found in the
disease is also localized to this chromosome.

Several lines of inquiry are likely to occupy the attention of investigators
in the months and years ahead. Certainly, attempts to use various strategies
for linkage analysis to locate the gene or genes responsible for autosomal
dominant cases of AD will go forward. Equally important are the efforts

to look at the disease process from the other end. One must start with the most specific and constant phenotypic abnormalities of the disease and study their genesis in a retrograde fashion, hoping ultimately to arrive at molecular alterations that may serve as candidates for the disease mutation or for an event rather proximal to it. The study of the genesis of neuritic plaques, with their involvement of axons and dendrites deriving from multiple neurotransmitter classes, is likely to remain a fruitful area of research. Understanding the nature of the regionally altered degradation of $\beta$APP in aged brain tissue and the role of "$\beta$-amyloid-associated" molecules such as $\alpha_1$-antichymotrypsin may lead to the identification of particular brain proteases that are responsible for this local processing. For example, we must determine why an arteriole bearing multiple amyloid deposits along its course in the cortical gray matter loses its perivascular amyloid when it penetrates into subcortical white matter. Of particular interest in the near future will be the gathering of further evidence that there are heightened and inappropriate trophic responses of cortical neurons in AD brain, leading perhaps to abnormal sprouting that contributes to the dysfunction of the complex circuitry of limbic and association cortices.

It is increasingly likely that the profound neuritic and perikaryal abnormalities observed in AD brain are the substrate, at least in part, of progressive cortical dysfunction in the patient. The growing opportunities for understanding the molecular events underlying these structural changes bode well for elucidating the pathogenesis, and ultimately the prevention, of this common and tragic disorder.

*Literature Cited*

Abraham, C. R., Selkoe, D. J., Potter, H. 1988. Immunochemical identification of the serine protease inhibitor $\alpha_1$-antichymotrypsin in the brain amyloid deposits of Alzheimer's disease. *Cell* 52: 487–501

Allsop, D., Landon, M., Kidd, M., Lowe, J. S., Reynolds, G. P., et al. 1986. Monoclonal antibodies raised against a subsequence of senile plaque core protein react with plaque cores, plaque periphery and cerebrovascular amyloid in Alzheimer's disease. *Neurosci. Lett.* 68: 252–56

Anderton, B. H., Breinburg, D., Downes, M. J., Green, P. J., Tomlinson, B. E., et al. 1982. Monoclonal antibodies show that neurofibrillary tangles and neurofilaments share antigenic determinants. *Nature* 298: 84–86

Bahmanyar, S., Higgins, G. A., Goldgaber, D., Lewis, D. A., Morrison, J. H., et al. 1987. Localization of amyloid $\beta$ protein messenger RNA in brains from patients with Alzheimer's disease. *Science* 237: 77–80

Bobin, S. A., Currie, J. R., Chen, M.-C., Iqbal, K., Miller, D. L., et al. 1987. Comparisons between the structures of cerebral vascular amyloid (CVA) and senile plaque core amyloid (SPCA) peptides found in Alzheimer's disease. *Fed. Proc. Am. Soc. Exp. Biol.* 46: 2137

Braak, H., Braak, E., Grundke-Iqbal, I., Iqbal, K. 1986. Occurrence of neuropil threads in the senile human brain and in Alzheimer's disease: A third location of paired helical filaments outside of neurofibrillary tangles and neuritic plaques. *Neurosci. Lett.* 65: 351–55

Brion, J. P., Couck, A. M., Passareiro, E.,

Flament-Durand, J. 1985a. Neurofibrillary tangles of Alzheimer's disease: an immunohistochemical study. *J. Submicrosc. Cytol.* 17: 89–96

Brion, J. P., van den Bosch de Aguilar, P., Flament-Durand, J. 1985b. In *Advances in Applied Neurological Science: Senile Dementia of the Alzheimer Type*, ed. J. Traber, W. H. Gispen, pp. 164–74. Berlin: Springer

Buell, S. J., Coleman, P. D. 1979. Dendritic growth in the aged human brain and failure of growth in senile dementia. *Science* 206: 854–56

Castano, E. M., Ghiso, J., Prelli, F., Gorevic, P. D., Migheli, A., et al. 1986. In vitro formation of amyloid fibrils from two synthetic peptides of different lengths homologous to Alzheimer's disease $\beta$-protein. *Biochem. Biophys. Res. Commun.* 141: 782–89

Cohen, M. L., Golde, T. E., Usiak, M. F., Younkin, L. H., Younkin, S. G. 1988. *In situ* hybridization of nucleus basalis neurons shows increased $\beta$-amyloid mRNA in Alzheimer disease. *Proc. Natl. Acad. Sci. USA* 85: 1227–31

Cork, L. C., Sternberger, N. H., Sternberger, L. A., Casanova, M. F., Struble, R. G., Price, D. L. 1986. Phosphorylated neurofilament antigens in neurofibrillary tangles in Alzheimer's disease. *J. Neuropathol. Exp. Neurol.* 45: 56–64

Delabar, J.-M., Goldgaber, D., Lamour, Y., Nicole, A., Huret, J.-L., et al. 1987. $\beta$-amyloid gene duplication in Alzheimer's disease and karyotypically normal Down Syndrome. *Science* 235: 1390–92

Dickson, D. W., Kress, Y., Crowe, A., Yen, S.-H. 1985. Monoclonal antibodies to Alzheimer's neurofibrillary tangles: 2. Demonstration of a common antigenic determinant between ANT and neurofibrillary degeneration in progressive supranuclear palsy. *Am. J. Pathol.* 120: 292–303

Divry, P. 1952. La Pathochimie generale et cellulaire des processus seniles et preseniles. *Proc. 1st Int. Congr. Neuropathol. (Rome)* 2: 313

Eidelberg, D., Sotrel, A., Joachim, C., Selkoe, D., Forman, A., et al. 1987. Adult onset Hallervorden-Spatz disease with neurofibrillary pathology. *Brain* 110: 993–1013

Eikelenboom, P., Stam, F. C. 1982. Immunoglobulins and complement factors in senile plaques: an immunoperoxidase study. *Acta Neuropathol.* 57: 239–42

Gambetti, P., Velasco, M. E., Dahl, D., Bignami, A., Roessmann, U., et al. 1980. Alzheimer neurofibrillary tangles: an immunohistochemical study. In *Aging of the Brain*, ed. L. Amaducci, A. N. Davison, P. Antuono. New York: Raven

Geddes, J. W., Monaghan, D. T., Cotman, C. W., Lott, I. T., Kim, R. C., et al. 1985. Plasticity of hippocampal circuitry in Alzheimer's disease. *Science* 230: 1179–81

Glenner, G. G. 1980. Amyloid deposits and amyloidosis. The $\beta$-fibrilloses. *N. Engl. J. Med.* 302: 1283–92

Glenner, G. G. 1983. Alzheimer's disease: multiple cerebral amyloidosis. See Katzman 1983, pp. 137–44

Glenner, G. G., Henry, J. H., Fujihara, S. 1981. Congophilic angiopathy in the pathogenesis of Alzheimer's degeneration. *Ann. Pathol.* 1: 120–29

Glenner, G. G., Wong, C. W. 1984a. Alzheimer's disease: initial report of the purification and characterization of a novel cerebrovascular amyloid protein. *Biochem. Biophys. Res. Commun.* 120: 885–90

Glenner, G. G., Wong, C. W. 1984b. Alzheimer's disease and Down's syndrome: sharing of a unique cerebrovascular amyloid fibril protein. *Biochem. Biophys. Res. Commun.* 122: 1131–35

Goedert, M., Wischik, C. M., Crowther, R. A., Walker, J. E., Klug, A. 1988. Cloning and sequencing of the cDNA encoding a core protein of the paired helical filament of Alzheimer disease: identification as the microtubule-associated protein tau. *Proc. Natl. Acad. Sci. USA* 85: 4051–55

Goldgaber, D., Lerman, M. I., McBride, O. W., Saffiotti, U., Gajdusek, D. C. 1987. Characterization and chromosomal localization of a cDNA encoding brain amyloid of Alzheimer's disease. *Science* 235: 877–80

Gonatas, N. K., Anderson, W., Evangelista, I. 1967. The contribution of altered synapses in the senile plaque: An electron microscopic study in Alzheimer's dementia. *J. Neuropathol. Exp. Neurol.* 26: 25–39

Gorevic, P., Goni, F., Pons-Estel, B., Alvarez, F., Peress, R., et al. 1986. Isolation and partial characterization of neurofibrillary tangles and amyloid plaque core in Alzheimer's disease: Immunohistological studies. *J. Neuropathol. Exp. Neurol.* 45: 647–64

Grundke-Iqbal, I., Johnson, A. B., Terry, R. D., Wisniewski, H. M., Iqbal, K. 1979. Alzheimer neurofibrillary tangles: antiserum and immunohistochemical staining. *Ann. Neurol.* 6: 532–37

Grundke-Iqbal, I., Iqbal, K., Quinlan, M., Tung, Y.-C., Zaidi, M. S., et al. 1986. Microtubule-associated protein tau: a component of Alzheimer paired helical filaments. *J. Biol. Chem.* 261: 6084–89

# 486 SELKOE

Guiroy, D. C., Miyazaki, M., Multhaup, G., Fischer, P., Garruto, R. M., et al. 1987. Amyloid of neurofibrillary tangles of Guamanian parkinsonism-dementia and Alzheimer disease share identical amino acid sequence. *Proc. Natl. Acad. Sci. USA* 84: 2073–77

Higgins, G. A., Lewis, D. A., Bahmanyar, S., Goldgaber, D., Gajdusek, D. C., et al. 1988. Differential regulation of amyloid-β protein mRNA expression within hippocampal neuronal subpopulations in Alzheimer disease. *Proc. Natl. Acad. Sci. USA* 85: 1297–1301

Ihara, Y., Nukina, N., Sugita, H., Toyokura, V. 1981. Staining of Alzheimer's neurofibrillary tangles with antiserum against 200K component of neurofilament. *Proc. Jpn. Acad.* 57: 152–56

Ihara, Y., Abraham, C., Selkoe, D. J. 1983. Antibodies to paired helical filaments in Alzheimer's disease do not recognize normal brain proteins. *Nature* 304: 727–30

Ihara, Y., Nukina, N., Miura, R., Ogawarra, M. 1986. Phosphorylated tau protein is integrated into paired helical filaments in Alzheimer's disease. *J. Biochem.* 99: 1807–10

Ihara, Y. 1987. Tau is a component of paired helical filaments (PHF). *J. Neurochem.* 48(Suppl.): S14B

Ihara, Y. 1988. Massive somatodendritic sprouting of cortical neurons in Alzheimer's disease. *Brain Res.* 459: 138–44

Iqbal, K., Zaidi, T., Thompson, C. H., Merz, P. A., Wisniewski, H. M. 1984. Alzheimer paired helical filaments: bulk isolation, solubility and protein composition. *Acta Neuropathol.* 62: 167–77

Ishii, T., Haga, S., Shimizu, F. 1975. Identification of components of immunoglobulins in senile plaques by means of fluorescent antibody technique. *Acta Neuropathol.* 32: 157–62

Joachim, C. L., Morris, J. H., Kosik, K. S., Selkoe, D. J. 1987a. Tau antisera recognize neurofibrillary tangles in a range of neurodegenerative disorders. *Ann. Neurol.* 22: 514–20

Joachim, C. L., Morris, J. H., Selkoe, D. J., Kosik, K. S. 1987b. Tau epitopes are incorporated into a range of lesions in Alzheimer's disease. *J. Neuropathol. Exp. Neurol.* 46: 611–22

Joachim, C. L., Duffy, L. K., Morris, J. H., Selkoe, D. J. 1988a. Protein chemical and immunocytochemical studies of meningovascular β-amyloid protein in Alzheimer's disease and normal aging. *Brain Res.* In press

Joachim, C. L., Morris, J. H., Selkoe, D. J. 1988b. Clinically diagnosed Alzheimer's disease: autopsy results in 150 cases. *Ann.*

*Neurol.* 24: 50–56

Kang, J., Lemaire, H.-G., Unterbeck, A., Salbaum, J. M., Masters, C. L., et al. 1987. The precursor of Alzheimer's disease amyloid A4 protein resembles a cell-surface receptor. *Nature* 325: 733–36

Katzman, R. 1983. *Banbury Report 15: Biological Aspects of Alzheimer's Disease,* ed. R. Katzman. New York: Cold Spring Harbor Lab.

Kidd, M. 1963. Paired helical filaments in electron microscopy of Alzheimer's disease. *Nature* 197: 192–93

Kidd, M. 1964. Alzheimer's disease—an electron microscopical study. *Brain* 87: 307–20

Kirschner, D. A., Abraham, C., Selkoe, D. J. 1986. X-ray diffraction from intraneuronal paired helical filaments and extraneuronal amyloid fibers in Alzheimer disease indicates cross-β conformation. *Proc. Natl. Acad. Sci. USA* 83: 503–7

Kirschner, D. A., Inouye, H., Duffy, L. K., Sinclair, A., Lind, M., et al. 1987. Synthetic peptide homologous to β protein from Alzheimer disease forms amyloid-like fibrils in vitro. *Proc. Natl. Acad. Sci. USA* 84: 6953–57

Kitaguchi, N., Takahashi, Y., Tokushima, Y., Shiojiri, S., Ito, H. 1988. Novel precursor of Alzheimer's disease amyloid protein shows protease inhibitory activity. *Nature* 331: 530–32

Kitamoto, T., Ogomori, K., Tateishi, J., Prusiner, S. B. 1987. Formic acid pretreatment enhances immunostaining of cerebral and systemic amyloids. *Lab. Invest.* 57: 230–36

Korenberg, J. R., West, R., Pulst, S.-M. 1988. The Alzheimer protein precursor gene maps to chromosome 21 sub-bands q21.15–q21.2. *Neurology* 38(Suppl. 1): 265

Kosik, K. S., Joachim, C. L., Selkoe, D. J. 1986. The microtubule-associated protein, tau, is a major antigenic component of paired helical filaments in Alzheimer's disease. *Proc. Natl. Acad. Sci. USA* 83: 4044–48

Kosik, K. S., Orecchio, L. D., Binder, L., Trojanowski, J. Q., Lee, V. M.-Y., Lee, G. 1988. Epitopes that span the tau molecule are shared with paired helical filaments. *Neuron.* In press

Kowall, N. W., Kosik, K. S. 1987. The cytoskeletal pathology of Alzheimer's disease is characterized by aberrant tau distribution. *Ann. Neurol.* 22: 639–43

Ksiesak-Reding, H., Dickson, D. W., Davis, P., Yen, S.-H. 1987. Recognition of tau epitopes by anti-neurofilament antibodies that bind to Alzheimer neurofibrillary tangles. *Proc. Natl. Acad. Sci. USA* 84: 3410–14

Kuzukara, S., Mori, H., Izumiyama, N., Yoshimura, M., Ihara, Y. 1988. Lewy bodies are ubiquitinated. A light- and electron-microscopic immunocytochemical study. *Acta Neuropathol.* 75: 345–53

Lee, G., Cowan, N., Kirschner, M. 1988. The primary structure and heterogeneity of tau protein from mouse brain. *Science* 239: 285–88

Love, S., Saitoh, T., Quijada, S., Cole, G. M., Terry, R. D. 1988. Alz-50, ubiquitin and tau immunoreactivity of neurofibrillary tangles, Pick bodies and Lewy bodies. *J. Neuropath. Exp. Neurol.* 47: 393–405

Mandybur, T. I. 1975. The incidence of cerebral amyloid angiopathy in Alzheimer's disease. *Neurology* 25: 120–26

Mann, D., Isiri, M. M. 1988. The site of the earliest lesions of Alzheimer's disease. *N. Engl. J. Med.* 318: 789

Masters, C. L., Beyreuther, K. 1987. Neuronal origin of cerebral amyloidogenic proteins: Their role in Alzheimer's disease and unconventional virus disease of the nervous system. In *Selective Neuronal Death, CIBA Found. Symp.* 126: 49–64. Chichester, UK: Wiley

Masters, C. L., Multhaup, G., Simms, G., Pottigiesser, J., Martins, R. N., et al. 1985a. Neuronal origin of a cerebral amyloid: neurofibrillary tangles of Alzheimer's disease contain the same protein as the amyloid of plaque cores and blood vessel. *EMBO J.* 4: 2757–63

Masters, C. L., Simms, G., Weinman, N. A., Multhaup, G., McDonald, B. L., et al. 1985b. Amyloid plaque protein in Alzheimer's disease and Down's syndrome. *Proc. Natl. Acad. Sci. USA* 82: 4245–49

Metuzals, J., Montpetit, V., Clapin, D. F. 1981. Organization of the neurofilamentous network. *Cell Tissue Res.* 214: 455–82

Miller, D. L., Currie, J. R., Iqbal, K., Potemska, A., Styles, J. 1988. Relationships among the cerebral amyloid peptides and their precursors. *Alzh. Dis. Assoc. Disorders* 2: 243 (Abstr.)

Miyakawa, T., Shimoji, A., Kuramoto, R., Higuchi, Y. 1982. The relationship between senile plaques and cerebral blood vessels in Alzheimer's disease and senile dementia: morphological mechanism of senile plaque production. *Virchows Arch. B* 40: 121–29

Mori, H., Kondo, J., Ihara, Y. 1987. Ubiquitin is a component of paired helical filaments in Alzheimer's disease. *Science* 235: 1641–44

Motte, J., Williams, R. S. 1987. Age related changes in the density and morphology

of plaques and neurofibrillary tangles in Down syndrome. *Mass. ADRC Sci. Session* (Abstr.), Vol. 2, Nov. 6

Nukina, N., Ihara, Y. 1985. Proteolytic fragments of Alzheimer's paired helical filaments. *J. Biochem.* 98: 1715–18

Nukina, N., Ihara, Y. 1986. One of the antigenic determinants of paired helical filaments is related to tau protein. *J. Biochem.* 99: 1541–44

Nukina, N., Kosik, K. S., Selkoe, D. J. 1987. Recognition of Alzheimer paired helical filaments by monoclonal neurofilament antibodies is due to crossreaction with tau protein. *Proc. Natl. Acad. Sci. USA* 84: 3415–19

Nukina, N., Kosik, K. S., Selkoe, D. J. 1988. The monoclonal antibody, Alz 50, recognizes tau protein in Alzheimer's disease brain. *Neurosci. Lett.* 87: 240–46

Perry, G., Friedman, R., Shaw, G., Chau, V. 1987. Ubiquitin is detected in neurofibrillary tangles and senile plaque neurites of Alzheimer disease brains. *Proc. Natl. Acad. Sci. USA* 84: 3033–36

Perry, G., Rizzuto, N., Autilio-Gambetti, L., Gambetti, P. 1985. Alzheimer's paired helical filaments contain cytoskeletal components. *Proc. Natl. Acad. Sci. USA* 82: 3916–20

Perry, G., Selkoe, D. J., Block, B. R., Stewart, D., Autilio-Gambetti, L., et al. 1986. Electron microscopic localization of Alzheimer neurofibrillary components recognized by an antiserum to paired helical filaments. *J. Neuropathol. Exp. Neurol.* 45: 161–68

Podlisny, M. B., Lee, G., Selkoe, D. J. 1987. Gene dosage of the amyloid-$\beta$ precursor protein in Alzheimer's disease. *Science* 238: 669–71

Ponte, P., Gonzalez-DeWhitt, P., Schilling, J., Miller, J., Hsu, D., et al. 1988. A new A4-amyloid mRNA contains a domain homologous to serine proteinase inhibitors. *Nature* 331: 525–27

Powers, J. M., Schlaepfer, W. W., Willingham, M. C., Hall, B. J. 1981. An immunoperoxidase study of senile cerebral amyloidosis with pathogenic considerations. *J. Neuropathol. Exp. Neurol.* 40: 592–612

Prelli, F., Castano, E. M., van Duinen, S. G., Bots, G. Th. A. M., Luyendij, W., et al. 1988. Different processing of Alzheimer's $\beta$-protein precursor in the vessel wall of patients with hereditary cerebral hemorrhage with amyloidosis-dutch type. *Biochem. Biophys. Res. Commun.* 151: 1150–55

Probst, A., Anderton, B. H., Ulrich, J., Kohler, R., Kahn, J., et al. 1983a. Pick's disease: an immunocytochemical study of neuronal changes. *Acta Neuropathol.* 60: 175–82

Probst, A., Basler, V., Bron, B., Ulrich, J. 1983b. Neuritic plaques in senile dementia of Alzheimer type: a Golgi analysis in the hippocampal region. *Brain Res.* 268: 249–54

Rasool, C. G., Abraham, C., Anderton, B. H., Haugh, M. C., Kahn, J., et al. 1984. Alzheimer's disease: immunoreactivity of neurofibrillary tangles with anti-neurofilament and anti-paired helical filament antibodies. *Brain Res.* 310: 249–60

Rasool, C. G., Selkoe, D. J. 1984. Alzheimer's disease: exposure of neurofilament immunoreactivity in SDS-insoluble paired helical filaments. *Brain Res.* 322: 194–98

Rasool, C. G., Selkoe, D. J. 1985. Sharing of specific antigens by degenerating neurons in Pick's disease and Alzheimer's disease. *N. Engl. J. Med.* 312: 700–5

Robakis, N. K., Ramakrishna, N., Wolfe, G., Wisniewski, H. M. 1987. Molecular cloning and characterization of a cDNA encoding the cerebrovascular and the neuritic plaque amyloid peptides. *Proc. Natl. Acad. Sci. USA* 84: 4190–94

Roberts, R., Gallagher, J., Spooner, E., Allen, T. D., Bloomfield, F., et al. 1988. Heparan-sulphate bound growth factors: a mechanism for stromal cell mediated hematopoiesis. *Nature* 332: 376–78

Rudelli, R., Strom, J. O., Welch, P. T., Ambler, M. W. 1982. Posttraumatic premature Alzheimer's disease: Neuropathologic findings and pathogenic considerations. *Arch. Neurol.* 39: 570–76

St. George-Hyslop, P. H., Tanzi, R. E., Polinsky, R. J., Haines, J. L., Nee, L., et al. 1987a. The genetic defect causing familial Alzheimer's disease maps on chromosome 21. *Science* 235: 885–89

St. George-Hyslop, P. H., Tanzi, R. E., Polinsky, R. J., Neve, R. L., Pollen, D., et al. 1987b. Absence of duplication of chromosome 21 genes in familial and sporadic Alzheimer's disease. *Science* 238: 664–66

Scheibel, M. E., Lindsay, R. D., Tomiyasu, U., Scheibel, A. B. 1976. Progressive dendritic changes in the aging human limbic system. *Exp. Neurol.* 53: 420–30

Scholz, W. 1938. Studien zur pathologie der hirngefasse II: die drusige entartung der hirnarterien und capillaren. *Z. Gesamte Neurol. Psychiatr.* 162: 694

Schubert, D., Schroeder, R., LaCorbiere, M., Saitoh, T., Cole, G. 1988. Amyloid β protein precursor is possibly a heparan sulfate proteoglycan core protein. *Science* 241: 223–26

Selkoe, D. J. 1984. The fibrous cyto-

pathology of degenerating neurons in Alzheimer's disease: an ultrastructural reappraisal. *Soc. Neurosci. Abstr.* 10: 813

Selkoe, D. J. 1986. Altered structural proteins in plaques and tangles: What do they tell us about the biology of Alzheimer's disease? *Neurobiol. Aging* 7: 425–32

Selkoe, D. J., Abraham, C. R. 1986. Isolation of paired helical filaments and amyloid fibers from human brain. *Methods Enzymol.* 134: 388–404

Selkoe, D. J., Abraham, C. R., Podlisny, M. B., Duffy, L. K. 1986. Isolation of low-molecular-weight proteins from amyloid plaque fibers in Alzheimer's disease. *J. Neurochem.* 146: 1820–34

Selkoe, D. J., Bell, D. S., Podlisny, M. B., Price, D. L., Cork, L. C. 1987. Conservation of brain amyloid proteins in aged mammals and humans with Alzheimer's disease. *Science* 235: 873–77

Selkoe, D. J., Ihara, Y., Salazar, F. J. 1982. Alzheimer's disease: insolubility of partially purified helical filaments in sodium dodecyl sulfate and urea. *Science* 215: 1243–45

Selkoe, D. J., Ihara, Y., Abraham, C., Rasool, C. G., McCluskey, A. H. 1983. Biochemical and immunocytochemical studies of Alzheimer paired helical filaments. See Katzman 1983, pp. 125–34

Selkoe, D. J., Podlisny, M. B., Joachim, C. L., Vickers, E. A., Lee, G., et al. 1988. β-Amyloid precursor of Alzheimer disease occurs as 110- to 135-kilodalton membrane-associated proteins in neural and nonneural tissues. *Proc. Natl. Acad. Sci. USA* 85: 7341–45

Shibayama, H., Kitoh, J. 1978. Electron microscopic structure of Alzheimer's neurofibrillary changes in a case of atypical senile dementia. *Acta Neuropathol.* 41: 229–34

Shirahama, T., Skinner, M., Westermark, P., Rubinow, A., Cohen, A. S., et al. 1982. Senile cerebral amyloid: prealbumin as a common constituent in the neuritic plaque, in the neurofibrillary tangles, and in the microangiopathic lesion. *Am. J. Pathol.* 197: 41–50

Shivers, B. D., Hilbich, C., Multhaup, G., Salbaum, M., Beyreuther, K., et al. 1988. Alzheimer's disease amyloidogenic glycoprotein: expression pattern in rat brain suggests a role in cell contact. *EMBO J.* 7: 1365–70

Sternberger, N. H., Sternberg, L. A., Ulrich, J. 1985. Aberrant neurofilament phosphorylation in Alzheimer's disease. *Proc. Natl. Acad. Sci. USA* 82: 4274–76

Struble, R. G., Cork, L. C., Whitehouse, P. J., Price, D. L. 1982. Cholinergic innervation in neuritic plaques. *Science* 216: 413–15

Struble, R. G., Powers, R. E., Casanova, M. F., Kitt, C. A., Brown, E. C., et al. 1987. Neuropeptidergic systems in plaques of Alzheimer's disease. *J. Neuropathol. Exp. Neurol.* 46: 567–84

Tagliavini, F., Giaccone, G., Frangione, B., Bugiani, O. 1988. Cortical pre-amyloid deposition in Alzheimer patients and normals. *J. Neuropathol. Exp. Neurol.* 47: 332

Tanzi, R. E., Bird, E. D., Latt, S. A., Neve, R. L. 1987a. The amyloid β protein gene is not duplicated in brains from patients with Alzheimer's disease. *Science* 238: 666–69

Tanzi, R. E., Gusella, J. F., Watkins, P. C., Bruns, G. A. P., St. George-Hyslop, P. H., et al. 1987b. Amyloid β-protein gene: c-DNA, mRNA distribution and genetic linkage near the Alzheimer locus. *Science* 235: 880–84

Tanzi, R. E., St. George-Hyslop, P. H., Haines, J. L., Polinsky, R. J., Nee, L., et al. 1987c. The genetic defect in familial Alzheimer's disease is not tightly linked to the amyloid β-protein gene. *Nature* 329: 156–57

Tanzi, R. E., McClatchey, A. I., Lamperti, E. D., Villa-Komaroff, L., Gusella, J. F., et al. 1988. Protease inhibitor domain encoded by an amyloid protein precursor mRNA associated with Alzheimer's disease. *Nature* 331: 528–30

Terry, R. D. 1963. The fine structure of neurofibrilary tangles in Alzheimer's disease. *J. Neuropathol. Exp. Neurol.* 22: 629–41

Terry, R. D., Gonatas, N. F., Weiss, M. 1964. Ultrastructural studies in Alzheimer's presenile dementia. *Am. J. Pathol.* 44: 269–97

Terry, R. D., Hansen, L. A., DeTeresa, R., Davies, P., Tobias, H., et al. 1987. Senile dementia of the Alzheimer type without neocortical neurofibrillary tangles. *J. Neuropathol. Exp. Neurol.* 46: 262–68

Uchida, Y., Ihara, Y., Tomonaga, M. 1988. Alzheimer's disease brain extract stimulates the survival of cerebral cortical neurons from neonatal rats. *Biochem. Biophys. Res. Commun.* 150: 1263–67

van Broeckhoven, C., Genthe, A. M., Vandenberghe, A., Horsthemke, B., Backhovens, H., et al. 1987. Failure of familial Alzheimer's disease to segregate with the A4-amyloid gene in several European families. *Nature* 329: 153–57

van Duinen, S. G., Castano, E. M., Prelli, F., Bots, G. T. A. B., Luyendijk, W., et al. 1987. Hereditary cerebral hemorrhage with amyloidosis in patients of Dutch

origin is related to Alzheimer disease. *Proc. Natl. Acad. Sci. USA* 84: 5991–94

Vanley, C. T., Aguilar, M. J., Kleinhenz, R. J., Lagios, M. D. 1981. Cerebral amyloid angiopathy. *Hum. Pathol.* 12: 609–16

Walker, L. C., Kitt, C. A., Cork, L. C., Struble, R. G., Dellovade, G. L., et al. 1988. Multiple transmitter systems contribute neurites to individual senile plaques. *J. Neuropathol. Exp. Neurol.* 47: 138–44

Wischik, C. M., Novak, M., Edwards, P. C., Klug, A., Tichelaar, W., et al. 1988a. Structural characterization of the core of the paired helical filament of Alzheimer disease. *Proc. Natl. Acad. Sci. USA* 85: 4884–88

Wischik, C. M., Novak, M., Thogersen, H. C., Edwards, P. C., Runswick, M. J., et al. 1988b. Isolation of a fragment of tau derived from the core of the paired helical filament of Alzheimer disease. *Proc. Natl. Acad. Sci. USA* 85: 4506–10

Wisniewski, H. M., Narang, M. K., Terry, R. D. 1976. Neurofibrillary tangles of paired helical filaments. *J. Neurol. Sci. 27:* 173–81

Wisniewski, H. M., Rabe, A., Wisniewski, K. E. 1988. Neuropathology and dementia in people with Down's syndrome. In *Molecular Neuropathology of Aging. Banbury Report*, ed. P. Davies, C. Finch, pp. 399–413. New York: Cold Spring Harbor Lab.

Wolozin, B., Davies, P. 1987. Alzheimer-related neuronal protein A68: specificity and distribution. *Ann. Neurol.* 22: 521–26

Wolozin, B. L., Pruchnicke, A., Dickson, D. W., Davies, P. 1986. A neuronal antigen in the brains of Alzheimer patients. *Science* 232: 648–50

Wong, C. W., Quaranta, V., Glenner, G. G. 1985. Neuritic plaques and cerebrovascular amyloid in Alzheimer disease are antigenically related. *Proc. Natl. Acad. Sci. USA* 82: 8729–32

Wood, J. G., Mirra, S. S., Pollock, N. J., Binder, L. I. 1986. Neurofibrillary tangles of Alzheimer's disease share antigenic determinants with the axonal microtubule-associated protein tau. *Proc. Natl. Acad. Sci. USA* 83: 4040–43

Yagashita, S., Itoh, T., Nan, W., Amano, N. 1981. Reappraisal of the fine structure of Alzheimer's neurofibrillary tangles. *Acta Neuropathol.* 54: 239–46

Yamada, T., Sasaki, H., Furuya, H., Miyata, T., Goto, I., et al. 1987. Complementary DNA for the mouse homolog of the human amyloid beta protein precursor. *Biochem. Biophys. Res. Commun.* 149: 665–71

Yamaguchi, H., Shoji, M., Harigaya, Y.,

Okamoto, K., Hirai, S. 1988. Alzheimer-type dementia: Diffuse type of senile plaques demonstrated by $\beta$-protein immunostaining. *Alzh. Dis. Assoc. Disorders* 2: 243

Yen, S.-H., Gaskin, F., Terry, R. D. 1981. Immunocytochemical studies of neurofibrillary tangles. *Am. J. Pathol.* 104: 77–89

Yen, S.-H., Kress, Y. 1983. The effect of chemical reagents. See Katzman 1983, pp. 155–65

Yen, S.-H., Crowe, A., Dickson, D. W. 1985. Monoclonal antibodies to Alzheimer's neurofibrillary tangles. I. Identification of polypeptides. *Am. J. Pathol.* 120: 282–91

Yoshimura, N. 1984. Evidence that paired helical filaments originate from neurofilaments. *Clin. Neuropathol.* 3: 22–27

*Ann. Rev. Neurosci. 1989. 12:491–516*

# EXTRACELLULAR MATRIX MOLECULES THAT INFLUENCE NEURAL DEVELOPMENT

## Joshua R. Sanes

Department of Anatomy and Neurobiology, Washington University School of Medicine, St. Louis, Missouri 63110

As developing neurons migrate, extend axons, and form synapses, much of their behavior is specific—that is, stereotypes and predictable. It is now clear that intrinsic neuronal predilections do not fully account for this specificity, but rather that extrinsic cues also influence neuronal choices. About ten years ago, experiments in several systems began to implicate the extracellular matrix (ECM) as an important source of this guidance: Contact with matrix was shown to influence neuronal migration, axonal growth, and synaptogenesis, as well as glial differentiation (reviewed in Sanes 1983). Once these phenomena had been documented, attention turned to a search for the ECM molecules that mediate them. Here, I summarize this recent work.

In reviewing very new and often preliminary data, I have tried to indicate some general themes that are emerging, but to refrain from drawing premature conclusions from a rapidly evolving body of work. It has seemed, too, more valuable to be comprehensive than to be critical at this stage, even though this precludes consideration of three areas that form the background for molecular analysis of neural ECM. First, I have cited few papers published before 1983, and refer the reader to the earlier review (Sanes 1983) for more detailed consideration of previous work. Second, even though most work on ECM in the nervous system relies on earlier studies of nonneural cells, I have omitted this relevant background. Recent reviews on the general biology and biochemistry of ECM include those of Akiyama & Yamada (1987), Buck & Horwitz (1987), Cunningham (1987), Hassell et al (1986), Hynes (1987), Mayne & Burgeson (1987), Martin & Timpl (1987), Ruoslahti (1988), Ruoslahti & Piersbacher (1987), and

491

0147–006X/89/0301–0491$02.00

Timpl & Dziadek (1986). Third, even though distinctions between "membrane-" and "matrix"-associated components of the cell surface are frequently arbitrary, I have tried to restrict this essay to the latter. Covault (1988), Edelman (1986), and Ekblom et al (1986) have recently considered cell-cell and cell-substratum (read ECM) interactions together.

## AXON OUTGROWTH

The earliest observations of growing axons, both in vivo and in vitro, indicated a "strange propensity of the nerve sprouts to adhere to supports or pre-established pathways" (Ramon y Cajal 1928). This form of guidance, variously called stereotropism, haptotropism, tactile adhesion, or contact sensibility, was long considered to be dominated by mechanical factors: Axons were channeled along structural inhomogeneities in their environment and tended "to follow passively the outlines of hard and smooth organs." In this context, the experiments of P. Letourneau (1975) were particularly influential. He coated tissue culture dishes with patterned substrata to which dissociated neurons adhered differentially, and demonstrated that axons grew preferentially along pathways of the more adhesive substance, even when the less adhesive substance was quite capable of supporting neurite outgrowth on its own. The notion that chemical interactions between axons and substrata-guided growth thus largely supplanted the previous focus on mechanical guidance.

A few years later, in analyzing the effects of nonneural cells on cultured neurons, F. Collins (1978, Collins & Garrett 1980, Collins & Lee 1984) separated material secreted by heart cells ("conditioned medium") into two active fractions: one that acted in solution to enhance neuronal survival, and another that bound to the substratum and then acted to induce neurite outgrowth. Similar substrate-attached factors were soon found in media conditioned by a variety of cell types (Berg, 1984). When applied to culture dishes in narrow stripes, the substrate-bound material proved capable of directing the course of axon outgrowth. Considering their results together with those of Letourneau, Collins & Garrett (1980) suggested that "during development the outgrowth of nerve fibers could be directed along pathways of the appropriate extracellular materials, which are released and deposited onto the extracellular matrix-substratum by neighboring embryonic cells."

Subsequently, considerable effort has gone into identifying the ECM molecules that guide elongating axons. Work has proceeded in three directions: testing the ability of known ECM components to promote neurite extension, identifying the active molecule in conditioned media, and documenting the distribution of candidate outgrowth-promoting molecules in vivo. All three approaches have converged on the conclusion that laminin, a noncollagenous glycoprotein present in most basal laminae (BLs), is an important promoter of neurite outgrowth.

First, neurons from a variety of sources and species extend neurites on substrata coated with purified laminin (Baron-Van Evercooren et al 1982, Lander et al 1983, Manthorpe et al 1983, Rogers et al 1983, Wewer et al 1983, Edgar et al 1984, Faivre-Bauman et al 1984, Liesi et al 1984a, Smalheiser et al 1984, Adler et al 1985, Hammarback et al 1985, 1986, Hopkins et al 1985, Davis et al 1985a, Unsicker et al 1985a,b, Engvall et al 1986). In several of these studies, laminin's ability to promote neurite outgrowth was compared to that of other ECM components, including collagens, proteoglycans, and fibronectin; in each case, laminin proved to be the superior substrate (e.g. Gundersen 1987a,b). In addition, laminin promotes outgrowth from some central neurons that are unresponsive to fibronectin or collagen (e.g. Rogers et al 1983), and permits short-term survival of nominally NGF-dependent cells in the absence of NGF (e.g. Baron-Van Evercooren et al 1982, Edgar et al 1984). In these respects, laminin appears to resemble conditioned media and to differ from other ECM molecules that have been tested.

Second, laminin turns out to be the major active molecule in neurite outgrowth–promoting media produced by a variety of cell types. Initial studies revealed that the active factor contained a heparan sulfate proteoglycan (Lander et al 1982), but it soon became apparent that laminin is also a component of the factor (Lander et al 1985a, Davis et al 1985b, Calof & Reichardt 1985, Dohrmann et al 1986). The laminin exists in a tight but non-covalent complex with heparan sulfate proteoglycan and, in some cases, with a chondroitin sulfate, proteoglycan, and a noncollagenous ECM glycoprotein, entactin (Lander et al 1985b, Davis et al 1987a). In each case, however, it appears that laminin accounts for most, if not all, of the neurite extension-promoting activity of the complex. In light of this clear-cut result, the reason for initially focusing on the proteoglycan is interesting: Antibodies to laminin that effectively blocked neurite outgrowth on substrata coated with pure laminin did not block outgrowth induced by the proteoglycan-laminin complex (Lander et al 1983, Manthorpe et al 1983, Edgar et al 1984, 1988). This observation remains unexplained and provocative. One possibility is that laminin in the complex differs somehow from the purified laminin used to generate the sera. Alternatively, the active site of the complexed laminin might be inaccessible to antibody; perhaps heparan sulfate proteoglycans on axonal membranes (e.g. Matthew et al 1985) can displace related proteoglycans in the complex, giving the axon privileged access to an active site that antibodies cannot reach. Yet another possibility is that neurites recognize an active site formed by a combination of laminin and proteoglycan, and are therefore not dependent on the binding site that isolated laminin presents; consistent with this notion is the finding that a monoclonal antibody that blocks outgrowth on factor-coated dishes (INO; Matthew & Patterson 1983) recognizes the complex but does not recognize either laminin or proteoglycan alone (Chiu et al 1986). Thus, although laminin is clearly crucial for the activity of cell-derived neurite outgrowth-promoting factors, its form and/or conformation are specialized in some way that has yet to be elucidated.

Third, immunohistochemical studies in vivo have shown that laminin is present along several pathways that growing axons follow. Thus, laminin immunoreactivity is present in developing rat muscle as axons invade (Chiu & Sanes 1984) and along the trajectories of trigeminal and spinal axons in chick embryos (Rogers et al 1986, Riggott & Moody 1987); in the latter cases, the authors note that laminin marks the sites of growing nerves more precisely than does fibronectin. Perhaps more telling, however, are studies on the regeneration of peripheral axons through denervated nerve trunks in adults. While axons degenerate following transection, the BL-coated Schwann cells that had ensheathed them survive (see Figure 1a), and growth cones advance along the inner surface of the Schwann cell BL as axons regenerate through the nerve (e.g. Scherer & Easter 1984). Even if the Schwann cells are killed by freezing, the acellular BL can support and guide axonal regeneration to some extent (Ide et al 1983, Ide 1986, 1987, Kuffler 1986; and see Keynes et al 1984, Ide 1984, Glasby et al 1986, and Davis et al 1987b,c, on the use of BLs as prosthetic devices to facilitate nerve regeneration). Schwann cell BLs contain laminin (Sanes 1982, Cornbrooks et al 1983), and the monoclonal antibody INO (which blocks neurite extension on the laminin-proteoglycan complex) reduces neurite outgrowth (although only by $\sim 20\%$) on sections of peripheral nerve in vitro (Sandrock & Matthew 1987a, Martini et al 1989) and in denervated irides in vivo (Sandrock & Matthew 1987b). Together, these results argue that laminin influences axon growth in vivo as well as in vitro.

A logical extension of work on laminin has been a search for its neuronal receptors. Here, three sets of studies have implicated three types of molecules. First, considerable attention has focused on the integrins, a class of cellular receptors for ECM molecules initially defined by two monoclonal antibodies (JG22 and CSAT) that inhibit the adhesion of muscle cells to ECM-derived substrata (Buck & Horwitz, 1987). These and other antibodies to integrin inhibit neurite extension on laminin, as well as on collagen, fibronectin, cell-derived neurite outgrowth-promoting factors, and some cell surfaces (Bozyczko & Horwitz 1986, Cohen et al

1986, 1987, Tomaselli et al 1986, 1987, Bixby et al 1987, Hall et al 1987). Interestingly, developmental changes in neuronal responsiveness to laminin (Cohen et al 1986, 1987) may reflect developmentally regulated (Hall et al 1987), target-dependent (Cohen 1987) changes in integrin isoforms. Second, several groups have made a systematic search for laminin-binding proteins in the membranes of neuroblastoma cells, using methods that had been developed to isolate a laminin receptor from muscle cells. Several laminin binding proteins distinct both from integrin and from the muscle-type receptor have been identified in this way (Douville et al 1987, Smalheiser & Schwartz 1987, Kleinman et al 1988), and antisera to two of them block neurite extension on laminin (Kleinman et al 1988). Third, investigating the involvement of carbohydrates in adhesive interactions, several groups have assayed neuronal attachment to laminin in the presence of carbohydrate-specific probes. In one group of studies, gangliosides and antibodies to gangliosides have been found to interfere with adhesion to laminin (Cheresh et al 1986, Laitinen et al 1987), and specific gangliosides (GD2 and GD3) shown to be complexed with an integrin-like receptor (Cheresh et al 1987). In a parallel effort, antibodies to an unusual carbohydrate ("L2" or "HNK-1"), which is associated with several adhesion molecules (Kruse et al 1985), have been shown to recognize integrin (Peshava et al 1987) and to block neurite outgrowth on laminin (Kunemund et al 1988, Dow et al 1988). In short, there is now evidence that neurons use more than one receptor to interact with laminin, and that carbohydrates can modulate the interaction.

Finally, the demonstrable potency of laminin in promoting axon outgrowth should not be taken to imply that other ECM components can be ignored. Fibronectin, collagen I, and collagen IV can all act as effective substrates for neurite outgrowth from peripheral neurons, and can also promote outgrowth from at least some central neurons (Akers et al 1981, Vlodavsky et al 1982, Hatten et al 1982, Carbonetto et al 1983, Thompson & Pelto 1982, Rauvala 1984). For fibronectin, the use of proteolytic fragments, domain-specific inhibitors, and synthetic peptides has permitted mapping of some regions of the molecule with which neurites interact, and has led to the suggestion that different neuronal types recognize different binding domains (Rogers et al 1985, 1987, Humphries et al 1988, Tobey et al 1985, Waite et al 1987, Mugnai et al 1988). Taking another tack, Schubert and colleagues have studied glycoprotein complexes called adherons, which are released by cultured cells and which mediate cell-substrate adhesion; several ECM molecules are present in adherons and may be involved in adhesion (Schubert & LeCorbiere 1985, Cole et al 1985, Cole & Glaser 1986, Berman et al 1987). Proteoglycans and their component glycosaminoglycan chains have generally not been found to promote neurite outgrowth, and may even inhibit outgrowth on other substrates (Carbonetto et al 1983, Manthorpe et al 1983, Akeson & Warren 1986), as expected from the result that laminin accounts for all of the activity of the laminin-proteoglycan complex. However, a recent report claims that antibodies to a heparan sulfate proteoglycan can inhibit outgrowth on the complex, and that isolated proteoglycan can induce neurite elongation (Hantaz-Ambroise et al 1987). Laminin and proteoglycan had different effects on neurons in these experiments, with laminin inducing neurite branching rapidly and the proteoglycan stimulating neurite elongation slowly; the failure of earlier studies to detect effects of the proteoglycan is ascribed to their use of short-term assays to measure neurite extension. More generally, of course, the choices that axons make *among* laminin-rich pathways must be guided by other, more specific cues; molecules that account for these choices remain to be identified.

# MIGRATION AND DIFFERENTIATION OF NEURAL CREST CELLS

The neural crest is a transitory aggregate of cells on the dorsolateral surface of the neural tube. Cells migrate from the crest along stereotyped pathways and then differentiate to from a variety of nonneural and neural derivatives; among the latter are the autonomic neurons of sympathetic, parasympathetic, and enteric ganglia, sensory neurons of dorsal root ganglia, chromaffin cells of the adrenal medulla, and Schwann cells (glia) of peripheral nerve trunks and ganglia. Because cells grafted from different regions of the crest generally migrate along pathways characteristic of their site of implantation (LeDouarin 1986), it is clear that cues provided along migratory pathways are important determinants of crest cell behavior. Early histological studies revealed that ECM is prominent in the migratory pathways and implicated various structures (e.g. BLs and interstitial granules) and molecules (e.g. glycosaminoglycans and fibronectin) in the guidance process (reviewed in Sanes 1983). More recent studies have detailed the arrangement of BLs that crest cells encounter (Sternberg & Kimber 1986b, Martins-Green & Erickson 1986, 1987, Payette et al 1988), documented the distribution of several ECM molecules along migratory pathways (Rogers et al 1986, Sternberg & Kimber 1986a, Duband & Thiery 1987, Pomeranz et al 1987), demonstrated that integrin is abundant on the surface of migrating crest cells (Duband et al 1986, Krotoski et al 1986) and compared the ability of isolated crest cells to migrate on various ECM-derived substrata in vitro (Rovasio et al 1983, Boucaut et al 1984, Bronner-Fraser 1985, Duband et al 1986, Perris & Johansson 1987, Tan et al 1987, Bilozur & Hay 1988).

An important extension of this work has been the attempt to perturb crest cell migration by injection of suitable blocking agents directly into chick embryos. In one such study Boucaut et al (1984) injected a synthetic peptide that competes with fibronectin for a binding site on its receptor, integrin (see Ruoslahti & Piersbacher 1987). This peptide blocked migration of crest cells on fibronectin-coated substrata and partially inhibited the emigration of cells from cranial neural crest in vivo; stymied crest cells formed ectopic aggregates within the neural tube of peptide-treated embryos. Now that subsequent work has shown that integrins recognize ECM molecules other than fibronectin (e.g. laminin), these results can be reinterpreted as implicting integrin rather than fibronectin per se in the migratory process. Consistent with this revised conclusion, Bronner-Fraser (1985, 1986b) has shown that each of two antibodies to integrin, but none of several control antibodies, also blocks migration of cranial neural crest cells and leads to formation of ectopic aggregates of cells in the neural tube. Other antibodies that perturb migration of cranial crest are INO, which recognizes the neurite-promoting laminin-proteoglycan complex, and HNK-1, which recognizes the carbohydrate epitope shared by several adhesion molecules (see above); in both of these cases, however, the defect is qualitatively different from that seen with the integrin-directed reagents, and may indicate interference at a relatively late stage of migration (Bronner-Fraser 1987, Bronner-Fraser & Lallier 1988).

Almost as revealing as the ability of these reagents to block the migration of cranial neural crest cells is their consistent inability to affect migration in the trunk region. In fact, coincident with the execution of the blocking experiments has come a reevaluation of the migratory pathways followed by trunk crest cells. It now appears, consistent with early results but at variance with some subsequent reports, that crest cells in the trunk migrate through the somites rather than along fibronectin-rich pathways between somites. Moreover, this invasion occurs primarily through the anterior or rostral half of each somite (or the sclerotome that arises from it); cells appear to be restricted from traversing the posterior half of each sclerotome (Rickmann et al 1985, Bronner-Fraser 1986a, Loring & Erickson 1987). In light of this finding, the symmetrical accumulations of laminin and fibronectin around somites seem unlikely to play a directive role in guiding crest migration in the trunk (Krotoski et al 1986, Krotoski & Bronner-Fraser 1987, Loring & Erickson 1987). On the other hand, asymmetrical distributions in somites have recently been reported for two ECM molecules that were derived from brain and are discussed below: tenascin (cytotactin) is concentrated in rostral half-somites and a chondroitin sulfate proteoglycan in caudal half-somites. These asymmetries do not result from crest migration, because they are apparent in somites isolated from contact with crest cells (Tan et al 1987, Mackie et al 1988). Tenascin and the proteoglycan now become obvious targets for attempts to perturb crest cell migration in the trunk.

ECM appears to influence the differentiation as well as the migration of neural crest cells. Thus, ECM components promote the adrenergic differentiation of cultured neural crest cells (Maxwell & Forbes 1987) and of the adrenal chromaffin cell, a crest derivative (Acheson et al 1986). Perhaps the best studied crest cell in this respect, however, is the Schwann cell. Schwann cells ensheath all axons in peripheral nerves, and form myelin around the larger-caliber axons. The Schwann cells also bear (and in large part synthesize) a BL that encircles the entire Schwann cell-axon unit (Figure 1a), and it now appears that at least some components of this BL must be present in order for myelination to occur. This conclusion stems from a variety of situations in vitro in which the abilities of Schwann cells to form BL and to myelinate are lost and regained together (see Bunge et al 1986 for review, and Carey et al 1986, 1987, Carey & Todd 1987, Eldridge et al 1987, 1989 for more recent work). Thus, although Schwann cells readily form BL and myelin when cultured in a rich medium, they contact axons but fail to form either BL or myelin when cultured: (a) in serum-free medium, (b) in the presence of an inhibitor of collagen secretion, cis-hydroxyproline, (c) in the presence of a proteoglycan synthesis inhibitor, $\beta$-xyloside, (d) with some neurons that are not generally myelinated, or (e) isolated from contact with a solid substratum. Most telling, addition of ascorbate (a cofactor for collagen synthesis), a BL-derived gel, or purified laminin to these deprived cultures promotes both BL deposition and myelin formation. Finally, in vivo, Schwann cells in the peripheral nerves of dystrophic (see Bunge et al 1986) and dysgenic (see Pincon Raymond et al 1987) mice show correlated defects in BL formation and myelination. Together, these results argue strongly that Schwann cells use one feature of their differentiated phenotype—a BL—to induce another—myelin.

## FORMATION OF NEUROMUSCULAR JUNCTIONS

Both pre- and postsynaptic membranes are highly specialized at the skeletal neuromuscular junction (Figure 1b). In the nerve terminal, synaptic vesicles cluster at membrane-associated densities to form "active zones" from which transmitter is released; far fewer vesicles and no active zones are found in preterminal portions of the motor axon. Postsynaptically, acetylcholine receptors, specific glycoconjugates, the neural cell adhesion molecule, and a variety of cytoskeletal elements are selectively associated with the infolded membrane that lies just beneath the nerve terminal; few of

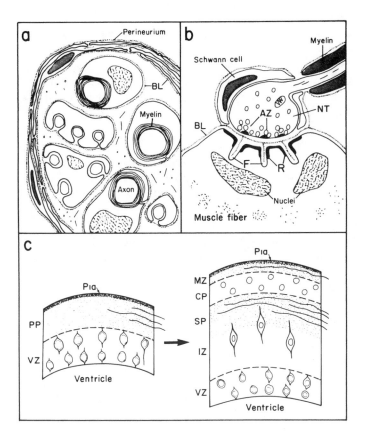

*Figure 1*  Some examples of ECM in the vertebrate nervous system. (*a*) A cross-sectioned peripheral nerve, showing the BL that ensheathes each axon-Schwann cell unit, the collagen and other fibrils that separate units, and the complex ECM of the perineurial cells that encircle each fascicle (see Bunge et al 1986). (*b*) A cross-sectioned skeletal muscle fiber, showing the BL that surrounds each fiber and passes between pre- and postsynaptic membranes at the neuromuscular junction. Active zones (AZ) in the nerve terminal (NT) lie opposite junctional folds (F) in the acetylcholine receptor (AChR)-rich postsynaptic membrane (see Sanes 1986). (*c*) Sagittal-sections of embryonic cerebral cortex, showing the distribution of fibronectin-immunoreactivity (*dots*) in the preplate (PP), a transient layer above the ventricular zone (VZ). As the cortical plate (CP) forms, it splits the PP into a subplate (SP) and a marginal zone (MZ), both of which are rich in fibronectin. Neurons migrate through the intermediate zone (IZ) and SP on their way from VZ to CP, and cortical afferents grow through the SP and MZ on their way to the cortex (see Stewart & Pearlman 1987, Chun & Shatz 1988).

these molecules and no folds appear extrasynaptically (see Salpeter 1987 for references).

The sheath of BL that encircles the entire muscle fiber extends through the synaptic cleft at the neuromuscular junction. Components of the synaptic portion of this BL play roles in organizing differentiation when neuromuscular junctions regenerate following damage to adult muscle. This was demonstrated by removing the cellular elements of the neuromuscular junction in ways that spared the BL, and then allowing either axons or myofibers to regenerate. When axons regenerated, they made nearly all of their contacts with BL sheaths at original synaptic sites, a specificity that was as marked as that observed when axons reinnervate denervated but undamaged muscle (Sanes et al 1978, Kuffler 1986). Furthermore, axons contacting synaptic BL differentiated into nerve terminals that appeared small but were normal by morphological, immunocytochemical, and functional criteria (Glicksman & Sanes 1983). Thus, components of BL are able to guide reinnervation of original synaptic sites and to organize the differentiation of axons into nerve terminals at those sites. Conversely, when myotubes regenerated within BL sheaths in the absence of axons or Schwann cells, they accumulated acetylcholine receptors, secreted acetylcholinesterase, and formed junctional folds in precise apposition to synaptic sites on the BL (Burden et al 1979, McMahan & Slater 1984, Anglister & McMahan 1985, Slater & Allen 1985, Hansen-Smith 1986). Thus, BL can trigger post- as well as presynaptic differentiation during regeneration.

Based on these results, several groups have searched for components of synaptic BL that might account for its organizing activities. The best characterized of these, and the only one known to affect synaptic differentiation, is agrin. McMahan and colleagues showed that material extracted from the ECM of *Torpedo* electric organ (a favorite tissue for biochemical studies, because its synapses resemble but are vastly more abundant than neuromuscular junctions) could induce clustering of acetylcholine receptors on cultured chick myotubes (Rubin & McMahan 1982, Godfrey et al 1984). The partially purified material was then used to prepare monoclonal antibodies that precipitated the active factor, dubbed agrin, and selectively strained synaptic BL (Fallon et al 1985, Reist et al 1987). These antibodies, in turn, identified a group of four polypeptides of $M_r$ $-70-150$ kD, all of which are recognized by anti-agrin, but only two of which are bioactive (Nitkin et al 1987, Smith et al 1987, Godfrey et al 1988a). The same proteins appear to induce accumulation of acetylcholinesterase and butyrylcholinesterase in addition to acetylcholine receptors (Wallace et al 1985, Wallace 1986, Nitkin et al 1987). Magill et al (1987) have found agrin-like immunoreactivity concentrated in motoneurons and suggest that motor neurons are a source of synaptic agrin. However, agrin is present in muscle before nerves arrive and is codistributed with acetylcholine receptors in myotubes that have developed aneurally (Fallon 1987 and personal communication, Godfrey et al 1988b). Furthermore, an independently derived set of anti-agrin antibodies do not stain motoneurons selectively (Godfrey et al 1988a). Based on their results, McMahan and colleagues have proposed that agrin-like molecules in synaptic BL induce receptor clustering when adult muscles regenerate, and perhaps during embryonic synaptogenesis as well.

Immunological and histochemical methods have led to the identification of several other components of synaptic BL that may also play roles in synaptogenesis. These can be summarized briefly:

1. Laminin, a heparan sulfate proteoglycan, collagen IV, and fibronectin are all present in both synaptic and extrasynaptic BL (Sanes 1982, Sanes & Chu 1983, Eldridge et al 1986). Axons can interact with all of these (see above), and laminin and collagen can induce AChR clustering in vitro (Vogel et al 1983, Kalcheim et al 1982a,b, see also Kalcheim et al 1985a,b). The widespread distribution of these molecules appears to disqualify them from participation in functions specific to synaptic BL. However, some monoclonal antibodies to laminin stain extrasynaptic but not synaptic BL (Gatchalian et al 1985), suggesting that this or other common BL molecules could be present in a distinctive form or conformation at the synapse.

2. Anderson & Fambrough (1983) and Sanes & Chiu (1983) have produced monoclonal antibodies that selectively stain synaptic BL in frog and rat muscle, respectively. The former antibodies recognize a heparan sulfate proteoglycan, and the latter a novel noncollagenous glycoprotein called JS1 that is homologous to laminin (based on cDNA sequence) and is adhesive to neurons (Hunter et al 1987, 1988, and in preparation). Both antigens are synthesized by muscle cells in the absence of neurons. These antibodies have been used to study the ways in which neuromuscular interactions regulate the accumulation of synaptic BL, in vivo and in vitro (Sanes & Lawrence 1983, Sanes et al 1984, Chiu & Sanes 1984, Anderson et al 1984, Anderson 1986). A general conclusion is that motoneurons appear to use a combination of activity-dependent and -independent mechanisms to induce the accumulation of synaptic BL components at synaptic sites; these specializations in the BL appear not to prefigure synaptic sites on embryonic myotubes, even though, once formed, such sites are clearly recognizable by regenerating axons.

3. Carlson et al (1986, Carlson & Wright 1987) have isolated a large chondroitin sulfate proteoglycan from *Torpedo* electric organ. This molecule appears to be associated with membranes in that it has a hydrophobic

domain and requires detergent to be solubilized; however, when the tissue is disrupted, most of the immunoreactive material is recovered in a matrix-rich fraction. Because of its large size and association with both membrane and matrix, Carlson has named the proteoglycan "Terminal Anchorage Protein 1" (TAP1), and proposed that it spans the synaptic cleft.

4. Barald et al (1987) have isolated an activity from rat synaptic BL that induces clustering of acetylcholine receptors but is apparently distinct from agrin.

5. Lectins specific for $\beta$-$N$-acetylgalactosamine-terminated glyco-conjugates selectively stain synaptic sites on muscle fibers from many vertebrate species, thus defining a synapse-specific carbohydrate (Sanes & Cheney 1982, Scott et al 1988). One of the molecules that bears this carbohydrate is the collagen-tailed form of acetylcholinesterase, which is itself anchored in synaptic BL, perhaps via attachements of its collagenous subunit (Lappin et al 1987) to a heparan sulfate proteoglycan (Brandan et al 1985, Brandan & Inestrosa 1986). While the functional significance of the carbohydrate moiety for acetylcholinesterase is unknown, other synapse-specific molecules, including a glycolipid, also bear $\beta$GalNAc-terminal sugars (Scott et al 1988), thus raising the possibility that $\beta$GalNAc plays some general role in synaptic differentiation.

Finally, consideration of the terrain that surrounds synaptic sites leads to the idea that ECM molecules may be involved in guiding axons to the patches of synaptic BL that they recognize. In that these sites occupy only $\sim 0.1\%$ of the muscle fiber surface, it is unlikely that regenerating axons would encounter them frequently by an unguided process of random search. Connective tissue sheaths of intramuscular nerves provide some guidance in this regard: axons generally regenerate through these sheaths, presumably along Schwann cell BL (see above), and are thereby delivered to regions of original synaptic sites (Scherer & Easter 1984, Kuffler 1986). However, axons also preferentially reinnervate original sites when they are forced to grow outside of perineurial sheaths, a finding that suggests the existence of a second intramuscular cue. In this regard, it is intriguing that three ECM components with which neurons interact—tenascin (J1; see below), fibronectin, and heparan sulfate proteoglycan—as well as a membrane-associated adhesion molecule, NCAM, all accumulate in interstitial areas surrounding denervated synaptic sites (Sanes et al 1986). All four of these molecules appear to be synthesized by fibroblasts that proliferate in perisynaptic areas following denervation (Gatchalian & Sanes 1987, Gatchalian et al 1989; see also Connor & McMahan 1987). When regenerating axons approach original synaptic sites, it is likely that they encounter these fibroblasts and possible that their growth is guided by the neuroactive ECM molecules that the fibroblasts secrete. Evidence favoring this suggestion is that cultured neurons grow longer neurites on fragments of denervated muscle taken from perisynaptic regions than on fragments taken from synapse-free regions (Covault et al 1987).

## ECM IN THE BRAIN

Students of the peripheral nervous system frequently voice the conviction that their work can provide insights and techniques that will be useful in understanding the experimentally less accessible CNS. Until recently, this appeared to be a shaky premise for studies of ECM: the brain's BLs are confined to meningeal surfaces and blood vessels, and it seemed to have no organized matrix. Some large extracellular spaces were reported to exist in embryonic brain, and were proposed to guide growing axons, but these appeared empty by conventional microscopy, and their very existence was suspected to be an artifact of the preparative techniques used. However, the introduction of fixatives that preserve and stain ECM (ruthenium red, Alcian blue, tannic acid) revealed that some of these extracellular spaces were filled with ECM-like granules and fibrils (Krayanek 1980, Nakanishi 1983, Bork et al 1987). More important, as antibodies have become available that recognize molecules clearly associated with ECM in the periphery, it has become apparent that many of these molecules are, in fact, present in the developing brain. These include fibronectin (Hatten et al 1982, Stewart & Pearlman 1987, Chun & Shatz 1988), laminin (Liesi et al 1984b, Liesi 1985a,b, Letourneau et al 1988, McLoon et al 1988), tenascin (Kruse et al 1985, Grumet et al 1985, Crossin et al 1986, Erickson & Taylor 1987), chondroitin sulfate proteoglycans (Aquino et al 1984a,b, Hoffman & Edelman 1987, Krusius et al 1987, Margolis et al 1987, Morris et al 1987), hyaluronectin (Delpech et al 1982, Bignami & Delpech 1985, Bignami & Dahl 1986, 1988), hyaluronic acid (Ripellino et al 1985), thrombospondin (O'Shea & Dixit 1987), JS1 (D. D. Hunter et al, in preparation), and agrin (Magill et al 1987). Some of these molecules (e.g. laminin and fibronectin) appear only transiently during embryogenesis, whereas others (e.g. tenascin and chondroitin sulfate proteoglycans) remain detectable in adult brain. In vitro, some of these molecules are synthesized by astrocytes (fibronectin: Price & Hynes 1985, Liesi et al 1986, Schmalenbach et al 1987; laminin: Liesi et al 1983, Wujek & Freese 1987, McLoon et al 1988, Ard & Bunge 1988; tenascin: Grumet et al 1985, Kruse et al 1985; JS1: D. D. Hunter et al, in preparation), whereas others are apparently produced by neurons (chondroitin sulfate proteoglycan: Hoffman & Edelman 1987; agrin: Magill et al 1987). Thus, in the space of just a few years, a rich and developmentally regulated ECM has been found in the brain.

What are these molecules doing? Here, speculation has been guided by the evidence, summarized above, that ECM mediates adhesion, guides cellular migrations, and promotes neurite outgrowth in the periphery. These processes are clearly crucial to morphogenesis in the CNS, and ECM molecules are able to promote adhesion and neurite extension by cultured central neurons (e.g. Akers et al 1981, Liesi et al 1984a, Smalheiser et al 1984, Adler et al 1985, Cohen et al 1987, Hall et al 1987; but see Kleitman et al 1988). It is therefore not surprising that patterns of immunoreactivity have been interpreted in light of results from peripheral systems.

The distribution of fibronectin in the developing cerebral cortex is particularly suggestive in this regard (Pearlman et al 1986, Stewart & Pearlman 1987, Chun & Shatz 1988; see Figure 1c). The first neurons born in the ventricular zone leave it to form the preplate beneath the pial surface. Later-born neurons migrate radially to form the cortical plate within the preplate. As the cortical plate expands, it splits the preplate into a marginal zone above and a subplate below; afferents from subcortical and contralateral areas enter the cortex through the marginal zone and subplate, invading the cortex proper only after a "waiting" period. Eventually, as lamination and synaptogenesis transform the cortical plate into the cortex, the marginal zone and subplate disappear. Immunohistochemistry in mouse and cat reveals that fibronectin is abundant in the preplate before afferents grow in, and that it is concentrated in the marginal zone and subplate but excluded from the cortical plate as these layers form. As the cortex matures, fibronectin is lost from the neuropil, remaining only in blood vessels and meninges. Fibronectin is thus appropriately distributed in space and time to influence the migrating neurons and ingrowing axons that join to form the cerebral cortex.

Laminin also appears transiently in the developing brain. Although it has not been studied in cortex, its distribution suggests that it might help to guide growing axons in subcortical areas. Thus, in rat, laminin-like immunoreactivity appears in the ventral longitudinal fasciculus ahead of the axons that enter the midbrain through this pathway (Letourneau et al 1988), and in the optic nerve ahead of ingrowing retinal axons (McLoon et al 1988). Furthermore, inter-species differences in laminin levels in the optic nerve correlate with the differing abilities of various animals to regenerate retinal ganglion cell axons following damage (Liesi 1985b, Hopkins et al 1985, McLoon 1986, Ford-Holevinski et al 1986, Gifto-christos & David 1988; but see Zak et al 1987). Thus, in mammals, which do not regenerate, laminin is absent from the adult nerve (save for blood vessels) and no new laminin appears following nerve transection. In contrast, laminin remains detectable in the optic nerves of adult fish and frogs, which do support regeneration, and laminin levels actually increase following transection of fish nerve. Together with the ability of many central neurons to extend axons on laminin in vitro (see above), these results have led to the suggestion that the absence of laminin is one of the factors that limits the regenerative ability of the adult mammalian CNS.

J1 is a glycoprotein that was isolated from mouse brain by Kruse et al (1985) on the basis of its ability to bind antibodies (L2, HNK-1; see above) that had previously been shown to recognize a carbohydrate shared by several other adhesion molecules. Grumet et al (1985; see also Jones et al 1988) used a similar strategy to isolate a neuron-glia adhesion molecule, which they called cytotactin, from chick brain. J1 and cytotactin are both present in a variety of peripheral tissues, where they are clearly associated with ECM (Crossin et al 1986, Sanes et al 1986). It now appears that cytotactin and the 200–220 kD forms of J1 are closely related proteins, and that both are similar, if not identical, to ECM-derived proteins that had been isolated from several sources and called myotendinous antigen, glioma-mesenchymal extracellular matrix protein, hexabrachion, and tenascin (Chiquet & Fambrough 1984, Chiquet-Ehrismann et al 1986, Erickson & Taylor 1987, Faissner et al 1988). Although their identity remains unproven, it is now justifiable to call all of these molecules tenascin. Antibodies to tenascin (J1 and cytotactin) inhibit neuron-glia adhesion in vitro (Kruse et al 1985, Grumet et al 1985, Hoffman & Edelman 1987) and may disrupt neuronal migration along glial cells in tissue slices (Chuong et al 1987; but see Antonicek et al 1987), although in other systems, tenascin (per se and as hexabrachion) appears to promote cell growth rather than cell adhesion (Chiquet-Ehrismann et al 1986, Erickson & Taylor 1987). In any event, the wide distribution of tenascin in embryonic brain (Crossin et al 1986), its restricted localization in adult brain (e.g. at nodes of Ranvier: ffrench-Constant et al 1986; see also Rieger et al 1986), and the neural regulation of its levels in the periphery (Sanes et al 1986) all make this molecule a subject of considerable current interest.

When tenascin is purified (as myotendinous antigen or cytotactin), it is found to be complexed with a large chondroitin sulfate proteoglycan (Chiquet & Fambrough 1984, Grumet et al 1985, Hoffman & Edelman 1987). When purified, this proteoglycan reassociates with tenascin (cyto-tactin) and blocks the binding of cytotactin-coated beads to neurons. These results, along with the observation that the chondroitin sulfate proteoglycan is made by and appears on neurons, suggests that it is a neuronal ligand for tenascin (Hoffman & Edelman 1987, Hoffman et al 1988). It seems, however, that tenascin and the proteoglycan are not each other's only ligands: tenascin can apparently also interact with fibronectin and the chondroitin sulfate proteoglycan can interact with laminin (Chiquet-Ehrismann et al 1986, Hoffman & Edelman 1987; see also Perris

& Johansson 1987), although it must be noted that many ECM molecules are promiscuous in their affinities for each other, and the physiological significance of each interaction will need to be determined separately. Finally, it is worth mentioning that immunological studies have identified two other chondroitin sulfate proteoglycans in the ECM of brain: one that appears extracellularly in embryonic brain, but intracellularly in adult brain (Aquino et al 1984b), and one that appears on the surface of interesting subsets of adult neurons, where its expression is modulated by variations in levels of neural activity (Zaremba & Hockfield 1987, Kalb & Hockfield 1988). The relationship, if any, of these chondroitin sulfate proteoglycans to the ligand for tenascin remains to be determined.

## ECM IN INSECTS

Although influences of ECM on neural development have been little studied in insects, it is clear that some axons navigate along—and presumably interact with—BLs during the formation of both peripheral and central nervous systems (Nardi 1983, Bastiani et al 1986). Very recent progress in isolating homologous molecules and their genes from *Drosophila* (laminin: Fessler et al 1987, Montell & Goodman 1988; collagen IV: Blumberg et al 1987; fibronectin: Gratecos et al 1988; integrin: Bogaert et al 1987) promises to make these interactions amenable to analysis by the powerful genetic techniques that are available for this species.

## CONCLUSION

About a decade ago, several aspects of the development of neurons and glia were shown to be influenced by interactions of cells with the extracellular matrices that they encounter. Over the past five years, remarkable progress has been made in identifying molecules whose distributions in vivo and/or activities in vitro make them candidate mediators of these effects. The tools are now available for a new attack on interactions that contribute importantly to the orderly construction of the nervous system.

*Literature Cited*

Acheson, A., Edgar, D., Timpl, R., Thoenen, H. 1986. Laminin increases both levels and activity of tyrosine hydroxylase in calf adrenal chromaffin cells. *J. Cell Biol.* 102: 151–59

Adler, R., Jerden, J., Hewitt, A. T. 1985. Responses of cultured neural retinal cells to substratum-bound laminin and other extracellular matrix molecules. *Dev. Biol.* 112: 100–14

Akers, R. M., Mosher, D. F., Lilien, J. E. 1981. Promotion of retinal neurite outgrowth by substratum-bound fibronectin. *Dev. Biol.* 86: 179–88

Akeson, R., Warren, S. L. 1986. PC12 adhesion and neurite formation on selected substrates are inhibited by some glycosaminoglycans and a fibronectin-derived tetrapeptide. *Exp. Cell Res.* 162: 347–62

Akiyama, S. K., Yamada, K. M. 1987. Fibronectin. *Advances in Enzymology and Related Areas of Molecular Biology*, ed. A. Meister, pp. 1–157. New York: Wiley

Anderson, M. J. 1986. Nerve-induced remodeling of muscle basal lamina during synaptogenesis. *J. Cell Biol.* 102: 863–77

Anderson, M. J., Fambrough, D. M. 1983. Aggregates of acetylcholine receptors are associated with plaques of a basal lamina heparan sulfate proteoglycan on the surface of skeletal muscle fibers. *J. Cell Biol.* 97: 1396–1411

Anderson, M. J., Klier, F. G., Tanguay, K. E. 1984. Acetylcholine receptor aggregation parallels the deposition of a basal lamina proteoglycan during development of the neuromuscular junction. *J. Cell Biol.* 99: 1769–84

Anglister, L., McMahan, U. J. 1985. Basal lamina directs acetylcholinesterase accumulation at synaptic sites in regenerating muscle. *J. Cell Biol.* 101: 735–43

Antonicek, H., Persohn, E., Schachner, M. 1987. Biochemical and functional characterization of a novel neuron-glia adhesion molecule that is involved in neuronal migration. *J. Cell Biol.* 104: 1587–95

Aquino, D. A., Margolis, R. U., Margolis, R. K. 1984a. Immunocytochemical localization of a chondroitin sulfate proteoglycan in nervous tissue. I. Adult brain, retina, and peripheral nerve. *J. Cell Biol.* 99: 1117–29

Aquino, D. A., Margolis, R. U., Margolis, R. K. 1984b. Immunocytochemical localization of a chondroitin sulfate proteoglycan in nervous tissue. II. Studies in developing brain. *J. Cell Biol.* 99: 1130–39

Ard, M. D., Bunge, R. P. 1988. Heparan sulfate proteoglycan and laminin production by astrocytes: Its relationship to differentiation and to neurite growth. *J. Neurosci.* 8: 2844–58

Barald, K. F., Phillips, G. D., Jay, J. C., Mizukami, I. F. 1987. A component in mammalian muscle synaptic basal lamina induces clustering of acetylcholine receptors. *Prog. Brain Res.* 71: 397–408

Baron-Van Evercooren, A., Kleinman, H. K., Ohno, S., Marangos, P., Schwartz, J. P., Dubois-Dalcq, M. E. 1982. Nerve growth factor, laminin, and fibronectin promote neurite growth in human fetal sensory ganglia cultures. *J. Neurosci. Res.* 8: 179–93

Bastiani, M. J., du Lac, S., Goodman, C. S. 1986. Guidance of neuronal growth cones in the grasshopper embryo. I. Recognition

of a specific axonal pathway by the pCC neuron. *J. Neurosci.* 6: 3518–31

Berg, D. K. 1984. New neuronal growth factors. *Ann. Rev. Neurosci.* 7: 149–70

Berman, P., Gray, P., Chen, E., Keyser, K., Ehrlich, D., Karten, H., LaCorbiere, M., Esch, F., Schubert, D. 1987. Sequence analysis, cellular localization, and expression of a neuroretina adhesion and cell survival molecule. *Cell* 51: 135–42

Bignami, A., Dahl, D. 1986. Brain-specific hyaluronate-binding protein: An immunohistological study with monoclonal antibodies of human and bovine central nervous system. *Proc. Natl. Acad. Sci. USA* 83: 3518–22

Bignami, A., Dahl, D. 1988. Expression of brain-specific hyaluronectin (BHN), a hyaluronate-binding protein, in dog postnatal development. *Exp. Neurol.* 99: 107–17

Bignami, A., Delpech, B. 1985. Extracellular matrix glycoprotein (hyaluronectin) in early cerebral development. Immunofluorescence study of the rat embryo. *Int. J. Dev. Neurosci.* 3: 301–7

Bilozur, M. E., Hay, E. D. 1988. Neural crest migration in 3D extracellular matrix utilizes laminin, fibronectin, or collagen. *Dev. Biol.* 125: 19–33

Bixby, J. L., Pratt, R. S., Lilien, J., Reichardt, L. F. 1987. Neurite outgrowth on muscle cell surfaces involves extracellular matrix receptors as well as $Ca^{2+}$-dependent and -independent cell adhesion molecules. *Proc. Natl. Acad. Sci. USA* 84: 2555–59

Blumberg, B., MacKrell, A. J., Olson, P. F., Kurkinen, M., Monson, J. M., Natzle, J. E., Fessler, J. H. 1987. Basement membrane procollagen IV and its specialized carboxyl domain are conserved in *Drosophila*, mouse and human. *J. Biol. Chem.* 262: 5947–50

Bogaert, T., Brown, N., Wilcox, M. 1987. The *Drosophila* PS2 antigen is an invertebrate integrin that, like the fibronectin receptor, becomes localized to muscle attachments. *Cell* 51: 929–40

Bork, T., Schabtach, E., Grant, P. 1987. Factors guiding optic fibers in developing *Xenopus* retina. *J. Comp. Neurol.* 264: 147–58

Boucaut, J.-C., Darribere, T., Poole, T. J., Aoyama, H., Yamada, K. M., Thiery, J. P. 1984. Biologically active synthetic peptides as probes of embryonic development: A competitive peptide inhibitor of fibronectin function inhibits gastrulation in amphibian embryos and neural crest cell migration in avian embryos. *J. Cell Biol.* 99: 1822–30

Bozyczko, D., Horwitz, A. F. 1986. The participation of a putative cell surface receptor for laminin and fibronectin in peripheral neurite extension. *J. Neurosci.* 6: 1241–51

Brandan, E., Maldonado, M., Garrido, J., Inestrosa, N. C. 1985. Anchorage of collagen-tailed acetylcholinesterase to the extracellular matrix is mediated by heparan sulfate proteoglycans. *J. Cell Biol.* 101: 985–92

Brandan, E., Inestrosa, N. C. 1986. The synaptic form of acetylcholinesterase binds to cell-surface heparan sulfate proteoglycans. *J. Neurosci. Res.* 15: 185–96

Bronner-Fraser, M. 1985. Alterations in neural crest migration by a monoclonal antibody that affects cell adhesion. *J. Cell Biol.* 101: 610–17

Bronner-Fraser, M. 1986a. Analysis of the early stages of trunk neural crest migration using monoclonal antibody HNK-1. *Dev. Biol.* 115: 44–55

Bronner-Fraser, M. 1986b. An antibody to a receptor for fibronectin and laminin perturbs cranial neural crest development in vivo. *Dev. Biol.* 117: 528–36

Bronner-Fraser, M. 1987. Perturbation of cranial neural crest migration by the HNK-1 antibody. *Dev. Biol.* 123: 321–31

Bronner-Fraser, M., Lallier, T. 1988. A monoclonal antibody against a laminin-heparan sulfate proteoglycan complex perturbs cranial neural crest migration in vivo. *J. Cell Biol.* 106: 1321–29

Buck, C. A., Horwitz, A. F. 1987. Cell surface receptors for extracellular matrix molecules. *Ann. Rev. Cell Biol.* 3: 179–205

Bunge, R. P., Bunge, M. B., Eldridge, C. F. 1986. Linkage between axonal ensheathment and basal lamina production by Schwann cells. *Ann. Rev. Neurosci.* 9: 305–28

Burden, S. J., Sargent, P. B., McMahan, U. J. 1979. Acetylcholine receptors in regenerating muscle accumulate at original synaptic sites in the absence of the nerve. *J. Cell Biol.* 82: 412–25

Calof, A. L., Reichardt, L. F. 1985. Response of purified chick motoneurons to myotube conditioned medium: Laminin is essential for the substratum-binding, neurite outgrowth-promoting activity. *Neurosci. Lett.* 59: 183–89

Carbonetto, S., Gruver, M. M., Turner, D. C. 1983. Nerve fiber growth in culture on fibronectin, collagen, and glycosaminoglycan substrates. *J. Neurosci.* 3: 2324–35

Carey, D. J., Todd, M. S. 1987. Schwann cell myelination in a chemically defined medium: Demonstration of a requirement for additives that promote Schwann cell

extracellular matrix formation. *Dev. Brain Res.* 32: 95–102

Carey, D. J., Todd, M. S., Rafferty, C. M. 1986. Schwann cell myelination: Induction by exogenous basement membrane-like extracellular matrix. *J. Cell Biol.* 102: 2254–63

Carey, D. J., Rafferty, C. M., Todd, M. S. 1987. Effects of inhibition of proteoglycan synthesis on the differentiation of cultured rat Schwann cells. *J. Cell Biol.* 105: 1013–21

Carlson, S. S., Caroni, P., Kelly, R. B. 1986. A nerve terminal anchorage protein from electric organ. *J. Cell Biol.* 103: 509–20

Carlson, S. S., Wight, T. N. 1987. Nerve terminal anchorage protein 1 (TAP-1) is a chondroitin sulfate proteoglycan: Biochemical and electron microscopic characterization. *J. Cell Biol.* 105: 3075–86

Cheresh, D. A., Pierschbacher, M. D., Herzig, M. A., Kalpana, M. 1986. Disialogangliosides GD2 and GD3 are involved in the attachment of human melanoma and neuroblastoma cells to extracellular matrix proteins. *J. Cell Biol.* 102: 688–96

Cheresh, D. A., Pytela, R., Pierschbacher, M. D., Klier, F. G., Ruoslahti, E., Reisfeld, R. A. 1987. An Arg-Gly-Asp-directed receptor on the surface of human melanoma cells exists in a divalent cation-dependent functional complex with the disialoganglioside GD2. *J. Cell Biol.* 105: 1163–73

Chiquet, M., Fambrough, D. M. 1984. Chick myotendinous antigen. II. A novel extracellular glycoprotein complex consisting of large disulfide-linked subunits. *J. Cell Biol.* 98: 1937–46

Chiquet-Ehrismann, R., Mackie, E. J., Pearson, C. A., Sakakura, T. 1986. Tenascin: An extracellular matrix protein involved in tissue interactions during fetal development and oncogenesis. *Cell* 47: 131–39

Chiu, A. Y., Matthew, W. D., Patterson, P. H. 1986. A monoclonal antibody that blocks the activity of a neurite regeneration-promoting factor: Studies on the binding site and its localization in vivo. *J. Cell Biol.* 103: 1383–98

Chiu, A. Y., Sanes, J. R. 1984. Differentiation of basal lamina in synaptic and extrasynaptic portions of embryonic rat muscle. *Dev. Biol.* 103: 456–67

Chun, J. J. M., Shatz, C. J. 1988. A fibronectin-like molecule is present in the developing cat cerebral cortex and is correlated with subplate neurons. *J. Cell Biol.* 106: 857–72

Chuong, C.-M., Crossin, K. L., Edelman, G. M. 1987. Sequential expression and differential function of multiple adhesion molecules during the formation of cerebellar cortical layers. *J. Cell Biol.* 104: 331–42

Cohen, J. 1987. Neurite-outgrowth on laminin by cultured retinal ganglion cells is regulated by the tectum. *Soc. Neurosci. Abstr.* 13: 1483

Cohen, J., Burne, J. F., Winter, J., Bartlett, P. 1986. Retinal ganglion cells lose response to laminin with maturation. *Nature* 322: 465–67

Cohen, J., Burne, J. F., McKinlay, C., Winter, J. 1987. The role of laminin and the laminin/fibronectin receptor complex in the outgrowth of retinal ganglion cell axons. *Dev. Biol.* 122: 407–18

Cole, G. J., Glaser, L. 1986. A heparin-binding domain from N-CAM is involved in neural cell-substratum adhesion. *J. Cell Biol.* 102: 403–12

Cole, G. J., Schubert, D., Glaser, L. 1985. Cell-substratum adhesion in chick neural retina depends upon protein-heparan sulfate interactions. *J. Cell Biol.* 100: 1192–99

Collins, F. 1978. Induction of neurite outgrowth by a conditioned-medium factor bound to the culture substratum. *Proc. Natl. Acad. Sci. USA* 75: 5210–13

Collins, F., Garrett, J. E. Jr. 1980. Elongating nerve fibers are guided by a pathway of material released from embryonic nonneuronal cells. *Proc. Natl. Acad. Sci. USA* 77: 6226–28

Collins, F., Lee, M. R. 1984. The spatial control of ganglionic neurite growth by the substrate-associated material from conditioned medium: An experimental model of haptotaxis. *J. Neurosci.* 4: 2823–29

Connor, E. A., McMahan, U. J. 1987. Cell accumulation in the junctional region of denervated muscle. *J. Cell Biol.* 104: 109–20

Cornbrooks, C. J., Carey, D. J., McDonald, J. A., Timpl, R., Bunge, R. P. 1983. In vivo and in vitro observations on laminin production by Schwann cells. *Proc. Natl. Acad. Sci, USA* 80: 3850–54

Covault, J. 1988. Molecular biology of cell adhesion in neural development. *Frontiers Mol. Biol.* In press

Covault, J., Cunningham, J. M., Sanes, J. R. 1987. Neurite outgrowth on cryostat sections of innervated and denervated skeletal muscle. *J. Cell Biol.* 105: 2479–88

Crossin, K. L., Hoffman, S., Grumet, M., Thiery, J.-P., Edelman, G. M. 1986. Site-restricted expression of cytotactin during development of the chicken embryo. *J. Cell Biol.* 102: 1917–30

Cunningham, L. W., Ed. 1987. Structural and contractile proteins. Part D, Extra-cellular matrix. *Methods in Enzymology,* Vol. 144. New York: Academic

Davis, G. E., Manthorpe, M., Varon, S. 1985a. Parameters of neuritic growth from ciliary ganglion neurons in vitro: Influence of laminin, Schwannoma polyornithine-binding neurite promoting factor and ciliary neuronotrophic factor. *Dev. Brain Res.* 17: 75–84

Davis, G. E., Manthorpe, M., Engvall, E., Varon, S. 1985b. Isolation and characterization of rat Schwannoma neurite-promoting factor: Evidence that the factor contains laminin. *J. Neurosci.* 5: 2662–71

Davis, G. E., Klier, G., Engvall, E., Cornbrooks, C., Varon, S., Manthorpe, M. 1987a. Association of laminin with heparan and chondroitin sulfate-bearing proteoglycans in neurite-promoting factor complexes from rat Schwannoma cells. *Neurochem. Res.* 12: 909–21

Davis, G. E., Engvall, E., Varon, S., Manthorpe, M. 1987b. Human amnion membrane as a substrtum for cultured peripheral and central nervous system neurons. *Dev. Brain Res.* 33: 1–10

Davis, G. E., Blaker, S. N., Engvall, E. Varon, S., Manthorpe, M., Gage, F. H. 1987c. Human amnion membrane serves as a substratum for growing axons in vitro and in vivo. *Science* 236: 1106–9

Delpech, A., Girard, N., Delpech, B. 1982. Localization of hyaluronectin in the nervous system. *Brain Res.* 245: 251–57

Dohrmann, U., Edgar, D., Sendtner, M., Thoenen, H. 1986. Muscle-derived factors that support survival and promote fiber outgrowth from embryonic chick spinal motor neurons in culture. *Dev. Biol.* 118: 209–21

Douville, P., Harvey, W., Carbonetto, S. 1987. Identification and purification of a high affinity laminin receptor from embryonic chick brain: Evidence for developmental regulation in the CNS. *Soc. Neurosci. Abstr.* 13: 1482

Dow, K. E., Mirski, S., Roder, J. C., Riopelle, R. J. 1988. Neuronal proteoglycans: Biosynthesis and funtional interaction with neurons in vitro. *J. Neurosci.* 8: 3278–89

Duband, J.-L., Rocher, S., Chen, W.-T., Yamada, K. M., Thiery, J. P. 1986. Cell adhesion and migration in the early vertebrate embryo: Location and possible role of the putative fibronectin receptor complex receptor complex. *J. Cell Biol.* 102: 160–78

Duband, J.-L., Thiery, J. P. 1987. Distribution of laminin and collagens during avian neural crest development. *Development* 101: 461–78

Edelman, G. M. 1986. Cell adhesion molecules in the regulation of animal form and tissue pattern. *Ann. Rev. Cell Biol.* 2: 81–116

Edgar, D., Timpl, R., Thoenen, H. 1984. The heparin-binding domain of laminin is responsible for its effects on neurite outgrowth and neuronal survival. *EMBO J.* 3: 1463–68

Edgar, D., Timpl, R., Thoenen, H. 1988. Structural requirements for the stimulation of neurite outgrowth by two variants of laminin and their inhibition by antibodies. *J. Cell Biol.* 106: 1299–1306

Ekblom, P., Vestweber, D., Kemler, R. 1986. Cell-matrix interactions and cell adhesion during development. *Ann. Rev. Cell Biol.* 2: 27–47

Eldridge, C. F., Sanes, J. R., Chiu, A. Y., Bunge, R. P., Cornbrooks, C. J. 1986. Basal lamina-associated heparan sulfate proteoglycan in the rat peripheral nervous system: Characterization and localization using monoclonal antibodies. *J. Neurocytol.* 15: 37–51

Eldridge, C. F., Bunge, M. B., Bunge, R. P., Wood, P. M. 1987. Differentiation of axon-related Schwann cells in vitro. I. Ascorbic acid regulates basal lamina assembly and myelin formation. *J. Cell Biol.* 105: 1023–34

Eldridge, C. F., Bunge, M. B., Bunge, R. P. 1989. Differentiation of axon-related Schwann cells in vitro. II. Control of myelin formation by basal lamina. *J. Neurosci.* In press

Engvall, E., Davis, G. E., Dickerson, K., Ruoslahti, E., Varon, S., Manthorpe, M. 1986. Mapping of domains in human laminin using monoclonal antibodies: Localization of the neurite-promoting site. *J. Cell Biol.* 103: 2457–65

Erickson, H. P., Taylor, H. C. 1987. Hexabrachion proteins in embryonic chicken tissues and human tumors. *J. Cell Biol.* 105: 1387–94

Faissner, A., Kruse, J., Chiquet-Ehrismann, R., Mackie, E. 1988. The high-molecular-weight J1 glycoproteins are immunochemically related to tenascin. *Differentiation* 37: 104–14

Faivre-Bauman, A., Puymirat, J., Loudes, C., Barret, A., Tixier-Vidal, A. 1984. Laminin promotes attachment and neurite elongation of fetal hypothalamic neurons grown in serum-free medium. *Neurosci. Lett.* 44: 83–89

Fallon, J. R. 1987. Localization of a synaptic organizing molecule in developing muscle. *Soc. Neurosci. Abstr.* 13: 374

Fallon, J. R., Nitkin, R. M., Reist, N. E., Wallace, B. G., McMahan, U. J. 1985. Acetylcholine receptor-aggregating factor is similar to molecules concentrated at neuromuscular junctions. *Nature* 315: 571–74

Fessler, L. I., Campbell, A. G., Duncan, K. G., Fessler, J. H. 1987. *Drosophila* laminin: Characterization and localization. *J. Cell Biol.* 105: 2383–91

ffrench-Constant, C., Miller, R. H., Kruse, J., Schachner, M., Raff, M. C. 1986. Molecular specialization of astrocyte processes at nodes of Ranvier in rat optic nerve. *J. Cell Biol.* 102: 844–52

Ford-Holevinski, T. S., Hopkins, J. M., McCoy, J. P., Agranoff, B. W. 1986. Laminin supports neurite outgrowth from explants of axotomized adult rat retinal neurons. *Dev. Brain Res.* 28: 121–26

Gatchalian, C., Chiu, A. Y., Sanes, J. R. 1985. Monoclonal antibodies to laminin that distinguish synaptic and extrasynaptic domains in muscle fiber basal lamina. *J. Cell Biol.* 101: 89a

Gatchalian, C. L., Sanes, J. R. 1987. Fibroblasts from denervated muscle synthesize NCAM, J1 and fibronectin. *Soc. Neurosci. Abstr.* 13: 375

Gatchalian, C. L., Schachner, M., Sanes, J. R. 1989. Synthesis of NCAM, J1, fibronectin and a heparan sulfate proteoglycan by fibroblasts that proliferate near synaptic sites in denervated muscle. *J. Cell Biol.* In press

Giftochristos, N., David, S. 1988. Laminin and heparan sulphate proteoglycan in the lesioned adult mammalian central nervous system, and their possible relationship to axonal sprouting. *J. Neurocytol.* 17: 385–97

Glasby, M. A., Gschmeissner, S. G., Hitchcock, R. J. I., Huang, C. L.-H. 1986. The dependence of nerve regeneration through muscle grafts in the rat on the availability and orientation of basement membrane. *J. Neurocytol.* 15: 497–510

Glicksman, M., Sanes, J. R. 1983. Development of motor nerve terminals formed in the absence of muscle fibers. *J. Neurocytol.* 12: 661–71

Godfrey, E. W., Nitkin, R. M., Wallace, B. G., Rubin, L. L., McMahan, U. J. 1984. Components of *Torpedo* electric organ and muscle that cause aggregation of acetylcholine receptors on cultured muscle cells. *J. Cell Biol.* 99: 615–27

Godfrey, E. W., Dietz, M. E., Morstad, A. L., Wallskog, P. A., Yorde, D. E. 1988. Acetylcholine receptor-aggregating proteins are associated with the extracellular matrix of many tissues in *Torpedo*. *J. Cell Biol.* In press

Godfrey, E. W., Siebenlist, R. E., Wallskog, P. A., Walters, L. M., Bolender, D. L., Yorde, D. E. 1988. Basement membrane components are concentrated in pre-muscle masses and at early acetylcholine receptor clusters in chick embryo hindlimb muscles. *Dev. Biol.* In press

Gratecos, D., Naidet, C., Astier, M., Thiery, J. P., Semeriva, M. 1988. *Drosophila* fibronectin: A protein that shared properties similar to those of its mammalian homologue. *EMBO J.* 7: 215–23

Grumet, M., Hoffman, S., Crossin, K. L., Edelman, G. M. 1985. Cytotactin, an extracellular matrix protein of neural and non-neural tissues that mediates glianeuron interaction. *Proc. Natl. Acad. Sci. USA* 82: 8075–79

Gundersen, R. W. 1987a. Response of sensory neurites and growth cones to patterned substrata of laminin and fibronectin in vitro. *Dev. Biol.* 121: 423–32

Gundersen, R. W. 1987b. Response of chick sensory growth cones to laminin and fibronectin: External morphology. *Soc. Neurosci. Abstr.* 13: 1484

Hall, D. E., Neugebauer, K. M., Reichardt, L. F. 1987. Embryonic neural retinal cell response to extracellular matrix proteins: Developmental changes and effects of the cell substratum attachment antibody (CSAT). *J. Cell Biol.* 104: 623–34

Hammarback, J. A., Palm, S. L., Furcht, L. T., Letourneau, P. C. 1985. Guidance of neurite outgrowth by pathways of substratum-adsorbed laminin. *J. Neurosci. Res.* 13: 213–20

Hammarback, J. A., Letourneau, P. C. 1986. Neurite extension across regions of low cell-substratum adhesivity: Implications for the guidepost hypothesis of axonal pathfinding. *Dev. Biol.* 117: 655–62

Hansen-Smith, F. M. 1986. Formation of acetylcholine receptor clusters in mammalian sternohyoid muscle regenerating in the absence of nerves. *Dev. Biol.* 118: 129–40

Hantaz-Ambroise, D., Vigny, M., Koenig, J. 1987. Heparan sulfate proteoglycan and laminin mediate two different types of neurite outgrowth. *J. Neurosci.* 7: 2293–2304

Hassell, J. R., Kimura, J. H., Hascall, V. C. 1986. Proteoglycan core protein families. *Ann. Rev. Biochem.* 55: 539–67

Hatten, M. E., Furie, M. B., Rifkin, D. B. 1982. Binding of developing mouse cerebellar cells to fibronectin: A possible mechanism for the formation of the external granular layer. *J. Neurosci.* 2: 1195–1206

Hoffman, S., Edelman, G. M. 1987. A proteoglycan with HNK-1 antigenic determinants is a neuron-associated ligand for cytotactin. *Proc. Natl. Acad. Sci. USA* 84: 2523–27

Hoffman, S., Crossin, K. L., Edelman, G. M.

1988. Molecular forms, binding functions, and developmental expression patterns of cytotactin and cytotactin-binding proteoglycan, an interactive pair of extracellular matrix molecules. *J. Cell Biol.* 106: 519–32

Hopkins, J. M., Ford-Holevinski, T. S., McCoy, J. P., Agranoff, B. W. 1985. Laminin and optic nerve regeneration in the goldfish. *J. Neurosci.* 5: 3030–38

Humphries, M. J., Akiyama, S. K., Komoriya, A., Olden, K., Yamada, K. M. 1989. Neurite extension of chicken peripheral nervous system neurons on fibronectin: Relative importance of specific adhesion sites in the central cell-binding domain and the alternatively-spliced type III connecting segment. *J. Cell Biol.* 106: 1289–97

Hunter, D. D., Merlie, J. P., Sanes, J. R. 1988. JS-1, a component of synaptic basal lamina at the neuromuscular junction. *Soc. Neurosci. Abstr.* 14: 894

Hunter, D. D., Sanes, J. R., Chiu, A. Y. 1987. An antigen concentrated in the basal lamina of the neuromuscular junction. *Soc. Neurosci. Abstr.* 13: 375

Hynes, R. O. 1987. Fibronectins: A family of complex and versatile adhesive glycoproteins derived from a single gene. *Harvey Lect.*, Ser. 81, pp. 133–52

Ide, C. 1984. Nerve regeneration through the basal lamina scaffold of the skeletal muscle. *Neurosci. Res.* 1: 379–91

Ide, C. 1986. Basal laminae and Meissner corpuscle regeneration. *Brain Res.* 384: 311–22

Ide, C. 1987. Role of extracellular matrix in the regeneration of a Pacinian corpuscle. *Brain Res.* 413: 155–69

Ide, C., Tohyama, K., Yokota, R., Nitatori, T., Onodera, S. 1983. Schwann cell basal lamina and nerve regeneration. *Brain Res.* 288: 61–75

Jones, F. S., Burgoon, M. P., Hoffman, S., Crossin, K. L., Cunningham, B. A., Edelman, G. M. 1988. A cDNA clone for cytotactin contains sequences similar to epidermal growth factor-like repeats and segments of fibronectic and fibrinogen. *Proc. Natl. Acad. Sci. USA* 85: 2186–90

Kalb, R., Hockfield, S. 1988. Molecular evidence for experience-dependent development of hamster motorneurons. *J. Neurosci.* 8: 2350–60

Kalcheim, C., Duksin, D., Vogel, Z. 1982a. Involvement of collagen in the aggregation of acetylcholine receptors on cultured muscle cells. *J. Biol. Chem.* 257: 12722–27

Kalcheim, C., Vogel, Z., Duksin, D. 1982b. Embryonic brain extract induces collagen biosynthesis in cultured muscle cells:

Involvement in acetylcholine receptor aggregation. *Proc. Natl. Acad. Sci. USA* 79: 3077–81

Kalcheim, C., Duksin, D., Bachar, E., Vogel, Z. 1985a. Collagen-stimulating factor from embryonic brain has ascorbate-like activity and stimulates prolyl hydroxylation in cultured muscle cells. *Eur. J. Biochem.* 146: 227–32

Kalcheim, C., Bachar, E., Duksin, D., Vogel, Z. 1985b. Ciliary ganglia and spinal cord explants release an ascorbate-like compound which stimulates proline hydroxylation and collagen formation in muscle cultures. *Neurosci. Lett.* 58: 219–24

Keynes, R. J., Hopkins, W. G., Huang, C. L. H. 1984. Regeneration of mouse peripheral nerves in degenerating skeletal muscle: Guidance by residual muscle fibre basement membrane. *Brain Res.* 295: 275–81

Kleinman, H. K., Ogle, R. C., Cannon, F. B., Little, C. C., Sweeney, T. M., Luckenbill-Edds, L. 1988. Laminin receptors for neurite formation. *Proc. Natl. Acad. Sci. USA* 85: 1282–86

Kleitman, N., Wood, P., Johnson, M. I., Bunge, R. P. 1988. Schwann cell surfaces but not extracellular matrix organized by Schwann cells support neurite outgrowth from embryonic rat retina. *J. Neurosci.* 8: 653–63

Krayanek, S. 1980. Structure and orientation of extracellular matrix in developing chick optic tectum. *Anat. Rec.* 197: 95–109

Krotoski, D., Bronner-Fraser, M. 1987. Distribution of laminin and integrin along trunk neural crest pathways in *Xenopus laevis. Soc. Neurosci. Abstr.* 13: 1637

Krotoski, D. M., Domingo, C., Bronner-Fraser, M. 1986. Distribution of a putative cell surface receptor for fibronectin and laminin in the avian embryo. *J. Cell Biol.* 103: 1061–71

Kruse, J., Keilhauer, G., Faissner, A., Timpl, R., Schachner, M. 1985. The J1 glycoprotein—a novel nervous system cell adhesion molecule of the L2/HNK-1 family. *Nature* 316: 146–48

Krusius, T., Reinhold, V. N., Margolis, R. K., Margolis, R. U. 1987. Structural studies on sialylated and sulphated O-glycosidic mannose-linked oligosaccharides in the chondroitin sulphate proteoglycan of brain. *Biochem. J.* 245: 229–34

Kuffler, D. P. 1986. Accurate reinnervation of motor end plates after disruption of sheath cells and muscle fibers. *J. Comp. Neurol.* 250: 228–35

Kunemund, V., Jungalwala, F. B., Fischer, G., Chou, D. K. H., Keilhauer, G., Schachner, M. 1988. The L2/HNK-1

carbohydrate of neural cell adhesion molecules is involved in cell interactions. *J. Cell Biol.* 106: 213–23

Laitinen, J., Lopponen, R., Merenmies, J., Rauvala, H. 1987. Binding of laminin to brain gangliosides and inhibition of laminin-neuron interaction by the gangliosides. *FEBS Lett.* 217: 94–100

Lander, A. D., Fujii, D. K., Gospodarowicz, D., Reichardt, L. F. 1982. Characterization of a factor that promotes neurite outgrowth: Evidence linking activity to a heparan sulfate proteoglycan. *J. Cell Biol.* 94: 574–85

Lander, A. D., Fujii, D. K., Reichardt, L. F. 1985a. Laminin is associated with the "neurite outgrowth-promoting factors" found in conditioned media. *Proc. Natl. Acad. Sci. USA* 82: 2183–87

Lander, A. D., Fujii, D. K., Reichardt, L. F. 1985b. Purification of a factor that promotes neurite outgrowth: Isolation of laminin and associated molecules. *J. Cell Biol.* 101: 898–913

Lander, A. D., Tomaselli, K., Calof, A. L., Reichardt, L. F. 1983. Studies on extracellular matrix components that promote neurite outgrowth. *Cold Spring Harbor Symp. Quant. Biol.* 48: 611–24

Lappin, R. I., Rubin, L. L., Lieberburg, I. M. 1987. Generation of subunit-specific antibody probes for *Torpedo* acetylcholinesterase: Cross-species reactivity and use in cell-free translations. *J. Neurobiol.* 18: 75–99

Le Douarin, N. M. 1986. Cell line segregation during peripheral nervous system ontogeny. *Science* 231: 1515–21

Letourneau, P. C. 1975. Cell-to-substratum adhesion and guidance of axonal elongation. *Dev. Biol.* 44: 92–101

Letourneau, P. C., Madsen, A. M., Palm, S. L., Furcht, L. T. 1989. Immunoreactivity for laminin in the developing ventral longitudinal pathway of the brain. *Dev. Biol.* 125: 135–44

Liesi, P. 1985a. Do neurons in the vertebrate CNS migrate on laminin? *EMBO J.* 4: 1163–70

Liesi, P. 1985b. Laminin-immunoreactive glia distinguish regenerative adult CNS systems from non-regenerative ones. *EMBO J.* 4: 2505–11

Liesi, P., Dahl, D., Vaheri, A. 1983. Laminin is produced by early rat astrocytes in primary culture. *J. Cell Biol.* 96: 920–24

Liesi, P., Dahl, D., Vaheri, A. 1984a. Neurons cultured from developing rat brain attach and spread preferentially to laminin. *J. Neurosci. Res.* 11: 241–51

Liesi, P., Kaakkola, S., Dahl, D., Vaheri, A. 1984b. Laminin is induced in astrocytes of adult brain by injury. *EMBO J.* 3: 683–86

Liesi, P., Kirkwood, T., Vaheri, A. 1986. Fibronectin is expressed by astrocytes cultured from embryonic and early postnatal rat brain. *Exp. Cell Res.* 163: 175–85

Loring, J. F., Erickson, C. A. 1987. Neural crest cell migratory pathways in the trunk of the chick embryo. *Dev. Biol.* 121: 220–36

Mackie, E. J., Tucker, R. P., Halfter, W., Chiquet-Ehrismann, R., Epperlein, H. H. 1988. The distribution of tenascin coincides with pathways of neural crest cell migration. *Development* 102: 237–50

Magill, C., Wallace, B. G., McMahan, U. J. 1987. Molecules similar to agrin are concentrated in motor neurons. *Soc. Neurosci. Abstr.* 13: 373

Manthorpe, M., Engvall, E., Ruoslahti, E., Longo, F. M., Davis, G. E., Varon, S. 1983. Laminin promotes neuritic regeneration from cultured peripheral and central neurons. *J. Cell Biol.* 97: 1882–90

Margolis, R. K., Ripellino, J. A., Goossen, B., Steinbrich, R., Margolis, R. U. 1987. Occurrence of the HNK-1 epitope (3-sulfoglucuronic acid) in PC12 pheochromocytoma cells, chromaffin granule membranes, and chondroitin sulfate proteoglycans. *Biochem. Biophys. Res. Commun.* 145: 1142–48

Martin, G. R., Timpl, R. 1987. Laminin and other basement membrane components. *Ann. Rev. Cell Biol.* 3: 57–85

Martini, R., Covault, J., Schachner, M., Sanes, J. R. 1989. Inhibition of neurite outgrowth on sciatic nerve by antibodies to the adhesion molecule L1. Submitted for publication

Martins-Green, M., Erickson, C. A. 1986. Development of neural tube basal lamina during neurulation and neural crest cell emigration in the trunk of the mouse embryo. *J. Embryol. Exp. Morph.* 98: 219–36

Martins-Green, M., Erickson, C. A. 1987. Basal lamina is not a barrier to neural crest cell emigration: Documentation by TEM and by immunofluorescent and immunogold labelling. *Development* 101: 517–33

Matthew, W. D., Greenspan, R. J., Lander, A. D., Reichardt, L. F. 1985. Immunopurification and characterization of a neuronal heparan sulfate proteoglycan. *J. Neurosci.* 5: 1842–50

Matthew, W. D., Patterson, P. H. 1983. The production of a monoclonal antibody that blocks the action of a neurite outgrowth-promoting factor. *Cold Spring Harbor Symp. Quant. Biol.* 48: 625–31

Maxwell, G. D., Forbes, M. E. 1987. Exogenous basement membrane-like matrix stimulate adrenergic development in avian neural crest cultures. *Development* 101: 767–76

Mayne, R., Burgeson, R. E., eds. 1987. *Structure and Function of Collagen Types.* New York: Academic

McLoon, S. C. 1986. Response of astrocytes in the visual system to Wallerian degeneration: An immunohistochemical analysis of laminin and glial fibrillary acidic protein (GFAP). *Exp. Neurol.* 91: 613–21

McLoon, S. C., McLoon, L. K., Palm, S. L., Furcht, L. T. 1988. Transient expression of laminin in the optic nerve of the developing rat. *J. Neurosci.* 8: 1981–90

McMahan, U. J., Slater, C. R. 1984. The influence of basal lamina on the accumulation of acetylcholine receptors at synaptic sites in regenerating muscle. *J. Cell Biol.* 98: 1453–73

Montell, D. J., Goodman, C. S. 1988. *Drosophila* substrate adhesion molecule: Sequence of laminin B1 chain reveals domains of homology with mouse. *Cell* 53: 463–73

Morris, J. E., Yanagishita, M., Hascall, V. 1987. Proteoglycans synthesized by embryonic chicken retina in culture: Composition and compartmentalization. *Arch. Biochem. Biophys.* 258: 206–18

Mugnai, G., Lewandowska, K., Carnemolla, B., Zardi, L., Culp, L. A. 1988. Modulation of matrix adhesive responses of human neuroblastoma cells by neighboring sequences in the fibronectins. *J. Cell Biol.* 106: 931–43

Nakanishi, S. 1983. Extracellular matrix during laminar pattern formation of neocortex in normal and reeler mutant mice. *Dev. Biol.* 95: 305–16

Nardi, J. B. 1983. Neuronal pathfinding in developing wings of the moth *Manduca sexta. Dev. Biol.* 95: 163–74

Nitkin, R. M., Smith, M. A., Magill, C., Fallon, J. R., Yao, Y.-M. M., Wallace, B. G., McMahan, U. J. 1987. Identification of agrin, a synaptic organizing protein from *Torpedo* electric organ. *J. Cell Biol.* 105: 2471–78

O'Shea, K. S., Dixit, V. M. 1987. Thrombospondin in the developing cerebellar cortex. *Soc. Neurosci. Abstr.* 13: 1114

Payette, R. F., Tennyson, V. M., Pomeranz, H. D., Pham, T. D., Rothman, T. P., Gershon, M. D. 1988. Accumulation of components of basal laminae: Association with the failure of neural crest cells to colonize the presumptive aganglionic bowel of ls/ls mutant mice. *Dev. Biol.* 125: 341–60

Pearlman, A. L., Kim, H. G., Schmitt, G. 1986. Early cortical afferents arrive after fibronectin-like immunoreactivity appears in their migratory pathway. *Soc. Neurosci. Abstr.* 12: 502

Perris, R., Johansson, S. 1987. Amphibian neural crest cell migration on purified extracellular matrix components: A chondroitin sulfate proteoglycan inhibits locomotion on fibronectin substrates. *J. Cell Biol.* 105: 2511–21

Pesheva, P., Horwitz, A. F., Schachner, M. 1989. Integrin, the cell surface receptor for fibronectin and laminin, is a member of the L2/HNK-1 family of adhesion molecules. *Neurosci. Lett.* 83: 303–6

Pincon-Raymond, M., Murawsky, M., Mege, R.-M., Rieger, F. 1987. Abnormal enwrapment of intramuscular axons by distal Schwann cells with defective basal lamina in the muscular dysgenic mouse embryo. *Dev. Biol.* 124

Pomeranz, H. D., Payette, R. F., Gershon, M. D. 1987. Migration of sacral neural crest cells to the avian bowel and ganglion of Remak: Relationship to laminin. *Soc. Neurosci. Abstr.* 13: 686

Price, J., Hynes, R. O. 1985. Astrocytes in culture synthesize and secrete a variant form of fibronectin. *J. Neurosci.* 5: 2205–11

Rauvala, H. 1984. Neurite outgrowth of neuroblastoma cells: Dependence on adhesion surface-cell surface interactions. *J. Cell Biol.* 98: 1010–16

Ramon y Cajal, S. 1928. *Degeneration and Regeneration of the Nervous System.* London: Hafner. (Reprinted 1968)

Reist, N. E., Magill, C., McMahan, U. J. 1987. Agrin-like molecules at synaptic sites in normal, denervated and damaged skeletal muscles. *J. Cell Biol.* 105: 2457–69

Rickmann, M., Fawcett, J. W., Keynes, R. J. 1985. The migration of neural crest cells and the growth of motor axons through the rostral half of the chick somite. *J. Embryol. Exp. Morph.* 90: 437–55

Rieger, F., Daniloff, J. K., Pincon-Raymond, M., Crossin, K. L., Grumet, M., Edelman, G. M. 1986. Neuronal cell adhesion molecules and cytotactin are colocalized at the node of ranvier. *J. Cell Biol.* 103: 379–91

Riggott, M. J., Moody, S. A. 1987. Distribution of laminin and fibronectin along peripheral trigeminal axon pathways in the developing chick. *J. Comp. Neurol.* 258: 580–96

Ripellino, J. A., Bailo, M., Margolis, R. U., Margolis, R. K. 1988. Light and electron microscopic studies on the localization of hyaluronic acid in developing rat cerebellum. *J. Cell Biol.* 106: 845–55

Ripellino, J. A., Klinger, M. M., Margolis, R. U., Margolis, R. K. 1985. The hyaluronic acid binding region as a specific probe for the localization of hyaluronic acid in tissue sections. *J. Histochem. Cytochem.* 33: 1060–66

Rogers, S. L., Letourneau, P. C., Palm, S. L., McCarthy, J., Furcht, L. T. 1983. Neurite extension by peripheral and central nervous system neurons in response to substratum-bound fibronectin and laminin. *Dev. Biol.* 98: 212–20

Rogers, S. L., McCarthy, J. B., Palm, S. L., Furcht, L. T., Letourneau, P. C. 1985. Neuron-specific interactions with two neurite-promoting fragments of fibronectin. *J. Neurosci.* 5: 369–78

Rogers, S. L., Edson, K. J., Letourneau, P. C., McLoon, S. C. 1986. Distribution of laminin in the developing peripheral nervous system of the chick. *Dev. Biol.* 113: 429–35

Rogers, S. L., Letourneau, P. C., Peterson, B. A., Furcht, L. T., McCarthy, J. B. 1987. Selective interaction of peripheral and central nervous system cells with two distinct cell-binding domains of fibronectin. *J. Cell Biol.* 105: 1435–42

Rovasio, R. A., Delouvee, A., Yamada, K. M., Timpl, R., Thiery, J.-P. 1983. Neural crest cell migration: Requirements for exogenous fibronectin and high cell density. *J. Cell Biol.* 96: 462–73

Rubin, L. L., McMahan, U. J. 1982. Regeneration of the neuromuscular junction: Steps toward defining the molecular basis of the interaction between nerve and muscle. In *Disorders of the Motor Unit,* ed. D. L. Schotland, pp. 187–96. New York: Wiley

Ruoslahti, E. 1988. Fibronectin and its receptors. *Ann. Rev. Biochem.* 57: 375–413

Ruoslahti, E., Pierschbacher, M. D. 1987. New perspectives in cell adhesion: RGD and integrins. *Science* 238: 491–97

Salpeter, M. M., ed. 1987. *The Vertebrate Neuromuscular Junction. Neurol. Neurobiol.,* Vol. 23. New York: Liss

Sandrock, A. W. Jr., Matthew, W. D. 1987a. Identification of a peripheral nerve neurite growth-promoting activity by development and use of an in vitro bioassay. *Proc. Natl. Acad. Sci. USA* 84: 6934–38

Sandrock, A. W. Jr., Matthew, W. D. 1987b. An in vitro neurite-promoting antigen functions in axonal regeneration in vivo. *Science* 237: 1605–8

Sanes, J. R. 1982. Laminin, fibronectin and collagen in synaptic and extrasynaptic portions of muscle fiber basement membrane. *J. Cell Biol.* 93: 442–51

Sanes, J. R. 1983. Roles of extracellular matrix in neural development. *Ann. Rev. Physiol.* 45: 581–600

Sanes, J. R. 1986. The extracellular matrix.

In *Myology*, ed. A. G. Engel, B. Q. Banker, Chap. 4. New York: McGraw-Hill

Sanes, J. R., Marshall, L. M., McMahan, U. J. 1978. Reinnervation of muscle fiber basal lamina after removal of myofibers. Differentiation of regenerating axons at original synaptic sites. *J. Cell Biol.* 78: 176–98

Sanes, J. R., Cheney, J. M. 1982. Lectin-binding reveals a synapse-specific carbohydrate in skeletal muscle. *Nature* 300: 646–47

Sanes, J. R., Chiu, A. Y. 1983. The basal lamina of the neuromuscular junction. *Cold Spring Harbor Symp. Quant. Biol.* 48: 667–78

Sanes, J. R., Lawrence, J. C. 1983. Activity-dependent accumulation of basal lamina by cultured rat myotubes. *Dev. Biol.* 97: 123–36

Sanes, J. R., Feldman, D. H., Cheney, J. M., Lawrence, J. C. 1984. Brain extract induces synaptic characteristics in the basal lamina of cultured myotubes. *J. Neurosci.* 4: 464–73

Sanes, J. R., Schachner, M., Covault, J. 1986. Expression of several adhesive macromolecules (N-CAM, L1, J1, NILE, uvomorulin, laminin, fibronectin, and a heparan sulfate proteoglycan) in embryonic, adult and denervated adult skeletal muscles. *J. Cell Biol.* 102: 420–31

Scherer, S. S., Easter, S. Jr. 1984. Degenerative and regenerative changes in the trochlear nerve of goldfish. *J. Neurocytol.* 13: 519–65

Schmalenbach, C., Matthiessen, H. P., Muller, H. W. 1987. Both fibronectin and laminin are neurite growth inducing components of rat astroglial conditioned medium. *Soc. Neurosci. Abstr.* 13: 1486

Schubert, D., LaCorbiere, M. 1985. Isolation of an adhesion-mediating protein from chick neural retina adherons. *J. Cell Biol.* 101: 1071–77

Scott, L. J. C., Bacou, F., Sanes, J. R. 1988. A synapse-specific carbohydrate at the neuromuscular junction: Association with both acetylcholinesterase and a glycolipid. *J. Neurosci.* 8: 932–44

Slater, C. R., Allen, E. G. 1985. Acetylcholine receptor distribution on regenerating mammalian muscle fibers at sites of mature and developing nerve-muscle junctions. *J. Physiol. Paris* 80: 238–46

Smalheiser, N. R., Crain, S. M., Reid, L. M. 1984. Laminin as a substrate for retinal axons in vitro. *Dev. Brain Res.* 12: 136–40

Smalheiser, N. R., Schwartz, N. B. 1987. Cranin: A laminin-binding protein of cell membranes. *Proc. Natl. Acad. Sci. USA* 84: 6457–61

Smith, M. A., Wallace, B. G., Yao, Y.-M.,

Schilling, J. W., Snow, P., McMahan, U. J. 1987. Purification and characterization of agrin. *Soc. Neurosci. Abstr.* 13: 374

Sternberg, J., Kimber, S. J. 1986a. Distribution of fibronectin, laminin and entactin in the environment of migrating neural crest cells in early mouse embryos. *J. Embryol. Exp. Morph.* 91: 267–82

Sternberg, J., Kimber, S. J. 1986b. The relationship between emerging neural crest cells and basement membranes in the trunk of the mouse embryo: A TEM and immunocytochemical study. *J. Embryol. Exp. Morph.* 98: 251–68

Stewart, G. R., Pearlman, A. L. 1987. Fibronectin-like immunoreactivity in the developing cerebral cortex. *J. Neurosci.* 7: 3325–33

Tan, S.-S., Crossin, K. L., Hoffman, S., Edelman, G. M. 1987. Asymmetric expression in somites of cytotactin and its proteoglycan ligand is correlated with neural crest cell distribution. *Proc. Natl. Acad. Sci. USA* 84: 7977–81

Thompson, J. M., Pelto, D. J. 1982. Attachment, survival and neurite extension of chick embryo retinal neurons on various culture substrates. *Dev. Neurosci.* 5: 447–57

Timpl, R., Dziadek, M. 1986. Structure, development and molecular pathology of basement membranes. *Int. Rev. Exp. Pathol.* 29: 1–112

Tobey, S. L., McClelland, K. J., Culp, L. A. 1985. Neurite extension by neuroblastoma cells on substratum-bound fibronectin's cell-binding fragment but not on the heparan sulfate-binding fragment, platelet factor-4. *Exp. Cell Res.* 158: 395–412

Tomaselli, K. J., Reichardt, L. F., Bixby, J. L. 1986. Distinct molecular interactions mediate neuronal process outgrowth on non-neuronal cell surfaces and extracellular matrices. *J. Cell Biol.* 103: 2659–72

Tomaselli, K. J., Damsky, C. H., Reichardt, L. F. 1987. Interactions of a neuronal cell line (PC12) with laminin, collagen IV, and fibronectin: Identification of integrin-related glycoproteins involved in attachment and process outgrowth. *J. Cell Biol.* 105: 2347–58

Unsicker, K., Skaper, S. D., Varon, S. 1985a. Developmental changes in the responses of rat chromaffin cells to neuronotrophic and neurite-promoting factors. *Dev. Biol.* 111: 425–33

Unsicker, K., Skaper, S. D., Davis, G. E., Manthorpe, M., Varon, S. 1985b. Comparison of the effects of laminin and the polyornithine-binding neurite promoting factor from RN22 Schwannoma cells on neurite regeneration from cultured newborn and adult rat dorsal root ganglion neurons. *Dev. Brain Res.* 17: 304–8

Vlodavsky, I., Levi, A., Lax, I., Fuks, Z., Schlessinger, J. 1982. Induction of cell attachment and morphological differentiation in a pheochromocytoma cell line and embryonal sensory cells by the extracellular matrix. *Dev. Biol.* 93: 285–300

Vogel, Z., Christian, C. N., Vigny, M., Bauer, H.-C., Sonderegger, P., Daniels, M. P. 1983. Laminin induces acetylcholine receptor aggregation on cultured myotubes and enhances the receptor aggregation activity of a neuronal factor. *J. Neurosci.* 3: 1058–58

Waite, K. A., Mugnai, G., Culp, L. A. 1987. A second cell-binding domain on fibronectin (RGDS-independent) for neurite extension of human neuroblastoma cells. *Exp. Cell Res.* 169: 311–27

Wallace, B. G. 1986. Aggregating factor from *Torpedo* electric organ induces patches containing cetylcholine receptors, acetylcholinesterase, and butyrylcholinesterase on cultured myotubes. *J. Cell Biol.* 102: 783–94

Wallace, B. G., Nitkin, R. M., Reist, N. E., Fallon, J. R., Moayeri, N. N., McMahan, U. J. 1985. Aggregates of acetylcholinesterase induced by acetylcholine receptor-aggregating factor. *Nature* 315: 574–77

Wewer, U., Albrechtsen, R., Manthorpe, M., Varon, S., Engvall, E., Ruoslahti, E. 1983. Human laminin isolated in a nearly intact, biologically active form from placenta by limited proteolysis. *J. Biol. Chem.* 258: 12654–60

Wujek, J. R., Freese, E. 1987. Neurite growth promoting factor(s) of astrocyte extracellular matrix: Identity and developmental regulation. *Soc. Neurosci. Abstr.* 13: 1484

Zak, N. B., Harel, A., Bawnik, Y., Benbasat, S., Vogel, Z., Schwartz, M. 1987. Laminin-immunoreactive sites are induced by growth-associated triggering factors in injured rabbit optic nerve. *Brain Res.* 408: 263–66

Zaremba, S., Hockfield, S. 1987. Monoclonal antibody CAT-301 recognizes a proteoglycan specific to the surface of subsets of mammalian CNS neurons. *Soc. Neurosci. Abstr.* 13: 1228

Ann. Rev. Neurosci. 1989. 12:517–34

# THE MACROGLIAL CELLS OF THE RAT OPTIC NERVE

*Robert H. Miller, Charles ffrench-Constant, and Martin C. Raff*

Medical Research Council Developmental Neurobiology Program, Biology Department, Medawar Building, University College London, London WC1E 6BT, England

## Introduction

Two major classes of macroglial cells are found in the vertebrate CNS: oligodendrocytes and astrocytes. Although it is known that oligodendrocytes form myelin sheaths (Bunge 1968, Peters & Vaughn 1970), the functions of astrocytes are less certain. Since they were first identified, a large number of diverse functions have been proposed for astrocytes in the developing and adult CNS, including the guidance of migrating neuronal cell bodies (Rakic 1971) and growth cones (Silver & Sapiro 1981, Silver et al 1982) during development, the induction of the formation of the blood brain barrier (Janzer & Raff 1987), the control of extracellular ion composition (Hertz 1981), and the formation of glial scars following injury (Maxwell & Kruger 1965, Vaughn & Pease 1970). To what extent these putative functions are carried out by distinct types of astrocytes, however, has remained uncertain.

Studies in the first half of this century defined two morphologically distinct populations of astrocytes in the vertebrate CNS (Cajal 1909, 1913). *Fibrous astrocytes* are found mainly in white matter, and have a stellate, process-bearing morphology; their cell bodies contain large numbers of fibrils (Weigert 1895). *Protoplasmic astrocytes* are found mainly in grey matter, and have sheet-like processes associated with neurons; their cell bodies contain relatively few fibrils. With the advent of electron microscopy, the astrocyte fibrils were identified as bundles of 10 nm intermediate filaments (Vaughn & Pease 1967, Mori & Leblond 1969), and

517

0147–006X/89/0301–0517$02.00

other ultrastructural features characteristic of both types of astrocytes were observed, including an electron-light cytoplasm, few microtubules, a large cell nucleus with a condensed rim of chromatin and numerous invaginations (Peters et al 1976).

Based on morphological and immunohistochemical observations of the developing vertebrate CNS, several schemes have been proposed describing the possible lineage relationship between astrocytes and other CNS glial cells. One proposal suggests that astrocytes develop from embryonic radial glial cells, which retract their radial processes and proliferate to form both fibrous and protoplasmic astrocytes (Schmechel & Rakic 1979, Privat & Rataboul 1986), and in some cases oligodendrocytes (Choi et al 1983, Choi & Kim 1985). Another is that a population of undifferentiated subpial glioblasts, not derived from radial glia, proliferate and differentiate into both astrocytes and oligodendrocytes (Fujita 1965). Lineage relationships, however, can rarely be established by simple descriptive studies, and thus the origin of astrocytes, and the lineage relationship between them and other cells of the CNS, have remained controversial.

Many studies of gliogensis in the vertebrate CNS have focused on developing white matter tracts, especially the rat optic nerve. The optic nerve has a number of advantages for studying glial cell development: It lacks neuronal cell bodies and is therefore one of the simplest parts of the CNS; it is a discrete axon tract, physically separate from the rest of the CNS; and the vast majority of its axons derive from retinal ganglion cells, and are myelinated and unbranched. Morphological studies have demonstrated three major classes of glial cells in the adult rat optic nerve: microglial cells and two classes of macroglial cells- astrocytes and oligodendrocytes. Astrocytes are the first macroglial cells to differentiate in the nerve (Skoff et al 1976b), first appearing between embryonic day 15 and 16 (E15-16); oligodendrocytes are not seen until after birth (Vaughn 1969, Kuwabara 1974, Skoff et al 1976a,b). Autoradiographic studies with $^3$H-thymidine have suggested that most astrocytes in the nerve are born by the end of the first postnatal week, whereas most oligodendrocytes are born during the first and second postnatal weeks (Skoff et al 1976a,b, Valat et al 1983). Attempts to deduce the lineage relationships between the two classes of macroglial cells from electron microscopic studies of normal and neonatally transected rat optic nerve have led to two conflicting hypotheses: One proposes that astrocytes and oligodendrocytes develop from different precursor cells in the neonatal nerve (Skoff et al 1976a,b); the other proposes that both astrocytes and oligodendrocytes develop from a common precursor cell (Privat et al 1981). Both hypotheses assumed that the astrocytes in the mature nerve are homogenous, and that all macroglial cells in the nerve develop from the optic stalk. Recent studies suggest that both of these assumptions may be wrong.

## Rat Optic Nerve Macroglial Cells in Culture

We have studied the differentiation and properties of perinatal rat optic nerve macroglial cells in culture, where we can control the cellular and fluid environment in which the cells develop. We have used conventional and monoclonal antibodies to identify and manipulate the specific types of glial cells and their precursors.

Cultures of perinatal rat optic nerve contain three distinct types of differentiated macroglial cells. *Oligodendrocytes* have a small cell body, a large number of branching processes, and are the only cells that are labeled on their surface by antibodies against galactocerebroside (GC) (Raff et al 1978, Ranscht et al 1982), the major glycolipid in myelin (Figure 1A). Thus we use the binding of anti-GC antibodies to define oligodendrocytes in cell suspensions and culture. In the electron microscope, these cells have a characteristic electron-dense cytoplasm containing few if any intermediate filaments but large numbers of microtubules (Raff et al 1983b). *Astrocytes* can be identified both in vitro and in vivo by the binding of antibodies against glial fibrillary acidic protein (GFAP) (Bignami et al 1972, Bignami & Dahl 1974, Pruss 1979), the major subunit of their intermediate (glial) filaments (Schachner et al 1977). Cultures of optic nerve cells contain two types of GFAP+ astrocytes, which differ in many of their properties (Raff et al 1983a). For simplicity we call them type-1 and type-2 astrocytes.

The majority of *type-1 astrocytes* in culture have a large cell body, and are flat, non–process-bearing, fibroblast-like cells (Figure 1B), which are labeled on their surface by a monoclonal antibody that reacts with a glycoprotein called Ran-2 (Rat neural antigen-2) (Bartlett et al 1981), but are not labeled by tetanus toxin or the monoclonal antibody A2B5 (Eisenbarth et al 1979), both of which bind to specific gangliosides (Van Heyningen 1963, Eisenbarth et al 1979). By contrast, most *type-2 astrocytes* in culture have a small cell body, a process-bearing morphology (Figure 1C), and are labeled on their surface by tetanus toxin and the A2B5 antibody (Raff et al 1983a); in short-term culture, they express little or no Ran-2 on their surface (Raff et al 1984). Whereas type-1 astrocytes are stimulated to proliferate in culture by epidermal growth factor (EGF) (Cohen 1962) and glial growth factor (GGF) (Lemke & Brockes 1984), type-2 astrocytes are not (Raff et al 1983a; M. Noble, personal communication). Some of the distinguishing properties of the three major types of macroglial cells in cultures of perinatal rat optic nerve cells are summarized in Table 1.

## Sequence of Macroglial Cell Development

Studies of cell suspensions prepared from optic nerves of different developmental ages indicate that the three macroglial cells differentiate on a

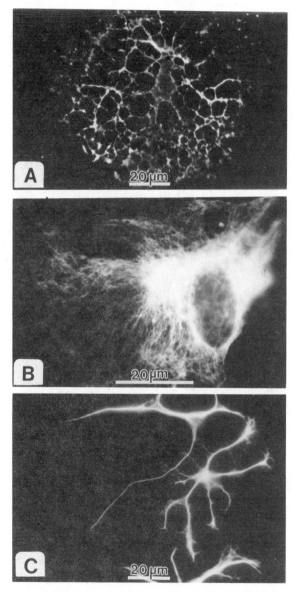

*Figure 1*    Immunofluorescence micrographs of the three major types of differentiated macro-glial cells in cultures of perinatal rat optic nerve. (*A*) An oligodendrocyte labeled on its surface with monoclonal antibody against galacerebroside. (*B*) A type-1 astrocyte labeled intracellularly with an antiserum against glial fibrillary acidic protein (GFAP). (*C*) A type-2 astrocyte labeled intracellularly with an antiserum against GFAP.

**Table 1** Some properties of the three major types of macroglial cells in the rat optic nerve

| | Type-1 astrocyte | Type-2 astrocyte | Oligodendrocyte | Ref. |
|---|---|---|---|---|
| Time of first appearance | E16 | P8–10 | P0 | Miller et al (1985) |
| Morphology in culture | Fibroblast-like | Process-bearing | Process-bearing | Raff et al (1978, 1983a) |
| Morphology in vivo | Sheet-like radial processes extending to pial surface and blood vessels | Cylindrical processes extending to nodes of Ranvier | Multidirectional processes extending to myelin | Miller et al (1989) |
| Antigenic phenotype in culture[a] | | | | |
| GFAP | + | + | − | Raff et al (1978) |
| GC | − | − | + | Raff et al (1978) |
| O4 | − | ± | + | Somner & Schahner (1981) |
| Ran-2 | + | − | − | Bartlett et al (1981), Raff et al (1984) |
| A2B5 | − | + | + → − | Raff et al (1983a, 1984, 1985) |
| HNK-1 (L2), NSP-4, J1 | − | + | + → − | ffrench-Constant & Raff (1986) |
| NG2 | − | + | + → − | Stallcup & Beasley (1987) |
| Responses to growth factors in vitro | | | | |
| EGF | + | − | − | Raff et al (1983a) |
| GGF | + | − | − | Raff et al (1983a), M. Noble, unpublished |

[a] Whereas GFAP and GC can be used as unambiguous markers of astrocytes and oligodendrocytes, respectively, this is not the case for the other antigens listed. A small proportion (<5%) of type-1-like astrocytes, for example, are A2B5+ and a small proportion (usually <5%) of type-2-like astrocytes are A2B5− in short-term cultures.

precise schedule. Type-1 astrocytes first appear at E16, oligodendrocytes around the day of birth (E21 = PO), and type-2 astrocytes in the second postnatal week (Miller et al 1985). The same three types of macroglial cells can be distinguished in cultures of embryonic rat brain, where, remarkably, the cells develop on the same schedule as they do in vivo (Abney et al 1981, Williams et al 1985). These findings suggest that the timing of glial cell differentiation depends on mechanisms that operate largely independently of CNS morphogenesis.

## Evidence for Two Macroglial Cell Lineages in Optic Nerve

Although both type-1 and type-2 astrocytes can display a variety of morphologies in vitro, depending on the culture conditions, there is no convincing evidence that in culture type-1 astrocytes can give rise to type-2 astrocytes, or vice versa (Raff et al 1983a). Instead, the two types of astrocytes seem to arise by two distinct lineages, at least in vitro.

Antibody labeling experiments have identified two non-overlapping populations of cells in cell suspensions prepared from E17 optic nerve: approximately 85% of the cells are Ran-2+, A2B5−, while about 10% are A2B5+, Ran-2− (Raff et al 1984). Pulse labeling studies, in which cells are first labeled on their surface with either anti-Ran-2 or A2B5 antibody and then allowed to differentiate in culture, demonstrate that the Ran-2+ population contains precursor cells for type-1 astrocytes but not for type-2 astrocytes or oligodendrocytes, whereas the A2B5+ population contains precursor cells for type-2 astrocytes and oligodendrocytes but not for type-1 astrocytes (Raff et al 1984). Furthermore, if cell suspensions of embryonic or postnatal optic nerve cells are treated with A2B5 antibody and complement before they are put into culture, no type-2 astrocytes or oligodendrocytes develop, although normal numbers of type-1 astrocytes are found in such cultures (Raff et al 1983b, 1984).

The direction of the differentiation of the A2B5+ cells depends on the culture medium. Quantitative experiments show that almost all of these cells in postnatal day 7 (P7) optic nerve develop into oligodendrocytes if cultured for three days in serum-free medium. However, almost all of them develop into type-2 astrocytes, if cultured for three days in 10% fetal calf serum (FCS) (Raff et al 1983b). Taken together, these results indicate that oligodendrocytes and type-2 astrocytes, but not type-1 astrocytes, develop in culture from a common A2B5+ progenitor cell. We call these bipotential cells *oligodendrocyte-type-2 astrocyte (O-2A) progenitor cells*.

The finding that O-2A progenitor cells develop into oligodendrocytes in serum-free medium, even when cultured on their own in microwells (Temple & Raff 1985), suggests that this is the constitutive developmental pathway for these cells. Type-2 astrocyte differentiation, by contrast, appears to require an inducing signal, such as the one in FCS; in vivo, however, the inducing signal seems to be a 25,000 dalton protein that appears relatively late in optic nerve development (Hughes & Raff 1987). There is evidence that both the proliferation of O-2A progenitor cells (Noble & Murray 1984) and the timing of their differentiation (Raff et al 1985) is controlled by specific protein signaling molecules secreted by type-1 astrocytes (Richardson et al 1988, L. E. Lilien and M. C. Raff, in preparation), as will be reviewed elsewhere.

## Evidence that the O-2A Progenitor Cells Migrate into the Developing Nerve from the Brain

A number of observations suggest that O-2A lineage cells in the rat optic nerve do not develop from optic stalk cells, but instead may be derived from cells that migrate into the nerve from the brain during development. The ability of oligodendrocytes or their precursors to migrate long distances in the CNS was initially suggested by transplantation studies. When small pieces of normal mouse brain were implanted in the brains of *shiverer* mutant mice, which are genetically unable to synthesize myelin basic protein (MBP) (Barbarese et al 1983, Roach et al 1983, 1985), MBP+ myelin was found long distances from the grafts (La Chapelle et al 1984). Time-lapse microcinephotographic studies of neonatal rat optic nerve cultures suggest that it is the progenitor cells rather than the oligodendrocytes that are migrating: In such cultures O-2A progenitor cells, which have a characteristic bipolar morphology (Temple & Raff 1986), migrate actively until they differentiate into oligodendrocytes, at which point locomotion stops (Small et al 1987).

There is indirect evidence that O-2A progenitor cells may migrate into the developing optic nerve from the brain. In E17 rats, O-2A progenitor cells are found almost exclusively at the chiasm-end of the nerve, and by birth, although some progenitors are found at the eye-end of the nerve, their numbers are still much smaller than at the chiasm-end. It is not until the second postnatal week that progenitor cells become evenly distributed along the length of the nerve (Small et al 1987). Type-1 astrocytes, however, develop first at the eye-end of the nerve, making it unlikely that the developmental gradient of O-2A progenitor cells reflects a gradient of differentiation along the nerve (Small et al 1987). These studies suggest that the neuroepithelial cells of the optic stalk are all destined to develop into type-1 astrocytes. The source of O-2A progenitor cells for the optic nerve is unknown. Small and co-workers (1987) have suggested that they may arise from cells at the base of the preoptic recess (which overlies the optic chiasm) that undergo a burst of proliferation around E16 in the rat.

If O-2A progenitor cells migrate down the developing nerve from the brain toward the eye, why do they not migrate into the retina and differentiate into oligodendrocytes (and type-2 astrocytes) and myelinate the retinal ganglion cell axons in the nerve fiber layer? Such myelination would severely impair vision. Several lines of evidence suggest that something in the lamina cribrosa region of the optic nerve (where the nerve pierces the sclera) acts as a barrier to prevent the migration of progenitor cells into the retina, a hypothesis originally proposed by Berliner (1931). First, no O-2A progenitor cells, oligodendrocytes, or type-2 astrocytes are found in the developing retina (ffrench-Constant et al 1988). Second, the lamina cribrosa region of the nerve is unmyelinated and contains very few oligodendrocytes (Hildebrand et al 1985) or type-2 astrocytes (ffrench-Constant et al 1988), suggesting that few O-2A progenitors enter it. Third, in rabbits, which lack a lamina cribrosa (Berliner 1931), the retina contains O-2A lineage cells (ffrench-Constant et al 1988), and the central part of the retina is myelinated by oligodendrocytes (Davis 1929, Berliner 1931, Narang & Wisniewsky 1977).

What is the nature of the barrier that prevents O-2A progenitor cells from entering the rat retina? One possibility is that it is a property of the specialized astrocytes in the region of the lamina cribrosa. In this region the astrocytes form a dense meshwork across the nerve, perpendicular to the axons, and their processes contain large numbers of glial filaments compared to astrocytes in more posterior regions of the nerve (Skoff et al 1986), similar to the astrocyte processes in a glial scar (Maxwell & Kruger 1965, Vaughn & Pease 1970). It is likely that the astrocytes in the lamina cribrosa region of the nerve are type-1 rather than type-2 astrocytes: They have a radial orientation, which is a characteristic of type-1 astrocytes in the rest of the nerve (see below), and they are present in the newborn rat (ffrench-Constant et al 1988), more than a week before the first type-2 astrocytes develop (Miller et al 1985).

It is not clear what makes the type-1 astrocytes in the lamina cribrosa region different from those elsewhere in the nerve. Perhaps the penetration of vascularized scleral connective tissue into the nerve in this region (Fine & Yanoff 1979), or the leakage of proteins from adjacent choroid blood vessels (Tso et al 1975, Flage 1977, Kistler & La Vail 1981), induces the astrocytes in this region to adopt a special character, much as CNS injury induces astrocytes to form a glial scar (Vaughn & Pease 1970, Miller et al 1986).

How might the type-1 astrocytes in the lamina cribrosa block O-2A progenitor cell migration? One possibility is that they form a mechanical barrier. Another is that they form a chemical barrier. In vitro studies, for example, have shown that type-1 astrocytes secrete a growth factor(s) that both stimulates O-2A progenitor cells to proliferate and inhibits their premature differentiation (Noble & Murray 1984, Raff et al 1985); it might be that type-1 astrocytes in the lamina cribrosa region do not secrete enough growth factor to maintain progenitor cell proliferation, and, as a consequence, the progenitor cells drop out of division, differentiate, and therefore can no longer migrate.

## Type-1 and Type-2 Astrocytes in the Adult Rat Optic Nerve

What do the two types of astrocytes seen in cultures of perinatal optic nerve correspond to in the intact adult optic nerve? Neither electron microscopic studies nor immunohistochemical staining with anti-GFAP antibodies of the adult optic nerve reveal two morphologically distinct types of astrocytes, possibly because of the complex cytoarchitecture of the nerve. When a modified Golgi impregnation technique (Stensaas & Stensaas 1968) is used, however, so that only a small number of the astrocytes is stained, two morphologically distinct types of astrocyte-like cells are readily seen (Miller et al 1989). The majority of impregnated astrocytes are found toward the periphery of the nerve and contribute to the glial-limiting membrane. They have large cell bodies close to the pial surface and predominantly radially oriented processes. The minority of impregnated astrocyte-like cells have smaller cell bodies, are distributed uniformly throughout the interior of the nerve, and have predominantly longitudinally oriented processes running parallel to the axons. When the same technique is used to study developing optic nerves, the first detectable astrocytes have predominantly radial processes extending to the pial surface, while astrocyte-like cells with predominantly longitudinally oriented processes are not seen until the second postnatal week, suggesting that these two types of cells correspond to type-1 and type-2 astrocytes in culture (Miller et al 1989).

In order to examine the two types of astrocytes in more detail, we injected horse radish peroxidase (HRP) into individual glial cells in freshly dissected adult optic nerves and then examined the injected cells by both light and electron microscopy (Miller et al 1989). Two morphologically distinct types of astrocytes are seen in such studies, equivalent to the two types of astrocytes seen in Golgi stained nerves. Most have large cell bodies located toward the periphery of the nerve and predominantly radially oriented processes. When the cell body is located at the pial surface, the processes run mainly toward the center of the nerve, where they terminate in endfeet on blood vessels. When the cell body is located in the interior of the nerve the processes run mainly toward the pial surface, where they contribute to the glial limiting membrane. Some cells send processes to both the interior and the pial surface, giving the cell a bipolar appearance. The minority type of astrocyte-like cells has a small cell body distributed uniformly throughout the interior of the nerve. It has fine, unbranching longitudinal processes, which run parallel to the axons and do not end on blood vessels or the pial surface. In the electron microscope, astrocytes with radial processes are seen either in linear arrays with other glial cells in the nerve, or as isolated cells between bundles of axons. Their processes form endfeet on blood vessels and at the pial surface, but do not contact nodes of Ranvier along myelinated axons. By contrast, astrocyte-like cells with predominantly longitudinal processes are always found in linear arrays with other glial cells. Their processes do not form endfeet on blood vessels or the pial surface, but instead form contacts with nodes of Ranvier (Miller et al 1989).

The findings with Golgi staining of developing nerves indicate that astrocyte-like cells with mainly longitudinal oriented processes in the rat optic nerve do not appear until the second postnatal week, suggesting that these cells are type-2 astrocytes. Strong support for this view comes from studies using several antibodies to stain cultures of neonatal optic nerve cells and ultra-thin and semi-thin frozen sections (Tokuyasu 1973) of adult optic nerve (ffrench-Constant et al 1986, ffrench-Constant & Raff 1986). The antibodies used in these studies included the monoclonal antibodies HNK-1 (Abo & Balch 1981), L2 (Kruse et al 1984), and anti-NSP-4 (Rougon et al 1983), and also polyclonal antibodies against the J1 glyco-proteins (Kruse et al 1985). All of these antibodies appear to recognize the same protein(s) (running between 160,000 and 180,000 daltons) in immunoblots of extracts of adult rat optic nerve (ffrench-Constant et al 1986). In culture, they label type-2 but not type-1 astrocytes (ffrench-Constant & Raff 1986), while in frozen sections, they label astrocyte processes associated with nodes of Ranvier, and also fine longitudinally oriented processes, at least some of which seemed to be continuous with the perinodal astrocyte processes (ffrench-Constant et al 1986, ffrench-Constant & Raff 1986). These findings initially led us to hypothesize that perinodal astrocyte processes are derived from type-2 astrocytes, which collaborate with oligodendrocytes to ensheath axons in white matter tracts (ffrench-Constant & Raff 1986).

Whereas Golgi and HRP-injection studies of adult optic nerve indicate that radially oriented astrocytes have multiple processes (Miller et al 1989), type-1 astrocytes in cultures of optic nerve are non–process-bearing (Raff et al 1983a). This apparent discrepency seems to be related to the absence of neurons in the optic nerve cultures: When optic nerve type-1 astrocytes are co-cultured with retinal neurons, the astrocytes adopt a process-bearing morphology very similar to that of the radial astrocytes in the intact nerve. Similar observations had been made previously with cerebellar astrocytes (Hatten 1985).

One set of our observations is difficult to reconcile with our current view of type-1 and type-2 astrocytes. In semi-thin frozen sections of adult optic nerve, the A2B5 monoclonal antibody (which labels the majority of type-2 astrocytes but very few type-1 astrocytes in cultures of perinatal rat optic nerve cells) stains some radially oriented astrocyte processes, and some astrocyte processes around blood vessels (Miller & Raff 1984), but not astrocyte processes associated with nodes of Ranvier (ffrench-Constant et al 1986). Because the A2B5 antibody did not label GFAP+ astrocytes in semi-thin frozen sections of developing optic nerve until the second post-natal week, we originally thought that the A2B5+ GFAP+ cells in such sections were type-2 astrocytes (Miller & Raff 1984). Subsequent results, however, with other antibodies that label most type-2 but not type-1 astrocytes in cultures of rat optic nerve (ffrench-Constant et al 1986, ffrench-Constant & Raff 1986), such as HNK-1, L1, anti-NSP-4, and anti-J1, taken together with the results obtained by Golgi staining and HRP injection, cast doubts on the reliability of A2B5 as a marker of type-2 astrocytes in tissue sections. One reason for the problem may be that whereas the A2B5 antibody has been used to label the surface of cells in suspension and culture (Raff et al 1983a), its staining in tissue sections is intracellular (Miller & Raff 1984). The significance of this subpopulation of GFAP+ cells that stains intracellularly with A2B5 antibody in tissue sections of rat optic nerve is uncertain; it is intriguing that such cells first appear in the second postnatal week (Miller et al 1985) and are not found in the unmyelinated lamina cribrosa region of the nerve (ffrench-Constant et al 1988), or in glial scars formed after optic nerve transection or after a stab lesion to the corpus callosum (Miller et al 1986). On the basis of A2B5 staining of sections of normal adult optic nerve and cerebral cortex, we originally suggested that type-1 and type-2 astrocytes might correspond to protoplasmic and fibrous astrocytes, respectively (Miller & Raff 1984); our subsequent observations have made this suggestion untenable.

## Macroglial Cell Function in the Optic Nerve

By myelinating the axons of the retinal ganglion cells, oligodendrocytes greatly increase the rate and efficiency of action potential propagation in the optic nerve. The functions of the astrocytes are less clear, but most of the functions that have been proposed seem to be carried out by type-1 astrocytes or their precursors. It is thought, for example, that the neuro-epithelial cells in the optic stalk guide the migration of the first retinal ganglion cell growth cones toward their target in the brainstem (Silver & Rutishauser 1984). These neuroepithelial cells are probably all precursor cells for type-1 astrocytes (Small et al 1987), as discussed above. Moreover, as we have seen, type-1 astrocytes form the glial limiting membrane (Miller et al 1989), and it seems likely that they provide a structural framework for the nerve.

Type-1 astrocytes also seem to play a crucial part in the formation of the blood brain barrier. In the CNS of higher vertebrates the endothelial cells that line the capillaries and venules are sealed together by tight junctions that make these vessels impermeable to the passive diffusion of most water soluble molecules in the blood (Reese & Karnovsky 1967). Transplantation experiments have shown that this property of the endothelial cells is not intrinsic to the cells but instead is induced by the environment of the CNS (Stewart & Wiley 1981). Since endfeet of type-1 astrocyte processes surround the blood vessels in the optic nerve, these astrocytes must be strong candidates for the crucial environmental factor that induces the endothelial cells to form tight vessels. Direct evidence that astrocytes are capable of inducing non-neural endothelial cells to form such tight vessels has been obtained by transplanting purified rat astrocytes onto the chorioallantoic membrane of a developing chick embryo (Janzer & Raff 1987). The chick endothelial cells that invade and vascularize the astrocyte aggregate are found to form tight blood vessels that are impermeable to the dye Evans blue. On the other hand, when fibroblasts are transplanted onto the chorioallantoic membrane, the vessels that form are leaky to the dye, suggesting that the ability to induce blood-brain properties in endothelial cells is a specific property of astrocytes. The astrocytes used in these experiments (purified from neonatal cerebral cortex) had the morphologic and antigenic properties of type-1 astrocytes, strongly supporting the idea that type-1 astrocytes are responsible for inducing the formation of the blood brain barrier.

One of the few established functions of astrocytes is the formation of glial scars following CNS injury (Maxwell & Kruger 1965, Vaughn & Pease 1970, Bignami & Dahl 1976, Fulcrand & Privat 1977). When the optic nerve is cut behind the eye in a newborn (David et al 1984) or adult rat (Miller et al 1986), and the astrocytes that form the glial scar are assessed 10–20 weeks after the transection by double labeling semi-thin frozen sections with A2B5 and anti-GFAP antibodies, almost all of the GFAP+ cells are A2B5−. On the basis of these findings we proposed that type-1 astrocytes may be mainly responsible for glial scar formation in the nerve (Miller et al 1986). Although A2B5 has subsequently been found not to be a reliable marker of type-2 astrocytes in tissue sections, there are other reasons for thinking that this proposal is likely to be correct. First, the astrocyte processes in adult transected optic nerves are not stained by anti-NSP-4 antibody (C. ffrench-Constant, unpublished observation), which stains type-2 astrocytes in neuron-free cultures and their processes in normal optic nerve (ffrench-Constant & Raff 1986). Second, within the first week following adult nerve transection, it is mainly the type-1 astrocytes that are stimulated to proliferate (Miller et al 1986). Third, the glial scar formed in an optic nerve looks the same whether the nerve is cut at birth or in adulthood (David et al 1984, Miller et al 1986), and yet no type-2 astrocytes are present at birth (Miller et al 1985), and they seem not to develop in neonatally transected nerves (David et al 1984).

Thus type-1 astrocytes appear to have multiple functions in the optic nerve. Type-2 astrocytes, on the other hand, seem to be specialized for servicing nodes of Ranvier (ffrench-Constant & Raff 1986, Miller et al 1989), but how they do so is unknown. In principle, they might help (a) control ion concentrations in the perinodal space, (b) provide nutrients for the nodal axon, and (c) set up and maintain the structure of the node (Waxman 1986). Whatever their function, it seems reasonable to view the O-2A cell lineage as specialized for myelinating axons and helping to construct nodes of Ranvier. The close lineage relationship between oligodendrocytes and type-2 astrocytes (Raff et al 1983b) thus makes functional sense.

Recently, Barres and her associates (1988) have examined the electro-physiological properties in both freshly isolated and cultured optic nerve glial cells and have found that each cell type has its own characteristic set of ion channels. Except for an inwardly rectifying $K+$ channel found in oligodendrocytes, which they hypothesize may play an important part in regulating the extracellular concentration of $K+$ in the nerve, the functions of the various ion channels are unknown.

## Astrocyte Heterogeneity in Other Regions of the CNS

Golgi staining of the glial cells in adult rat spinal cord also reveals two morphologically distinct types of astrocytes in the peripheral white matter (Liuzzi & Miller 1987). As in the optic nerve, the majority of the astrocytes have predominantly radially oriented processes and share many charac-teristics with type-1 astrocytes in the optic nerve: They are the first astro-cytes to appear in the cord; they send processes to the pial surface of the cord (Liuzzi & Miller 1987); and following cord transection they form a dense glial scar distal to the point of transection (Barrett et al 1984). A minority population of astrocyte-like cells develop relatively late. These cells have a similar morphology to type-2 astrocytes in the optic nerve. They have relatively small cell bodies, which are found in linear arrays with oligodendrocytes, and predominantly longitudinally oriented pro-cesses that do not contact the pial surface (Liuzzi & Miller 1987). Although intracellular HRP injection studies have not been attempted in the spinal cord, it seems likely that these longitudinal processes interact with axons at nodes of Ranvier.

Thus it seems likely that the two types of astrocyte found in the optic nerve are a general feature of white matter tracts in the CNS. Consistent with this view is the finding that cultures of perinatal rat cerebrum (Williams et al 1985) and cerebellum (Levi et al 1986a) contain two types of astrocytes that are very similar to type-1 and type-2 astrocytes in cultures of perinatal rat optic nerve cells. Moreover, cells with the properties of O-2A progenitor cells have been identified in cultures of rat cerebrum (Beahr et al 1987) and cerebellum (Levi et al 1986b).

Clearly, however, other types of astrocytes or astrocyte-like cells exist in addition to the two found in white matter tracts (Schachner 1982). The rat retina, for example, which contains no white matter, has no O-2A lineage cells (ffrench-Constant et al 1988), but does have two types of macroglial cells. Müller cells span the full thickness of the retina and have feet on both retinal surfaces, whereas type-1-like astrocytes are confined to the nerve fiber layer (Büssow 1980). Recent evidence indicates that the Müller cells develop from the same neuroepithelial cells that give rise to retinal neurons (Turner & Cepko 1987), whereas the type-1 astrocytes migrate in from the optic nerve head (Watanabe & Raff 1988). The lineage relationships of the various other astrocytes and astrocyte-like cells in the CNS, such as Bergmann glia in the cerebellum and the astrocytes in the cerebral cortex, remain to be determined.

## Phylogeny of Macroglial Cells

The ontogeny of macroglial cells in the rat optic nerve (type-1 astrocytes followed by oligodendrocytes followed by type-2 astrocytes) seems to recapitulate phylogeny. In the lamprey spinal cord, for example, which lacks myelin, radial glial cells are the only glia (Bullock 1977). In the myelinated newt and frog spinal cord there are oligodendrocytes and a single type of radial astrocyte that spans the thickness of the cord. In the frog spinal cord these astrocytes have regionally specialized radial processes (Miller & Liuzzi 1986): In grey matter they contain few intermediate filaments and branch extensively, whereas in white matter they contain large numbers of intermediate filaments and have few branches, suggesting that the processs perform different functions in the two regions (Miller & Liuzzi 1986). Two morphologically distinct populations of white matter astrocytes are first seen in reptiles and birds, where the degree of myelination in the CNS is also greatly increased compared to amphibians (F. J. Liuzzi and R. H. Miller, unpublished observation). As CNS complexity increased during evolution, it seems that the requirement for more types of functionally specialized glial cells increased as well.

## Conclusions

The relative simplicity of the optic nerve makes it an attractive part of the CNS in which to study glial cell development and function. Recent studies on the rat optic nerve in vivo and in vitro have provided new insights into the lineage relationships and functions of the three classes of macroglial cells in the nerve. Type-1 astrocytes develop first arising from resident optic stalk cells. They extend radial processes to the pial surface and to blood vessels, providing a structural framework for the nerve and inducing endothelial cells to form the blood-brain barrier. Later, oligodendrocytes and then type-2 astrocytes develop from bipotential progenitor cells, which are thought to migrate into the nerve from the brain. These two glial cells collaborate in ensheathing axons: oligodendrocytes make myelin, while type-2 astrocytes extend longitudinal processes that contact nodes of Ranvier. It seems likely that the same principles of macroglial cell development and function operate in other white matter tracts in the mammalian CNS.

*Literature Cited*

Abney, E., Bartlett, P. F., Raff, M. C. 1981. Astrocytes, ependymal cells and oligodendrocytes develop on schedule in dissociated cell cultures of embryonic brain. *Dev. Biol.* 83: 301–10

Abo, T, Balch, C. M. 1981. A differentiation antigen of human NK and K cells identified by a monoclonal antibody (HNK-1). *J. Immunol.* 127: 1024–29

Barbarese, E., Neilson, M. L., Carson, J. H. 1983. The effect of the *shiverer* mutation on myelin basic protein expression on homozygous and heterozygous mouse brain. *J. Neurochem.* 40: 1680–86

Barres, B. A., Chun, L. L. Y., Corey, D. P. 1988. Ion channel expression by white matter glia: 1. Type-2 astrocytes and oligodendrocytes. *Glia* 1: 10–30

Barrett, C. P., Donati, E. J., Guth, L. 1984. Differences between adult and neonatal rats in their astroglial response to spinal cord injury. *Exp Neurol.* 84: 374–85

Bartlett, P. F., Noble, M., Pruss, R. M., Raff, M. C., Williams, C. A. 1981. Rat neural antigen-2 (Ran-2): a cell surface antigen on astrocytes, ependymal cells, Müller cells and leptomeninges defined by a monoclonal antibody. *Brain Res.* 204: 339–51

Beahr, T., Movotny, E., Barker, J., McMorris, S. A., Dubois-Dalacq, M. 1987. O-2A progenitor cells sorted from postnatal rat brain closely resemble their optic nerve counterpart. *J. Cell Biol.* 105: 318A

Berliner, M. L. 1931. Cytological studies on the retina 1. Normal coexistance of oligodendroglia and myelinated nerve fibers. *Arch Opthal* 6: 740–51

Bignami, A., Eng, L. F., Dahl, D., Uyeda, C. T. 1972. Localization of the glial fibrillary acidic protein in astrocytes by immunofluorescence. *Brain Res.* 43: 429–35

Bignami, A, Dahl, D. 1974. Astrocyte specific protein and radial glia in the cerebral cortex of new born rat. *Nature* 252: 55–56

Bignami, A, Dahl, D. 1976. The astroglial response to stabbing. Immunofluorescence studies with antibodies to astrocyte-specific protein (GFA) in mammalian and submammalian vertebrates. *Neuropathol. Appl. Neurobiol.* 2: 99–110

Bullock, T. H. 1977. *Introduction to Nervous Systems*, pp 435–46. San Francisco: Freeman

Bunge, R. P. 1968. Glial cells and the central myelin sheath. *Physiol. Rev.* 48: 197–251

Büssow, H. 1980. The astrocytes in the retina and optic nerve head of mammals: A special glia for the ganglion cell axons. *Cell Tiss. Res.* 206: 367–78

Cajal, S. R. 1909. *Histologie du Systeme Nerveux de l'Homme et des Vertebres*, pp. 230–52. Paris: Maloine

Cajal, S. R. 1913. Sobre un nuevo proceder de impregnacion de la neuroglia y sus resultados en los centros nerviosos del hombre y animales. *Trab. Lab. Invest. Biol. Univ. Madrid* 11: 219–37

Choi, B. H., Kim, R. C. 1985. Expression of glial fibrillary acidic protein by immature oligodendroglia and its implications. *J. Neuroimmunol.* 8: 215–35

Choi, B. H., Kim, R. C., Lapham, L. W. 1983. Do radial glia give rise to both astroglial and oligodendroglial cells? *Dev. Brain Res.* 8: 119–30

Cohen, S. 1962. Isolation of a mouse submaxillary gland protein accelerating incisor eruption and eyelid opening in the newborn animal. *J. Biol. Chem.* 237: 1555–62

David, S., Miller, R. H., Patel, R., Raff, M. C. 1984. Effects of neonatal transection on glial cell development in the rat optic nerve: Evidence that the oligodendrocyte-type-2 astrocyte cell lineage depends on axons for its survival. *J. Neurocytol.* 13: 961–74

Davis, F. A. 1929. The anatomy and histology of the eye and orbit of the rabbit. *Trans. Am. Ophthal. Soc.* 27: 402–41

Eisenbarth, G. S., Walsh, F. S., Nirenberg, M. 1979. Monoclonal antibody to a plasma membrane antigen of neurons. *Proc. Nat. Acad. Sci. USA.* 76: 4913–17

ffrench-Constant, C., Miller, R. H., Kruse, J., Schachner, M., Raff, M. C. 1986. Molecular specialization of astrocyte processes at nodes of Ranvier in rat optic nerve. *J. Cell Biol.* 102: 844–52

ffrench-Constant, C., Miller, R. H., Burne, J. F., Raff, M. C. 1988. Evidence that migratory oligodendrocyte-type-2 astrocyte (O-2A) progenitor cells are kept out of the rat retina by a barrier at the eye-end of the optic nerve. *J. Neurocytol.* 17: 13–25

ffrench-Constant, C. and Raff, M. C. 1986. The oligodendrocyte-type-2 astrocyte cell lineage is specialized for myelination. *Nature* 323: 335–38

Fine, B. S., Yanoff, M. 1979. *Ocular Histology: A Text and Atlas*, pp 273–86. Hagerstown: Harper & Row

Flage, T. 1977. Permeability properties of the tissues in the optic nervehead of the rabbit and monkey: An ultrastructural study. *Acta Ophthalmol.* 55: 652–64

Fujita, S. 1965. An autoradiographic study on the origin and fate of the sub-pial glioblast in the embryonic chick spinal cord. *J. Comp. Neuro.* 124: 51–60

Fulcrand, J., Privat, A. 1977. Neuroglial reactions secondary to Wallerian degeneration in the optic nerve of the postnatal rat: ultrastructural and quantitative study. *J. Comp. Neurol.* 176: 189–224

Hatten, M. E. 1985. Neuronal regulation of astroglial morphology and proliferation in vitro. *J. Cell Biol.* 100: 384–96

Hertz, L. 1981. Functional interactions between astrocytes and neurons. In *Glial and Neuronal Cell Biology*, ed. S. Fedoroff, pp. 45–58. New York: Liss

Hildebrand, E., Remahl, S., Waxman, S. G. 1985. Axo-glial relations in the retina-optic nerve junction of the adult rat: Electron-microscopic observations. *J. Neurocytol.* 14: 597–617

Hughes, S. M., Raff, M. C. 1987. An inducer protein may control the timing of fate switching in a bipotential glial progenitor

cell in the rat optic nerve. *Development* 101: 157–67

Janzer, R. C., Raff, M. C. 1987. Astrocytes induce blood brain barrier properties in endothelial cells. *Nature* 325: 253–57

Kistler, H. B., LaVail, J. H. 1981. Penetration of horseradish peroxidase into the optic nerve after vitreal or vascular injections in the developing chick. *Invest. Ophthalmol. Visual Sci.* 20: 705–16

Kruse, J., Keilhauer, G., Faissner, A., Timpl, R., Schachner, M. 1985. The J1 glycoprotein—a novel nervous system cell adhesion molecule of the L2/HNK-1 family. *Nature* 316: 146–48

Kruse, J., Mailhammer, R., Wernecke, H., Faissner, A., Sommer, I., Goridis, C., Schachner, M. 1984. Neural cell adhesion molecules and myelin associated glycoprotein share a common carbohydrate moiety recognised by monoclonal antibodies L2 and HNK-1. *Nature.* 311: 153–55

Kuwabara, T. 1974. Development of the optic nerve of the rat. *Invest. Ophthalmol.* 13: 732–45

La Chapelle, F., Gumple, M., Baluac, M., Jacque, C., Due, P., Baumann, N. 1984. Transplantion of CNS fragments into the brain of *shiverer* mutant mice: Extensive myelination by implanted oligodendrocytes. 1. Immunological studies. *Dev. Neurosci.* 6: 325–34

Lemke, G. E., Brockes, J. P. 1984. Identification and purification of glial growth factor. *J. Neurosci.* 4: 75–83

Levi, G., Gallo, V., Wilkins, G. P., Ghen, J. 1986a. Astrocyte subpopulations and glial precursors in rat cerebellar cell cultures. *Adv. Biosci.* 61: 21–30

Levi, G., Gallo, V., Ciotti, M. T. 1986b. Bipotential precursors of putative fibrous astrocytes and oligodendrocytes in rat cerebellar cultures express distinct surface features and "neuron-like" $\gamma$-aminobutyric acid transport. *Proc. Nat. Acad. Sci. USA* 83: 1504–8

Liuzzi, K. J., Miller, R. H. 1987. Radially oriented astrocytes in the normal adult rat spinal cord. *Brain Res.* 403: 385–88

Maxwell, D. S., Kruger, L. 1965. The fine structure of astrocytes in the cerebral cortex and their response to focal injury produced by heavy ionising particles. *J. Cell Biol.* 25: 141–57

Miller, R. H., Abney E. R., David, S., ffrench-Constant, C., Lindsay, R., Patel, R., Stone, J., Raff, M. C. 1986. Is reactive gliosis a property of a distinct subpopulation of astrocytes? *J. Neurosci.* 6: 22–29

Miller R. H., David, S., Patel R., Abney E. R., Raff, M. C. 1985. A quantitative immunohistochemical study of macroglial cell development in the rat optic nerve: *In vivo* evidence for two distinct glial cell lineages. *Dev. Biol.* 111: 35–41

Miller, R. H., Fulton, B. P., Raff, M. C. 1989. A novel type of astrocytes contributes to the structure of nodes of Ranvier in rat optic nerve. *Eur. J. Neurosci.* In press

Miller, R. H., Liuzzi, F. J. 1986. Regional specialization of the radial glial cells of the adult frog spinal cord. *J. Neurocytol.* 15: 187–96

Miller, R. H., Raff, M. C. 1984. Fibrous and potoplasmic astrocytes are biochemically and developmentally distinct. *J Neurosci.* 4: 585–92

Mori, S., Leblond, C. P. 1969. Electron microscopic features and proliferation of astrocytes in the corpus callosum of the rat. *J. Comp. Neurol.* 137: 197–205

Narang, H. K., Wisniewski, H. M. 1977. The sequence of myelination in the epiretinal portion of the optic nerve in the rabbit. *Neuropath. Appl. Neurobiol.* 3: 15–27

Noble, M., Murray, K. 1984. Purified astrocytes promote the *in vitro* division of a bipotential progenitor cell. *EMBO J.* 3: 2243–47

Peters, A., Palay, S. L., de F. Webster, H. 1976. *The Fine Structure of the Nervous System: The Neurones and the Supporting Cells.* Philadelphia: Saunders

Peters, A., Vaughn J. E. 1970. Morphology and development of the myelin sheath. In *Myelination*, ed. A. Davison, A. Peters, pp. 3–79. Springfield, Ill: Thomas

Privat, A., Rataboul, P. 1986. Fibrous and protoplasmic astrocytes. In *Astrocytes*, ed. S. Fedoroff, A. Vernadakis, 1: 105–26. New York: Academic

Privat, A., Valat, J., Fulcrand, J. 1981. Proliferation of neuroglial cell lines in the degenerating optic nerve of young rats. A radioautographic study. *J. Neuropath. Exp. Neurol.* 40: 46–60

Pruss, R. 1979. Thy-1 antigen on astrocytes in long term cultures of rat central nervous system. *Nature* 280: 688–90

Raff, M. C., Abney, E. R., Cohen, J., Lindsay, R., Noble, M. 1983a. Two types of astrocytes in cultures of developing rat white matter: Differences in morphology, surface gangliosides, and growth characteristics. *J. Neurosci.* 3: 1289–1300

Raff, M. C., Abney, E. R., Fok-Seang, J. 1985. Reconstitution of a developmental clock *in vitro*: A critical role for astrocytes in the timing of oligodendrocyte differentiation. *Cell* 42: 61–69

Raff, M. C., Abney E. R., Miller, R. H. 1984. Two glial cell lineages diverge prenatally in the rat optic nerve. *Dev. Biol.* 106: 53–60

Raff, M. C., Miller, R. H., Noble, M. 1983b. A glial progenitor cell that develops *in vitro* into an astrocyte or an oligodendrocyte depending on the culture medium. *Nature* 303: 390–96

Raff, M. C., Mirsky, R., Fields, K. L., Lisak, R. P., Dorfman, S. H., Silberberg, D. H., Gregson, N. A., Liebowitz, S., Kennedy, M. C. 1978. Galactocerebroside is a specific cell surface antigenic marker for oligodendrocytes in culture. *Nature* 274:813–16

Rakic, P. 1971. Neuron-glial relationship during granule cell migration in developing cerebellar cortex. A Golgi and electronmicroscopic study in *Maccacus rhesus. J. Comp. Neurol.* 141: 238–312

Ranscht, B., Clapshaw, P. A., Price, J., Noble, M., Seifert, W. 1982. Development of oligodendrocytes and Schwann cells studied with a monoclonal antibody against galactocerebroside. *Proc. Natl. Acad. Sci. USA* 79: 2709–13

Reese, T. S., Karnovsky, M. J. 1967. Fine structural localization of a blood brain barrier to exogenous peroxidase. *J. Cell Biol.* 34: 207–17

Richardson, W. D., Pringle, N., Mosley, M., Westermark, B., Dubois-Daleq, M. 1988. A role for platelet derived growth factor in normal gliogenesis in the central nervous system. *Cell* 53: 309–19

Roach, A., Boylan, K., Horvath, S., Prusiner, S. B., Hood, L. E. 1983. Characterisation of clones of cDNA representing rat myelin basic protein: Absence of expression in brain of shiverer mutant mice. *Cell* 34: 799–806

Roach, A., Takahashi, N., Provitcheva, D., Hood, L. E. 1985. Chromosomal mapping of mouse basic protein gene and structure and transcription of the partially deleted gene in shiverer mutant mice. *Cell* 42: 149–55

Rougon, G., Hirsch, M. R., Hirn, M., Guenet, J. L., Goridis, C. 1983. Monoclonal antibody to neural cell surface protein: Identification of a glycoprotein family of restricted cellular localization. *Neuroscience* 10: 511–20

Schachner, M. 1982. Immunological analysis of cellular heterogeneity in the cerebellum. In *Neuroimmunology*, ed., J. Brockes, pp. 215–50. New York: Plenum

Schachner, M., Hedley-Whyte, E. T., Hsu, D. W., Schoonmaker, G., Bigmami, A. 1977. Ultrastructural localization of glial fibrillary acidic protein in mouse cerebellum by immunoperoxidase labelling. *J. Cell Biol.* 75: 67–73

Schmechel, D. E., Rakic, P. 1979. A Golgi study of radial glial cells in developing monkey telenchephalon: Morphogensis and transformation into astrocytes. *Anat. Embryol.* 156: 115–52

Silver, J., Lorenz, S. E., Wahlsten, D., Coughlin, J. 1982. Axonal guidance during the development of the great cerebral commissures: Descriptive and experimental studies in vivo on the role of preformed glial pathways. *J. Comp Neurol.* 210: 10–29

Silver, J., Sapiro, J. 1981. Axonal guidance during development of the optic nerve: The role of pigmented epithelia and other intrinsic factors. *J. Comp. Neurol.* 202: 521–38

Silver, J., Rutishauser, U. 1984. Guidance of optic axons in vivo by a preformed adhesive pathway on neuroepithelial endfeet. *Dev. Biol.* 106: 485–99

Skoff, R., Price, D., Stocks, A. 1976a. Electron microscopic autoradiographic studies of gliogensis in rat optic nerve. 1. Cell proliferation. *J. Comp. Neurol.* 169: 291–312

Skoff, R., Price, D., Stocks, A. 1976b. Electron microscopic autoradiographic studies of gliogensis in rat optic nerve. 2. Time of origin. *J. Comp. Neurol.* 169: 313–33

Skoff, R., Knapp, P. E., Bartlett, W. P. 1986. Astrocyte diversity in the optic nerve: A cytoarchitectural study. In *Astrocytes*, ed. S. Fedoroff, A. Vernadakis, 1: 269–91. New York: Academic

Small, R. K., Riddle, P., Noble, M. 1987. Evidence for migration of oligodendrocyte-type-2 astrocyte progenitor cells into the developing rat optic nerve. *Nature* 328: 155–57

Sommer, I., Schachner, M. 1981. Monoclonal antibodies (01-04) for oligodendrocyte cell surfaces: An immunological study in the central nervous system. *Dev. Biol.* 83: 311–27

Stallcup. W., Beasley, L. 1987. Bipotential glial precursor cells of the optic nerve express the NG2 proteoglycan. *J. Neurosci.* 7: 1737–44

Stensaas, L. J., Stensaas, S. S. 1968. Astrocytic neuroglial cells, oligodendroglia and microgliacytes in the spinal cord of the toad. 1. Light microscopy. *Z. Zellforsch.* 86: 473–89

Stewart, P. A., Wiley, M. J. 1981. Developing nervous tissue induces formation of blood-brain barrier characteristics in invading endothelial cells: A study using quail-chick transplant chimeras. *Dev. Biol.* 84: 183–92

Temple, S., Raff, M. C. 1985. Differentiation of a bipotential glial progenitor cell in single cell microwell. *Nature* 313: 223–25

Temple, S., Raff, M. C. 1986. Clonal analysis of oligodendrocyte development in culture: Evidence for a developmental clock that counts cell divisions. *Cell* 44: 773–79

Tokuyasu, K. T. 1973. A technique for ultracryotomy of cell suspensions and tissues. *J. Cell Biol.* 37: 551–65

Tso, M. O. M., Shih, C. Y., McLean, M. I. W. 1975. Is there a blood brain barrier at the optic nerve head? *Arch. Ophthalmol.* 93: 815–25

Turner, D. L., Cepko, C. L. 1987. A common progenitor for neuron and glia persists in the rat retina late in development. *Nature* 328: 131–36

Van Heyningen, W. E. 1963. The fixation of tetanus toxin, strychnine, serotonin and other substances by gangliosides. *J. Gen. Microbiol.* 31: 375–87

Valat, J., Privat, A., Fulcrand, J. 1983. Multiplication and differentiation of glial cells in the optic nerve of the postnatal rat. *Anat. Embryol.* 167: 335–46

Vaughn, J. E. 1969. An electron microscope analysis of gliogensis in rat optic nerves. *Z. Zellforsch.* 94: 293–324

Vaughn, J. E., Pease, D. C. 1967. Electron microscopy of classically stained astrocytes. *J. Comp. Neurol.* 131: 143–53

Vaughn, J. E., Pease, D. C. 1970. Electron microscopic studies of Wallerian degeneration in rat optic nerves 11. Astrocytes, oligodendrocytes and adventitial cells. *J. Comp. Neurol.* 140: 207–26

Watanabe, T., Raff, M. C. 1988. Retinal astrocytes are immigrants from the optic nerve. *Nature* 332: 834–37

Waxman, S. G. 1986. The astrocyte as a component of the node of Ranvier. *Trends Neurosci.* 9: 250–53

Weigert, F. 1895. *Beitrage zur Kenntnis der normalen menschlichen Neuroglia.* Frankfurt am Main: Weisbrod

Williams, B. P., Abney, E. R., Raff, M. C. 1985. Macroglial cell development in embryonic rat brain: Studies using monoclonal antibodies, fluorescence activated cell sorting and cell culture. *Dev. Biol.* 112: 126–34

# SUBJECT INDEX

## A

Absorbance, 228-31
Acetylcholine
  retina and, 205-19
Acetylcholine receptors
  desensitization of, 91
  $Ca^{2+}$ ions and, 91
  dendritic spines and, 91-92
  second messenger process
    and, 91
  neuromuscular junctions and
    formation of, 498
Acoustic startle response
  insects and, 355-65
  behavioral aspects of, 362
  crickets, 355, 361-64
  green lacewings, 355, 360-
    61
  moths, 355, 360-61
  praying mantises, 355, 364-
    65
  neural basis of, 362-64
Adenosinemonophosphate (See
  cAMP)
Adrenal chromaffin cells
  catecholamine and, 415
Adrenal demedullation, 259
Adrenal gland
  catecholamine biosynthesis
    and, 415-16
Adrenal medulla
  catecholamine biosynthesis
    and, 415-16
Adrenergic nerves
  catecholamine and, 415
Adrenergic receptors
  amino terminus of, 75-76
  carboxy tail of, 76-79
  cytoplasmic regions of
    Xenopus oocytes and, 76
  gene structure of, 79-81
  guanine nucleotide regulator
    proteins and, 67-79
  homologous desensitization
    and, 78-79
  membrane-spanning domains
    of, 73-75
  protein sequences of, 73-79
  protein structure and, 69-72
  structure of, 67-81
β-Adrenergic receptor kinase
  phosphorylation of $\alpha_2$-AR
    and, 78
  phosphorylation of $\beta_2$-AR
    and, 77

phosphorylation of rhodopsin
  and, 77
Alz-50, 471
Alzheimer's disease
  amyloid angiopathy in, 464-
    67
  β-amyloid protein in, 466-
    84
  animal model of, 264
  cholinergic forbrain neuclei
    lesions and, 264
  cerebral microvessels in, 463
  cholinergic function and, 262
  Down's syndrome and, 463-
    64
  familial, 481-82
  chromosome 21 and, 481
  genetic linkage analysis
    and, 481
  intraneural and extraneural
    proteinaceous filaments
    in, 463
  morphological changes
    accompanying, 464-70
  neurofibrillary degeneration
    and, 464, 472
  neurofibrillary tangles and,
    464
  neuritic plaques and, 464
  paired helical filaments in,
    465, 467-74
2-Amino-4-phosphonobutyrate
  retina and, 209-19
γ-Aminobutyric acid (GABA)
  retina and, 205-19
  directional selectivity in,
    206, 218-19
  memory storage and, 274
Amnesia
  opiate antagonists and, 266
  retrograde, 255, 268
Amygdala
  memory storage and, 266,
    270-72, 277
Amyloid
  Alzheimer's disease and, 465-
    67
Amyloid angiopathy, 464-67
β-Amyloid protein
  Alzheimer's disease and, 466-
    67, 474-84
  biochemistry of, 474-81
β-Amyloid protein precursor
  Alzheimer's disease and, 474-
    84
Anticonvulsant

AG2 spider toxin, 411
Antidromic stimulation, 429
Apoliopoprotein E, 107
Area 7a
  eye position signals and, 381
Astrocytes
  development of, 518
  fibrous, 517
  mitogens for
    epidermal growth factor,
      111
    fibroblast growth factors,
      111
    insulin, 111
    insulin-like growth factors I
      and II, 111
    protoplasmic, 517
  type I, 519-31
  type II, 519-31
Astrocytic products
  apoliopoprotein E, 107
  basic fibroblast growth factor,
    1114-16
  laminin, 108-10
  nexin, 107-8
  S100b, 108
Autoactive neurons, 190
Axonal growth
  basal laminae and, 493-96
  collagen and, 493-95
    collagen I, 495
    collagen IV, 495
  entactin and, 493
  extracellular matrix and, 491-
    96
  fibronectin and, 493
  laminin and, 493-96
  molecular correlates of
    actin and, 129
    CSAT proteins, 146
    c-src, 145-46
    GAP-24, 145
    GAP-43 (a.k.a. B-50/F1/
      pp46/P57), 129-49
    JG22 proteins, 147
    LDL receptor, 147
    NCAM, 147
    neurofilament subunits and,
      129
    tubulin and, 129, 144-45
  proteoglycan-laminin complex
    and, 493
Axonal regeneration, 128-30
  abortive, 131
  GAP-43 (a.k.a. B-50/F1/pp46/
    P-57) and, 130-36

535

# CUMULATIVE INDEXES

## CONTRIBUTING AUTHORS, VOLUMES 8–12

544

# CHAPTER TITLES, VOLUMES 8–12